Lecture Notes in Computer Science 6196

Commenced Publication in 1973
Founding and Former Series Editors:
Gerhard Goos, Juris Hartmanis, and Jan van Leeuwen

My T. Thai Sartaj Sahni (Eds.)

Computing and Combinatorics

16th Annual International Conference, COCOON 2010
Nha Trang, Vietnam, July 19-21, 2010
Proceedings

Volume Editors

My T. Thai
University of Florida, CSE Building, Room 566
P.O. Box 116120, Gainesville, Florida 32611-6120, USA
E-mail: mythai@cise.ufl.edu

Sartaj Sahni
University of Florida, CSE Building, Room 305
P.O. Box 116120, Gainesville, Florida 32611-6120, USA
E-mail: sahni@cise.ufl.edu

Library of Congress Control Number: 2010929275

CR Subject Classification (1998): F.2, C.2, G.2, F.1, E.1, I.3.5

LNCS Sublibrary: SL 1 – Theoretical Computer Science and General Issues

ISSN 0302-9743
ISBN-10 3-642-14030-0 Springer Berlin Heidelberg New York
ISBN-13 978-3-642-14030-3 Springer Berlin Heidelberg New York

springer.com

Printed in Germany

Typesetting: Camera-ready by author, data conversion by Scientific Publishing Services, Chennai, India
Printed on acid-free paper 06/3180

Preface

The papers in this volume were selected for presentation at the 16th Annual International Computing and Combinatorics Conference (COCOON 2010), held during July 19–21, 2010 in Nha Trang, Vietnam. Previous meetings of this conference were held in Singapore (2002), Big Sky (2003), Jeju Island (2004), Kunming (2005), Taipei (2006), Alberta (2007), Dalian (2008) and New York (2009).

COCOON 2010 provided a forum for researchers working in the areas of algorithms, theory of computation, computational complexity, and combinatorics related to computing. In all, 133 papers were submitted from 40 countries and regions, of which 54 were accepted. Authors of the submitted papers were from Australia (10), Bangladesh (11), Belgium (1), Canada (23), Chile (1), China (20), Colombia (1), Czech Republic (6), Denmark (1), France (25), France(1), Germany (13), Greece (2), Hong Kong (7), Hungary (2), India (18), Indonesia (8), Islamic Republic of Iran (2), Ireland (1), Israel (6), Italy (6), Japan (31), Republic of Korea(4), Malaysia (1), The Netherlands (2), New Zealand (2), Norway (3), Pakistan (1), Poland (1), Portugal (1), Russian Federation (3), Singapore (6), Slovakia (1), Spain (7), Sweden (2), Taiwan (19), Thailand (2), UK (2), USA (44), and Vietnam (15).

The submitted papers were evaluated by an international Technical Program Committee (TPC). Each paper was evaluated by at least three TPC members, with possible assistance of the external referees, as indicated by the referee list found in the proceedings. Some of these 54 accepted papers will be selected for publication in a special issue of *Algorithmica*, *Journal of Combinatorial Optimization*, and *Discrete Mathematics, Algorithms, and Application* under the standard refereeing procedure. In addition to the selected papers, the conference also included two invited presentations by Manuel Blum (Carnegie Mellon University) and Oscar H. Ibarra (University of California Santa Barbara).

We are extremely thankful to all the TPC members, each of whom reviewed about 14 papers despite a very tight schedule. The time and effort spent per TPC member were tremendous. By the same token, we profusely thank all the external referees who helped review the submissions. The TPC and external referees not only helped select a strong program for the conference, but also gave very informative feedback to the authors of all submitted papers. We also thank the local Organizing Committee for their contribution to making the conference a success. Many thanks are due to the two invited speakers and all the authors who submitted papers for consideration, all of whom contributed to the quality of COCOON 2010.

July 2010

My T. Thai
Sartaj Sahni

Organization

Executive Committee

Conference and Program Chair	My T. Thai (University of Florida) Sartaj Sahni (University of Florida)
Local Organizing Committee Chair	Anh-Vu Duc-Dinh (HCM University of Technology)
Local Organizing Committee Members	Thoai Nam (HCM University of Technology) Thuan Dinh Nguyen (Nha Trang University) Nguyen-Phan Nguyen-Thai (VNPT Khanh Hoa)

Technical Program Committee

Tetsuo Asano	Japan Advanced Institute of Science and Technology, Japan
Mike Atallah	Purdue University, USA
Nikhil Bansal	IBM Research, USA
Therese Biedl	University of Waterloo, Canada
Amit Chakrabarti	Dartmouth College, USA
Chandra Chekuri	University of Illinois at Urbana-Champaign, USA
Siu-Wing Cheng	Hong Kong University of Science and Technology, Hong Kong
Peter Eades	University of Sydney, Australia
Uriel Feige	Weizmann Institute of Science, Israel
Greg N. Frederickson	Purdue University, USA
Raffaele Giancarlo	University of Palermo, Italy
Ashish Goel	Stanford University, USA
Carsten Gutwenger	Dortmund University of Technology, Germany
MohammadTaghi Hajiaghayi	AT&T Research, USA
Pinar Heggernes	University of Bergen, Norway
Wen-Lian Hsu	Academia Sinica, Taiwan
Hiro Ito	Kyoto University, Japan
Valentine Kabanets	Simon Fraser University, Canada
Sanjeev Khanna	University of Pennsylvania, USA
James R. Lee	University of Washington, USA
Ming Li	University of Waterloo, Canada
Xiang-Yang Li	Illinois Institute of Technology, USA
Prabhat K. Mahanti	University of New Brunswick, CA
Greg Plaxton	University of Texas at Austin, USA

Sartaj Sahni	University of Florida, USA, Co-chair
Daniel Stefankovic	University of Rochester, USA
Cliff Stein	Columbia University, USA
Martin Strauss	University of Michigan, USA
Chaitanya Swamy	University of Waterloo, Canada
My T. Thai	University of Florida, USA, Co-chair
Chris Umans	California Institute of Technology, USA
Lisa Zhang	Bell Laboratories, Alcatel-Lucent, USA
Yun Zhang	Pioneer Hi-Bred International Inc., USA

External Referees

Christian Komusiewicz
Peyman Afshani
Mohamed Aly
Yuichi Asahiro
Dave Bacon
M. Bateni
Hoda Bidkhori
Punya Biswal
Melissa Chase
Markus Chimani
Elad Cohen
Graham Cormode
Chiara Epifanio
Jiri Fiala
Mathew Francis
Takuro Fukunaga
Philippe Gambette
Anna Gilbert
Petr Golovach
Anupam Gupta
Walter Gutjahr
Toru Hasunuma
Hiroshi Hirai
Takashi Horiyama
Tsan-sheng Hsu
Toshimasa Ishii
Volkan Isler
David Johnson
Marcin Kaminski
Karsten Klein
Jochen Koenemann
Dieter Kratsch
Sara Krehbiel
Amit Kumar
Biagio Lenzitti
Vahid Liaghat
Dan Luu
Bin Ma
Hamid Mahini
Sven Mallach
Evangelos Markakis
Jiri Matousek
Daniel Meister
Ankur Moitra
Hiroki Morizumi
Mitsuo Motoki
Ashwin Nayak
Jesper Nederlof
Hirotaka Ono
Shayan Oveis Gharan
Charis Papadopoulos
Ian Post
Anup Rao
Saket Saurabh
Marinella Sciortino
Yaoyun Shi
Quentin Stout
Zoya Svitkina
Takeyuki Tamura
Dimitrios Thilikos
Akihiro Uejima
Cesare Valenti
Erik Jan Van Leeuwen
Kasturi Varadarajan
Hoi-Ming Wong
Yulai Xie
Yuichi Yoshida
M. Zadimoghaddam
Bernd Zey

Table of Contents

Invited Talks

Understanding and Inductive Inference 1
Manuel Blum

Computing with Cells: Membrane Systems 2
Oscar H. Ibarra

Complexity and Inapproximability

Boxicity and Poset Dimension 3
Abhijin Adiga, Diptendu Bhowmick, and L. Sunil Chandran

On the Hardness against Constant-Depth Linear-Size Circuits 13
Chi-Jen Lu and Hsin-Lung Wu

A k-Provers Parallel Repetition Theorem for a version of No-Signaling Model 23
Ricky Rosen

The Curse of Connectivity: t-Total Vertex (Edge) Cover 34
Henning Fernau, Fedor V. Fomin, Geevarghese Philip, and Saket Saurabh

Counting Paths in VPA Is Complete for $\#\mathsf{NC}^1$ 44
Andreas Krebs, Nutan Limaye, and Meena Mahajan

Depth-Independent Lower Bounds on the Communication Complexity of Read-Once Boolean Formulas 54
Rahul Jain, Hartmut Klauck, and Shengyu Zhang

Approximation Algorithms

Multiplying Pessimistic Estimators: Deterministic Approximation of Max TSP and Maximum Triangle Packing 60
Anke van Zuylen

Clustering with or without the Approximation 70
Frans Schalekamp, Michael Yu, and Anke van Zuylen

A Self-stabilizing 3-Approximation for the Maximum Leaf Spanning Tree Problem in Arbitrary Networks 80
Sayaka Kamei, Hirotsugu Kakugawa, Stéphane Devismes, and Sébastien Tixeuil

Approximate Weighted Farthest Neighbors and Minimum Dilation Stars 90
John Augustine, David Eppstein, and Kevin A. Wortman

Approximated Distributed Minimum Vertex Cover Algorithms for Bounded Degree Graphs 100
Yong Zhang, Francis Y.L. Chin, and Hing-Fung Ting

Graph Theory and Algorithms

Maximum Upward Planar Subgraph of a Single-Source Embedded Digraph 110
Aimal Rextin and Patrick Healy

Triangle-Free 2-Matchings Revisited 120
Maxim Babenko, Alexey Gusakov, and Ilya Razenshteyn

The Cover Time of Deterministic Random Walks 130
Tobias Friedrich and Thomas Sauerwald

Finding Maximum Edge Bicliques in Convex Bipartite Graphs 140
Doron Nussbaum, Shuye Pu, Jörg-Rüdiger Sack, Takeaki Uno, and Hamid Zarrabi-Zadeh

A Note on Vertex Cover in Graphs with Maximum Degree 3 150
Mingyu Xiao

Computing Graph Spanners in Small Memory: Fault-Tolerance and Streaming 160
Giorgio Ausiello, Paolo G. Franciosa, Giuseppe F. Italiano, and Andrea Ribichini

Factorization of Cartesian Products of Hypergraphs 173
Alain Bretto and Yannick Silvestre

Graph Drawing and Coloring

Minimum-Segment Convex Drawings of 3-Connected Cubic Plane Graphs (Extended Abstract) 182
Sudip Biswas, Debajyoti Mondal, Rahnuma Islam Nishat, and Md. Saidur Rahman

On Three Parameters of Invisibility Graphs 192
Josef Cibulka, Jan Kynčl, Viola Mészáros, Rudolf Stolař, and Pavel Valtr

Imbalance Is Fixed Parameter Tractable 199
Daniel Lokshtanov, Neeldhara Misra, and Saket Saurabh

The Ramsey Number for a Linear Forest versus Two Identical Copies of Complete Graphs ... 209
I W. Sudarsana, Adiwijaya, and S. Musdalifah

Computational Geometry

Optimal Binary Space Partitions in the Plane ... 216
Mark de Berg and Amirali Khosravi

Exact and Approximation Algorithms for Geometric and Capacitated Set Cover Problems ... 226
Piotr Berman, Marek Karpinski, and Andrzej Lingas

Effect of Corner Information in Simultaneous Placement of K Rectangles and Tableaux ... 235
Shinya Anzai, Jinhee Chun, Ryosei Kasai, Matias Korman, and Takeshi Tokuyama

Detecting Areas Visited Regularly ... 244
Bojan Djordjevic and Joachim Gudmundsson

Tile-Packing Tomography Is $\mathbb{NP}$-hard ... 254
Marek Chrobak, Christoph Dürr, Flavio Guíñez, Antoni Lozano, and Nguyen Kim Thang

The Rectilinear k-Bends TSP ... 264
Vladimir Estivill-Castro, Apichat Heednacram, and Francis Suraweera

Tracking a Generator by Persistence ... 278
Oleksiy Busaryev, Tamal K. Dey, and Yusu Wang

Auspicious Tatami Mat Arrangements ... 288
Alejandro Erickson, Frank Ruskey, Mark Schurch, and Jennifer Woodcock

Automata, Logic, Algebra and Number Theory

Faster Generation of Shorthand Universal Cycles for Permutations ... 298
Alexander Holroyd, Frank Ruskey, and Aaron Williams

The Complexity of Word Circuits ... 308
Xue Chen, Guangda Hu, and Xiaoming Sun

On the Density of Regular and Context-Free Languages ... 318
Michael Hartwig

Extensions of the Minimum Cost Homomorphism Problem ... 328
Rustem Takhanov

The Longest Almost-Increasing Subsequence 338
Amr Elmasry

Universal Test Sets for Reversible Circuits (Extended Abstract) 348
Satoshi Tayu, Shota Fukuyama, and Shuichi Ueno

Approximate Counting with a Floating-Point Counter 358
Miklós Csűrös

Network Optimization and Scheduling Algorithm

Broadcasting in Heterogeneous Tree Networks 368
Yu-Hsuan Su, Ching-Chi Lin, and D.T. Lee

Contention Resolution in Multiple-Access Channels: k-Selection in Radio Networks 378
Antonio Fernández Anta and Miguel A. Mosteiro

Online Preemptive Scheduling with Immediate Decision or Notification and Penalties 389
Stanley P.Y. Fung

Computational Biology and Bioinformatics

Discovering Pairwise Compatibility Graphs 399
Muhammad Nur Yanhaona, Md. Shamsuzzoha Bayzid, and Md. Saidur Rahman

Near Optimal Solutions for Maximum Quasi-bicliques 409
Lusheng Wang

Fast Coupled Path Planning: From Pseudo-Polynomial to Polynomial 419
Yunlong Liu and Xiaodong Wu

Constant Time Approximation Scheme for Largest Well Predicted Subset 429
Bin Fu and Lusheng Wang

On Sorting Permutations by Double-Cut-and-Joins 439
Xin Chen

A Three-String Approach to the Closest String Problem 449
Zhi-Zhong Chen, Bin Ma, and Lusheng Wang

A $2k$ Kernel for the Cluster Editing Problem 459
Jianer Chen and Jie Meng

Data Structure and Sampling Theory

On the Computation of 3D Visibility Skeletons 469
Sylvain Lazard, Christophe Weibel, Sue Whitesides, and Linqiao Zhang

The Violation Heap: A Relaxed Fibonacci Like Heap 479
Amr Elmasry

Threshold Rules for Online Sample Selection 489
Eric Bach, Shuchi Chawla, and Seeun Umboh

Heterogeneous Subset Sampling 500
Meng-Tsung Tsai, Da-Wei Wang, Churn-Jung Liau, and Tsan-sheng Hsu

Cryptography, Security, Coding and Game Theory

Identity-Based Authenticated Asymmetric Group Key Agreement Protocol 510
Lei Zhang, Qianhong Wu, Bo Qin, and Josep Domingo-Ferrer

Zero-Knowledge Argument for Simultaneous Discrete Logarithms 520
Sherman S.M. Chow, Changshe Ma, and Jian Weng

Directed Figure Codes: Decidability Frontier 530
Michał Kolarz

Author Index 541

Understanding and Inductive Inference

Manuel Blum

Computer Science Department
Carnegie Mellon University
Pittsburgh, PA 15213, USA
mblum@cs.cmu.edu

Abstract. This talk will be about different kinds of understanding, including especially that provided by deductive and inductive inference. Examples will be drawn from Galileo, Kepler, and Fermat. Applications will be given to the physicists problem of inferring the laws of nature, and to the restricted case of inferring sequences in Sloanes Encyclopedia of Integer Sequences.

While there is much theory and understanding of deductive inference, there is relatively little understanding of the process of inductive inference. Inductive inference is about hypothesizing, hopefully inferring, the laws of physics from observation of the data. It is what Kepler did in coming up with his three laws of planetary motion.

Once laws are in place, deductive inference can be used to derive their consequences, which then uncovers much of the understanding in physics.

This talk will define inductive inference, state some of its open problems, and give applications to a relatively simpler but still largely unexplored task, namely, inferring a large natural class of algorithmically generated integer sequences that includes (as of this writing) 20in Sloanes Encyclopedia of integer sequences.

M.T. Thai and S. Sahni (Eds.): COCOON 2010, LNCS 6196, p. 1, 2010.

Computing with Cells: Membrane Systems

Oscar H. Ibarra

Department of Computer Science
University of California
Santa Barbara, CA 93106, USA
ibarra@cs.ucsb.edu

Abstract. Membrane computing is a part of the general research effort of describing and investigating computing models, ideas, architectures, and paradigms from processes taking place in nature, especially in biology. It identifies an unconventional computing model, namely a P system, which abstracts from the way living cells process chemical compounds in their compartmental structure. P systems are a class of distributed maximally parallel computing devices of a biochemical type. We give a brief overview of the area and present results that answer some interesting and fundamental questions concerning issues such as determinism versus nondeterminism, membrane and alphabet-size hierarchies, various notions of parallelism, reachability, computational complexity. We also look at neural-like systems, called spiking neural P systems. Such systems incorporate the idea of spiking neurons in membrane computing. We study various classes and characterize their computing power and complexity.

M.T. Thai and S. Sahni (Eds.): COCOON 2010, LNCS 6196, p. 2, 2010.

Boxicity and Poset Dimension*

Abhijin Adiga, Diptendu Bhowmick, and L. Sunil Chandran

Department of Computer Science and Automation, Indian Institute of Science, Bangalore–560012, India
abhijin@csa.iisc.ernet.in, diptendubhowmick@gmail.com, sunil@csa.iisc.ernet.in

Abstract. Let G be a simple, undirected, finite graph with vertex set $V(G)$ and edge set $E(G)$. A k-dimensional box is a Cartesian product of closed intervals $[a_1, b_1] \times [a_2, b_2] \times \cdots \times [a_k, b_k]$. The *boxicity* of G, $\text{box}(G)$ is the minimum integer k such that G can be represented as the intersection graph of k-dimensional boxes, i.e. each vertex is mapped to a k-dimensional box and two vertices are adjacent in G if and only if their corresponding boxes intersect. Let $\mathcal{P} = (S, P)$ be a poset where S is the ground set and P is a reflexive, anti-symmetric and transitive binary relation on S. The dimension of $\mathcal{P}$, $\dim(\mathcal{P})$ is the minimum integer t such that P can be expressed as the intersection of t total orders.

Let $G_{\mathcal{P}}$ be the *underlying comparability graph* of $\mathcal{P}$. It is a well-known fact that posets with the same underlying comparability graph have the same dimension. The first result of this paper links the dimension of a poset to the boxicity of its underlying comparability graph. In particular, we show that for any poset $\mathcal{P}$, $\text{box}(G_{\mathcal{P}})/(\chi(G_{\mathcal{P}}) - 1) \leq \dim(\mathcal{P}) \leq 2\text{box}(G_{\mathcal{P}})$, where $\chi(G_{\mathcal{P}})$ is the chromatic number of $G_{\mathcal{P}}$ and $\chi(G_{\mathcal{P}}) \neq 1$.

The second result of the paper relates the boxicity of a graph G with a natural partial order associated with its *extended double cover*, denoted as G_c. Let $\mathcal{P}_c$ be the natural height-2 poset associated with G_c by making A the set of minimal elements and B the set of maximal elements. We show that $\frac{\text{box}(G)}{2} \leq \dim(\mathcal{P}_c) \leq 2\text{box}(G) + 4$.

These results have some immediate and significant consequences. The upper bound $\dim(\mathcal{P}) \leq 2\text{box}(G_{\mathcal{P}})$ allows us to derive hitherto unknown upper bounds for poset dimension. In the other direction, using the already known bounds for partial order dimension we get the following: (1) The boxicity of any graph with maximum degree Δ is $O(\Delta \log^2 \Delta)$ which is an improvement over the best known upper bound of $\Delta^2 + 2$. (2) There exist graphs with boxicity $\Omega(\Delta \log \Delta)$. This disproves a conjecture that the boxicity of a graph is $O(\Delta)$. (3) There exists no polynomial-time algorithm to approximate the boxicity of a bipartite graph on n vertices with a factor of $O(n^{0.5-\epsilon})$ for any $\epsilon > 0$, unless $NP = ZPP$.

Keywords: Boxicity, partial order, poset dimension, comparability graph, extended double cover.

---

* This work was supported by DST grant SR/S3/EECE/62/2006 and Infosys Technologies Ltd., Bangalore, under the "Infosys Fellowship Award".

M.T. Thai and S. Sahni (Eds.): COCOON 2010, LNCS 6196, pp. 3–12, 2010.

1 Introduction

1.1 Boxicity

A k-box is a Cartesian product of closed intervals $[a_1, b_1] \times [a_2, b_2] \times \cdots \times [a_k, b_k]$. A k-box representation of a graph G is a mapping of the vertices of G to k-boxes in the k-dimensional Euclidean space such that two vertices in G are adjacent if and only if their corresponding k-boxes have a non-empty intersection. The *boxicity* of G denoted $\text{box}(G)$, is the minimum integer k such that G has a k-box representation. Boxicity was introduced by Roberts [19]. Cozzens [8] showed that computing the boxicity of a graph is NP-hard. This was later strengthened by Yannakakis [26] and finally by Kratochvìl [17] who showed that determining whether boxicity of a graph is at most two itself is NP-complete.

It is easy to see that a graph has boxicity at most 1 if and only if it is an *interval graph*, i.e. each vertex of the graph can be associated with a closed interval on the real line such that two intervals intersect if and only if the corresponding vertices are adjacent. By definition, boxicity of a complete graph is 0. Let G be any graph and G_i, $1 \leq i \leq k$ be graphs on the same vertex set as G such that $E(G) = E(G_1) \cap E(G_2) \cap \cdots \cap E(G_k)$. Then we say that G is the *intersection graph* of G_i s for $1 \leq i \leq k$ and denote it as $G = \bigcap_{i=1}^{k} G_i$.

Lemma 1. Roberts [19]: *The boxicity of a non-complete graph G is the minimum positive integer b such that G can be represented as the intersection of b interval graphs. Moreover, if $G = \bigcap_{i=1}^{m} G_i$ for some graphs G_i then* $\text{box}(G) \leq \sum_{i=1}^{m} \text{box}(G_i)$.

1.2 Poset Dimension

A *partially ordered set* or *poset* $\mathcal{P} = (S, P)$ consists of a non-empty set S, called the *ground set* and a reflexive, antisymmetric and transitive binary relation P on S. A *total order* is a partial order in which every two elements are comparable. It essentially corresponds to a permutation of elements of S. A *height-2 poset* is one in which every element is either a minimal element or a maximal element. A *linear extension* L of a partial order P is a total order which satisfies ($x \leq y$ in $P \Rightarrow x \leq y$ in L). A *realizer* of a poset $\mathcal{P} = (S, P)$ is a set of linear extensions of P, say $\mathcal{R}$ which satisfy the following condition: for any two distinct elements x and y, $x < y$ in P if and only if $x < y$ in L, $\forall L \in \mathcal{R}$. The *poset dimension* of $\mathcal{P}$ (sometimes abbreviated as dimension of $\mathcal{P}$) denoted by $\dim(\mathcal{P})$ is the minimum integer k such that there exists a realizer of P of cardinality k. Poset dimension was introduced by Dushnik and Miller [10]. Clearly, a poset is one-dimensional if and only if it is a total order. Pnueli et al. [18] gave a polynomial time algorithm to recognize dimension 2 posets. In [26] Yannakakis showed that it is NP-complete to decide whether the dimension of a poset is at most 3. For more references and survey on dimension theory of posets see Trotter [23,24]. Recently, Hegde and Jain [16] showed that it is hard to design an approximation algorithm for computing the dimension of a poset.

A simple undirected graph G is a comparability graph if and only if there exists some poset $\mathcal{P} = (S, P)$, such that S is the vertex set of G and two vertices are adjacent in G if and only if they are comparable in $\mathcal{P}$. We will call such a poset an *associated poset* of G. Likewise, we refer to G as the *underlying comparability graph* of $\mathcal{P}$. Note that for a height-2 poset, the underlying comparability graph is a bipartite graph with partite sets A and B, with say A corresponding to minimal elements and B to maximal elements. For more on comparability graphs see [15]. It is easy to see that there is a unique comparability graph associated with a poset, whereas, there can be several posets with the same underlying comparability graph. However, Trotter, Moore and Sumner [25] proved that posets with the same underlying comparability graph have the same dimension.

2 Our Main Results

The results of this paper are the consequence of our attempts to bring out some connections between boxicity and poset dimension. As early as 1982, Yannakakis had some intuition regarding a possible connection between these problems when he established the NP-completeness of both poset dimension and boxicity in [26]. But interestingly, no results were discovered in the last 25 years which establish links between these two notions. Perhaps the researchers were misled by some deceptive examples such as the following one: Consider a complete graph K_n where n is even and remove a perfect matching from it. The resulting graph is a comparability graph and the dimension of any of its associated posets is 2, while its boxicity is $n/2$.

First we state an upper bound and a lower bound for the dimension of a poset in terms of the boxicity of its underlying comparability graph.

Theorem 1. *Let $\mathcal{P} = (V, P)$ be a poset such that $\dim(\mathcal{P}) > 1$ and $G_{\mathcal{P}}$ its underlying comparability graph. Then, $\dim(\mathcal{P}) \leq 2\text{box}(G_{\mathcal{P}})$.*

Theorem 2. *Let $\mathcal{P} = (V, P)$ be a poset and let $\chi(G_{\mathcal{P}})$ be the chromatic number of its underlying comparability graph $G_{\mathcal{P}}$ such that $\chi(G_{\mathcal{P}}) > 1$. Then, $\dim(\mathcal{P}) \geq \frac{\text{box}(G_{\mathcal{P}})}{\chi(G_{\mathcal{P}})-1}$.*

Note that if $\mathcal{P}$ is a height-2 poset, then $G_{\mathcal{P}}$ is a bipartite graph and therefore $\chi(G_{\mathcal{P}}) = 2$. Thus, from the above results we have the following:

Corollary 1. *Let $\mathcal{P} = (V, P)$ be a height-2 poset and $G_{\mathcal{P}}$ its underlying comparability graph. Then, $\text{box}(G_{\mathcal{P}}) \leq \dim(\mathcal{P}) \leq 2\text{box}(G_{\mathcal{P}})$.*

Definition 1. *The extended double cover of G, denoted as G_c is a bipartite graph with partite sets A and B which are copies of $V(G)$ such that corresponding to every $u \in V(G)$, there are two vertices $u_A \in A$ and $u_B \in B$ and $\{u_A, v_B\}$ is an edge in G_c if and only if either $u = v$ or u is adjacent to v in G.*

We prove the following lemma relating the boxicity of G and G_c.

Lemma 2. *Let G be any graph and G_c its extended double cover. Then,*

$$\frac{\text{box}(G)}{2} \leq \text{box}(G_c) \leq \text{box}(G) + 2.$$

Let $\mathcal{P}_c$ be the natural height-2 poset associated with G_c, i.e. the elements in A are the minimal elements and the elements in B are the maximal elements. Combining Corollary 1 and Lemma 2 we have the following theorem:

Theorem 3. *Let G be a graph and $\mathcal{P}_c$ be the natural height-2 poset associated with its extended double cover. Then,* $\frac{\dim(\mathcal{P}_c)}{2} - 2 \leq \text{box}(G) \leq 2\dim(\mathcal{P}_c)$ *and therefore* $\text{box}(G) = \Theta(\dim(\mathcal{P}_c))$.

2.1 Consequences

New upper bounds for poset dimension: Our results lead to some hitherto unknown bounds for poset dimension. Some general bounds obtained in this manner are listed below:

1. It is proved in [5] that for any graph G, boxicity of G is at most tree-width (G) + 2. Applying this bound in Theorem 1 it immediately follows that, for a poset $\mathcal{P}$, $\dim(\mathcal{P}) \leq 2\,\text{tree-width}\,(G_{\mathcal{P}}) + 4$.
2. The *threshold dimension* of a graph G is the minimum number of *threshold graphs* such that G is the edge union of these graphs. For more on threshold graphs and threshold dimension see [15]. Cozzens and Halsey [9] proved that $\text{box}(G) \leq \text{threshold-dimension}(\overline{G})$, where $\overline{G}$ is the complement of G. From this it follows that $\dim(\mathcal{P}) \leq 2\ \text{threshold-dimension}(\overline{G_{\mathcal{P}}})$.
3. In [2] it is proved that $\text{box}(G) \leq \left\lfloor \frac{\text{MVC}(G)}{2} \right\rfloor + 1$, where MVC$(G)$ is the cardinality of the *minimum vertex cover* of G. Therefore, we have $\dim(\mathcal{P}) \leq \text{MVC}\,(G_{\mathcal{P}}) + 2$.

Some more interesting results can be obtained if we restrict $G_{\mathcal{P}}$ to belong to certain subclasses of graphs.

4. Scheinerman [20] proved that the boxicity of outer planar graphs is at most 2 and later Thomassen [21] proved that the boxicity of planar graphs is at most 3. Therefore, it follows that $\dim(\mathcal{P}) \leq 4$ if $G_{\mathcal{P}}$ is outer planar and $\dim(\mathcal{P}) \leq 6$ if $G_{\mathcal{P}}$ is planar.
5. Bhowmick and Chandran [1] proved that boxicity of an AT-free graph is at most its chromatic number. Hence, $\dim(\mathcal{P}) \leq 2\chi(G_{\mathcal{P}})$, if $G_{\mathcal{P}}$ is AT-free.
6. The boxicity of a *d-dimensional hypercube* is $O(d/\log(d))$ [6]. Therefore, if $G_{\mathcal{P}}$ is a height-2 poset whose comparability graph is a *d-dimensional hypercube*, then from Corollary 1 we have $\dim(\mathcal{P}) = O(d/\log(d))$.
7. Chandran et al. [3] recently proved that chordal bipartite graphs have arbitrarily high boxicity. From Corollary 1 it follows that height-2 posets whose underlying comparability graph are chordal bipartite can have arbitrarily high dimension.

Improved upper bound for boxicity based on maximum degree: Füredi and Kahn [14] proved the following

Lemma 3. *Let $\mathcal{P}$ be a poset and Δ be the maximum degree of $G_{\mathcal{P}}$. There exists a constant c such that* $\dim(\mathcal{P}) < c\Delta(\log \Delta)^2$.

From Lemma 2 and Corollary 1 we have $\text{box}(G) \leq 2\text{box}(G_c) \leq 2\dim(\mathcal{P}_c)$, where G_c is the extended double cover of G. Note that by construction $\Delta(G_c) = \Delta(G) + 1$. On applying the above lemma, we have

Theorem 4. *For any graph G having maximum degree Δ there exists a constant c' such that* $\text{box}(G) < c'\Delta(\log \Delta)^2$.

This result is an improvement over the previous upper bound of $\Delta^2 + 2$ by Esperet [12].

Counter examples to the conjecture of [4]: Chandran et al. [4] conjectured that boxicity of a graph is $O(\Delta)$. We use a result by Erdős, Kierstead and Trotter [11] to show that there exist graphs with boxicity $\Omega(\Delta \log \Delta)$, hence disproving the conjecture. Let $\mathbb{P}(n,p)$ be the probability space of height-2 posets with n minimal elements forming set A and n maximal elements forming set B, where for any $a \in A$ and $b \in B$, $\text{Prob}(a < b) = p(n) = p$. They proved the following:

Theorem 5. [11] *For every $\epsilon > 0$, there exists $\delta > 0$ so that if $(\log^{1+\epsilon} n)/n < p < 1 - n^{-1+\epsilon}$, then, $\dim(\mathcal{P}) > (\delta pn \log(pn))/(1 + \delta p \log(pn))$ for almost all $\mathcal{P} \in \mathbb{P}(n,p)$.*

When $p = 1/\log n$, for almost all posets $\mathcal{P} \in \mathbb{P}(n, 1/\log n)$, $\Delta(G_{\mathcal{P}}) < \delta_1 n/\log n$ and by the above theorem $\dim(\mathcal{P}) > \delta_2 n$, where δ_1 and δ_2 are some positive constants (see [24] for a discussion on the above theorem). From Theorem 1, it immediately implies that for almost all $\mathcal{P} \in \mathbb{P}(n, 1/\log n)$, $\text{box}(G_{\mathcal{P}}) \geq \frac{\dim(\mathcal{P})}{2} > \delta' \Delta(G_{\mathcal{P}}) \log \Delta(G_{\mathcal{P}})$ for some positive constant δ', hence proving the existence of graphs with boxicity $\Omega(\Delta \log \Delta)$.

Approximation hardness for the boxicity of bipartite graphs: Hegde and Jain [16] proved the following

Theorem 6. *There exists no polynomial-time algorithm to approximate the dimension of an n-element poset within a factor of $O(n^{0.5-\epsilon})$ for any $\epsilon > 0$, unless $NP = ZPP$.*

This is achieved by reducing the *fractional chromatic number problem* on graphs to the poset dimension problem. In addition they observed that a slight modification of their reduction will imply the same result for even height-2 posets. From Corollary 1, it is clear that for any height-2 poset $\mathcal{P}$, $\dim(\mathcal{P}) = \Theta(\text{box}(G_{\mathcal{P}}))$. Suppose there exists an algorithm to compute the boxicity of bipartite graphs with approximation factor $O(n^{0.5-\epsilon})$, for some $\epsilon > 0$, then, it is clear that the same algorithm can be used to compute the dimension of height-2 posets with approximation factor $O(n^{0.5-\epsilon})$, a contradiction. Hence,

Theorem 7. *There exists no polynomial-time algorithm to approximate the boxicity of a bipartite graph on n-vertices with a factor of $O(n^{0.5-\epsilon})$ for any $\epsilon > 0$, unless $NP = ZPP$.*

3 Notations

Let $[n]$ denote $\{1, 2, \ldots, n\}$ where n is a positive integer. For any graph G, let $V(G)$ and $E(G)$ denote its vertex set and edge set respectively. If G is undirected, for any $u, v \in V(G)$, $\{u, v\} \in E(G)$ means u is adjacent to v and if G is directed, $(u, v) \in E(G)$ means there is a directed edge from u to v. Whenever we refer to an AB bipartite (or co-bipartite) graph, we imply that its vertex set can be partitioned into non-empty sets A and B where both these sets induce independent sets (cliques respectively).

In a poset $\mathcal{P} = (S, P)$, the notations aPb, $a \leq b$ in P and $(a, b) \in P$ are equivalent and are used interchangeably. $G_{\mathcal{P}}$ denotes the underlying comparability graph of $\mathcal{P}$. Given an AB bipartite graph G, the natural poset associated with G with respect to the bipartition is the poset obtained by taking A to be the set of minimal elements and B to be the set of maximal elements. In particular, if G_c is the extended double cover of G, we denote by $\mathcal{P}_c$ the natural associated poset of G_c.

Suppose I is an interval graph. Let f_I be an *interval representation* for I, i.e. it is a mapping from the vertex set to closed intervals on the real line such that for any two vertices u and v, $\{u, v\} \in E(I)$ if and only if $f_I(u) \cap f_I(v) \neq \emptyset$. Let $l(u, f_I)$ and $r(u, f_I)$ denote the left and right end points of the interval corresponding to the vertex u respectively. In this paper, we will never consider more than one interval representation for an interval graph. Therefore, we will simplify the notations to $l(u, I)$ and $r(u, I)$. Further, when there is no ambiguity about the graph under consideration and its interval representation, we simply denote the left and right end points as $l(u)$ and $r(u)$ respectively.

4 Proof of Theorem 1

Let $\text{box}(G_{\mathcal{P}}) = k$. Note that since $\dim(\mathcal{P}) > 1$, $G_{\mathcal{P}}$ cannot be a complete graph and therefore $k \geq 1$. Let $\mathcal{I} = \{I_1, I_2, \ldots, I_k\}$ be a set of interval graphs such that $G_{\mathcal{P}} = \bigcap_{i=1}^{k} I_i$. Now, corresponding to each I_i we will construct two total orders L_i^1 and L_i^2 such that $\mathcal{R} = \{L_i^j | i \in [k] \text{ and } j \in [2]\}$ is a realizer of $\mathcal{P}$.

Let $I \in \mathcal{I}$ and f_I be an interval representation of I. We will define two partial orders P_I and $\overline{P}_I$ as follows: $\forall a \in V$, (a, a) belongs to P_I and $\overline{P}_I$ and for every non-adjacent pair of vertices $a, b \in V$ with respect to I,

$$\left.\begin{array}{l}(a, b) \in P_I \\ (b, a) \in \overline{P}_I\end{array}\right\} \text{ if and only if } r(a, f_I) < l(b, f_I).$$

Partial orders constructed in the above manner from a collection of intervals are called *interval orders* (See [24] for more details). It is easy to see that $\overline{I}$ (the complement of I) is the underlying comparability graph of both P_I and $\overline{P}_I$.

Let G_1 and G_2 be two directed graphs with vertex set V and edge set $E(G_1) = (P \cup P_I) \setminus \{(a,a)|a \in V\}$ and $E(G_2) = (P \cup \overline{P}_I) \setminus \{(a,a)|a \in V\}$ respectively. Note that from the definition it is obvious that there are no directed loops in G_1 and G_2.

Lemma 4. *G_1 and G_2 are acyclic directed graphs.*

Proof. We will prove the lemma for G_1 – a similar proof holds for G_2. First of all, since it can be assumed that none of the graphs in $\mathcal{I}$ are complete graphs, $P_I \neq \emptyset$. Suppose P_I is a total order, i.e. if P is an antichain, then it is clear that $E(G_1) = P_I$ and therefore G_1 is acyclic. Henceforth, we will assume that P_I is not a total order. Suppose G_1 is not acyclic. Let $C = \{(a_0, a_1), (a_1, a_2), \ldots, (a_{t-2}, a_{t-1}), (a_{t-1}, a_0)\}$ be a shortest directed cycle in G_1.

First we will show that $t > 2$. If $t = 2$, then there should be $a, b \in V$ such that $(a,b), (b,a) \in E(G_1)$. Since P is a partial order, (a,b) and (b,a) cannot be simultaneously present in P. The same holds for P_I. Thus, without loss of generality we can assume that $(a,b) \in P$ and $(b,a) \in P_I$. But if $(a,b) \in P$, then, a and b are adjacent in $G_{\mathcal{P}}$ and thus adjacent in I. Then clearly the intervals of a and b intersect and therefore $(b,a) \notin P_I$, a contradiction.

Now, we claim that two consecutive edges in C cannot belong to P (or P_I). Suppose there do exist such edges, say (a_i, a_{i+1}) and (a_{i+1}, a_{i+2}) which belong to P (or P_I) (note that the addition is modulo t). Since P (or P_I) is a partial order, it implies that $(a_i, a_{i+2}) \in P$ (or P_I) and as a result we have a directed cycle of length $t-1$, a contradiction to the assumption that C is a shortest directed cycle. Therefore, the edges of C alternate between P and P_I. It also follows that $t \geq 4$.

Without loss of generality we will assume that $(a_1, a_2), (a_3, a_4) \in P_I$. We claim that $\{(a_1, a_2), (a_3, a_4)\}$ is an induced poset of P_I. First of all a_2 and a_3 are not comparable in P_I as they are comparable in P. If either $\{a_1, a_3\}$ or $\{a_2, a_4\}$ are comparable, then we can demonstrate a shorter directed cycle in G_1, a contradiction. Finally we consider the pair $\{a_1, a_4\}$. If $t = 4$, then they are not comparable as they are comparable in P while if $t \neq 4$ and if they are comparable, then, it would again imply a shorter directed cycle, a contradiction. Hence, $\{(a_1, a_2), (a_3, a_4)\}$ is an induced subposet. In the literature such a poset is denoted as $\mathbf{2} + \mathbf{2}$ where $+$ refers to *disjoint sum* and $\mathbf{2}$ is a two-element total order. Fishburn [13] has proved that a poset is an interval order if and only if it does not contain a $\mathbf{2} + \mathbf{2}$. This implies that P_I is not an interval order, a contradiction. We have therefore proved that there cannot be any directed cycles in G_1. In a similar way we can show that G_2 is an acyclic directed graph. □

Since G_1 and G_2 are acyclic, we can construct total orders, say L^1 and L^2 using *topological sort* on G_1 and G_2 such that $P \cup P_I \subseteq L^1$ and $P \cup \overline{P}_I \subseteq L^2$ (For more details on topological sort, see [7]). For each I_i, we create linear extensions L_i^1 and L_i^2 of P as described above. It is easy to verify that $\mathcal{R} = \{L_i^j | i \in [k], j \in [2]\}$ is a realizer of $\mathcal{P}$. Hence proved.

Tight Example for Theorem 1: Consider the *crown* poset S_n^0: a height-2 poset with n minimal elements $a_1, a_2, \ldots, a_n$ and n maximal elements $b_1, b_2, \ldots, b_n$

and $a_i < b_j$, for $j = i+1, i+2, \ldots, i-1$, where the addition is modulo n. Its underlying comparability graph is the bipartite graph obtained by removing a perfect matching from the complete bipartite graph $K_{n,n}$. The dimension of this poset is n (see [22,24]) while the boxicity of the graph is $\lceil \frac{n}{2} \rceil$ [2].

5 Proof of Theorem 2

We will prove that $\text{box}(G_{\mathcal{P}}) \leq (\chi(G_{\mathcal{P}}) - 1)\dim(\mathcal{P})$. Let $(\chi(G_{\mathcal{P}}) - 1) = p$, $\dim(\mathcal{P}) = k$ and $\mathcal{R} = \{L_1, \ldots, L_k\}$ a realizer of $\mathcal{P}$. Now we color the vertices of $G_{\mathcal{P}}$ as follows: For a vertex v, if γ is the size of a longest chain in $\mathcal{P}$ such that v is its maximum element, then we assign color γ to it. This is clearly a proper coloring scheme since if two vertices x and y are assigned the same color, say γ and $x < y$, then it implies that the size of a longest chain in which y occurs as the maximum element is at least $\gamma + 1$, a contradiction. Also, this is a minimum coloring because the maximum number that gets assigned to any vertex equals the size of a longest chain in $\mathcal{P}$, which corresponds to the clique number of $G_{\mathcal{P}}$.

Now we construct pk interval graphs $\mathcal{I} = \{I_{ij} | i \in [p], j \in [k]\}$ and show that $G_{\mathcal{P}}$ is an intersection graph of these interval graphs. Let Π_j be the *permutation induced* by the total order L_j on $[n]$, i.e. xL_jy if and only if $\Pi_j^{-1}(x) < \Pi_j^{-1}(y)$. The following construction applies to all graphs in $\mathcal{I}$ except I_{pk}. Let $I_{ij} \in \mathcal{I} \setminus \{I_{pk}\}$. We assign the point interval $\left[\Pi_j^{-1}(v), \Pi_j^{-1}(v)\right]$ for all vertices v colored i. For all vertices v colored $i' < i$, we assign $\left[\Pi_j^{-1}(v), n+1\right]$ and for those colored $i' > i$, we assign $\left[0, \Pi_j^{-1}(v)\right]$. The interval assignment for the last interval graph I_{pk} is as follows: for all vertices v colored $p + 1 = \chi(G_{\mathcal{P}})$ we assign the point interval $\left[\Pi_k^{-1}(v), \Pi_k^{-1}(v)\right]$ and for the rest of the vertices we assign the interval $\left[\Pi_k^{-1}(v), n+1\right]$. The proof of $G_{\mathcal{P}} = \bigcap_{I \in \mathcal{I}} I$ is omitted due to space constraints.

A complete k-partite graph serves as a tight example for Theorem 2. While its associated poset has dimension 2, its boxicity and chromatic number are k. However, it would be interesting to see if there are posets of higher dimension for which Theorem 2 is tight.

6 Boxicity of the Extended Double Cover

In this section, we will give an outline of the proof of Lemma 2. First, we define the associated co-bipartite graph.

Definition 2. *Let H be an AB bipartite graph. The associated co-bipartite graph of H, denoted by H^* is the graph obtained by making the sets A and B cliques, but keeping the set of edges between vertices of A and B identical to that of H, i.e. $\forall u \in A, v \in B$, $\{u, v\} \in E(H^*)$ if and only if $\{u, v\} \in E(H)$.*

The associated co-bipartite graph H^* is not to be confused with the complement of H (i.e. $\overline{H}$) which is also a co-bipartite graph. The following Lemma is stated without proof due to space constraints:

Lemma 5. *Let H be an AB bipartite graph and H^* its associated co-bipartite graph. If H^* is a non-interval graph, then*

$$\frac{\text{box}(H^*)}{2} \leq \text{box}(H) \leq \text{box}(H^*).$$

If H^* is an interval graph, then $\text{box}(H) \leq 2$.

6.1 Proof of Lemma 2

Definition 3. *(Canonical interval representation of a co-bipartite interval graph:) Let I be an AB co-bipartite interval graph. A canonical interval representation of I satisfies: $\forall u \in A$, $l(u) = l$ and $\forall u \in B$ $r(u) = r$, where the points l and r are the leftmost and rightmost points respectively of the interval representation.*

It is easy to verify that such a representation exists for every co-bipartite interval graph.

$\text{box}(G_c) \leq \text{box}(G) + 2$*:* Let $\text{box}(G) = k$ and $G = I_1 \cap I_2 \cap \ldots \cap I_k$ where I_is are interval graphs. For each I_i, we construct interval graphs I'_i with vertex set $V(G_c)$ as follows: Consider an interval representation for I_i. For every vertex u in I_i, we assign the interval of u to u_A and u_B in I'_i. Let I'_{k+1} and I'_{k+2} be interval graphs where (1) all vertices in A are adjacent to all the vertices in B (2) In I'_{k+1}, A induces a clique and B induces an independent set while in I'_{k+2} it is the other way round. It is easy to verify that $G_c = I'_1 \cap I'_2 \cap \ldots \cap I'_{k+2}$.

$\text{box}(G) \leq 2\text{box}(G_c)$*:* We will assume without loss of generality that $|V(G)| > 1$. This implies G_c is not a complete graph and therefore $\text{box}(G_c) > 0$. Let us consider the associated co-bipartite graph of G_c, i.e. G^*_c. We will show that $\text{box}(G) \leq \text{box}(G^*_c)$ and the required result follows from Lemma 5. Let $\text{box}(G^*_c) = p$ and $G^*_c = J_1 \cap J_2 \cap \ldots \cap J_p$ where J_is are interval graphs. Let us assume canonical interval representation for each J_i (recall Definition 3). Corresponding to each J_i, we construct an interval graph J'_i with vertex set $V(G)$ as follows: The interval for any vertex u is the intersection of the intervals of u_A and u_B, i.e. $l(u, J'_i) = l(u_B, J_i)$ and $r(u, J'_i) = r(u_A, J_i)$. Note that since u_A and u_B are adjacent in J_i, their intersection is non-empty. It is easy to verify that $G = \bigcap_{i=1}^{p} J'_i$.

References

1. Bhowmick, D., Chandran, L.S.: Boxicity and cubicity of asteroidal triple free graphs. Disc. Math. 310, 1536–1543 (2010)
2. Chandran, L.S., Das, A., Shah, C.D.: Cubicity, boxicity, and vertex cover. Disc. Math. 309, 2488–2496 (2009)
3. Chandran, L.S., Francis, M.C., Mathew, R.: Chordal bipartite graphs with high boxicity. In: Japan conference on computation geometry and graphs, Kanazawa (2009)

4. Chandran, L.S., Francis, M.C., Sivadasan, N.: Boxicity and maximum degree. J. Combin. Theory Ser. B 98(2), 443–445 (2008)
5. Chandran, L.S., Sivadasan, N.: Boxicity and treewidth. J. Combin. Theory Ser. B 97(5), 733–744 (2007)
6. Chandran, L.S., Sivadasan, N.: The cubicity of hypercube graphs. Disc. Math. 308(23), 5795–5800 (2008)
7. Cormen, T.H., Leiserson, C.E., Rivest, R.L., Stein, C.: Introduction to algorithms. MIT press, Cambridge (2001)
8. Cozzens, M.B.: Higher and multi-dimensional analogues of interval graphs, Ph.D thesis, Department of Mathematics, Rutgers University, New Brunswick, NJ (1981)
9. Cozzens, M.B., Halsey, M.D.: The relationship between the threshold dimension of split graphs and various dimensional parameters. Disc. Appl. Math. 30, 125–135 (1991)
10. Dushnik, B., Miller, E.W.: Partially ordered sets. Amer. J. Math 6(3), 600–610 (1941)
11. Erdős, P., Kierstead, H., Trotter, W.T.: The dimension of random ordered sets. Random structures and algorithms 2, 253–275 (1991)
12. Esperet, L.: Boxicity of graphs with bounded degree. European J. Combin. 30(5), 1277–1280 (2009)
13. Fishburn, P.C.: Intransitive indifference with unequal indifference intervals. J. Math. Psych. 7, 144–149 (1983)
14. Füredi, Z., Kahn, J.: On the dimensions of ordered sets of bounded degree. Order 3, 15–20 (1986)
15. Golumbic, M.C.: Algorithmic Graph Theory and Perfect Graphs. Academic Press, New York (1980)
16. Hegde, R., Jain, K.: The hardness of approximating poset dimension. Electronic Notes on Discrete Mathematics 29, 435–443 (2007)
17. Kratochvil, J.: A special planar satisfiability problem and a consequence of its NPcompleteness. Disc. Appl. Math. 52, 233–252 (1994)
18. Pnueli, A., Lempel, A., Even, S.: Transitive orientation of graphs and identification of permutation graphs. Canad. J. Math 23, 160–175 (1971)
19. Roberts, F.S.: On the boxicity and cubicity of a graph. In: Recent Progresses in Combinatorics, pp. 301–310. Academic Press, New York (1969)
20. Scheinerman, E.R.: Intersection classes and multiple intersection parameters, Ph.D. thesis, Princeton University (1984)
21. Thomassen, C.: Interval representations of planar graphs. J. Combin. Theory Ser. B 40, 9–20 (1986)
22. Trotter, W.T.: Dimension of the crown S_n^K. Disc. Math. 8, 85–103 (1974)
23. Trotter, W.T.: Combinatorics and partially ordered sets: Dimension Theory. The Johns Hopkins University Press, Baltimore (1992)
24. Trotter, W.T.: Graphs and partially ordered sets: recent results and new directions. In: Surveys in graph theory, San Fransisco, CA (1995), Congr. Numer. 116 (1996)
25. Trotter, W.T., Moore, J.I., Sumner, D.P.: The dimension of a comparability graph. Proc. Amer Math Soc. 60, 35–38 (1976)
26. Yannakakis, M.: The complexity of the partial order dimension problem. SIAM J. Alg. Disc. Math. 3(3), 351–358 (1982)

On the Hardness against Constant-Depth Linear-Size Circuits

Chi-Jen Lu[1] and Hsin-Lung Wu[2]

[1] Institute of Information Science, Academia Sinica, Taipei, Taiwan
cjlu@iis.sinica.edu.tw
[2] Department of Computer Science and Information Engineering, National Taipei University, Taipei, Taiwan
hsinlung@mail.ntpu.edu.tw

Abstract. The notion of average-case hardness is a fundamental one in complexity theory. In particular, it plays an important role in the research on derandomization, as there are general derandomization results which are based on the assumption that average-case hard functions exist. However, to achieve a complete derandomization, one usually needs a function which is extremely hard against a complexity class, in the sense that any algorithm in the class fails to compute the function on $1/2 - 2^{-\Omega(n)}$ fraction of its n-bit inputs. Unfortunately, lower bound results are very rare and they are only known for very restricted complexity classes, and achieving such extreme hardness seems even more difficult. Motivated by this, we study the hardness against linear-size circuits of constant depth in this paper. We show that the parity function is extremely hard for them: any such circuit must fail to compute the parity function on at least $1/2 - 2^{-\Omega(n)}$ fraction of inputs.

1 Introduction

One of the main research topics in theoretical computer science is to understand the power of randomness in computation. A major open question is the BPP versus P question, which asks if all randomized polynomial-time algorithms can be derandomized into deterministic ones. To derandomize BPP, a standard way is to construct the so-called pseudorandom generators (PRG), which stretch a short random seed into a long string that looks random to circuits of polynomial size. As shown by Nisan and Wigderson [5], to have BPP = P, it suffices to have such a PRG with a seed of logarithmic length, which exists under some average-case hardness assumption. The assumption is that there exists a function in $\mathsf{DTIME}(2^{O(n)})$ such that any circuit of size 2^{cn}, for some constant c, must fail to compute the function on $1/2 - 2^{-\Omega(n)}$ fraction of the inputs. After a series of works [1,3,7,6,8], it is now known that this average-case hardness assumption follows from a seemly weaker worst-case hardness assumption: any such circuit must fail to compute the function on at least one input. However, no one can prove this assumption, and in fact proving lower bounds remains a notoriously difficult task in theoretical computer science. So far, lower bound results are only known for very restricted complexity classes.

M.T. Thai and S. Sahni (Eds.): COCOON 2010, LNCS 6196, pp. 13–22, 2010.

On the other hand, in addition to the class BPP, people are also interested in derandomizing other randomized complexity classes. Thus, people have also been looking for hard functions in other complexity classes in order to build PRG against them. We say that a function has hardness δ against a complexity class if any algorithm in the class fails to compute the function on δ fraction of the inputs. Using the approach of Nisan and Wigderson [5], to construct a PRG with a logarithmic-length seed, one needs a function which has hardness $1/2 - 2^{-\Omega(n)}$ against the complexity class. Unfortunately, such hardness results are very rare. The celebrated result of Håstad [2] only achieves a slight weaker hardness bound: the parity function has hardness $1/2 - 2^{-\Omega((n/s)^{1/(d-1)})}$ against any circuit of depth d and size at most 2^s. Consequently, the best PRG against AC^0 (constant-depth polynomial-size) circuits currently known requires a seed of poly-logarithmic, instead of logarithmic, length. In fact, even with depth $d = 2$, Håstad's result can only guarantee a hardness of $1/2 - 2^{-\Omega(n/\log n)}$. We are not aware of any previous result which can prove a hardness of $1/2 - 2^{-\Omega(n)}$ for an explicit function against constant-depth circuits, and existing techniques do not seem powerful enough to achieve this.

As a step toward this direction, we would like to start from a more restricted class of constant-depth circuits. In this paper, we consider constant-depth circuits of linear size. They contain a special type of depth-two circuits known as read-once DNFs, disjunctive normal form formulas in which each input variable is read at most once, and more generally, they also contains DNFs in which each input variable is read only a constant number of times. Note that even against read-once DNFs, which have depth two and size $O(n)$, Håstad's lower bound still only gives a hardness of $1/2 - 2^{-\Omega(n/\log n)}$ for the parity function.

We show that our goal in fact can be achieved, by proving a hardness of $1/2 - 2^{-\Omega(n)}$ for constant-depth linear-size circuits. More precisely, our main result is the following, which we will prove in Section 3.

Theorem 1. *For any constant-depth linear-size circuit* $C : \{0,1\}^n \to \{0,1\}$,

$$\Pr_{x \in \{0,1\}^n} [C(x) \neq \mathsf{PARITY}(x)] \geq 1/2 - 2^{-\Omega(n)}.$$

Our approach follows roughly that of Håstad's, using random restrictions to repeatedly simplify the circuit and reduce the circuit depth. However, to prove a stronger lower bound, we need a more sophisticated way of applying restrictions, which we borrow from [4]. In fact, our Theorem 1 is mainly inspired by the work of [4].

Furthermore, for the special cases of read-once DNFs and read-twice DNFs, we also provide a completely different approach and derive tighter bounds with explicit constants. More precisely, we have the following two theorems, which we will prove in Section 4 and Section 5, respectively.

Theorem 2. *For any read-once* DNF $C : \{0,1\}^n \to \{0,1\}$,

$$\Pr_{x \in \{0,1\}^n} [C(x) \neq \mathsf{PARITY}(x)] \geq 1/2 - 2^{-n}.$$

Theorem 3. *For any read-twice* DNF $C : \{0,1\}^n \to \{0,1\}$,

$$\Pr_{x \in \{0,1\}^n} [C(x) \neq \mathsf{PARITY}(x)] \geq 1/2 - 2^{-(2/15)n}.$$

Note that if we ignore the constant factors, the two theorems above are just special cases of Theorem 1. However, we feel that the completely different proofs we use to prove them may have independent interest and may find applications elsewhere. Moreover, the constants we derive for the two theorems are tighter than those from Theorem 1, and in fact the hardness $1/2 - 2^{-n}$ we prove in Theorem 2 is almost the highest hardness one can have.

2 Preliminaries

For a set S, let $|S|$ denote the number of elements in it. For any $n \in \mathbb{N}$, let $[n]$ denote the set $\{1, \ldots, n\}$. We will consider the parity function $\mathsf{PARITY} : \{0,1\}^n \to \{0,1\}$, defined as $\mathsf{PARITY}(x_1, \cdots, x_n) = x_1 \oplus \cdots \oplus x_n$. In fact, as typically in complexity theory, we will consider a sequence of the parity functions as well as a sequence of circuits on n input bits, for $n \in \mathbb{N}$, but to simplify our presentation, we will not try to make this explicit.

Let $\mathsf{AND}(x_1, \cdots, x_n)$ and $\mathsf{OR}(x_1, \cdots, x_n)$ denote the functions $x_1 \wedge \cdots \wedge x_n$ and $x_1 \vee \cdots \vee x_n$ respectively. A variable x_i and its negation $\neg x_i$ are called literals. A monomial is an AND of literals, and the size of a monomial is defined as the number of literals in it. A disjunctive normal form formula, abbreviated as DNF, is a formula expressed as an OR of monomials.

A DNF is called read-once if every variable (including its negation) appears at most once in the formula, while it is called read-twice if every variable (including its negation) appears at most twice. For a DNF $C = \bigvee_{i \in [m]} C_i$ and for $I \subseteq [m]$, let C_I denote the formula $\bigwedge_{i \in I} C_i$. Let $\textsc{Var}(C_I)$ and $\textsc{Lit}(C_I)$ denote the sets of variables and literals appearing in C_I, respectively. For example, if $C_I = \neg x_1 \wedge x_2 \wedge \neg x_3$, then $\textsc{Var}(C_I) = \{x_1, x_2, x_3\}$ and $\textsc{Lit}(C_I) = \{\neg x_1, x_2, \neg x_3\}$. We say that C_I covers a variable if the variable appears in C_I, and we say that C_I covers C if C_I covers all the variables of C.

We will also consider circuits of depth more than two, which consist of AND/OR gates of unbounded fan-in and fan-out and have variables and have literals as their inputs. We measure the size of a circuit as the number of AND/OR gates it has and the depth of a circuit as the number of gates on the longest path from an input bit to the output gate.

We will follow Håstad's idea [2] of using random restrictions to simplify constant-depth circuits. A restriction ρ on a variable set V is a mapping from V to $\{0, 1, \star\}$, and we say that a variable $x_i \in V$ is fixed by ρ if $\rho(x_i) \in \{0,1\}$ while we say that it remains free if $\rho(x_i) = \star$. Moreover, we say that a circuit is killed by a restriction if the resulting circuit after the restriction becomes a constant function. We will use the following type of random restrictions.

Definition 1. *For $\tau \in [0,1]$, let $\mathcal{R}_\tau$ denote the distribution over restrictions on a set of variables such that a restriction ρ sampled from $\mathcal{R}_\tau$ maps each variable independently to*

$$\begin{cases} \star \text{ with probabilty } \tau, \\ 0 \text{ with probabilty } (1-\tau)/2, \\ 1 \text{ with probabilty } (1-\tau)/2. \end{cases}$$

Finally, we will need the following lemma, which is known as the principle of inclusion and exclusion.

Lemma 1. *Let $B_1, \cdots, B_m$ be subsets of some ground set, and for any $I \subseteq [m]$, let B_I denote the set $\bigcap_{i \in I} B_i$. Then we have $\left|\bigcup_{i \in [m]} B_i\right| = \sum_{\emptyset \neq I \subseteq [m]} (-1)^{|I|+1} |B_I|$.*

3 Proof of Theorem 1

Consider any circuit of depth $d = O(1)$ and size $O(n)$ on a set V of n input variables. Recall that we measure the size of the circuit by its number of AND/OR gates, and we allow a gate to have an unbounded fan-out so that it can be read by more than one other gates. Our goal is to show that the circuit must disagree with the parity function on at least $1/2 - 2^{-\Omega(n)}$ fraction of the inputs. We would like to follow the approach of Håstad [2], which uses random restrictions to reduce the circuit depth repeatedly. However, since Håstad considered constant-depth circuits of polynomial size, he applied random restrictions which leave only $o(n)$ variables free, and consequently he can only show a fraction $1/2 - 2^{-o(n)}$ of disagreement. To show our result, we must have $\Omega(n)$ free variables left, so a more careful way of restricting input variables is needed, which we borrow from [4].

Before we start, let us convert the circuit into the following form. First, we arrange its gates into d layers, in which gates at the same layer are of the same type and receive inputs only from the layer below them. This can be achieved by adding at most d dummy gates of fan-in one for each gate, so the resulting circuit still has linear size. Assume without loss of generality that the bottom layer consists of OR gates. For our purpose later, we modify the circuit further by adding a layer of n AND gates of fan-in one as the bottom layer, one for each input variable, so that now the circuit depth becomes $d+1$ and the circuit size remains linear, while the bottom two layers form a collection of DNFs. Let us denote this new depth $d+1$ circuit by C, and we will work on this circuit C from now on. Note that it satisfies the following property.

Definition 2. *We call a layered circuit an (α, β)-circuit if each of its bottom gates has fan-in at most α and each variable is read a total of at most β times by all the bottom gates.*

Our key lemma is the following, which generalizes a lemma in [4], and we will prove it later in Subsection 3.1.

Lemma 2. *Consider any (α, β)-circuit of size $m = O(n)$ on a set V of n input variables, with constant α, β. Suppose we apply a random restriction ρ from $\mathcal{R}_{1/2}$ on V, let $B_{\alpha,\beta}(\rho)$ be the event that there exists some $V' \subseteq V$ of size $\Omega(n)$ such that every bottom DNF or CNF of the circuit which is not killed by ρ has at most $O(1)$ free variables in V'. Then we have $\Pr_\rho[\neg B_{\alpha,\beta}(\rho)] \leq 2^{-\Omega(n)}$.*

Remark 1. As mentioned in Section 2, we actually consider a sequence of circuits on n input variables, for $n \in \mathbb{N}$, in the lemma above, so all the constants there are with respect to this parameter n.

We would like to use this lemma repeatedly in d iterations to reduce the circuit depth from $d+1$ down to 2. Initially, we have an (α_1, β_1)-circuit $C_1 = C$ of size $O(n)$ and depth $d+1$ on a set $V_1 = V$ of n input variables, with $\alpha_1 = \beta_1 = 1$, which is why we added a layer of n AND gates of fan-in one to the original circuit to obtain the circuit C. Then we will proceed in d iterations as follows.

In iteration i, for i from 1 to $d-1$, we start with an (α_i, β_i)-circuit C_i of depth $d+2-i$ on a variable set V_i, and we do the following. Let us assume that the bottom two layers of C_i consists of DNFs; the case of CNFs can be handled similarly, by switching between AND gates and OR gates. First, we apply a random restriction $\rho_{i,1}$ from $\mathcal{R}_{1/2}$ on the variable set V_i of C_i. Let us call $\rho_{i,1}$ good if the event $B_{\alpha_i,\beta_i}(\rho_{i,1})$ in Lemma 2 happens, so that there exists a subset $V_i' \subseteq V_i$ of size $\Omega(|V_i|) = \Omega(n)$ such that each bottom DNF which is not killed by $\rho_{i,1}$ has only a constant number of free variables in V_i'. Now consider any good $\rho_{i,1}$. Let us fix all those variables in $V_i \setminus V_i'$ by applying a random restriction $\rho_{i,2}$ from $\mathcal{R}_0$ on them. Then for any $\rho_{i,2}$, any bottom DNF in the resulting circuit depends only on a constant number of free variables, so we can convert each bottom DNF into a CNF of constant size and merge it with the AND gate above it. As a result, we reduce the circuit depth by one, while still having $\Omega(n)$ free variables. Note that now the bottom layer of the new circuit consists of $O(n)$ OR gates, each with a constant fan-in, because $O(n)$ such CNFs were produced, each consisting of a constant number of such OR gates. At this point, we may still not be able to apply Lemma 2 again, since each free variable may appear in more than a constant number of these OR gates. Fortunately, a Markov inequality shows that only a constant fraction of the variables can appear in more than some constant β_{i+1} number of the bottom OR gates. Let us fix those variables by applying a random restriction $\rho_{i,3}$ from $\mathcal{R}_0$ on them, and this completes iteration i. In summary, if $\rho_{i,1}$ is good, then for any $\rho_{i,2}, \rho_{i,3}$, we end up with an $(\alpha_{i+1}, \beta_{i+1})$-circuit C_{i+1} of depth $d+1-i$ and size $O(n)$, for some constant $\alpha_{i+1}, \beta_{i+1}$, on a set V_{i+1} of $\Omega(n)$ free variables.

Now suppose $\rho_{i,1}$ is good for every iteration $i \in [d-1]$. Then after $d-1$ iterations, we obtain an (α_d, β_d)-circuit C_d of depth 2 and size $O(n)$, for some constant α_d, β_d, on a set V_d of $\Omega(n)$ free variables. We can apply a lower bound result of Håstad [2] to show that C_d disagrees with the parity function on at least $1/2 - 2^{-\Omega(n)}$ fraction of the inputs. In fact, we can also prove such a lower bound by repeating our process for one more iteration, with random restrictions $\rho_{d,1}, \rho_{d,2}, \rho_{d,3}$ (actually, $\rho_{d,3}$ is not needed). Again, call the random restriction $\rho_{d,1}$ good if the event $B_{\alpha_d,\beta_d}(\rho_{d,1})$ in Lemma 2 happens. When this happens, for any $\rho_{d,2}, \rho_{d,3}$, we obtain a circuit which depends only on a constant number of variables in a set V_{d+1} of $\Omega(n)$ free variables.

Thus, after d iterations, we have $3d$ restrictions and let $\boldsymbol{\rho}$ denote the resulting restriction by composing all these restrictions together. Call $\boldsymbol{\rho}$ good if $\rho_{i,1}$ is good for every $i \in [d]$. We have shown that when $\boldsymbol{\rho}$ is good, there are

$\Omega(n)$ free variables left but the resulting circuit, which we denote as $C_{\boldsymbol{\rho}}$, depends only on a constant number of them. Thus, when $\boldsymbol{\rho}$ is good, we have $\Pr_z [C_{\boldsymbol{\rho}}(z) \neq \mathsf{PARITY}_{\boldsymbol{\rho}}(z)] = 1/2$, where $\mathsf{PARITY}_{\boldsymbol{\rho}}$ denotes the parity function on V restricted by $\boldsymbol{\rho}$. For our convenience, let us also imagine a restriction $\rho_{0,1}$ on V before we start which leaves every variable free, and we always call it good. According to Lemma 2, we know that if $\rho_{0,1}, \rho_{1,1}, \ldots, \rho_{i-1,1}$ are all good, then the probability that $\rho_{i,1}$ is not good is at most $2^{-\Omega(n)}$. This implies that $\Pr_{\boldsymbol{\rho}} [\boldsymbol{\rho} \text{ is not good}] \leq d \cdot 2^{-\Omega(n)} = 2^{-\Omega(n)}$.

Finally, observe that the distribution of sampling an input x uniformly from $\{0,1\}^n$ is identical to that of sampling the restriction $\boldsymbol{\rho}$ first and then setting its free variables to a random binary string z of length $|\boldsymbol{\rho}^{-1}(\star)|$. As a result, we have $\Pr_x [C(x) \neq \mathsf{PARITY}(x)] = \Pr_{\boldsymbol{\rho},z} [C_{\boldsymbol{\rho}}(z) \neq \mathsf{PARITY}_{\boldsymbol{\rho}}(z)] \geq \Pr_{\boldsymbol{\rho}} [\boldsymbol{\rho} \text{ is good}] \cdot \Pr_{\boldsymbol{\rho},z} [C_{\boldsymbol{\rho}}(z) \neq \mathsf{PARITY}_{\boldsymbol{\rho}}(z) \mid \boldsymbol{\rho} \text{ is good}] \geq \left(1 - 2^{-\Omega(n)}\right) \cdot 1/2 = 1/2 - 2^{-\Omega(n)}$. This proves Theorem 1.

3.1 Proof of Lemma 2

Let us first consider the case that the bottom layer of the circuit consists of AND gates so that we can see the bottom two layers of the circuit as a collection Φ of DNFs. These DNFs in Φ have the same set V of input variables and they may share some bottom AND gates.

Let A denote the event that $|\rho^{-1}(\star)| \geq 0.4n$, and let B denote the event that there exists some $S \subseteq V$ of size at most $\ell = 0.1n$ such that any DNF in Φ which is not killed by ρ has at most r free variables outside of S, for some large enough constant r. It is easy to see that $B_{\alpha,\beta}(\rho)$ happens if $A \wedge B$ happens, by letting $V' = \rho^{-1}(\star) \setminus S$, which implies that $\Pr_\rho [\neg B_{\alpha,\beta}(\rho)] \leq \Pr_\rho [\neg A \vee \neg B] \leq \Pr_\rho [\neg A] + \Pr_\rho [\neg B]$. By Chernoff's bound, we know that $\Pr_\rho [\neg A] \leq 2^{-\Omega(n)}$. Next, we bound the probability $\Pr_\rho [\neg B]$. For this, we need the following.

Claim. Suppose $\neg B$ happens. Then there exists a subset $\Psi \subseteq \Phi$ of size at most $\lceil \ell/(r+1) \rceil$ reading at least ℓ input variables such that no DNF in Ψ is set to 1 by ρ.

Proof. Suppose $\neg B$, so that for any $S \subseteq V$ of size at most ℓ, there is a DNF in Φ which is not killed by ρ and has at least $r+1$ free variables outside of S. Then following [4], we can find the subset $\Psi \subseteq \Phi$ needed by the claim as follows. Starting with $\Psi = \emptyset$ and $S = \emptyset$, if there is any $\phi \in \Phi$ which is not set to 1 by ρ and has at least $r+1$ free variables outside of S, we update Ψ and S by including ϕ in Ψ and including those free variables of ϕ in S, and then we repeat the process. Note that we can always find such a ϕ as long as $|S| \leq \ell$, and this process will stop within $\lceil \ell/(r+1) \rceil$ steps. The final collection Ψ is what we want.

Consider any fixed $\Psi \subseteq \Phi$. Note that if none of the DNF in Ψ is set to 1 by ρ, then none of the bottom AND gates in Ψ is set to 1 by ρ. Although these AND gates may share input variables, we claim that there is actually a large subset of them which share no input variables. This is because Ψ has at least ℓ input variables, each read by at most β bottom AND gates, and each bottom AND gate in Ψ reads at most α input variables. Thus, Ψ must have at least $\ell/(\alpha\beta)$ bottom AND gates which share no input variables, and hence the probability that none of its bottom AND gates is set to 1 by ρ is at most $(1-4^{-\alpha})^{\ell/(\alpha\beta)} \leq e^{-\ell/(\alpha\beta 4^{\alpha})}$. Then by a union bound, over Ψ of size at most $\lceil \ell/(r+1)\rceil$, we have $\Pr_{\rho}[\neg B] \leq \sum_{j=1}^{\lceil \ell/(r+1)\rceil}\binom{m}{j}e^{-\ell/(\alpha\beta 4^{\alpha})} \leq 2^{-\Omega(n)}$, for a large enough constant r. As a result, we have $\Pr_{\rho}[\neg B_{\alpha,\beta}(\rho)] \leq \Pr_{\rho}[\neg A] + \Pr_{\rho}[\neg B] \leq 2^{-\Omega(n)}$, which proves Lemma 2 for the case that Φ is a collection of DNFs. For the case that Φ is a collection of CNFs, the proof is almost identical. The only difference is that now we bound the probability that none of the bottom OR gates is set to 0 by ρ, and it is not hard to see that the same analysis also works.

4 Proof of Theorem 2

Note that to prove that $\Pr_{x\in\{0,1\}^n}[C(x) \neq \mathsf{PARITY}(x)] \geq 1/2 - \delta$, for $0 \leq \delta \leq 1/2$, it suffices to prove that

$$\Pr_{x\in\{0,1\}^n}[C(x) = \mathsf{PARITY}(x)] - \Pr_{x\in\{0,1\}^n}[C(x) \neq \mathsf{PARITY}(x)] \leq 2\cdot\delta,$$

which becomes $(p_{1,1}+p_{0,0}) - (p_{1,0}+p_{0,1}) \leq 2\cdot\delta$ by using the notation that $p_{a,b} = \Pr_{x\in\{0,1\}^n}[C(x) = a \wedge \mathsf{PARITY}(x) = b]$, for $a, b \in \{0,1\}$. Since PARITY outputs 1 for exactly one half of its inputs, we have $p_{1,0}+p_{0,0} = 1/2 = p_{0,1}+p_{1,1}$, which implies that $(p_{1,1}+p_{0,0})-(p_{1,0}+p_{0,1}) = 2(p_{1,1}-p_{1,0})$, and our goal becomes that of proving $p_{1,1} - p_{1,0} \leq \delta$.

Next, we will provide a generic bound of $p_{1,1} - p_{1,0}$ for general DNFs, which will be specialized later in this and the next section to read-once and read-twice DNFs, respectively. Consider any general DNF $C = \bigvee_{i\in[m]} C_i$, where $C_1, \cdots, C_m$ are its monomials. Recall that for $I \subseteq [m]$, $C_I = \bigwedge_{i\in I} C_i$. For $b \in \{0,1\}$, let us consider the following set:

- $B_I^b = \{x \in \{0,1\}^n : C_I(x) = 1 \wedge \mathsf{PARITY}(x) = b\}$.

For $I = \{i\}$, we will simply write B_i^b for $B_{\{i\}}^b$, and observe that $B_I^b = \bigcap_{i\in I} B_i^b$. Then we have

$$\begin{aligned} p_{1,1} - p_{1,0} &= \Pr_{x\in\{0,1\}^n}\left[x \in \bigcup_{i\in[m]} B_i^1\right] - \Pr_{x\in\{0,1\}^n}\left[x \in \bigcup_{i\in[m]} B_i^0\right] \\ &= 2^{-n}\cdot\left(\left|\bigcup_{i\in[m]} B_i^1\right| - \left|\bigcup_{i\in[m]} B_i^0\right|\right) \\ &= 2^{-n}\cdot\sum_{\emptyset\neq I\subseteq[m]} (-1)^{|I|+1}\left(|B_I^1| - |B_I^0|\right), \end{aligned} \tag{1}$$

where the last equality follows from Lemma 1. It remains to bound the last equation above, for which we need the following.

Claim. For any $I \subseteq [m]$, $|B_I^1| - |B_I^0| = 0$ if C_I does not cover C, and $\big||B_I^1| - |B_I^0|\big| \leq 1$ otherwise.

Proof. First, suppose that C_I does not cover C. Note that for any x such that $C_I(x) = 1$, we also have $C_I(x') = 1$ for x' obtained from x by flipping the first variable not covered by C_I. On the other hand, since x and x' differ in exactly one bit, we have $\mathsf{PARITY}(x) \neq \mathsf{PARITY}(x')$. This implies the existence of a bijection between B_I^1 and B_I^0, so we have $|B_I^1| - |B_I^0| = 0$.

Next, suppose that C_I covers C. This means that all the variables or their negation appear in C_I, while C_I is the conjunction of them. Thus, we have $|B_I^1|, |B_I^0| \leq 1$ and $\big||B_I^1| - |B_I^0|\big| \leq 1$.

Applying the bound in this claim to the equation in (1), we have $p_{1,1} - p_{1,0} \leq 2^{-n} \cdot |\{I \subseteq [m] : C_I \text{ covers } C\}|$, which implies the following lemma.

Lemma 3. *For any general* DNF $C = \bigvee_{i \in [m]} C_i$, *we have*

$$\Pr_{x \in \{0,1\}^n} [C(x) \neq \mathsf{PARITY}(x)] \geq 1/2 - 2^{-n} \cdot |\{I \subseteq [m] : C_I \text{ covers } C\}|.$$

Note that the above lemma holds for any general (not necessarily read-once or read-twice) DNF. Now consider any read-once DNF C, and note that C can only be covered by at most one C_I, with $I = [m]$. Then Theorem 2 follows immediately from Lemma 3.

5 Proof of Theorem 3

To prove the theorem, we would like to apply Lemma 3 in the previous section. Consider any read-twice DNF C, and we would like to bound the number of I's such that C_I covers C. First note that we can assume without loss of generality that C is not a constant function because otherwise we have $\Pr_{x \in \{0,1\}^n} [C(x) \neq \mathsf{PARITY}(x)] = 1/2$. Next, we need the following lemma for transforming C into some standard form, which we will prove in Subsection 5.1.

Lemma 4. *Any non-constant read-twice* DNF C *can be transformed into an equivalent read-twice* DNF $\bigvee_{i \in [m]} C_i$ *such that for some* $t \leq m$ *the following two conditions hold:*

- *For any* $i \leq t$, $|\mathrm{VAR}(C_i)| = 1$ *and* C_i *is not covered by* C_j *for any* $j \neq i$.
- *For any* $j > t$, $|\mathrm{VAR}(C_j)| \geq 2$.

By this lemma, we will consider C in its standard form $\bigvee_{i \in [m]} C_i$ from now on. Let $J = [m] \setminus [t]$. It is easy to see that for $I \subseteq [m]$, if C_I covers C, then $[t] \subseteq I$, which implies that $|\{I \subseteq [m] : C_I \text{ covers } C\}| = \big|\{I \subseteq J : C_I \text{ covers } \bigvee_{i \in J} C_i\}\big|$. This means that it suffices to focus on the covering of $\bigvee_{i \in J} C_i$. Let us drop those t monomials of size one, and let us also use C to denote the resulting circuit and m the resulting number of monomials (or simply assume that $t = 0$).

Now all the monomials of C have size at least two, and we have $m \leq n$ because $2m \leq \sum_{i\in[m]} |\text{VAR}(C_i)| = \sum_{k\in[n]} |\{i \in [m] : x_k \in \text{VAR}(C_i)\}| \leq 2n$.

To bound the number of I's such that C_I covers C, we bound the probability over a randomly selected I such that C_I covers C. Here we select $I \subseteq [m]$ in the way that each $i \in [m]$ is included in I independently with probability 1/2. Our key lemma is the following, which we will prove in Subsection 5.2.

Lemma 5. *Given any read-twice* DNF $C = \bigvee_{i\in[m]} C_i$ *such that* $|\text{VAR}(C_i)| \geq 2$ *for any* $i \in [m]$, *we have* $\Pr_{I\subseteq[m]} [C_I \textit{ covers } C] \leq 2^{-(4/15)m+(2/15)n}$.

By Lemma 5 and Lemma 3, we have $\Pr_{x\in\{0,1\}^n} [C(x) \neq \mathsf{PARITY}(x)] \geq 1/2 - 2^{-n} \cdot \left(2^m \cdot 2^{-(4/15)m+(2/15)n}\right) = 1/2 - 2^{-(13/15)n+(11/15)m} \geq 1/2 - 2^{-(2/15)n}$, where the last line uses the fact that $m \leq n$. This proves Theorem 3.

5.1 Proof of Lemma 4

First, we rearrange the monomials of C and remove any duplicate monomial (which clearly does not change the functionality of C) to get a DNF $\bigvee_{i\in[\bar{m}]} C_i$, for some $\bar{m} \in \mathbb{N}$, such that for some t, we have $|\text{VAR}(C_i)| = 1$ for any $i \leq t$ and $|\text{VAR}(C_i)| \geq 2$ for any $i > t$.

Note that for any distinct $i, j \leq t$, we can assume that $C_i \neq \neg C_j$ because otherwise C would be a constant (always one) function. Therefore we have that for any $i \leq t$, C_i is not covered by C_j for any distinct $j \leq t$. It remains to process those C_j's, for $j > t$, so that they do not cover any C_i, for $i \leq t$.

Consider any C_j, with $j > t$, which by definition has $|\text{VAR}(C_j)| \geq 2$. While C_j covers some C_i with $i \leq t$, we do the following according to two cases:

(1) If $C_i \in \text{LIT}(C_j)$, we simply discard C_j from C. This does not change the functionality of C because for any input x such that $C_j(x) = 1$, the resulting formula still evaluates to $C_i(x) = 1$.

(2) If $\neg C_i \in \text{LIT}(C_j)$, we remove the literal $\neg C_i$ from C_j. This again does not change the functionality of C, because for any input x on which the initial and the resulting C_j evaluate differently, we must have $C_i(x) = 1$, which implies that the initial and the resulting DNFs both evaluate to 1.

When C_j does not cover any C_i with $i \leq t$, we keep C_j in the resulting DNF and process the next remaining monomial. Note that after we process all the monomials, the resulting DNF satisfies the two conditions of the lemma. Moreover, it is a read-twice DNF with the same functionality as the initial one.

5.2 Proof of Lemma 5

For $k \in [n]$, let A_k denote the event, over a randomly selected $I \subseteq [m]$, that the variable x_k is covered by C_I. By definition,

$$\Pr_{I\subseteq[m]} [C_I \text{ covers } C] = \Pr_{I\subseteq[m]} \left[\bigwedge_{k\in[n]} A_k \right].$$

Note that for any $k \in [n]$, we have $\Pr_{I \subseteq [m]}[A_k] \le 1 - (1/2)^2 = 3/4$, because each variable is covered by at most two monomials and each monomial is selected independently with probability 1/2. We would like to show the existence of a large set $V \subseteq [n]$ such that the events A_k's for $k \in V$ are all independent from each other, as this would give a small bound for $\Pr_{I \subseteq [m]}\left[\bigwedge_{k \in [n]} A_k\right] \le \prod_{k \in V} \Pr_{I \subseteq [m]}[A_k] \le (3/4)^{|V|}$.

We form the set V by adding variables to it in the following iterative way, and we also use another set R to keep track of the remaining variables:

– Initially, we have $V = \emptyset$ and $R = [n]$. Then while there exists some $k \in R$, we include k in V, and for any monomial C_i which covers the variable x_k, we remove from R any j such that the variable x_j is covered by C_i.

Note that after the loop terminates, we have at most $2|V|$ monomials which cover all the occurrences of each variable x_k with $k \in V$ and cover each variable x_j with $j \in [n] \setminus V$ at least once. Since each variable can be covered at most twice and each monomial has size at least two, there can be at most $(n - |V|)/2$ monomials left, and thus $2|V| + (n - |V|)/2 \ge m$. This implies that $|V| \ge (2/3)m - (1/3)n$, and we have $\Pr_{I \subseteq [m]}[C_I \text{ covers } C] \le (3/4)^{2m/3 - n/3} \le 2^{-(4/15)m + (2/15)n}$.

References

1. Babai, L., Fortnow, L., Nisan, N., Wigderson, A.: BPP has subexponential time simulations unless EXPTIME has publishable proofs. Computational Complexity 3(4), 307–318 (1993)
2. Håstad, J.: Computational limitations for small depth circuits. Ph.D. thesis, MIT Press (1986)
3. Impagliazzo, R., Wigderson, A.: P=BPP if E requires exponential circuits: derandomizing the XOR lemma. In: Proceedings of the 29th ACM Symposium on Theory of Computing, pp. 220–229 (1997)
4. Miltersen, P.B., Radhakrishnan, J., Wegener, I.: On converting CNF to DNF. Theoretical Computer Science 347(1-2), 325–335 (2005)
5. Nisan, N., Wigderson, A.: Hardness vs. randomness. Journal of Computer and System Sciences 49(2), 149–167 (1994)
6. Shaltiel, R., Umans, C.: Simple extractors for all min-entropies and a new pseudorandom generator. In: Proceedings of the 42nd Annual IEEE Symposium on Foundations of Computer Science, pp. 648–657 (2001)
7. Sudan, M., Trevisan, L., Vadhan, S.: Pseudorandom generators without the XOR lemma. Journal of Computer and System Sciences 62(2), 236–266 (2001)
8. Umans, C.: Pseudo-random generators for all hardnesses. Journal of Computer and System Sciences 67(2), 419–440 (2003)

A K-Provers Parallel Repetition Theorem for a Version of No-Signaling Model

Ricky Rosen

Department of Computer Science, Tel-Aviv University,
Tel-Aviv 69978, Israel

Abstract. The parallel repetition theorem states that for any two provers one round game with value at most $1-\epsilon$ (for $\epsilon < 1/2$), the value of the game repeated n times in parallel is at most $(1-\epsilon^3)^{\Omega(n/\log s)}$ where s is the size of the answers set [Raz98],[Hol07]. It is not known how the value of the game decreases when there are three or more players. In this paper we address the problem of the error decrease of parallel repetition game for k-provers where $k > 2$. We consider a special case of the No-Signaling model and show that the error of the parallel repetition of k provers one round game, for $k > 2$, in this model, decreases exponentially depending only on the error of the original game and on the number of repetitions. There were no prior results for k-provers parallel repetition for $k > 2$ in any model.

1 Introduction

In a k provers one round game there are k *provers* and a *verifier*. The verifier selects randomly $(x_1, \ldots, x_k) \in X_1 \times \cdots \times X_k$, a question for each prover, according to some distribution $\mathrm{P}_{X_1 \ldots X_k}$ where X_i is the questions set of prover i where $i \in 1, \ldots, k$. Each prover knows only the question addressed to her, prover 1 knows only x_1 and prover i knows only x_i. The provers cannot communicate during the transaction. The provers send their answers to the verifier, $a_i \in A_i$ where A_i is the answers set of prover i. The verifier evaluates an acceptance predicate $V(x_1, \ldots, x_k, a_1, \ldots, a_k)$ and accepts or rejects based on the outcome of the predicate. The acceptance predicate as well as the distribution of the questions are known in advance to the provers. The provers answer the questions according to a strategy. The strategy of the provers is also called a protocol. The type of the protocol determines the type of model. In the *classic model* the provers' strategy is a k tuple of functions, one for each prover, where each function is from her questions to her answers, i.e., for all i, $f_{a_i} : X_i \to A_i$. We also call this model MIP$(k, 1)$. In the *No-Signaling model* the provers' strategy is a k tuple of functions, one for each prover, where each function is a random function from the questions of **all** the provers to her answers but in a way that does not reveal any information about the other provers' questions. We denote this model by $MIP^{\diamond}(k, 1)$. The no signaling condition ensures that the answer of each prover given its question is independent of the questions of the other provers (but not of the other provers answers). This definition is the natural generalization of the

M.T. Thai and S. Sahni (Eds.): COCOON 2010, LNCS 6196, pp. 23–33, 2010.

no-signaling model of two provers one round game to no signaling model of k provers one round game. However, in this paper we generalize the no-signaling model in a different way. In this variation, no $k-1$ provers together can signal to the remaining prover. In the standard model one also demands that no prover can signal to any composite system of at most $k-1$ provers. To get a sense of such games, let us consider a generalization of the CHSH [CHSH69], [PR94] game for 3 players. Let us denote the three provers by Alice, Bob and Charlie. The verifier chooses uniformly at random three questions $x, y, z \in \{0,1\}$ and sends x to Alice, y to Bob and z to Charlie. The provers sends their answers, denoted by a, b, c and the verifier accepts if and only if $x \wedge y \wedge z = a \oplus b \oplus c$. We will consider the following strategy: if $x \wedge y \wedge z = 0$ then (a, b, c) is distributed uniformly over $\{(0,0,0),(1,0,1),(1,1,0),(0,1,1)\}$. If $x \wedge y \wedge z = 1$ then (a, b, c) is distributed uniformly over $\{(1,0,0),(0,1,0),(0,0,1),(1,1,1)\}$. This is a no-signaling strategy since for every question of Alice, knowing her answer to that question does not reveal any information about Bob's or Charlie's questions. Similar argument apply to Bob and Charlie.

The *value* of the game is the maximum of the probability that the verifier accepts, where the maximum is taken over all the provers strategies. More formally, the value of the game is:

$$\max_{f_{a_1},\ldots,f_{a_k}} \mathbb{E}_{x_1,\ldots,x_k}\left[V\left(x_1,\ldots,x_k, f_{a_1}(x_1),\ldots,f_b(x_k)\right)\right]$$

where the expectation is taken with respect to the distribution $P_{X_1 \ldots X_k}$. We denote the value of the game in the classical model by $w(G)$ and the value of the game in the No-Signaling model by $\omega^{\diamondsuit}(\mathrm{G})$.

Roughly speaking, the *n-fold parallel repetition* of the game, denoted by $(\mathrm{G}^{\otimes n})$, is playing the game n times in parallel, i.e. the verifier sends n questions to each prover and receives n answers from each prover. The verifier evaluates the acceptance predicate on each game and accepts if and only if all the predicates were true. Obviously $\omega^{\diamondsuit}(\mathrm{G}^{\otimes n}) \leq \omega^{\diamondsuit}(\mathrm{G})$ but one might expect that $\omega^{\diamondsuit}(\mathrm{G}^{\otimes n}) = (\omega^{\diamondsuit}(\mathrm{G}))^n$. However, although the verifier treats each game of the n games independently, the provers may not. The answer of each question addressed to a prover may depend on all the questions addressed to that prover. There are examples for the MIP$(2,1)$ model [For89], [FL92], [Raz98], [FV02], [Hol07], and for the $MIP^{\diamondsuit}(2,1)$ model [Hol07] for which the $\omega^{\diamondsuit}(\mathrm{G}^{\otimes n}) = (\omega^{\diamondsuit}(\mathrm{G}))$ for small n. Raz provided an example for MIP$(k,1)$ where $\omega^{\diamondsuit}(\mathrm{G}^{\otimes k}) = (\omega^{\diamondsuit}(\mathrm{G}))$.

Another model of k-provers one round game is the *quantum model*, denote by MIP$^*(k,1)$, in which the provers have a joint quantum state and so the function of each prover is from the questions of all the provers to her answers but only for such functions that can be implemented by an entangled quantum state.

Clearly, the value of a game in the MIP$(k,1)$ model is less or equal to the value of the game in the $\omega^*(\mathrm{G})$ model which is less or equal to the value of the game in $MIP^{\diamondsuit}(k,1)$ model. However, this bound does not imply a bound for the parallel repetition of the MIP$(k,1)$ model since there are games for which the value of the game in the MIP$(k,1)$ model is strictly less than 1 but the value of the game in $MIP^{\diamond}(k,1)$ model is 1, therefore implying only the trivial bound. If the value of the game in $MIP^{\diamond}(k,1)$ model is strictly less than 1 then we obtain a non trivial bound on the value of the game for the MIP$(k,1)$ model.[1]

Many fundamental questions related to the MIP$(2,1)$ model have been answered by now, but there was no known upper bounds on MIP$(k,1)$ for $k > 2$ in any model. However, it was conjectured that the error decreases exponentially. Although this model of no-signaling is not the natural no-signaling model, it allows us to obtain a first upper bound result for the MIP$(k,1)$ for $k > 2$ for special types of games[2] and it raises explicitly the questions for what models one can look at more than 2 provers.

1.1 Related Work

A series of papers deals with the nature of the error decrease of parallel repetition game [CCL92], [Fei91], [FK94], [Raz98], [Hol07], [Rao08]. The breakthrough result was done by Raz's [Raz98]. In his celebrated result, Raz proved that the error of MIP$(2,1)$ decreases exponentially depending also on the size of the answer set. Holenstein [Hol07] lately simplified this result and revealed new insights on the nature of the problem. Holenstein also improved the constants to give a tighter bound on the error of the n parallel repetition. Holenstein also proved that in the $MIP^{\diamond}(2,1)$ model, the error decreases exponentially and does not depend on the size of the answer set.

2 Our Results

A 3-provers one round No-Signaling game

$$\mathrm{G} = (X, Y, Z, A, B, C, P_{XYZ}, Q)$$

is an object consisting on three finite question sets X, Y, Z, a probability measure P_{XYZ} on XYZ

$$\mathrm{P}_{XYZ} : X \times Y \times Z \to \mathbb{R}^{+},$$

answer sets A, B, C and an acceptance predicate

$$Q : X \times Y \times Z \times A \times B \times C \to \{0,1\}.$$

A No-Signaling protocol is a set of three No-Signaling functions: the function

$$p_1 : X \times Y \times Z \times R \to A,$$

the function

$$p_2 : X \times Y \times Z \times R \to B$$

[1] It decreases exponentially and does not depend on the size of the answers support.

[2] Namely, games where the value of the game played according to a no-signaling protocol is strictly less than 1.

and the function:

$$p_3 : X \times Y \times Z \times R \to C.$$

A function

$$p_1 : X \times Y \times Z \times R \to A$$

is a No-Signaling function if:

$\forall x \in X, a \in A, y, y' \in Y, z, z' \in Z \ \Pr_R[p_1(x, y, z, R) = a] = \Pr_R[p_1(x, y', z', R) = a].$

For k provers, a function p_i is no-signaling if:

$\forall x \in X_i, a \in A_i, x_j, x'_j \in X_j \text{ for } j \neq i \ \Pr_R[p_i(x_1, \ldots x_{i-1}, x, x_{i+1}, \ldots x_k, R) = a] =$

$\Pr_R[p_i(x'_1, \ldots x'_{i-1}, x, x'_{i+1}, \ldots x'_k, R) = a].$

Definition 2.1 (value of the Game). *The* value of the game *is defined as:*

$\omega^{\diamond}(\mathrm{G}) = \max_{p_1, p_2, p_3} \mathbb{E}_{XYZR}[Q(X, Y, Z, p_1(X, Y, Z, R), p_2(X, Y, Z, R), p_3(X, Y, Z, R))]$

where the expectation is taken with respect to P_{XYZ} *and the maximum is taken over all No-Signaling protocols.*

The n-fold parallel repetition 3-provers game $\mathrm{G}^{\otimes n}$ consists of the sets X^n, Y^n, Z^n, a probability measure on those sets, $P^{\otimes n}_{X^n Y^n Z^n}$ where

$$P^{\otimes n}_{X^n Y^n Z^n}(x^n, y^n, z^n) = \prod_{i=1}^{n} P_{XYZ}(x_i, y_i, z_i)$$

for $x^n \in X^n,\ y^n \in Y^n,\ z^n \in Z^n$. It also consists of the sets A^n, B^n, C^n and an accepting predicate $Q^{\otimes n}$ where

$$Q^{\otimes n}(x^n, y^n, z^n, a^n, b^n, c^n) = \bigwedge_{i=1}^{n} Q(x_i, y_i, z_i, a_i, b_i, c_i)$$

for $x^n \in X^n,\ y^n \in Y^n,\ z^n \in Z^n, a^n \in A^n,\ b^n \in B^n,\ c^n \in C^n$. The n-fold parallel repetition for k-provers game (for every $k > 3$) is defined in the obvious way and therefore omitted.

For a game G^n and a strategy (p_1, p_2, p_3) we define:

$$P^{\otimes n}_{X^n Y^n Z^n A^n B^n C^n}(x^n, y^n, z^n, a^n, b^n, c^n) \triangleq$$
$$P^{\otimes n}_{X^n Y^n Z^n}(x^n, y^n, z^n) \cdot \Pr[p_1(x^n) = a^n \ \wedge \ p_2(y^n) = b^n \ \wedge \ p_3(y^n) = c^n]$$

We can now present our theorem:

Theorem 2.2. *For every* $k \geq 2$*, every positive integer* n*, all games* $\mathrm{G} = (X, Y, Z, A, B, C, P_{XYZ}, Q)$*, played in the* $\mathrm{MIP}^{\diamond}(k, 1)$ *model satisfy:*

$$\omega^{\diamond}(\mathrm{G}^{\otimes n}) \leq \left(1 - \frac{(1 - \omega^{\diamond}(\mathrm{G}))^2}{100(1 + 4k)^2}\right)^n$$

Holenstein [Hol07] proved the theorem for $k = 2$. For simplicity, we will prove the theorem for $k = 3$ and generalize it for $k > 3$ in the full version of the paper.

3 Preliminaries

We denote an n-dimensional vector by superscripts n, e.g., $A^n = (A_1, \dots, A_n)$ where A_i is its ith entry and $A^{-i} = (A_1, \dots, A_{i-1}, A_{i+1}, \dots, A_n)$. The statistical difference between two probability distributions P and Q, defined over then same sample space Ω, is $\|\mathrm{P} - \mathrm{Q}\|_1 = 1/2 \sum_{x \in \Omega} |\mathrm{P}(x) - \mathrm{Q}(x)|$.

Definition 3.1 (Divergence). *We define the Kullback Leibler divergence, also called the* informational divergence *or simply* divergence. *Let* P *and* Q *be two probability measures defined on the same sample space, Ω. The divergence of* P *with respect to* Q *is:*

$$\mathcal{D}(\mathrm{P}\|\mathrm{Q}) = \sum_{x \in \Omega} \mathrm{P}(x) \log \frac{\mathrm{P}(x)}{\mathrm{Q}(x)}$$

where $0 \log \frac{0}{0}$ *is defined to be* 0 *and* $p \log \frac{p}{0}$ *where* $p \neq 0$ *is defined to be* ∞.

Vaguely speaking, we could think of the divergence as a way to measure the information we gained by knowing that a random variable is distributed according to P rather than Q. This indicates how *far* Q is from P, if we don't gain much information then the two distributions are very *close* in some sense.

4 Proof Sketch

We closely follow Holenstein's proof [Hol07] and generalize it for three provers one round No-Signaling game. Furthermore, in full version of the paper, we generalization the result for every $k \geq 3$ provers.

Fixing a No-Signaling strategy (p_1, p_2, p_3), we define W_i to be the probability of winning game i. Using this notation the parallel repetition is an upper bound for $\Pr[W_1 \wedge W_2 \ldots \wedge W_n]$. Since

$$\Pr[W_1 \wedge \ldots \wedge W_n] = \Pr[W_{i_1}] \Pr[W_{i_2} | W_{i_1}] \cdots \Pr[W_{i_n} | W_{i_1} W_{i_2} \cdots W_{i_{n-1}}]$$

we will upper bound $\Pr[W_{i_m} | W_{i_1} \ldots W_{i_{m-1}}]$ for every $m \leq n^\delta$ (where δ is just some small constant).

We will show that conditioning on the event of winning all $m-1$ games, i.e., conditioning on $W_{i_1} \ldots W_{i_{m-1}}$, there is a game i_m on which the probability of winning that game is at most $\omega^\diamond(\mathrm{G}) + \epsilon$. The way to do so, is by showing that given a No-Signaling strategy for the game $\mathrm{G}^{\otimes n}$ and conditioned on winning $m-1$ games, the provers can use this strategy to obtain a No-Signaling strategy for G. To obtain a strategy for G we will need two lemmas. The first lemma, Lemma 5.2 shows that fixing a strategy and conditioning on the event of winning $m-1$ games, there is a coordinate i_m for which $P^{\otimes n}_{X^n Y^n Z^n | W_{i_1} \ldots W_{i_{m-1}}}(x^n, y^n, z^n)$ projected on the i_mth coordinate is similar to P_{XYZ}. This lemma is Claim 5.1 in Raz's paper [Raz98] and Lemma 5 in Holenstein's paper [Hol07]. The second lemma, Lemma 5.4 shows that we can obtain a No-Signaling strategy to play the i_mth coordinate from a No-Signaling strategy for $\mathrm{G}^{\otimes n}$ after conditioning on the event $W_{i_1} \wedge \ldots \wedge W_{i_{m-1}}$ (after conditioning we may not obtain a No-Signaling function but we show that there is a No-Signaling function which is very close to it.). This is a generalization of Lemma 23 in Holenstein's paper [Hol07]. We then conclude that the provers can not win the game i_m with probability greater than $\omega^{\diamond}(\mathrm{G}) + \epsilon$.

5 Technical Lemmas

Lemma 5.1 (RazHolenstein). *Let* $\mathrm{P}_{U^\ell} = \mathrm{P}_{U_1}\mathrm{P}_{U_2}\cdots\mathrm{P}_{U_\ell}$ *be a probability distribution over* U^ℓ *(i.e.,* P_{U^ℓ} *is a product distribution over* $U_1, \ldots U_\ell$*) and let* W *be some event, then the following holds:*

$$\sum_{i=1}^{\ell} \left\| \mathrm{P}_{U_i|W} - \mathrm{P}_{U_i} \right\| \leq \sqrt{\ell \cdot \log \frac{1}{\Pr[W]}}$$

For completeness, we include the proof of the lemma in full version of the paper.

We need the following corollary of Lemma 5.1.

Corollary 5.2. *Let* $\mathrm{P}_{TU^\ell} = \mathrm{P}_T\mathrm{P}_{U_1|T}\mathrm{P}_{U_2|T}\cdots\mathrm{P}_{U_\ell|T}$ *and let* W *be some event, then the following holds:*

$$\sum_{i=1}^{\ell} \left\| \mathrm{P}_{TU_i|W} - \mathrm{P}_{T|W}\mathrm{P}_{U_i|T} \right\| \leq \sqrt{\ell \cdot \log \frac{1}{\Pr[W]}}$$

For the proof see full version of the paper.

For the second lemma we will need to use the following proposition.

Proposition 5.3. *Let* P_{RST} *be a probability distribution over* RST *and let* $\widetilde{\mathrm{P}}_R$ *be some distribution over* R*. There exists a distribution* Q_{RST} *satisfying the following:*

$$\|\mathrm{Q}_{RST} - \mathrm{P}_{RST}\| \leq \|\mathrm{P}_R - \widetilde{\mathrm{P}}_R\| \quad (1)$$

$$\|\mathrm{Q}_R - \widetilde{\mathrm{P}}_R\| = 0 \quad (2)$$

$$\|\mathrm{Q}_S - \mathrm{P}_S\| = 0 \quad (3)$$

$$\|\mathrm{Q}_T - \mathrm{P}_T\| = 0 \quad (4)$$

The proof is following closely Holenstein's [Hol07] proof of Lemma 22.

Proof: We show a finite process for changing P_{RST} into Q_{RST} with the desired properties. Let $H = \{r \mid \mathrm{P}_R(r) > \widetilde{\mathrm{P}}_R(r)\}$ and let $L = \{r \mid \mathrm{P}_R(r) < \widetilde{\mathrm{P}}_R(r)\}$. Until $|H| \neq 0$, fix some $r \in H$ and $r' \in L$ and fix s, t such that $\mathrm{P}_{RST}(r, s, t) > 0$. Define,

$$\alpha := \min\{\mathrm{P}_{RST}(r, s, t), |\mathrm{P}_R(r) - \widetilde{\mathrm{P}}_R(r)|, |\mathrm{P}_R(r') - \widetilde{\mathrm{P}}_R(r')|\}$$

and set $\mathrm{P}_{RST}(r, s, t) := \mathrm{P}_{RST}(r, s, t) - \alpha$ and $\mathrm{P}_{RST}(r', s, t) := \mathrm{P}_{RST}(r', s, t) + \alpha$. The process only decreases $\|\mathrm{P}_R - \widetilde{\mathrm{P}}_R\|$ and therefore Condition 1 holds. By definition, we continue the process until Condition 2 holds. Since in every iteration, the process decreases some fixed (r, s, t) by α and increases (r', s, t) by α , for the same s, t then Condition 3 and Condition 4 are satisfied. ■

Lemma 5.4. *Let* P_{XYZRST} *be a probability distribution over* $XYZRST$ *and let* $\bar{\mathrm{P}}_{XYZ}$ *be some distribution over* XYZ. *If*

$$\|\bar{\mathrm{P}}_{XYZ}\mathrm{P}_{R|X} - \mathrm{P}_{XYZR}\| \leq \epsilon_1 \tag{5}$$
$$\|\bar{\mathrm{P}}_{XYZ}\mathrm{P}_{S|Y} - \mathrm{P}_{XYZS}\| \leq \epsilon_2 \tag{6}$$
$$\|\bar{\mathrm{P}}_{XYZ}\mathrm{P}_{T|Z} - \mathrm{P}_{XYZT}\| \leq \epsilon_3 \tag{7}$$

then, there exists a conditional distribution $\mathrm{Q}_{RST|X=x,Y=y,Z=z}$ *satisfying the following:*

$$\|\bar{\mathrm{P}}_{XYZ}\mathrm{Q}_{RST|XYZ} - \mathrm{P}_{XYZRST}\| \leq \min\{\epsilon_1, \epsilon_2, \epsilon_3\} + 2(\epsilon_1 + \epsilon_2 + \epsilon_3) \tag{8}$$
$$\|\mathrm{Q}_{R|X=x,Y=y,Z=z} - \mathrm{Q}_{R|X=x}\| = 0 \tag{9}$$
$$\|\mathrm{Q}_{S|X=x,Y=y,Z=z} - \mathrm{Q}_{S|Y=y}\| = 0 \tag{10}$$
$$\|\mathrm{Q}_{T|X=x,Y=y,Z=z} - \mathrm{Q}_{T|Z=z}\| = 0 \tag{11}$$

Proof: For fixed x, y, z we apply Lemma 5.3 three times. We first apply the lemma on $\mathrm{P}_{RST|X=x,Y=y,Z=z}$ and $\mathrm{P}_{R|X=x}$ to obtain $\hat{\mathrm{Q}}_{RST|X=x,Y=y,Z=z}$ such that

$$\|\hat{\mathrm{Q}}_{RST|X=x,Y=y,Z=z} - \mathrm{P}_{RST|X=x,Y=y,Z=z}\| \leq |\mathrm{P}_{R|X=x,Y=y,Z=z} - \mathrm{P}_{R|X=x}\|$$
$$\|\hat{\mathrm{Q}}_{R|X=x,Y=y,Z=z} - \mathrm{P}_{R|X=x}\| = 0$$
$$\|\hat{\mathrm{Q}}_{S|X=x,Y=y,Z=z} - \mathrm{P}_{S|X=x,Y=y,Z=z}\| = 0$$
$$\|\hat{\mathrm{Q}}_{T|X=x,Y=y,Z=z} - \mathrm{P}_{T|X=x,Y=y,Z=z}\| = 0$$

Applying Lemma 5.3 on $\hat{\mathrm{Q}}_{RST|X=x,Y=y,Z=z}$ and $\mathrm{P}_{S|Y=y}$ and combining the previous result we obtain $\tilde{\mathrm{Q}}_{RST|X=x,Y=y,Z=z}$ such that

$$\|\tilde{\mathrm{Q}}_{RST|X=x,Y=y,Z=z} - \hat{\mathrm{Q}}_{RST|X=x,Y=y,Z=z}\| \leq \|\mathrm{P}_{S|X=x,Y=y,Z=z} - \mathrm{P}_{S|Y=y}\|$$
$$\|\tilde{\mathrm{Q}}_{R|X=x,Y=y,Z=z} - \mathrm{P}_{R|X=x}\| = 0$$
$$\|\tilde{\mathrm{Q}}_{S|X=x,Y=y,Z=z} - \mathrm{P}_{S|Y=y}\| = 0$$
$$\|\tilde{\mathrm{Q}}_{T|X=x,Y=y,Z=z} - \hat{\mathrm{Q}}_{T|X=x,Y=y,Z=z}\| = 0$$

Apply Lemma 5.3 again, this time on $\tilde{Q}_{RST|X=x,Y=y,Z=z}$ and $\tilde{\mathrm{Q}}_{T|Z=z}$, we obtain $\mathrm{Q}_{RST|X=x,Y=y,Z=z}$ such that

$$\|\mathrm{Q}_{RST|X=x,Y=y,Z=z} - \tilde{\mathrm{Q}}_{RST|X=x,Y=y,Z=z}\| \leq \|\mathrm{P}_{T|X=x,Y=y,Z=z} - \mathrm{P}_{T|Z=z}\|$$
$$\|\mathrm{Q}_{R|X=x,Y=y,Z=z} - \mathrm{P}_{R|X=x}\| = 0$$
$$\|\mathrm{Q}_{S|X=x,Y=y,Z=z} - \mathrm{P}_{S|Y=y}\| = 0$$
$$\|\mathrm{Q}_{T|X=x,Y=y,Z=z} - \mathrm{P}_{T|Z=z}\| = 0$$

From Condition 5, Condition 6 and Condition 7 we obtain: $\|\bar{\mathrm{P}}_{XYZ} - \mathrm{P}_{XYZ}\| \leq \min\{\epsilon_1, \epsilon_2, \epsilon_3\}$

Hence,

$$\begin{aligned}
&\|\bar{\mathrm{P}}_{XYZ}\mathrm{Q}_{RST|XYZ} - \mathrm{P}_{XYZRST}\| \\
&\leq \min\{\epsilon_1, \epsilon_2, \epsilon_3\} + \|\mathrm{P}_{XYZ}\mathrm{Q}_{RST|XYZ} - \mathrm{P}_{XYZRST}\| \\
&= \min\{\epsilon_1, \epsilon_2, \epsilon_3\} + \sum_{xyz} \mathrm{P}_{XYZ}(x,y,z)\|\mathrm{Q}_{RST|X=x,Y=y,Z=z} - \mathrm{P}_{RST|X=x,Y=y,Z=z}\| \\
&\leq \min\{\epsilon_1, \epsilon_2, \epsilon_3\} + \sum_{xyz} \mathrm{P}_{XYZ}(x,y,z)\cdot \\
&\quad (\|\mathrm{P}_{R|X=x,Y=y,Z=z} - \mathrm{P}_{R|X=x}\| + \|\mathrm{P}_{S|X=x,Y=y,Z=z} - \mathrm{P}_{S|Y=y}\| + \|\mathrm{P}_{T|X=x,Y=y,Z=z} - \mathrm{P}_{T|Z=z}\|) \\
&\leq \min\{\epsilon_1, \epsilon_2, \epsilon_3\} + \|\mathrm{P}_{XYZR} - \mathrm{P}_{XYZ}\mathrm{P}_{R|X}\| + \|\mathrm{P}_{XYZS} - \mathrm{P}_{XYZ}\mathrm{P}_{S|Y}\| + \|\mathrm{P}_{XYZS} - \mathrm{P}_{XYZ}\mathrm{P}_{T|Z}\| \\
&\leq \min\{\epsilon_1, \epsilon_2, \epsilon_3\} + 2\,(\epsilon_1 + \epsilon_2 + \epsilon_3)
\end{aligned}$$

■

6 Parallel Repetition Theorem

Proposition 6.1. *For every game* $\mathrm{G} = (X, Y, Z, A, B, C, P_{XYZ}, Q)$ *fix any No-Signaling strategy for the n-fold parallel repetition game,* $\mathrm{G}^{\otimes n}$*, and let* $W = W_{i_1} \wedge \ldots \wedge W_{i_{m-1}}$ *be the event of winning on some fixed* $i_1, \ldots, i_{m-1}$ *coordinates with respect to that strategy. There exists a coordinate* i_m *such that*

$$\Pr[W_{i_m}|W] \leq \omega^{\diamond}(\mathrm{G}) + 13\sqrt{\frac{1}{n-m}\log\left(\frac{1}{\Pr[W]}\right)}$$

Proof: Assume without loss of generality that $i_1, \ldots, i_{m-1} \in \{n-m+1, \ldots, n\}$ (the last m coordinates). Let $T = (X^n, A^n)$ and $U^{n-m} = (Y^{n-m}, Z^{n-m})$ then from Lemma 5.2 we obtain:

$$\sum_{i=1}^{n-m} \left\|\mathrm{P}_{X^nA^nY_iZ_i|W} - \mathrm{P}_{X^nA^n|W}\mathrm{P}_{Y_iZ_i|X^nA^n}\right\| \leq \sqrt{(n-m)\log\frac{1}{\Pr[W]}}.$$

Since the strategy is No-Signaling:

$$\sum_{i}^{n-m} \left\|\mathrm{P}_{X^nA^nY_iZ_i|W} - \mathrm{P}_{X^nA^n|W}\mathrm{P}_{Z_iY_i|X_i}\right\| \leq \sqrt{(n-m)\log\frac{1}{\Pr[W]}}.$$

This can be written as:

$$\sum_{i}^{n-m} \left\|\mathrm{P}_{X^nA^nY_iZ_i|W} - \mathrm{P}_{X_i|W}\mathrm{P}_{X^{-i}A^n|X_i,W}\mathrm{P}_{Z_iY_i|X_i}\right\| \leq \sqrt{(n-m)\log\frac{1}{\Pr[W]}}. \tag{12}$$

Applying Lemma 5.1 with $U^{n-m} = X^{n-m}$ we obtain:

$$\sum_{i}^{n-m} \left\|\mathrm{P}_{X_i|W} - \mathrm{P}_{X_i}\right\| \leq \sqrt{(n-m)\log\frac{1}{\Pr[W]}} \tag{13}$$

Combining Equation 12 and Equation 13 yields:

$$\sum_{i}^{n-m} \|\mathrm{P}_{X^n A^n Y_i Z_i|W} - \mathrm{P}_{X_i}\mathrm{P}_{X^{-i}A^n|X_i,W}\mathrm{P}_{Z_i Y_i|X_i}\| \leq 2\sqrt{(n-m)\log\frac{1}{\Pr[W]}}$$

This can be written as:

$$\sum_{i}^{n-m} \|\mathrm{P}_{X^n A^n Y_i Z_i|W} - \mathrm{P}_{X_i Y_i Z_i}\mathrm{P}_{X^{-i}A^n|X_i,W}\| \leq 2\sqrt{(n-m)\log\frac{1}{\Pr[W]}}$$

Since $\mathrm{P}_{X_i Y_i Z_i} = \mathrm{P}_{XYZ}$,

$$\sum_{i}^{n-m} \|\mathrm{P}_{X^n A^n Y_i Z_i|W} - \mathrm{P}_{XYZ}\mathrm{P}_{X^{-i}A^n|X_i,W}\| \leq 2\sqrt{(n-m)\log\frac{1}{\Pr[W]}} \tag{14}$$

Similar argument for $T = (Y^n, B^n)$ and $U^{n-m} = (X^{n-m}, Z^{n-m})$ yields:

$$\sum_{i}^{n-m} \|\mathrm{P}_{Y^n B^n X_i Z_i|W} - \mathrm{P}_{XYZ}\mathrm{P}_{Y^{-i}B^n|Y_i,W}\| \leq 2\sqrt{(n-m)\log\frac{1}{\Pr[W]}} \tag{15}$$

And for $T = (Z^n, C^n)$ and $U^{n-m} = (X^{n-m}, Y^{n-m})$ yields:

$$\sum_{i}^{n-m} \|\mathrm{P}_{Z^n C^n X_i Y_i|W} - \mathrm{P}_{XYZ}\mathrm{P}_{Z^{-i}C^n|Z_i,W}\| \leq 2\sqrt{(n-m)\log\frac{1}{\Pr[W]}} \tag{16}$$

By Lemma 5.4, there exists a distribution $\mathrm{Q}_{X^{-i}A^n Y^{-i}B^n Y^{-i}C^n|X_i=x,Y_i=y,Z_i=z}$ which can be implemented by a No-Signaling function and satisfy:

$$\sum_{i=1}^{n-m} \|\mathrm{P}_{XYZ}\mathrm{Q}_{X^{-i}A^n Y^{-i}B^n Y^{-i}C^n|XYZ} - \mathrm{P}_{XYZX^{-i}A^n Y^{-i}B^n Y^{-i}C^n|W}\| \leq 13\sqrt{(n-m)\log\left(\frac{1}{\Pr[W]}\right)}.$$

Therefore, there exists a coordinate i_m such that

$$\|\mathrm{P}_{XYZ}\mathrm{Q}_{X^{-i_m}A^n Y^{-i_m}B^n Y^{-i_m}C^n|XYZ} - \mathrm{P}_{XYZX^{-i_m}A^n Y^{-i_m}B^n Y^{-i_m}C^n|W}\| \leq 13\sqrt{\frac{1}{n-m}\log\left(\frac{1}{\Pr[W]}\right)}.$$

Hence, given input (X, Y, Z) for the game G, Alice, Bob and Charlie can use the No-Signaling Strategy for $\mathrm{G}^{\otimes n}$ to obtain a No-Signaling strategy for G. The provers play G in the i_m coordinate and answer $A_{i_m}, B_{i_m}, C_{i_m}$ (and ignoring the redundant information). Thus there is a coordinate i_m on which

$$\Pr[W_{i_m}|W] \leq \omega^{\diamondsuit}(\mathrm{G}) + 13\sqrt{\frac{1}{n-m}\log\left(\frac{1}{\Pr[W]}\right)}$$

■

Theorem 6.2. *For all games* $\mathrm{G} = (X, Y, Z, A, B, C, P_{XYZ}, Q)$ *and any positive integer* n,

$$\omega^{\diamondsuit}(\mathrm{G}^{\otimes n}) \leq \left(1 - \frac{(1-\omega^{\diamondsuit}(\mathrm{G}))^2}{4000}\right)^n$$

Proof: Fix any strategy for the n-folded parallel repetition game $\mathrm{G}^{\otimes n}$ and let $q_m = \Pr[W_{i_1} \wedge \ldots \wedge W_{i_m}]$. By Proposition 6.1 we obtain:

$$q_{m+1} \leq q_m \cdot \left(\omega^{\diamondsuit}(\mathrm{G}) + 13\sqrt{\frac{1}{n-m} \log\left(\frac{1}{q_m}\right)}\right)$$

We show by induction that $q_{m+1} \leq \left(\frac{1+\omega^{\diamondsuit}(\mathrm{G})}{2}\right)^{m+1}$ for any $m+1 \leq \frac{(1-\omega^{\diamondsuit}(\mathrm{G}))(n-m)}{1000}$. If $q_m \leq \left(\frac{1+\omega^{\diamondsuit}(\mathrm{G})}{2}\right)^{m+1}$ then the claim immediately follows. If $q_m > \left(\frac{1+\omega^{\diamondsuit}(\mathrm{G})}{2}\right)^{m+1}$ then,

$$\log\left(\frac{1}{q_m}\right) \leq \log\left(\frac{1+\omega^{\diamondsuit}(\mathrm{G})}{2}\right)^{-(m+1)}.$$

Since for any $\alpha, \beta < 1$, $(1-\alpha)^b \leq 1 - \alpha\beta$, then $(1-\frac{1}{2})^{1-\omega^{\diamondsuit}(\mathrm{G})} \leq (1 - \frac{1-\omega^{\diamondsuit}(\mathrm{G})}{2})$ and we obtain that

$$\log\left(\frac{1+\omega^{\diamondsuit}(\mathrm{G})}{2}\right)^{-(m+1)} =$$

$$\log\left(1 - \frac{1-\omega^{\diamondsuit}(\mathrm{G})}{2}\right)^{-(m+1)} \leq \log\left(\frac{1}{2}^{-(m+1)(1-\omega^{\diamondsuit}(\mathrm{G}))}\right) = (m+1)(1-\omega^{\diamondsuit}(\mathrm{G}))$$

Thus for any m satisfying $m + 1 \leq \frac{(1-\omega^{\diamondsuit}(\mathrm{G}))(n-m)}{1000}$,

$$q_{m+1} \leq q_m \cdot \left(\omega^{\diamondsuit}(\mathrm{G}) + 13\sqrt{\frac{1}{n-m}(m+1)(1-\omega^{\diamondsuit}(\mathrm{G}))}\right) \leq q_m \cdot \left(\frac{1+\omega^{\diamondsuit}(\mathrm{G})}{2}\right)$$

combining the induction hypothesis we obtain $q_m \cdot \left(\frac{1+\omega^{\diamondsuit}(\mathrm{G})}{2}\right) \leq \left(\frac{1+\omega^{\diamondsuit}(\mathrm{G})}{2}\right)^{m+1}$. Taking $m = n\frac{(1-\omega^{\diamondsuit}(\mathrm{G}))}{2000}$ we get,

$$q_m \leq \left(1 - \frac{1-\omega^{\diamondsuit}(\mathrm{G})}{2}\right)^{\frac{n(1-\omega^{\diamondsuit}(\mathrm{G}))}{2000}} \leq \left(1 - \frac{(1-\omega^{\diamondsuit}(\mathrm{G}))^2}{4000}\right)^n$$

■

Acknowledgement

We would like to express my deepest gratitude to Oded Regev and Ran Raz for very valuable discussions on this subject and for their comments on the previous version of this paper.

References

[CCL92] Cai, J.-Y., Condon, A., Lipton, R.J.: On games of incomplete information. In: Choffrut, C., Lengauer, T. (eds.) STACS 1990. LNCS, vol. 415, Springer, Heidelberg (1990)

[CHSH69] Clauser, J.F., Horne, M.A., Shimony, A., Holt, R.A.: Proposed experiment to test local hidden-variable theories. Phys. Rev. Lett. 23(15), 880–884 (1969)

[CT06] Cover, T.M., Thomas, J.A.: Elements of information theory, 2nd edn. Wiley-Interscience/John Wiley & Sons, Hoboken/ NJ (2006)

[Fei91] Feige, U.: On the success probability of the two provers in one-round proof systems. In: Structure in Complexity Theory Conference, pp. 116–123 (1991)

[FK94] Feige, U., Kilian, J.: Two prover protocols: low error at affordable rates. In: Proceedings of the 26th Annual Symposium on the Theory of Computing, pp. 172–183. ACM Press, New York (May 1994)

[FL92] Feige, U., Lovász, L.: Two-prover one-round proof systems: their power and their problems (extended abstract). In: STOC '92: Proceedings of the twenty-fourth annual ACM symposium on Theory of computing, pp. 733–744. ACM Press, New York (1992)

[For89] Fortnow, L.J.: Complexity - theoretic aspects of interactive proof systems. Technical Report MIT-LCS//MIT/LCS/TR-447, Department of Mathematics, Massachusetts Institute of Technology (1989)

[FV02] Feige, U., Verbitsky, O.: Error reduction by parallel repetition—a negative result. Combinatorica 22(4), 461–478 (2002)

[FV02] Feige, U., Verbitsky, O.: Error reduction by parallel repetition—a negative result. Combinatorica 22(4), 461–478 (2002)

[Hol07] Holenstein, T.: Parallel repetition: simplifications and the no-signaling case. In: STOC '07: Proceedings of the 39th Annual ACM Symposium on Theory of Computing (2007)

[PR94] Popescu, S., Rohrlich, D.: Nonlocality as an axiom for quantum theory. In: Foundations of Physics, vol. 24(3), pp. 379–385 (1994)

[Rao08] Rao, A.: Parallel repetition in projection games and a concentration bound. In: STOC (2008)

[Raz98] Raz, R.: A parallel repetition theorem. SIAM J. Comput. 27(3), 763–803 (1998)

The Curse of Connectivity: t-Total Vertex (Edge) Cover

Henning Fernau[1], Fedor V. Fomin[2], Geevarghese Philip[3], and Saket Saurabh[3]

[1] Universität Trier FB 4—Abteilung Informatik, 54286 Trier, Germany
fernau@uni-trier.de

[2] Department of Informatics, University of Bergen, 5020 Bergen, Norway
fomin@ii.uib.no

[3] The Institute of Mathematical Sciences, Taramani, Chennai 600 113, India
{gphilip,saket}@imsc.res.in

Abstract. We investigate the effect of certain natural connectivity constraints on the parameterized complexity of two fundamental graph covering problems, namely k-VERTEX COVER and k-EDGE COVER. Specifically, we impose the additional requirement that each connected component of a solution have at least t vertices (resp. edges from the solution), and call the problem t-TOTAL VERTEX COVER (resp. t-TOTAL EDGE COVER). We show that

- both problems remain fixed-parameter tractable with these restrictions, with running times of the form $\mathcal{O}^*\left(c^k\right)$ for some constant $c > 0$ in each case;
- for every $t \geq 2$, t-TOTAL VERTEX COVER has no polynomial kernel unless the Polynomial Hierarchy collapses to the third level;
- for every $t \geq 2$, t-TOTAL EDGE COVER has a linear vertex kernel of size $\frac{t+1}{t}k$.

1 Introduction

k-VERTEX COVER and k-EDGE COVER are two related combinatorial problems that exhibit contrasting behavior in terms of their solvability. In the k-VERTEX COVER problem, or k-VC for short, we are given as input a graph G and a positive integer k, and are asked if there exists a set S of at most k vertices in G such that every edge in G is adjacent to at least one of the vertices in S; such an S is called a *vertex cover* of G. In k-EDGE COVER, or k-EC for short, we are again given as input a graph G and a positive integer k, and are asked if there exists a set S of at most k *edges* in G such that for each vertex v in G, there is at least one edge $e = \{u, v\} \in S$; such an S is called an *edge cover* of G.

k-VC is one of the first problems that Karp showed to be NP-complete [18], and is one of the six basic NP-complete problems chosen for special mention by Garey and Johnson as being at the "core" of known NP-complete problems [15, Section 3.1]. k-VC and its variants have been extensively investigated from the point of view of various algorithmic paradigms, including approximation and parameterized algorithms. In particular, k-VC is considered to be the drosophila of the field

M.T. Thai and S. Sahni (Eds.): COCOON 2010, LNCS 6196, pp. 34–43, 2010.

of parameterized complexity, where a large set of results of different kinds have been obtained on the problem and its variants [2,8,12,14,16,19,20,22]. In sharp contrast, k-EC has long been known to be solvable in polynomial time [24].

In this paper we investigate the parameterized complexity of variants of these problems where additional connectivity constraints are imposed on the solution S. More specifically, for each integer $t \geq 1$ we define variants of the two problems, named t-TOTAL VERTEX COVER and t-TOTAL EDGE COVER. In t-TOTAL VERTEX COVER, or t-TVC for short, (resp. t-TOTAL EDGE COVER, or t-TEC for short), we ask whether there is a vertex cover (resp. edge cover) S of the input graph such that each connected component of the subgraph induced on S contains at least t vertices (resp. at least t edges from S). These problems were introduced by Fernau and Manlove [12] who initiated the study of the parameterized complexity of these problems. We significantly improve their results and obtain several new results: in particular, we complete the picture on how even the slightest connectivity requirement dramatically changes the complexity of these problems.

Our results. As noted above, k-EC has been known to be solvable in polynomial time for over half a century. It was recently observed [12] that the least possible connectivity requirement on S, namely that each connected component of the graph induced on S have at least 2 edges, makes the problem NP-hard.

We show a similar result for t-TVC, not in the context of classical complexity, but in the realm of parameterized complexity. It is a well-known result in parameterized complexity that k-VC has a $2k$-sized vertex kernel. We show that adding a connectivity constraint results in a dramatic change in kernelizability: for any fixed $t \geq 2$, t-TVC has no polynomial-size kernel unless the Polynomial Hierarchy (PH) collapses to the third level, which is deemed unlikely in complexity theory. We complement the no-polynomial-kernel result with results on the fixed-parameter tractability of t-TVC and t-TEC when parameterized by the solution size. More specifically, we show the following:

- t-TVC can be solved in $\mathcal{O}\left(16.1^{k+\mathcal{O}(\log^2 k)} \times n^{\mathcal{O}(1)}\right)$ time. To obtain these results we combine the classical result of Otter [25] on the number of unlabeled trees with a modification of the color-coding technique of Alon et al. [1].
- t-TEC has a vertex kernel of size $\frac{t+1}{t}k$,
- t-TEC can be solved in $\mathcal{O}\left(2^{\frac{t+1}{t}k+\mathcal{O}(\sqrt{k})} \times n^{\mathcal{O}(1)}\right)$ time. To obtain this result, we combine kernelization techniques with a classical result of Hardy and Ramanujan [17] and the Fast Fourier Transform [21].

These results significantly improve earlier work on these problems.

Due to space constraints, proofs of results labeled with a [$\star$] have been deferred to a full version of the paper. Likewise, we omit the definitions of standard notions from graph theory and parameterized complexity theory, and refer the reader to the excellent expositions by Diestel [10] and Flum and Grohe [13], respectively.

2 Total Vertex Covers

For each positive integer t, we defined the problem t-TVC above. For $t = 1$, t-TOTAL VERTEX COVER is the k-VERTEX COVER problem, and for $t = k$ it becomes the k-CONNECTED VERTEX COVER problem, or k-CVC for short; these are two classical NP-complete problems [15, Problem GT1]. For $2 \leq t \leq k$, t-TVC has been shown to be NP-hard by reduction from k-VC [12, Theorem 3]; we give an alternate proof of NP-hardness in Claim 1 below.

Kernelizability. We first investigate the kernelizability of t-TVC. Recall that for $t = 1$, t-TVC is just k-VC, and for $t = k$ it becomes k-CVC. The former problem has a vertex kernel of size at most $2k$ [7], and the latter problem does not have a kernel of size k^c for any constant c unless PH collapses to the third level [11]. It turns out that this change in polynomial kernelizability occurs at the smallest value of t possible.

Theorem 1. *For each fixed $t \geq 2$, t-TVC has no kernel of size bounded by k^c, for any fixed constant c, unless* PH *collapses to the third level.*

To prove this, we need a few notions and results from the recently developed theory of kernel lower bounds [5,6,11]. We start off by defining a new problem:

RED-BLUE DOMINATING SET (RBDS)
Input: An undirected bipartite graph $G = (R \uplus B, E)$, and a positive integer k.
Parameter: $k + |B|$
Question: Does there exist a set $D \subseteq R$ of at most k vertices of G such that every $v \in B$ is adjacent to some $u \in D$ (i.e., D is a dominating set of B)?

We will use RBDS as a source problem to show a kernel lower bound for t-TVC. Towards this end we state the following fact.

Fact 1. [11, Theorem 2] RBDS *parameterized by $(k + |B|)$ does not admit a polynomial kernel unless* PH *collapses to the third level.*

Let $\Pi \subseteq \Sigma^* \times \mathbb{N}$ be a parameterized problem, and let $1 \notin \Sigma$ be a new symbol. The *derived classical problem* associated with Π is $\{x1^k | (x, k) \in \Pi\}$. Let P and Q be parameterized problems. We say that P is *polynomial time and parameter reducible* to Q, written $P \leq_{Ptp} Q$, if there exists a polynomial time computable function $f : \Sigma^* \times \mathbb{N} \to \Sigma^* \times \mathbb{N}$, and a polynomial $p : \mathbb{N} \to \mathbb{N}$, and for all $x \in \Sigma^*$ and $k \in \mathbb{N}$, if $f((x, k)) = (x', k')$, then $(x, k) \in P$ if and only if $(x', k') \in Q$, and $k' \leq p(k)$. We call f a polynomial parameter transformation (or a PPT) from P to Q. This notion of a reduction is useful in showing kernel lower bounds because of the following theorem:

Fact 2. [6, Theorem 3] *Let P and Q be parameterized problems whose derived classical problems are P^c, Q^c, respectively. Let P^c be* NP*-complete, and $Q^c \in$* NP. *Suppose there exists a PPT from P to Q. Then, if Q has a polynomial kernel, then P also has a polynomial kernel.*

Theorem 1. *For each fixed $t \geq 2$, t-TVC has no kernel of size bounded by k^c, for any fixed constant c, unless PH collapses to the third level* [1].

Proof. We begin by noting that by a simple reduction from the NP-complete SET COVER problem [15], the derived classical problem corresponding to RBDS is NP-complete. Also, the derived classical problem corresponding to t-TVC is evidently in NP. Now suppose t-TVC has a polynomial kernel, and that there is a PPT from RBDS to t-TVC. Then by Fact 2, RBDS has a polynomial kernel, and hence by Fact 1 PH collapses to the third level. Thus t-TVC does not have polynomial kernel unless PH collapses to the third level. Hence to prove the theorem, it suffices to show that there is a PPT from RBDS to t-TVC. We now give such a reduction.

Given an instance $(G = (R \uplus B, E), k)$ of RBDS, we construct an instance of t-TVC as follows: If B contains isolated vertices then (G, k) is a NO instance, and in this case we construct a trivial NO instance of t-TVC. Otherwise, we add a distinct path of length (number of edges) $t-1$ starting from each vertex $v \in B$. Thus, if $t = 2$, then we attach a new, distinct pendant vertex w_i to each $v_i \in B$; if $t = 3$, then we add a path (v_i, u_i^1, w_i) to each $v_i \in B$. In general, for $t \geq 2$, we add a path $(v_i, u_i^1, u_i^2, \dots u_i^{t-2}, w_i)$ to each $v_i \in B$, where the vertices u_i^j and w_i are all new and distinct: see Fig. 1 for an illustration. We call the resulting graph H and $(H, k + (t-1)|B|)$ is the constructed instance of t-TVC.The following claim completes our proof:

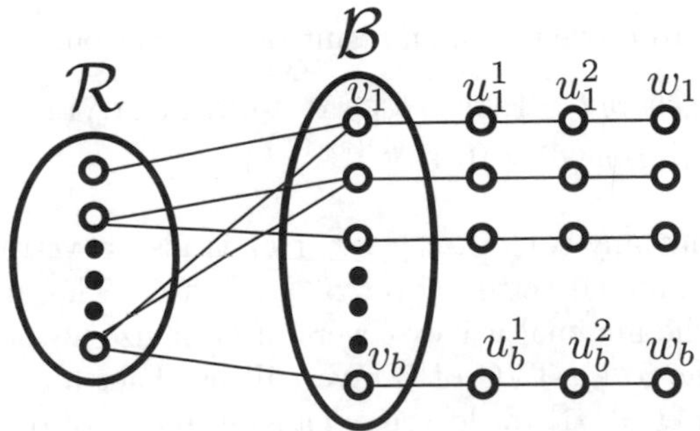

Fig. 1. The reduced instance for $t = 4$

Claim 1. [$\star$] Let $(G = (R \uplus B, E), k)$ be an instance of RBDS, and t a fixed positive integer. Let H be the graph constructed from G as described above. Then (G, k) is a YES instance of RBDS if and only if $(H, k + (t-1)|B|)$ is a YES instance of t-TVC. □

Fixed Parameter Tractability. We now investigate the fixed-parameter tractability of t-TVC. Two special cases of the problem, for the two extreme values $t = 1$ (k-VC) and $t = k$ (k-CVC), have been studied extensively from the perspective of

[1] We note in passing that it was wrongly claimed in [12] that the problem has a kernel of size $\mathcal{O}(k(k+t))$.

parameterized algorithms. The k-VC problem is perhaps the most well-studied problem in parameterized algorithmics. After a long series of improvements, the current fastest FPT algorithm for this problem runs in time $\mathcal{O}^*(1.2738^k)$ [8]. Similarly k-CVC also has a history of improvements, and the current fastest FPT algorithm for this problem runs in time $\mathcal{O}^*(2.7606^k)$ [22]. We show in this section that t-TVC is FPT parameterized by the solution size k, and give an $\mathcal{O}^*(c^k)$ time algorithm.

Let $G = (V, E)$ be the input graph. If $|V| \leq k$, then we can solve the problem in polynomial time by checking if each component of G has at least t vertices. Also, deleting isolated vertices does not affect the solution. Hence we assume without loss of generality that $|V| > k$, and that G has no isolated vertices. We start with a structural claim which is useful later.

Claim 2. [$\star$] Let $G = (V, E), |V| > k$ be a graph without any isolated vertices. Then G has a t-TVC of size at most k iff G has a t-TVC of size exactly k.

We use the following fact in the proof of the next lemma:

Fact 3. [4,25] *The number of unlabeled trees on k vertices is at most 2.96^k. Moreover, all non-isomorphic unlabeled trees on k vertices can be enumerated in time $\mathcal{O}(2.96^k k^c)$ for some constant c independent of k.*

Lemma 1. [$\star$] *All unlabeled* forests *on k vertices are enumerable in $\mathcal{O}^*(2.96^k)$.*

Now we are ready to prove the main result of this section.

Theorem 2. *For every $t \geq 1$, the t-TOTAL VERTEX COVER problem is in FPT, and can be solved in time $\mathcal{O}^*\left(16.1^{k+\mathcal{O}(\log^2 k)}\right)$.*

Proof. Observe that any t-TVC, say S, of G is also a vertex cover of G and hence contains a minimal vertex cover $S' \subseteq S$ of G. The idea of our proof is to enumerate all the minimal vertex covers of G of size at most k and then try to expand each one to a t-TVC of G. We will use Fact 3 and the color-coding technique of Alon et al. [1] to do the expansion phase of our algorithm. More precisely, our algorithm is based on the following claim.

Claim 3. [$\star$] A graph $G = (V, E)$ has a t-TVC of size k if and only if there exists a minimal vertex cover C of G of size at most k, and a subset $T \subseteq V \setminus C$ of size $k - |C|$, such that there exists a forest F on k vertices which is isomorphic to a spanning subgraph of $G[C \cup T]$, and in which each connected component has at least t vertices.

For the algorithm we essentially mimic the claim. First we enumerate all inclusion-minimal vertex covers of G of size at most k. This can be done in time $\mathcal{O}^*(2^k)$ by a simple 2-way branching on edges—for every edge at least one of its endpoints should be in any vertex cover. For each such vertex cover C, we do the following:

1. Color each $v \in C$ with a distinct color from $\{1, 2, \ldots, |C|\}$.
2. Let $l = k - |C|$. Color the vertices of the independent set $G[V \setminus C]$ uniformly at random with l new colors.

Let S be a fixed t-TVC of G of size at most k, if there exists one. Define a "good" coloring of V to be a coloring in which the vertices in $S \setminus C$ are all distinctly colored. The above procedure will yield a good coloring of V with probability $l!/l^l \geq e^{-l}$.

Next, we iterate through all unlabeled forests on k vertices, and check if at least one of these forests is isomorphic to a spanning forest $\mathcal{F}$ of $G[S]$, where each connected component of $\mathcal{F}$ has at least t vertices. By Lemma 1, we can iterate through all such forests in $\mathcal{O}^*\left(2.96^k\right)$ time. To check if a given forest F on k vertices is isomorphic to a witness for $G[S]$, we do the following:

1. We check if there is at least one tree in F that has less than t vertices. If yes, then we reject F.
2. Next we check if there is a colorful subgraph (one in which each vertex has a distinct color) isomorphic to F in the colored graph obtained above. Since F is of treewidth at most 1, this can be done in $\mathcal{O}\left(2^k \cdot k \cdot n^2\right)$ time [3, Corollary 12]. If such a subgraph is present, then F satisfies the requirements of Claim 3, and so we return YES. Otherwise we reject F.

The expected running time of the algorithm is

$$\mathcal{O}^*\left(\sum_{l=0}^{k} 2^{k-l} \times e^l \times 2.96^k \times 2^k\right) = \mathcal{O}^*\left((2 \times 2.96 \times 2)^k \times \sum_{l=0}^{k}\left(\left(\frac{e}{2}\right)^l\right)\right) = \mathcal{O}^*\left(16.1^k\right).$$

To obtain a deterministic algorithm we have to replace the randomized step of the algorithm, that is, where we color the vertices of $G[V \setminus C]$ uniformly at random by l colors, with a deterministic procedure. This is done by making use of (n, l, l)*-perfect hash families.* An (n, l, l)-perfect hash family, $\mathcal{H}$, is a set of functions from $\{1, \ldots, n\}$ to $\{1, \ldots, l\}$ such that for every subset $S \subseteq \{1, \ldots, n\}$ of size l there exists a function $f \in \mathcal{H}$ such that f is injective on S. That is, for all $i, j \in S$, $f(i) \neq f(j)$. There exists a construction of (n, l, l)-perfect hash family of size $\mathcal{O}(e^l \cdot l^{\mathcal{O}(\log l)} \cdot \log n)$ and one can produce this family in time linear in the output size [23]. Using this instead of a random coloring, we obtain the desired deterministic algorithm. The run time of the derandomized algorithm is $\mathcal{O}^*\left(16.1^{k+\mathcal{O}\left(\log^2 k\right)}\right)$. □

3 Total Edge Covers

For each positive integer t, the parameterized t-TOTAL EDGE COVER problem was defined in the Introduction. For $t = 1$, t-TEC is the k-EC problem, which is solvable in polynomial time [24] while for $t \geq 2$, the problem is NP-complete [12, Theorem 3]. We use the following fact to obtain an equivalent formulation of the problem.

Fact 4. (From the proof of [12, Theorem 16]) *In any connected graph G with n vertices, and for any $t < n$, there exists a minimal t-TEC, say S, of G such that the graph $G(S)$ induced by the edge set S is acyclic.*

Lemma 2. [$\star$] *Given a graph $G = (V, E)$, the t-TOTAL EDGE COVER problem is equivalent to the following problem: does there exist a partition of the vertex set V into q parts $V_1, \ldots, V_q$, for some q, such that (i) $G[V_i]$ is connected, (ii) $|V_i| \geq t+1$ for each $1 \leq i \leq q$, and (iii) $\sum_{i=1}^{q}(|V_i|-1) \leq k$?*

Kernelizability. We use this reformulation to improve on a known simple vertex kernel of size at most $2k$ for t-TOTAL EDGE COVER [12].

Theorem 3. *t-TEC admits a vertex kernel of size $\frac{t+1}{t}k$.*

Proof. Let $(G = (V, E), k)$ be a YES instance of t-TEC. Hence, by Lemma 2 there exists a partition of V into q parts of the kind stated in Lemma 2. Now $|V_i| \geq t+1 \implies |V_i| - 1 \geq t \implies \sum_{i=1}^{q}(|V_i|-1) \geq qt$. By Lemma 2, $\sum_{i=1}^{q}(|V_i|-1) \leq k$, and so $qt \leq k$, and $q \leq \frac{k}{t}$. $\sum_{i=1}^{q}(|V_i|-1) \leq k \implies \sum_{i=1}^{q}|V_i| \leq k + q \leq k + \frac{k}{t} = \frac{t+1}{t}k$, and so G has at most $\frac{t+1}{t}k$ vertices. $\square$

Corollary 1. *If t-TEC has an exact exponential time algorithm that runs in $\mathcal{O}^*\left(c^{f(|V|)}\right)$ time on an input instance $(G = (V, E), k)$ for some function $f()$, then the problem has an FPT algorithm that runs in $\mathcal{O}^*\left(c^{f\left(\frac{t+1}{t}k\right)}\right)$ time.*

Fixed Parameter Tractability. We now present an exact exponential-time algorithm for the problem, with running time $\mathcal{O}^*\left(2^{n+\mathcal{O}(\sqrt{n})}\right)$ where n is the number of vertices in the input graph. By Corollary 1, this yields an FPT algorithm for the problem with running time $\mathcal{O}^*\left(c^k\right)$ for some fixed constant c. This is a significant improvement over the $\mathcal{O}^*\left((2k)^{2k}\right)$ bound obtained earlier [12].

Let $(G = (V, E), k)$ be an input instance of t-TOTAL EDGE COVER. First, we enumerate all unordered partitions of n. By the Hardy-Ramanujan asymptotic formula, n has at most $2^{\mathcal{O}(\sqrt{n})}$ unordered partitions [17]. The partitions of n can be generated with constant average delay [26], and so we can enumerate all unordered partitions of n in $2^{\mathcal{O}(\sqrt{n})}$ time.

For each partition of n as $n = n_1 + n_2 + \cdots + n_q; 1 \leq q \leq n$, we check if there exists a partition of V into q parts $V_1, \ldots, V_q$ such that (i) $|V_i| = n_i$ for $1 \leq i \leq q$, and (ii) the partition satisfies the conditions of Lemma 2.

To do these checks, we construct the q lists

$$L_i = \{V' \subseteq V \mid |V'| = n_i \text{ and } G[V'] \text{ is connected}\}$$

for $1 \leq i \leq q$. For $1 \leq i \leq q$ we compute the polynomial

$$P_i = \sum_{V' \in L_i} x^{\chi(V')}$$

where x is a formal variable and $\chi(V')$ is the (binary number represented by the) characteristic vector of $V' \subseteq V$. That is, let $V = \{v_1, v_2, \ldots, v_n\}$. Then $\chi(V')$ is a bit vector with $|V|$ bits where, for $1 \leq j \leq |V|$, the jth bit of $\chi(V')$ is 1 if and only if $v_j \in V'$. We treat $\chi(V')$ as a binary number in all computations. The lists L_i and the polynomials P_i can be computed in $\mathcal{O}^*(2^n)$ time, by enumerating all subsets of V. We now compute the product $Q = P_1 \times P_2 \times \cdots \times P_q$ in the given order, with a small modification: given the partial product Q_i of the first i terms, we first compute $Q_i \times P_{i+1}$. Then we delete all those terms αx^β in $Q_i \times P_{i+1}$ where (the binary representation of) β does not contain exactly $\sum_{j=1}^{i+1} n_j$ 1s, and set Q_{i+1} to be the resulting polynomial. This pruning operation ensures that the partial product Q_i, for $1 \leq i \leq q$, represents exactly those sets of size $\sum_{j=1}^{i} n_j$ that can be obtained by taking the union of one set each from $L_1, L_2, \ldots, L_i$. It is easy to see that the product $Q_q = Q$ is non-zero if and only if there exists a partition of V into q parts satisfying the required conditions.

The degree of each polynomial involved in the multiplications is at most $2^{|V|} - 1 = 2^n - 1$, and so, using the Fast Fourier Transform, we can multiply two of these polynomials in $\mathcal{O}(2^n \log 2^n) = \mathcal{O}(n2^n)$ time [9, Chapter 30]. We have to perform at most $q \leq n$ such multiplications to compute Q, and so given the P_is we can compute Q in $\mathcal{O}(n^2 2^n) = \mathcal{O}^*(2^n)$ time. The running time of this algorithm is thus $2^{\mathcal{O}(\sqrt{n})} \times (\mathcal{O}^*(2^n) + \mathcal{O}^*(2^n)) = \mathcal{O}^*\left(2^{n+\mathcal{O}(\sqrt{n})}\right)$, and so we have:

Theorem 4. *t-TOTAL EDGE COVER can be solved in $\mathcal{O}^*\left(2^{n+\mathcal{O}(\sqrt{n})}\right)$ time, where n is the number of vertices in the input graph.*

From this theorem and Corollary 1 we get:

Theorem 5. *t-TEC can be solved in $\mathcal{O}^*(2^{\frac{t+1}{t}k+\mathcal{O}(\sqrt{k})})$ time.*

The above algorithm uses exponential space (e.g., for constructing the lists L_i). An approach similar to the one used in Section 2 results in an FPT algorithm that runs in polynomial space:

Theorem 6. [$\star$] *t-TEC can be solved in time $\mathcal{O}^*\left(2^{O(k)}\right)$ time in polynomial space.*

4 Conclusion

We investigated the parameterized complexity of two problems, namely t-TOTAL VERTEX COVER and t-TOTAL EDGE COVER, obtained by imposing certain connectivity constraints on two classical problems, namely k-VERTEX COVER and k-EDGE COVER, respectively. We showed that for any $t \geq 2$ the t-TOTAL VERTEX COVER problem has no polynomial-size kernel unless PH collapses. We also showed that both these problems have FPT algorithms that run in time $\mathcal{O}^*\left(c^k\right)$ for different constants c.

One direction of future research would be to examine the effect of such connectivity constraints on other parameterized graph problems. Another would be to try to improve the base c of the exponent of the running times that we obtained for t-TOTAL VERTEX COVER and t-TOTAL EDGE COVER.

References

1. Alon, N., Yuster, R., Zwick, U.: Color-coding. Journal of the Association for Computing Machinery 42(4), 844–856 (1995)
2. Amini, O., Fomin, F.V., Saurabh, S.: Implicit branching and parameterized partial cover problems (extended abstract). In: Hariharan, R., Mukund, M., Vinay, V. (eds.) IARCS Annual Conference on Foundations of Software Technology and Theoretical Computer Science, FSTTCS 2008, Bangalore, India, December 9-11. LIPIcs, vol. 2, pp. 1–12. Schloss Dagstuhl–Leibniz-Zentrum fuer Informatik (2008)
3. Amini, O., Fomin, F.V., Saurabh, S.: Counting subgraphs via homomorphisms. In: Albers, S., Marchetti-Spaccamela, A., Matias, Y., Nikoletseas, S., Thomas, W. (eds.) ICALP 2009. LNCS, vol. 5555, pp. 71–82. Springer, Heidelberg (2009)
4. Beyer, T., Hedetniemi, S.M.: Constant time generation of rooted trees. SIAM Journal on Computing 9(4), 706–712 (1980)
5. Bodlaender, H.L., Downey, R.G., Fellows, M.R., Hermelin, D.: On problems without polynomial kernels (extended abstract). In: Aceto, L., Damgård, I., Goldberg, L.A., Halldórsson, M.M., Ingólfsdóttir, A., Walukiewicz, I. (eds.) ICALP 2008, Part I. LNCS, vol. 5125, pp. 563–574. Springer, Heidelberg (2008)
6. Bodlaender, H.L., Thomassé, S., Yeo, A.: Kernel bounds for disjoint cycles and disjoint paths. In: Fiat, A., Sanders, P. (eds.) ESA 2009. LNCS, vol. 5757, pp. 635–646. Springer, Heidelberg (2009)
7. Chen, J., Kanj, I.A., Jia, W.: Vertex Cover: Further observations and further improvements. Journal of Algorithms 41(2), 280–301 (2001)
8. Chen, J., Kanj, I.A., Xia, G.: Improved parameterized upper bounds for Vertex Cover. In: Královič, R., Urzyczyn, P. (eds.) MFCS 2006. LNCS, vol. 4162, pp. 238–249. Springer, Heidelberg (2006)
9. Cormen, T.H., Leiserson, C.E., Rivest, R.L., Stein, C.: Introduction to Algorithms, 2nd edn. The MIT Press, Cambridge (2001)
10. Diestel, R.: Graph Theory, 3rd edn. Springer, Heidelberg (2005)
11. Dom, M., Lokshtanov, D., Saurabh, S.: Incompressibility through Colors and IDs. In: Albers, S., Marchetti-Spaccamela, A., Matias, Y., Nikoletseas, S., Thomas, W. (eds.) ICALP 2009. LNCS, vol. 5555, pp. 378–389. Springer, Heidelberg (2009)
12. Fernau, H., Manlove, D.F.: Vertex and edge covers with clustering properties: Complexity and algorithms. Journal of Discrete Algorithms 7, 149–167 (2009)
13. Flum, J., Grohe, M.: Parameterized Complexity Theory. In: Texts in Theoretical Computer Science. An EATCS Series, Springer, Heidelberg (2006)
14. Fomin, F.V., Lokshtanov, D., Raman, V., Saurabh, S.: Subexponential algorithms for partial cover problems. In: Kannan, R., Kumar, K.N. (eds.) IARCS Annual Conference on Foundations of Software Technology and Theoretical Computer Science (FSTTCS 2009). LIPIcs, vol. 4, pp. 193–201. Schloss Dagstuhl–Leibniz-Zentrum fuer Informatik (2009)
15. Garey, M.R., Johnson, D.S.: Computers and Intractability: A Guide to the Theory of NP–Completeness. Freeman, San Francisco (1979)
16. Guo, J., Niedermeier, R., Wernicke, S.: Parameterized complexity of vertex cover variants. Theory of Computing Systems 41(3), 501–520 (2007)
17. Hardy, G.H., Ramanujan, S.: Asymptotic formulae in combinatory analysis. Proceedings of the London Mathematical Society 17 (1918)
18. Karp, R.M.: Reducibility among combinatorial problems. In: Miller, R.E., Thatcher, J.W. (eds.) Complexity of Computer Communications, pp. 85–103 (1972)

19. Kneis, J., Langer, A., Rossmanith, P.: Improved upper bounds for partial vertex cover. In: Broersma, H., Erlebach, T., Friedetzky, T., Paulusma, D. (eds.) WG 2008. LNCS, vol. 5344, pp. 240–251. Springer, Heidelberg (2008)
20. Kneis, J., Mölle, D., Richter, S., Rossmanith, P.: Intuitive algorithms and t-vertex cover. In: Asano, T. (ed.) ISAAC 2006. LNCS, vol. 4288, pp. 598–607. Springer, Heidelberg (2006)
21. Knuth, D.E.: The Art of Computer Programming, 3rd edn. Seminumerical Algorithms, vol. 3. Addison-Wesley, Reading (1998)
22. Mölle, D., Richter, S., Rossmanith, P.: Enumerate and Expand: Improved algorithms for Connected Vertex Cover and Tree Cover. Theory of Computing Systems 43(2), 234–253 (2008)
23. Naor, M., Schulman, L.J., Srinivasan, A.: Splitters and near-optimal derandomization. In: 36th Annual Symposium on Foundations of Computer Science (FOCS '95), pp. 182–193. IEEE Computer Society Press, Los Alamitos (1995)
24. Norman, R.Z., Rabin, M.O.: An algorithm for a minimum cover of a graph. Proceedings of the American Mathematical Society 10, 315–319 (1959)
25. Otter, R.: The number of trees. Annals of Mathematics 49(3), 583–599 (1948)
26. Zoghbi, A., Stojmenovic, I.: Fast algorithms for generating integer partitions. International Journal of Computer Mathematics 70, 319–332 (1998)

Counting Paths in VPA Is Complete for $\#\mathsf{NC}^1$

Andreas Krebs[1], Nutan Limaye[2], and Meena Mahajan[3]

[1] University of Tübingen, Germany
mail@krebs-net.de
[2] Tata Institute of Fundamental Research, Mumbai 400 005, India
nutan@tcs.tifr.res.in
[3] The Institute of Mathematical Sciences, Chennai 600 113, India
meena@imsc.res.in

Abstract. We give a $\#\mathsf{NC}^1$ upper bound for the problem of counting accepting paths in any fixed visibly pushdown automaton. Our algorithm involves a non-trivial adaptation of the arithmetic formula evaluation algorithm of Buss, Cook, Gupta, Ramachandran ([9]). We also show that the problem is $\#\mathsf{NC}^1$ hard. Our results show that the difference between $\#\mathsf{BWBP}$ and $\#\mathsf{NC}^1$ is captured exactly by the addition of a visible stack to a nondeterministic finite-state automata.

1 Introduction

We investigate the complexity of the following problem: Fix any visibly pushdown automata V. Given a word w over the input alphabet of V, compute the number of accepting paths that V has on w. We show that this problem is complete for the counting class $\#\mathsf{NC}^1$.

The class $\#\mathsf{NC}^1$ was first singled out for systematic study in [10], and has been studied from many different perspectives; see [1,5,7,10]. It consists of functions from strings to numbers that can be computed by arithmetic circuits (using the operations $+$ and $\times$ and the constants $0, 1$) of polynomial size and logarithmic depth. Equivalently, these functions compute the number of accepting proof trees in a Boolean NC^1 circuit (a polynomial size logarithmic depth circuit over $\vee$ and $\wedge$). It is known that characteristic functions of Boolean NC^1 languages can be computed in $\#\mathsf{NC}^1$ and that functions in $\#\mathsf{NC}^1$ can be computed in deterministic logspace. It is also known that functions in $\#\mathsf{NC}^1$ can be computed by Boolean circuits of polynomial size and $O(\log n \log^* n)$ depth; that is, almost in Boolean NC^1. An analogue of Barrington's celebrated thereom ([4]) stating that Boolean NC^1 equals languages accepted by families of bounded-width branching programs BWBP almost goes through here: functions computed by arithmetic BWBP, denoted $\#\mathsf{BWBP}$, are also computable in $\#\mathsf{NC}^1$, and $\#\mathsf{NC}^1$ functions are expressible as the difference of two $\#\mathsf{BWBP}$ functions. All attempts so far to remove this one subtraction and place $\#\mathsf{NC}^1$ in $\#\mathsf{BWBP}$ have failed.

A nice characterization of $\#\mathsf{BWBP}$, extending Barrington's result for the Boolean case, is in terms of branching programs over monoids, and yields the following ([10]): there is a fixed NFA (nondeterministic finite-state automaton)

M.T. Thai and S. Sahni (Eds.): COCOON 2010, LNCS 6196, pp. 44–53, 2010.

N such that any function f in #BWBP can be reduced to counting accepting paths of N. In particular, $f(x)$ equals the number of accepting paths of N on a word $g(x)$ that is a projection of x (each letter in $g(x)$ is either a constant or depends on exactly one letter of x) and is of size polynomial in the length of x. There has been no similar characterization of #NC^1 so far (though there is a characterization of its closure under subtraction GapNC^1, using integer matrices of constant dimension). Our result does exactly this; the hardness proof shows that there is a fixed VPA (visibly pushdown automaton) V such that any function f in #NC^1 can be reduced via projections to counting accepting paths of V, and the algorithm shows that any #VPA function (the number of accepting paths in any VPA) can be computed in #NC^1. Thus, the difference (if any) between #BWBP and #NC^1, which is known to vanish with one subtraction, is captured exactly by the extension of NFA to VPA.

What exactly are visibly pushdown automata? These are pushdown automata (PDA) with certain restrictions on their transition functions. (They are also called input-driven automata, in some of the older literature. See [15,6,11,3].) There are no ϵ moves. The input alphabet is partitioned into call, return and internal letters. On a call letter, the PDA must push a symbol onto its stack, on a return letter it must pop a symbol, and on an internal move it cannot access the stack at all. While this is a severe restriction, it still allows VPA to accept non-regular languages (for example $\{a^n b^n \mid n \geq 0\}$). At the same time, VPA are less powerful than all PDA; they cannot even check if a string has an equal number of a's and b's. In fact, due to the visible nature of the stack, membership testing for VPA is significantly easier than for general PDA; it is known to be in Boolean NC^1 ([11]). In other words, as far as membership testing is concerned, VPA are no harder than NFA.

However, the picture changes where counting accepting paths is concerned. An obvious upper bound on #VPA functions is the upper bound for #PDA functions. It is tempting to speculate that since membership testing for VPA and NFA have the same complexity, so does counting. This was indeed claimed, erroneously, in [13]; the subsequent version in [14] retracted this claim and showed an upper bound of LogDCFL for #VPA functions. In this paper, we improve this upper bound to #NC^1, and show via a hardness proof that improving it to #BWBP or #NFA would imply #BWBP= #NC^1. Our main results are:

Theorem 1. #VPA $\subseteq$ #NC^1.
For every fixed VPA V*, there is a family of polynomial-size logarithmic depth bounded fanin circuits over* $+$ *and* $\times$ *that computes, for each word* w*, the number of accepting paths of* V *on* w.

Theorem 2. #NC^1 $\leq$ #VPA.
There is a fixed VPA V *such that for any function family* $\{f_n\}$ *in* #NC^1*, there is a uniform reduction (via projections)* π *such that for each word* w*,* $f(w)$ *equals the number of accepting paths of* V *on* $\pi(w)$.

Combining the two results, we see that branching programs over VPA characterize #NC^1 functions. Using notation from [14,10]:

Corollary 1. $\#\mathsf{NC}^1 = \#\mathsf{BP\text{-}VPA}$

Here is a high-level description of how we achieve the upper bound. In [8], Buss showed that Boolean formulas can be evaluated in NC^1. In [9], Buss, Cook, Gupta, and Ramachandran extended this to arithmetic formulas over semi-rings. We use the algorithm of [9], but not as a black-box. We show that counting paths on a word in a VPA can be written as a formula in a new algebra which is not a semi-ring. However, the crucial way in which semi-ring properties are used in [9] is to assert that for any specified position (called scar) in the formula, the final value is a linear function of the value computed at the scar. We show that this property holds even over our algebra, because of the behaviour of VPA. Thus the strategy of [9] for choosing scar positions can still be used to produce a logarithmic depth circuit, where each gate computes a constant number of operations over this algebra. We also note that in our algebra, basic operations are in fact computable in $\#\mathsf{NC}^0$ (constant-depth constant fanin circuits over $+$ and $\times$). Thus the circuit produced above can be implemented in $\#\mathsf{NC}^1$.

The rest of this paper is organised as follows. In Section 2 we set up the basic notations required for our main result. In Section 3 we present an overview of the arithmetic formula evaluation algorithm of [9], highlighting the points where we will make changes. Section 4 describes our adaptation of this algorithm, placing #VPA in $\#\mathsf{NC}^1$. In Section 5, we show that #VPA functions are hard for $\#\mathsf{NC}^1$.

2 Preliminaries

Definition 1 (Visibly Pushdown Automaton). *A visibly pushdown automaton on finite words over a tri-partitioned alphabet $\Sigma = \Sigma_c \cup \Sigma_r \cup \Sigma_i$ is a tuple $V = (Q, Q_I, \Gamma, \delta, Q_F)$ where Q is a finite set of states, $Q_I \subseteq Q$ is a set of initial states, Γ is a finite stack alphabet that contains a special bottom-of-stack symbol $\perp$, δ is the transition function $\delta \subseteq (Q \times \Sigma_c \times Q \times \Gamma) \cup (Q \times \Sigma_r \times \Gamma \times Q) \cup (Q \times \Sigma_i \times Q)$, and $Q_F \subseteq Q$ is a set of final states. The letters in Σ_c, Σ_r, and Σ_i are called call letters, return letters, and internal letters, respectively.*

The technical definition in [3] also allows pop moves on an empty stack, with a special bottom-of-stack marker $\perp$. However, we will not need such moves because of *well-matchedness*, discussed later below.

Definition 2 (#VPA). *A function $f : \Sigma^* \longrightarrow \mathbb{N}$ is said to be in* #VPA *if there is a* VPA *V over the alphabet Σ such that for each $w \in \Sigma^*$, $f(w)$ is exactly the number of accepting paths of V on w.*

Without loss of generality, we can assume that there are no internal letters. (If there are, then design another VPA where each internal letter ν is replaced by a string $\alpha\beta$ where α is a new call letter with a new stack symbol and β is a new return letter.)

We say that a string is *well-matched* if the VPA never sees a return letter when its stack is empty, and at the end of the word its stack is empty. It can be assumed that all strings are well-matched; in [14], there is a conversion from

VPA V to VPA V', and a reduction (computable in NC^1) from inputs w of V to inputs w' of V' such that the number of accepting paths is preserved.

Theorem 1 places #VPA functions in $\#\mathsf{NC}^1$. But $\#\mathsf{NC}^1$ is a class of functions from $\{0,1\}^*$ to $\mathbb{N}$, while VPA may have arbitrary input alphabets. Using standard terminology (see, eg, [2,10]), we assume that the leaves of the $\#\mathsf{NC}^1$ circuits are labeled by predicates of the form $[w_i = a]$.

For hardness, we use the notion of reductions via projections.

Definition 3 (Projections, [10]). *A function $f : \Sigma^* \longrightarrow \Delta^*$ is a projection if for each $x \in \Sigma^*$, each letter in $f(x)$ is either a constant or depends on exactly one bit of x.*

For a definition of uniformity in projections, see [10].

3 An Overview of the BCGR Algorithm

The algorithm of [9] for arithmetic formula evaluation over commutative semirings $(S, +, \cdot)$ builds upon Brent's recursive evaluation method [7], but uses a pebbling game to make the construction uniform and oblivious. The recursive strategy is as follows:

For formula ϕ let A be the value of ϕ, and for a position j within it, let ϕ_j be the formula rooted at j and let A_j be the value of ϕ_j. Let $A(j, X)$ be the function corresponding to the formula ϕ with the sub-formula ϕ_j replaced by the indeterminate X. We say ϕ is scarred at j. Then $A(j, X) = B \cdot X + C$, and to compute the value A, we recursively determine B, C, and the correct value of X (that is, A_j). The recursive procedure ensures that there is at most one scar at each stage. Thus while considering $A(j, X)$, the next scar is always chosen to be an ancestor of ϕ_j. It is shown in [9] that there is a way of choosing scars such that the recursion terminates in $O(\log |\phi|)$ rounds. The main steps of the algorithm of [9] can thus be stated as follows:

1. Convert the given formula ϕ' to an equivalent formula represented in post-fix form, with the longer operand of each operator appearing first in the expression. Call this PLOF (Post-Fix Longer Operand First). Pad the formula with a unary identity operator, if necessary, so that its length is a power of 2. Let ϕ be the resulting formula.
2. Construct an $O(\log |\phi|)$ depth fanin 3 "circuit" $\mathcal{C}$, where each "gate" of $\mathcal{C}$ is a constant-size program, or a block, associated with a particular sub-formula. The top-most block is associated with the entire formula. A block associated with interval g of length $4m$ has as its three children the blocks associated with the prefix interval g_1, the centred interval g_2, and the suffix interval g_3, each of length $2m$. Each interval has upto 9 designated positions or sub-formulas, and the block computes the values of the sub-formulas rooted at these positions. The set of these positions will contain all possible scar positions that are good (that can lead to an $O(\log 4m)$ depth recursion).

3. Describe Boolean NC^1 circuitry that determines the designated positions within each block. This circuitry depends only on the position of the block within $\mathcal{C}$, and on the letters appearing in the associated interval, not on the values computed so far.
4. Using this Boolean circuitry along with the values at the designated positions in the children g_i, compute the values at designated positions for the interval g using $\#\mathsf{NC}^0$ circuits. Plug in this $\#\mathsf{NC}^0$ circuit for each block of $\mathcal{C}$ to get a $\#\mathsf{NC}^1$ circuit.

To handle the non-commutative case, add an operator $\cdot'$. In conversion to PLOF, if the operands of $\cdot$ have to be switched, then replace the operator by $\cdot'$. The actual operator is correctly applied within the $\#\mathsf{NC}^0$ block at the last step.

4 Adaption of the BCGR Algorithm

The algorithm in the previous section works for any non-commutative ring. Unfortunately we were not able to find a non-commutative ring in which we can compute the value of a given VPA. So we will define an algebraic structure that uses constant size matrices as its elements and has two operations $\otimes$ and $\odot$, but which is not a ring. Then we show that the algorithm of the previous section can be modified to work for our algebraic structure, and give the precise differences.

Let $V = (Q, Q_I, \Gamma, \delta, Q_F)$ be a fixed VPA and let $q = |Q|$.

In describing the NC^1 algorithm for membership testing in VPA, Dymond [11] constructed a formula using operators Ext and $\circ$ described below. In [14] it was shown that this formula can also evaluate the number of paths (and the number was computed using a deterministic auxiliary logspace pushdown machine which runs in polynomial time). Essentially, the formula builds up a $q \times q$ matrix $\hat{M}_w$ for the input word w by building such matrices for well-matched subwords. The (i,j)th entry of $\hat{M}_w$ gives the number of paths from state q_i to state q_j on reading w. (Due to well-matchedness, the stack contents are irrelevant for this number.) For the zero-length word $w = \epsilon$, the matrix is the identity matrix $\hat{I}$. The unary $\mathrm{Ext}_{\alpha\beta}$ operator computes, for any word w, the matrix $\hat{M}_{\alpha w \beta}$ from the matrix $\hat{M}_w$. The binary $\circ$ operator computes the matrix $\hat{M}_{ww'}$ from the matrices $\hat{M}_w$ and $\hat{M}_{w'}$. The formula over these matrices can be obtained from the input word in NC^1 (and in fact, even in TC^0, see [14,12]). The leaves of the formula all carry the identity matrix $\hat{I} = \hat{M}_\epsilon$.

Lemma 1 ([11,14]). *Fix a* VPA V. *For every well-matched word, there is a formula over* $\mathrm{Ext}_{\alpha\beta}, \circ$, *constructible in* NC^1, *which computes the* $q \times q$ *matrix* $\hat{M}_w$. *The* (i,j)*th entry of* $\hat{M}_w$ *is the number of paths from* q_i *to* q_j *while reading* w.

The unary operators $\mathrm{Ext}_{\alpha\beta}$ used above are functions mapping $\mathbb{N}^{q\times q}$ to $\mathbb{N}^{q\times q}$. From the definition of the operators,
$[\mathrm{Ext}_{\alpha\beta}(M)]_{ij} = \sum_{kl} M_{kl} \cdot |\{X \mid q_i \xrightarrow{\alpha} q_k X; \;\; q_l X \xrightarrow{\beta} q_j\}|$

it is easy to see that these functions are linear operators on $\mathbb{N}^{q\times q}$. The linear operators of $\mathbb{N}^{q\times q}$ can be written as $q\times q$ matrices with $q\times q$ matrices as entries, or simply as matrices of size $q^2 \times q^2$. So we can represent every unary $\mathrm{Ext}_{\alpha\beta}$ operator as a $q^2 \times q^2$ matrix. However, the binary $\circ$ operator works with $q \times q$ matrices. To unify these two sizes, we embed the $q \times q$ matrices from Lemma 1 into $q^2 \times q^2$ matrices in a particular way.

We require that for a matrix M at some position in Dymond's formula, our corresponding matrix M satisfies the following: $\hat{M}_{ij} = M_{(ij)(oo)}$ for all indices o. (The values at $M_{(ij)(kl)}$ when $k \neq l$ are not important.)

The operators to capture $\mathrm{Ext}_{\alpha\beta}$ and $\circ$ are defined as follows.

Definition 4. *Let $\mathbb{M}$ be the family of $q^2 \times q^2$ matrices over $\mathbb{N}$.*

1. *The matrix I is defined as the "pointwise" identity matrix, i.e. $I_{(ij)(kl)} = 1$ if $i = j \wedge k = l$ and $I_{(ij)(kl)} = 0$ otherwise.*
2. *For each well-matched string $\alpha\beta$ of length 2 (i.e. $\alpha \in \Sigma_c, \beta \in \Sigma_r$), the matrix $EXT^{\alpha\beta} \in \mathbb{M}$ is the matrix corresponding to $\mathrm{Ext}_{\alpha\beta}$ and is defined as $EXT^{\alpha\beta}_{(ij)(kl)} = [\mathrm{Ext}_{\alpha\beta}(E_{kl})]_{ij}$, where E_{kl} is a $q \times q$ matrix with a 1 at position (k,l) and zeroes everywhere else.*
3. *The operator $\otimes : \mathbb{M} \longrightarrow \mathbb{M}$ is matrix multiplication, i.e. $M = S \otimes T \iff M_{(ij)(kl)} = \sum_{u,v} S_{(ij)(uv)} T_{(uv)(kl)}$.*
4. *The operator $\odot : \mathbb{M} \longrightarrow \mathbb{M}$ is a "point-wise" matrix multiplication: $M = S \odot T \iff M_{(ij)(kl)} = \sum_{u} S_{(iu)(kl)} T_{(uj)(kl)}$.*
5. *The algebraic structure $\mathbb{V}'$ is defined as follows. $\mathbb{V}' = (\mathbb{M}, \otimes, \odot, I, \{EXT^{\alpha\beta}\}_{\alpha\in\Sigma_c,\beta\in\Sigma_r})$.*

Now we change Dymond's formula into a formula that works over the algebra $\mathbb{V}'$ and produces a matrix in $\mathbb{M}$. The $\circ$ operator that represented concatenation is now replaced by the new $\odot$ operator. Each unary $\mathrm{Ext}_{\alpha\beta}$ operator with argument ψ is replaced by the sub-formula $EXT^{\alpha\beta} \otimes \psi$. Each leaf matrix $\hat{I}$ is replaced by I. See Example 1 below.

Example 1. Let $w = \alpha\alpha\beta\alpha\alpha\beta\beta\beta$ be word over $\{\alpha, \beta\}$, where α is a call letter and β is a return letter. Then Dymond's formula is

$$\mathrm{Ext}_{\alpha\beta}(\ \mathrm{Ext}_{\alpha\beta}(\hat{I})\ \circ\ \mathrm{Ext}_{\alpha\beta}(\ \mathrm{Ext}_{\alpha\beta}(\hat{I})\ )\ ).$$

The formula we obtain over $\mathbb{V}'$ is given below

$$(EXT^{\alpha\beta} \otimes ((EXT^{\alpha\beta} \otimes I) \odot (EXT^{\alpha\beta} \otimes (EXT^{\alpha\beta} \otimes I)))).$$

Lemma 2. *Let w be a well-matched word, and let $\hat{\phi}$ be the corresponding formula from Lemma 1. Let ϕ be the formula over $\mathbb{V}'$ obtained by changing $\hat{\phi}$ as described above. Then, for each sub-formula $\hat{\psi}$ of $\hat{\phi}$ and corresponding ψ of ϕ, if $\hat{\psi}$ computes $\hat{P} \in \mathbb{N}^{q\times q}$ and ψ computes $P \in \mathbb{M}$, then:*

$$\forall i, j, o \in [q] \quad \hat{P}_{(ij)} = P_{(ij)(oo)}$$

From Lemma 2, and since we do only a syntactic replacement at each node of Dymond's formula, we conclude:

Lemma 3. *For every well-matched word w, there is a formula over $\mathbb{V}'$, constructible in NC^1, which computes a $q^2 \times q^2$ matrix M satisfying the following: For each i, j, the number of paths from q_i to q_j while reading w is $M_{(ij)(oo)}$ for all o.*

For using the template of [9], we need to convert our formula to PLOF format. But our operators $\otimes$ and $\odot$ are not commutative. We handle this exactly as in [9], extending the algebra to include the antisymmetric operators. We also need that the length of the formula is a power of 2. In order to handle this, we introduce a unary identity operator $\ominus(S)$.

Definition 5. *The algebraic structure $\mathbb{V}$ is the extension of the structure $\mathbb{V}'$ to include the operators $\otimes'$, $\odot'$, and $\ominus$, defined as follows.*
$S \otimes' T = T \otimes S$; $S \odot' T = T \odot S$, $\ominus(S) = S$.

Since the conversion to PLOF as outlined in [9] does not depend on the semantics of the structure, we can do the same here. Further, since our formula has a special structure (at each $\otimes$ node, the left operand is a leaf of the form $EXT^{\alpha\beta}$), we can rule out using some operators.

Lemma 4. *Given a formula over $\mathbb{V}$ in infix notation as constructed in Lemma 3, there is a formula over $\mathbb{V}$ in postfix longest operator first PLOF form, constructible in NC^1, which computes the same matrix. The formula over $\mathbb{V}$ in PLOF form will not make use of the operator $\otimes$.*

In the following we do not want to allow arbitrary formulas but only formulas that describe the run of the VPA V.

Definition 6. *Let ϕ be a valid formula in $\mathbb{V}$, and let ϕ_j be a sub-formula of ϕ appearing as a prefix in the PLOF representation of ϕ. If we replace ϕ_j by an indeterminate X and obtain a formula ψ over $\mathbb{V}[X]$, we call ψ a formula with a left-most scar. In a natural way a formula with a left-most scar represents a function $f : \mathbb{M} \to \mathbb{M}$ which we call the value of ψ.*

We say a formula/formula with a left-most scar over $\mathbb{V}$ is valid if it is obtained by Lemma 3, and converted to PLOF as in Lemma 4, and then possibly scarred at a prefix sub-formula. In the following we will only consider valid formulas. Following the algorithm we need to show that we can write every valid formula with a left-most scar as a fixed expression. From Lemma 4 it follows that these have the following form.

Lemma 5. *Let ψ be a valid formula in PLOF over $\mathbb{V}$ with a left-most scar X. Then ψ is of the form X, or $\sigma\ominus$, or $\sigma EXT^{\alpha\beta}\otimes'$, or $\sigma\tau\odot$, or $\sigma\tau\odot'$, where σ is a valid formula over $\mathbb{V}$ with a left-most scar X, $\alpha \in \Sigma_c$, $\beta \in \Sigma_r$, and τ is a valid formula.*

Let ψ be a formula over an algebraic structure $\mathbb{V}$ with a left-most scar X. Then the value of this formula with a scar in general can be represented by a function $f : \mathbb{M} \to \mathbb{M}$. In our case the situation is much simpler; we show that all the functions that occur in our construction can be represented by functions of the form $f(X) = B \cdot X$, where $B \in \mathbb{M}$ is an element of our structure, *i.e.* a $q^2 \times q^2$ matrix of the natural numbers. (By the definition of $\otimes$ we know that $B \otimes X$ is also given by matrix multiplication $B \cdot X$, but still we use the $\cdot$ operator here to distinguish between the different uses of the semantical equivalent expressions.) Actually the situation is a bit more technical. Since we are only interested in computing the values in the "diagonal", we only need to ensure that these values are computed correctly. In the following lemma we show that the algebraic structure does allow us to represent the computation of these values succinctly. In Lemma 7 we will show that these succinct representations can be computed as required.

Lemma 6. *Let ψ be a valid formula in PLOF over $\mathbb{V}$ with a left-most scar X, and let $f : \mathbb{V} \to \mathbb{V}$ be the value of ψ. Then there is an element $B \in \mathbb{M}$ such that $f(X)_{(ij)(oo)} = (B \cdot X)_{(ij)(oo)}$ for all $i, j, o \in [q]$.*

Since we are able to represent every scarred formula by a constant size matrix, we can apply a conversion similar to the BCGR algorithm and end up with a circuit of logarithmic depth. The only thing that remains is to show that computations within the blocks can be computed by $\#\mathsf{NC}^0$ circuits, so that the total circuit will be in $\#\mathsf{NC}^1$. Note that each block is expected to compute for its associated formula the element $B \in \mathbb{M}$ guaranteed by Lemma 6, assuming the corresponding elements at designated sub-formulas have been computed.

The construction of the blocks is very restricted by the BCGR algorithm. We know that there are only two major cases that happen:

- Either we need to compute a formula ψ with a scar X from the value of the formula ψ with a scar Y and the value of the formula Y with the scar X (here X can be the empty scar),
- or we need to compute a formula ψ with operand one of $\otimes'$ $\odot$, $\odot'$, $\ominus$, and a scar X, where the left argument contains the scar X and we are already given the value of the left argument with the scar X and the value of the right argument.

Hence we need to show that we can compute the operators in our algebraic structure in constant depth, as well as the computations for the matrix representations of our functions as in the proof of Lemma 6.

Lemma 7. *Let $S, T \in \mathbb{M}$. There are $\#\mathsf{NC}^0$ circuits that compute the matrix P, where P is defined in any one of the following ways: $P = S \cdot T$, $P = S \otimes' T$, $P = S \odot T$, $P = S \odot' T$, $P_{(ij)(kl)} = \sum_m S_{(im)(kl)} T_{(mj)(11)}$, or $P_{(ij)(kl)} = \sum_m S_{(im)(11)} T_{(mj)(kl)}$.*

Hence we can replace the blocks by $\#\mathsf{NC}^0$ circuits and obtain a $\#\mathsf{NC}^1$ circuit. But this implies Theorem 1.

5 Hardness

In this section we will show that there is a hardest function in $\#\mathsf{NC}^1$ and that it can be computed by a VPA. Let $\Sigma = \{[,], \bar{]}, +, \times, 0, 1\}$. We will define a function $f : \Sigma^* \to \mathbb{N}$ that is $\#\mathsf{NC}^1$ hard under projections and is computable by a VPA. Informally speaking, the following language will represent equations with $+, \times$ over the natural numbers, where the operation $+$ is always bracketed by $[$ and $\bar{]}]$. The 2 closing brackets are necessary for the technical reasons.

Definition 7. *Define L to be the smallest language containing 0, 1, and uv and $[u + v\bar{]}]$ for each $u, v \in L$, Also, define the function $f : \Sigma^* \to \mathbb{N}$ as: $f(0) = 0; f(1) = 1$; for $u, v \in L$, $f([u + v\bar{]}]) = f(u) + f(v)$ and $f(uv) = f(u) \cdot f(v)$; $f(u) = 0$ for $u \notin L$.*

Please note that the value of $f(w)$ does not depend on the decomposition of w since multiplication on natural numbers is associative.

Lemma 8. *Computing the value of $f(w)$ on words $w \in L$ is $\#\mathsf{NC}^1$ hard under projections.*

We will design a fixed VPA V. The idea for this VPA is to compute inductively the value of f as paths from one state q_C to q_C, while always keeping a single path from another state q_I to q_I. This allows us to temporarily store a value in the computation, and with the stack of the VPA we use this as a stack for storing values.

We will first give the definition of the VPA and then prove that it computes f on all words of L. It will have a tripartitioned input alphabet $\Sigma = \Sigma_c \cup \Sigma_r \cup \Sigma_i$ where $\Sigma_c = \{[, +\}$, $\Sigma_r = \{\bar{]},]\}$, and $\Sigma_i = \{0, 1\}$.

We let $V = \{\{q_C, q_I\}, \{q_C\}, \{T_1, T_2\}, \delta, \{q_C\}\}$, where the transition function is defined as :

Σ_i	q_C	q_I
1	q_C	q_I
0	$-$	q_I

Σ_c	q_C	q_I
$[$	$q_C T_1, q_I T_1$	$q_I T_2$
$+$	$q_I T_1$	$q_C T_1, q_I T_2$

Σ_r	$q_C T_1$	$q_I T_1$	$q_C T_2$	$q_I T_2$
$\bar{]}$	q_C	q_C	$-$	q_I
$]$	q_C	$-$	$-$	q_I

We now show that this VPA actually computes f.

Lemma 9. *For all $w \in L$, the number of accepting paths of the* VPA *V is exactly computed by $f(w)$.*

It is also easy to see that L itself can be recognized by an VPA (with the same partition into call and return letters). Since f is 0 outside L, hence there is another VPA that computes f on all of Σ^*. This establishes Theorem 2.

References

1. Allender, E.: Arithmetic circuits and counting complexity classes. In: Krajicek, J. (ed.) Complexity of Computations and Proofs. Quaderni di Matematica, vol. 13, pp. 33–72. Seconda Universita di Napoli (2004); An earlier version appeared in the Complexity Theory Column, SIGACT News 28(4), 2–15 (December 1997)

2. Allender, E., Jiao, J., Mahajan, M., Vinay, V.: Non-commutative arithmetic circuits: depth reduction and size lower bounds. Theoretical Computer Science 209, 47–86 (1998)
3. Alur, R., Madhusudan, P.: Visibly pushdown languages. In: STOC, pp. 202–211 (2004)
4. Barrington, D.A.: Bounded-width polynomial size branching programs recognize exactly those languages in NC^1. Journal of Computer and System Sciences 38, 150–164 (1989)
5. Ben-Or, M., Cleve, R.: Computing algebraic formulas using a constant number of registers. SIAM Journal on Computing 21, 54–58 (1992)
6. Von Braunmuhl, B., Verbeek, R.: Input-driven languages are recognized in log n space. In: Ésik, Z. (ed.) FCT 1993. LNCS, vol. 710, pp. 40–51. Springer, Heidelberg (1993)
7. Brent, R.P.: The parallel evaluation of general arithmetic expressions. Journal of the ACM 21, 201–206 (1974)
8. Buss, S.: The Boolean formula value problem is in ALOGTIME. In: STOC, pp. 123–131 (1987)
9. Buss, S., Cook, S., Gupta, A., Ramachandran, V.: An optimal parallel algorithm for formula evaluation. SIAM Journal of Computation 21(4), 755–780 (1992)
10. Caussinus, H., McKenzie, P., Thérien, D., Vollmer, H.: Nondeterministic NC^0 computation. Journal of Computer and System Sciences 57, 200–212 (1998); Preliminary version in Proceedings of the 11th IEEE Conference on Computational Complexity, pp. 12–21 (1996)
11. Dymond, P.W.: Input-driven languages are in log n depth. Information processing letters 26, 247–250 (1988)
12. Limaye, N., Mahajan, M., Meyer, A.: On the complexity of membership and counting in height-deterministic pushdown automata. In: Hirsch, E.A., Razborov, A.A., Semenov, A., Slissenko, A. (eds.) CSR 2008. LNCS, vol. 5010, pp. 240–251. Springer, Heidelberg (2008)
13. Limaye, N., Mahajan, M., Raghavendra Rao, B.V.: Arithmetizing classes around NC^1 and L. In: Thomas, W., Weil, P. (eds.) STACS 2007. LNCS, vol. 4393, pp. 477–488. Springer, Heidelberg (2007)
14. Limaye, N., Mahajan, M., Raghavendra Rao, B.V.: Arithmetizing classes around NC^1 and L. Theory of Computing Systems 46(3), 499–522 (2010)
15. Mehlhorn, K.: Pebbling mountain ranges and its application to DCFL recognition. In: de Bakker, J.W., van Leeuwen, J. (eds.) ICALP 1980. LNCS, vol. 85, pp. 422–432. Springer, Heidelberg (1980)

Depth-Independent Lower Bounds on the Communication Complexity of Read-Once Boolean Formulas

Rahul Jain[1], Hartmut Klauck[2], and Shengyu Zhang[3]

[1] Centre for Quantum Technologies and Department of Computer Science, National University of Singapore
rahul@comp.nus.edu.sg

[2] Centre for Quantum Technologies and SPMS, Nanyang Technological University
hklauck@gmail.com

[3] Department of Computer Science and Engineering, The Chinese University of Hong Kong
syzhang@cse.cuhk.edu.hk

Abstract. We show lower bounds of $\Omega(\sqrt{n})$ and $\Omega(n^{1/4})$ on the randomized and quantum communication complexity, respectively, of all n-variable read-once Boolean formulas. Our results complement the recent lower bound of $\Omega(n/8^d)$ by Leonardos and Saks [LS09] and $\Omega(n/2^{O(d \log d)})$ by Jayram, Kopparty and Raghavendra [JKR09] for randomized communication complexity of read-once Boolean formulas with depth d.

We obtain our result by "embedding" either the Disjointness problem or its complement in any given read-once Boolean formula.

1 Introduction

A *read-once Boolean formula* $f : \{0,1\}^n \to \{0,1\}$ is a function which can be represented by a Boolean formula involving AND and OR such that each variable appears, possibly negated, at most once in the formula. An *alternating* AND-OR *tree* is a layered tree in which each internal node is labeled either AND or OR and the leaves are labeled by variables; each path from the root to the any leaf alternates between AND and OR labeled nodes. It is well known (see eg. [HW91]) that given a read-once Boolean formula $f : \{0,1\}^n \to \{0,1\}$ there exists a unique alternating AND-OR tree, denoted T_f, with n leaves labeled by input Boolean variables $z_1, \ldots, z_n$, such that the output at the root, when the tree is evaluated according to the labels of the internal nodes, is equal to $f(z_1 \ldots z_n)$. Given an alternating AND-OR tree T, let f_T denote the corresponding read-once Boolean formula evaluated by T.

Let $x, y \in \{0,1\}^n$ and let $x \wedge y, x \vee y$ represent the *bit-wise* AND, OR of the strings x and y respectively. For $f : \{0,1\}^n \to \{0,1\}$, let $f^\wedge : \{0,1\}^n \times \{0,1\}^n \to \{0,1\}$ be given by $f^\wedge(x,y) = f(x \wedge y)$. Similarly let $f^\vee : \{0,1\}^n \times \{0,1\}^n \to \{0,1\}$ be given by $f^\vee(x,y) = f(x \vee y)$. Recently Leonardos and Saks [LS09],

M.T. Thai and S. Sahni (Eds.): COCOON 2010, LNCS 6196, pp. 54–59, 2010.

investigated the *two-party randomized communication complexity* with constant error, denoted $\mathsf{R}(\cdot)$, of $f^\wedge, f^\vee$ and showed the following. (Please refer to [KN97] for familiarity with basic definitions in communication complexity.)

Theorem 1 ([LS09]). *Let $f : \{0,1\}^n \to \{0,1\}$ be a read-once Boolean formula such that T_f has depth d. Then*

$$\max\{\mathsf{R}(f^\wedge), \mathsf{R}(f^\vee)\} \geq \Omega(n/8^d).$$

In the theorem, the depth of a tree is the number of edges on a longest path from the root to a leaf. Independently, Jayram, Kopparty and Raghavendra [JKR09] proved randomized lower bounds of $\Omega(n/2^{O(d \log d)})$ for general read-once Boolean formulas and $\Omega(n/4^d)$ for a special class of "balanced" formulas.

It follows from results of Snir [Sni85] and Saks and Wigderson [SW86] (via a generic simulation of trees by communication protocols [BCW98]) that for the read-once Boolean formula with their canonical tree being a *complete binary* alternating AND-OR trees, the randomized communication complexity is $O(n^{0.753\ldots})$, the best known so far. However in this situation, the results of [LS09, JKR09] do not provide any lower bound since $d = \log_2 n$ for the complete binary tree. We complement their result by giving universal lower bounds that do not depend on the depth. Below $\mathsf{Q}(\cdot)$ represents the two-party *quantum communication complexity.*

Theorem 2. *Let $f : \{0,1\}^n \to \{0,1\}$ be a read-once Boolean formula. Then,*

$$\max\{\mathsf{R}(f^\wedge), \mathsf{R}(f^\vee)\} \geq \Omega(\sqrt{n}).$$

$$\max\{\mathsf{Q}(f^\wedge), \mathsf{Q}(f^\vee)\} \geq \Omega(n^{1/4}).$$

Remark

1. Note that the maximum in Theorem 1 and 2 is necessary since for example if f is the AND of the n input bits then it is easily seen that $\mathsf{R}(f^\wedge)$ is 1.
2. The randomized lower bound in the above theorem is easy to observe for balanced trees, as is also remarked in [LS09].
3. We obtain our result by "embedding" either the Disjointness problem or its complement in any given read-once Boolean formula. This simple idea was also used by Zhang [Zha09] and independently by Sherstov [She10] to show some relations between decision tree complexity and communication complexity.

2 Proofs

In this section we show the proof of Theorem 2. We start with the following definition.

Definition 1 (Embedding). *We say that a function*

$$g_1 : \{0,1\}^r \times \{0,1\}^r \to \{0,1\}$$

can be embedded *into a function* $g_2 : \{0,1\}^t \times \{0,1\}^t \to \{0,1\}$, *if there exist maps* $h_a : \{0,1\}^r \to \{0,1\}^t$ *and* $h_b : \{0,1\}^r \to \{0,1\}^t$ *such that* $\forall x, y \in \{0,1\}^r$, $g_1(x,y) = g_2(h_a(x), h_b(y))$.

It is easily seen that if g_1 can be embedded into g_2 then the communication complexity of g_2 is at least as large as that of g_1.

Let us define the *Disjointness* problem $\mathsf{DISJ}_n : \{0,1\}^n \times \{0,1\}^n \to \{0,1\}$ as $\mathsf{DISJ}_n(x,y) = \bigwedge_{i=1,\dots,n}(x_i \vee y_i)$ (where the usual negation of the variables is left out for notational simplicity). Similarly the *Non-Disjointness* problem $\mathsf{NDISJ}_n : \{0,1\}^n \times \{0,1\}^n \to \{0,1\}$ is defined as $\mathsf{NDISJ}_n(x,y) = \bigvee_{i=1,\dots,n}(x_i \wedge y_i)$. We shall also use the following well-known lower bounds.

Fact 1 ([KS92, Raz92]). $\mathsf{R}(\mathsf{DISJ}_n) = \Omega(n), \mathsf{R}(\mathsf{NDISJ}_n) = \Omega(n)$.

Fact 2 ([Raz03]). $\mathsf{Q}(\mathsf{DISJ}_n) = \Omega(\sqrt{n}), \mathsf{Q}(\mathsf{NDISJ}_n) = \Omega(\sqrt{n})$.

Recall that for the given read-once Boolean formula $f : \{0,1\}^n \to \{0,1\}$ its canonical tree is denoted T_f. We have the following lemma which we prove in Section 2.1.

Lemma 3. *1. Let T_f have its last layer consisting only of* AND *gates. Let m_0 be the largest integer such that DISJ_{m_0} can be embedded into $f^\vee$ and m_1 be the largest integer such that NDISJ_{m_1} can be embedded into $f^\vee$. Then $m_0 m_1 \geq n$.*

2. Let T_f have its last layer consisting only of OR *gates. Let m_0 be the largest integer such that DISJ_{m_0} can be embedded into $f^\wedge$ and m_1 be the largest integer such that NDISJ_{m_1} can be embedded into $f^\wedge$. Then $m_0 m_1 \geq n$.*

With this lemma, we can prove the lower bounds on $\max\{\mathsf{R}(f^\wedge), \mathsf{R}(f^\vee)\}$ and $\max\{\mathsf{Q}(f^\wedge), \mathsf{Q}(f^\vee)\}$ as follows. For an arbitrary read-once formula f with n variables, consider the sets of leaves

$$L_{odd} = \{\text{leaves in } T_f \text{ on odd levels}\}, L_{even} = \{\text{leaves in } T_f \text{ on even levels}\}$$

At least one of the two sets has size at least $n/2$; without loss of generality, let us assume that it is L_{odd}. Depending on whether the root is AND or OR, this set consists only of AND gates or OR gates, corresponding to case 1 or 2 in Lemma 3. Then by the lemma, either $\mathsf{DISJ}_{\sqrt{n/2}}$ or $\mathsf{NDISJ}_{\sqrt{n/2}}$ can be embedded in f (by setting the leaves in L_{even} to 0's). By Fact 1 and 2, we get the lower bounds in Theorem 2.

2.1 Proof of Lemma 3

We shall prove the first statement; the second statement follows similarly. We first prove the following claim.

Claim 1. *For an n-leaf* ($n > 2$) *alternating* AND-OR *tree* T *such that all its internal nodes just above the leaves have exactly two children (denoted the* x*-child and the* y*-child), let* $s(T)$ *denote the number of such nodes directly above the leaves. Let* $m_0(T)$ *be the largest integer such that* DISJ_{m_0} *can be embedded into* f_T *and* $m_1(T)$ *be the largest integer such that* NDISJ_{m_1} *can be embedded into* f_T*. Then* $m_0(T)m_1(T) \geq s(T)$*.*

Proof. The proof is by induction on depth d of T. When $n > 2$, the condition of the tree makes $d > 1$, so the base case is $d = 2$.

Base Case $d = 2$**:** In this case T consists either of the root labeled AND with $s(T)$ (fan-in 2) children labeled ORs or it consists of the root labeled OR with $s(T)$ (fan-in 2) children labeled ANDs. We consider the former case and the latter follows similarly. In the former case f_T is clearly $\mathsf{DISJ}_{s(T)}$ and hence $m_0(T) = s(T)$. Also $m_1(T) \geq 1$ as follows. Let us choose the first two children v_1, v_2 of the root. Further choose the x child of v_1 and the y child of v_2 which are kept free and the values of all other input variables are set to 0. It is easily seen that the function (of input bits x, y) now evaluated is NDISJ_1. Hence $m_0(T)m_1(T) \geq s(T)$.

Induction Step $d > 2$**:** Assume the root is labeled AND (the case when the root is labeled OR follows similarly). Let the root have r children $v_1, \ldots, v_r$ which are labeled OR and let the corresponding subtrees be $T_1, \ldots, T_r$ rooted at $v_1, \ldots, v_r$ respectively. Without loss of generality let the first r' (with $0 \leq r' \leq r$) of these trees be of depth 1 in which case the corresponding $s(\cdot) = 0$. It is easily seen that

$$s(T) = r' + \left(\sum_{i=r'+1}^{r} s(T_i) \right) .$$

For $i > r'$, we have from the induction hypothesis that $m_1(T_i)m_0(T_i) \geq s(T_i)$.

It is clear that $m_0(T) \geq \sum_{i=1}^{r} m_0(T_i)$, since we can simply combine the Disjointness instances of the subtrees. Also we have

$$m_1(T) \geq \max\{m_1(T_{r'+1}), \ldots, m_1(T_r), 1\},$$

because we can either take any one of the subtree instances (and set all other inputs to 0), or at the very least can pick a pair of x, y leaves (as in the base case above) and fix the remaining variables appropriately to realize a single AND gate which amounts to embedding NDISJ_1. Now,

$$\begin{aligned} m_0(T)m_1(T) &\geq \left(\sum_{i=1}^{r} m_0(T_i) \right) \cdot (\max\{m_1(T_1), \ldots, m_1(T_r), 1\}) \\ &\geq r' + \left(\sum_{i=r'+1}^{r} m_0(T_i)m_1(T_i) \right) \\ &\geq r' + \left(\sum_{i=r'+1}^{r} s(T_i) \right) = s(T) . \end{aligned}$$

Now we prove Lemma 3: Let us view $f^\vee : \{0,1\}^{2n} \to \{0,1\}$ as a read-once Boolean formula, with input (x,y) of $f^\vee$ corresponding to the x- and y- children of the internal nodes just above the leaves. Note that in this case $T_{f^\vee}$ satisfies the conditions of the above claim and $s(T_{f^\vee}) = n$. Hence the proof of the first statement in Lemma 3 finishes.

3 Concluding Remarks

1. The randomized communication complexity varies between $\Theta(n)$ for the Tribes_n function (a read-once Boolean formula whose canonical tree has depth 2) [JKS03] and $O(n^{0.753\ldots})$ for functions corresponding to completely balanced AND-OR trees (which have depth $\log n$). It will probably be hard to prove a generic lower bound much larger than $\sqrt{n}$ for all read-once Boolean formulas $f : \{0,1\}^n \to \{0,1\}$, since the best known lower bound on the randomized query complexity of every read-once Boolean formula is $\Omega(n^{.51})$ [HW91] and communication complexity lower bounds immediately imply slightly weaker query complexity lower bounds (via the generic simulation of trees by communication protocols [BCW98]).
2. Ambainis et al. [ACR+07] show how to evaluate any alternating AND-OR tree T with n leaves by a quantum query algorithm with slightly more than $\sqrt{n}$ queries; this also gives the same upper bound for the communication complexity of $\max\{\mathsf{Q}(f_T^\wedge), \mathsf{Q}(f_T^\vee)\}$. On the other hand, it is easily seen that the *parity* of n bits can be computed by a formula of size $O(n^2)$ involving AND, OR. Therefore it is easy to show that the function *Inner Product modulo* 2 i.e. the function $\mathsf{IP}_m : \{0,1\}^m \times \{0,1\}^m \to \{0,1\}$ given by $\mathsf{IP}_m(x,y) = \sum_{i=1}^m x_i y_i \bmod 2$, with $m = \sqrt{n}$ can be reduced to the evaluation of an alternating AND-OR tree of size $O(n)$ and logarithmic depth. Since it is known that $\mathsf{Q}(\mathsf{IP}_{\sqrt{n}}) = \Omega(\sqrt{n})$ [CvDNT99], we get an example of an alternating AND-OR tree T with n leaves and $\log n$ depth such that $\mathsf{Q}(f_T^\wedge) = \Omega(\sqrt{n})$. Since the same lower bound also holds for shallow trees such as OR, hence $\Theta(\sqrt{n})$ might turn out to be the correct bound on $\max\{\mathsf{Q}(f_T^\wedge), \mathsf{Q}(f_T^\vee)\}$ for all alternating AND-OR trees T with n leaves regardless of the depth.

Acknowledgments. Research of Rahul Jain and Hartmut Klauck is supported by the internal grants of the Centre for Quantum Technologies (CQT), which is funded by the Singapore Ministry of Education and the Singapore National Research Foundation. Research of Shengyu Zhang is partially supported by the Hong Kong grant RGC-419309, and partially by CQT for his research visit.

References

[ACR+07] Ambainis, A., Childs, A., Reichardt, B., Spalek, R., Zhang, S.: Any AND-OR formula of size n can be evaluated in time $n^{1/2+o(1)}$ on a quantum computer. In: Proceedings of the 48th Annual IEEE Symposium on Foundations of Computer Science, pp. 363–372 (2007)

[BCW98] Buhrman, H., Cleve, R., Wigderson, A.: Quantum vs. classical communication and computation. In: Proceedings of the Thirtieth Annual ACM Symposium on the Theory of Computing (STOC), pp. 63–68 (1998)

[CvDNT99] Cleve, R., van Dam, W., Nielsen, M., Tapp, A.: Quantum entanglement and the communication complexity of the inner product function. In: Williams, C.P. (ed.) QCQC 1998. LNCS, vol. 1509, pp. 61–74. Springer, Heidelberg (1999)

[HW91] Heiman, R., Wigderson, A.: Randomized vs. deterministic decision tree complexity for read-once boolean functions. In: Proceedings of the 6th Structures in Complexity Theory Conference, pp. 172–179 (1991)

[JKR09] Jayram, T.S., Kopparty, S., Raghavendra, P.: On the communication complexity of read-once ac^0 formula. In: Proceedings of the 24th Annual IEEE Conference on Computational Complexity, pp. 329–340 (2009)

[JKS03] Jayram, T.S., Kumar, R., Sivakumar, D.: Two applications of information complexity. In: Proceedings of the thirty-fifth annual ACM symposium on Theory of computing (STOC), pp. 673–682 (2003)

[KN97] Kushilevitz, E., Nisan, N.: Communication Complexity. Cambridge University Press, Cambridge (1997)

[KS92] Kalyanasundaram, B., Schnitger, G.: The probabilistic communication complexity of set intersection. SIAM Journal on Discrete Mathematics 5(4), 545–557 (1992); Earlier version in Structures '87

[LS09] Leonardos, N., Saks, M.: Lower bounds on the randomized communication complexity of read-once functions. In: Proceedings of the 24th Annual IEEE Conference on Computational Complexity, pp. 341–350 (2009)

[Raz92] Razborov, A.: On the distributional complexity of disjointness. Theoretical Computer Science 106(2), 385–390 (1992)

[Raz03] Razborov, A.: Quantum communication complexity of symmetric predicates. Izvestiya of the Russian Academy of Science, mathematics 67(1), 159–176 (2003), quant-ph/0204025

[She10] Sherstov, A.: On quantum-classical equivalence for composed communication problems. Quantum Information and Computation 10(5-6), 435–455 (2010)

[Sni85] Snir, M.: Lower bounds for probabilistic linear decision trees. Theoretical Computer Science 38, 69–82 (1985)

[SW86] Saks, M., Wigderson, A.: Probabilistic boolean decision trees and the complexity of evaluating game trees. In: Proceedings of the 27th Annual IEEE Symposium on Foundations of Computer Science, pp. 29–38 (1986)

[Zha09] Zhang, S.: On the tightness of the Buhrman-Cleve-Wigderson simulation. In: Proceedings of the 20th International Symposium on Algorithms and Computation, pp. 434–440 (2009)

Multiplying Pessimistic Estimators: Deterministic Approximation of Max TSP and Maximum Triangle Packing

Anke van Zuylen*

Institute for Theoretical Computer Science, 1-208 FIT Building, Tsinghua University, Beijing 100084, China
anke@tsinghua.edu.cn

Abstract. We give a generalization of the method of pessimistic estimators [14], in which we compose estimators by multiplying them. We give conditions on the pessimistic estimators of two expectations, under which the product of the pessimistic estimators is a pessimistic estimator of the product of the two expectations. This approach can be useful when derandomizing algorithms for which one needs to bound a certain probability, which can be expressed as an intersection of multiple events; using our method, one can define pessimistic estimators for the probabilities of the individual events, and then multiply them to obtain a pessimistic estimator for the probability of the intersection of the events. We apply this method to give derandomizations of all known approximation algorithms for the maximum traveling salesman problem and the maximum triangle packing problem: we define simple pessimistic estimators based on the analysis of known randomized algorithms and show that we can multiply them to obtain pessimistic estimators for the expected weight of the solution. This gives deterministic algorithms with better approximation guarantees than what was previously known.

Keywords: derandomization, approximation algorithms, pessimistic estimators, maximum traveling salesman problem, maximum triangle packing.

1 Introduction

In this paper, we consider two NP-hard maximization problems: the maximum traveling salesman problem and the maximum triangle packing problem. For these two problems, the best known approximation algorithms are randomized algorithms: the solution they output has an *expected* weight that is at least a certain factor of the optimum. Practically, one may prefer deterministic approximation algorithms to randomized approximation algorithms, because the

* This work was supported in part by the National Natural Science Foundation of China Grant 60553001, and the National Basic Research Program of China Grant 2007CB807900, 2007CB807901.

M.T. Thai and S. Sahni (Eds.): COCOON 2010, LNCS 6196, pp. 60–69, 2010.

approximation guarantee of a randomized algorithm only says something about the expected quality of the solution; it does not tell us anything about the variance in the quality. Theoretically, it is a major open question whether there exist problems that we can solve in randomized polynomial time, for which there is no deterministic polynomial time algorithm. The algorithms for the problems we consider and their analysis get quite complicated, so complicated, even, that two papers on maximum triangle packing contained subtle errors, see [11,2]. A direct derandomization of these algorithms remained an open problem. In fact, Chen et al. [5] claim in the conference version of their paper on the maximum triangle packing problem that their "randomized algorithm is too sophisticated to derandomize". We show that this is not the case: our derandomizations are quite simple, and are a direct consequence of the analysis of the randomized algorithms, a simple lemma and the right definitions.

Two powerful tools in the derandomization of randomized algorithms are the method of conditional expectation [8] and the method of pessimistic estimators, introduced by Raghavan [14]. We explain the idea of the method of conditional expectation and pessimistic estimators informally; we give a formal definition of pessimistic estimators in Section 2. Suppose we have a randomized algorithm for which we can prove that the expected outcome has a certain property, say, it will be a tour of length at least L. The method of conditional expectation does the following: it iterates over the random decisions by the algorithm, and for each such decision, it evaluates the expected length of the tour conditioned on each of the possible outcomes of the random decision. By the definition of conditional expectation, at least one of these outcomes has expected length at least L; we fix this outcome and continue to the next random decision. Since we maintain the invariant that the conditional expected length is at least L, we will finish with a deterministic solution of length at least L. It is often difficult to exactly compute the conditional expectations one needs to achieve this. A pessimistic estimator is a lower bound on the conditional expectation that is efficiently computable and can be used instead of the true conditional expectation.

In the algorithms we consider in this paper, it would be sufficient for obtaining a deterministic approximation algorithm to define pessimistic estimators for the probability that an edge occurs in the solution returned by the algorithm. Since the algorithms are quite involved, this seems non-trivial. However, the analysis of the randomized algorithm does succeed in bounding this probability, by defining several events for each edge, where the probability of the edge occurring in the algorithm's solution is exactly the probability of the intersection of these events. With the right definitions of events, defining pessimistic estimators for the probabilities of these individual events turns out to be relatively straightforward. We therefore give conditions under which we can multiply pessimistic estimators to obtain a pessimistic estimator for the expectation of the product of the variables they estimate (or, in our case, the probability of the intersection of the events). This allows us to leverage the analysis of the randomized algorithms for the two problems we consider and gives rise to quite simple derandomizations. In the case of the algorithms for the maximum triangle packing problem [10,5], we also simplify a part of the analysis that is used for both the original randomized algorithm and our derandomization, thus making the analysis more concise.

The strength of our method is that it makes it easier to see that an algorithm can be derandomized using the method of pessimistic estimators, by "breaking up" the pessimistic estimator for a "complicated" probability, into a product of pessimistic estimators for "simpler" probabilities. This is useful if, as in the applications given in this paper, the analysis of the randomized algorithm also bounds the relevant probability by considering it as the probability of an intersection of simpler events, or put differently, the randomized analysis bounds the relevant probability by (repeated) conditioning.

1.1 Results and Related Work

It is straightforward to show, using linearity of expectation, that the sum of pessimistic estimators that estimate the expectation of two random variables, say Y and Z respectively, is a pessimistic estimator of the expectation of $Y+Z$. The product of these pessimistic estimators is not necessarily a pessimistic estimator of $\mathbb{E}[YZ]$, and we are not aware of any research into conditions under which this relation does hold, except for a lemma by Widgerson and Xiao [16]. In their work, a pessimistic estimator is an upper bound (whereas for our applications, we need estimators that are lower bounds on the conditional expectation), and they show that if we have pessimistic estimators for the probability of events σ and τ respectively, then the sum of the estimators is a pessimistic estimator for the probability of $\sigma \cap \tau$. Because we need estimators that are lower bounds, this result does not hold for our setting. The conditions we propose for our setting are quite straightforward and by no means exhaustive: the strength of these conditions is that they are simple to check for the two applications we consider. In fact, our framework makes it easier to prove the guarantee for the randomized algorithms for the maximum triangle packing problem as well.

Using our approach, we show how to obtain deterministic approximation algorithms that have the same performance guarantee as the best known randomized approximation algorithms for the maximum traveling salesman problem and the maximum triangle packing problem. In the maximum traveling salesman problem (max TSP) we are given an undirected weighted graph, and want to find a tour of maximum weight that visits all the vertices exactly once. We note that we do not assume that the weights satisfy the triangle inequality; if we do assume this, better approximation guarantees are known [12]. Max TSP was shown to be APX-hard by Barvinok et al. [1]. Hassin and Rubinstein [9] give a randomized approximation algorithm for max TSP with expected approximation guarantee $(\frac{25}{33} - \varepsilon)$ for any constant $\varepsilon > 0$. Chen, Okamoto and Wang [3] propose a derandomization, by first modifying the randomized algorithm and then using the method of conditional expectation, thus achieving a slightly worse guarantee of $(\frac{61}{81} - \varepsilon)$. In addition, Chen and Wang [6] propose an improvement of the Hassin and Rubinstein algorithm, resulting in a randomized algorithm with guarantee $(\frac{251}{331} - \varepsilon)$.

In the maximum triangle packing problem, we are again given an undirected weighted graph, and we want to find a maximum weight set of disjoint 3-cycles. The maximum triangle packing problem cannot be approximated to within 0.9929 unless $P = NP$ as was shown by Chleík and Chlebíková [7]. Hassin and Rubinstein [10] show a randomized approximation algorithm with expected approximation guarantee $(\frac{43}{83} - \varepsilon)$ for any constant $\varepsilon > 0$. Chen, Tanahashi and Wang [5] improve this algorithm to give a randomized algorithm with improved guarantee of $(0.5257 - \varepsilon)$. A recent result of Tanahashi and Chen [4], which has not yet appeared, shows an improved approximation ratio for the related maximum 2-edge packing problem. They show how to derandomize their new algorithm using the method of pessimistic estimators. The derandomization is quite involved and many parts of the proof requires a tedious case analysis. Their method also implies a deterministic algorithm with guarantee $(0.518 - \varepsilon)$ for the maximum triangle packing problem. Using our method, we can give a simpler argument to show their result.

2 Pessimistic Estimators

We use the following definition of a pessimistic estimator.

Definition 1 (Pessimistic Estimator). *Given random variables $X_1, \ldots, X_r$, and a function $Y = f(X_1, \ldots, X_r)$, a pessimistic estimator ϕ with guarantee y for $\mathbb{E}[Y]$ is a set of functions $\phi^{(0)}, \ldots, \phi^{(r)}$ where $\phi^{(\ell)}$ has domain $\mathcal{X}_\ell := \{(x_1, \ldots, x_\ell) : \mathbb{P}[X_1 = x_1, \ldots, X_\ell = x_\ell] > 0\}$ for $\ell = 1, 2, \ldots, r$ such that $\phi^{(0)} \geq y$ and:*

- *(i)* $\mathbb{E}[Y | X_1 = x_1, \ldots, X_r = x_r] \geq \phi^{(r)}(x_1, \ldots, x_r)$ *for any* $(x_1, \ldots, x_r) \in \mathcal{X}_r$*;*
- *(ii)* $\phi^{(\ell)}(x_1, \ldots, x_\ell)$ *is computable in deterministic polynomial time for any* $\ell \leq r$ *and* $(x_1, \ldots, x_\ell) \in \mathcal{X}_\ell$*;*
- *(iii)* $\mathbb{E}[\phi^{(\ell+1)}(x_1, \ldots, x_\ell, X_{\ell+1})] \geq \phi^{(\ell)}(x_1, \ldots, x_\ell)$ *for any* $\ell \leq r$ *and* $(x_1, \ldots, x_\ell) \in \mathcal{X}_\ell$*;*
- *(iv) we can efficiently enumerate all* $x_{\ell+1}$ *such that* $(x_1, \ldots, x_{\ell+1}) \in \mathcal{X}_{\ell+1}$ *for any* $\ell < r$ *and* $(x_1, \ldots, x_\ell) \in \mathcal{X}_\ell$ *.*

In this paper, we are concerned with using pessimistic estimators to derandomize approximation algorithms (for maximization problems). The random variables $X_1, \ldots, X_r$ represent the outcomes of the random decisions made by the algorithm, and Y is the objective value of the solution returned by the algorithm.

Lemma 2. *Given random variables $X_1, \ldots, X_r$, a function $Y = f(X_1, \ldots, X_r)$, and a pessimistic estimator ϕ for $\mathbb{E}[Y]$ with guarantee y, we can find $(x_1, \ldots, x_r)$ such that*

$$\mathbb{E}[Y | X_1 = x_1, \ldots, X_r = x_r] \geq y.$$

Proof. By conditions (ii)-(iv) in Definition 1, we can efficiently find x_1 such that $\phi^{(1)}(x_1) \geq \phi^{(0)} \geq y$, and by repeatedly applying these conditions, we find $(x_1, \ldots, x_r)$ such that $\phi^{(r)}(x_1, \ldots, x_r) \geq y$. Then, by condition (i) in Definition 1, it follows that $\mathbb{E}[Y | X_1 = x_1, \ldots, X_r = x_r] \geq y$. □

We make a few remarks about Definition 1. It is sometimes required that condition (i) holds for all $\ell = 0, \ldots, r$. However, the condition we state is sufficient. Second, note that we do not require that the support of the random variables is of polynomial size. This fact may be important, see for example Section 3. Also note that we can combine (ii), (iii) and (iv) into a weaker condition, namely that for any $\ell < r$ and $(x_1, \ldots, x_\ell) \in \mathcal{X}_\ell$ we can efficiently find $x_{\ell+1}$ such that $\phi^{(\ell+1)}(x_1, \ldots, x_\ell, x_{\ell+1}) \geq \phi^{(\ell)}(x_1, \ldots, x_\ell)$. Our stricter definition will make it easier to define under which conditions we can combine pessimistic estimators.

The following lemma is easy to prove.

Lemma 3. *If Y, Z are both functions of $X_1, \ldots, X_r$ and ϕ is a pessimistic estimator for $\mathbb{E}[Y]$ with guarantee y and θ is a pessimistic estimator for $\mathbb{E}[Z]$ with guarantee z, then $\phi + \theta$ is a pessimistic estimator for $\mathbb{E}[Y + Z]$ with guarantee $y + z$. Also, for any input parameter $w \in \mathbb{R}^{\geq 0}$, $w\phi$ is a pessimistic estimator of $\mathbb{E}[wY]$ with guarantee wy.*

In our applications, the input is an undirected graph $G = (V, E)$ with edge weights $w_e \geq 0$ for $e \in E$. Let Y_e be an indicator variable that is one if edge e is present in the solution returned by the algorithm, and suppose we have pessimistic estimators ϕ_e with guarantee y_e for $\mathbb{E}[Y_e] = \mathbb{P}[Y_e = 1]$ for all $e \in E$. By the previous lemma, $\sum_{e \in E} w_e \phi_e$ is a pessimistic estimator with guarantee $\sum_{e \in E} w_e y_e$ for $\mathbb{E}[\sum_{e \in E} w_e Y_E]$, the expected weight of the solution returned by the algorithm. Hence such pessimistic estimators are enough to obtain a deterministic solution with weight $\sum_{e \in E} w_e y_e$.

Rather than finding a pessimistic estimator for $\mathbb{E}[Y_e]$ directly, we will define pessimistic estimators for (simpler) events such that $\{Y_e = 1\}$ is the intersection of these events. We now give two sufficient conditions on pessimistic estimators so that we can multiply them and obtain a pessimistic estimator of the expectation of the product of the underlying random variables. We note that our conditions are quite specific and that the fact that we can multiply the estimators easily follows from them. The strength of the conditions is in the fact that they are quite easy to check for the pessimistic estimators that we consider.

Definition 4 (Product of Pessimistic Estimators). *Given random variables $X_1, \ldots, X_r$, and two pessimistic estimators ϕ, θ for the expectation of $Y = f(X_1, \ldots, X_r)$, $Z = g(X_1, \ldots, X_r)$ respectively, we define the product of ϕ and θ, denoted by $\phi \cdot \theta$, as*

$$\phi \cdot \theta^{(\ell)}(x_1, \ldots, x_\ell) = \phi^{(\ell)}(x_1, \ldots, x_\ell) \cdot \theta^{(\ell)}(x_1, \ldots, x_\ell) \text{ for } \ell = 1, \ldots, r.$$

Definition 5 (Uncorrelated Pessimistic Estimators). *Given random variables $X_1, \ldots, X_r$, and two pessimistic estimators ϕ, θ for the expectation of functions $Y = f(X_1, \ldots, X_r)$, $Z = g(X_1, \ldots, X_r)$ respectively, we say ϕ and θ are uncorrelated, if for any $\ell < r$ and $(x_1, \ldots, x_\ell) \in \mathcal{X}_\ell$, the random variables $\phi^{(\ell+1)}(x_1, \ldots, x_\ell, X_{\ell+1})$ and $\theta^{(\ell+1)}(x_1, \ldots, x_\ell, X_{\ell+1})$ are uncorrelated.*

Definition 6 (Conditionally Nondecreasing Pessimistic Estimators). *Given random variables $X_1, \ldots, X_r$, and two pessimistic estimators ϕ, θ for the* *expectation of functions* $Y = f(X_1, \ldots, X_r)$, $Z = g(X_1, \ldots, X_r)$ *respectively, we say that* ϕ is nondecreasing conditioned on θ *if* θ *only takes on non-negative values and*

$$\mathbb{E}\left[\phi^{(\ell+1)}(x_1, \ldots, x_\ell, X_{\ell+1}) \;\middle|\; \theta^{(\ell+1)}(x_1, \ldots, x_\ell, X_{\ell+1}) = c\right] \geq \phi^{(\ell)}(x_1, \ldots, x_\ell)$$

for any $\ell < r$ *and* $(x_1, \ldots, x_\ell) \in \mathcal{X}_\ell$ *and* $c > 0$ *such that* $\mathbb{P}\left[\theta^{(\ell+1)}(x_1, \ldots, x_\ell, X_{\ell+1}) = c\right] > 0$.

The proof of the following two lemmas is straightforward, and is omitted due to space constraints.

Lemma 7. *Given two uncorrelated pessimistic estimators as in Definition 5 which take on only non-negative values, their product is a pessimistic estimator of* $\mathbb{E}[YZ]$ *with guarantee* $\phi^{(0)}\theta^{(0)}$.

Lemma 8. *Given two pessimistic estimators as in Definition 6, their product is a pessimistic estimator of* $\mathbb{E}[YZ]$ *with guarantee* $\phi^{(0)}\theta^{(0)}$.

Before we sketch how to use Definition 1 and Lemmas 2, 3, 7 and 8 to obtain deterministic algorithms for the max TSP and maximum triangle packing problems, we make the following remark.

Remark 9 (Random subsets instead of random variables). In the algorithms we consider, the random decisions take the form of random subsets, say $S_1, \ldots, S_r$, of the edge set E. We could define random variables $X_1, \ldots, X_r$ which take on n-digit binary numbers to represent these decisions, but to bypass this extra step, we allow $X_1, \ldots, X_r$ in the definition of pessimistic estimators to represent either random variables or random subsets of a fixed ground set.

3 Maximum Traveling Salesman Problem

In the maximum traveling salesman problem (max TSP), we are given a complete undirected graph $G = (V, E)$ with edge weights $w(e) \geq 0$ for $e \in E$. Our goal is to compute a Hamiltonian circuit (tour) of maximum edge weight. For a subset of edges E', we denote by $w(E') = \sum_{e \in E'} w(e)$. We let OPT denote the value of the optimal solution. A *cycle cover* (also known as a 2-factor) of G is a subgraph in which each vertex has degree 2. A maximum cycle cover is a cycle cover of maximum weight and can be computed in polynomial time (see, for instance, Chapter 10 of [13]). A *subtour* on G is a subgraph that can be extended to a tour by adding edges, i.e., each vertex has degree at most 2 and there are no cycles on a strict subset of V.

Both the algorithm of Hassin and Rubinstein [9] and the algorithm by Chen and Wang [6] start with an idea due to Serdyukov [15]: Compute a maximum weight cycle cover $\mathcal{C}$ and a maximum weight matching W of G. Note that the weight of the edges in $\mathcal{C}$ is at least OPT and the weight of W is at least $\frac{1}{2}OPT$. Now, move a subset of the edges from $\mathcal{C}$ to the matching W so that we get two subtours. Extend the two subtours to tours; at least one of the tours has weight at least $\frac{3}{4}OPT$.

The algorithms of Hassin and Rubinstein [9] and Chen and Wang [6] use the same idea, but are more complicated, especially the algorithm in [6]. Due to space constraints, we will only sketch the algorithm by Hassin and Rubinstein [9] here. The algorithm constructs three tours instead of two; the first tour is deterministic and we refer the reader to [9] for its construction. The second and third tours are constructed by a randomized algorithm which is inspired by Serdyukov's algorithm [15]: a subset of the edges from $\mathcal{C}$ is moved from $\mathcal{C}$ to W in such a way that we obtain two subtours. However, to deal with the case when a large part of the optimal tour consists of edges with endpoints in different cycles in $\mathcal{C}$, a third set of edges M is computed, which is a maximum weight matching on edges that connect nodes in different cycles in $\mathcal{C}$. In the last step of the algorithm, the algorithm adds as many edges from M as possible to the remaining edges of $\mathcal{C}$ while maintaining that the graph is a subtour. Finally, the two constructed subtours are extended to tours.

The part of the algorithm that is difficult to derandomize is the choice of the edges that are moved from $\mathcal{C}$ to W. These edges are chosen so that, in expectation, a significant fraction of the edges from M can be added in the last step. We will call a vertex *free* if one of the two edges adjacent to it in $\mathcal{C}$ is chosen to be moved to W. Note that if we add all edges of M for which both endpoints are free to the remaining edges in $\mathcal{C}$, then we may create new cycles. By the definition of M, such cycles must each contain at least two edges from M. Hence we know that we can add at least half the (weight of the) edges of M for which both endpoints are free without creating cycles. We therefore try to choose edges to move from $\mathcal{C}$ to W in order to maximize the weight of the edges in M for which both endpoints are free.

We first give the following lemma, which states the resulting approximation guarantee if the weight of the edges in M for which both endpoints are free is at least $\frac{1}{4}w(M)$. The proof of this lemma is implied by the proof of Theorem 5 in Hassin and Rubinstein [9].

Lemma 10 ([9]). *Given a weighted undirected graph $G = (V, E)$, let W a maximum weight matching on G and $\mathcal{C} = \{C_1, \ldots, C_r\}$ be a maximum weight cycle cover. Let M be a maximum weight matching on edges that connect nodes in different cycles in $\mathcal{C}$. If we can find nonempty subsets $X_1, \ldots, X_r$, where $X_i \subseteq C_i$ such that*

- *$(V, W \cup \bigcup_{i=1}^{r} X_i)$ is a subtour;*
- *the weight of the edges from M for which both endpoints are "free" (adjacent to some edge in $\bigcup_{i=1}^{r} X_i$) is at least $\frac{1}{4}w(M)$;*

then there exists a $(\frac{25}{33} - \varepsilon)$-approximation algorithm for max TSP.

Hassin and Rubinstein [9] succeed in finding random sets $X_1, \ldots, X_r$ such that the second condition holds in expectation: The algorithm considers the cycles in $\mathcal{C}$ one by one, and when considering cycle C_i in $\mathcal{C}$, they give a (deterministic) procedure (see the proof of Lemma 1 in [9]) for computing two non-empty candidate sets of edges in C_i, each of which may be added to $(V, W \cup \bigcup_{\ell=1}^{i-1} X_\ell)$ so that the resulting graph is a subtour. The candidate sets are such that each vertex in C_i is adjacent to at least one of the two candidate sets. Now, X_i is equal to one of these two candidate sets, each with probability 1/2. The two candidate sets from C_i depend on the edges $X_1, \ldots, X_{i-1}$ which were chosen to be moved from $C_1, \ldots, C_{i-1}$, and it is hence not clear if one can efficiently determine the probability distribution of the set of edges from C_i that will be moved to W. Fortunately, we do not need to know this distribution when using the appropriate pessimistic estimators.

Lemma 11. *There exists a deterministic* $(\frac{25}{33} - \varepsilon)$*-approximation algorithm for max TSP.*

Proof. For each vertex $v \in V$, we define the function Y_v which is 1 if v is adjacent to an edge in $\bigcup_{i=1}^r X_i$, and 0 otherwise. Let C_i be the cycle in $\mathcal{C}$ that contains v, and note that Y_v is a function of $X_1, \ldots, X_i$. We define a pessimistic estimator ϕ_v with guarantee 1/2 for the probability that Y_v is one. No matter which edges are in $X_1, \ldots, X_{i-1}$, the two candidate sets in C_i always have the property that each vertex in C_i is adjacent to at least one of the two candidate sets. We can therefore let $\phi_v(x_1, \ldots, x_\ell) = \frac{1}{2}$ if $\ell < i$, and we define it to be 0 or 1 depending on whether v has degree one if we remove the edges corresponding to $X_i = x_i$ from C_i if $i \geq \ell$. Clearly, ϕ_v satisfies properties (i)-(iii) in Definition 1 with respect to $\mathbb{P}\big[Y_v = 1\big]$. For (iv), note that given $X_1 = x_1, \ldots, X_{i-1} = x_{i-1}$, there are only two possible sets x_{i+1} (however, it is not necessarily the case that the random variable X_i has polynomial size support if we do not condition on $X_1, \ldots, X_{i-1}$).

Now, note that for an edge $\{u, v\} \in M$, u and v are in different cycles of the cycle cover. Therefore, for any $\ell < r$, either $\phi_u^{(\ell+1)}(x_1, \ldots, x_\ell, X_{\ell+1})$ or $\phi_v^{(\ell+1)}(x_1, \ldots, x_\ell, X_{\ell+1})$ is non-stochastic. Hence ϕ_u and ϕ_v are uncorrelated pessimistic extimators, and by Lemma 7, $\phi_u \cdot \phi_v$ is a pessimistic estimator with guarantee $\frac{1}{4}$ for $\mathbb{P}\big[Y_u = 1, Y_v = 1\big]$ which is exactly the probability that both endpoints of $\{u, v\}$ are free. By Lemmas 2, 3, we can use these pessimistic estimators to find subsets $x_1, \ldots, x_r$ that satisfy the conditions in Lemma 10. □

4 Maximum Triangle Packing

In the maximum triangle packing problem, we are again given an undirected graph $G = (V, E)$ with weights $w(e) \geq 0$ for $e \in E$. The goal is to find a maximum weight triangle packing, i.e., a maximum weight set of disjoint 3-cycles. We assume without loss of generality that $|V| = 3n$: otherwise, we can try all possible ways of removing one or two vertices so that the remaining number of vertices is a multiple of 3.

The algorithms for the maximum triangle packing problem by Hassin and Rubinstein [10] and Chen et al. [5] are similar in spirit to the algorithms for the max TSP. Due to space constraints, we again only sketch the ideas for the algorithm in [10]. The algorithm computes three triangle packings, and the first two are computed by deterministic algorithms. The third packing is computed by computing a subtour of large weight; note that we can find a triangle packing of weight equal to at least 2/3 times the weight of any subtour.

The subtour is constructed in a similar way as in the previous section. First, a cycle cover $\mathcal{C}$ is computed; each cycle in the cycle cover has at most $1 + 1/\varepsilon$ edges and the weight of the cycle cover is large (at least $(1-\varepsilon)$ times the weight of the maximum cycle cover). Also, a matching M is computed on the edges for which the endpoints are in different cycles of $\mathcal{C}$. Now, we remove a non-empty subset of the edges from each cycle in the cycle cover, and subsequently add edges from M for which both endpoints are "free", i.e. have degree zero or one. Denote the graph we obtain by G'. Note that G' may contain cycles, but each cycle contains at least two edges from M. To get a subtour, we choose one edge from M in each cycle (uniformly at random) and delete it.

The algorithm of Hassin and Rubinstein [10] removes edges from the cycles in the cycle cover by a random procedure in such a way that

(1) the expected weight of the edges removed from cycle $C \in \mathcal{C}$ is $\frac{1}{3}w(C)$ if $|C| = 3$, and $\frac{1}{4}w(C)$ otherwise;
(2) the probability that v becomes free (i.e. at least one edge in $\mathcal{C}$ that is adjacent to v is deleted) is at least $\frac{1}{2}$;
(3) the probability that $e \in M$ is added to G' (i.e. both endpoints of e are free) is at least $\frac{1}{4}$;
(4) conditioned on the event that e is in G', the probability that e is in a cycle in G' containing at most two other edges from M is less than the probability that e is not in a cycle in G'.

Note that (4) ensures that the probability that e is deleted from G', conditioned on both of its endpoints being free, is at most $\frac{1}{4}$. Hence, (3) and (4) ensure that the expected weight of the edges from M in the subtour is at least $\frac{1}{4} \times \frac{3}{4}w(M)$.

In the full version of this paper, we show how to define pessimistic estimators for the probability that each edge in $\mathcal{C} \cup M$ is in the subtour. For an edge in $\mathcal{C}$, this is straightforward. For edge $e = \{u, v\}$ in M, this probability is the probability of the intersection of the events that both u and v are free, and the event that edge e is not deleted from a cycle. We therefore define separate pessimistic estimators for the probabilities of these events: an estimator ϕ_u and ϕ_v for the probability that u and v, respectively, are free (each with guarantee $\frac{1}{2}$), and an estimator ϕ_{q_e} for the probability that an edge e is in the subtour, conditioned on the event that it was added to G' (with guarantee $\frac{3}{4}$). As in the previous section, the estimators ϕ_u, ϕ_v are uncorrelated for $e = \{u, v\} \in M$. Therefore, their product is a pessimistic estimator for the probability that e is added to G'. The estimator ϕ_{q_e}, that we define, turns out to be correlated with estimator $\phi_u \cdot \phi_v$. However, it is not hard to show that ϕ_{q_e} is nondecreasing conditioned on $\phi_u \cdot \phi_v$. Hence by Lemma 8, the product of $\phi_u \cdot \phi_v$ and ϕ_{q_e} is a pessimistic estimator for the probability that e is in the subtour. Hence we can find a deterministic algorithm that finds a subtour of weight at least as large as the expected weight of the tour by Hassin and Rubinstein's algorithm.

In the full version of this paper, we also apply our method to the randomized algorithms of Chen and Wang [6] and Chen et al. [5]. Finally, we give a simple proof of fact (4) listed above from the analysis of [10], which does not involve checking many cases as in the original proof.

Acknowledgements. The author thanks Bodo Manthey for pointing out that no direct derandomizations were known for the algorithms considered here.

References

1. Barvinok, A., Johnson, D.S., Woeginger, G.J., Woodroofe, R.: The maximum traveling salesman problem under polyhedral norms. In: Bixby, R.E., Boyd, E.A., Ríos-Mercado, R.Z. (eds.) IPCO 1998. LNCS, vol. 1412, pp. 195–201. Springer, Heidelberg (1998)
2. Chen, Z.-Z.: Personal Communication (2009)
3. Chen, Z.-Z., Okamoto, Y., Wang, L.: Improved deterministic approximation algorithms for Max TSP. Inform. Process. Lett. 95(2), 333–342 (2005)
4. Chen, Z.-Z., Tanahashi, R.: A deterministic approximation algorithm for maximum 2-path packing. To appear in IEICE Transaction
5. Chen, Z.-Z., Tanahashi, R., Wang, L.: An improved randomized approximation algorithm for maximum triangle packing. Discrete Appl. Math. 157(7), 1640–1646 (2009); Preliminary version appeared in AAIM '08
6. Chen, Z.-Z., Wang, L.: An improved randomized approximation algorithm for Max TSP. J. Comb. Optim. 9(4), 401–432 (2005)
7. Chlebík, M., Chlebíková, J.: Approximation hardness for small occurrence instances of NP-hard problems. In: Petreschi, R., Persiano, G., Silvestri, R. (eds.) CIAC 2003. LNCS, vol. 2653, pp. 152–164. Springer, Heidelberg (2003)
8. Erdős, P., Spencer, J.: Probabilistic methods in combinatorics. Academic Press, New York (1974); Probability and Mathematical Statistics, vol. 17
9. Hassin, R., Rubinstein, S.: Better approximations for max TSP. Inform. Process. Lett. 75(4), 181–186 (2000)
10. Hassin, R., Rubinstein, S.: An approximation algorithm for maximum triangle packing. Discrete Appl. Math. 154(6), 971–979 (2006); Preliminary version appeared in ESA 2004
11. Hassin, R., Rubinstein, S.: Erratum to: An approximation algorithm for maximum triangle packing. Discrete Appl. Math. 154(6), 971–979 (2006); Discrete Appl. Math., 154(18), 2620 (2006)
12. Kowalik, L., Mucha, M.: Deterministic 7/8-approximation for the metric maximum tsp. Theoretical Computer Science 410, 5000–5009 (2009); Preliminary version appeard in APPROX-RANDOM '08
13. Lovász, L., Plummer, M.D.: Matching theory. North-Holland Mathematics Studies, vol. 121. North-Holland Publishing Co, Amsterdam (1986); Annals of Discrete Mathematics, 29
14. Raghavan, P.: Probabilistic construction of deterministic algorithms: approximating packing integer programs. J. Comput. System Sci. 37(2), 130–143 (1988)
15. Serdyukov, A.I.: An algorithm with an estimate for the travelling salesman problem of the maximum. Upravlyaemye Sistemy 89(25), 80–86 (1984)
16. Wigderson, A., Xiao, D.: Derandomizing the Ahlswede-Winter matrix-valued Chernoff bound using pessimistic estimators, and applications. Theory of Computing 4(1), 53–76 (2008)

Clustering with or without the Approximation*

Frans Schalekamp[1], Michael Yu[2], and Anke van Zuylen[1]

[1] ITCS, Tsinghua University
{frans,anke}@tsingua.edu.cn
[2] MIT
mikeyu@mit.edu

Abstract. We study algorithms for clustering data that were recently proposed by Balcan, Blum and Gupta in SODA'09 [4] and that have already given rise to two follow-up papers. The input for the clustering problem consists of points in a metric space and a number k, specifying the desired number of clusters. The algorithms find a clustering that is provably close to a target clustering, provided that the instance has the "$(1+\alpha,\varepsilon)$-property", which means that the instance is such that all solutions to the k-median problem for which the objective value is at most $(1+\alpha)$ times the optimal objective value correspond to clusterings that misclassify at most an ε fraction of the points with respect to the target clustering. We investigate the theoretical and practical implications of their results.

Our main contributions are as follows. First, we show that instances that have the $(1+\alpha,\varepsilon)$-property and for which, additionally, the clusters in the target clustering are large, are easier than general instances: the algorithm proposed in [4] is a constant factor approximation algorithm with an approximation guarantee that is better than the known hardness of approximation for general instances. Further, we show that it is NP-hard to check if an instance satisfies the $(1+\alpha,\varepsilon)$-property for a given (α,ε); the algorithms in [4] need such α and ε as input parameters, however. We propose ways to use their algorithms even if we do not know values of α and ε for which the assumption holds. Finally, we implement these methods and other popular methods, and test them on real world data sets. We find that on these data sets there are no α and ε so that the dataset has both $(1+\alpha,\varepsilon)$-property and sufficiently large clusters in the target solution. For the general case, we show that on our data sets the performance guarantee proved by [4] is meaningless for the values of α,ε such that the data set has the $(1+\alpha,\varepsilon)$-property. The algorithm nonetheless gives reasonable results, although it is outperformed by other methods.

* This work was supported in part by the National Natural Science Foundation of China Grant 60553001, and the National Basic Research Program of China Grant 2007CB807900, 2007CB807901. Part of this work was done while the second author was visiting the Institute for Theoretical Computer Science at Tsinghua University in the summer of 2009.

M.T. Thai and S. Sahni (Eds.): COCOON 2010, LNCS 6196, pp. 70–79, 2010.

1 Introduction

Clustering is an important problem which has applications in many situations where we try to make sense of large amounts of data, such as in biology, marketing, information retrieval, et cetera. A common approach is to infer a distance function on the data points based on the observations, and then to try to find the correct clustering by solving an optimization problem such as the k-median problem, k-means problem or min-sum clustering problem. Unfortunately, these three optimization problems are all NP-hard, hence we do not expect to find algorithms that find the optimal solution in polynomial time. Research has therefore focused on finding good heuristics (such as for example the popular k-means++ algorithm [1]), exact methods (see for example [8]), and approximation algorithms: polynomial time algorithms that come with a guarantee β that the returned solution has objective value at most β times the optimum value. Research into approximation algorithms for these three clustering problems has produced a large number of papers that demonstrate approximation algorithms as well as lower bounds on the best possible guarantee. However, in many cases there is still a gap between the best known approximation algorithm and the best known lower bound.

In a recent paper, Balcan, Blum and Vempala [5] (see also Balcan, Blum and Gupta [4]) observe the following: The optimization problems that we try to solve are just proxies for the real problem, namely, finding the "right" clustering of the data. Hence, if researchers try so hard to find better approximation algorithms, that must mean that we believe that this will help us find clusterings that are closer to the target clustering. More precisely, Balcan, Blum and Vempala [5] turn this implicit belief into the following explicit assumption: there exist $\alpha > 0, \varepsilon > 0$ such that any solution with objective value at most $(1 + \alpha)$ times the optimum value misclassifies at most an ε fraction of the points (with respect to the unknown target clustering). We will call this the $(1 + \alpha, \varepsilon)$-property. By making this implicit assumption explicit, Balcan et al. [4] are able to show that, given (α, ε) such that the $(1 + \alpha, \varepsilon)$-property holds, quite simple algorithms will give a clustering that misclassifies at most an $O(\varepsilon)$-fraction of the points. In the case when the clusters in the target clustering are "large" (where the required size is a function of ε/α), they give an algorithm that misclassifies at most an ε fraction of the points. In the general case, they give an algorithm that misclassifies at most an $O(\varepsilon/\alpha)$-fraction. They do not need better approximation algorithms for the k-median problem to achieve these results: in fact, [4] shows that finding a $(1 + \alpha)$-approximation algorithm does not become easier if the instance satisfies the $(1 + \alpha, \varepsilon)$-property.

These results seem quite exciting, because they allow us to approximate the target clustering without approximating the objective value of the corresponding optimization problem. As Balcan et al. [4] point out, especially if approximating the objective to within the desired accuracy is hard, we have no choice but to "bypass" the objective value if we want to approximate the target clustering.

However, it is not immediately clear how useful these results are in practice. A first concern is that the algorithms need parameters α and ε such that the instance satisfies the $(1+\alpha, \varepsilon)$-property. The paper by Balcan et al. [4] gives no suggestions on how a practitioner can find such α and ε. And of course an interesting question is whether these new algorithms outperform previously known methods in approximating the target clustering, if we do know α and ε, especially in the case when α is smaller than the guarantee of the best known approximation algorithm.

In this paper, we set out to investigate practical and theoretical implications of the algorithms in Balcan et al. [4] We now briefly describe our contributions.

1.1 Our Contributions

We focus on the case when the optimization problem we need to solve is the k-median problem. Our main theoretical contribution is a proof that the algorithm for "large clusters" given by Balcan et al. [4] is in fact an approximation algorithm with a guarantee $1 + \frac{1}{1/2+5/\alpha}$. One could argue that the algorithm of Balcan et al. [4] is most interesting when $\alpha \leq 2$ (since for $\alpha > 2$ one can use the algorithm of Arya et al. [2] to obtain the claimed result), and hence in those cases their algorithm has an approximation guarantee of at most $\frac{4}{3}$. However, for the general case of the k-median problem, there is a hardness of approximation of $1 + \frac{1}{e}$ [11]. As $1 + \frac{1}{e} \approx 1.37$ is larger than $\frac{4}{3}$, this means that these instances are provably easier than the general class of instances. We note that Balcan et al. [4] show that approximating the k-median objective does *not* become easier if we are guaranteed that the instance has the $(1+\alpha, \varepsilon)$-property. We show that it *does* become easier for those instances that have the $(1+\alpha, \varepsilon)$-property and for which the clusters in the target clustering are "large".

For the general case, we show that it is NP-hard to check whether a data sets satisfies the $(1+\alpha, \varepsilon)$-property for a given α, ε; however, knowledge of such parameters is necessary to run the algorithm of Balcan et al. [4].

We implement the algorithms of Balcan et al. [4] and compare the results to the outcome of previously known methods for various real world data sets. We show how to efficiently run the algorithms for all possible values of the parameters (α, ε) (regardless of whether the assumption holds for the pair of values), and suggest a heuristic for choosing a good solution among the generated clusterings. The algorithm for "large clusters" fails to find a solution on any of our instances and for any value of the parameters α, ε. The algorithm for the general case, however, does return reasonably good solutions. We compare these results to other methods, and find that they are reasonable, but that there are other methods, both heuristics and approximation algorithms, which are significantly better both in terms of approximating the target clustering and approximating the k-median objective.

We also show how to enumerate all values of α, ε for which the $(1+\alpha, \varepsilon)$-property holds, which we note are not practical as they need or calculate the optimal k-median solution. We find that indeed our data sets never satisfy the $(1+\alpha, \varepsilon)$-property *and* the large clusters assumption. For the general case, we find that the proven guarantee on the misclassification of $O(\varepsilon/\alpha)$ is greater than one.

1.2 Related Work

Due to space constraints, we focus our discussion on research that is similar in spirit to our work, in the sense that it restricts the input space and uses these additional assumptions to obtain improved algorithmic results.

Balcan, Blum and Vempala [5] study the problem of approximating an unknown target clustering given a distance metric on the points. They identify properties of the distance metric that allow us to approximate an unknown target clustering. One of the properties they define is the $(1+\alpha, \varepsilon)$-property which is exploited by Balcan, Blum and Gupta [4] to find clusterings that are provably close to the target clustering. We describe their work in more detail in the next section. Two follow-up papers have extended their results in two directions: Balcan, Röglin and Teng [7] consider the setting where all but a γ-fraction of the points have the $(1+\alpha, \varepsilon)$-property and this γ-fraction of points is adversarially chosen. Balcan and Braverman [6] improve the results in [4] if the goal is to approximate the target clustering and the input satisfies the $(1+\alpha, \varepsilon)$-property with respect tot the min-sum objective.

Ostrovsky, Rabani, Schulman and Swamy [15] identify natural properties under which variants of Lloyd's algorithm are guaranteed to quickly find near-optimal solutions to the k-means problem. The recent paper of Bilu and Linial [10] gives polynomial time algorithms for a certain class of inputs to the max-cut problem, which they call "stable" instances. There are strong similarities between [4] and [10]: both approaches define classes of inputs, for which they can give algorithms that perform better than what is possible for general instances. In fact, it is possible to show that the $(1+\alpha, \varepsilon)$-property implies a stability property in similar vein to the stability property defined in [10]. Theorem 1 in this paper shows that the class of inputs defined by $(1+\alpha, \varepsilon)$-property combined with the assumption that the clusters of the target clustering are large, is easier to approximate than general instances of the k-median problem.

2 Problem Definition

2.1 k-Median Problem

In the k-median problem, we are given a set of elements X and a distance function $d : X \times X \to \mathbb{R}^{\geq 0}$ which forms a metric (i.e., d satisfies the triangle inequality), a subset of elements $V \subset X$ that need to be covered and a parameter $k \in \mathbb{N}$. We denote $|V| = n$. The goal is to choose k cluster centers $v_1, \ldots, v_k \in X$ so as to minimize $\sum_{u \in V} \min_{i=1,\ldots,k} d(u, v_i)$.

We denote by OPT the optimum ojective value of a given instance, and we say an algorithm is an β-approximation algorithm for the k-median problem if for any instance it is guaranteed to output cluster centers $v_1, \ldots, v_k \in X$ so that $\sum_{u \in V} \min_{i=1,\ldots,k} d(u, v_i) \leq \beta OPT$.

2.2 Setting of Balcan, Blum and Gupta

In the setting considered by Balcan et al. [4], an instance also includes an unknown target clustering, i.e. a partition $C_1^*, \ldots, C_k^*$ of V. We say a clustering $C_1, \ldots, C_k$ is ε-close to the target clustering, if there exists a permutation π such that $\frac{1}{n}\sum_{i=1}^{k} |C_i^* \backslash C_{\pi(i)}| \leq \varepsilon$. The misclassification is defined as the smallest ε such that the clustering is ε-close to the target clustering.

When clustering data into k clusters, an often used approach is to define a distance function on the data based on observations, and to solve an optimization problem (for example, the k-median problem) to obtain a clustering. Balcan et al. [4] argue that the quest for better approximation algorithms thus implies a belief that better approximations will result in solutions that are closer to the unknown target clustering. They formalize this implicit belief into the following property.

Definition 1 ($(1+\alpha,\varepsilon)$-property). *An instance satisfies the $(1+\alpha,\varepsilon)$-property, if any k-median solution with objective value at most $(1+\alpha)OPT$ is ε-close to the target clustering.*

Balcan et al. [4] propose and analyze two algorithms: the first one is for instances that both satisfy the $(1+\alpha,\varepsilon)$-property and additionally are such that the clusters in the target clustering are of size at least $(3+10/\alpha)\varepsilon n+2$. This algorithm needs α, ε and OPT as inputs. (There is a way to get around having to give OPT as an input, as is shown in [4].) The algorithm is guaranteed to return a clustering that is ε-close to the target clustering.

The second algorithm is for instances that (just) satisfy the $(1+\alpha,\varepsilon)$-property. This algorithm is remarkably simple, but also assumes knowledge of α, ε and OPT. This algorithm for the less restricted input space is guaranteed to return a clustering that is $O(\varepsilon/\alpha)$-close to the target clustering. Instead of OPT an approximate value can be used, at the cost of a deteriorating guarantee: If the value is guaranteed to be at most βOPT, then the misclassification guarantee becomes $O(\varepsilon\beta/\alpha)$. In the remainder of this paper, we will use the abbreviations BBGlarge and BBGgeneral to refer to these algorithms.

Instead of working with the $(1+\alpha,\varepsilon)$-property directly, it is usually easier to work with a weaker property that is implied by the $(1+\alpha,\varepsilon)$-property (as shown in Lemma 3.1 in [4]). We will refer to this as the "weak $(1+\alpha,\varepsilon)$-property". Note that this weaker property does not depend on a target clustering.

Definition 2 (weak $(1+\alpha,\varepsilon)$-property if all clusters in the target clustering have more than $2\varepsilon n$ points). *In the optimal k-median solution, there are at most εn points for which the second closest center is strictly less than $\alpha OPT/(\varepsilon n)$ farther than the closest center.*

Definition 3 (weak $(1+\alpha,\varepsilon)$-property). *In the optimal k-median solution, there are at most $6\varepsilon n$ points for which the second closest center is strictly less than $\alpha OPT/(2\varepsilon n)$ farther than the closest center.*

3 Theoretical Aspects of the BBG Algorithms

We now show that in fact the algorithm BBGlarge succeeds in finding a solution which not only has a classification error of at most ε, but is also an approximately optimal k-median solution.

Theorem 1. *If the k-median instance satisfies the $(1+\alpha,\varepsilon)$-property and each cluster in the target clustering has size at least $(3+10/\alpha)\varepsilon n+2$, then the algorithm for large clusters proposed by Balcan et al. [4] gives a $(1+1/(1/2+5/\alpha))$-approximation algorithm for k-median clustering.*

Proof (sketch). In the proof of the fact that the algorithm for large clusters only misclassifies an ϵ fraction of the points, Balcan et al. [4] distinguish two types of points; red points, and non-red (green or yellow) points. They show that the only points that are potentially misclassified are the red points, and there are only ϵn red points.

To show the approximation result, we first repeat the final step of their algorithm: Given clusters $C_1,\ldots,C_k$, we compute for each point x the index $j(x)$ of the cluster that minimizes the median distance from x. The new clustering is obtained by letting $C'_j=\{x : j(x)=j\}$. It follows from the analysis of Balcan et al. [4] that the new clustering again only misclassifies red points. Moreover, we know that for each misclassified red point x, (say in cluster i instead of cluster j), there is a large number of "good reasons" why we made this mistake: pairs of non-red points $y(i), y(j)$ where $y(i)$ is in cluster i and $y(j)$ is in cluster j, and x is closer to $y(i)$ than to $y(j)$. In order to prove an approximation guarantee, we need to bound (or "charge") the difference between the distance from x to its cluster center in our solution and in the optimal solution. By using the triangle inequality, it is not hard to show that we can charge this difference against the sum of the distances from $y(i)$ and $y(j)$ to their respective cluster centers. Since $y(i)$ and $y(j)$ are non-red, and hence correctly clustered, we charge against a piece of the optimal solution. And, since there are "many" pairs $y(i), y(j)$, we only need to charge each piece a "small" number of times, where small turns out to be $1/(1/2+5/\alpha)$. We refer the reader to the full version [16] for further details. □

We note that, although we need to know values of α,ε to get the guaranteed bound on misclassification, we do not need it to obtain the approximation result in Theorem 1: It is not hard to show that we can try all relevant values of α and ε in polynomial time, and get the approximation result by returning the solution with smallest objective value.

We finally remark that the algorithms proposed by Balcan et al. [4] are most interesting for instances with a $(1+\alpha,\varepsilon)$-property with $\alpha \leq 2$: if $\alpha > 2$ then we could then find an ε-close clustering by running a $(3+2/p)$-approximation algorithm [2] for sufficiently large p. Hence for those α for which the Balcan et al. algorithms are interesting, we have shown that the large clusters algorithm gives a $\frac{4}{3}$-approximation algorithm.

Our second result in this section is to show that verifying if an instance has the $(1+\alpha,\varepsilon)$-property for a given α,ε is NP-hard. The proof is a reduction from max k-cover, and is given in the appendix of the full version of this paper [16].

Lemma 1. *It is NP-hard to verify whether an instance has the (weak)* $(1+\alpha,\varepsilon)$*-property for a given* α,ε.

We should remark two things about Lemma 1. First of all, in our proof we need to choose $\alpha \approx \varepsilon/n^3$. In that case, the guarantee given by Balcan et al. [4] is $O(\varepsilon/\alpha) = O(n^3)$, hence this does not constitute an interesting case for their algorithm. Second, our lemma does not say that it is NP-hard to find *some* α,ε for which the $(1+\alpha,\varepsilon)$-property holds. However, we do not know how to find such α,ε efficiently.

4 Practical Aspects of the BBG Algorithms

4.1 Data Sets

We use two popular sets of data to test the algorithms, and compare their outcomes to other methods. We use the pmed data sets from the OR-Library [9] to investigate whether the methods proposed by Balcan et al. [4] give improved performance on commonly used k-median data sets compared to known algorithms in either misclassification, objective value or performance relative to the running time. These instances are distance based but do not have a ground truth clustering. Note that the $(1+\alpha,\varepsilon)$-property implies that the optimal k-median clustering is ε-close to the target clustering (whatever the target clustering is), hence we can assume that the optimal k-median clustering is the target clustering while changing the misclassification of any solution with respect to the target by at most ε.

The second data sets we use come from the University of California, Irivine (UCI) Machine Learning Repository [3]. For these data sets a ground truth clustering is known and given. The data sets we use have only numeric attributes and no missing values. To get a distance functions, we first apply a "z-transform" on each of the dimensions, i.e., for each attribute we normalize the values to have mean 0 and standard deviation 1. Next, we calculate the Euclidean distance between each pair of points. We note that it may be possible to define distance functions that give better results in terms of approximating the target clustering. This is not within the scope of this paper, as we are only interested in comparing the performance of different algorithms for a given distance function.

4.2 Implementing the BBG Algorithms

Balcan et al. [4] do not discuss how to find values of α and ε to use. Indeed, Lemma 1 gives an indication that this may be far from trivial. It is not hard to realize however, that by varying α and ε, both the algorithms can generate only a polynomial number of different outcomes: In the case of BBGgeneral, α

and ε are used only to determine a threshold graph, for which there are only $O(n^2)$ possibilities. In the case of BBGlarge, a second graph is formed, which connects vertices that have at least b neighbors in the threshold graph, where b is a function of α and ε. This leads to a total of $O(n^3)$ possible outcomes for the BBGlarge algorithm. We therefore propose bypassing the fact that we do not know which values of α and ε to use by iterating over all possible outcomes. By being careful in the implementation, it is possible to get reasonably efficient algorithms; we refer the interested reader to the full version of this paper [16] for more details.

For the algorithm for the large clusters case, we found a somewhat surprising outcome: it did not return any clustering on any instance! This means that there exists no α, ε such that the data satisfied the $(1+\alpha, \varepsilon)$-property *and* the clusters in the target clustering had size at least $(3 + 10/\alpha)\varepsilon n + 2$.

For the algorithm for the general case, our implementation ideas reduce the number of solutions to consider from $O(n^2)$ to a much smaller number; at most 5% of the maximum possible number of solutions $n(n-1)/2$. Our next challenge is how to choose a good solution, that is close to the target clustering. A natural choice is to choose the outcome $C_1, \ldots, C_k$ with the lowest k-median objective value defined as $\sum_{i=1}^{k} \min_{c \in X} \sum_{v \in C_i} d(c, v)$. The following lemma shows that unfortunately the k-median objective is not always a reliable indicator of the best solution. The proof is given in the full version of this paper [16].

Lemma 2. *For a given instance, let δ be the misclassification of the solution with lowest k-median objective value among all outcomes obtained by the BBGgeneral algorithm for all threshold graphs. Then $\delta \neq O(\varepsilon/\alpha)$, even if the instance has the weak $(1 + \alpha, \varepsilon)$-property.*

We tried several other criteria for choosing a solution, which are inspired by the analysis of Balcan et al. [4]. None of these criteria is guaranteed to choose a solution with small misclassification, but some of them, including the k-median objective value, seem to work quite well in practice. Since the k-median objective value is quick to evaluate, we chose this criterion for our experimental comparison: on average, the misclassification of the best solution among all solutions generated is 6 percentage points lower than the misclassification of the solution with the best k-median objective value.

4.3 Verifying the $(1 + \alpha, \varepsilon)$-Property for These Data Sets

We found that all instances of our data sets do not lie in the restricted input space for which the BBG algorithm for large clusters is designed. The values of (α, ε) for which the weak $(1 + \alpha, \varepsilon)$-property for the general case (Definition 3) holds are such that the exact guarantee proved by Balcan et al., which is $(25+40/\alpha)\varepsilon$, is much larger than 1. However, ε/α was itself is always less than 1, so an $O(\varepsilon/\alpha)$ guarantee is meaningful for smaller constants. For more discussion we refer to the full version of this paper [16].

5 Comparison to Other Methods

We compare the quality and running time of the BBG algorithms to various heuristics and approximation algorithms for the k-median problem. More specifically, we implemented the following algorithms in MATLAB: the primal-dual algorithm proposed by Jain and Vazirani [13]; the primal-dual algorithm proposed by Jain, Mahdian, Markakis, Saberi and Vazirani [12]; Lloyd's algorithm [14]; k-means++ by Arthur and Vassilvitskii [1]; two variants of Local Search [2].

Due to space constraints, we refer the reader to the full version of this paper [16] for an overview of the outcome of the experiments. Although we found in the previous section that the theoretical guarantees of the BBGgeneral algorithm are meaningless for our data sets, it is clear from our experiments that the algorithm does give reasonable clusterings, and it is fast, even when checking all threshold graphs. However, for our data sets, other algorithms clearly outperform the BBGgeneral algorithm. In terms of overall performance, Local Search, which chooses random improving moves, is superior, both in terms of k-median objective and closeness to the target clustering. The algorithm of Jain et al. [12] is second in terms of performance, followed by kmeans++.

6 Conclusion and Open Problems

In this paper, we investigate theoretical and practical aspects of a new approach to clustering proposed by Balcan et al. [4]. We show that the assumption needed for their strongest result (the "large" clusters case) defines a set of "easy" instances: instances for which we can approximate the k-median objective to within a smaller ratio than for general instances. Our practical evaluations show that our instances do not fall into this category. For the algorithm for the general case, we give some theoretical justification that it may be hard to find the values of parameters α, ε that are needed as input. We show how to adapt the algorithm so we do not need to know these parameters, but this approach does not come with any guarantees on the misclassification. In our experimental comparison, the performance is reasonable but some existing methods are significantly better.

An interesting direction to evaluate the pratical performance of the algorithms by Balcan et al. [4] would be to test them on "easy" instances, i.e. instances for which the $(1+\alpha, \varepsilon)$-property holds for values ε, α for which ε/α is small, perhaps by identifying a small set of points whose removal ensures that this is the case, which was studied by Balcan, Röglin and Teng [7].

Theoretically, our results also raise the question whether it is possible to show an approximation guarantee for the algorithms for instances that satisfy the $(1 + \alpha, \varepsilon)$-property and for which the target clustering has "large" clusters that were proposed for other objective functions, namely k-means and min-sum k-clustering, by Balcan et al. [4] and Balcan and Braverman [6]. For the general case, more research into exploiting this property may lead to algorithms which outperform existing methods. On the other hand, it would be interesting to have a lower bound on the misclassification of any (reasonable) algorithm when given an α, ε, such that the $(1+\alpha, \varepsilon)$-property holds. In particular, it would be interesting to know if the dependence on ε/α in either the guarantee on the misclassification or in the minimum cluster size is unavoidable.

Finally, an interesting direction is to find (other) classes of inputs defined by natural properties for which one can give algorithms that perform better than what is possible for the general class of inputs, both for the k-median problem and other optimization problems.

References

1. Arthur, D., Vassilvitskii, S.: k-means++: the advantages of careful seeding. In: SODA '07: 18th Annual ACM-SIAM Symposium on Discrete Algorithms, pp. 1027–1035 (2007)
2. Arya, V., Garg, N., Khandekar, R., Meyerson, A., Munagala, K., Pandit, V.: Local search heuristics for k-median and facility location problems. SIAM J. Comput. 33(3), 544–562 (2004)
3. Asuncion, A., Newman, D.: UCI machine learning repository (2007), `http://www.ics.uci.edu/~mlearn/MLRepository.html`
4. Balcan, M.-F., Blum, A., Gupta, A.: Approximate clustering without the approximation. In: SODA '09: 19th Annual ACM -SIAM Symposium on Discrete Algorithms, pp. 1068–1077 (2009)
5. Balcan, M.-F., Blum, A., Vempala, S.: A discriminative framework for clustering via similarity functions. In: STOC 2008: 40th Annual ACM Symposium on Theory of Computing, pp. 671–680 (2008)
6. Balcan, M.-F., Braverman, M.: Finding low error clusterings. In: COLT 2009: 22nd Annual Conference on Learning Theory (2009)
7. Balcan, M.-F., Röglin, H., Teng, S.-H.: Agnostic clustering. In: Gavaldà, R., Lugosi, G., Zeugmann, T., Zilles, S. (eds.) ALT 2009. LNCS, vol. 5809, pp. 384–398. Springer, Heidelberg (2009)
8. Beasley, J.E.: A note on solving large p-median problems. European Journal of Operational Research 21(2), 270–273 (1985)
9. Beasley, J.E.: OR-Library p-median - uncapacitated (1985), `http://people.brunel.ac.uk/~mastjjb/jeb/orlib/pmedinfo.html`
10. Bilu, Y., Linial, N.: Are stable instances easy. In: ICS 2010: The First Symposium on Innovations in Computer Science, pp. 332–341 (2010)
11. Gupta, A.: Personal Communication (2009)
12. Jain, K., Mahdian, M., Markakis, E., Saberi, A., Vazirani, V.V.: Greedy facility location algorithms analyzed using dual fitting with factor-revealing LP. J. ACM 50(6), 795–824 (2003)
13. Jain, K., Vazirani, V.V.: Approximation algorithms for metric facility location and k-median problems using the primal-dual schema and Lagrangian relaxation. J. ACM 48(2), 274–296 (2001)
14. Lloyd, S.: Least squares quantization in PCM. IEEE Transactions on Information Theory 28(2), 129–137 (1982)
15. Ostrovsky, R., Rabani, Y., Schulman, L.J., Swamy, C.: The effectiveness of Lloyd-type methods for the k-means problem. In: FOCS '06:47th Annual IEEE Symposium on Foundations of Computer Science, pp. 165–176 (2006)
16. Schalekamp, F., Yu, M., van Zuylen, A.: Clustering with or without the approximation, `http://www.itcs.tsinghua.edu.cn/~frans/pub/ClustCOCOON.pdf`

A Self-stabilizing 3-Approximation for the Maximum Leaf Spanning Tree Problem in Arbitrary Networks

Sayaka Kamei[1,⋆], Hirotsugu Kakugawa[2,⋆⋆], Stéphane Devismes[3], and Sébastien Tixeuil[4,⋆⋆⋆]

[1] Dept. of Information Engineering, Hiroshima University, Japan
s-kamei@se.hiroshima-u.ac.jp
[2] Dept. of Computer Science, Osaka University, Japan
kakugawa@ist.osaka-u.ac.jp
[3] Université Joseph Fourier, Grenoble I, France
Stephane.Devismes@imag.fr
[4] LIP6 UMR 7606, Université Pierre et Marie Curie, France
Sebastien.Tixeuil@lip6.fr

Abstract. The maximum leaf spanning tree (MLST) is a good candidate for constructing a virtual backbone in self-organized multihop wireless networks, but is practically intractable (NP-complete). Self-stabilization is a general technique that permits to recover from catastrophic transient failures in self-organized networks without human intervention. We propose a fully distributed self-stabilizing approximation algorithm for the MLST problem on arbitrary topology networks. Our algorithm is the first self-stabilizing protocol that is specifically designed for the construction of an MLST. It improves other previous self-stabilizing solutions both for generality (arbitrary topology graphs *vs.* unit disk graphs or generalized disk graphs, respectively) and for approximation ratio, as it guarantees the number of its leaves is at least $1/3$ of the maximum one. The time complexity of our algorithm is $O(n^2)$ rounds.

1 Introduction

Multihop wireless ad hoc or sensor networks have neither fixed physical infrastructure nor central administration. They typically operate in a self-organizing manner that permits them to autonomously construct routing and communication primitives that are used by higher level applications. The construction of virtual backbones infrastructures usually makes use of graph related properties over the graph induced by communication capabilities (*i.e.* nodes represent machines, and edges represent the ability for two machines within wireless range to communicate) of the network. For example, a connected dominating set (CDS) is a good candidate for a virtual backbone since it

⋆ This work is supported in part by a Grant-in-Aid for Young Scientists ((B)22700074) of JSPS.

⋆⋆ This work is supported in part by Kayamori Foundation of Informational Science Advancement, a Grant-in-Aid for Scientific Research ((B)20300012) of JSPS, and "Global COE (Centers of Excellence) Program" of the Ministry of Education, Culture, Sports, Science and Technology, Japan.

⋆⋆⋆ This work is supported in part by ANR projects SHAMAN and ALADDIN.

M.T. Thai and S. Sahni (Eds.): COCOON 2010, LNCS 6196, pp. 80–89, 2010.

guarantees reachability of every node yet preserves energy. The maximum leaf spanning tree (MLST) problem consists in constructing a spanning tree with the maximum number of leaves. Finding the MLST is tantamount to finding the minimum CDS: let $G = (V,E)$ be a graph and $cds(G)$ be the size of the minimum CDS of G, then the number of leaves of the MLST of G is $|V| - cds(G)$ [1].

One of the most versatile techniques to ensure forward recovery of distributed systems and networks is that of *self- stabilization* [2]. A distributed algorithm is self-stabilizing if after faults and attacks hit the system and place it in some arbitrary global state, the system recovers from this catastrophic situation without external (*e.g.* human) intervention in finite time. As self-stabilization makes no hypothesis about the nature or the extent of the faults (self-stabilization only deals with the effect of the faults), it can also be used to deal with other transient changes while the network is being operated (topology change, message loss, spontaneous resets, etc.).

1.1 Related Works

In [3], Galbiati *et al.* proved that the MLST problem is MAX-SNP-hard, *i.e.*, there exists $\varepsilon > 0$ such that finding approximation algorithm[1] with approximation ratio $1+\varepsilon$ is NP-hard. In [1], Solis-Oba proposed a 2-approximation algorithm, and in [4], Lu *et al.* proposed a 3-approximation algorithm. Note that none of those algorithms [1,4] is distributed, not to mention self-stabilizing.

Spanning tree construction is one of the main studied problems in self-stabilizing literature. One of the main recent trends in this topic is to provide self-stabilizing protocols for constrained variants of the spanning tree problem, *e.g.* [5], [6], [7], etc. None of those metrics give any guarantee on the number of leaves.

In [8], Guha *et al.* showed that the existence of an algorithm for finding the minimum CDS with approximation ratio α implies the existence of an algorithm of the MLST problem with approximation ratio 2α. In turn, there exist self-stabilizing approximation algorithms for finding the minimum CDS. In [9], Kamei *et al.* proposed a self-stabilizing 7.6-approximation algorithm for the minimum CDS problem in unit disk graphs, *i.e.*, this algorithm is also an approximation algorithm for the MLST problem with approximation ratio 15.2 (by [8]). However, this algorithm [9] does not guarantee any approximation ratio in general topology networks. The subsequent work of Raei *et al.* [10] proposed a self-stabilizing $20\lfloor \ln R/\ln(2\cos(\pi/5)) \rfloor$-approximation algorithm in generalized disk graphs where $R = r_{max}/r_{min}$ and r_{max} (resp. r_{min}) is the maximum (resp. minimum) transmission range. This algorithm is an approximation for the MLST problem with approximation ratio $40\lfloor \ln R/\ln(2\cos(\pi/5)) \rfloor$. Again, this algorithm [10] does not guarantee any approximation ratio in general topology networks.

1.2 Our Contribution and Outline of This Paper

We propose a fully distributed self-stabilizing approximation algorithm for the MLST problem on arbitrary topology networks. Its time complexity is $O(n^2)$ rounds. To our

[1] An approximation algorithm for the MLST problem is an algorithm that guarantees approximation ratio $|T_{opt}|/|T_{alg}|$, where $|T_{alg}|$ is the number of leaves obtained by the approximation algorithm in the worst case and $|T_{opt}|$ is the number of leaves of the optimal solution.

knowledge, our algorithm is the first self-stabilizing protocol that is specifically designed for the construction of MLST. It improves over previous self-stabilizing derivations both for generality (arbitrary topology graphs *vs.* unit disk graphs [9] (resp. generalized disk graphs [10])) and for approximation ratio (3 *vs.* 15.2 [9] (resp. $40\lfloor \ln R / \ln(2\cos(\pi/5)) \rfloor$ [10])).

The improved approximation ratio permits to improve significantly the load and the energy consumed by the virtual backbone. The improved generality on the communication graph enables our scheme to be useful even in networks that cannot be modeled by (generalized) disk graphs (such as wired networks).

This paper is organized as follows. In Section 2, we formally describe the system model and the distributed MLST problem. In Section 3, we present our self-stabilizing approximation algorithm for the distributed MLST problem, and prove the correctness and analyze the time complexity of the proposed algorithm. Details of proofs are omitted because of limitation of space, and they will appear in the full paper. Concluding remarks can be found in Section 4.

2 Preliminaries

Let $V = \{P_1, P_2, ..., P_n\}$ be a set of n processes and $E \subseteq V \times V$ be a set of bidirectional communication links in a distributed system. Each links is an unordered pair of distinct processes. Then, the topology of the distributed system is represented as an undirected graph $G = (V, E)$. We assume that G is connected and simple. In this paper, we use "graphs" and "distributed systems" interchangeably. We assume that each process has unique identifier. By P_i, we denote the identifier of process P_i for each process P_i.

We call *subgraph* of G any graph $G' = (V', E')$ such that $V' \subseteq V$, $E' \subseteq E$, and $\forall P_i, P_j$, $(P_i, P_j) \in E' \Rightarrow P_i, P_j \in V'$. By N_i, we denote the set of neighboring processes of P_i. For each process P_i, the set N_i is assumed to be a constant. We define the *degree* of P_i as the number of its neighbors. The degree of P_i in the subgraph G' is the number of edges of E' incident to P_i. We assume that the maximum degree of G is at least 3. We define the *distance* between P_i and P_j as the number of the edges of the shortest path between them.

A set of local variables defines the local state of a process. By Q_i, we denote the local state of each process $P_i \in V$. A tuple of the local state of each process $(Q_1, Q_2, ..., Q_n)$ forms a *configuration* of a distributed system. Let Γ be a set of all configurations.

As communication model, we assume that each process can read the local state of neighboring processes. This model is called the *state reading model*. Although a process can read the local state of neighboring processes, it cannot update them; it can only update its local state.

We say that P_i is *privileged* in a configuration γ if and only if at least one of the conditions of the algorithm is true and P_i must change the value of its variables in γ. An atomic step of each process P_i consists of following three sub-steps: (1) read the local states of all neighbors and evaluate the conditions of the algorithm, (2) compute the next local state, and (3) update the local state.

Executions of processes are scheduled by an external (virtual) scheduler called *daemon*. That is, the daemon decides which processes to execute in the next step. Here, we assume a *distributed weakly fair daemon*. *Distributed* means that, at each step, the daemon selects an arbitrary non-empty set of privileged processes, and each selected processes executes the atomic step in parallel. *Weakly fair* means that every continuously privileged process will be eventually executed.

For any configuration γ, let γ' be any configuration that follows γ. Then, we denote this transition relation by $\gamma \rightarrow \gamma'$. For any configuration γ_0, a *computation* E starting from γ_0 is a maximal (possibly infinite) sequence of configurations $E = \gamma_0, \gamma_1, \gamma_2, \ldots$ such that $\gamma_t \rightarrow \gamma_{t+1}$ for each $t \geq 0$.

Definition 1 (Self-Stabilization). *Let Γ be a set of all configurations. A system S is* self-stabilizing *with respect to Λ such that $\Lambda \subseteq \Gamma$ if and only if it satisfies the following two conditions:*

- *Convergence: Starting from an arbitrary configuration, a configuration eventually becomes one in Λ, and*
- *Closure: For any configuration $\lambda \in \Lambda$, any configuration γ that follows λ is also in Λ as long as the system does not fail.*

Each $\gamma \in \Lambda$ is called a legitimate *configuration. Conversely, any configuration that is not legitimate is said* illegitimate. □

A *spanning tree* $T = (V', E')$ is any acyclic connected subgraph of G such that $V' = V$ and $E' \subseteq E$. A *leaf* of a spanning tree is any process of degree one. Generally, the MLST problem is defined as follows.

Definition 2. *The* maximum leaf spanning tree *is a spanning tree whose number of leaves is maximum.* □

We consider solving the MLST problem in distributed systems in this paper. We assume that each process does not know global information of the network. Under this assumption, we defined the distributed MLST problem as follows.

Definition 3. *Let $G = (V, E)$ be a graph that represents a distributed system. Then, the* distributed maximum leaf spanning tree problem *is defined as follows.*

- *Each process P_i must select a neighbor on G or itself (if the father of P_i is P_i, then P_i is a* root*) as its father on a spanning tree T_{ml} and output it, and*
- *The spanning tree T_{ml} is a maximum leaf spanning tree of G.* □

3 Proposed Algorithm

Our algorithm SSMLST is based on the sequential approximation algorithm in [4].

We call *tree* any subgraph T of G that has no cycle and more than one process. We construct disjoint trees $T_1, T_2, \cdots$, where $T_i = (V_i, E_i)$, $V = V_1 \cup V_2 \cup \cdots$, $|V_i| > 1$, and $V_i \cap V_j = \emptyset$ for any i and j. We call *forest* any set of trees $\{T_1, T_2, \cdots\}$. Note that some process P_i can be alone and does not join the forest, *i.e.*, $S_i = (\{P_i\}, \emptyset)$, in this case P_i is called *singleton*.

Let $d_k(G)$ be the set of nodes that have degree k on G, and let $\bar{d}_k(G)$ be the set of nodes that have degree at least k on G.

Definition 4. *([4]) Let T be a tree of G. If $\bar{d_3}(T)$ is not empty and every node in $d_2(T)$ is adjacent in T to exactly two nodes in $\bar{d_3}(T)$, then let T be* leafy tree. *Let $T_1, T_2, \cdots$ be disjoint trees on G. If each $T_1, T_2, \cdots$ is leafy, then $F = \{T_1, T_2, \cdots\}$ is a* leafy forest. *If F is not a subgraph of any other leafy forest of G, then we call F* maximal leafy forest. □

In [4], Lu *et al.* showed the following theorem.

Theorem 1. *([4]) Let F be a maximal leafy forest of G, and let T_{ml} be a spanning tree of G such that F is a subgraph of T_{ml}. Let T_{span} be any spanning tree of G. Then, $|d_1(T_{ml})| \geq |d_1(T_{span})|/3$.*

Our algorithm SSMLST first constructs a maximal leafy forest (MLF) of G, and then, it constructs a spanning tree T_{ml} of G that is a supergraph of the MLF. Hence, T_{ml} is an approximation of the MLST with ratio 3 according to Theorem 1.

The proposed algorithm is a fair composition [11] of four layers:

1. In the first layer, each process P_i computes its degree D_i on G and the maximum couple $MAX = (D_0, P_0)$ of degree D_0 and ID P_0 on G, where $(D_i, P_i) > (D_j, P_j) \equiv [D_i > D_j \vee (D_i = D_j \wedge P_i > P_j)]$ for each process P_i and P_j. For this layer, we can use a self-stabilizing leader election algorithm for arbitrary networks, for example [12] (The time complexity of [12] is $O(n)$ rounds.). In such an algorithm, the process with the minimum or the maximum ID is elected as a leader. It is modified to elect the process with the maximum value (D_0, P_0) for our purpose.
2. The second layer SSMLF (see subsection 3.1) computes an MLF on G.
3. The third layer SSTN (see subsection 3.2) modifies the cost of each link based on the MLF.
4. The last layer computes a minimum cost spanning tree T_{ml} based on the costs computed by SSTN. Such costs make T_{ml} includes the MLF. For this layer, we can use one of existing self-stabilizing algorithms, *e.g.*, [6] (The time complexity of [6] is $O(n^2)$ rounds.).

3.1 The Second Layer: Construction of the Maximal Leafy Forest

We now propose a self-stabilizing algorithm called SSMLF that constructs of a maximal leafy forest (MLF) of G. The formal description of SSMLF is shown in Fig. 1.

Each process P_i computes the following outputs:

- $root_i$ is set to $\emptyset$ if P_i is a singleton. Otherwise, P_i belongs to some tree T and $root_i$ is set to the couple (D_r, P_r), where P_r is the root of the tree of P_i.
- $father_i$ is set to the couple (D_j, P_j) where P_j is the father of P_i. If P_i is neither a root nor a singleton, then $P_j \in N_i$. In this case, we say that "*P_j is a father of P_i*" and "*P_i is a child of P_j*". In either cases (P_i is a singleton or the root of its tree), $P_j = P_i$. Note that, in SSMLF, each process P_i distinguishes its incident link to $father_i$ as its parent-link in its tree.
- $rank_i$ is the distance from P_i to the root of its tree.
- $MaxChildren_i$: the *expected number of children* of P_i in its tree. (We shall explain that later.)

Constant (Input)
N_i: the set of neighbors of P_i on G.
D_i: the degree of P_i on G (an output from the first layer).
MAX: the maximum couple of degree and ID on G (an output from the first layer).
Variable (Output)
$root_i = (D_r, P_r)(\emptyset \leq root_i \leq MAX)$: P_r is the root of tree T to which P_i belongs.
$father_i$: the couple (D_j, P_j) of the father P_j of P_i on T.
$rank_i$: the distance from the root to P_i on T.
$MaxChildren_i$: the number of children and child candidates on T.
Macro
$MaxRoot_i = \max\{root_j, (D_i, P_i) \mid P_j \in N_i\}$
$MinRank_i = \begin{cases} -1 & \text{(In case that } MaxRoot_i = (D_i, P_i)) \\ \min\{rank_j \mid P_j \in N_i \wedge root_j = MaxRoot_i\} & \text{(otherwise)} \end{cases}$
$CCand_i =$
$\{P_j \in N_i \mid root_j = \emptyset \vee root_j < MaxRoot_i \vee (root_j = MaxRoot_i \wedge rank_j > MinRank_i + 2)\}$
$CountMaxChildren_i = \begin{cases} D_i & \text{(In case that } (D_i, P_i) = MAX) \\ |\{P_j \in N_i \mid father_j = (D_i, P_i)\}| + |CCand_i| & \text{(otherwise)} \end{cases}$
$FCand_i = \{P_j \in N_i \mid rank_j + 1 \leq n \wedge (MaxChildren_j \geq 3 \vee$
$(MaxChildren_j = 2 \wedge father_j \neq (D_j, P_j) \wedge root_j > (D_j, P_j)))\}$
Algorithm for process P_i:
do *forever*{
if $(root_i > MAX)$ {
    $root_i := \emptyset;$
} elseif $(MaxChildren_i \neq CountMaxChildren_i)$ {
    $MaxChildren_i := CountMaxChildren_i;$
/* For the roots. */
} elseif $(MaxChildren_i \geq 3 \wedge MaxRoot_i = (D_i, P_i))$ {
    $root_i := (D_i, P_i);\ rank_i := 0;\ father_i := (D_i, P_i);$
/* For other nodes in tree.*/
} elseif $(\exists P_j \in FCand_i, [root_j = MaxRoot_i \wedge rank_j = MinRank_i])$ {
    $root_i := MaxRoot_i;$
    $rank_i := MinRank_i + 1;$
    $father_i := \max\{(D_j, P_j) \mid P_j \in FCand_i \wedge root_j = root_i \wedge rank_j = rank_i - 1\};$
} elseif $(MinRank_i + 1 \leq n \wedge MaxChildren_i \geq 2 \wedge MaxRoot_i \neq (D_i, P_i))$ {
    $root_i := MaxRoot_i;$
    $rank_i := MinRank_i + 1;$
    $father_i := \max\{(D_j, P_j) \mid P_j \in N_i \wedge root_j = root_i \wedge rank_j = rank_i - 1\};$
} elseif $(\exists P_j \in FCand_i, [root_j = MaxRoot_i])$ {
    $root_i := MaxRoot_i;$
    $rank_i := \min\{rank_j \mid P_j \in FCand_i \wedge root_j = root_i\} + 1;$
    $father_i := \max\{(D_j, P_j) \mid P_j \in FCand_i \wedge root_j = root_i \wedge rank_j = rank_i - 1\};$
} elseif $(|FCand_i| \geq 1)$ {
    $root_i := \max\{root_j \mid P_j \in FCand_i\};$
    $rank_i := \min\{rank_j \mid P_j \in FCand_i \wedge root_j = root_i\} + 1;$
    $father_i := \max\{(D_j, P_j) \mid P_j \in FCand_i \wedge root_j = root_i \wedge rank_j = rank_i - 1\};$
} else {
/* For singleton.*/
    $root_i := \emptyset;\ rank_i := 0;\ father_i := (D_i, P_i);$
}
}

Fig. 1. SSMLF: Self-stabilizing algorithm for construction of the maximal leafy forest

In the following explanations, we call *large tree* any tree rooted at a process with a large couple of degree on G and ID. Also, we call a *child-candidate* of process P_i the neighbor of P_i that may become a child of P_i in the future, *e.g.*, singleton, process belonging to a tree that is not larger than the tree of P_i, or process belonging to the tree of P_i that can minimize their height in the tree by changing its father to P_i. The *expected number of children* is the number of its children and child-candidates. Each process P_i counts the number of expected number of children to make the tree leafy, and joins a tree as large as possible. That is, if P_i is the root and its tree is larger than the one of its neighbors, then all its neighbors will join the tree of P_i.

According to Definition 4, SSMLF constructs of a maximal leafy forest (MLF) of G by assigning its outputs following Definitions 5 and 6.

Definition 5. *Let $sdeg_i$ be the degree of process P_i in its tree, i.e., $sdeg_i \equiv |\{P_j \in N_i \mid father_j = (D_i, P_i)\}| + |\{P_j \in N_i \mid father_i = (D_j, P_j)\}|$.* □

Definition 6. *Let $T_k = (V_k, E_k)$ be a tree rooted at a process P_r where E_k is a set of links represented by the value of $father_i$ of each process P_i of $V_k \subseteq V$ and $sdeg_r \geq 3$ holds. Consider each process P_i such that $sdeg_i = 2$ on T_k. If $sdeg_f \geq 3$ and $sdeg_j \geq 3$ where $father_i = (D_f, P_f)$ and $father_j = (D_i, P_i)$, then T_k is leafy tree. If each tree $T_1, T_2, \cdots$ is disjoint and leafy on G, then the set $\{T_1, T_2, \cdots\}$ is a leafy forest F. If F is not a subgraph of any other leafy forest, then F is maximal leafy forest.* □

In order to evaluate its output variables, a process P_i uses several macros:

- *$MaxRoot_i$* returns the largest value in the *root*-variables of P_i and its neighbors.
- *$CCand_i$* returns the set of child-candidates of P_i.
- *$CountMaxChildren_i$* returns the expected number of children of P_i.
- *$FCand_i$* returns the *father candidates* of P_i, that is, the neighbors that P_i can choose in order to make its tree leafy, *i.e.*, process P_j such that $rank_j$ is not obviously inconsistent and that has a chance of holding $sdeg_j \geq 3$.
- *$MinRank_i$* returns the *rank*-value of the (current or future) father of P_i in its tree. (If P_i is the root of its tree, then $MinRank_i$ returns -1 so that P_i sets $rank_i$ to 0.)

Now, we give more details about SSMLF. Consider a process P_i. In Lines 1-2, if the value of $root_i$ is obviously inconsistent, *i.e.*, $root_i > MAX$, then P_i resets its value to $\emptyset$. In Lines 3-4, P_i updates $MaxChildren_i$, if necessary. Then, if $root_i$ and $MaxChildren_i$ are correctly evaluated, P_i must choose its status among *root*, *internal node or leaf* of a tree and *singleton*, and updates its variables $root_i$, $rank_i$, and $father_i$ in consequence. That is what it does in Lines 5-24. Below, we detail these lines:

- In Lines 5-6, if $MaxChildren_i \geq 3$ and P_i can become a root of large tree, then P_i becomes a root.
- In Lines 7-10, P_i selects a process P_j such that the distance from the root to P_j is the minimum on a largest tree as $father_i$ only if P_j has a chance of holding $sdeg_j \geq 3$. That is, $P_j \in FCand_i$.
- In Lines 11-14, if P_i has a chance of holding $sdeg_i \geq 3$, then P_i selects a process P_j such that the distance from the root to P_j is the minimum on a largest tree as $father_i$ even if $sdeg_j < 3$, in order to make the tree leafy. By the condition of $MinRank_i + 1 \leq n$, P_i does not select a process P_j such that the value of $rank_j$ is obviously inconsistent.

- In Lines 15-18, if P_i does not have a chance of holding $sdeg_i \geq 3$, P_i selects a process P_j on a largest tree as $father_i$ only if P_j has a chance of holding $sdeg_j \geq 3$ by the condition $P_j \in FCand_i$.
- In Lines 19-22, if P_i cannot belong to a largest tree, P_i belongs to other tree.
- In Line 24, if P_i cannot belong to any tree, P_i becomes a singleton.

To show the correctness, we must first define the set of legitimate configurations of SSMLF. Such a definition is given in Definition 7 below:

Definition 7. *A configuration of* SSMLF *is legitimate if and only if each process P_i satisfies the following conditions:*

- *The connection by $father_i$ represents the maximal leafy forest.*
- *If P_i is on a tree T of the maximal leafy forest, then the value of $root_i$ of P_i represents the root of T.*
- *If P_i is not on any tree of the maximal leafy forest, then P_i is called a* singleton, *$father_i = (D_i, P_i)$ and $root_i = \emptyset$.*

By Λ_f, we denote a set of legitimate configuration. □

Theorem 2. *The algorithm* SSMLF *is self-stabilizing with respect to Λ_f.*

Proof Outline. The first part of the proof consists in proving that any terminal configuration of SSMLF is legitimate. So, in the following, we consider a configuration γ' where no process is privileged. In γ', each connected component T represented by the value of $father_i$ is a tree or a singleton on G. Additionally, there exists no process such that the value of $rank_i$ is obviously inconsistent in γ', *i.e.*, for each process P_i, $rank_i \leq n-1$ holds. Then, each tree is rooted at a process P_i such that $father_i = root_i = (D_i, P_i)$ and $sdeg_i = MaxChildren_i = D_i$, and P_0 such that $(D_0, P_0) = MAX$ is a root of a tree. Additionally, each singleton P_i holds $father_i = (D_i, P_i)$, $root_i = \emptyset$ and $sdeg_i = 0$. Other processes P_i selects a process P_j such that $sdeg_j \geq 3$ as its father if $sdeg_i < 3$, *i.e.*, each tree T represented by the values of *fathers* is leafy, and the leafy forest is maximal.

The second part of the proof consists in proving that eventually the configuration becomes γ' by the algorithm SSMLF.

For any configuration γ and any computation starting from γ, each value of $root_i$ of each process P_i on G eventually become smaller than or equal to MAX, and P_0 such that $MAX = (D_0, P_0)$ decides the values of its each variable as a root of a tree and never change. After that, some processes P_i eventually form the tree $T_0 = (V_0, E_0)$ rooted at P_0 and never change the values of their each variable.

Let $G^1 = (V^1, E^1)$ be an induced subgraph by $V \setminus V_0$. Let $(D_{\bar{1}}, P_{\bar{1}})$ be the maximum couple among the processes on G^1 which are not neighboring to any process in V_0. If $D_{\bar{1}} \geq 3$, $P_{\bar{1}}$ eventually fixes the values of its variables to be a root forever. After that, some processes P_i eventually form the tree $T_1 = (V_1, E_1)$ rooted at $P_{\bar{1}}$ and stop changing the values of their variables.

By repeating this discussion, we have a series of trees $T_0 = (V_0, E_0), T_1 = (V_1, E_1), \cdots,$ $T_k = (V_k, E_k)$, where each process in $V_0, V_1, \cdots, V_k$ fixes the values of all its variables. Let G^k be an induced subgraph by $V \setminus \{V_0 \cup V_1 \cup .. \cup V_k\}$. If $D_{\bar{k}} < 3$ holds where the

Constant (Input)
N_i: the set of neighbors on G.
$root_i$: ID of the root of tree T to which P_i belongs on the MLF (an output from the second layer).
$father_i$: ID of the father of P_i in T (an output from the second layer).
Variable (Output)
$W(P_i)[P_j]$: new cost of the edge between P_i and $P_j \in N_i$.
Algorithm for process P_i:

```
do forever{
    if (∃P_j ∈ N_i[root_i ≠ root_j ∧ W(P_i)[P_j] ≠ 1]){
        W(P_i)[P_j] := 1; /* For Inter-tree edge */
    } elseif (∃P_j ∈ N_i[root_i = root_j ∧ (father_i ≠ P_j ∧ father_j ≠ P_i) ∧ W(P_i)[P_j] ≠ ∞]){
        W(P_i)[P_j] := ∞; /* For Intra-tree edge */
    } elseif (∃P_j ∈ N_i[root_i = root_j ∧ (father_i = P_j ∨ father_j = P_i) ∧ W(P_i)[P_j] ≠ 0]){
        W(P_i)[P_j] := 0; /* For Tree edge */
    }
}
```

Fig. 2. SSTN: Self-stabilizing algorithm for transforming the network

maximum couple on G^k among the processes which are not neighboring to any process in $\{V_0 \cup V_1 \cup \cdots \cup V_k\}$, processes on G^k cannot form any leafy tree, and they become singletons. Then, no process is privileged. □

Theorem 3. *The time complexity of algorithm* SSMLF *is* $O(n^2)$ *rounds.*

3.2 The Third Layer: Modification of Edge Cost

The third layer algorithm SSTN computes a network cost from the MLF computed by the second layer. In the fourth layer, the minimum spanning tree is computed based on this cost. The minimum spanning tree is the approximate solution of the MLST problem. Then, the edges in any tree must include the minimum spanning tree, and the *intra-tree* edges such that both endpoints are in the same tree must not be in the minimum spanning tree. Additionally, we would like to make the number of the connector edges between each tree as small as possible. The connector edges are selected from *inter-tree* edges such that its endpoints are not in the same tree.

Formal description of this layer SSMLF is shown in Fig. 2.

Definition 8. *A configuration of* SSTN *is legitimate if and only if each edge satisfies the following three conditions.*

- *If e is a tree edge, then* $W(P_i)[P_j]$ *is* 0,
- *If e is an intra-tree edge, then* $W(P_i)[P_j]$ *is* ∞, *and*
- *If e is an inter-tree edge, then* $W(P_i)[P_j]$ *is* 1.

By Λ_t, *we denote a set of legitimate configuration.* □

Theorem 4. *The algorithm* SSTN *is self-stabilizing with respect to* Λ_t. *The time complexity of algorithm* SSTN *is* $O(n)$ *rounds.*

Each of the four layers stabilize in at most $O(n^2)$ rounds, hence we can conclude:

Theorem 5. *The algorithm* SSMLST *is a self-stabilizing approximation algorithm for the MLST problem with approximation ratio 3. Its time complexity is* $O(n^2)$ *rounds.*

4 Conclusion

In this paper, we proposed a self-stabilizing distributed approximation algorithm for the MLST problem in arbitrary topology networks with approximation ratio 3. However, there exists a sequential solution [1] proposed by Solis-Oba that has approximation ratio 2. Investigating the trade-off between approximation ratio and complexity of the self-stabilizing mechanism to achieve it is an immediate future work. Also, we would like to mention the importance to complement the self-stabilizing abilities of a distributed algorithm with some additional *safety* properties that are guaranteed when the permanent and intermittent failures that hit the system satisfy some conditions.

References

1. Solis-Oba, R.: 2-approximation algorithm for finding a spanning tree with maximum number of leaves. In: Bilardi, G., Pietracaprina, A., Italiano, G.F., Pucci, G. (eds.) ESA 1998. LNCS, vol. 1461, pp. 441–452. Springer, Heidelberg (1998)
2. Tixeuil, S.: Self-stabilizing Algorithms. Chapman & Hall/CRC Applied Algorithms and Data Structures. In: Algorithms and Theory of Computation Handbook, 2nd edn. Taylor & Francis, Abington (2009)
3. Galbiati, G., Maffioli, F., Morzenti, A.: A short note on the approximability of the maximum leaves spanning tree problem. Information Processing Letters 52(1), 45–49 (1994)
4. Lu, H.I., Ravi, R.: Approximating maximum leaf spanning trees in almost linear time. Journal of algorithms 29, 132–141 (1998)
5. Blin, L., Potop-Butucaru, M., Rovedakis, S.: Self-stabilizing minimum-degree spanning tree within one from the optimal degree. In: Proceedings of the 23th IEEE International Parallel and Distributed Processing Symposium (2009)
6. Blin, L., Potop-Butucaru, M.G., Rovedakis, S., Tixeuil, S.: A new self-stabilizing minimum spanning tree construction with loop-free property. In: Keidar, I. (ed.) DISC 2009. LNCS, vol. 5805, pp. 407–422. Springer, Heidelberg (2009)
7. Butelle, F., Lavault, C., Bui, M.: A uniform self-stabilizing minimum diameter tree algorithm. In: Helary, J.-M., Raynal, M. (eds.) WDAG 1995. LNCS, vol. 972, pp. 257–272. Springer, Heidelberg (1995)
8. Guha, S., Khuller, S.: Approximation Algorithms for Connected Dominating Sets. Algorithmica 20, 347–387 (1998)
9. Kamei, S., Kakugawa, H.: A self-stabilizing distributed approximation algorithm for the minimum connected dominating set. In: Proceedings of the 9th IPDPS Workshop on Advances in Parallel and Distributed Computational Models, p. 224 (2007)
10. Raei, H., Tabibzadeh, M., Ahmadipoor, B., Saei, S.: A self-stabilizing distributed algorithm for minimum connected dominating sets in wireless sensor networks with different transmission ranges. In: Proceedings of the 11th International Conference on Advanced Communication Technology, pp. 526–530 (2009)
11. Dolev, S.: Self-Stabilization. The MIT Press, Cambridge (2000)
12. Datta, A.K., Larmore, L.L., Vemula, P.: Self-stabilizing leader election in optimal space. In: Kulkarni, S., Schiper, A. (eds.) SSS 2008. LNCS, vol. 5340, pp. 109–123. Springer, Heidelberg (2008)

Approximate Weighted Farthest Neighbors and Minimum Dilation Stars

John Augustine[1], David Eppstein[2], and Kevin A. Wortman[3]

[1] School of Physical & Mathematical Sciences
Nanyang Technological University
jea@ics.uci.edu

[2] Computer Science Department
University of California, Irvine
Irvine, CA 92697, USA
eppstein@ics.uci.edu

[3] Computer Science Department
California State University, Fullerton
Fullerton, CA 92831, USA
kwortman@fullerton.edu

Abstract. We provide an efficient reduction from the problem of querying approximate multiplicatively weighted farthest neighbors in a metric space to the unweighted problem. Combining our techniques with core-sets for approximate unweighted farthest neighbors, we show how to find approximate farthest neighbors that are farther than a factor $(1-\varepsilon)$ of optimal in time $O(\log n)$ per query in D-dimensional Euclidean space for any constants D and ε. As an application, we find an $O(n \log n)$ expected time algorithm for choosing the center of a star topology network connecting a given set of points, so as to approximately minimize the maximum dilation between any pair of points.

1 Introduction

Data structures for proximity problems such as finding closest or farthest neighbors or maintaining closest or farthest pairs in sets of points have been a central topic in computational geometry for a long time [2]. Due to the difficulty of solving these problems exactly in high dimensions, there has been much work on approximate versions of these problems, in which we seek neighbors whose distance is greater than a factor $(1-\varepsilon)$ of optimal [4]. In this paper, we consider the version of this problem in which we seek to approximately answer farthest neighbor queries for point sets with multiplicative weights. That is, we have a set of points p_i, each with a weight $w(p_i)$, and for any query point q we seek to approximate $\max_i w(p_i)\, d(p_i, q)$. That problem can be solved exactly by constructing the weighted farthest neighbor Voronoi diagram, which Lee and Wu [6] show how to construct in $O(n \log^2 n)$ time. The query phase is a simple point location problem that takes $O(\log n)$ time.

We provide the following new results:

– We describe in Theorem 3 a general reduction from the approximate weighted farthest neighbor query problem to the approximate unweighted farthest neighbor

M.T. Thai and S. Sahni (Eds.): COCOON 2010, LNCS 6196, pp. 90–99, 2010.

query problem. In any metric space, suppose that there exists a data structure that can answer unweighted approximate farthest neighbor queries (farther than a factor $(1-\varepsilon)$ of optimal) for a given n-item point set in query time $Q(n,\varepsilon)$, space $S(n,\varepsilon)$, and preprocessing time $P(n,\varepsilon)$. Then our reduction provides a data structure for answering $(1-\varepsilon)$-approximate weighted farthest neighbor queries in time $O((\frac{1}{\varepsilon}\log\frac{1}{\varepsilon})Q(n,\varepsilon/2))$ per query, that uses space $O((\frac{1}{\varepsilon}\log\frac{1}{\varepsilon})S(n,\varepsilon/2))$, and can be constructed in preprocessing time $O((\frac{1}{\varepsilon}\log\frac{1}{\varepsilon})P(n,\varepsilon/2))$. The results are summarized in Table 1.

- We apply core-sets [1] to find a data structure for the approximate unweighted farthest neighbor query problem in $\mathbb{R}^D$, for any constant D, with query time $O(\varepsilon^{\frac{1-D}{2}})$, space $O(\varepsilon^{\frac{1-D}{2}})$, and preprocessing time $O(n+\varepsilon^{\frac{3}{2}-D})$ (Theorem 5). Applying our reduction results in a data structure for the approximate weighted farthest neighbor query problem in $\mathbb{R}^D$, with preprocessing time $O(\frac{n}{\varepsilon}\log\frac{1}{\varepsilon}+\varepsilon^{\frac{1}{2}-D}\log\frac{1}{\varepsilon})$, query time $O(\varepsilon^{\frac{-1-D}{2}}\log\frac{1}{\varepsilon})$, and space $O(\varepsilon^{\frac{-1-D}{2}}\log\frac{1}{\varepsilon})$ (Corollary 6). Table 1 summarizes the results.
- As a motivating example for our data structures, we consider the problem of finding a star-topology network for a set of points in $\mathbb{R}^D$, having one of the input points as the hub of the network, and minimizing the maximum dilation of any network path between any pair of points. By results of Eppstein and Wortman [5], this problem can be solved exactly in time $O(n2^{\alpha(n)}\log^2 n)$ in the plane, and $O(n^2)$ in any higher dimension. By using our data structure for approximate weighted farthest neighbor queries, we find in Corollary 16 a solution to this problem having dilation within a $(1+\varepsilon)$ factor of optimal, for any constant ε and constant dimension, in expected time $O(n\log n)$. More generally, as shown in Theorem 15, we can approximately evaluate the dilations that would be achieved by using every input point as the hub of a star topology, in expected $O(n\log n)$ total time.

Table 1. Summary of results for farthest neighbor problems. For the sake of brevity, $O(\cdot)$ notations are omitted.

Weighted or unweighted	unweighted	weighted	unweighted	weighted
Metric Space	metric space	metric space	$\mathbb{R}^D$	$\mathbb{R}^D$
Preprocessing time	$P(n,\varepsilon)$	$\frac{1}{\varepsilon}\log\frac{1}{\varepsilon}P(n,\varepsilon/2)$	$n+\varepsilon^{3/2-D}$	$(n/\varepsilon)\log\frac{1}{\varepsilon}+\varepsilon^{1/2-D}\log\frac{1}{\varepsilon}$
Space	$S(n,\varepsilon)$	$\frac{1}{\varepsilon}\log\frac{1}{\varepsilon}S(n,\varepsilon/2)$	$\varepsilon^{\frac{1-D}{2}}$	$\varepsilon^{\frac{-1-D}{2}}\log\frac{1}{\varepsilon}$
Query time	$Q(n,\varepsilon)$	$\frac{1}{\varepsilon}\log\frac{1}{\varepsilon}Q(n,\varepsilon/2)$	$\varepsilon^{\frac{1-D}{2}}$	$\varepsilon^{\frac{-1-D}{2}}\log\frac{1}{\varepsilon}$

2 Problem Definition

We first define the *weighted farthest neighbor query problem* in a metric space, and then extend its definition to relevant variants and special cases. Let $\mathcal{M}=(S,d)$ be a metric space. For any two points p and q in S, $d(p,q)$ denotes the distance between them. We assume we are given as input a finite set $P \subset S$; we denote by p_i the points in P. In addition, we are given positive weights on each point of P, which we denote by a function $w : P \mapsto \mathbb{R}^+$. We wish to preprocess the point set such that for any query point $q \in S$, we can quickly produce a point $f \in P$ that maximizes the weighted distance from q to r. More precisely, f should be the point in P that maximizes (or approximately maximizes) the weighted distance $d_w(q, p_i) = w(p_i) \cdot d(q, p_i)$ from the query point q. If we restrict the weights of all points to be equal, then this problem simplifies to the *unweighted farthest neighbor query problem.*

Finding the exact farthest neighbor to a query point can be computationally expensive. Hence, we are interested in approximate solutions to this problem, with sufficient guarantees that our algorithm produces a reasonably distant point in the weighted sense. To achieve this, we define the *approximate weighted farthest neighbor query problem*, wherein we are given an input of the same type as the weighted farthest neighbor query problem, and wish to preprocess the point set so that we can efficiently produce an $(1-\varepsilon)$-approximate weighted farthest point. That is, we wish to find a point $r \in P$ such that $d_w(q, r) \geq (1-\varepsilon) d_w(q, p_i)$ for some predefined ε and all $p_i \in P$. A similar definition holds for the *approximate unweighted farthest neighbor query problem* as well. When the metric space under consideration is Euclidean, we call the problem the *Euclidean approximate weighted farthest neighbor query problem.*

Without loss of generality, we can assume that $0 < w(p) \leq 1$ for all $p \in P$, and that exactly one point p_1 has weight 1. The first assumption can be made to hold by dividing all weights by the largest weight; this leaves approximations to the weighted farthest neighbor unchanged. The second assumption can be made to hold by arbitrarily breaking ties among contenders for the maximum-weight point.

3 Reduction from Weighted to Unweighted

First, let us suppose that we have a family of algorithms $\mathcal{A}_\varepsilon$ for the approximate unweighted farthest neighbor query problem defined over any metric space $\mathcal{M} = (S, d)$. $\mathcal{A}_\varepsilon$ consists of two components, one for preprocessing and the other for querying. For an input instance of the approximate unweighted farthest neighbor query problem, let $f \in P$ maximize the distance between our query point q and f. Using $\mathcal{A}_\varepsilon$, we can find an $(1-\varepsilon)$-approximate farthest point r such that $(1-\varepsilon)d(q, f) \leq d(q, r)$. The running time of $\mathcal{A}_\varepsilon$ has two components, namely preprocessing time and query time denoted $P(n, \varepsilon)$, and $Q(n, \varepsilon)$ respectively. It has a space requirement of $S(n, \varepsilon)$. Our real concern is the weighted version of the problem and in this section, we provide a data structure to reduce an instance of the approximate weighted farthest neighbor query problem such that it can be solved by invoking A_ε. We call this family of reduction algorithms $\mathcal{R}_\varepsilon$.

The preprocessing and querying components of $\mathcal{R}_\varepsilon$ are provided in Procedures 1 and 2 respectively. For convenience, we separate out the points in S into two subsets S_1 and S_2. The first subset S_1 contains all points in P with weights in $[\varepsilon/2, 1]$, and S_2 contains $P \setminus S_1$. We use different approaches for the two subsets producing three candidate approximate farthest neighbors. Our final result is the farther of the three candidates in the weighted sense. We now show that the algorithm is correct both when $f \in S_1$ and when $f \in S_2$.

Procedure 1. Preprocessing for the approximate unweighted farthest neighbor problem

1: **INPUT:** Set of points P with $n = |P|$ and a distance metric d. A fixed error parameter ε.
2: **Let** $S_1 \subseteq P$ be points with weights in $[\varepsilon/2, 1]$ and $S_2 = P \setminus S_2$.
{Farthest point is in S_1.}
3: **Partition** S_1 into buckets $B_0, B_1, \ldots$ such that each B_i contains all points in S_1 whose weights are in $((\varepsilon/2)(1+\varepsilon/2)^{i-1}, (\varepsilon/2)(1+(\varepsilon/2))^i]$.
4: **Preprocess** each bucket by calling the preprocessing component of A_ε instance with error parameter $\varepsilon/2$.
{Farthest point is in S_2.}
5: **Assign** p_1 to be point in P with highest weight (assumed to be 1).
6: **Assign** $r_2 = \arg\max_{p \in S_2} d_w(q, p)$.

Case $f \in S_1$:

In this case, we only consider points in S_1. Therefore, we can assume that the weights are within a factor of $2/\varepsilon$ of each other. We partition the points set S_1 into buckets such that each bucket B_i contains all points in S_1 with weight in $[(\varepsilon/2)(1+\varepsilon/2)^{(i-1)}, (\varepsilon/2)(1+\varepsilon/2)^i)$. The number of buckets needed to cover S_1 is $\Theta((1/\varepsilon)\ln(1/\varepsilon))$.

Our data structure consists of separate instances (consisting of both the preprocessing and querying components) of A_ε with the error parameter $\varepsilon/2$ for each bucket B_i. In addition, each B_i is preprocessed by the preprocessing instance of A_ε. To query our data structure for the $(1-\varepsilon/2)$-approximate farthest neighbor $k_i \in B_i$ from q, we run the querying instance of $\mathcal{A}_\varepsilon$ on each preprocessed bucket B_i. Our candidate approximate weighted farthest neighbor for this case will be the $r_1 = \arg\max_{k_i} d(q, k_i)$.

Lemma 1. *If $f \in S_1$, then $d_w(q, r_1) \geq (1-\varepsilon)d_w(q, f)$, where f is the exact weighted farthest point and r_1 is as defined above.*

Proof. Let us just consider bucket B_i whose weights are between some $(\varepsilon/2)(1+\varepsilon/2)^{(i-1)}$ and $(\varepsilon/2)(1+\varepsilon/2)^i$. Let f_i be the exact weighted farthest point from q in B_i. Within the bucket B_i, k_i is an $(1-\varepsilon/2)$-approximate farthest point if we don't consider the weights, i.e., $(1-\varepsilon/2)d(q, f_i) \leq d(q, k_i)$. However, when we consider the weighted distance, $w(k_i)$ and $w(f_i)$ differ by a factor of at most $(1+\varepsilon/2)$ and hence our inequality for the weighted case becomes

$$\frac{1-\varepsilon/2}{1+\varepsilon/2} d_w(q, f_i) < (1-\varepsilon)d_w(q, f_i) \leq d_w(q, k_i)$$

since $\frac{1-\varepsilon/2}{1+\varepsilon/2} > (1-\varepsilon)$ for $\varepsilon > 0$. Considering over all values of i,

$$(1-\varepsilon)d_w(q, f) = \max_i (1-\varepsilon)d_w(q, f_i) < \max_i d_w(q, k_i) = d_w(q, r_1).$$

□

Case $f \in S_2$:

In this case, we generate two candidate points. The first is p_1, the point with the greatest weight (normalized to 1). The second candidate is $r_2 = \arg\max_{p \in S_2} d_w(p_1, p)$. Intuitively, if p_1 is not an $(1-\varepsilon)$-approximate weighted farthest neighbor of q, then it is close enough to q to ensure that r_2 will be a $(1-\varepsilon)$-approximate weighted farthest neighbor of q. Lemma 2 formalizes this intuition.

Lemma 2. *If $f \in S_2$, and p_1 is not an $(1-\varepsilon)$-approximate weighted farthest point from q, then $d_w(q,r_2) \geq (1-\varepsilon)d_w(q,f)$.*

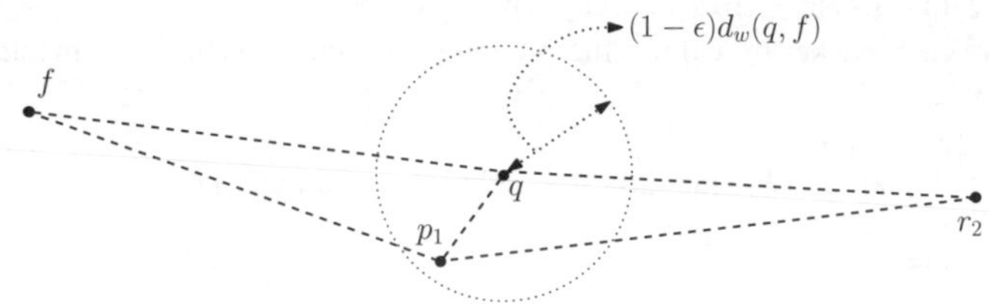

Fig. 1. Figure illustrating Lemma 2

Proof. Given how we choose r_2, $d_w(p_1,r_2) \geq d_w(p_1,f)$. Consider the ball centered at q with radius $(1-\varepsilon)d_w(q,f)$ (shown in Figure 1). Clearly, p_1 is inside this ball, because

$$d(q,p_1) = d_w(q,p_1) < (1-\varepsilon)d_w(q,f). \tag{1}$$

Applying triangle inequality on $\triangle p_1 f q$, we get $d(p_1,f) \geq d(q,f) - d(q,p_1)$ and multiplying throughout by $w(f) \leq (\varepsilon/2)$, we get

$$d_w(p_1,f) \geq d_w(q,f) - (\varepsilon/2)d(q,p_1). \tag{2}$$

Similarly, applying triangle inequality on $\triangle p_1 q r_2$, we get $d(q,p_1)+d(q,r_2) \geq d(p_1,r_2)$ and multiplying throughout by $w(r_2) \leq (\varepsilon/2)$, we get

$$(\varepsilon/2)d(q,p_1) + d_w(q,r_2) \geq d_w(p_1,r_2). \tag{3}$$

Applying Equation 2 and Equation 3 in $d_w(p_1,r_2) \geq d_w(p_1,f)$, we get

$$(\varepsilon/2)d(q,p_1) + d_w(q,r_2) \geq d_w(q,f) - (\varepsilon/2)d(q,p_1).$$

Rearranging and substituting Equation 1, we get $d_w(q,r_2) \geq d_w(q,f) - \varepsilon(1-\varepsilon)d_w(q,f) \geq (1-\varepsilon)d_w(q,f)$. □

The preprocessing and querying components (Procedures 1 and 2 respectively) are presented in pseudocode format. Each procedure has two parts to it corresponding to the two cases when $f \in S_1$ and $f \in S_2$. In Procedure 1, we first partition S_1 into buckets and preprocess them. Secondly, we compute p_1 and r_2 according to the requirements of Lemma 2. In Procedure 2, we choose the farthest neighbor r_1 of q in S_1 by querying each bucket and choosing the farthest from the pool of results. Our final $(1-\varepsilon)$-approximate farthest neighbor r of q is the weighted farthest point in $\{r_1, r_2, p_1\}$ from q. The following theorem follows.

Theorem 3. *The family of reduction algorithms $\mathcal{R}_\varepsilon$ answer $(1-\varepsilon)$-approximate weighted farthest neighbor queries in time $O((\frac{1}{\varepsilon}\log\frac{1}{\varepsilon})Q(n,\varepsilon/2))$ per query, and the data structure can be constructed in preprocessing time $O((\frac{1}{\varepsilon}\log\frac{1}{\varepsilon})P(n,\varepsilon/2))$ with a space requirement of $O((\frac{1}{\varepsilon}\log\frac{1}{\varepsilon})S(n,\varepsilon/2))$.*

Procedure 2. Query step of the approximate unweighted farthest neighbor problem

```
INPUT: Query point q, B_i for all 1 ≤ i ≤ 1/ε ln(1/ε), and S_2.
{Farthest point is in S_1.}
Assign r_1 ← q
for each B_i do
    k_i ← A_(ε/2)(B_i)
    if d_w(q, r_1) ≤ d_w(q, k_i) then
        r_1 ← k_i
    end if
end for
{Farthest point in S_2 is subsumed in following line.}
Return r ← arg max_{r_1, r_2 p_1} (d_w(q, r_1), d_w(q, r_2), d_w(q, p_1))
```

4 Euclidean Approximate Unweighted Farthest Neighbor Queries

In this section, our problem definition remains intact except for the restriction that S is $\mathbb{R}^D$, where D is a constant, with the Euclidean distance metric. Notice that our points are unweighted and we are seeking a family of approximation algorithms $\mathcal{A}_\varepsilon$ that we assumed to exist as a black box in the previous section. Let $f \in P$ maximize the distance between our query point q and points in S. For a given ε, we are asked to find an $(1-\varepsilon)$-approximate farthest point r such that $d(q,r) \geq (1-\varepsilon)d(q,f)$. We can solve this using the ε-kernel technique surveyed in [1]. Let u be the unit vector in some direction. We denote the directional width of P in direction u by $w(u,P)$ and is given by

$$w(u,P) = \max_{s \in P} \langle u,s \rangle - \min_{s \in P} \langle u,s \rangle,$$

where $\langle u,s \rangle$ is the dot product of u and s. An ε-kernel is a subset K of P such that for all unit directions u,

$$(1-\varepsilon)w(u,P) \leq w(u,K).$$

It is now useful to state a theorem from [3, 1] that provides us an ε-kernel in time linear in n and polynomial in $1/\varepsilon$ with the dimension D appearing in the exponent.

Theorem 4. *Given a set P of n points in $\mathbb{R}^D$ and a parameter $\varepsilon > 0$, one can compute an ε-kernel of P of size $O(1/\varepsilon^{(D-1)/2})$ in time $O(n + 1/\varepsilon^{D-(3/2)})$.*

Consider points s_1 and s_2 in P and k_1 and k_2 in K that maximize the directional widths in the direction u, which is the unit vector in the direction of $\overrightarrow{qf}$.

$$(1-\varepsilon)w(u,P) = (1-\varepsilon)\langle u,s_1 \rangle - (1-\varepsilon)\langle u,s_2 \rangle \leq \langle u,k_1 \rangle - \langle u,k_2 \rangle.$$

Now, $\langle u,k_2 \rangle \geq \langle u,s_2 \rangle$, because otherwise, $w(u,P)$ can be maximized further. Hence with some substitution, we get $(1-\varepsilon)\langle u,s_1 \rangle \leq \langle u,k_1 \rangle$.

The point f maximizes the left hand side in the direction of $\overrightarrow{qf}$ because it is the farthest point from q. Therefore, the above inequality suggests that $k_1 \in K$ is an $(1-\varepsilon)$-approximate farthest neighbor of q. This implies that the point farthest from q obtained by sequentially searching the ε-kernel K will be an $(1-\varepsilon)$-approximate farthest neighbor of q. Hence, we can state the following theorem.

Theorem 5. *Given a set P of n points in $\mathbb{R}^D$ and a parameter $\varepsilon > 0$, there exists a family of algorithms $\mathcal{A}_\varepsilon$ that answers the $(1-\varepsilon)$-approximate unweighted farthest neighbor query problem in preprocessing time $O(n + e^{\frac{3}{2}-D})$, and the query time and space requirements are both $O(\varepsilon^{\frac{1-D}{2}})$.*

Combining this theorem with our general reduction from the weighted to the unweighted problem gives us the following corollary.

Corollary 6. *Given a set P of n points in $\mathbb{R}^D$, for any constant D, and given a weight function $w : P \mapsto \mathbb{R}^+$ and a parameter $\varepsilon > 0$, there exists a family of algorithms that answers the $(1-\varepsilon)$-approximate weighted farthest neighbor query problem in preprocessing time $O(\frac{n}{\varepsilon}\log\frac{1}{\varepsilon} + \varepsilon^{\frac{1}{2}-D}\log\frac{1}{\varepsilon})$; both query time and space required are in $O(\varepsilon^{\frac{-D-1}{2}}\log\frac{1}{\varepsilon})$.*

5 Constrained Minimum Dilation Stars

The *dilation* between two vertices v and w of a Euclidean graph is defined as the ratio of the weight of the shortest path from v to w, divided by the direct distance between v and w. The dilation of the entire graph is defined as the greatest dilation between any pair of vertices in the graph. A *star* is a connected graph with exactly one internal vertex, called its *center*. Any collection of n points admits n possible stars, since any individual point may serve as a star center. In this section we consider the problem of computing the dilation of all of these n stars. A solution to this problem provides the foundation for a solution to the problem of choosing an optimal center: simply search for the point whose corresponding dilation value is smallest.

Eppstein and Wortman [5] considered the problem of selecting star centers that are optimal with respect to dilation. They showed that for any set of n points $P \subset \mathbb{R}^D$, there exists a set C of $O(n)$ pairs of points such that the worst pair for any center $c \in \mathbb{R}^D$ is contained in C. They give an $O(n\log n)$-time algorithm for constructing C, and go on to consider the problem of computing the center that admits a star with minimal dilation. They present an $O(n\log n)$ expected-time algorithm for the case when c may be any point in $\mathbb{R}^D$ and D is an arbitrary constant, and an $O(n2^{\alpha(n)}\log^2 n)$ expected-time algorithm for the case when c is constrained to be one of the input points and $D = 2$.

These results imply that the dilation of all n stars with centers from P may be computed in $O(n^2)$ time: construct C, then for each $c \in P$ and $(v,w) \in C$, evaluate the dilation of the path $\langle v,c,w\rangle$. In this section we improve on this time bound through approximation. We will show that a $(1-\varepsilon)$-approximation of the dilation of all n stars may be computed in $O(n\log n)$ expected time, and hence an approximately optimal center $c \in P$ may be identified in $O(n\log n)$ expected time.

Our approximation algorithm first uses the results of [5] to compute the optimal $c_{OPT} \in \mathbb{R}^D$. Note that c_{OPT} may or may not be in P, and in general will not be. Our algorithm then partitions P into two subsets: the $k \in O(1)$ points nearest to c_{OPT}, and the other $n-k$ points. We say that a star center $c \in P$ is *k-low* if it is one of the k points nearest to c_{OPT}, or *k-high* otherwise. The dilation values for all the k-low centers are computed exactly in $O(n\log n)$ time using a combination of known techniques.

The dilation between any two points $p_i, p_j \in P$ through any k-high center c may be approximated by a weighted distance from c to the centroid of p_i and p_j. Hence the dilation of the star centered on any k-high c may be approximated by the distance from c to the centroid farthest from c. We use the data structure described in Section 4 to answer each of these $n-k$ weighted farthest-neighbor queries in $O(\log n)$ time, which makes the overall running time of the approximation algorithm $O(n \log n)$.

Definition 7. *Let G be a Euclidean star with vertices V and center c. The dilation between any $v, w \in V \setminus \{c\}$ is $\delta(c,v,w) = \frac{d(v,c)+d(w,c)}{d(v,w)}$, where $d(x,y)$ is the Euclidean distance $|xy|$ between x and y. The dilation of the entire star is $\Delta(c) = \max_{v,w \in V \setminus \{c\}} \delta(c,v,w)$.*

Definition 8. *Let*

- *$P = \langle p_0, \ldots, p_{n-1} \rangle$ be a set of n input points from $\mathbb{R}^D$ for some constant D, each with the potential to be the center of a star with vertices P,*
- *$c_{OPT} \in \mathbb{R}^D$ be the point minimizing Δ, i.e. $\Delta(c_{OPT}) \le \Delta(x)\ \forall x \in \mathbb{R}^D$,*
- *$S = \langle s_0, \ldots, s_{n-1} \rangle$ be the sequence formed by sorting P by distance from c_{OPT},*
- *$\varepsilon > 0$ be a constant parameter,*
- *$\Gamma = 2/\varepsilon - 1$,*
- *k be a constant depending only on Γ,*
- *$L = \{s_i \in S \mid 0 \le i < k\}$ be the k-low centers,*
- *and $H = P \setminus L$ be the k-high centers.*

We require the following claim, which is proved in [5]:

Claim 9. *Let c be the center of a Euclidean star in $\mathbb{R}^D$ for $D \in O(1)$ having vertices V. If $\Delta(c) \le \Gamma$ for some constant Γ, then there exists a constant ρ_Γ such that for any integer i, the D-dimensional annulus centered on c with inner radius ρ_Γ^i and outer radius ρ_Γ^{i+1} contains only $O(1)$ points from V.*

Lemma 10. *For any Γ there exists a constant k depending only on Γ such that any center $s_i \in S$ with $i \ge k$ has dilation $\Delta(s_i) \ge \Gamma$.*

Proof. We first consider the case when $\Delta(c_{OPT}) > \Gamma$. By definition $\Delta(s_i) \ge \Delta(c_{OPT})$ for any $s_i \in S$, so we have that every $\Delta(s_i) > \Gamma$ regardless of the value of k. We now turn to the case when $\Delta(c_{OPT}) \le \Gamma$. Define ρ_Γ as in Claim 9, and let $A_j = \{x \in \mathbb{R}^D \mid \rho_\Gamma^j \le d(x, c_{OPT}) \le \rho_\Gamma^{j+1}\}$ be the jth annulus centered on c_{OPT}. Define $l = 1 + \lceil \log_{\rho_\Gamma}(\Gamma - 1) \rceil$, and suppose $s_i \in A_j$. Let v, w be two input points that lie in the annulus A_{j-l}. By the definition of l we have $\rho_\Gamma^{l-1} - 1 \ge \Gamma$.

$$\frac{2(\rho_\Gamma^{j-1} - \rho_\Gamma^{j-l})}{2\rho_\Gamma^{j-l}} \ge \Gamma.$$

Since $v \in A_{j-l}$, we have $d(s_i, v) \ge (\rho_\Gamma^{j-1} - \rho_\Gamma^{j-l})$, and similarly $d(s_i, w) \ge (\rho_\Gamma^{j-1} - \rho_\Gamma^{j-l})$. Further, $d(v,w) \le 2\rho_\Gamma^{j-l}$. So by substitution

$$\frac{(d(s_i,v)) + (d(s_i,w))}{(d(v,w))} \ge \Gamma.$$

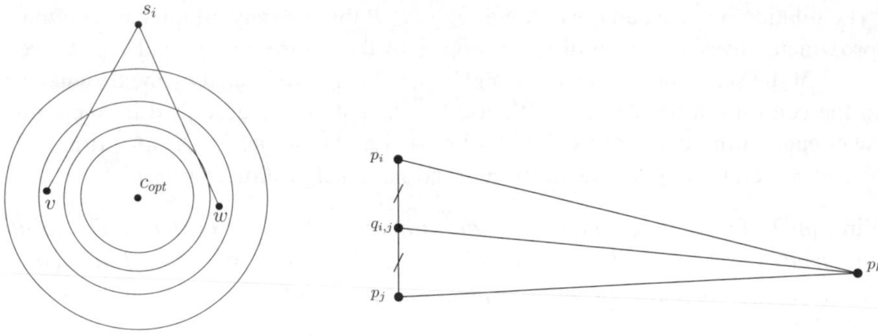

Fig. 2. Planar annuli containing the points s_i, v, and w

Fig. 3. Example points p_i, p_j, $q_{i,j}$, and c

Therefore, $\Delta(s_i) \geq \Gamma$, so the lemma holds if the annuli $A_j, A_{j-1}, \ldots, A_{j-l}$ contain at least k points. We can ensure this by selecting k such that the points $s_{i-k}, \ldots, s_i$ necessarily span at least $l+1$ annuli. By Claim 9, there exists some $m \in O(1)$ such that any annulus contains no more than m input points. So we set $k \geq (l+1)m$. The observation that l depends only on Γ completes the proof. □

Corollary 11. *For any $c \in H$, $\Delta(c) \geq \Gamma$; and $|L| \in O(1)$.*

Lemma 12. *The set L, as well as the quantity $\Delta(c)$ for every $c \in L$, may be computed in $O(n \log n)$ expected time. (Proof is deferred to the full version of the paper.)*

Lemma 13. *If $c \in H$, $p_i, p_j \in P$ be the pair of points such that $\delta(c, p_i, p_j) = \Delta(c)$, and $d(p_i, c) \geq d(p_j, c)$, then $d(p_j, c) \geq (1-\varepsilon) d(p_i, c)$.*

Proof. By the assumption $c \in H$ and Corollary 11, $\delta(c, p_i, p_j) = \frac{d(p_i,c)+d(p_j,c)}{d(p_i,p_j)} \geq \Gamma$; by the triangle inequality, $d(p_i, p_j) \geq d(p_i, c) - d(p_j, c)$, so

$$\frac{d(p_i,c)+d(p_j,c)}{(d(p_i,c)-d(p_j,c))} \geq \Gamma$$

With some rearranging, $d(p_j, c) \geq \frac{\Gamma-1}{\Gamma+1} d(p_i, c)$. By the definition of Γ, we get $d(p_j, c) \geq \frac{(2/\varepsilon-1)-1}{(2/\varepsilon-1)+1} d(p_i, c) \geq \frac{2/\varepsilon-2}{2/\varepsilon} d(p_i, c) \geq (1-\varepsilon) d(p_i, c)$. □

Lemma 14. *Define $q_{i,j} = \frac{1}{2}(p_i + p_j)$ and $w_{i,j} = \frac{2}{d(p_i,p_j)}$; then for any $c \in H$ and corresponding $p_i, p_j \in P$ such that $\delta(c, p_i, p_j) = \Delta(c)$, $w_{i,j} \cdot d(q_{i,j}, c) \geq (1-\varepsilon) \cdot \delta(c, p_i, p_j)$. (Proof is deferred to the full version of the paper.)*

Theorem 15. *A set of n values $\hat{\Delta}(p_i)$ subject to $(1-\varepsilon)\Delta(p_i) \leq \hat{\Delta}(p_i) \leq \Delta(p_i)$ may be computed in $O(n \log n)$ expected time.*

Proof. By Lemma 12 it is possible to separate the k-low and k-high centers, and compute the exact dilation of all the k-low centers, in $O(n \log n)$ expected time. Section 4 describes a data structure that can answer approximate weighted farthest neighbor queries within a factor of $(1-\varepsilon)$ in $O(1)$ time after $O(n)$ preprocessing. As shown in Lemma 14, the dilation function δ may be approximated for the k-high centers up to a factor of $(1-\varepsilon)$ using weights $w_{i,j} = \frac{2}{d(p_i,p_j)}$. So for each k-high p_i, $\Delta(p_i)$ may be approximated by the result of a weighted farthest neighbor query. Each of these $|H| = n-k$ queries takes $O(1)$ time, for a total expected running time of $O(n \log n)$. □

Corollary 16. *Let* $\Delta_{OPT} = \min_{p_i \in P} \Delta(p_i)$. *Then a point* $\hat{c} \in P$ *satisfying* $\Delta_{OPT} \leq \Delta(\hat{c}) \leq (1+\varepsilon)\Delta_{OPT}$ *may be identified in* $O(n \log n)$ *expected time.*

Acknowledgment. We are grateful to the anonymous reviewers of a previous version of the paper for their useful suggestions.

References

1. Agarwal, P.K., Har-Peled, S., Varadarajan, K.R.: Geometric approximation via coresets. In: Goodman, J.E., Pach, J., Welzl, E. (eds.) Combinatorial and Computational Geometry. MSRI Publications, vol. 52, pp. 1–30. Cambridge University Press, Cambridge (2005)
2. Arya, S., Mount, D.: Computational geometry: Proximity and location. In: Mehta, D.P., Sahni, S. (eds.) Handbook of Data Structures and Applications, pp. 63-1–63-22. CRC Press, Boca Raton (2005)
3. Chan, T.M.: Faster core-set constructions and data stream algorithms in fixed dimensions. In: Proc. 20th Symp. Computational Geometry, pp. 152–159. ACM Press, New York (2004)
4. Duncan, C., Goodrich, M.T.: Approximate geometric query structures. In: Mehta, D.P., Sahni, S. (eds.) Handbook of Data Structures and Applications, pp. 26-1–26-17. CRC Press, Boca Raton (2005)
5. Eppstein, D., Wortman, K.A.: Minimum dilation stars. In: Proc. 21st Symp. Computational Geometry, pp. 321–326. ACM Press, New York (June 2005)
6. Lee, D.T., Wu, V.B.: Multiplicative weighted farthest neighbor Voronoi diagrams in the plane. In: Proceedings of the International Workshop on Discrete Mathematics and Algorithms, pp. 154–168 (1993)

Approximated Distributed Minimum Vertex Cover Algorithms for Bounded Degree Graphs

Yong Zhang[1,2], Francis Y.L. Chin[2,⋆], and Hing-Fung Ting[2,⋆⋆]

[1] College of Mathematics and Computer Science, Hebei University, China
[2] Department of Computer Science, The University of Hong Kong, Hong Kong
{yzhang,chin,hfting}@cs.hku.hk

Abstract. In this paper, two distributed algorithms for the minimum vertex cover problem are given. In the unweighted case, we propose a 2.5-approximation algorithm with round complexity $O(\Delta)$, where Δ is the maximal degree of G, improving the previous 3-approximation result with the same round complexity $O(\Delta)$. For the weighted case, we give a 4-approximation algorithm with round complexity $O(\Delta)$.

1 Introduction

Minimum vertex cover (MVC in short) is a fundamental and classical problem in computer science. For a graph $G = (V, E)$, we say a subset $S \subseteq V$ is a *vertex cover* if any edge $e \in E$ is incident to at least one vertex in S. The problem of MVC is to find a vertex cover S such that the number of vertices in S is minimum. If each vertex $u \in V$ has a weight $w(u)$, this problem is turned to be the weighted version of minimum vertex cover (WMVC in short). The objective of the WMVC is to find a vertex cover $S \subseteq V$ such that the total weight of vertices in S is minimized, i.e., $\min w(S)$, where $w(S) = \sum_{u \in S} w(u)$.

In this paper, we study the distributed algorithms for both the unweighted and weighted versions of the minimum vertex cover problem. In the distributed model, each vertex is a server and can do some computation on its local information. Each vertex only knows its local information, i.e., the property of itself (e.g., weight) and which vertices are its neighbors. Each vertex can exchange information with its neighbors. We assume the exchange of information is done synchronously. At each round, a vertex can send/receive messages to/from its neighbors. In this model, each vertex does not know the topology of G and the number of vertices in V. The complexity of distributed algorithm is measured by the the number of rounds to achieve the expected result (e.g., a vertex cover).

Related Works:
Minimum vertex cover is a very well studied problem in theoretical computer science. Both the unweighted and weighted versions are NP-hard [3]. There are

⋆ Research supported by HK RGC grant HKU-7113/07E and the William M.W. Mong Engineering Research Fund.

⋆⋆ Research supported by HK RGC grant HKU-7171/08E.

M.T. Thai and S. Sahni (Eds.): COCOON 2010, LNCS 6196, pp. 100–109, 2010.

no polynomial algorithm can achieve the minimum vertex cover unless $P = NP$, thus, most studies focus on the approximation algorithms for this problem. In 2005, Karakostas [6] gave a $(2 - \Theta(1/\sqrt{\log n}))$-approximation algorithm, which is the best known upper bound. The current lower bound of the approximation ratio is $10\sqrt{5} - 21 > 1.36$ [2].

Distributed computing for finding a vertex cover is also well studied during these years. In the unweighted version, a 2-approximation algorithm is a straightforward extension of the algorithm in [5] for finding the maximal matching in $O(\log^4 n)$ rounds. Note that all vertices in the maximal matching 2-approximate the minimum vertex cover. Recently, another 2-approximation algorithm was proposed in [1] with round complexity $O(\Delta^2)$, where Δ is the maximum degree of the graph. With less number of rounds, a 3-approximation $O(\Delta)$-rounds algorithm was given in [8]. For the weighted version, Khuller et al [7] gave a $(2 + \varepsilon)$-approximation in $O(\log n \log \frac{1}{\varepsilon})$ rounds. A randomized 2-approximation algorithm was given by Grandoni et al [4], the expected number of rounds is $O(\log n + \log \hat{W})$, where $\hat{W}$ is the average weight of all vertices.

Our Contributions:
In this paper, we study the distributed algorithms for both the unweighted and weighted versions of the minimum vertex cover problem. In the unweighted case, we improve the previous 3-approximation result and give a 2.5-approximation algorithm with the complexity $O(\Delta)$ (Section 2); in the weighted case, we give a 4-approximation algorithm with $O(\Delta)$ rounds (Section 3).

2 Algorithm for MVC

In this section, we give a distributed 2.5-approximation algorithm for finding the minimum vertex cover in graph $G = (V, E)$. The high level idea is described as follows. Find a subgraph $\hat{G} = (\hat{V}, \hat{E})$ of G, such that (1) the degree of each vertex in $\hat{V}$ is at least one and at most two, thus, each connected component in $\hat{G}$ is either a cycle or a path; and (2) each edge in G is incident to at least one vertex in $\hat{V}$, thus, $\hat{V}$ is a vertex cover of G. Let $\{P_1, P_2, ..., P_k\}$ denote a partition of the connected components in $\hat{G}$, and any two connected components in different partitions are disjoint. In each partition, define the *local approximation ratio* to be the ratio between the number of vertices in the partition and the number of vertices selected into the minimum vertex cover. In partition P_i, let p_i and o_i be the number of vertices selected into the vertex cover by our algorithm (all vertices in this partition) and by the optimum algorithm, respectively. Since the partition of connected components are disjoint each other, the size of the minimum vertex cover for G is at least $\sum_{i=1}^{k} o_i$. Thus, the approximation ratio of our algorithm is at most $\sum_{i=1}^{k} p_i / \sum_{i=1}^{k} o_i \leq \max p_i/o_i$.

Our algorithm has two phases: the *partition phase* and the *modification phase*. In the partition phase, similar to the above description, we find a subgraph $G' = (V', E')$ of G such that each connected component in G' contains only one path or cycle. Consider a connected component P_i with p_i vertices,

- if P_i is a cycle
  P_i contains p_i edges, to cover these edges, any vertex cover, including the minimum vertex cover, must select at least $\lceil p_i/2 \rceil$ vertices from P_i. Since V' contains all vertices in P_i, the local approximation ratio for P_i w.r.t. V' is $p_i/\lceil p_i/2 \rceil$, which is at most 2.
- If P_i is a path
  P_i contains $p_i - 1$ edges, to cover these edges, any vertex cover must select at least $\lceil (p_i - 1)/2 \rceil$ vertices. Since V' contains all vertices in P_i, the local approximation ratio for P_i w.r.t. V' is $p_i/\lceil (p_i - 1)/2 \rceil$.

Combine the above two cases, the approximation ratio of such algorithm is at most $\max_i p_i/\lceil (p_i - 1)/2 \rceil$. In the worst case, consider a path with two edges (3 vertices), the minimum vertex cover may contain only one vertex, but the algorithm may select all these three vertices, the local approximation ratio is 3. Intuitively, the longer the path, the lower the ratio. Therefore, we have a modification phase for handling the length-2 paths in G'.

2.1 The Partition Phase

This part is similar to the algorithm in [8], and is the background of the modification phase, we describe this phase briefly.

- Each vertex sends requests to its neighbors one by one, until it receives an accept message, or all neighbors reject its requests.
- Each vertex must accept one incoming request if it has not received any requests before. After that, this vertex will reject all the other incoming requests.

From the above request-accept algorithm, each vertex accepts at most one neighbor's request, and receives at most one neighbor's accept message. We can construct the graph $G' = (V', E')$ as follows: V' contains such vertices that either accepts the request from one of its neighbors, or receives an accept message from one of its neighbors. An edge (u, v) is in E' if u accepts an request from v, or vise versa. From the definition of G', we can see that G' is a subgraph of G. Polishchuk and Suomela proved the following theorem in [8].

Theorem 1. *V' is a vertex cover of G and the size of V' is at most 3 times to the size of the minimum vertex cover. The round complexity of the above algorithm is $O(\Delta)$.*

Now we give a tight example for the partition phase. Consider the graph G is a combination of some length-2 paths, as shown in Figure 1. According to the above algorithm, the middle vertex of each length-2 path may accept the request from upper vertex, and the lower vertex may accept the request from middle vertex. Thus, all vertices are selected into the vertex cover, but the minimum vertex cover only contains the middle vertex of each path.

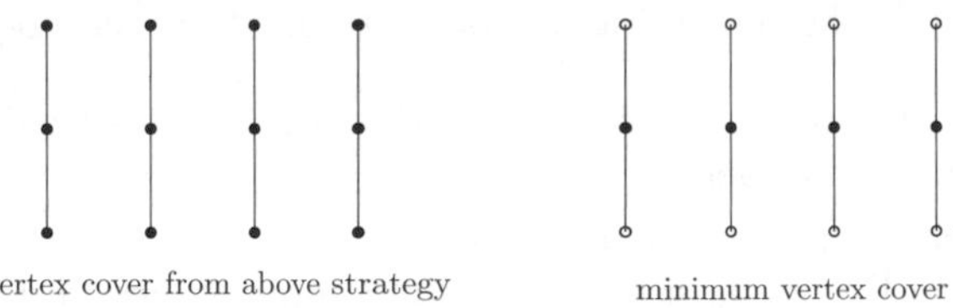

Fig. 1. Tight example of the partition phase (dark vertices are selected into the vertex cover)

2.2 The Modification Phase

From last subsection, if all vertices in V' are selected into the vertex cover, the approximation ratio is 3.

In $G' = (V', E')$, each vertex receives at most one accept message and accepts at most one of its neighbors' requests, the degree of each vertex in G' is at most 2. Thus, each connected component in G' is either a path or a cycle. Note that V' is a vertex cover of G, any edge in E contains at least one vertex in V'.

From the previous analysis, the approximation ratio after the partition phase is at most $p_i/\lceil(p_i-1)/2\rceil$, which is at most 2.5 if $p_i \neq 3$. This gives us an heuristic to improve the approximation ratio by modifying the length-2 paths in G'.

In the modification phase, the components with lengths not equal to 2 remain (Step 1 of the modification phase); some length-2 paths are extended to length-3 paths (Step 2 of the modification phase); some terminals of some length-2 paths are removed from G' (Step 3 of the modification phase). After the modification phase, the selected vertices is still a vertex cover.

Now we describe the modification phase step by step.

Step 1: Select all vertices in some connected component with length 1, or more than 2 into V'', then remove all these vertices and edges incident to them from G.

Fact 1. *For any vertex in graph G', checking whether it is in a connected component of length not equal to 2 can be done in constant rounds.*

Proof. In G', each vertex knows its degree. The checking procedure starts from each vertex with degree one. For example, the degree of u is one, u sends a message $(u, 0)$ to its neighbor. When vertex v receives a message (u, i) from one of its neighbors, if v's degree is 2, v sends the message $(u, i+1)$ to its another neighbor; otherwise, v's degree is one, it sends the message $(u, i+1)$ back to its only neighbor.

For vertex u with degree one, if it receives a message $(u, 1)$, or it does not receive any feedback message after 4 rounds, u can confirm that it is not in a length-2 path. For any vertex with degree 2, if it receives messages $(u, 0)$ and $(u, 2)$, this vertex must be in a length-2 path; and must not otherwise. □

After Step 1, all the remaining components in G' are length-2 paths.

Step 2: Each terminal of the remaining (length-2) paths sends requests to its neighbors in G one by one.

- Vertex in V' rejects all requests.
- Vertex in $V - V'$ must accept one incoming request if it has not received any requests before. After that, this vertex will reject all other requests.
- Each terminal will not sends requests if it gets an accept message, i.e., each terminal can be accepted at most once.

  If terminal u is accepted by vertex v, add (u, v) to this connected component, the length of this path is increased to be 3. Such component must be disjoint with other components because each vertex in $V - V'$ can accept at most one request. After such request-accept strategy,
- select all vertices in such components with extended length into V''.
- remove all these vertices and edges incident to them from G.

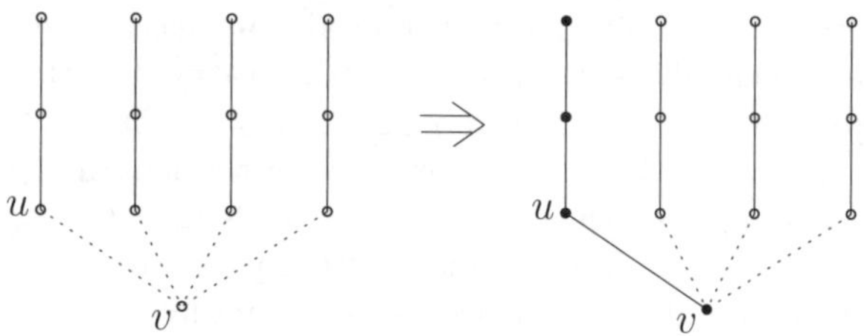

Fig. 2. Example of step 2

For example, consider the graphs as shown in Figure 2. The left part shows the structure before Step 2. In G, there are several edges connecting a vertex v and some terminals of length-2 paths in G', such that v does not belong to any component in G'. According to step 2 of this strategy, these terminals may send requests to v, and v must accept one of them. W.l.o.g., v accepts the request from u and rejects all the other requests. Thus, the length-2 path including vertex u can be extended to v. By such modification, the extended path is still disjoint with other components in G', and the length of this path is increased to be 3.

Since each terminal sends at most Δ requests, the total number of rounds in this step is at most $O(\Delta)$.

Step 3: Each terminal of the remaining (length-2) paths sends requests to its neighbors in G one by one.

- Each vertex in V'' and each non-terminal in the remaining length-2 paths must reject all requests.
- Each terminal in the remaining length-2 paths must accept one incoming request if it does not receive any requests before. After that, this vertex will reject all the other requests.

- Each terminal will not send any request if it gets an accept message, i.e., each terminal can be accepted at most once.

  Select all vertices, except those terminals rejected by all its neighbors, into V''.

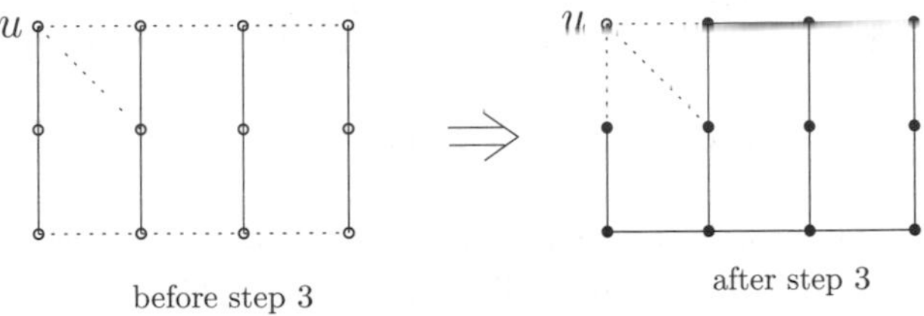

Fig. 3. Example of step 3

For example, consider the graphs as shown in Figure 3. The left part shows the structure before Step 3, there are several length-2 paths remained after step 2. After processing step 3, some terminals accept a request from one of its neighbors, or receive an accept message from one of its neighbor. Some terminals are rejected by all its neighbors, e.g., vertex u. In this example, we select all vertices except u since any edge adjacent to u must be adjacent to some selected vertex.

For each vertex pair (u, v), there exists an edge in G connecting them after step 3 if either u accepts a request from v or vise versa, as shown in Figure 3. Similar to the above analysis, the total number of rounds in this step is at most $O(\Delta)$.

Theorem 2. *The selected vertex set V'' is a vertex cover of G, the size of V'' is at most 2.5 times to the size of the minimum vertex cover. The round complexity of this algorithm is $O(\Delta)$.*

Proof. Consider an edge (u, v), from the partition phase, at least one vertex is selected into V', w.l.o.g., $u \in V'$. In the modification phase, if u is selected in step 1, 2, or 3, the edge (u, v) is covered. Otherwise, suppose u is not selected in any step. From the description of modification phase, u must be a terminal of a length-2 path. We claim that any vertex in $V - V'$ connected to u must be selected in step 2. Note that the vertices in $V - V'$ will not belong to any connected component after the partition phase. If there exists a vertex $v \in V - V'$ and (u, v) is an edge in G, u must have sent a request to v and v should have accepted this request if it does not receive any other requests. Since u is not selected, v must have rejected the request from u and thus v has been selected in step 2. In step 3, since u is not selected, all its requests must have been rejected. That means all its neighbors must have been selected in some step. By considering all cases, for any edge (u, v), either u or v must be contained in the vertex cover. Thus, V'' is a vertex cover.

Now let us consider the local approximation ratio for the partitions of the connected components step by step. In step 1, the first partition contains all components whose sizes are either 1 or strictly larger than 2. Thus, the local approximation ratio of each component in this partition is at most 2.5. In step 2, the length of each connected component with extended length is 3, thus, the local approximation ratios for these components are at most 2. Note that in step 3, some terminals of some length-2 paths are not selected, i.e., the lengths of these paths are reduced to be 1. For these paths with reduced length, the optimum algorithm must select at least one vertex, thus, the local approximation ratio for them are at most 2.

In step 3, consider the last partition includes all remaining length-2 paths. We select all vertices in these paths into the vertex cover. Since each terminal of such paths must either accept one request, or receive one accept message. There exist edges in G corresponding to every pair of request-accept, and such request-accept pair must be between some terminals of some length-2 paths. Therefore, only selecting the middle vertex of each path cannot cover all edges. Suppose there are k length-2 paths in the remaining length-2 paths. The minimum vertex cover contains one vertex in x paths, and contains at least two vertices in $k - x$ paths. Thus, the minimum vertex cover selects at least $x + 2(k - x) = 2k - x$ vertices, while our algorithm selects $3k$ vertices. We say two length-2 paths in this partition are adjacent if some terminals in these two paths form a request-accept pair, i.e., there exists an edge between the terminals of these two paths. For each path with 1 selected vertex in the minimum cover, each of the two terminals must connect to some terminal of its adjacent length-2 path, which must contain at least 2 selected vertices. For each path with at least 2 selected vertices in the minimum vertex cover, since these two selected vertices may be terminals and each terminal may be contained in two request-accept pairs, the number of adjacent length-2 paths is at most 4, which may contain one selected vertex in the minimum vertex cover. Thus, we have $x \leq 4(k - x)$, and in this partition, the local approximation ratio is at most $3k/(2k - x) \leq 2.5$.

By considering all these cases, the local approximation ratio for each component in the partition in each step is at most 2.5. Thus, the size of the selected vertex cover is at most 2.5 times to the size of the minimum vertex cover.

Now we analyze the complexity of the strategy. In the partition phase, the round complexity is $O(\Delta)$. Checking the length of a connected component is not 2 can be done in constant time. Since the degree of each vertex is at most Δ, step 1 can be done in $O(\Delta)$ rounds. From the description of step 2 and 3, the round complexity of these two steps is $O(\Delta)$ too. Therefore, the round complexity of our 2.5-approximation algorithm is $O(\Delta)$. □

3 Algorithm for WMVC

In this section, we propose a distributed algorithm to approximate the weighted vertex cover. The intuition idea can be regarded as a generalization of the distributed 3-approximation algorithm in [8], where each vertex may accept one request or receive one accept message from one of its neighbors. In the weighted case, we still implement the request-accept strategy. For each vertex u, if it accepts the request from vertex v, the weight of u will be decreased. On the other hand, if vertex u receives an accept message, the weight will be decreased too. During the processing of the algorithm, the weights of vertices are decreased step by step. After the termination of the algorithm, we select vertex u into the vertex cover if $w(u) \leq 0$. At each step, each vertex may send requests or accept/reject messages to its neighbors. To specify the messages, each request from u is $(req, w(u))$; if u rejects the request from v, u will send $(rej, -1)$ to v; otherwise, each accept message is associated with a positive number, e.g., $(acc, x > 0)$. Let $w^o(u)$ be the original weight of vertex u, i.e., in the initial state, $w^o(u) = w(u)$.

Before describing the algorithm, we give a simple example to illustrate the decreasing of weights. Consider the graphs as shown in Figure 4, the left graph contains four vertices with weights 5, 1, 1, and 0.5 respectively. According to the request-accept strategy, the upper vertex sends request to the first lower vertex, and each lower vertex sends request to the upper vertex. All these four requests will be accepted and the weights of vertices will be decreased accordingly. The upper vertex accepts three requests and receives one accept message from vertices with weights 1 and 0.5, the weight of this vertex is decreased by 3.5. Similarly, the final weights of the lower vertices are -1, 0, and 0 respectively.

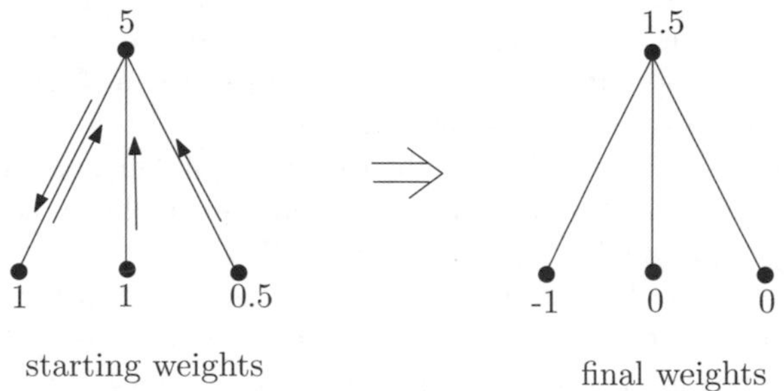

Fig. 4. Weights of vertices are decreased according to the algorithm

Suppose step 1 is the first step and each round contains two continuous steps. Now we describe the algorithm on the i-th round for each vertex $u \in V$.

At step $2i - 1$:
u receives message x from its $(i-1)$-th neighbor v.

- if $x = (rej, -1)$, u is rejected by v;
- if $x = (acc, y > 0)$, u is accepted by v, then set $w(u) = w(u) - y$;

If $w(u) > 0$, u sends request message $(req, w(u))$ to its i-th neighbor.

At step $2i$:
u receives requests from v_1, v_2, ..., v_k,

- if $w(u) \geq \sum_{i=1}^{k} w(v_i)$, u sends accept message $(acc, w(v_i))$ to each v_i and set $w(u) = w(u) - \sum_{i=1}^{k} w(v_i)$;
- otherwise, assume $\sum_{i=1}^{\ell} w(v_i) \leq w(u)$ and $\sum_{i=1}^{\ell+1} w(v_i) > w(u)$, u sends accept message $(acc, w(v_i))$ to each v_i $(1 \leq i \leq \ell)$; sends accept message (acc, x) to $v_{\ell+1}$, where $x = w(u) - \sum_{i=1}^{\ell} w(v_i)$; sends reject message $(rej, -1)$ to $v_{\ell+2}$, ..., v_k; then set $w(u) = 0$.

Theorem 3. *After the termination of the algorithm, the vertex set $S = \{u \in V | w(u) \leq 0\}$ is a vertex cover.*

Proof. This is proved by contradiction. Assume that after the termination of the algorithm, there exists an edge (u, v) such that $w(u) > 0$ and $w(v) > 0$. From the above algorithm, since $w(u) > 0$, u must send a request to v at some time. If v accepts u's request, either u's or v's weight will be decreased to 0. Otherwise, v rejects u's request, that means v's weight is already decreased to be 0. Contradiction! □

To upper bound the total weights of selected vertices, we can assign weights to edges according to the above algorithm. Firstly, set the initial weights of all edges to be 0. If an accept message (acc, x) is sent through (u, v), set the edge weight $p_{(u,v)} = x$, where x is a positive value contained in the accept message. Note that in each round, there may be two accept messages sent through an edge (see the example in Figure 4), in this case, two positive values are associated with these two messages, set the edge weight to be the higher one. After processing the algorithm, each edge is assigned a non-negative weight. Now we analyze the total weight of edges incident to the vertex in S.

Consider vertex $u \in S$, suppose $w(u)$ is decreased to non-positive in the i-th round. From round 1 to round $i - 1$, u may send accept messages to some neighbors, and receive some accept messages from some neighbors. At each round, u may send more than one accept messages but receives at most one accept message. All these accept messages may assign weights to the edges incident to u, and the total assigned weights is no more than the original weight $w^o(u)$. Let $w'(u)$ denote the decreased weight of vertex u in the first $i - 1$ rounds, i.e., $w(u) = w^o(u) - w'(u)$ before the i-th round. Then in the i-th round, u may send some accept messages to some of its neighbors, or receives an accept message from one of its neighbors. From the rules of assigning weights on edges, the total assigned weights of those edges through which u sends accept message is at most $w^o(u) - w'(u)$; the weight of the edge through which u receives an accept message is at most $w^o(u) - w'(u)$ too. Note that at most two messages may be sent through an edge incident to u, we have the following lemma:

Lemma 2. *For each vertex $u \in S$, the total assigned weights on edges incident to u is at least $w^o(u)$ and at most $2 \cdot w^o(u)$.*

If we assign weight p_e to each edge $e \in E$ and satisfying $\sum_{(u,v)} p_{(u,v)} \leq w(u)$ for each $u \in V$, we have the following fact.

Fact 3. *$\sum_{e \in E} p_e \leq w(S)$ for any vertex cover S.*

Proof. $\sum_{e \in E} p_e \leq \sum_{u \in S} \sum_{(u,v)} p_{(u,v)} \leq \sum_{u \in S} w(u) = w(S)$. The first inequality holds because S is a vertex cover set. □

Let p_e denote the weight assigned to edge $e \in E$, let $p'_e = p_e/2$ for each $e \in E$, we have

$$w(S) = \sum_{u \in S} w^o(u) \leq \sum_{u \in S} \sum_{(u,v)} p_{(u,v)} = 2 \cdot \sum_{u \in S} \sum_{(u,v)} p'_{(u,v)}.$$

From Lemma 2, we know that for any $u \in V$, $\sum_{(u,v)} p'_{(u,v)} \leq w^o(u)$, then from Fact 3, the total weights of p'_e for all edges is no more than the total weights of any vertex cover, including the minimum vertex cover S^*. Thus,

$$w(S) \leq 2 \cdot \sum_{u \in S} \sum_{(u,v)} p'_{(u,v)} \leq 2 \cdot \sum_{u \in V} \sum_{(u,v)} p'_{(u,v)} = 4 \cdot \sum_{e \in E} p'_e \leq 4 \cdot w(S^*)$$

The third equality holds because each edge is incident to two vertices and counted twice. Therefore, we have the following conclusion:

Theorem 4. *Our algorithm 4-approximates the minimum weighted vertex cover with round complexity $O(\Delta)$.*

Proof. From previous analysis, we know the algorithm is 4-approximation. From the description of algorithm, the round complexity is $O(\Delta)$. □

References

1. Åstrand, M., Floréen, P., Polishchuk, V., Rybicki, J., Suomela, J., Uitto, J.: A Local 2-Approximation Algorithm for the Vertex Cover Problem. In: Keidar, I. (ed.) DISC 2009. LNCS, vol. 5805, pp. 191–205. Springer, Heidelberg (2009)
2. Dinur, I., Safra, S.: The importance of being biased. In: Proceedings of the ACM Symposium on the Theory of Computing (STOC 2002), pp. 33–42. ACM, New York (2002)
3. Garey, M.R., Johnson, D.S.: Computers and Intractability. A Guide to the Theory of NP-Completeness. Freeman, New York
4. Grandoni, F., Könemann, J., Panconesi, A.: Distributed Weighted Vertex Cover via Maximal Matchings. ACM Transactions on Algorithms 5(1)
5. Hańćkowiak, M., Karoński, M., Panconesi, A.: On the distributed complexity of computing maximal matchings. SIAM J. Discrete Math. 15(1), 41–57
6. Karakostas, G.: A better approximation ratio for the vertex cover problem. In: Caires, L., Italiano, G.F., Monteiro, L., Palamidessi, C., Yung, M. (eds.) ICALP 2005. LNCS, vol. 3580, pp. 1043–1050. Springer, Heidelberg (2005)
7. Khuller, S., Vishkin, U., Young, N.: Aprimal-dual parallel approximation technique applied to weighted set and vertex cover. J. Algorithms 17(2), 280–289 (1994)
8. Polishchuk, V., Suomela, J.: A simple local 3-approximation algorithm for vertex cover. Information Processing Letters 109, 642–645 (2009)

Maximum Upward Planar Subgraph of a Single-Source Embedded Digraph

Aimal Rextin[1] and Patrick Healy[2]

[1] Department of Computing,
National University of Science Technology, Pakistan
aimal.rextin@seecs.edu.pk

[2] Computer Science Department,
University of Limerick, Ireland
patrick.healy@ul.ie

Abstract. We show how to compute a maximum upward planar single-source subgraph of a single-source embedded DAG G_ϕ. We first show that finding a maximum upward planar subgraph of a single-source embedded digraph is NP-complete. We then give a new characterization of upward planar single-source digraphs. We use this characterization to present an algorithm that computes a maximum upward planar single-source subgraph of a single-source embedded DAG. This algorithm takes $O(n^4)$ time in the worst case and $O(n^3)$ time on average.

1 Introduction

A drawing of a graph is *readable* if it allows humans to easily understand the graph's structure. Generally, a drawing is readable if the vertices of the graph are laid in a regular form and the edges are easily distinguished. It is difficult for humans to distinguish edges if a large number intersect or *cross*. Ideally, we like no edge crossings, but this is impossible for non-planar graphs. Hence, it is important that a drawing has minimum number of edges crossings [7, 14]. Moreover in a drawing of a digraph G, we want the edges to point in one direction because this allows humans to easily comprehend the directional information of G. Hence, a digraph is readable if it has a drawing with no edge crossings and all edges point in one direction; we call such digraphs as *upward planar*.

It is NP-hard to test if a digraph G is upward planar [10], hence upward planarity testing is either done for a fixed embedding [2, 8], or for special classes of digraphs like single-source digraphs [11, 3], series-parallel digraphs [9], and outer planar digraphs [13]. A *planar embedding* is an equivalence class of planar drawings for a graph G, such that each drawing of this class has the same circular order of edges around each vertex of G. A graph G with a planar embedding ϕ is denoted by G_ϕ and we call it an *embedded* digraph. Let $G'_{\phi'}$ be a subgraph of G_ϕ with embedding ϕ'. We say that $G'_{\phi'}$ is *compatible* with G_ϕ if the circular order of edges for a vertex $v \in G'_{\phi'}$ is obtained by deleting zero or more edges from $v \in G_\phi$.

M.T. Thai and S. Sahni (Eds.): COCOON 2010, LNCS 6196, pp. 110–119, 2010.

It is reasonable to say that a non-upward planar digraph G is more readable if we draw its largest possible subgraph in an upward planar fashion. This corresponds to finding the maximum upward planar subgraph of G. A subgraph $\tilde{G}$ of the digraph G is an *upward planar subgraph* of G if $\tilde{G}$ is upward planar. An upward planar subgraph $\tilde{G}$ is a *maximum upward planar subgraph* if there is no other upward planar subgraph of G with more edges than $\tilde{G}$. Binucci, Didimo, and Giordano [6,5] showed that finding even the maximum upward planar subgraph of a planarly embedded digraph is NP-complete.

Theorem 1 (Binucci, Didimo, and Giordano [6,5]). *Finding the maximum upward planar subgraph of a planarly embedded digraph is NP-complete.*

One can try to find the maximum upward planar subgraph of an embedded single-source digraph. However, the following theorem states that this problem is also NP-complete. The proof of the following theorem is analogous to the proof of Binucci *et al.* [6,4].

Theorem 2. *Computing the maximum upward planar subgraph of a single-source embedded digraph is NP-complete.*

We therefore focus on computing a *maximum upward planar single-source subgraph of a single-source embedded digraph* G_ϕ. Let s be the unique source of G_ϕ. Computing such a subgraph is useful because the single-source of G_ϕ represents a special vertex, for example many digraphs generated by project management applications have a special "start" node. A general upward planar subgraph computation algorithm might result in a subgraph $\tilde{G}_{\tilde{\phi}}$ having more than one source. A generic upward planar drawing algorithm can end up drawing these sources at the same level as s in an upward planar drawing of $\tilde{G}_{\tilde{\phi}}$, i.e. s loses its distinction.

Let G_ϕ be a single-source embedded graph, such that the source s is a cut-vertex. Let $G^1_{\phi^1}, G^2_{\phi^2}, \ldots, G^k_{\phi^k}$ be the blocks of G_ϕ wrt to s. Moreover, we let $\tilde{G}^i_{\tilde{\phi}^i}$ be a maximum upward planar single-source subgraph of the block $G^i_{\phi^i}$, where $1 \le i \le k$. The maximum upward planar single-source subgraph of G_ϕ is $\tilde{G}^1_{\tilde{\phi}^1} \cup \tilde{G}^2_{\tilde{\phi}^2} \cup \cdots \cup \tilde{G}^k_{\tilde{\phi}^k}$. Hence, in the rest of this paper we assume that the source s is not a cut-vertex of G_ϕ.

In Section 2, we introduce a special spanning tree of a single-source digraph called a directed spanning tree. In Section 3, we use directed spanning trees to identify edge configurations that cannot be present in a single-source upward planar embedded digraph. In Section 4, we present our algorithm to compute a maximum upward planar single-source subgraph $G'_{\phi'}$ of G_ϕ, such that $G'_{\phi'}$ is compatible with G_ϕ.

1.1 Preliminaries

In a digraph, a *source* vertex has only outgoing edges, a *sink* vertex has only incoming edges. A planar drawing Γ divides the plane into non-overlapping regions called *faces*; the unique unbounded region is called the *external face* and each bounded region is called an *internal face*. The *facial boundary* of a face f is the path we get by traversing the boundary of f in the clockwise (counterclockwise) direction when f is an internal (the external) face. All drawings of a graph with the same embeddings will have the same set of facial boundaries. An embedded digraph G_ϕ is *bimodal* if $\phi(v)$ can be partitioned into consecutive incoming and outgoing edges for every vertex $v \in G_\phi$; bimodality is a necessary condition for upward planarity [2].

We now present an upward planarity characterization of a single-source embedded digraph G_ϕ. However, we first define what we mean by angles and a face-sink graph. In an embedded digraph G_ϕ, an *angle* is a triplet $\langle e_1, v, e_2\rangle$ such that the edges are incident to the vertex v, and edge e_1 is immediately before edge e_2 in $\phi(v)$. A vertex v is incident to the angle $\langle e, v, e\rangle$ when e is the only edge incident to v. If e_1 and e_2 point to v then it is a *sink-switch*, and if e_1 and e_2 point away from v then it is a *source-switch* [8]. The *face-sink graph* F of G_ϕ is an undirected graph, whose vertices are the faces of G_ϕ and the vertices of G_ϕ incident to a sink-switch. There is an edge (f, v) in F if face f is incident to a sink-switch $\langle e_1, v, e_2\rangle$ in G_ϕ. Bertolazzi *et al.* [3] presented an $O(n)$-time algorithm to test the upward planarity of a single-source embedded digraph G_ϕ.

Theorem 3 (Bertolazzi *et al.* [3]). *Let G_ϕ be a planar embedded digraph with a single-source s. G_ϕ is upward planar with face h as the external face if and only if the following conditions are satisfied: the face-sink graph F of G_ϕ is a forest; F has exactly one tree $\hat{T}$ with no internal vertices; while all other trees have exactly one internal vertex; $\hat{T}$ contains the node corresponding to the face h and s is incident to face h in G_ϕ.*

2 Directed Spanning Trees

Let G be a digraph with a unique source s and let G_ϕ be a planar embedding of G. A *directed spanning tree (DST)* T is a spanning tree of G with a directed path from s to every other vertex u. Every single-source digraph has at least one directed spanning tree because $s \rightsquigarrow u$ for every vertex $u \in G$. Directed spanning trees are used in the next section to identify edge configurations that cannot be present in an upward planar embedded single-source digraph.

A *tree-path* in a DST T is a directed path from s to a leaf of T. Let Π_1 and Π_2 be two tree-paths of T. The *disjoint subpath* of Π_1 with respect to Π_2 is the subpath of Π_1 that is not shared by Π_2; it is denoted by $\Pi_1 \setminus \Pi_2$. The edges $E(G) \setminus E(T)$ are called the *non-tree edges* of G wrt T. An edge $e = (u, v)$ is a *forward edge* if both u and v are on the same tree-path and $u \rightsquigarrow v$. An edge $e = (u, v)$ is a *back edge* if both u and v are on the same tree-path and $v \rightsquigarrow u$. An edge $e = (u, v)$ is a *cross edge* if $u \in \Pi_1 \setminus \Pi_2$ and $v \in \Pi_2 \setminus \Pi_1$.

The DST $T_{\phi'}$ is a *compatible embedded directed spanning tree* ($CEDST$) if an embedded DST $T_{\phi'}$ is compatible with G_ϕ. Theorem 3 implies that a compatible embedded DST $T_{\phi'}$ of G_ϕ is always upward planar. A *compatible strongly embedded directed spanning tree* ($CSEDST$) is an equivalence class of planar drawings of T obtained by deleting the edges $E(G) \setminus E(T)$ from a planar drawing of G_ϕ with a given external face. We say that a $CSEDST$ *realizes* the given external face. In other words, a compatible strongly embedded DST $T_{\phi'}$ realizes a given external face and $T_{\phi'}$ is compatible with G_ϕ. Note that a compatible strongly embedded DST is always upward planar. We assume that a strongly embedded DST is compatible with G_ϕ unless specified otherwise.

Let Γ be a drawing of G_ϕ with a face f at the exterior and let s be incident to f. Let $L(s)$ represent the circular list of edges incident to s in the clockwise direction. Since we are assuming that s is not a cut-vertex, there are at least two faces incident to s in Γ. In Γ, if we circumtraverse s by starting from inside a face $g \neq f$, we will ultimately reach the first edge $\hat{e}_1$ incident to f. Similarly, if we circumtraverse s by starting from inside f, we reach the last edge $\hat{e}_2$ incident to f. The ordered list $I_s(f) = \{e_1^i, e_2^i, \ldots, e_k^i\}$ denotes the edges of $L(s)$ that lie between $\hat{e}_1$ and $\hat{e}_2$ in the clockwise direction; $I_s(f)$ is said to be *inside the face* f. The ordered list $O_s(f) = \{e_1^o, e_2^o, \ldots, e_l^o\}$ denotes the edges of $L(s)$ that lie between $\hat{e}_2$ and $\hat{e}_1$ in the clockwise direction; $O_s(f)$ is said to be *outside the face* f. There are no other edges in $L(s)$ except $O_s(f)$, $I_s(f)$, and $\{\hat{e}_1, \hat{e}_2\}$. The following lemma implies that $I_s(f)$ is always empty.

Lemma 4. *If s is not a cut-vertex in G_ϕ, then $I_s(f) = \emptyset$ for every external face f.*

Let $T_{\phi'}$ be a compatible strongly embedded DST that realizes a face f. We let $\overrightarrow{L}(s)$ denote a *linearization* of $L(s)$, such that the first edge of $\overrightarrow{L}(s)$ is incident to f. Note that $\overrightarrow{L}(s)$ represents the left to right ordering of the edges incident to s in an upward planar drawing of $T_{\phi'}$ that realizes a face f. We have $\overrightarrow{L}(s) = \hat{e}_2, e_1^o, e_2^o, \ldots, e_l^o, \hat{e}_1$ as the unique linearization of $L(s)$ because $I_s(f) = \emptyset$, while $\hat{e}_1$ and $\hat{e}_2$ are incident to f.

A compatible strongly embedded DST $T_{\phi'}$ with a given linearization, $\overrightarrow{L}(s)$, induces a total order on the tree-paths of $T_{\phi'}$. This ordering is denoted by $\prec$. Intuitively, this ordering means that in an upward planar drawing of $T_{\phi'}$, a tree-path Π_1 is drawn to the left of a tree-path Π_2 if $\Pi_1 \prec \Pi_2$ and it is drawn to the right of Π_2 if $\Pi_2 \prec \Pi_1$. We now define this ordering more formally. Let v be the last common vertex of the tree-paths Π_1 and Π_2 when we traverse Π_1 and Π_2 from s to the leaves of Π_1 and Π_2. Moreover, let e_{Π_1} and e_{Π_2} be the first edges after v in Π_1 and Π_2 respectively. We have the following two cases:

- **$v = s$:** In this case, $\Pi_1 \prec \Pi_2$ if e_{Π_1} comes before e_{Π_2} in $\overrightarrow{L}(s)$, otherwise $\Pi_2 \prec \Pi_1$.
- **$v \neq s$:** Note that v has only one incoming edge incident to v in T; let us denote it by e. If we traverse $L(v)$ in clockwise direction by starting from e, then $\Pi_1 \prec \Pi_2$ if e_{Π_1} comes before e_{Π_2} during this traversal otherwise $\Pi_2 \prec \Pi_1$.

We now show how to determine if $\Pi_1 \prec \Pi_2$ or $\Pi_2 \prec \Pi_1$ in constant time given a vertex $u \in \Pi_1 \setminus \Pi_2$ and a vertex $v \in \Pi_2 \setminus \Pi_1$. Our method requires a preprocessing step that takes $O(n)$ time using a simple modified `DFS` subroutine. We associate a number called *path ID* with every node of a strongly embedded

DST $T_{\phi'}$ with a given linearization. The path ID represents the left-most path that goes through a vertex wrt to a given linearization. Note that $\Pi_1 \prec \Pi_2$ if $ID(u) < ID(v)$.

We now define the two *sides* for an internal vertex u in a tree-path Π of a strongly embedded *DST* $T_{\phi'}$ with linearization $\overrightarrow{L}(s)$. Intuitively, a side of u corresponds to a geometrical side of u in an upward planar drawing of $T_{\phi'}$. These two sides are defined by how one traverses the external face. Recall from Section 1.1 that we traverse an external face in the counterclockwise direction. Assume that e_1 enters u and e_2 leaves u, such that $e_1 \in E(T)$ and $e_2 \in E(T)$. There are two angles incident to u in Π: $\langle e_1, u, e_2\rangle$ and $\langle e_2, u, e_1\rangle$. We call $\langle e_1, u, e_2\rangle$ the *right side angle* and we call $\langle e_2, u, e_1\rangle$ the *left side angle* . Note that the leaves and the source s do not have right and left sides.

3 Forbidden Configurations

In this section, we show that a strongly embedded single-source digraph G_ϕ is upward planar if and only if certain edge configurations are absent in a strongly embedded *DST* $T_{\phi'}$ with linearization $\overrightarrow{L}(s)$. We first define these *forbidden configurations* and then give a characterization of upward planarity wrt these forbidden configurations. Figure 1 illustrate two of these forbidden configurations.

Definition 5 (Forbidden Configuration 1 (FC1)). *Let $T_{\phi'}$ be a strongly embedded DST with linearization $\overrightarrow{L}(s)$. We let Π_1 and Π_2 represent two distinct tree-paths of $T_{\phi'}$. We assume that u is a vertex of Π_1 and v is an internal vertex of Π_2. A cross edge $e = (u, v)$ forms a forbidden configuration FC1 wrt to $T_{\phi'}$ if one of the following is true: $\Pi_1 \prec \Pi_2$ and e enters v from the right side (called FC1-a); and $\Pi_2 \prec \Pi_1$ and e enters v from the left side (called FC1-b).*

Definition 6 (Forbidden Configuration 2 (FC2)). *Let $T_{\phi'}$ be a strongly embedded DST with linearization $\overrightarrow{L}(s)$. We let Π_1, Π_2, and Π_3 be three distinct tree-paths of $T_{\phi'}$. Assume that u is a vertex of Π_1, w is a vertex of Π_2, and v is a vertex of Π_3. The cross edges (u, v) and (v, w) form a forbidden configuration FC2 if one of the following is true:$\Pi_1 \prec \Pi_2 \prec \Pi_3$ (called FC2-a); and $\Pi_3 \prec \Pi_2 \prec \Pi_1$ (called FC2-b).*

We now present our main result (Theorem 7). It gives a characterization for strongly embedded single-source upward planar digraphs. This characterization does not lead to an upward planarity testing algorithm that performs better than the known upward planarity testing algorithms for single-source digraphs [11,3]. However, we can use this characterization to compute a maximum upward planar single-source subgraph of an embedded single-source digraph.

Theorem 7. *Let G_ϕ be a strongly embedded single-source digraph with f as the external face. Let $T_{\phi'}$ denote a strongly embedded DST of G_ϕ that realizes f. We assume that $T_{\phi'}$ has the linearization $\overrightarrow{L}(s)$. The strongly embedded G_ϕ is upward planar if and only if it has no forbidden configuration wrt $T_{\phi'}$.*

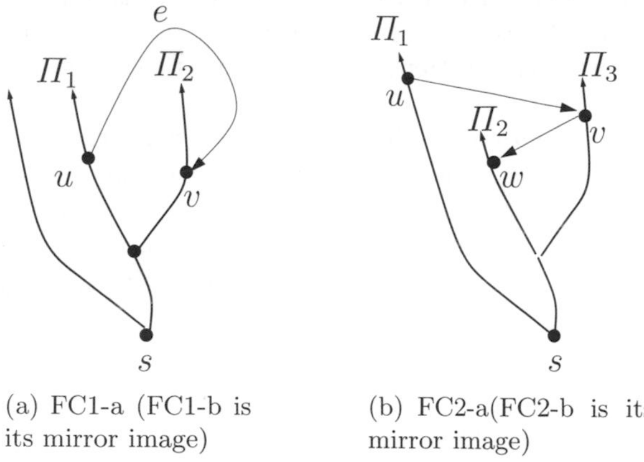

(a) FC1-a (FC1-b is its mirror image)

(b) FC2-a(FC2-b is its mirror image)

Fig. 1. Forbidden Configurations

4 Algorithm

In this section, we will present an algorithm that computes a maximum upward planar single-source subgraph of a strongly embedded single-source DAG G_ϕ. This algorithm is based on Theorem 7. We repeat this process for all possible external faces and choose the upward planar subgraph that retains the maximum number of edges.

We can compute a *DST* T by running a `Left-First-DFS` from s, where *`Left-First-DFS`* is a modification of the simple DFS algorithm. When running `Left-First-DFS`, let $v \neq s$ be a vertex that we have reached in the graph traversal. Let e be the tree-edge connecting v to its predecessor. The next vertex that we will visit will be the first unvisited vertex when we circumtraverse $L(v)$ in the clockwise direction by starting from e.

`Left-First-DFS` will take $O(n)$ time because G is planar and we visit each edge in G only once. Since G_ϕ is a DAG, there are no back-edges with respect to a DST that is computed via `Left-First-DFS`. Once `Left-First-DFS` is complete, we get a strongly embedded *DST* by ordering the edges out of s according to the unique linearization for the given external face. Let us represent the strongly embedded *DST* computed by using the `Left-First-DFS` by $\hat{T}_{\hat{\phi}}$.

Lemma 8. *In $\hat{T}_{\hat{\phi}}$, we cannot have a cross edge from a tree-path Π_1 to a tree-path Π_2 such that $\Pi_1 \prec \Pi_2$. Hence, there can neither be an FC1-a edge nor an FC2 edge-pair in $\hat{T}_{\hat{\phi}}$.*

A naive approach to computing a maximum upward planar subgraph of G_ϕ with a given external face would be to first compute $\hat{T}_{\hat{\phi}}$ and then remove all FC1 edges. However, different *DST*s can have different numbers of FC1 edges, hence we cannot assume that $\hat{T}_{\hat{\phi}}$ will have the minimum number of FC1 edges. We say that a strongly embedded *DST* is a *min-FC1 DST* if there is no other strongly embedded DST with the same external face with fewer FC1 edges. One can try to compute all possible DSTs, but as we will show now the number of possible DSTs is exponential. Hence, the only practical option left is to convert $\hat{T}_{\hat{\phi}}$ into a min-FC1 DST in polynomial time.

We define *transformation* as the operation of converting a DST $T_{\phi'}$ into a new DST $T'_{\phi''}$. This is done by making a co-tree edge $e' = (w, v)$ wrt to $T_{\phi'}$ a tree edge and making a tree edge $e = (u, v)$ of $T_{\phi'}$ a co-tree edge. That is $V(T'_{\phi''}) = V(T_{\phi'})$ and $E(T'_{\phi''}) = E(T_{\phi'}) \setminus \{e\} \cup \{e'\}$. We denote this operation by $transform(T_{\phi'}, e, e')$. We can convert $\hat{T}_{\hat{\phi}}$ into $T_{\phi'}$ by applying $transform(\hat{T}_{\hat{\phi}}, (v_0, v_1), (u_1, v_1))$. Note that we can obtain a new tree from $\hat{T}_{\hat{\phi}}$ by choosing any subset of the co-tree edges as tree-edges and making the same number of tree-edges of $\hat{T}_{\hat{\phi}}$ co-tree edges of the new DST. Since there are $O(n)$ co-tree edges, there are $O(2^n)$ possible DSTs.

Let $T_{\phi'}$ be a strongly embedded DST and let $T'_{\phi''}$ be the embedded DST by applying $transform(T_{\phi'}, e, e')$. We say that $transform(T_{\phi'}, e, e')$ *creates* an FC1 edge $\tilde{e}$ in $T'_{\phi''}$ if $\tilde{e}$ is not an FC1 edge in $T_{\phi'}$ and it is an FC1 edge in $T'_{\phi''}$. Similarly, we say that $transform(T_{\phi'}, e, e')$ *destroys* an FC1 edge $\tilde{e}$ in $T'_{\phi''}$ if $\tilde{e}$ is an FC1 edge in $T_{\phi'}$ and it is not FC1 edge in $T'_{\phi''}$. We say that $transform(T_{\phi'}, e, e')$ *reduces* the number of FC1 edges if the number of FC1 edges that are destroyed by the transform operation is more than the number FC1 edges that are created.

We now consider when FC1 edges are created or destroyed. An obvious observation one can make is that an FC1 edge e is created (resp. destroyed) when e has to circumtraverse (resp. no longer has to circumtraverse) a portion of the tree after the transform operation. This is only possible if a portion of the tree changes its location relative to the other vertices. We now identify how a tree changes after a transform operation. This will help to identify the cases when an FC1 edge is created or destroyed. Let us define $S(T_{\phi'}, v)$ as the subgraph of $T_{\phi'}$ that is induced by the vertices reachable from a vertex v. Let $T_{\phi'}$ be a strongly embedded DST with a tree-edge $e = (u, v)$ and let $e' = (w, v)$ be a co-tree edge wrt $T_{\phi'}$. Let us apply $transform(T_{\phi'}, e, e')$ to get an embedded DST $T'_{\phi''}$. One can see that only the vertices of $S(T_{\phi'}, v)$ move either left or right in the new tree because it was reachable in $T_{\phi'}$ from e but in $T'_{\phi''}$ it is reachable from e'. We say that vertices v_1 and v_2 change their relative positions after a transform operation if v_1 was left (resp. right) of v_2 before the transform operation is applied and it is right (resp. left) of v_2 after the transform operation is applied.

Property 9. If an FC1 edge $\tilde{e} = (v_1, v_2)$ is created or destroyed as a result of applying $transform(T_{\phi'}, e, e')$, then the relative locations of v_1 and v_2 in the DST $T'_{\phi''}$ change after the transform operation is applied.

Although Property 9 gives us a necessary condition of when an FC1 edge is created or destroyed, it is not a sufficient condition. The following lemma identifies necessary and sufficient conditions for FC1 edges to be created and destroyed.

Lemma 10. *Let $T_{\phi'}$ be a strongly embedded DST. Let $e = (u, v)$ be a tree edge of $T_{\phi'}$ and let $e' = (w, v)$ be a co-tree edge of $T_{\phi'}$. An FC1 edge $\tilde{e} = (v_1, v_2)$ is created or destroyed as a result of applying $transform(T_{\phi'}, e, e')$ if and only if the following conditions are all satisfied: exactly one end-vertex of $\tilde{e}$ is in $S(T_{\phi'}, v)$ and the other end-vertex is not in $S(T_{\phi'}, v)$; the edge $\tilde{e}$ is in the area $\mathcal{A}$, where $\mathcal{A}$ is the area enclosed by $s \rightsquigarrow v$, $s \rightsquigarrow w$, and the edge e'; e' is an FC1 edge.*

The following theorem characterizes when our algorithm should stop.

Theorem 11. *A transform operation cannot be applied to reduce the number of FC1 edges in a strongly embedded DST $T_{\phi'}$ of G_ϕ if and only if $T_{\phi'}$ is a min-FC1 DST of G_ϕ for the given external face.*

Algorithm 1 presents our suggested algorithm. In each iteration of the while-loop, we reduce the number of FC1 edges. Since there are $O(n)$ FC1 edges in the initial DST $\hat{T}_{\hat{\phi}}$, the total number of iterations of the while-loop will be at most $O(n)$. There are $O(n)$ co-tree edges, hence the for-loop will be executed $O(n)$ times in each iteration of the while-loop. Counting the number of FC1 edges in the current DST $T^c_{\phi^c}$ takes linear time by using path-IDs. Hence the overall time complexity of Algorithm 1 is $O(n^3)$.

Algorithm 1. `Max Upward Planar Single-Source Subgraph`

Input: The initial strongly embedded DST $\hat{T}_{\hat{\phi}}$
Output: A strongly embedded maximum upward planar subgraph $G'_{\phi'}$ of G_ϕ
1: Let $T^c_{\phi^c}$ denote the current strongly embedded DST
2: $T^c_{\phi^c} \Leftarrow \hat{T}_{\hat{\phi}}$
3: **while** 1 **do**
4:   $c \Leftarrow$ number of FC1 edges in $T^c_{\phi^c}$
5:   Traverse $T^c_{\phi^c}$ in DFS fashion.
6:   **for all** co-tree edge $e' = (w, v)$ encountered at a vertex v **do**
7:     Let e be the tree-edge of $T^c_{\phi^c}$ with head v
8:     $T_{\phi'} = transform(T^c_{\phi^c}, e, e')$
9:     $c' \Leftarrow$ number of FC1 edges in $T_{\phi'}$
10:     **if** $c' < c$ **then**
11:       $T^c_{\phi^c} \Leftarrow T_{\phi'}$; $c = c'$
12:       **GOTO** Line 4
13:     **else**
14:       $T^c_{\phi^c} = transform(T_{\phi'}, e', e)$ //convert back to the original $T^c_{\phi^c}$
15:     **end if**
16:   **end for**
17:   **return** $T^c_{\phi^c}$
18: **end while**

Algorithm 1 will terminate because the number of FC1 edges is reduced in each iteration and when further reduction is not possible then the algorithm terminates. We can see from Theorem 11 that the DST computed by Algorithm 1 is a min-FC1 DST. Algorithm 1 will not create an FC2 edge-pair is created as shown below.

Lemma 12. *Let* $\hat{T}_{\hat{\phi}}$ *be the initial strongly embedded DST. While running Algorithm 1, the current strongly embedded DST* $T^c_{\phi^c}$ *will never have an FC2 edge-pair.*

Theorem 13. *A maximum upward planar subgraph* $G'_{\phi'}$ *for a strongly embedded single-source digraph* G_ϕ *is obtained by removing all FC1 edges wrt to the DST* $T_{\phi'}$ *computed by Algorithm 1.*

Computing a maximum upward planar single-source subgraph for a strongly embedded digraph takes $O(n^3)$ time. We find the maximum upward planar single-source subgraph for every possible external face and then choose the external face that retains the maximum edges. Since, there are $O(n)$ possible external faces, the maximum upward planar single-source subgraph of an embedded single-source digraph with a fixed external face can be computed in $O(n^4)$ time. However, we only need to consider the $deg(s)$ faces that are incident to s. The average degree of a planar graph is less than or equal to $6 - \frac{12}{|V|}$ because $\sum deg(v) = 2|E|$ for any graph and $|E| \leq 3|V| - 6$ for any planar graph. Hence, the average degree of a simple planar graph is less than 6. This means that the average running time of our algorithm is $O(n^3)$.

5 Conclusion

We have shown that we can compute a maximum upward planar single-source subgraph of an embedded single-source embedded digraph in $O(n^4)$ worst case time and $O(n^3)$ average time. This work naturally leads to the following open questions: a) can we improve the running time of this algorithm? and, b) is there a class of embedded digraphs that admits polynomial time computation of maximum upward planar subgraphs?

References

1. Abbasi, S., Healy, P., Rextin, A.: An improved upward planarity testing algorithm and related applications. In: Das, S., Uehara, R. (eds.) WALCOM 2009. LNCS, vol. 5431, pp. 334–344. Springer, Heidelberg (2009)
2. Bertolazzi, P., Battista, G.D., Liotta, G., Mannino, C.: Upward drawings of triconnected digraphs. Algorithmica 12(6), 476–497 (1994)
3. Bertolazzi, P., Battista, G.D., Mannino, C., Tamassia, R.: Optimal upward planarity testing of single-source digraphs. SIAM J. Comput. 27(1), 132–169 (1998)
4. Binucci, C., Didimo, W., Giordano, F.: On the complexity of finding maximum upward planar subgraph of an embedded planar digraph. Technical Report RT001-07, University of Perugia (Febuary 2007)
5. Binucci, C., Didimo, W., Giordano, F.: Maximum upward planar subgraphs of embedded planar digraphs. In: Hong, S.-H., Nishizeki, T., Quan, W. (eds.) GD 2007. LNCS, vol. 4875. Springer, Heidelberg (2008)
6. Binucci, C., Didimo, W., Giordano, F.: Maximum upward planar subgraphs of embedded planar digraphs. Computational Geometry: Theory and Applications 41(3), 230–246 (2008)

7. Battista, G.D., Eades, P., Tamassia, R., Tollis, I.G.: Graph Drawing: Algorithms for the Visualization of Graphs. Prentice Hall, Englewood Cliffs (1999)
8. Didimo, W.: Computing upward planar drawings using switch-regularity heuristics. In: Vojtáš, P., Bieliková, M., Charron-Bost, B., Sýkora, O. (eds.) SOFSEM 2005. LNCS, vol. 3381, pp. 117–126. Springer, Heidelberg (2005)
9. Didimo, W., Giordano, F., Liotta, G.: Upward spirality and upward planarity testing. In: Healy, P., Nikolov, N.S. (eds.) GD 2005. LNCS, vol. 3843, pp. 117–128. Springer, Heidelberg (2006)
10. Garg, A., Tamassia, R.: On the computational complexity of upward and rectilinear planarity testing. SIAM J. Comput. 31(2), 601–625 (2001)
11. Hutton, M.D., Lubiw, A.: Upward planar drawing of singlesource acyclic digraphs. SIAM J. Comput. 25(2), 291–311 (1996)
12. Hutton, M.: Upward planar drawing of single source acyclic digraphs. Master's thesis, University of Waterloo (1990)
13. Papakostas, A.: Upward planarity testing of outerplanar DAGs. In: Tamassia, R., Tollis, I.G. (eds.) GD 1994. LNCS, vol. 894, pp. 298–306. Springer, Heidelberg (1995)
14. Sugiyama, K., Tagawa, S., Toda, M.: Methods for visual understanding of hierarchical system structures. IEEE Transactions on Systems, Man and Cybernetics 1, 109–125 (1981)

Triangle-Free 2-Matchings Revisited

Maxim Babenko*, Alexey Gusakov, and Ilya Razenshteyn

Moscow State University
max@adde.math.msu.su,
agusakov@gmail.com,
ilyaraz@gmail.com

Abstract. A *2-matching* in an undirected graph $G = (VG, EG)$ is a function $x\colon EG \to \{0, 1, 2\}$ such that for each node $v \in VG$ the sum of values $x(e)$ on all edges e incident to v does not exceed 2. The *size* of x is the sum $\sum_e x(e)$. If $\{e \in EG \mid x(e) \neq 0\}$ contains no triangles then x is called *triangle-free*.

Cornuéjols and Pulleyblank devised a combinatorial $O(mn)$-algorithm that finds a triangle free 2-matching of maximum size (hereinafter $n := |VG|$, $m := |EG|$) and also established a min-max theorem.

We claim that this approach is, in fact, superfluous by demonstrating how their results may be obtained directly from the Edmonds–Gallai decomposition. Applying the algorithm of Micali and Vazirani we are able to find a maximum triangle-free 2-matching in $O(m\sqrt{n})$-time. Also we give a short self-contained algorithmic proof of the min-max theorem.

Next, we consider the case of regular graphs. It is well-known that every regular graph admits a perfect 2-matching. One can easily strengthen this result and prove that every d-regular graph (for $d \geq 3$) contains a perfect triangle-free 2-matching. We give the following algorithms for finding a perfect triangle-free 2-matching in a d-regular graph: an $O(n)$-algorithm for $d = 3$, an $O(m + n^{3/2})$-algorithm for $d = 2k$ ($k \geq 2$), and an $O(n^2)$-algorithm for $d = 2k + 1$ ($k \geq 2$).

1 Introduction

1.1 Basic Notation and Definitions

We shall use some standard graph-theoretic notation throughout the paper. For an undirected graph G we denote its sets of nodes and edges by VG and EG, respectively. For a directed graph we speak of arcs rather than edges and denote the arc set of G by AG. A similar notation is used for paths, trees, and etc. Unless stated otherwise, we do not allow loops and parallel edges or arcs in graphs. An undirected graph is called *d-regular* (or just *regular* if the value of d is unimportant) if all degrees of its nodes are equal to d. A subgraph of G induced by a subset $U \subseteq VG$ is denoted by $G[U]$.

* Supported by RFBR grant 09-01-00709-a.

M.T. Thai and S. Sahni (Eds.): COCOON 2010, LNCS 6196, pp. 120–129, 2010.

1.2 Triangle-Free 2-Matchings

Definition 1. *Given an undirected graph G, a* 2-matching *in G is a function $x \colon EG \to \{0,1,2\}$ such that for each node $v \in VG$ the sum of values $x(e)$ on all edges e incident to v does not exceed 2.*

A natural optimization problem is to find, given a graph G, a *maximum* 2-matching x in G, that is, a 2-matching of maximum *size* $||x|| := \sum_e x(e)$. When $||x|| = |VG|$ we call x *perfect.*

If $\{e \mid x(e) = 1\}$ partitions into a collection of node-disjoint circuits of odd length then x is called *basic.* Applying a straightforward reduction one can easily see that for each 2-matching there exists a basic 2-matching of the same or larger size (see [CP80, Theorem 1.1]). From now on we shall only consider basic 2-matchings x.

One may think of a basic 2-matching x as a collection of node disjoint *double edges* (each contributing 2 to $||x||$) and *odd length circuits* (where each edge of the latter contributes 1 to $||x||$). See Fig. 1.2(a) for an example.

Computing the maximum size $\nu_2(G)$ of a 2-matching in G reduces to finding a maximum matching in an auxiliary bipartite graph obtained by splitting the nodes of G. Therefore, the problem is solvable in $O(m\sqrt{n})$-time with the help of Hopcroft–Karp's algorithm [HK73] (hereinafter $n := |VG|$, $m := |EG|$). A simple min-max relation is known (see [Sch03, Th. 6.1.4] for an equivalent statement):

Theorem 1. $\nu_2(G) := \min_{U \subseteq VG} (|VG| + |U| - \mathrm{iso}(G - U))$.

Here $\nu_2(G)$ is the maximum size of a 2-matching in G, $G - U$ denotes the graph obtained from G by removing nodes U (i.e. $G[VG - U]$) and $\mathrm{iso}(H)$ stands for the number of isolated nodes in H. The reader may refer to [Sch03, Ch. 30] and [LP86, Ch. 6] for a survey.

Let $\mathrm{supp}(x)$ denote $\{e \in EG \mid x(e) \neq 0\}$. The following refinement of 2-matchings was studied by Cornuéjols and Pulleyblank [CP80] in connection with the Hamilton cycle problem:

Definition 2. *Call a 2-matching x* triangle-free *if supp(x) contains no triangle.*

They investigated the problem of finding a maximum size triangle-free 2-matching, devised a combinatorial algorithm, and gave an $O(n^3)$ estimate for its running time. Their algorithm initially starts with $x := 0$ and then performs a sequence of *augmentation steps* each aiming to increase $||x||$. Totally, there are $O(n)$ steps and a more careful analysis easily shows that the step can be implemented to run in $O(m)$ time. Hence, in fact the running time of their algorithm is $O(mn)$.

The above algorithm also yields a min-max relation as a by-product. Denote the maximum size of a triangle-free 2-matching in G by $\nu_2^3(G)$.

Definition 3. *A* triangle cluster *is a connected graph whose edges partition into disjoint triangles such that any two triangles have at most one node in common and if such a node exists, it is an articulation point of the cluster. (See Fig. 1.2(b) for an example.)*

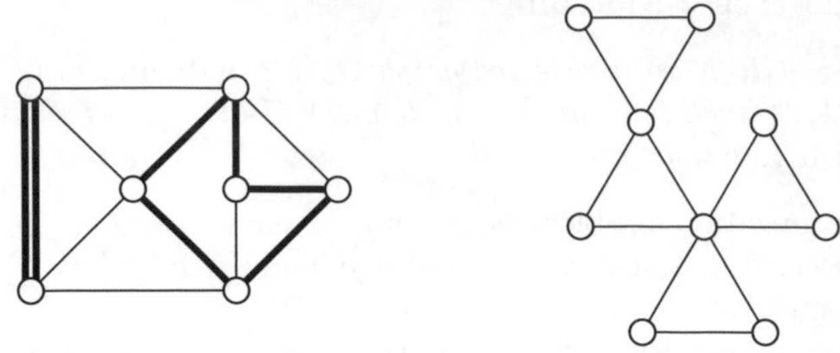

Fig. 1. (a) A perfect basic 2-matching. (b) A triangle cluster.

Let cluster(H) be the number of the connected components of H that are triangle clusters.

Theorem 2. $\nu_2^3(G) := \min_{U \subseteq VG} (|VG| + |U| - \text{cluster}(G - U))$.

One may notice a close similarity between Theorem 2 and Theorem 1.

1.3 Our Contribution

The goal of the present paper is to devise a faster algorithm for constructing a maximum triangle-free 2-matching. We give a number of results that improve the above-mentioned $O(mn)$ time bound.

Firstly, let G be an arbitrary undirected graph. We claim that the direct augmenting approach of Cornuéjols and Pulleyblank is, in fact, superfluous. In Section 2 we show how one can compute a maximum triangle-free 2-matching with the help of the Edmonds–Gallai decomposition [LP86, Sec. 3.2]. The resulting algorithm runs in $O(m\sqrt{n})$ time (assuming that the maximum matching in G is computed by the algorithm of Micali and Vazirani [MV80]). Also, this approach directly yields Theorem 2.

Secondly, there are some well-known results on matchings in regular graphs.

Theorem 3. *Every 3-regular bridgeless graph has a perfect matching.*

Theorem 4. *Every regular bipartite graph has a perfect matching.*

The former theorem is usually credited to Petersen while the second one is an easy consequence of Hall's condition.

Theorem 5 (Cole, Ost, Schirra [COS01]). *There exists a linear time algorithm that finds a perfect matching in a regular bipartite graph.*

Theorem 4 and Theorem 5 imply the following:

Corollary 1. *Every regular graph has a perfect 2-matching. The latter 2-matching can be found in linear time.*

In Section 3 we consider the analogues of Corollary 1 with 2-matchings replaced by triangle-free 2-matchings. We prove that every d-regular graph ($d \geq 3$) has a perfect triangle-free 2-matching. This result gives a simple and natural strengthening to the non-algorithmic part of Corollary 1.

As for the complexity of finding a perfect 2-matching in a d-regular graph it turns out heavily depending on d. The ultimate goal is a linear time algorithm but we are only able to fulfill this task for $d = 3$. The case of even d ($d \geq 4$) turns out reducible to $d = 4$, so the problem is solvable in $O(m + n^{3/2})$ time by the use of the general algorithm (since $m = O(n)$ for 4-regular graphs). The case of odd d ($d \geq 5$) is harder, we give an $O(n^2)$-time algorithm, which improves the general time bound of $O(m\sqrt{n})$ when $m = \omega\left(n^{3/2}\right)$.

2 General Graphs

2.1 Factor-Critical Graphs, Matchings, and Decompositions

We need several standard facts concerning maximum matchings (see [LP86, Ch. 3] for a survey). For a graph G, let $\nu(G)$ denote the maximum size of a matching in G and $\mathrm{odd}(H)$ be the number of connected components of H with an odd number of vertices.

Theorem 6 (Tutte–Berge). $\nu(G) = \min_{U \subseteq VG} \frac{1}{2}\left(|VG| + |U| - \mathrm{odd}(G - U)\right)$.

Definition 4. *A graph G is* factor-critical *if for any $v \in VG$, $G - v$ admits a perfect matching.*

Theorem 7 (Edmonds–Gallai). *Consider a graph G and put*

$$\begin{aligned} D &:= \{v \in VG \mid \text{there exists a maximum size matching missing } v\}, \\ A &:= \{v \in VG \mid v \text{ is a neighbor of } D\}, \\ C &:= VG - (A \cup D). \end{aligned}$$

Then $U := A$ achieves the minimum in the Tutte–Berge formula, and D is the union of the odd connected components of $G[VG - A]$. Every connected component of $G[D]$ is factor-critical. Any maximum matching in G induces a perfect matching in $G[C]$ and a matching in $G[VG - C]$ that matches all nodes of A to distinct connected components of $G[D]$.

We note that once a maximum matching M in G is found, an Edmonds–Gallai decomposition of G can be constructed in linear time by running a search for an M-augmenting path. Most algorithms that find M yield this decomposition as a by-product. Also, the above augmenting path search may be adapted to produce an *odd ear decomposition* of every odd connected component of $G[VG - A]$:

Definition 5. *An* ear decomposition *$G_0, G_1, \ldots, G_k = G$ of a graph G is a sequence of graphs where G_0 consists of a single node, and for each $i = 0, \ldots, k-1$, G_{i+1} obtained from G_i by adding the edges and the intermediate nodes of an ear. An* ear *of G_i is a path P_i in G_{i+1} such that the only nodes of P_i belonging to G_i are its (possibly coinciding) endpoints. An ear decomposition with all ears having an odd number of edges is called* odd.

The next statement is widely-known and, in fact, comprises a part of the blossom-shrinking approach to constructing a maximum matching.

Lemma 1. *Given an odd ear decomposition of a factor-critical graph G and a node $v \in VG$ one can construct in linear time a matching M in G that misses exactly the node v.*

Finally, we classify factor-critical graphs depending on the existence of a perfect triangle-free 2-matching. The proof of the next lemma is implicit in [CP83] and one can easily turn it into an algorithm:

Lemma 2. *Each factor-critical graph G is either a triangle cluster or has a perfect triangle-free 2-matching x. Moreover, if an odd ear decomposition of G is known then these cases can be distinguished and x (if exists) can be constructed in linear time.*

2.2 The Algorithm

For the sake of completeness, we first establish an upper bound on the size of a triangle-free 2-matching.

Lemma 3. *For each $U \subseteq VG$, $\nu_2^3(G) \le |VG| + |U| - \mathrm{cluster}(G - U)$.*

Proof
Removing a single node from a graph G may decrease $\nu_2^3(G)$ by at most 2. Hence, $\nu_2^3(G) \le \nu_2^3(G-U) + 2|U|$. Also, $\nu_2^3(G-U) \le (|VG| - |U|) - \mathrm{cluster}(G - U)$ since every connected component of $G - U$ that is a triangle cluster lacks a perfect triangle-free 2-matching. Combining these inequalities, one gets the desired result. □

The next theorem both gives an efficient algorithm a self-contained proof of the min-max formula.

Theorem 8. *A maximum triangle-tree 2-matching can be found in $O(m\sqrt{n})$ time.*

Proof
Construct an Edmonds–Gallai decomposition of G, call it (D, A, C), and consider odd ear decompositions of the connected components of $G[D]$. As indicated earlier, the complexity of this step is dominated by finding a maximum matching M in G. The latter can be done in $O(m\sqrt{n})$ time (see [MV80]).

The matching M induces a perfect matching M_C in $G[C]$. We turn M_C into double edges in the desired triangle-free 2-matching x by putting $x(e) := 2$ for each $e \in M_C$.

Next, we build a bipartite graph H. The nodes in the upper part of H correspond to the components of $G[D]$, the nodes in the lower part of H are just the nodes of A. There is an edge between a component C and a node v in H if and only if there is at least one edge between C and v in G. Let us call the components that are triangle clusters *bad* and the others *good.* Consider another bipartite graph H' formed from H by dropping all nodes (in the upper part) corresponding to good components.

The algorithm finds a maximum matching $M_{H'}$ in H' and then augments it to a maximum matching M_H in H. This is done in $O(m\sqrt{n})$ time using Hopcroft–Karp algorithm [HK73]. It is well-known that an augmentation can only increase the set of matched nodes, hence every bad component matched by $M_{H'}$ is also matched by M_H and vice versa. From the properties of Edmonds–Gallai decomposition it follows that M_H matches all nodes in A.

Each edge $e \in M_H$ corresponds to an edge $\widetilde{e} \in EG$, we put $x(\widetilde{e}) := 2$.

Finally, we deal with the components of $G[D]$. Let C be a component that is matched (in M_H) by, say, an edge $e_C \in M_H$. As earlier, let $\widetilde{e}_C$ be the preimage of e_C in G. Since C is factor-critical, there exists a matching M_C in C that misses exactly the node in C covered by $\widetilde{e}_C$. We find M_C in linear time (see Lemma 1) and put $x(e) := 2$ for each $e \in M_C$.

As for the unmatched components, we consider good and bad ones separately. If an unmatched component C is good, we apply Lemma 2 to find (in linear time) and add to x a perfect triangle-free 2-matching in C. If C is bad, we employ Lemma 1 and find (in linear time) a matching M_C in C that covers all the nodes expect for an arbitrary chosen one and set $x(e) := 2$ for each $e \in M_C$.

The running time of the above procedure is dominated by constructing the Edmonds–Gallai decomposition of G and finding matchings $M_{H'}$ and M_H. Clearly, it is $O(m\sqrt{n})$.

It remains to prove that x is a maximum triangle-free 2-matching. Let n_{bad} be the number of bad components in $G[D]$. Among these components, let k_{bad} be matched by $M_{H'}$ (and, hence, by M_H). Then $||x|| = |VG| - (n_{\text{bad}} - k_{\text{bad}})$. From König–Egervary theorem (see, e.g., [LP86]) there exists a vertex cover L in H' of cardinality k_{bad} (i.e. a subset $L \subseteq VH'$ such that each edge in H' is incident to at least one node in L). Put $L = L_A \cup L_D$, where L_A are the nodes of L belonging to the lower part of H and L_D are the nodes from the upper part. The graph $G - L_A$ contains at least $n_{\text{bad}} - |L_D|$ components that are triangle clusters. (They correspond to the uncovered nodes in the upper part of H'. Indeed, these components are only connected to L_A in the lower part.) Hence, putting $U := L_A$ in Lemma 3 one gets $\nu_2^3(G) \le |VG| + |L_A| - (n_{\text{bad}} - |L_D|)$ $|VG| + |L| - n_{\text{bad}} = |VG| - (n_{\text{bad}} - k_{\text{bad}}) = ||x||$. Therefore, x is a maximum triangle-free 2-matching, as claimed.

3 Regular Graphs

3.1 Existence of a Perfect Triangle-Free 2-Matching

Theorem 9. *Let G* [illegible]

and let k be the number of triangles in H. One has $|VH| = 2k + 1$. Each node of H is incident to either d or $d - 1$ edges. Let q_i denote the number of nodes of degree $d - 1$ in H. Since $|EH| = 3k$ it follows that $(2k + 1)d - 6k - q_i = d + (2d - 6)k - q_i \geq d - q_i$ edges of G connect H to U. Totally, the nodes in U have at least $\sum_{i=1}^{t}(d - q_i) \geq td - q$ incident edges. On the other hand, each node of U has the degree of at most d, hence $td - q \leq |U|d$ therefore $t - |U| \leq q/d$. By the min-max formula (see Theorem 2) this implies the desired bound. □

Corollary 2. *Every d-regular graph ($d \geq 3$) has a perfect triangle-free 2-matching.*

3.2 Cubic Graphs

For $d = 3$ we speed up the general algorithm ultimately as follows:

Theorem 10. *A a perfect triangle-free 2-matching in a 3-regular graph can be found in linear time.*

Proof

Consider a 3-regular graph G. First, we find an arbitrary inclusion-wise maximal collection of node-disjoint triangles $\Delta_1, \ldots, \Delta_k$ in G. This is done in linear time by performing a local search at each node $v \in VG$. Next, we contract $\Delta_1, \ldots, \Delta_k$ into *composite* nodes $z_1, \ldots, z_k$ and obtain another 3-regular graph G' (note that G' may contain multiple parallel edges).

Construct a bipartite graph H' from G' as follows. Every node $v \in VG'$ is split into a pair of nodes v^1 and v^2. Every edge $\{u, v\} \in EG'$ generates edges $\{u^1, v^2\}$ and $\{v^1, u^2\}$ in H'. There is a natural surjective many-to-one correspondence between perfect matchings in H' and perfect 2-matchings in G'. Applying the algorithm of Cole, Ost and Schirra [COS01] to H' we construct a perfect 2-matching x' in G' in linear time. As usual, we assume that x' is basic, in particular x' contains no circuit of length 2 (i.e. $\mathrm{supp}(x')$ contains no pair of parallel edges).

Our final goal is to expand x' into a perfect triangle-free 2-matching x in G. The latter is done as follows. Consider an arbitrary composite node z_i obtained by contracting Δ_i in G. Suppose that a double edge e of x' is incident to z_i in G'. We keep the preimage of e as a double edge of x and add another double edge connecting the remaining pair of nodes in Δ_i. See Fig. 2(a).

Next, suppose that x' contains an odd-length circuit C' passing through z_i. Then, we expand z_i to Δ_i and insert an additional pair of edges to C'. Note that the length of the resulting circuit C is odd and is no less than 5. See Fig. 2(b).

Clearly, the resulting 2-matching x is perfect. But why is it triangle-free? For sake of contradiction, suppose that Δ is a triangle in $\mathrm{supp}(x)$. Then, Δ is an odd circuit in x' and no node of Δ is composite. Hence, Δ is a triangle disjoint from $\Delta_1, \ldots, \Delta_k$ — a contradiction. □

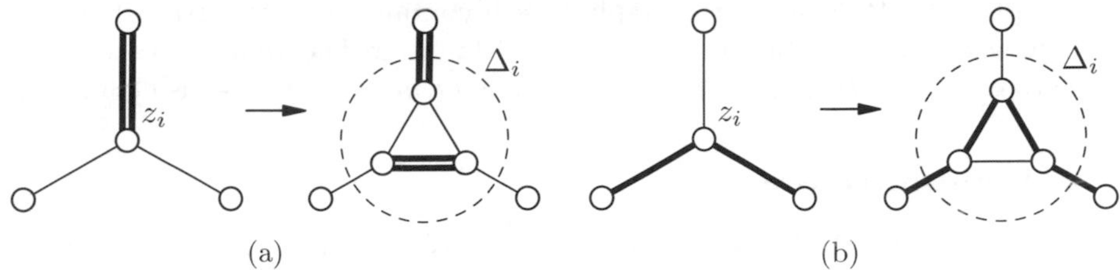

Fig. 2. Uncontraction of z_i

Combining the above connection between triangle-free 2-matchings in G and 2-matchings in G' with the result of Voorhoeve [Voo79] one can prove the following:

Theorem 11. *There exists a constant $c > 1$ such that every 3-regular graph G contains at least c^n perfect triangle-free 2-matchings.*

3.3 Even-Degree Graphs

To find a perfect triangle-free 2-matching in a $2k$-regular graph G ($k \geq 2$) we replace it by a 4-regular spanning subgraph and then apply the general algorithm.

Lemma 4. *For each $2k$-regular ($k \geq 1$) graph G there exists and can be found in linear time a 2-regular spanning subgraph.*

Proof
Since the degrees of all nodes in G are even, EG decomposes into a collection of edge-disjoint circuits. This decomposition takes linear time. For each circuit C from the above decomposition we choose an arbitrary direction and traverse C in this direction turning undirected edges into directed arcs. Let $\overrightarrow{G}$ denote the resulting digraph. For each node v exactly k arcs of $\overrightarrow{G}$ enter v and exactly k arcs leave v.

Next, we construct a bipartite graph H from $\overrightarrow{G}$ as follows: each node $v \in \overrightarrow{G}$ generates a pair of nodes $v^1, v^2 \in VH$, each arc $(u, v) \in A\overrightarrow{G}$ generates an edge $\{u^1, v^2\} \in EH$. The graph H is k-regular and, hence, contains a perfect matching M (which, by Theorem 5, can be found in linear time). Each edge of M corresponds to an arc of $\overrightarrow{G}$ and, therefore, to an edge of G. Clearly, the set of the latter edges forms a 2-regular spanning subgraph of G. □

Theorem 12. *A perfect triangle-free 2-matching in a d-regular graph ($d = 2k$, $k \geq 2$) can be found in $O(m + n^{3/2})$ time.*

Proof
Consider an undirected $2k$-regular graph G. Apply Lemma 4 and construct find a 2-regular spanning subgraph H_1 of G. Next, discard the edges of H_1 and reapply Lemma 4 thus obtaining another 2-regular spanning subgraph H_2 (here we use that $k \geq 2$). Their union $H := (VG, EH_1 \cup EH_2)$ is a 4-regular spanning

subgraph of G. By Corollary 2 graph H still contains a perfect triangle-free 2-matching x, which can be found by the algorithm from Theorem 8. It takes $O(m)$ time to construct H and $O(n^{3/2})$ time, totally $O(m + n^{3/2})$ time, as claimed. □

3.4 Odd-Degree Graphs

The case $d = 2k + 1$ ($k \geq 2$) is more involved. We extract a spanning subgraph H of G whose node degrees are 3 and 4. A careful choice of H allows us to ensure that the number of nodes of degree 3 is $O(n/d)$. Then, by Theorem 9 subgraph H contains a nearly-perfect triangle-free 2-matching. The latter is found and then augmented to a perfect one with the help of the algorithm from [CP80]. More details follow.

Lemma 5. *There exists and can be found in linear time a spanning subgraph H of graph G with nodes degrees equal to 3 and 4. Moreover, at most $O(n/d)$ nodes in H are of degree 3.*

Proof

Let us partition the nodes of G into pairs (in an arbitrary way) and add $n/2$ *virtual* edges connecting these pairs. The resulting graph G' is $2k + 2$-regular. (Note that G' may contain multiple parallel edges.)

Our task is find a 4-regular spanning subgraph H' of G' containing at most $O(n/d)$ virtual edges. Once this subgraph is found, the auxiliary edges are dropped creating $O(n/d)$ nodes of degree 3 (recall that each node of G is incident to exactly one virtual edge).

Subgraph H' is constructed by repeatedly pruning graph G'. During this process graph G' remains d'-regular for some even d' (initially $d' := d + 1$).

At each pruning step we first examine d'. Two cases are possible. Suppose d' is divisible by 4, then a *large* step is executed. The graph G' is decomposed into a collection of edge-disjoint circuits. In each circuit, every second edge is marked as *red* while others are marked as *blue*. This red-blue coloring partitions G' into a pair of spanning $d'/2$-regular subgraphs. We replace G' by the one containing the smallest number of virtual edges. The second case (which leads to a *small* step) applies if d' is not divisible by 4. Then, with the help of Lemma 4 a 2-regular spanning subgraph is found in G'. The edges of this subgraph are removed from G', so d' decreases by 2.

The process stops when d' reaches 4 yielding the desired subgraph H'. Totally, there are $O(\log d)$ large (and hence also small) steps each taking time proportional to the number of remaining edges. The latter decreases exponentially, hence the total time to construct H' is linear.

It remains to bound the number of virtual edges in H'. There are exactly $t := \lfloor \log_2 (d + 1)/4 \rfloor$ large steps performed by the algorithm. Each of the latter decreases the number of virtual edges in the current subgraph by at least a factor of 2. Hence, at the end there are $O(n/2^t) = O(n/d)$ virtual edges in H', as required. □

Theorem 13. *A perfect triangle-free 2-matching in a d-regular graph* ($d = 2k+1$, $k \geq 2$) *can be found in* $O(n^2)$ *time.*

Proof
We apply Lemma 5 and construct a subgraph H in $O(m)$ time. Next, a maximum triangle-free 2-matching x is found in H, which takes $O(|EH| \cdot |VH|^{1/2}) = O(n^{3/2})$ time. By Theorem 9 the latter 2-matching obeys $n - ||x|| = O(n/d)$. To turn x into a perfect triangle-free 2-matching in G we apply the algorithm from [CP80] and perform $O(n/d)$ augmentation steps. Each step takes $O(m)$ time, so totally the desired perfect triangle-free 2-matching is constructed in $O(m + n^{3/2} + mn/d) = O(n^2)$ time. □

Acknowledgements

The authors are thankful to Andrew Stankevich for fruitful suggestions and helpful discussions.

References

[COS01] Cole, R., Ost, K., Schirra, S.: Edge-coloring bipartite multigraphs in $O(E \log D)$ time. Combinatorica 21, 5–12 (2001)

[CP80] Cornuéjols, G., Pulleyblank, W.R.: Perfect triangle-free 2-matchings. Mathematical Programming Studies 13, 1–7 (1980)

[CP83] Cornuéjols, G., Pulleyblank, W.R.: Critical graphs, matchings and tours or a hierarchy of relaxations for the travelling salesman problem. Combinatorica 3(1), 35–52 (1983)

[HK73] Hopcroft, J.E., Karp, R.M.: An $n^{5/2}$ algorithm for maximum matchings in bipartite graphs. SIAM Journal on Computing 2(4), 225–231 (1973)

[LP86] Lovász, L., Plummer, M.D.: Matching Theory. Akadémiai Kiadó, North Holland (1986)

[MV80] Micali, S., Vazirani, V.: An $O(\sqrt{|V|}.|E|)$ algorithm for finding maximum matching in general graphs. In: Proc. 21st IEEE Symp. Foundations of Computer Science, pp. 248–255 (1980)

[Sch03] Schrijver, A.: Combinatorial Optimization. Springer, Berlin (2003)

[Voo79] Voorhoeve, M.: A lower bound for the permanents of certain (0, 1)-matrices. Nederl. Akad. Wetensch. Indag. Math. 41(1), 83–86 (1979)

The Cover Time of Deterministic Random Walks

Tobias Friedrich[1] and Thomas Sauerwald[2]

[1] Max-Planck-Institut für Informatik, Campus E1.4, 66123 Saarbrücken, Germany
[2] Simon Fraser University, Burnaby B.C. V5A 1S6, Canada

Abstract. The rotor router model is a popular deterministic analogue of a random walk on a graph. Instead of moving to a random neighbor, the neighbors are served in a fixed order. We examine how fast this "deterministic random walk" covers all vertices (or all edges). We present general techniques to derive upper bounds for the vertex and edge cover time and derive matching lower bounds for several important graph classes. Depending on the topology, the deterministic random walk can be asymptotically faster, slower or equally fast compared to the classical random walk.

1 Introduction

We examine the cover time of a simple deterministic process known under various names such as "rotor router model" or "Propp machine." It can be viewed as an attempt to derandomize random walks on graphs $G = (V, E)$. In the model each vertex $x \in V$ is equipped with a "rotor" together with a fixed sequence of the neighbors of x called "rotor sequence." While a particle (chip, coin, ...) performing a random walk leaves a vertex in a random direction, the deterministic random walk always goes in the direction the rotor is pointing. After a particle is sent, the rotor is updated to the next position of its rotor sequence. We examine how fast this model covers all vertices and/or edges, when one particle starts a walk from an arbitrary vertex.

1.1 Deterministic Random Walks

The idea of rotor routing appeared independently several times in the literature. First under the name "Eulerian walker" [28], then as "edge ant walk" [33] and later as "whirling tour" [16]. Around the same time it was also popularized by James Propp and analyzed by Cooper and Spencer [10] who called it the "Propp machine." Later the term "deterministic random walk" was established in Doerr *et al.* [11, 14]. For brevity, we omit the "random" and just refer to "deterministic walk."

Cooper and Spencer [10] showed the following remarkable similarity between the expectation of a random walk and a deterministic walk with cyclic rotor sequences: If an (almost) arbitrary distribution of particles is placed on the vertices of an infinite grid $\mathbb{Z}^d$ and does a simultaneous walk in the deterministic walk model, then at all times and on each vertex, the number of particles deviates

M.T. Thai and S. Sahni (Eds.): COCOON 2010, LNCS 6196, pp. 130–139, 2010.

from the expected number the standard random walk would have gotten there, by at most a constant. This constant is precisely known for the cases $d = 1$ [11] and $d = 2$ [14]. It is further known that there is no such constant for infinite trees [12]. Levine and Peres [24, 25] also extensively studied a related model called internal diffusion-limited aggregation for deterministic walks.

As in these works, our aim is to understand random walk and their deterministic counterpart from a theoretical viewpoint. However, we would like to mention that the rotor router mechanism also led to improvements in applications. With a random initial rotor direction, the quasirandom rumor spreading protocol broadcasts faster in some networks than its random counterpart [15]. A similar idea is used in quasirandom load balancing [20].

We consider our model of a deterministic walk based on rotor routing to be a simple and canonic derandomization of a random walk which is not tailored for search problems. On the other hand, there is a vast literature on local deterministic agents/robots/ants patrolling or covering all vertices or edges of a graph (*e.g.* [21, 23, 30, 32, 33]). For instance, Cooper *et al.* [9] studied a model where the walk uses adjacent edges which have been traversed the smallest number of times. However, all of these models are more specialized and require additional counters/identifiers/markers/pebbles on the vertices or edges of the explored graph.

1.2 Cover Time of Random Walks

In his survey, Lovász [26] mentions three important measures of a random walk: cover time, hitting time, and mixing time. These three (especially the first two) are closely related. Here we will mainly concentrate on the *cover time*, which is the expected number of steps to visit every node. The study of the cover time of random walks on graphs was initiated in 1979. Motivated by the space-complexity of the s–t-connectivity problem, Aleliunas *et al.* [2] showed that the cover time is upper bounded by $\mathcal{O}(|V|\,|E|)$ for any graph. For regular graphs, Feige [18] gave an improved upper bound of $\mathcal{O}(|V|^2)$ for the cover time. Broder and Karlin [4] proved several bounds which rely on the spectral gap of the transition matrix. Their bounds imply that the cover time on a regular expander graph is $\Theta(|V| \log |V|)$. In addition, many papers are devoted to the study of the cover time on special graphs such as hypercubes [1], random graphs [7], random regular graphs [6], random geometric graphs [8] etc. A general lower bound of $(1 - o(1))\,|V| \ln |V|$ for any graph was shown by Feige [17].

A natural variant of the cover time is the so-called *edge cover time*, which measures the expected number of steps to traverse all edges. Amongst other results, Zuckerman [35, 36] proved that the edge cover time of general graphs is at least $\Omega(|E| \log |E|)$ and at most $\mathcal{O}(|V|\,|E|)$.

1.3 Cover Time of Deterministic Walks (Our Results)

For the case of a cyclic rotor sequence the edge cover time is known to be $\Theta(|E| \operatorname{diam}(G))$ (see Yanovski *et al.* [34] for the upper and Bampas *et al.* [3] for the lower bound). It is further known that there are rotor sequences such that

Table 1. Comparison of the vertex cover time of random and deterministic walk on different graphs ($n = |V|$)

Graph class G	Vertex cover time $\mathsf{VC}(G)$ of the random walk	Vertex cover time $\widetilde{\mathsf{VC}}(G)$ of the deterministic walk
k-ary tree, $k = \mathcal{O}(1)$	$\Theta(n \log^2 n)$ [36, Cor. 9]	$\Theta(n \log n)$ (Thm. 4.2 and 3.13)
star	$\Theta(n \log n)$ [36, Cor. 9]	$\Theta(n)$ (Thm. 4.1)
cycle	$\Theta(n^2)$ [26, Ex. 1]	$\Theta(n^2)$ (Thm. 4.3 and 3.11)
lollipop graph	$\Theta(n^3)$ [26, Thm. 2.1]	$\Theta(n^3)$ (Thm. 4.4 and 3.14)
expander	$\Theta(n \log n)$ [4, Cor. 6], [31]	$\Theta(n \log n)$ (Thm. 4.5, Cor. 3.7)
two-dim. torus	$\Theta(n \log^2 n)$ [36, Thm. 4], [5, Thm. 6.1]	$\Theta(n^{1.5})$ (Thm. 4.6 and 3.11)
d-dim. torus ($d \geqslant 3$)	$\Theta(n \log n)$ [36, Cor. 12], [5, Thm. 6.1]	$\mathcal{O}(n^{1+1/d})$ (Thm. 3.11)
hypercube	$\Theta(n \log n)$ [1, p. 372], [27, Sec. 5.2]	$\Theta(n \log^2 n)$ (Thm. 4.7 and 3.12)
complete	$\Theta(n \log n)$ [26, Ex. 1]	$\Theta(n^2)$ (Thm. 4.1 and 3.10)

the edge cover time is precisely $|E|$ [28]. We allow arbitrary rotor sequences and present three techniques to upper bound the edge cover time based on the local divergence (Thm. 3.5), expansion of the graph (Thm. 3.6), and a corresponding flow problem (Thm. 3.9). With these general theorems it is easy to prove upper bounds for expanders, complete graphs, torus graphs, hypercubes, k-ary trees and lollipop graphs. In addition we show in Section 4 that these bounds can be matched by very canonical rotor sequences. Though a general lower bound of $\Omega(|E| \operatorname{diam}(G))$ was shown in [3], we believe that the study of these canonical rotor sequences is of independent interest. Unfortunately, all proofs had to be omitted to meet the page limit. They will be given in the full version of the paper.

It is not our aim to prove superiority of the deterministic walk, but it is instructive to compare our results for the vertex and edge cover time with the respective bounds of the random walk. Tables 1 and 2 group the graphs in three classes depending whether random or deterministic walk is faster. In spite of the strong adversary (as the order of the rotors is completely arbitrary), the deterministic walk is surprisingly efficient. It is known that the edge cover time of random walks can be asymptotically larger than its vertex cover time. Somewhat unexpectedly, this is not the case for the deterministic walk. To highlight this issue, let us consider hypercubes and complete graph. For these graphs, the vertex cover time of the deterministic walk is larger while the edge cover time is smaller (complete graph) or equal (hypercube) compared to the random walk.

2 Models and Preliminaries

2.1 Random Walks

We consider weighted random walks on finite connected graphs $G = (V, E)$. For this, we assign every pair of vertices $u, v \in V$ a weight $c(u, v) \in \mathbb{N}_0$ (rational weights can be handled by scaling) such that $c(u, v) = c(v, u) > 0$ if $\{u, v\} \in E$ and $c(u, v) = c(v, u) = 0$ otherwise. This defines transition probabilities $\mathbf{P}_{u,v} := c(u, v)/c(u)$ with $c(u) := \sum_{w \in V} c(u, w)$. So, whenever a random

Table 2. Comparison of the edge cover time of random and deterministic walk on different graphs ($n = |V|$)

Graph class G	**Edge cover time $\mathsf{EC}(G)$ of the random walk**	**Edge cover time $\widetilde{\mathsf{EC}}(G)$ of the deterministic walk**
k-ary tree, $k = \mathcal{O}(1)$	$\Theta(n \log^2 n)$ [36, Cor. 9]	$\Theta(n \log n)$ (Thm. 4.2 and 3.13)
star	$\Theta(n \log n)$ [36, Cor. 9]	$\Theta(n)$ (Thm. 4.1)
complete	$\Theta(n^2 \log n)$ [35, 36]	$\Theta(n^2)$ (Thm. 4.1 and 3.10)
expander	$\Theta(n \log n)$ [35, 36]	$\Theta(n \log n)$ (Thm. 4.5, Cor. 3.7)
cycle	$O(n^2)$ [26, Ex. 1]	$O(n^2)$ (Thm. 4.3 and 3.11)
lollipop graph	$\Theta(n^3)$ [26, Thm. 2.1], [35, Lem. 2]	$\Theta(n^3)$ (Thm. 4.4 and 3.14)
hypercube	$\Theta(n \log^2 n)$ [35, 36]	$\Theta(n \log^2 n)$ (Thm. 4.7 and 3.12)
two-dim. torus	$\Theta(n \log^2 n)$ [35, 36]	$\Theta(n^{1.5})$ (Thm. 4.6 and 3.11)
d-dim. torus ($d \geqslant 3$)	$\Theta(n \log n)$ [35, 36]	$\mathcal{O}(n^{1+1/d})$ (Thm. 3.11)

walk is at a vertex u it moves to a vertex v in the next step with probability $\mathbf{P}_{u,v}$. Moreover, note that for all $u, v \in V$, $c(u,v) = c(v,u)$ while $\mathbf{P}_{u,v} \neq \mathbf{P}_{v,u}$ in general. This defines a time-reversible, irreducible, finite Markov chain $X_0, X_1, \ldots$ with transition matrix $\mathbf{P}$. The t-step probabilities of the walk can be obtained by taking the t-th power of $\mathbf{P}^t$. In what follows, we prefer to use the term weighted random walk instead of Markov chain to emphasize the limitation to rational transition probabilities.

It is intuitively clear that a random walk with large weights $c(u,v)$ is harder to approximate deterministically with a simple rotor sequence. To measure this, we use $c_{\max} := \max_{u,v \in V} c(u,v)$. An important special case is the *unweighted random walk* with $c(u,v) \in \{0,1\}$ for all $u, v \in V$ on a simple graph. In this case, $\mathbf{P}_{u,v} = 1/\deg(u)$ for all $\{u,v\} \in E$, and $c_{\max} = 1$. Our general results hold for weighted (random) walks. However, the derived bounds for specific graphs are only stated for unweighted walks. With *random walk* we mean unweighted random walk and if a random walk is allowed to be weighted we will emphasize this. For weighted and unweighted random walks we define for a graph G,

- cover time: $\mathsf{VC}(G) = \max_{u \in V} \mathbf{E}\left[\min\left\{t \geqslant 0 \colon \bigcup_{\ell=0}^{t} \{X_\ell\} = V\right\} \mid X_0 = u\right]$,
- edge cover time:
  $\mathsf{EC}(G) = \max_{u \in V} \mathbf{E}\left[\min\left\{t \geqslant 0 \colon \bigcup_{\ell=1}^{t} \{X_{\ell-1}, X_\ell\} = E\right\} \mid X_0 = u\right]$.

The (edge) cover time of a graph class $\mathcal{G}$ is the maximum of the (edge) cover times of all graphs of the graph class. Observe that $\mathsf{VC}(\mathcal{G}) \leqslant \mathsf{EC}(G)$ for all graphs G. For vertices $u, v \in V$ we further define

- (expected) hitting time: $\mathsf{H}(u,v) = \mathbf{E}\left[\min\left\{t \geqslant 0 \colon X_t = v\right\} \mid X_0 = u\right]$,
- stationary distribution: $\pi_u = c(u) / \sum_{w \in V} c(w)$.

2.2 Deterministic Random Walks

We define weighted deterministic random walks (or short: weighted deterministic walks) based on rotor routers as introduced by Holroyd and Propp [22]. For a weighted random walk, we define the corresponding weighted deterministic walk as follows. We use a tilde ($\widetilde{\ }$) to mark variables related to the deterministic walk. To each vertex u we assign a rotor sequence $\widetilde{s}(u) = (\widetilde{s}(u,1), \widetilde{s}(u,2), \ldots, \widetilde{s}(u,\widetilde{d}(u))) \in V^{\widetilde{d}(u)}$ of arbitrary length $\widetilde{d}(u)$ such that the number of times a neighbor v occurs in the rotor sequence $\widetilde{s}(u)$ corresponds to the transition probability to go from u to v in the weighted random walk, that is, $\mathbf{P}_{u,v} = |\{i \in [\widetilde{d}(u)] \colon \widetilde{s}(u,i) = v\}|/\widetilde{d}(u)$ with $[\widetilde{d}(u)] := \{1, \ldots, \widetilde{d}(u)\}$. For a weighted random walk, $\widetilde{d}(u)$ is a multiple of the lowest common denominator of the transition probabilities from u to its neighbors. For the standard random walk, a corresponding canonical deterministic walk would be $\widetilde{d}(u) = \deg(u)$ and a permutation of the neighbors of u as rotor sequence $\widetilde{s}(u)$. As the length of the rotor sequences crucially influences the performance of a deterministic walk, we set $\widetilde{\kappa} := \max_{u \in V} \widetilde{d}(u)/\deg(u)$ (note that $\widetilde{\kappa} \geqslant 1$). The set V together with $\widetilde{s}(u)$ and $\widetilde{d}(u)$ for all $u \in V$ defines the deterministic walk, sometimes abbreviated **D**. Note that every deterministic walk has a unique corresponding random walk while there are many deterministic walks corresponding to one random walk.

We also assign to each vertex u an integer $\widetilde{r}_t(u) \in [\widetilde{d}(u)]$ corresponding to a rotor at u pointing to $\widetilde{s}(u, \widetilde{r}_t(u))$ at step t. A *rotor configuration* C describes the rotor sequences $\widetilde{s}(u)$ and initial rotor directions $\widetilde{r}_0(u)$ for all vertices $u \in V$. At every time step t the walk moves from $\widetilde{x}_t$ in the direction of the current rotor of $\widetilde{x}_t$ and this rotor is incremented[1] to the next position according to the rotor sequence $\widetilde{s}(\widetilde{x}_t)$ of $\widetilde{x}_t$. More formally, for given $\widetilde{x}_t$ and $\widetilde{r}_t(\cdot)$ at time $t \geqslant 0$ we set $\widetilde{x}_{t+1} := s(\widetilde{x}_t, \widetilde{r}_t(\widetilde{x}_t))$, $\widetilde{r}_{t+1}(\widetilde{x}_t) := \widetilde{r}_t(\widetilde{x}_t) \bmod \widetilde{d}(\widetilde{x}_t) + 1$, and $\widetilde{r}_{t+1}(u) := \widetilde{r}_t(u)$ for all $u \neq \widetilde{x}_t$. Let $\mathcal{C}$ be the set of all possible rotor configurations (that is, $\widetilde{s}(u)$, $\widetilde{r}_0(u)$ for $u \in V$) of a corresponding deterministic walk for a fixed weighted random walk (and fixed rotor sequence length $\widetilde{d}(u)$ for each $u \in V$). Given a rotor configuration $C \in \mathcal{C}$ and an initial location $\widetilde{x}_0 \in V$, the vertices $\widetilde{x}_0, \widetilde{x}_1, \ldots \in V$ visited by a deterministic walk are completely determined. For deterministic walks we define for a graph G and vertices $u, v \in V$,

- deterministic cover time:
  $\widetilde{\mathsf{VC}}(G) = \max_{\widetilde{x}_0 \in V} \max_{C \in \mathcal{C}} \min \left\{ t \geqslant 0 \colon \bigcup_{\ell=0}^{t} \{\widetilde{x}_\ell\} = V \right\}$,
- deterministic edge cover time:
  $\widetilde{\mathsf{EC}}(G) = \max_{\widetilde{x}_0 \in V} \max_{C \in \mathcal{C}} \min \left\{ t \geqslant 0 \colon \bigcup_{\ell=1}^{t} \{\widetilde{x}_{\ell-1}, \widetilde{x}_\ell\} = E \right\}$,
- hitting time: $\widetilde{\mathsf{H}}(u,v) = \max_{C \in \mathcal{C}} \min \{ t \geqslant 0 \colon \widetilde{x}_t = u, \widetilde{x}_0 = v \}$.

Note that the definition of the deterministic cover time takes the *maximum* over all possible rotor configurations, while the cover time of a random walk takes the *expectation* over the random decisions. Also, $\widetilde{\mathsf{VC}}(G) \leqslant \widetilde{\mathsf{EC}}(G)$ for all graphs G. We further define for fixed configurations $C \in \mathcal{C}$, $\widetilde{x}_0$, and vertices $u, v \in V$,

- number of visits to vertex u: $\widetilde{N}_t(u) = \left| \{ 0 \leqslant \ell \leqslant t \colon \widetilde{x}_\ell = u \} \right|$,
- number of traversals of a directed edge $u \to v$: $\widetilde{N}_t(u \to v) = \left| \{ 1 \leqslant \ell \leqslant t \colon (\widetilde{x}_{\ell-1}, \widetilde{x}_\ell) = (u,v) \} \right|$.

[1] In this respect we slightly deviate from the model of Holroyd and Propp [22] who first increment the rotor and then move the chip, but this change is insignificant here.

2.3 Graph-Theoretic Notation

We consider finite, connected graphs $G = (V, E)$. Unless stated differently, $n := |V|$ is the number vertices and $m := |E|$ the number of undirected edges. By δ and Δ we denote the minimum and maximum degree of the graph, respectively. For a pair of vertices $u, v \in V$, we denote by $\operatorname{dist}(u, v)$ their distance, i.e., the length of a shortest path between them. For a vertex $u \in V$, let $\Gamma(u)$ denote the set of all neighbors of u. More generally, for any $k \geqslant 1$, $\Gamma^k(u)$ denotes the set of vertices v with $\operatorname{dist}(u, v) = k$. For any subsets $S, T \subseteq V$, $E(S)$ denotes the set of edges with one endpoint in S and $E(S, T)$ denotes the edges $\{u, v\}$ with $u \in S$ and $v \in T$. As a walk is something directed, we also have to argue about directed edges though our graph G is undirected. In slight abuse of notation, for $\{u, v\} \in E$ we might also write $(u, v) \in E$ or $(v, u) \in E$.

3 Upper Bounds on the Deterministic Cover Times

Very recently, Holroyd and Propp [22] proved that several natural quantities of the weighted deterministic walk as defined in Section 2.2 concentrate around the respective expected values of the corresponding weighted random walk. To state their result formally, we set for a vertex $v \in V$,

$$K(v) := \max_{u \in V} \mathsf{H}(u, v) + \frac{1}{2}\left(\frac{\widetilde{d}(v)}{\pi_v} + \sum_{i,j \in V} \widetilde{d}(i)\, \mathbf{P}_{i,j}\, |\mathsf{H}(i, v) - \mathsf{H}(j, v) - 1|\right).$$

Theorem 3.1 ([22, Thm. 4]). *For all weighted deterministic walks, all vertices $v \in V$, and all times t, $\left|\pi_v - \widetilde{N}_t(v)/t\right| \leqslant K(v)\, \pi_v / t$.*

Roughly speaking, Theorem 3.1 states that the proportion of time spent by the weighted deterministic walk concentrates around the stationary distribution for all configurations $C \in \mathcal{C}$ and all starting points $\widetilde{x}_0$. To quantify the hitting or cover time with Theorem 3.1, we choose $t = K(v) + 1$ to get $\widetilde{N}_t(v) > 0$. To get a bound for the edge cover time, we choose $t = 3K(v)$ and observe that then $\widetilde{N}_t(v) \geqslant 2\pi_v K(v) > \widetilde{d}(v)$. This already shows the following corollary.

Corollary 3.2. *For all weighted deterministic walks,*

$$\begin{aligned} \widetilde{\mathsf{H}}(u, v) &\leqslant K(v) + 1 && \text{for all } u, v \in V, \\ \widetilde{\mathsf{VC}}(G) &\leqslant \max_{v \in V} K(v) + 1, \\ \widetilde{\mathsf{EC}}(G) &\leqslant 3 \max_{v \in V} K(v). \end{aligned}$$

One obvious question that arises from Theorem 3.1 and Corollary 3.2 is how to bound the value $K(v)$. While it is clear that $K(v)$ is polynomial in n (provided that $c_{\max}$ and $\widetilde{\kappa}$ are polynomially bounded), it is not clear how to get more precise upper bounds. A key tool to tackle the difference of hitting times in $K(v)$ is the following elementary lemma, where in case of a periodic walk the sum is taken as a Cesáro summation. The proof of this lemma and the proofs of all following results is omitted due to space limitations.

Lemma 3.3. *For all weighted random walks and all vertices* $i, j, v \in V$, $\sum_{t=0}^{\infty} \left(\mathbf{P}_{i,v}^t - \mathbf{P}_{j,v}^t\right) = \pi_v \left(\mathsf{H}(j,v) - \mathsf{H}(i,v)\right)$.

To analyze weighted random walks, we use the notion of local divergence which is an important quantity in the analysis of load balancing algorithms [19, 29].

Definition 3.4. *The local divergence of a weighted random walk is* $\Psi(\mathbf{P}) := \max_{v \in V} \Psi(\mathbf{P}, v)$, *where* $\Psi(\mathbf{P}, v)$ *is the local divergence w.r.t. to a vertex* $v \in V$ *defined as* $\Psi(\mathbf{P}, v) := \sum_{t=0}^{\infty} \sum_{\{i,j\} \in E} \left|\mathbf{P}_{i,v}^t - \mathbf{P}_{j,v}^t\right|$.

Using Corollary 3.2 and Lemma 3.3, we can prove the following bound on the hitting time of a deterministic walk.

Theorem 3.5. *For all deterministic walks and all vertices* $v \in V$,

$$K(v) \leqslant \max_{u \in V} \mathsf{H}(u,v) + \frac{\widetilde{\kappa}\, c_{\max}}{\pi_v} \Psi(\mathbf{P}, v) + 2m\, \widetilde{\kappa}\, c_{\max}.$$

Note that Theorem 3.5 is more general than just giving an upper bound for hitting and cover times via Corollary 3.2. It can be useful in the other directions, too. To give a specific example, we can apply the result of Theorem 4.7 that $\widetilde{\mathsf{EC}}(G) = \Omega(n \log^2 n)$ for hypercubes and $\max_{u,v} \mathsf{H}(u,v) = \mathcal{O}(n)$ (cf. [26]) to Theorem 3.5 and obtain a lower bound of $\Omega(n \log^2 n)$ on the local divergence of hypercubes.

3.1 Upper Bound Depending on the Expansion

We now derive an upper bound for $\widetilde{\mathsf{EC}}(G)$ that depends on the expansion properties of G. Let $\lambda_2(\mathbf{P})$ be the second-largest eigenvalue in absolute value of $\mathbf{P}$.

Theorem 3.6. *For all graphs* G, $\widetilde{\mathsf{EC}}(G) = \mathcal{O}\big(\frac{\Delta}{\delta} \frac{n}{1-\lambda_2(\mathbf{P})} + n\, \widetilde{\kappa}\, \frac{\Delta}{\delta} \frac{\Delta \log n}{1-\lambda_2(\mathbf{P})}\big)$.

Here, we call a graph with constant maximum degree an expander graph, if $1/(1-\lambda_2(\mathbf{P})) = \mathcal{O}(1)$ (equivalently, we have for all subsets $X \subseteq V, 1 \leqslant |X| \leqslant n/2$, $|E(X, X^c)| = \Omega(|X|)$ (cf. [13, Prop. 6])). Using Theorem 3.6, we immediately get the following upper bound on $\widetilde{\mathsf{EC}}(G)$ for expanders.

Corollary 3.7. *For all expander graphs,* $\widetilde{\mathsf{EC}}(G) = \mathcal{O}(\widetilde{\kappa}\, n \log n)$.

3.2 Upper Bound by Flows

We relate the edge cover time of the unweighted random walk to the optimal solution of the following flow problem.

Definition 3.8. *Consider the flow problem where a distinguished source node s sends a flow amount of 1 to each other node in the graph. Then $f_s(i,j)$ denotes the load transferred along edge $\{i,j\}$ (note $f_s(i,j) = -f_s(j,i)$) such that $\sum_{\{i,j\}\in E} f_s(i,j)^2$ is minimized.*

Theorem 3.9. *For all graphs G,*

$$\widetilde{\mathsf{EC}}(G) = \mathcal{O}\Big(\frac{\Delta}{\delta}\max_{u,v\in V}\mathsf{H}(u,v) + \Delta\, n\, \widetilde{\kappa} + \widetilde{\kappa}\,\Delta \max_{s\in V}\sum_{\{i,j\}\in E}|f_s(i,j)|\Big)$$

where f_s is the flow with source s according to Definition 3.8.

3.3 Upper Bounds for Common Graphs

We now demonstrate how to apply above general results to obtain upper bounds for the edge cover time of the deterministic walk for many common graphs. As the general bounds Theorems 3.5, 3.6 and 3.9 all have a linear dependency on $\widetilde{\kappa}$, the following upper bounds can be also stated depending on $\widetilde{\kappa}$. However, for clarity we assume $\widetilde{\kappa} = \mathcal{O}(1)$ here.

Theorem 3.10. *For complete graphs, $\widetilde{\mathsf{EC}}(G) = \mathcal{O}(n^2)$.*

Theorem 3.11. *For d-dimensional torus graphs ($d \geqslant 1$ constant), $\widetilde{\mathsf{EC}}(G) = \mathcal{O}(n^{1+1/d})$.*

Theorem 3.12. *For hypercubes, $\widetilde{\mathsf{EC}}(G) = \mathcal{O}(n\log^2 n)$.*

Theorem 3.13. *For k-ary trees ($k \geqslant 2$ constant), $\widetilde{\mathsf{EC}}(G) = \mathcal{O}(n\log n)$.*

Theorem 3.14. *For lollipop graphs, $\widetilde{\mathsf{EC}}(G) = \mathcal{O}(n^3)$.*

4 Lower Bounds on the Deterministic Cover Time

We first give a general lower bound of $\Omega(m)$ on the deterministic cover time for all graphs. Afterwards, for all graphs examined in Section 3.3 for which this general bound is not tight we present stronger lower bounds which match their respective upper bounds.

Theorem 4.1. *For all graphs, $\widetilde{\mathsf{VC}}(G) \geqslant m - \delta$.*

Theorem 4.2. *For k-ary trees ($k \geqslant 2$ constant), $\widetilde{\mathsf{VC}}(G) = \Omega(n\log n)$.*

Theorem 4.3. *For cycles, $\widetilde{\mathsf{VC}}(G) = \Omega(n^2)$.*

Theorem 4.4. *For lollipop graphs, $\widetilde{\mathsf{VC}}(G) = \Omega(n^3)$.*

Theorem 4.5. *There are expander graphs with $\widetilde{\mathsf{VC}}(G) = \Omega(n\log n)$.*

Theorem 4.6. *For two-dimensional torus graphs, $\widetilde{\mathsf{VC}}(G) = \Omega(n^{3/2})$.*

Theorem 4.7. *For hypercubes, $\widetilde{\mathsf{VC}}(G) = \Omega(n\log^2 n)$.*

References

[1] Aldous, D.: On the time taken by random walks on finite groups to visit every state. Zeitschrift für Wahrscheinlichkeitstheorie und verwandte Gebiete, 361–374 (1983)

[2] Aleliunas, R., Karp, R., Lipton, R., Lovász, L., Rackoff, C.: Random walks, universal traversal sequences, and the complexity of maze problems. In: 20th Annual IEEE Symposium on Foundations of Computer Science (FOCS '79), pp. 218–223 (1979)

[3] Bampas, E., Gasieniec, L., Hanusse, N., Ilcinkas, D., Klasing, R., Kosowski, A.: Euler tour lock-in problem in the rotor-router model. In: Keidar, I. (ed.) DISC 2009. LNCS, vol. 5805, pp. 423–435. Springer, Heidelberg (2009)

[4] Broder, A., Karlin, A.: Bounds on the cover time. Journal of Theoretical Probability 2(1), 101–120 (1989)

[5] Chandra, A., Raghavan, P., Ruzzo, W., Smolensky, R., Tiwari, P.: The electrical resistance of a graph captures its commute and cover times. Computational Complexity 6(4), 312–340 (1997)

[6] Cooper, C., Frieze, A.: The cover time of random regular graphs. SIAM Journal of Discrete Mathematics 18(4), 728–740 (2005)

[7] Cooper, C., Frieze, A.: The cover time of the giant component of a random graph. Random Structures & Algorithms 32(4), 401–439 (2008)

[8] Cooper, C., Frieze, A.: The cover time of random geometric graphs. In: 19th Annual ACM-SIAM Symposium on Discrete Algorithms (SODA '09) , pp. 48–57 (2009)

[9] Cooper, C., Ilcinkas, D., Klasing, R., Kosowski, A.: Derandomizing random walks in undirected graphs using locally fair exploration strategies. In: Albers, S., Marchetti-Spaccamela, A., Matias, Y., Nikoletseas, S., Thomas, W. (eds.) ICALP 2009. LNCS, vol. 5556, pp. 411–422. Springer, Heidelberg (2009)

[10] Cooper, J., Spencer, J.: Simulating a random walk with constant error. Combinatorics, Probability & Computing 15, 815–822 (2006)

[11] Cooper, J., Doerr, B., Spencer, J., Tardos, G.: Deterministic random walks on the integers. European Journal of Combinatorics 28(8), 2072–2090 (2007)

[12] Cooper, J., Doerr, B., Friedrich, T., Spencer, J.: Deterministic random walks on regular trees. In: 19th Annual ACM-SIAM Symposium on Discrete Algorithms (SODA '08), pp. 766–772 (2008)

[13] Diaconis, P., Saloff-Coste, L.: Comparison theorems for reversible Markov chains. Annals of Applied Probability 3(3), 696–730 (1993)

[14] Doerr, B., Friedrich, T.: Deterministic random walks on the two-dimensional grid. Combinatorics, Probability & Computing 18(1-2), 123–144 (2009)

[15] Doerr, B., Friedrich, T., Sauerwald, T.: Quasirandom rumor spreading. In: 19th Annual ACM-SIAM Symposium on Discrete Algorithms (SODA '08), pp. 773–781 (2008)

[16] Dumitriu, I., Tetali, P., Winkler, P.: On playing golf with two balls. SIAM J. Discrete Math. 16(4), 604–615 (2003)

[17] Feige, U.: A tight lower bound for the cover time of random walks on graphs. Random Structures & Algorithms 6(4), 433–438 (1995)

[18] Feige, U.: Collecting coupons on trees, and the cover time of random walks. Computational Complexity 6(4), 341–356 (1997)

[19] Friedrich, T., Sauerwald, T.: Near-perfect load balancing by randomized rounding. In: 41st Annual ACM Symposium on Theory Progarmming (STOC'09), pp. 121–130 (2009)

[20] Friedrich, T., Gairing, M., Sauerwald, T.: Quasirandom load balancing. In: 21st Annual ACM-SIAM Symposium on Discrete Algorithms (SODA '10), pp. 1620–1629 (2010)

[21] Gasieniec, L., Pelc, A., Radzik, T., Zhang, X.: Tree exploration with logarithmic memory. In: 18th Annual ACM-SIAM Symposium on Discrete Algorithms (SODA '07) , pp. 585–594 (2007)

[22] Holroyd, A.E., Propp, J.: Rotor walks and markov chains (2000), arXiv:0904.4507

[23] Korf, R.E.: Real-time heuristic search. Artif. Intell. 42(2-3), 189–211 (1990)

[24] Levine, L., Peres, Y.: Spherical asymptotics for the rotor-router model in $\mathbb{Z}^d$. Indiana Univ. Math. J. 57, 431–450 (2008)

[25] Levine, L., Peres, Y.: Strong spherical asymptotics for rotor-router aggregation and the divisible sandpile. Potential Analysis 30, 1–27 (2009)

[26] Lovász, L.: Random walks on graphs: A survey. Combinatorics, Paul Erdös is Eighty 2, 1–46 (1993)

[27] Palacios, J.L.: Expected hitting and cover times of random walks on some special graphs. Random Structures & Algorithms 5(1), 173–182 (1994)

[28] Priezzhev, V.B., Dhar, D., Dhar, A., Krishnamurthy, S.: Eulerian walkers as a model of self-organized criticality. Phys. Rev. Lett. 77, 5079–5082 (1996)

[29] Rabani, Y., Sinclair, A., Wanka, R.: Local divergence of Markov chains and the analysis of iterative load balancing schemes. In: 39th Annual IEEE Symposium on Foundations of Computer Science (FOCS '98), pp. 694–705 (1998)

[30] Reingold, O.: Undirected connectivity in log-space. J. ACM 55(4) (2008)

[31] Rubinfeld, R.: The cover time of a regular expander is $\mathcal{O}(n \log n)$. Inf. Process. Lett. 35(1), 49–51 (1990)

[32] Wagner, I.A., Lindenbaum, M., Bruckstein, A.M.: Smell as a computational resource - a lesson we can learn from the ant. In: Israeli Symposium on Theory of Computing and Systems (ISTCS '96), pp. 219–230 (1996)

[33] Wagner, I.A., Lindenbaum, M., Bruckstein, A.M.: Distributed covering by ant-robots using evaporating traces. IEEE Transactions on Robotics and Automation 15(5), 918–933 (1999)

[34] Yanovski, V., Wagner, I.A., Bruckstein, A.M.: A distributed ant algorithm for efficiently patrolling a network. Algorithmica 37(3), 165–186 (2003)

[35] Zuckerman, D.: On the time to traverse all edges of a graph. Inf. Process. Lett. 38(6), 335–337 (1991)

[36] Zuckerman, D.: A technique for lower bounding the cover time. SIAM Journal on Discrete Mathematics 5(1), 81–87 (1992)

Finding Maximum Edge Bicliques in Convex Bipartite Graphs*

Doron Nussbaum[1], Shuye Pu[2], Jörg-Rüdiger Sack[1], Takeaki Uno[3], and Hamid Zarrabi-Zadeh[1]

[1] School of Computer Science, Carleton University, Ottawa, Ontario K1S 5B6, Canada
{nussbaum,sack,zarrabi}@scs.carleton.ca

[2] Program in Molecular Structure and Function, Hospital for Sick Children, 555 University Avenue, Toronto, Ontario M5G 1X8, Canada
shuyepu@sickkids.ca

[3] National Institute of Informatics, 2-1-2 Hitotsubashi, Chiyoda-ku, Tokyo 101-8430, Japan
uno@nii.ac.jp

Abstract. A bipartite graph $G = (A, B, E)$ is convex on B if there exists an ordering of the vertices of B such that for any vertex $v \in A$, vertices adjacent to v are consecutive in B. A complete bipartite subgraph of a graph G is called a biclique of G. In this paper, we study the problem of finding the maximum edge-cardinality biclique in convex bipartite graphs. Given a bipartite graph $G = (A, B, E)$ which is convex on B, we present a new algorithm that computes the maximum edge-cardinality biclique of G in $O(n \log^3 n \log \log n)$ time and $O(n)$ space, where $n = |A|$. This improves the current $O(n^2)$ time bound available for the problem.

1 Introduction

In this paper, we study the following optimization problem in graph theory: given a bipartite graph $G = (A, B, E)$, find a complete bipartite subgraph of G with two vertex sets $S \subseteq A$ and $T \subseteq B$ such that $|S| \times |T|$ is maximized. This problem is called *maximum edge biclique*, as opposed to the *maximum vertex biclique* problem in which the objective is to maximize $|S| + |T|$. Besides applications to molecular biology [2,5,19,26], maximum edge biclique has applications to manufacturing optimization [6], formal concept analysis [11], and conjunctive clustering [21]. Our motivation for studying this problem came from an application to analyzing DNA microarray data as illustrated in Figure 1.

While maximum vertex biclique is solvable in polynomial time, the maximum edge biclique problem in general bipartite graphs is known to be NP-complete [22]. Indeed, it is hard to approximate the maximum edge biclique in general bipartite graphs to within a factor of n^δ, for some $\delta > 0$ [9,14] (see also [25] for corresponding inapproximability results in weighted version of the

* Research supported by NSERC and SUN Microsystems.

M.T. Thai and S. Sahni (Eds.): COCOON 2010, LNCS 6196, pp. 140–149, 2010.

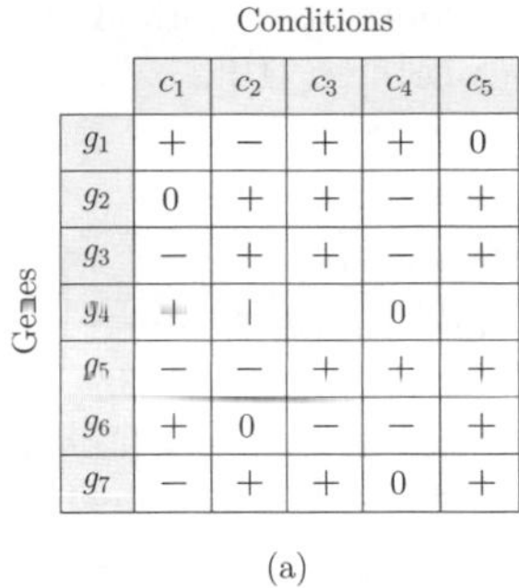

Conditions

Genes	c_1	c_2	c_3	c_4	c_5
g_1	+	−	+	+	0
g_2	0	+	+	−	+
g_3	−	+	+	−	+
g_4	+	|		0	
g_5	−	−	+	+	+
g_6	+	0	−	−	+
g_7	−	+	+	0	+

(a)

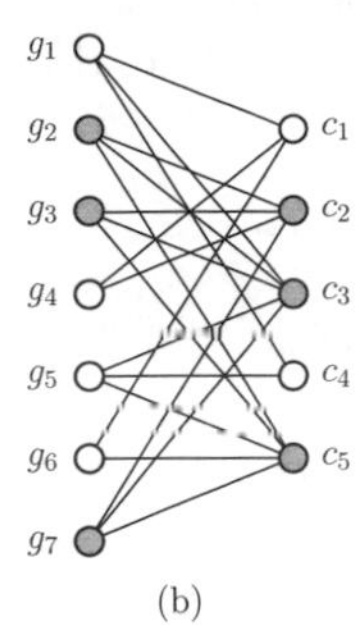

(b)

Fig. 1. (a) An exemplary DNA microarray data matrix. Each entry $[i, j]$ of the matrix represents a significant increase or a significant decrease (shown by + and −, respectively) in the expression level of the gene g_i under the condition c_j. The symbol $\emptyset$ represents no significant change. In this example the subset of genes $\{g_2, g_3, g_7\}$ and the subset of conditions $\{c_2, c_3, c_5\}$ form a bicluster. (b) A bipartite graph corresponding to the significant increases in the microarray data. Vertices forming the maximum edge biclique in this graph are shown in gray.

problem). However, for special subclasses of bipartite graphs, such as chordal bipartite graphs and convex graphs, polynomial-time algorithms for the maximum edge biclique problem are available using algorithms that enumerate all maximal bicliques in a given graph [1,7,8,12,16].

Convex bipartite graphs introduced by Glover [13] naturally arise in several industrial and scheduling applications, and a number of efficient algorithms have been developed on this graph class for problems such as maximum matching, maximum independent set, and minimum feedback vertex set (see e.g. [4,17,18,23,24]). A bipartite graph $G = (A, B, E)$ is called convex (on B) if there exists an ordering of the vertices of B such that, for any vertex $v \in A$, vertices adjacent to v are consecutive in B. The motivation for studying convex bipartite graphs in the context of analyzing DNA microarray data is that a linear ordering of genes exists naturally in several forms, such as chronological ordering in the course of evolution, and spatial ordering on chromosomes.

All the existing algorithms for solving the maximum edge biclique problem are based on enumerating all maximal bicliques in the input graph (see e.g. [1,8,16]). It is known that the number of maximal bicliques in a convex bipartite graph with n vertices is $O(n^2)$ [1]. Indeed, it is not hard to construct convex bipartite graphs that have $\Theta(n^2)$ maximal bicliques. Therefore, the existing algorithms for solving the maximum edge biclique problem have a running time of $\Omega(n^2)$ on convex bipartite graphs.

In this paper, we show that the maximum edge biclique problem can be solved more efficiently on convex bipartite graphs by using a pruning technique that avoids enumerating all the maximal bicliques. More precisely, we present a new algorithm that, given a convex bipartite graph $G = (A, B, E)$ with the corresponding convex ordering on B, computes the maximum edge biclique of G in

$O(n \log^3 n \log\log n)$ time and $O(n)$ space, where $n = |A|$. This improves the current $O(n^2)$ time bound available for the problem [16].

2 Preliminaries

A graph G is *bipartite*, if its set of vertices can be partitioned into two disjoint sets A and B such that every edge of G connects a vertex in A to a vertex in B. We denote such a bipartite graph by $G = (A, B, E)$, where E is the set of edges of G. A complete bipartite subgraph of a graph G is called a *biclique* of G. Given a biclique C of G, we refer to the number of edges in C by the *size* of C, and denote it by $|B|$. Moreover, for a vertex v of G, we denote the set of vertices adjacent to v by $N(v)$.

Let $G = (A, B, E)$ be a bipartite graph. An ordering $\prec$ of B has the *adjacency property* if for every vertex $a \in A$, $N(a)$ consists of vertices that are consecutive (i.e., form an interval) in the ordering $\prec$ of B. A bipartite graph $G = (A, B, E)$ is *convex* if there is an ordering of A or B that fulfills the adjacency property.

For convex graphs, there are linear-time recognition algorithms that output the corresponding orderings on the vertex sets in linear time [3,15,20]. Throughout this paper, we assume that the input graph is convex on B, and the vertices of A and B are labeled with integers $1, 2, \ldots$ in the same order imposed by the adjacency property. Figure 2(a) shows an examples of a convex bipartite graph.

In this paper, we denote by $[a \,..\, b]$ the set of integers that lie between two integers a and b, including both. Such a set is called an *integer interval*. Given an integer interval $I = [a \,..\, b]$, the *size* of I, denoted by $|I|$, is the number of integers, $b - a + 1$, contained in I.

3 Problem Transformation

In order to solve the maximum edge biclique problem, we first transform it from a graph theoretical problem to a geometric problem, and then provide an efficient algorithm for solving the geometric problem in Section 4. The main problem considered in this paper is the following:

Problem 1 (Maximum Edge Biclique). Given a convex bipartite graph $G = (A, B, E)$ with $|A| = n$ and $|B| = k$, find a biclique of G that has the maximum number of edges.

We transform the maximum edge biclique problem to a variant of the point dominance problem in the plane. Given two points $p, q \in \mathbb{R}^2$, we say that q is *dominated* by p if $q_x \leqslant p_x$ and $q_y \geqslant p_y$ (in other words, if q lies in a rectangle whose bottom-right corner is at p). Let S be a set of n points in the grid $[1 \,..\, k] \times [1 \,..\, k]$. We refer to each point of S as a *token*. For a grid point (i, j), we define the *dominance number* of (i, j) w.r.t. S to be $\text{DOM}(i, j) = |\{(x, y) \in S : (x, y)$ is dominated by $(i, j)\}|$, and define the *magnitude* of (i, j) to be $\text{MAG}(i, j) = \text{DOM}(i, j) \times (j - i + 1)$. We call the gird point maximizing $\text{MAG}(i, j)$ the *maximum point* of the grid.

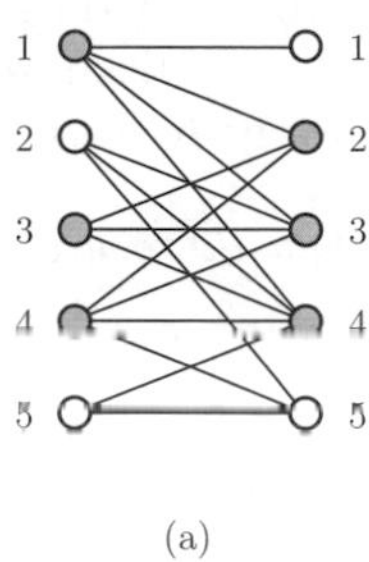

(a)

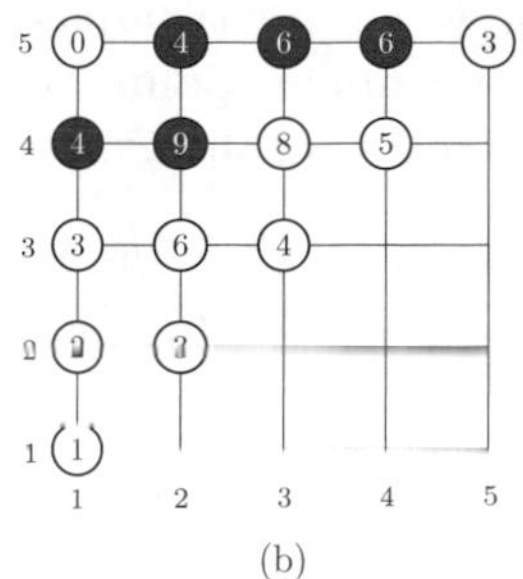

(b)

Fig. 2. (a) A convex bipartite graph. (b) The corresponding grid. Tokens are shown in black. The number inside each grid point denotes its magnitude.

Problem 2 (Maximum Point). Given a set S of n tokens on a $k \times k$ grid, find a grid point (i, j) that maximizes $\text{MAG}(i, j)$.

Lemma 1. *Problem 1 is equivalent to Problem 2.*

Proof. By the convexity of G, $N(a)$ is an integer interval in $[1 \mathinner{..} k]$ for each vertex $a \in A$. Let π be a function that maps each integer interval $[i \mathinner{..} j] \subseteq [1 \mathinner{..} k]$ to a grid point (i, j) on the grid $[1 \mathinner{..} k] \times [1 \mathinner{..} k]$. For each vertex $a \in A$, we define $\pi(a) \equiv \pi(N(a))$. Let $S = \{\pi(a) : a \in A\}$. We show that finding the maximum edge biclique in G is equivalent to solving the maximum point problem on the set S (see Figure 2).

The key observation is that for any pair of integer intervals I and R, $R \subseteq I$ if and only if $\pi(I)$ is dominated by $\pi(R)$. This is because $[i \mathinner{..} j] \subseteq [i' \mathinner{..} j']$ if and only if $i' \leqslant i$ and $j' \geqslant j$ which means that (i', j') is dominated by (i, j). Let $R = [i \mathinner{..} j] \subseteq [1 \mathinner{..} k]$ represent a set of subsequent vertices in B. Define $A_R = \{a \in A : R \subseteq N(a)\}$. Every vertex in A_R is connected to every vertex in R. Therefore, $A_R \times R$ defines a biclique C_R of G with $|A_R| \times |R|$ edges. Moreover, $|A_R| = |\{a \in A : R \subseteq N(a)\}| = |\{a \in A : \pi(N(a)) \text{ is dominated by } \pi(R)\}| = |\{a \in A : \pi(a) \text{ is dominated by } (i, j)\}| = \text{DOM}(i, j)$ w.r.t. S. Therefore, $|C_R| = |A_R| \times |R| = \text{DOM}(i, j) \times (j - i + 1) = \text{MAG}(i, j)$, and thus, finding a clique of maximum size in G is equivalent to finding a grid point with maximum magnitude. □

Note that $\text{MAG}(i, j)$ is less than or equal to zero for $j < i$. Therefore, to find the maximum point, we only need to consider grid points (i, j) with $j \geqslant i$.

4 The Algorithm

Let $S = \{(x_1, y_1), \ldots, (x_n, y_n)\}$ be a set of n tokens on a $k \times k$ grid. It is easy to find the maximum point in this grid by a simple scan over the grid points. If the number of tokens is small compared to the grid size, then tokens are sparsely scattered across the grid. An immediate question is whether we can compute the maximum magnitude without necessarily visiting all the grid points.

In this section, we answer this question affirmatively by providing an algorithm that finds the maximum point by examining only a small subset (namely, a subquadratic number) of the grid points. We start by the following observation.

Observation 1. *If (x, y) is the maximum point, then there are tokens (x_i, y_i) and (x_j, y_j) in S such that $x_i = x$ and $y_j = y$.*

Proof. Suppose by contradiction that there is no token in S such that $x_i = x$. Let $Q \subseteq S$ be the set of tokens dominated by (x, y), and let $q = (x', y')$ be the token with the largest x-coordinate in Q (ties are broken arbitrarily). Obviously, $\text{DOM}(x', y) = \text{DOM}(x, y)$, as there is no token in the rectangle $[x' + 1, x] \times [y, k]$. Moreover, $x' < x$ by our selection of q. Therefore, $\text{DOM}(x', y) \times (y - x' + 1) > \text{DOM}(x, y) \times (y - x + 1)$, which means that $\text{MAG}(x', y) > \text{MAG}(x, y)$, contradicting the assumption that (x, y) is the maximum point. Similarly, if there is no token with $y_j = y$, we get into a similar contradiction. □

Observation 1 enables us to restricts the candidates for the maximum point to (x_i, y_j) for some $1 \leqslant i, j \leqslant n$. Let $P = \{x : x = x_i \text{ for some } i\}$ and $Q = \{y : y = y_j \text{ for some } j\}$. Let $s = |P|$ and $t = |Q|$. We denote the elements in P in increasing order by $(p_1, \ldots, p_s)$, and the elements in Q in increasing order by $(q_1, \ldots, q_t)$. The maximum point can be now found via a simple scan over the grid points in $P \times Q$ in $O(n + s \times t) = O(n^2)$ time. (Note that this time bound is independent of the size of the original grid, k.) In the following, we show that this $O(n^2)$ bound can be further improved, using a more clever pruning of the candidate points.

Let $\text{D}(i, j) \equiv \text{DOM}(p_i, q_j)$ be the dominance number, and $\mu(i, j) \equiv \text{MAG}(p_i, q_j)$ be the magnitude of the grid point (p_i, q_j) in our refined grid. The problem is to find a pair (i, j) for which $\mu(i, j)$ is maximum. We define $\text{GAP}(i, j, j') = \mu(i, j') - \mu(i, j)$ for $j' \geqslant j$. For example, in the grid shown in Figure 3, $\text{GAP}(3, 5, 7) = -4$ and $\text{GAP}(5, 5, 7) = 4$.

Lemma 2. *For indices $i < i'$ and $j < j'$, $\text{GAP}(i, j, j') \leqslant \text{GAP}(i', j, j')$ if there is no token at (x, y) such that $p_i < x \leqslant p_{i'}$ and $q_j \leqslant y < q_{j'}$.*

Proof. Omitted due to space constraints. To appear in the full version. □

For a pair of indices j and j' ($j < j'$), we say that a *flip* occurs between j and j' at index $i > 1$ if either

$$\text{GAP}(i - 1, j, j') \geqslant 0 \quad \text{and} \quad \text{GAP}(i, j, j') < 0,$$

or

$$\text{GAP}(i - 1, j, j') < 0 \quad \text{and} \quad \text{GAP}(i, j, j') \geqslant 0.$$

A flip satisfying the former condition is called a *type-1 flip*, and a *type-2 flip* otherwise. For example, in the grid shown in Figure 3, there is a type-2 flip between $j = 3$ and $j' = 5$ at $i = 2$, and there is a type-1 flip between $j = 5$ and $j' = 6$ at index $i = 3$.

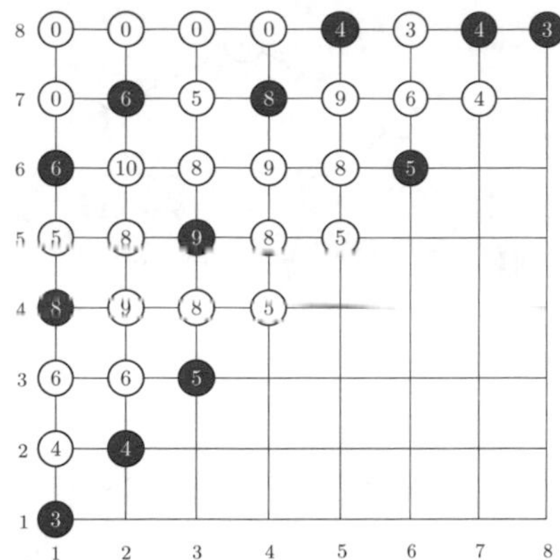

Fig. 3. A refined grid with tokens shown in black. The number on each grid point denotes its magnitude.

Lemma 3. *At any index i, a type-1 flip can occur between j and j' $(j < j')$ only if there is a token at (p_i, q_h) for some $j \leqslant h < j'$.*

Proof. This is a direct corollary of Lemma 2. □

Our idea for finding the maximum point is to construct a binary tree that maintains the maximum magnitude for the points having the same x-coordinate, and update the tree by using flips to find the overall maximum. The algorithm constructs a balanced binary tree B whose leaves are indices $1, \ldots, t$ in increasing order. For an internal node x of B, we denote the set of leaves in the subtree rooted at x by $\text{DES}(x)$, the left child of x by $l(x)$ (corresponding to smaller indices), and the right child of x by $r(x)$ (corresponding to larger indices). Note that $\text{DES}(x)$ forms an interval of indices. We denote the largest index among $\text{DES}(x)$ by $u(x)$, and the smallest index by $b(x)$. For an index i, we define $\mathfrak{J}_i^x$ to be the index j maximizing $\mu(i,j)$ among all indices in $\text{DES}(x)$. Ties are broken by choosing the largest index. (See Figure 4.) The algorithm iteratively increases i one by one, and at each step i, updates $\mathfrak{J}_i^x$ when $\mathfrak{J}_{i-1}^x \neq \mathfrak{J}_i^x$.

Observation 2. *$\mathfrak{J}_{i-1}^x \neq \mathfrak{J}_i^x$ only if either* (a) $\mathfrak{J}_{i-1}^{l(x)} \neq \mathfrak{J}_i^{l(x)}$, (b) $\mathfrak{J}_{i-1}^{r(x)} \neq \mathfrak{J}_i^{r(x)}$, *or* (c) *there is a flip between $\mathfrak{J}_{i-1}^{l(x)}$ and $\mathfrak{J}_{i-1}^{r(x)}$ at index i.*

By this observation, cases (a) and (b) need a descendant of x which involves case (c). Therefore, to find updates for all nodes in B, we only have to find occurrences of case (c). In the example shown in Figure 4, for the node labeled z we have $\mathfrak{J}_1^z \neq \mathfrak{J}_2^z$ because there is a type-1 flip between leaves 7 and 8 at $i = 2$. Moreover, for the root of the tree, r, we have $\mathfrak{J}_1^r \neq \mathfrak{J}_2^r$ because a type-2 flip occurs between leaves 4 and 6 at $i = 2$.

In each step i, we say that a type-1 *event* (resp., a type-2 event) occurs at a node x, if there is type-1 (resp., type-2) flip between $\mathfrak{J}_i^{l(x)}$ and $\mathfrak{J}_i^{r(x)}$ at index i. We also say that a type-0 event occurs at a node x in step i if there is a token

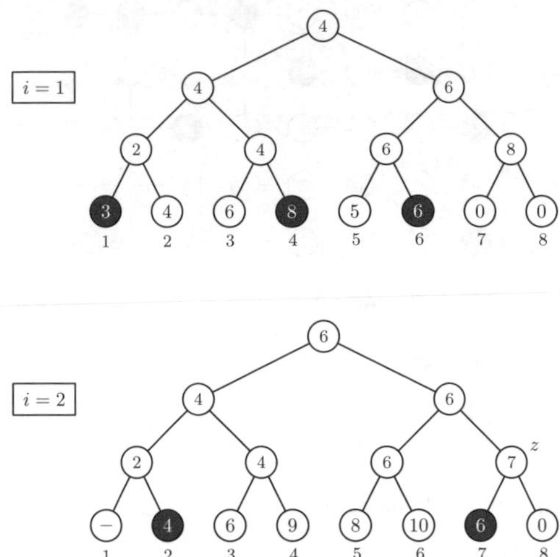

Fig. 4. Binary trees corresponding to the first two columns of the grid depicted in Figure 3. Numbers in the leaves are magnitudes. For each internal node x, the number shown in x is $\mathfrak{J}_i^x$.

at (p_i, q_j) for some $\mathfrak{J}_i^{l(x)} \leqslant j < \mathfrak{J}_i^{r(x)}$. By Lemma 3, any type-1 event is also a type-0 event. Therefore, we only need to consider type-0 and type-2 events.

To find events in each step efficiently, we keep the earliest coming event for each internal node. More precisely, for each node x and index i, we define the *next event* of x by the smallest index i' such that $i' > i$ and either a type-0 or a type-2 event occurs at x in step i'. If no such index i' exists, we say that x has no next flip. In each i-th iteration, we maintain the next event for each internal node, and thus, we can find all nodes at which an event occurs in each step by looking only at the next events.

The complete procedure for computing the maximum point is presented in Algorithm 1. In this algorithm, a node x called an *update node* if an event occurs at some descendant of x.

Algorithm 1. FINDMAXPOINT(S)

```
Initialize:
    build a binary tree B with leaves 1, ..., t
    set 𝔍_1^ℓ = ℓ for each leaf ℓ of B
    mark all internal nodes of B as update nodes for step i = 1
for i = 1 to s do
    for each update node x at step i in the bottom-up order do
        compute μ(i, 𝔍_i^{l(x)}) and μ(i, 𝔍_i^{r(x)})
        update 𝔍_i^x
        compute the next event of x
return max_{i=1}^{s} μ(i, 𝔍_i^{root(B)})
```

Lemma 4. *The total number of update nodes during the execution of Algorithm 1 is $O(n \log^2 n)$.*

Proof. The number of ancestors of a leaf is $O(\log n)$, thus the number of type-0 events is $O(n \log n)$ in total during the execution of the algorithm. We show below that the same bound applies to the number of type-2 events.

Consider the sequence of events occurring at a node x. We show that in this sequence, there is at most one type-2 event between any two consecutive type-0 events. Fix a node x, and consider two consecutive type-0 events occurring at x, say in steps a and b. Since no other type-0 event occurs in between a and b, by Lemma 3 there is no token at (p_i, q_j) for all $a < i < b$ and $b(x) \leqslant j < u(x)$.

Let c be the first step, $a < c < b$, at which a type-2 event occurs at x. We prove that for all subsequent steps i, $c < i < b$, no type-2 event can occur at x. The proof is based on the following claim:

Claim. For all $c < i < b$, $\mu(i, \mathfrak{J}_i^{l(x)}) \leqslant \mu(i, \mathfrak{J}_i^{r(x)})$.

To show this, fix an i such that $c < i < b$. We have

$$\mu(c, \mathfrak{J}_i^{l(x)}) \;\leqslant\; \mu(c, \mathfrak{J}_c^{l(x)}) \;\leqslant\; \mu(c, \mathfrak{J}_c^{r(x)}), \tag{1}$$

where the right-hand inequality holds because a type-2 flip has occurred at x in step c, and the left-hand inequality holds because $\mathfrak{J}_c^{l(x)}$ points to the maximum leaf in $\text{DES}(l(x))$ in step c. Using (1) and Lemma 2 we get

$$\mu(i, \mathfrak{J}_i^{l(x)}) \;\leqslant\; \mu(i, \mathfrak{J}_c^{r(x)}), \tag{2}$$

because there is no token at (p_h, q_j) for all $c < h \leqslant i$ and $b(x) \leqslant j < u(x)$. Using (2) and the fact that $\mu(i, \mathfrak{J}_c^{r(x)}) \leqslant \mu(i, \mathfrak{J}_i^{r(x)})$, we obtain the claim statement.

It thus follows that the number of type-2 events does not exceed the number of type-0 events, and therefore, we have $O(n \log n)$ events in total. Since each event can be involved in at most $O(\log n)$ update nodes, the total number of update nodes during the execution of the algorithm is $O(n \log^2 n)$. □

Theorem 1. *Algorithm 1 solves the maximum point problem for a set of n tokens on a grid in $O(n \log^3 n \log\log n)$ time and $O(n)$ space.*

Proof. The correctness of the algorithm follows directly from the fact that at each step i, $\mathfrak{J}_i^{root(B)}$ maintains the location of the maximum point in column i, and therefore, line 7 of the algorithm returns the maximum point in the whole grid. The running time of Algorithm 1 is dominated by the time needed by the two for-loops. The computation of the magnitude $\mu(i, j)$ in line 4 involves computing $\text{D}(i, j)$ which can be done in $O(\log\log n)$ time using an orthogonal range search on the refined grid [10]. For any node x, the computation of the next type-0 event involves finding the earliest coming token in the rectangle

$[i \mathinner{..} s] \times [\mathfrak{I}_i^{l(x)} \mathinner{..} \mathfrak{I}_i^{r(x)}]$, and the computation of the next type-2 event involves a binary search on index i with the computation of $O(\log n)$ magnitudes, both can be done in $O(\log n \log\log n)$ time. By Lemma 4, the two for-loops together iterate $O(n \log^2 n)$ times. Therefore, the running time of the algorithm is $O(n \log^3 n \log\log n)$ in total. The binary tree has $O(n)$ nodes and each node requires memory of constant size to keep the maximum index and the next event. The space complexity is therefore $O(n)$. Note that the initialization step involves sorting tokens by their integer coordinates, constructing the binary tree B, and initializing the data structure for orthogonal range search, all of which can be done in $O(n)$ time and $O(n)$ space. The proof of the theorem is thus complete. □

The following is a direct corollary of Lemma 1 and Theorem 1.

Theorem 2. *Given a convex bipartite graph $G = (A, B, E)$, the maximum edge biclique of G can be computed in $O(n \log^3 n \log\log n)$ time and $O(n)$ space, where $n = |A|$.*

5 Discussion and Conclusions

In this paper, we presented an efficient algorithm for solving the maximum edge biclique problem in convex bipartite graphs in $O(n \log^3 n \log\log n)$ time. The objective function used in our algorithm was $\text{DOM}(i,j) \times (j-i+1)$. The algorithm works as long as the monotonicity of the GAP function is preserved. Therefore, we can generalize the objective function to any arbitrary function of the form $f(\text{DOM}(i,j)) \times g(j-i)$ such that f is monotone increasing in $\text{DOM}(i,j)$, and g is monotone increasing in j.

Better running times can be obtained for special subclasses of convex bipartite graphs. In particular, we have shown that by reducing the maximum edge biclique problem to the problem of finding the largest-area rectangle inside a simple polygon with certain properties, we can solve the maximum edge biclique problem in biconvex graphs and bipartite permutation graphs in $O(n\alpha(n))$ and $O(n)$ time, respectively, where $n = \min(|A|, |B|)$, and $\alpha(n)$ is the slowly growing inverse of the Ackermann function. Details will appear in the full version.

Some problems remain open. An immediate problem is whether we can improve the running time of the algorithm presented in this paper by removing (some of) the logarithmic factors. Finding a better algorithm for the maximum edge biclique problem in chordal bipartite graphs (which is a direct superclass of convex bipartite graphs) is another interesting open problem.

References

1. Alexe, G., Alexe, S., Crama, Y., Foldes, S., Hammer, P.L., Simeone, B.: Consensus algorithms for the generation of all maximal bicliques 145(1), 11–21 (2004)
2. Ben-Dor, A., Chor, B., Karp, R., Yakhini, Z.: Discovering local structure in gene expression data: the order-preserving submatrix problem. J. Comput. Biol. 10(3-4), 373–384 (2003)

3. Booth, K.S., Lueker, G.S.: Testing for the consecutive ones property, interval graphs, and graph planarity using PQ-tree algorithms. J. Comput. Syst. Sci. 13(3), 335–379 (1976)
4. Brodal, G., Georgiadis, L., Hansen, K., Katriel, I.: Dynamic matchings in convex bipartite graphs, pp. 406–417 (2007)
5. Chen, Y., Church, G.: Biclustering of expression data. In: Proc. 8th Internat. Conf. Intelligent Systems for Molecular Biology, pp. 93–103 (2000)
6. Dawande, M., Keskinocak, P., Swaminathan, J.M., Tayur, S.: On bipartite and multipartite clique problems 41(2), 388–403 (2001)
7. Dias, V.M., de Figueiredo, C.M., Szwarcfiter, J.L.: Generating bicliques of a graph in lexicographic order. Theoretical Comput. Sci. 337(1-3), 240–248 (2005)
8. Dias, V.M., de Figueiredo, C.M., Szwarcfiter, J.L.: On the generation of bicliques of a graph 155(14), 1826–1832 (2007)
9. Feige, U.: Relations between average case complexity and approximation complexity, pp. 534–543 (2002)
10. Fries, O., Mehlhorn, K., Näher, S., Tsakalidis, A.: A $\log\log n$ data structure for three-sided range queries. Inf. Process. Lett. 25(4), 269–273 (1987)
11. Ganter, B., Wille, R.: Formal Concept Analysis, Mathematical Foundations. Springer, Berlin (1996)
12. Gély, A., Nourine, L., Sadi, B.: Enumeration aspects of maximal cliques and bicliques 157(7), 1447–1459 (2009)
13. Glover, F.: Maximum matching in a convex bipartite graph 14, 313–316 (1967)
14. Goerdt, A., Lanka, A.: An approximation hardness result for bipartite clique. In: Technical Report 48, Electronic Colloquium on Computation Complexity (2004)
15. Habib, M., McConnell, R., Paul, C., Viennot, L.: Lex-BFS and partition refinement, with applications to transitive orientation, interval graph recognition and consecutive ones testing. Theoretical Comput. Sci. 234(1-2), 59–84 (2000)
16. Kloks, T., Kratsch, D.: Computing a perfect edge without vertex elimination ordering of a chordal bipartite graph. Inf. Process. Lett. 55(1), 11–16 (1995)
17. Liang, Y.D., Chang, M.: Minimum feedback vertex sets in cocomparability graphs and convex bipartite graphs 34(5), 337–346 (1997)
18. Lipski, W., Preparata, F.P.: Efficient algorithms for finding maximum matchings in convex bipartite graphs and related problems 15(4), 329–346 (1981)
19. Madeira, S.C., Oliveira, A.L.: Biclustering algorithms for biological data analysis: A survey 1(1), 24–45 (2004)
20. Meidanis, J., Porto, O., Telles, G.P.: On the consecutive ones property 88(1-3), 325–354 (1998)
21. Mishra, N., Ron, D., Swaminathan, R.: On finding large conjunctive clusters. In: Proc. 16th Annu. Conf. Computational Learning Theory, pp. 448–462 (2003)
22. Peeters, R.: The maximum edge biclique problem is NP-complete 131(3), 651–654 (2003)
23. Soares, J., Stefanes, M.: Algorithms for maximum independent set in convex bipartite graphs 53(1), 35–49 (2009)
24. Steiner, G., Yeomans, J.S.: A linear time algorithm for maximum matchings in convex, bipartite graphs. Comput. Math. Appl. 31(12), 91–96 (1996)
25. Tan, J.: Inapproximability of maximum weighted edge biclique and its applications. In: Agrawal, M., Du, D.-Z., Duan, Z., Li, A. (eds.) TAMC 2008. LNCS, vol. 4978, pp. 282–293. Springer, Heidelberg (2008)
26. Tanay, A., Sharan, R., Shamir, R.: Discovering statistically significant biclusters in gene expression data. Bioinformatics 18(Supplement 1), S136–S144 (2002)

A Note on Vertex Cover in Graphs with Maximum Degree 3

Mingyu Xiao*

School of Computer Science and Engineering
University of Electronic Science and Technology of China
Chengdu 610054, China
myxiao@gmail.com

Abstract. We show that the k-Vertex Cover problem in degree-3 graphs can be solved in $O^*(1.1616^k)$ time, which improves previous results of $O^*(1.1940^k)$ by Chen, Kanj and Xia and $O^*(1.1864^k)$ by Razgon. In this paper, we will present a new way to analyze algorithms for the problem. We use $r = k - \frac{2}{5}n$ to measure the size of the search tree, and then get a simple $O(1.6651^{k-\frac{2}{5}n_0})$-time algorithm, where n_0 is the number of vertices with degree ≥ 2 in the graph. Combining this result with fast algorithms for the Maximum Independent Set problem in degree-3 graphs, we improve the upper bound for the k-Vertex Cover problem in degree-3 graphs.

Keywords: Parameterized Algorithms, Vertex Cover, Independent Set, Sparse Graphs.

1 Introduction

Given a graph $G = (V, E)$ with n vertices and m edges, and an integer k, a *vertex cover* in G is a set of vertices such that each edge has at least one endpoint in it, and the k-*Vertex Cover* problem is to decide if G has a vertex cover of size at most k. A graph with maximum degree 3 is called a degree-3 graph. We may use VC and VC-3 to denote the k-Vertex Cover problem in general graphs and degree-3 graphs respectively.

The k-Vertex Cover problem is one of the most extensively studied problems in the line of research on exact and parameterized algorithms for NP-hard problems. There is a long list of continuous improvements on the combinatorial explosion with respect to the size k of the vertex cover when solving the problem exactly. The currently best exponential bound is $O^*(1.2738^k)$ due to Chen *et al.* [1].

One of the most important cases in the k-Vertex Cover problem is the problem in low-degree graphs. In fact, many effective algorithms in the literature are based on careful analysis on the problem in low-degree graphs [1,2]. To a certain

* This work was supported in part by National Natural Science Foundation of China Grant No. 60903007.

M.T. Thai and S. Sahni (Eds.): COCOON 2010, LNCS 6196, pp. 150–159, 2010.

extent, the k-Vertex Cover problems in low-degree graphs are the bottleneck for improving the algorithm for the problem in general graphs. It is not surprising that the k-Vertex Cover problems in low-degree graphs have been extensively studied [2,3,4]. This situation is also true for algorithms for a closely related problem—the *Maximum Independent Set* problem (to find a maximum size set of vertices in a graph such that no edge has both endpoints in the set).

In this paper, we concentrate on VC 3 and present an $O^*(1.6651^{k-\frac{2}{5}n_0})$-time algorithm for the problem, where n_0 is the number of vertices of degree ≥ 2 in the graph. In fact, there is a technique to reduce vertices of degree ≤ 1 directly, we can simply assume that the graph has no vertices of degree ≤ 1 and then $n_0 = n$. Obviously, when k is close to n, our algorithm is effective. Otherwise, we can use fast algorithms for finding a maximum independent set in the graph and then get a minimum vertex cover by taking all the remaining vertices. For instance, if $n \geq 1.7659k$, we use our algorithm in this paper to solve the problem in $O^*(1.6651^{k-\frac{2}{5}n}) = O^*(1.1616^k)$ time, else we use the $O^*(1.0885^n)$-time algorithm presented in [5] to find a maximum independent set in degree-3 graphs. Then we can always solve VC-3 in $O^*(1.1616^k)$ time, which improves previous results of $O^*(1.1940^k)$ by Chen, Kanj and Xia [3] and $O^*(1.1864^k)$ by Razgon [4]. In this paper, we use a modified O notation that suppresses all polynomially bounded factors. For two functions f and g, we write $f(n) = O^*(g(n))$ if $f(n) = g(n)n^{O(1)}$.

Like many exact and parameterized algorithms for solving NP-hard problems, our algorithm will use a branch and reduce paradigm to construct a search tree. Each node of the search tree corresponds to a sub instance of the problem and the exponential part of the time complexity of the algorithm is roughly equivalent to the size of the search tree. There are two basic ways to reduce the size of the search tree: effectively enumerate feasible partial solutions and efficiently reduce the substance sizes. To do so, we will use some reduction and branching rules. The running time analysis leads to a linear recurrence with respect to a measure for each branching rule that can be solved by using standard techniques. For graph algorithms, the number of vertices or edges in the graph is a natural measure. For parameterized algorithms, the parameter k is a natural measure.

In this paper, we will use a combination of vertex number and vertex cover size in the graph as a measure to estimate the size of our search tree. We note that if an n-vertex graph has maximum degree 3, minimum degree ≥ 2 and a vertex over of size k, then $k \geq \frac{2}{5}n$ (see Corollary 1). Motivated by this, we analyze our algorithm by using $r = k - \frac{2}{5}n$ as a measure and explain how efficiently r is reduced in each sub instance. The main idea of our algorithm and its analysis are simple enough to describe here:

If there are any degree-1 or degree-2 vertices, we apply some simple reduction rules to deal with them until the graph becomes a 3-regular graph or has a vertex of degree ≥ 4. Then we branch by either including some vertices in the vertex cover or excluding some vertices from the vertex cover. To analyze the algorithm, we let $C(r)$ denote the worst-case size of the search tree in our algorithm when the parameter of the graph is r. When the maximum degree of the graph is greater than 3, we can always branch with recurrence relation $C(r) \leq C(r - \frac{4}{5}) + C(r - \frac{11}{5})$. When the graph is a 3-regular graph, we can branch on a vertex with recurrence relation $C(r) \leq C(r - \frac{3}{5}) + C(r - \frac{7}{5})$, and before branching we can reduce r by at least $\frac{2}{5}$. Considering all these together, we know that in the worst case we can branch with $C(r) \leq C(r-1) + C(r - \frac{9}{5})$. By solving the recurrence, we get $C(r) = O(1.6651^r)$.

2 Preliminaries

2.1 Structural Properties

The following structural properties are simple and intuitive, but crucial for our analysis.

Lemma 1. *If an n-vertex graph has maximum degree d, minimum degree ≥ 2, and a vertex cover of size k, then*

$$k \geq \frac{2}{2+d}n. \tag{1}$$

Proof. Let S be a vertex cover in the graph $G = (V, E)$ and $k = |S|$. We consider the edges with one endpoint in S and one in $V - S$. Clearly, $V - S$ is an independent set and each vertex in $V - S$ has degree ≥ 2. Then there are at least $2|V - S| = 2(n - k)$ edges between S and $V - S$. On the other hand, each vertex in S has degree $\leq d$. Then

$$2(n-k) \leq dk.$$

Therefore, we get (1). ∎

Corollary 1. *If an n-vertex graph has maximum degree* 3, *minimum degree* ≥ 2, *and a vertex cover of size k, then*

$$k \geq \frac{2}{5}n. \tag{2}$$

We will use $r = k - \frac{2}{5}n$ as a measure to analyze our algorithm for VC-3. When $r \leq 0$, the problem can be solved in polynomial time, and in each step of our algorithm r will decrease. Therefore, we get a bounded search tree.

In fact, there are many good kernelization results on the k-Vertex Cover problem that show some better relations between k and n [6,7]. After executing a kernelization algorithm in polynomial time, we can get bound $k \geq \frac{1}{2}n$, which is tighter than (1) and (2). In this paper, we do not use $r' = k - \frac{1}{2}n$ as a measure, because the parameter r' may increase when we apply the kernelization algorithms.

2.2 Simple Reduction Rules

Before explaining our algorithm, we introduce some reduction rules to deal with degree-1 and degree-2 vertices and some other local structures. Most of these reduction rules are well-known in the literature.

Folding a degree-1 vertex
When there is a degree-1 vertex v, we can put the unique neighbor u of v into the vertex cover and reduce k by 1 and remove u (also all the edges incident on it and degree-0 vertices) from the graph.

Removing a dominated vertex
We say that vertex v dominates *vertex u, if $N[v] \subseteq N[u]$, where $N[v] = N(v) \cup \{v\}$ is the set of vertices with distance at most 1 from v. If v dominates u, we can put u into the vertex cover and reduce k by 1 and remove u.*

In fact, folding a degree-1 vertex is a special case of removing a dominated vertex, we distinguish them in this paper just for the analysis purpose.

Folding a degree-2 vertex
Let v be a degree-2 vertex that does not dominate any other vertices, and u and w the two neighbors of v. Then, we can construct a new graph by removing v, u and w from G and introducing a new vertex s that is adjacent to all neighbors of u and w in G (except the removed vertex v), and reduce k by 1.

Note that in definition of the folding a degree-2 vertex operation, we assume that v does not dominate any other vertices, and then u and w are not adjacent to each other. In our algorithm, when we fold a degree-2 vertex, all dominated vertices have been removed.

The above three reduction rules are frequently used in the literature [1,8]. We omit the proofs for correctness here. Note that after folding a degree-2 vertex, it is possible to create a vertex of degree ≥ 4. When a vertex of degree ≥ 4 is created in our algorithm, we will deal with it immediately.

2.3 General Crown Decomposition

Besides the above simple reduction rules, we still use crown reduction rules to reduce the input size and simplify our analysis. Crown reduction rules are effective and frequently used techniques to design parameterized algorithms and kernelization algorithms. It was firstly introduced by Chor, Fellows and Juedes [9] and has applications in many problems [10]. In a crown decomposition we can simply select some vertices into the vertex cover directly. In this paper we will use a *general crown decomposition* and call the crown decomposition introduced in [9] the *proper crown decomposition.*

Definition 1. *A general crown decomposition of graph G is a partition of the vertex set of G into three sets, H, C and J such that the following conditions hold:*

(1) $H \neq \emptyset$ and $C \neq \emptyset$,
(2) H is an independent set, and
(3) C is a cutset of H and J, i.e. there are no edges between H and J.

Definition 2. *A proper crown decomposition of graph G is a crown decomposition such that C is matched into H, i.e., there is an injective assignment $m : C \longrightarrow H$ such that $\forall c \in C$, c is adjacent to $m(c)$.*

Lemma 2. *Let G be a graph. If G admits a proper crown decomposition (H, C, J), then G has a vertex cover of size k if and only if G' has a vertex cover of size k', where $G' = G - H - C$ and $k' = k - |C|$.*

By Lemma 2 we can reduce a proper crown decomposition in a k-Vertex Cover instance by removing H and C from the graph and putting C into the vertex cover. This reduction rule can be regarded as a generalization of the degree-1 vertex reduction rule. Here arises the problem how to effectively determine if a graph has a proper crown decomposition.

Lemma 3. *If graph $G = (V, E)$ admits a general crown decomposition (H, C, J) such that $|C| \leq |H|$, then a proper crown decomposition (H', C', J') with $H' \subseteq H$ for G can be found in time $O(|V| + |E|)$.*

The lemma for the case $|C| < |H|$ was proved in [9,11]. In fact, the lemma still holds for $|C| = |H|$. One can prove the result by slightly modifying the constructive proofs in [9,11].

Chen, Kanj and Xia have also studied a crown structure in [1]. They have proved the following lemma and called the corresponding operation *general folding operation.*

Lemma 4. *Let (H, C, J) be a general crown decomposition of graph G, if $|C| = |H| + 1$ and for each $\emptyset \neq S \subseteq H$ we have $|N(S)| > |S| + 1$, then*

(1) when the graph induced by C is not an independent set, G has a vertex cover of size k if and only if G' has a vertex cover of size k', where $G' = G - H - C$ and $k' = k - |C|$.
(2) when the graph induced by C is an independent set, G has a vertex cover of size k if and only if G' has a vertex cover of size $k - |H|$, where G' is the graph obtained from G by shrinking $H \cup C$ into a single vertex and removing all self-loops and parallel edges.

As mentioned in [1], a general crown decomposition described in Lemma 4 can be found by using the $O(|V|+|E|)$-time algorithm presented in [9,11]. By putting above together, we get

Lemma 5. *Let (H, C, J) be a general crown decomposition of graph $G = (V, E)$, if $|C| = |H| + 1$, then we can use $O(|V| + |E|)$ time to find a proper crown decomposition (H', C', J') of G such that $\emptyset \neq H' \subseteq H$ or prove that (H, C, J) is a general crown decomposition satisfying the condition in Lemma 4.*

By Lemma 3, Lemma 4 and Lemma 5, we know that when there is a general crown decomposition (H, C, J) such that $|C| \leq |H| + 1$, we can use some crown reduction rules to reduce the graph without branching. For convenience, we will call such kind of general crown decompositions *foldable crown decompositions.*

In our algorithm, we do not need to find out all foldable crown decompositions before branching. We only check whether a vertex set V' *implies* a foldable crown decomposition when putting V' into the vertex cover in a branching operation. Assume that in a subbranch V' is put into the vertex cover and removed from the graph. Then if any dominated vertices are created in the remaining graph, we will further remove them until no such vertices exist. We consider a general crown decomposition (H, C, J) of G created by the above operations, where $C \supseteq V'$ is the set of vertices being put into the vertex cover and H the set of degree-0 vertices created after removing C from G. We say that V' *implies* a foldable crown decomposition, if $|C| \leq |H| + 1$ and $H \neq \emptyset$.

3 The Simple Algorithm

In our algorithm, we will branch on a vertex by including it in the vertex cover or excluding it from the vertex cover. If a vertex is included in the vertex cover, we delete it (together with all incident edges and degree-0 vertices), if a vertex is excluded from the vertex cover, we put the neighbor set of it in the vertex cover and delete the neighbor set. Before branching, we also require the graph satisfying the following **Degree Conditions**: either (i) the graph is a 3-regular graph, or (ii) the graph has a vertex v of degree ≥ 4 such that neighbors of v are either degree-3 or degree-4 vertices, and all degree-3 neighbors of v are not adjacent to any vertex of degree ≤ 2, and all vertices other than $N(v) \cup \{v\}$ are of degree ≤ 3. For convenience, we will call the vertex v in the above definition the *central vertex*. We also say any vertex in a 3-regular graph a *central vertex*. The Degree Conditions will be used to simplify the analysis and avoid high-degree vertices after branching. It is easy to see that we can iteratively fold degree-2 vertices in the graph to get a graph satisfying the Degree Conditions.

1. Fold degree-1 vertices until no such kind of vertices exist.
2. If $k < \frac{2}{5}n$, halt.
3. Remove dominated vertices and fold degree-2 vertices until the graph satisfying the Degree Conditions.
4. Pick up a central vertex v.
5. If $\{v\}$ or $N(v)$ implies a foldable crown decomposition, reduce the graph without branching.
6. Branch on v by including it in the vertex cover or excluding it from the vertex cover.

Fig. 1. The Main Steps of Our Algorithm for Vertex Cover

Our algorithm is presented in Fig. 1. Then correctness of the algorithm follows from the following observations. After branching on a central vertex in a graph satisfying the Degree Conditions, the graph in each subbranch has no vertex of degree ≥ 4, and once vertices of degree 1 are created, the algorithm will reduce them in Step 1. Then we can use $r = k - \frac{2}{5}n$ as a measure to analyze the algorithm. We will show that r will not increase when we apply reduction rules (see Lemma 6) and r will decrease after branching in Step 6 and reducing degree-1 vertices in Step 1. Then the algorithm will always stop. To analyze the running time, we need to show how efficiently r can be reduced in each operation in the algorithm.

Lemma 6. *Let G be a graph with maximum degree ≤ 3 and minimum degree ≥ 2. Then after applying any of the simple reduction rules or crown reduction rules, the parameter $r = k - \frac{2}{5}n$ will not increase.*

Proof. When a reduction rule creates new vertex of high degree, the parameter r will not increase. Otherwise, some vertices will be selected into the vertex cover. Let C be the set of vertices being put into the vertex cover in the operations, S the set of degree-0 vertices created after removing C, and $G' = G - C - S$ the remaining graph. Clearly, in G', the number of vertices is $n' = n - |C \cup S|$ and the parameter k' equals $k - |C|$. Since each vertex in S is of degree ≥ 2 and each vertex in C is of degree ≤ 3 (if some vertices of degree ≥ 4 are created when folding a degree-2 vertex or folding a crown decomposition of case 2 in Lemma 4, we simply unfold them), we get $2|S| \leq 3|C|$. Then in G',

$$r' = k' - \frac{2}{5}n' = k - |C| - \frac{2}{5}n + \frac{2}{5}(|C| + |S|) = r + \frac{1}{5}(2|S| - 3|C|) \leq r.$$

∎

Lemma 7. *After folding a degree-2 vertex, we can reduce parameter r by $\frac{1}{5}$ without creating any degree-0 or degree-1 vertices.*

Since after executing the folding operation, the number of vertices will decrease by 2 and k will decrease by 1, then r will decrease by $1 - \frac{2}{5} \times 2 = \frac{1}{5}$. It is easy to see that after executing the operation, no degree-0 or degree-1 vertices will be created, but a vertex of higher degree may be created.

We assume that the initial graph satisfying: (1) having maximum degree ≤ 3 and minimum degree 2; (2) having two nonadjacent degree-2 vertices.

Lemma 8. *Let v be a central vertex in a graph satisfying the Degree Conditions and $N(v)$ does not imply a foldable crown decomposition, then the graph after removing $N(v)$ and iteratively folding degree-1 vertices has two nonadjacent degree-2 vertices.*

Proof. We consider the decomposition (H, C, J) of G, where C is the set of vertices being put into the vertex cover and H the set of degree-0 vertices created after removing C from G. Clearly, the graph $G' = G - H - C$ has not any degree-1 vertices. Since $|H| \leq |C| - 2$ and all vertices in C are of degree ≥ 2, there are at least four edges between C and J. Then after removing $H \cup C$, G' has at least four degree-2 vertices and at least one pair of them are nonadjacent. ∎

It is also easy to see that if the graph after removing v has not two nonadjacent degree-2 vertices, the only case is that v is a vertex of degree ≥ 4.

Now we are ready to analyze the running time of the algorithm. In Step 6, our algorithm will branch on a central vertex v by including it in the vertex cover or excluding it from the vertex cover. First, we look at the branch where v is included in the vertex cover. Since the graph satisfies the Degree Conditions and v is a central vertex, we know that no degree-1 vertices will be created after deleting v. In this branch, n decreases by 1 and k decreases by 1, and then r decreases by $1 - \frac{2}{5} = \frac{3}{5}$. In the branch where v is excluded from the vertex cover ($N[v]$ is deleted), we consider two cases. Case 1: no degree-1 vertex is created after deleting $N(v)$. For this case, n decreases by $|N(v)| + 1$ and k decreases by $|N(v)|$, and then r decreases by $|N(v)| - \frac{2}{5}(|N(v)|+1) = \frac{3}{5}|N(v)| - \frac{2}{5}$. Case 2: some degree-1 vertices are created after deleting $N(v)$. For this case, our algorithm will execute Step 1 to remove them. Let (H, C, J) be the decomposition of G, where C is the set of vertices being put into the vertex cover in the two steps (Step 6 and Step 1) and H the set of degree-0 vertices created after removing C from G. Then we have $|H| \leq |C| - 2$, otherwise $N(v)$ implies a foldable crown decomposition and the graph will be reduced in Step 5. Note that in the operations n decreases by $|H| + |C|$ and k decreases by $|C|$. Totally, r decreases by

$$|C| - \frac{2}{5}(|H| + |C|) \geq |C| - \frac{2}{5}(2|C| - 2) = \frac{1}{5}|C| + \frac{4}{5}, \tag{3}$$

where $|C| \geq |N(v)| + 1 \geq 5$. Furthermore, we know that for this case, some neighbor of v is a degree-4 vertex since the graph satisfies the Degree Conditions.

If v is a vertex of degree ≥ 4, and no degree-1 vertex is created after deleting $N(v)$. Then we can branch by reducing r by either $\frac{3}{5}$ or $(\frac{3}{5} \times 4 - \frac{2}{5}) = 2$. Note that before branching, we need to fold a degree-2 vertex to get the degree-4 vertex v and reduce r by $\frac{1}{5}$ in the operation. Thus, we get the following recurrence relation

$$C(r) \leq C(r - \frac{1}{5} - \frac{3}{5}) + C(r - \frac{1}{5} - 2) = C(r - \frac{4}{5}) + C(r - \frac{11}{5}). \tag{4}$$

If v is a vertex of degree ≥ 4, and some degree-1 vertex is created after deleting $N(v)$. Then we can branch by reducing r by either $\frac{3}{5}$ or $(\frac{1}{5} \times 5 + \frac{4}{5}) = \frac{9}{5}$ by (3). Before branching, we reduce r by at least $\frac{2}{5}$ to get two vertices of degree ≥ 4 (v and a neighbor of v). Then get the recurrence relation

$$C(r) \leq C(r - \frac{2}{5} - \frac{3}{5}) + C(r - \frac{2}{5} - \frac{9}{5}) = C(r - 1) + C(r - \frac{11}{5}). \tag{5}$$

If v is a degree-3 vertex, then Case 2 in the branch where $N(v)$ is removed will not happen. We can branch by reducing r by either $\frac{3}{5}$ or $(\frac{3}{5} \times 3 - \frac{2}{5}) = \frac{7}{5}$. We will show that we can reduce r by at least $\frac{2}{5}$ (amortized cost) before branching on v. Therefore, we can branch with recurrence relation

$$C(r) \leq C(r - \frac{2}{5} - \frac{3}{5}) + C(r - \frac{2}{5} - \frac{7}{5}) = C(r - 1) + C(r - \frac{9}{5}). \tag{6}$$

To show that, we consider the following three cases. Case I: the graph has two nonadjacent degree-2 vertices before branching v. For this case, we can reduce r by at least $\frac{2}{5}$ by folding degree-2 vertices directly. Case II: the graph has only two adjacent degree-2 vertices before branching v. By Lemma 8 and the discussion below Lemma 8, we know that the graph is obtained by removing a vertex v' of degree ≥ 4 from G', where v' has at least two neighbors of degree 4 in G'. Note that in the analysis of branching on v' we only use at most one degree-4 neighbor of v', then we can 'save' cost $\frac{1}{5}$ (from another degree-4 neighbor of v') as *credit* to pay when branching on v. We can also reduce r by $\frac{1}{5}$ by folding the two adjacent degree-2 vertices. Totally, we can reduce r by at least $\frac{2}{5}$ before branching. Case III: the graph has at most one degree-2 vertex before branching v. By Lemma 8 and the discussion below Lemma 8, we know that the graph is obtained by removing a vertex v' of degree ≥ 4 from G', where v' has at least three neighbors of degree 4 in G'. Then we can get credit cost at least $\frac{2}{5}$ from the neighbors of v'. Therefore, for any case, we can reduce r by $\frac{2}{5}$ before branching.

Since (6) covers (4) and (5), and $C(r) = O(1.6651^r)$ satisfies (6), we get

Theorem 1. *The k-Vertex Cover problem in an n-vertex graph with maximum degree* 3 *and minimum degree* ≥ 2 *can be solved in* $O^*(1.6651^{k-\frac{2}{5}n})$ *time.*

Next, we combine our algorithm with some fast algorithms for the Maximum Independent Set problem in degree-3 graphs to get the final running time bound for solving the k-Vertex Cover problem in degree-3 graphs.

Clearly, if the Maximum Independent Set problem in an n-vertex degree-3 graph can be solved in $O^*(c^n)$ time, then the k-Vertex Cover problem in an n-vertex graph with maximum degree 3 and minimum degree ≥ 2 can be solved in $O^*(\min(c^n, 1.6651^{k-\frac{2}{5}n}))$ time. There are many fast algorithms for the Maximum Independent Set problem in degree-3 graphs [12,4,5]. We will use the result $c \leq 1.0885$ introduced in [5]. Suppose $n = t \cdot k$ and let $1.0885^n = 1.6651^{k-\frac{2}{5}n}$, then we get $t = 1.7659$ and $1.0885^n = 1.6651^{k-\frac{2}{5}n} = 1.1616^k$.

Given a degree-3 graph, we first fold all the degree-1 vertices. Then when $n \geq 1.7659 \cdot t$, we apply the algorithm presented in the paper; when $n < 1.7659 \cdot t$, we use the simple $O^*(1.0885^n)$-time algorithm presented in [5] to find a minimum vertex cover. Therefore,

Theorem 2. *The k-Vertex Cover problem in degree-*3 *graphs can be solved in* $O^*(1.1616^k)$ *time.*

Our result improves the previous results of $O^*(1.1940^k)$ by Chen, Kanj and Xia [3] and $O^*(1.1864^k)$ by Razgon [4].

4 Concluding Remarks

In this paper, we have presented simple and improved algorithms for the k-Vertex Cover problem in degree-3 graphs. To analyze them, we adopt a combination parameter as a measure of the size of the search tree. The combination parameter catches some intrinsic properties of the graph. Measuring it may lead to simple and improved algorithms. Experience tells us that when k is close to $\frac{n}{2}$, the k-Vertex Cover problem usually becomes more complicated and harder to solve. If we use this measure to analyze algorithms, we may improve the upper bound for the worst cases and break the bottlenecks of previous algorithms.

References

1. Chen, J., Kanj, I.A., Xia, G.: Simplicity is beauty: Improved upper bounds for vertex cover. Technical Report TR05-008, DePaul University, Chicago IL (2005)
2. Chen, J., Kanj, I.A., Jia, W.: Vertex cover: Further observations and further improvements. J. Algorithms 41(2), 280–301 (2001)
3. Chen, J., Kanj, I.A., Xia, G.: Labeled search trees and amortized analysis: Improved upper bounds for NP-hard problems. Algorithmica 43(4), 245–273 (2005); A preliminary version appeared in ISAAC 2003
4. Razgon, I.: Faster computation of maximum independent set and parameterized vertex cover for graphs with maximum degree 3. J. of Discrete Algorithms 7(2), 191–212 (2009)
5. Xiao, M.: A simple and fast algorithm for maximum independent set in 3-degree graphs. In: Rahman, M. S., Fujita, S. (eds.) WALCOM 2010. LNCS, vol. 5942, pp. 281–292. Springer, Heidelberg (2010)
6. Nemhauser, G.L., Trotter, L.E.: Vertex packings: Structural properties and algorithms. Mathematical Programming 8(1), 232–248 (1975)
7. Abu-Khzamy, F.N., Collinsy, R.L., Fellows, M.R., Langstony, M.A., Sutersy, W.H., Symons, C.T.: Kernelization algorithms for the vertex cover problem: Theory and experiments. In: Proceedings of the 6th workshop on algorithm engineering and experiments (ALENEX '04), pp. 62–68 (2004)
8. Fomin, F.V., Grandoni, F., Kratsch, D.: Measure and conquer: a simple $O(2^{0.288n})$ independent set algorithm. In: SODA, pp. 18–25. ACM Press, New York (2006)
9. Chor, B., Fellows, M., Juedes, D.W.: Linear kernels in linear time, or how to save k colors in $o(n^2)$ steps. In: Hromkovic, J., Nagl, M., Westfechtel, B. (eds.) WG 2004. LNCS, vol. 3353, pp. 257–269. Springer, Heidelberg (2004)
10. Dehne, F., Fellows, M., Rosamond, F., Shaw, P.: Greedy localization, iterative compression and modeled crown reductions: new fpt techniques, an improved algorithm for set splitting and a novel 2k kernelization for vertex cover. In: Downey, R.G., Fellows, M.R., Dehne, F. (eds.) IWPEC 2004. LNCS, vol. 3162, pp. 271–280. Springer, Heidelberg (2004)
11. Fellows, M.R.: Blow-ups, win/win's, and crown rules: Some new directions in fpt. In: Bodlaender, H.L. (ed.) WG 2003. LNCS, vol. 2880, pp. 1–12. Springer, Heidelberg (2003)
12. Bourgeois, N., Escoffier, B., Paschos, V.T.: An $O^*(1.0977^n)$ exact algorithm for max independent set in sparse graphs. In: Grohe, M., Niedermeier, R. (eds.) IWPEC 2008. LNCS, vol. 5018, pp. 55–65. Springer, Heidelberg (2008)

Computing Graph Spanners in Small Memory: Fault-Tolerance and Streaming*

Giorgio Ausiello[1], Paolo G. Franciosa[2], Giuseppe F. Italiano[3], and Andrea Ribichini[1]

[1] Dipartimento di Informatica e Sistemistica, Sapienza Università di Roma, via Ariosto 25, I-00185 Roma, Italy
{ausiello,ribichini}@dis.uniroma1.it

[2] Dipartimento di Statistica, Probabilità e Statistiche Applicate, Sapienza Università di Roma, piazzale Aldo Moro 5, I-00185 Roma, Italy
paolo.franciosa@uniroma1.it

[3] Dipartimento di Informatica, Sistemi e Produzione, Università di Roma "Tor Vergata", via del Politecnico 1, 00133 Roma, Italy
italiano@disp.uniroma2.it

Abstract. Let G be an undirected graph with m edges and n vertices. A spanner of G is a subgraph which preserves approximate distances between all pairs of vertices. An f-vertex fault-tolerant spanner is a subgraph which preserves approximate distances, under the failure of any set of at most f vertices. The contribution of this paper is twofold: we present algorithms for computing fault-tolerant spanners, and propose streaming algorithms for computing spanners in very small internal memory. In particular, we give algorithms for computing f-vertex fault-tolerant (3,2)- and (2,1)-spanners of G with the following bounds: our (3,2)-spanner contains $O(f^{4/3}n^{4/3})$ edges and can be computed in time $\tilde{O}(f^2m)$, while our (2,1)-spanner contains $O(fn^{3/2})$ edges and can be computed in time $\tilde{O}(fm)$. Both algorithms improve significantly on previously known bounds.

Assume that the graph G is presented as an input stream of edges, which may appear in any arbitrary order, and that we do not know in advance m and n. We show how to compute efficiently (3,2)- and (2,1)-spanners of G, using only very small internal memory and a slow access external memory device. Our spanners have asymptotically optimal size and the I/O complexity of our algorithms for computing such spanners is optimal up to a polylogarithmic factor. Our f-vertex fault-tolerant (3,2)- and (2,1)-spanners can also be computed efficiently in the same computational model described above.

1 Introduction

A spanner of an undirected graph is a subgraph on the same vertex set which preserves approximate distances between all pairs of vertices. More formally,

* Partially supported by the Italian Ministry of Education, University and Research (MIUR) under Project AlgoDEEP.

M.T. Thai and S. Sahni (Eds.): COCOON 2010, LNCS 6196, pp. 160–172, 2010.

given $\alpha \geq 1$ and $\beta \geq 0$, an (α, β)-spanner of a graph $G(V, E)$ is a subgraph $S(V, E')$ of G such that $d_S(x, y) \leq \alpha \cdot d_G(x, y) + \beta$ for each pair of vertices $x, y \in V$, where $d_G(x, y)$ represents the distance between x and y in graph G. If $\alpha = 1$ the spanner is called a *purely additive* spanner, while it is called a *purely multiplicative* spanner if $\beta = 0$.

The interest towards graph spanners arises mainly from their sparsity: approximating distances by a multiplicative factor, or even by simply an additive term, allows one to reduce substantially the space requirements. Indeed, it is known (see [17]) that any graph with n vertices contains a $(k, k-1)$-spanner with $O(n^{1+1/k})$ edges. There are several algorithms for computing sparse spanners both in the static and the dynamic case, i.e., while maintaining a spanner under insertions and deletions of edges (see e.g., [2, 4, 6–8, 11, 15, 17, 19–21]). A modular framework for building various families of spanners is due to Pettie [18].

Another interesting notion introduced in the literature is the concept of fault-tolerant spanner: given an integer $f \geq 1$, an f-vertex fault-tolerant spanner of a graph is a subgraph that preserves distances, within the same distortion, under the failure of any set of at most f vertices. More formally, an *f-vertex fault-tolerant (α, β)-spanner* of $G(V, E)$ is a graph $S(V, E')$, with $E' \subseteq E$, such that for each subset of vertices $F \subseteq V$, with $|F| \leq f$, and for each pair of vertices $x, y \in V \setminus F$ we have $d_{S \setminus F}(x, y) \leq \alpha \cdot d_{G \setminus F}(x, y) + \beta$, where by $G \setminus F$ we mean the subgraph of $G(V, E)$ induced by vertex set $V \setminus F$. An f-edge fault-tolerant spanner is defined analogously, under the failure of at most f edges.

Due to their sparsity, graph spanners appear to be useful especially when dealing with very large and dense graphs. In this scenario, it is quite natural to assume that the input graph must be read from an external device and that it cannot be stored in internal memory. In addition to the classical external memory model by Vitter [22], in which data stored on external devices is transferred into/from internal memory in blocks of fixed size, the more restrictive *data streaming* model has also been proposed [16]. In the data streaming model, it is assumed that input data is scanned as a sequence of values in arbitrary order, and that the storage available to the algorithm is smaller than the input size. Thus, the algorithm is not allowed to access randomly the input data. The complexity of a data streaming algorithm is measured in terms of the number of passes over the input sequence (in some applicative scenarios, only one pass is allowed), the amount of internal storage used (always smaller than the input size) and the time needed to process the input. In the case of graph problems, an algorithm that requires $\Omega(n)$ internal storage, where n is the number of vertices, is often called a *semi-streaming* algorithm.

State of the art: Several algoritms have been proposed in the literature for computing graph spanners in the data streaming model [3, 5, 12–14]. Given a graph with n vertices, the spanners computed by all these algorithms have $\Omega(n^{1+1/k})$ edges, where $2k+1$ is the desired multiplicative stretch factor. Slightly better stretch/size trade-offs can be obtained in case of small stretch spanners: the $(k, k-1)$-spanners in [3], for $k \in \{2, 3\}$, have $O(n^{1+1/k})$ edges. An extensive study of the experimental behavior of the above algorithms is in [1]. In particular, experimental results show that small stretch spanners (i.e., spanners obtained with small values of k) provide the best performance in terms of stretch/size trade-offs, and that spanners of larger stretch are not likely to be of practical value.

The only attempt at computing spanners using a small internal memory (i.e., smaller than the spanner size) is due to Ausiello *et al.* [3], where an external memory variant of the algorithm for (2,1)-spanners and (3,2)-spanners is described. The algorithm works in the well known external memory model with block transfer [22], and reads the graph in a single pass from the input stream. That algorithm requires optimal $O(1/B)$ amortized block transfers for processing each edge in the stream, where B is the size of each external memory block[1], and works in $\Theta(n)$ internal memory and $O\left(\frac{n^{4/3}}{B}\right)$ external memory blocks. All the above algorithms are in fact *semi-streaming* algorithms, since they require that all the information about vertices be stored in internal memory.

While fault-tolerant spanners for geometric graphs have been studied in the last decade, they have only recently been introduced for general graphs by Chechik *et al.* [10]. Their construction builds an f-vertex fault-tolerant $(2k-1, 0)$-spanner containing, with high probability, $O(f^2 k^{f+1} \cdot n^{1+1/k} \log^{1-1/k} n)$ edges. The total time required to compute the spanner is $O(f^2 k^{f+1} \cdot n^{3+1/k} \log^{1-1/k} n)$.

Our results: The contribution of this paper is twofold: we introduce practical algorithms for computing fault-tolerant small stretch spanners for general graphs, which improve substantially on previously known bounds [10] and propose, for the first time, streaming algorithms for computing spanners in very small (i.e., even as small as $\Theta(B)$) internal memory. We assume that the input is read in a single pass on a data stream, and the computation can exploit only a very small internal memory plus an external memory device from/to which data are transferred in blocks of size B, like in the classical EM model [22]. We measure the performance of our algorithms in terms of the number of block transfers. Since we are interested in dealing with very large graphs, and some of today's applications are required to run even on small computing devices (such as mobile devices and smart phones), it seems crucial to use a limited amount of internal memory, and to store most of the information needed on slow access external memory devices.

Let $G = (V, E)$ be an undirected unweighted graph, with m edges and n vertices. Our streaming algorithms compute (3,2)- and (2,1)-spanners of G having asymptotically optimal size: the (3,2)-spanner contains $O(n^{4/3})$ edges, while the (2,1)-spanner contains $O(n^{3/2})$ edges. Both algorithms work with an internal memory of size $s = \Theta(\max(\log n, B))$ plus an external memory proportional to the size of the spanner, and require a single pass on the input stream, in which graph edges may appear in any arbitrary order. The I/O complexity of our algorithms is $\tilde{O}(m/B)$ in the worst case, and hence it is optimal up to a polylogarithmic factor. Our algorithms do not require advance knowledge of the

[1] As it is usually assumed with external memory algorithms, external memory is organized in blocks, each containing B words, where a word has size $\Theta(\log n)$.

number n of vertices or the number m of edges in the graph. While processing the input stream, the algorithms build an optimal spanner on the portion of the graph examined so far: in all the complexity bounds, n and m refer to the number of vertices and edges currently known to the algorithm. We remark that data stored on external memory are accessed only by sequential scans and sorting, and thus the algorithms are expected to be very efficient in practice.

We also provide algorithms for computing f-vertex fault-tolerant (3,2)- and (2,1)- spanners of general graphs, which can be built in the same computational model described above. The (3,2)-spanner contains $O(f^{4/3}n^{4/3})$ edges and can be computed with $\tilde{O}\left(f^2 m/B\right)$ I/O's in the worst case. The (2,1)-spanner contains $O(fn^{3/2})$ edges and can be computed with $\tilde{O}\left(fm/B\right)$ I/O's in the worst case. In the classical RAM model, the computations of our f-vertex fault-tolerant (3,2)- and (2,1)-spanners require respectively only time $\tilde{O}\left(f^2 m\right)$ and $\tilde{O}\left(fm\right)$. The corresponding fault-tolerant spanners of Chechik *et al.* [10] achieve the following bounds: their fault-tolerant (5,0)-spanner contains $O(f^2 3^{f+1} \cdot n^{4/3} \log^{2/3} n)$ edges with high probability and can be computed in time $\tilde{O}(f^2 3^{f+1} \cdot n^{10/3})$, while their fault-tolerant (3,0)-spanner contains $O(f^2 2^{f+1} \cdot n^{3/2} \log^{1/2} n)$ edges with high probability and can be computed in time $\tilde{O}(f^2 2^{f+1} \cdot n^{7/2})$. Thus, our algorithms improve significantly the bounds in [10] in the case of small stretch spanners.

For lack of space, in this extended abstract we only describe the algorithms for computing (3,2)-spanners and f-vertex fault-tolerant (3,2)-spanners, which are more involved than the corresponding algorithms for (2,1)-spanners.

2 Computing Graph Spanners with Small Internal Memory

The algorithm builds on a graph clustering method described in [2], which is recalled in Section 2.1. In Sections 2.2 and 2.3 we show how to exploit this clustering to compute spanners in very small internal memory.

2.1 Clustering the Graph

We define the *neighborhood* of a vertex v as $N(v) = \{v\} \cup \{x \mid (v, x) \in E\}$. A *cluster* is a set of vertices contained in $N(v)$; vertex v is the *center* of the cluster. Note that, according to this definition, the center of a cluster is not necessarily contained in the cluster itself. Each vertex may belong to at most one cluster; a vertex is said to be *clustered* if it belongs to a cluster, and it is said to be *free* otherwise. The size of a cluster is given by the number of vertices contained in the cluster. Given a vertex v in graph $G(V, E)$ and a subset of edges $E' \subseteq E$, we define $\mathit{free}(v, E')$ as the set of all free vertices w such that $((v, w) \in E') \vee (w = v)$. Obviously, for any $E' \subseteq E$, we have that $\mathit{free}(v, E') \subseteq N(v)$. A *clustering* of a graph is simply a set of disjoint clusters. Given a clustering of the graph, the following three kinds of edges can be defined:

cluster edges: for each cluster C, where v is the center of C, all edges joining v to each vertex in $C \setminus \{v\}$;

free edges: for each free vertex v, all the edges incident to v;
bridge edges: for each pair of clusters C_1, C_2, one arbitrary edge joining a vertex in C_1 and a vertex in C_2.

As shown in [2], the subgraph induced by the union cluster edges, free edges and bridge edges is a (3,2)-spanner of the original graph. Let $\vartheta = n^{1/3}$ be a clustering threshold. If each cluster contains at least ϑ vertices, by computing clusters in a greedy fashion until no more clusters of size at least ϑ can be created, we obtain a (3,2)-spanner having only $O(n^{4/3})$ edges.

2.2 Data Structures

We assume that the graph is given as a stream of edges, in arbitrary order. In the following, we denote by n the number of vertices currently known by the algorithm. Our algorithm processes edges in batches, where each batch E_i contains $\Theta(n)$ edges. In each batch, the algorithm checks whether either new clusters of size at least $n^{1/3}$ can be built, or the size of existing clusters can be increased by at least $n^{1/3}$ vertices. Free edges in the batch which are not used to create or to extend a cluster are appended to a set of edges (named **CandidateFree** in the sequel). When the size of **CandidateFree** exceeds $2 \cdot n^{4/3}$, its edges are reconsidered by the algorithm so that they may be used again to contribute to the clustering process. This allows us to build a cluster centered on v also in the case that edges (v, u), with u free, are far apart in the input stream. Edges in the batch whose endpoints are contained in two different clusters are considered as possible bridges between those clusters, and appended to a set named **CandidateBridges**. When the size of **CandidateBridges** exceeds $n^{4/3}$, it will be reconsidered by the algorithm in order to identify one bridge for each cluster pair (by discarding duplicate bridges).

We maintain the following information in external memory:

VertexStatus: the set of all vertices encountered so far in the input stream. Each vertex is marked as either free or clustered, and we also mark cluster centers. For clustered vertices we store the label of the corresponding cluster. The status of each vertex may change from *free* to *clustered* while processing an edge batch; those changes have to be taken into account in the remaining part of the batch and in subsequent batches. This is done by means of an appropriate data structure, as explained in Section 2.3.

CandidateFree: a set containing all the edges that have already been processed and currently identified as free edges, i.e., they have not been discarded and have not yet contributed to the clustering.

CandidateBridges: a set of edges, each joining a pair of currently created clusters. The same pair of clusters might be temporarily connected by more than one edge in this set.

ClusterEdges: edges joining each clustered vertex to the center of its cluster.

During the execution of the algorithm, a (3,2)-spanner of the graph currently observed is given by the union of **ClusterEdges**, **CandidateFree** and **CandidateBridges**.

Throughout the algorithm, we maintain the following invariants:

- **CandidateFree** and **CandidateBridges** have both size $O(n^{4/3})$;
- for each pair of clusters, if any edges joining them have been read from the input stream, then at least one of those edges is contained in **CandidateBridges** or in **CandidateFree**.

We now show how to process a generic edge batch E_i. Each undirected edge (x, y) in the batch read from the input stream is duplicated as a pair of symmetric directed edges (x, y) and (y, x). We proceed as follows:

Step 1: In order to determine the type of each edge in E_i, we sort edges in the batch according to their first endpoint and determine whether each first endpoint is currently free or clustered, by scanning **VertexStatus** in parallel with the batch. The same is done for the second endpoint. The batch is finally sorted again according to the first endpoint.
For each edge $(x, y) \in E_i$, if x and y belong to the same cluster, then edge (x, y) is deleted from E_i, while if x and y belong to different clusters, then (x, y) is moved from E_i to the set **CandidateBridges**.

Step 2: We consider each sequence of free edges $(v, x_1), (v, x_2), \ldots, (v, x_k)$ in E_i, and we check whether there are at least $n^{1/3}$ free vertices among $v, x_1, x_2, \ldots, x_k$. If this is the case, then if v is the center of an already created cluster, we add all its free neighbors to this cluster; otherwise, we create a new cluster centered in v containing all those free vertices. Edges joining newly clustered vertices to center v are appended to **ClusterEdges**. The edges joining v to vertices that are in different clusters are appended to the set **CandidateBridges**.
Otherwise, if the sequence $(v, x_1), (v, x_2), \ldots, (v, x_k)$ in E_i does not contain at least $n^{1/3}$ free endpoints, these edges are added to the set **CandidateFree**.

Step 3: The new status of vertices clustered during Step 2 is saved in **VertexStatus**. These changes have to be taken into account while processing the remaining part of batch E_i. This is done as described in Section 2.3.

Whenever the set **CandidateFree** contains more than $2 \cdot n^{4/3}$ edges, it is sorted according to their first endpoints and processed again as described in Steps 1, 2 and 3 above.

We also delete duplicate bridges, i.e., bridges joining the same pair of clusters, from **CandidateBridges** whenever its size exceeds $n^{4/3}$. This is done by sorting lexicographically **CandidateBridges** according to the cluster labels of the edge endpoints, and by keeping only a single bridge for each pair of clusters.

2.3 Updating Vertices in Current Batch

The main difficulty in the algorithm described above is to update efficiently the status of edge endpoints in the current batch while a new cluster is created or more vertices are added to an already existing cluster. To accomplish this task, during the processing of a batch we insert newly clustered vertices into a hierarchy $\mathbf{VertexStatus}_i$ ($\mathbf{VS}_i$, for short), following the ideas of Bentley and Saxe [9]. More precisely:

- $\mathbf{VS}_0$ contains at most $n^{1/3}$ vertices;
- each $\mathbf{VS}_i$, for $1 \leq i \leq k$ where k is the smallest integer such that $2^k n^{1/3} \geq n$, is either empty or contains a number of vertices $|\mathbf{VS}_i|$ such that $2^{i-1} n^{1/3} < |\mathbf{VS}_i| \leq 2^i n^{1/3}$;
- each $\mathbf{VS}_i$, for $0 \leq i \leq k$, is ordered according to the vertex label.

Vertices in $\mathbf{VS}_i$'s are maintained under insertion of a set V_c of new vertices as follows:

- let ℓ be the smallest index such that $|V_c| + \sum_{i=0}^{\ell} |\mathbf{VS}_i| \leq 2^\ell n^{1/3}$;
- merge and sort V_c and $\mathbf{VS}_i$, $0 \leq i \leq \ell$, into $\mathbf{VS}_\ell$.
- set all $\mathbf{VS}_i$ to $\emptyset$, $0 \leq i < \ell$.

At the beginning of each batch, all $\mathbf{VS}_i$'s are merged into the smallest level that can contain all clustered vertices, and all the endpoints in the current batch are updated as described in Step 1 of the algorithm.

Every time some vertices are clustered, the change in their status must be reflected into the remaining part of the current batch. Let ℓ be the index of the highest level modified in the **VertexStatus** hierarchy due to the insertion: we update only the endpoints' status of the next $|\mathbf{VS}_\ell|$ edges in the current batch, and set a marker of *level* ℓ at the end of the updated portion of the batch. While processing the batch, any time a marker of level k is found, the endpoints' status of the next $|\mathbf{VS}_k|$ edges in the current batch are updated according to vertices in $\mathbf{VS}_k$.

3 Correctness and Complexity

In this section, we show that the algorithm described in Section 2 computes a (3,2)-spanner of the original graph having $O(n^{4/3})$ edges and analyze its performance.

Theorem 1. *The algorithm described in Section 2 computes a (3,2)-spanner of the original graph.*

Proof. (Sketch) Given a graph $G = (V, E)$ and a clustering C of G, it is shown in [2] that the graph $S_C = (V, E_C)$, where E_C is the union of cluster edges, free edges and bridge edges as defined in Section 2, associated to clustering C, is a (3-2)-spanner of G. By definition of (α, β)-spanner, any graph $S' = (V, E')$, with $E' \supseteq E_C$, is still a (3-2)-spanner of G.

We show that our algorithm computes a supergraph $G'(V, E')$ of a (3,2)-spanner $S_C(V, E_C)$ of $G(V, E)$ by showing that all edges deleted from the graph are contained in $E \setminus E_C$. In particular, all edges in $E \setminus E'$ are either:

a) edges joining two vertices in the same cluster, both different from the cluster center, or
b) edges with endpoints in clusters C_1 and C_2 such that there is already a similar edge in E'.

The algorithm (in Step 1) deletes an edge e while processing a batch if and only if its endpoints are already in the same cluster; these edges are thus in set a) above. Other edges are either kept in the spanner or moved into the set of candidate bridges. The set of candidate bridges is periodically processed, and only duplicated bridges between the same pair of clusters are deleted; these deleted edges are thus in set b) above.

We must also show that edges are correctly classified as free or as candidate bridges. The status of each edge endpoint in the current batch is correctly read from **VertexStatus** at the beginning of the batch processing. If a vertex v is included into a cluster while processing the batch, then v is inserted into $\mathbf{VS}_i$ for some i, and this update is immediately reflected into the next $|\mathbf{VS}_i|$ edges in the batch. The new status of v is propagated forward in the batch when the marker of level i is encountered, unless v is moved to some $\mathbf{VS}_j$, with $j > i$. In this case, changes in $\mathbf{VS}_j$ are reflected into the batch edges as well, and eventually propagated when the level j marker is met. □

Theorem 2. *The (3,2)-spanner computed by our algorithm contains $O(n^{4/3})$ edges.*

Proof. (Sketch) The (3,2)-spanner is given by the union of cluster edges, free edges and candidate bridges.

- Each cluster edge joins a clustered vertex to its center. Thus, the number of cluster edges is at most n.
- Free edges are only saved if a vertex does not have more than $n^{1/3}$ free neighbors, so the number of free edges is $O(n^{4/3})$.
- Set **CandidateBridges** is processed every time its size exceeds $n^{4/3}$. After the processing, at most one bridge for each pair of clusters is maintained, so there are $O(n^{4/3})$ bridge edges in the spanner. □

Theorem 3. *Our algorithm computes a (3,2)-spanner in $\tilde{O}\left(\frac{m}{B}\right)$ I/O's, using $O\left(\frac{n^{4/3}}{B}\right)$ external memory blocks and $O(\log n)$ internal memory.*

Proof. (Sketch) The following operations are performed in external memory:

Process a batch from the input stream: labelling endpoints requires sorting the batch a constant number of times, plus a constant number of sequential scans of **VertexStatus** and of the batch in parallel. This is done in $O(Sort(n))$ I/O's for each of the $O(m/n)$ batches, plus the time needed for updating the status of vertices (see below).

Maintain $\mathbf{VS}_i$'s under insertion of vertices: each clustered vertex is inserted in the data structure only once, and possibly moved from $\mathbf{VS}_0$ to $\mathbf{VS}_1$, and so on to $\mathbf{VS}_k$, with $k = \left\lceil \frac{2}{3}\log_2 n \right\rceil$. Each level $\mathbf{VS}_i$ is filled and merged $O\left(\frac{n^{2/3}}{2^i}\right)$ times, requiring $O(Sort(2^i \cdot n^{1/3}))$ I/O's each time. This gives a total of $O(\log n \cdot Sort(n))$ I/O's.

Update the status of endpoints in current batch: each time a $\mathbf{VS}_i$ is modified, clustered vertices in $\mathbf{VS}_i$ are reflected into the next $|\mathbf{VS}_i|$ edges in current batch. The same is done when a marker of level i is found in the current batch. If the propagation occurs in the whole batch, it costs $O\left(\frac{n}{2^i n^{1/3}} Sort(2^i n^{1/3})\right) = O(Sort(n))$ I/O's for each level, giving overall $O(\log n \cdot Sort(n))$ I/O's for each batch in which at least $n^{1/3}$ vertices are clustered. This may occur in at most $\min\{m/n, n^{2/3}\}$ batches, giving overall $\tilde{O}(m/B)$ I/O's.

Examine CandidateFree: each time **CandidateFree** contains more than $2 \cdot n^{4/3}$ edges, we examine this set in batches of size $\Theta(n)$, moving into **CandidateBridges** edges having clustered endpoints, and clustering free vertices when possible. At the end of this process, at most $n^{1/3}$ free edges for each vertex may remain into **CandidateFree**, otherwise their endpoints would be clustered. Hence, the whole process is performed at most $\frac{m}{n^{4/3}}$ times, each time requiring to process $O(n^{1/3})$ batches. This requires overall $\tilde{O}(m/B)$ I/O's.

Discard edges in excess from CandidateBridges: There can be at most $n^{2/3}$ clusters, so at most $n^{4/3}/2$ bridges are required. Hence, each time **CandidateBridges** is processed, at least half of the edges are discarded. This means that the processing is performed $O(\frac{m}{n^{4/3}})$ times, each time requiring $O(Sort(n^{4/3}))$ I/O's, for a total of $\tilde{O}(m/B)$ I/O's.

All the above operations can be performed using $\Theta(\max\{\log n, B\})$ internal memory. □

4 Computing *f*-Vertex Fault-Tolerant Spanners

Our algorithms can be adapted to build fault-tolerant (2,1)- and (3,2)-spanners, improving significantly the bounds in [10] in the case of small stretch spanners. Due to space constraints, in this extended abstract we focus on fault-tolerant (3,2)-spanners only.

The main changes to the algorithm consist in allowing the same vertex to belong to up to $f+1$ different clusters, and in joining the same pair of clusters by up to $f \cdot (f+1)+1$ bridges. Clusters are no longer disjoint, but each vertex can be the center of at most one cluster. Moreover, to keep under control the number of edges in the spanner, the size of each cluster is required to be at least $f^{4/3}n^{1/3}$, as illustrated in Theorem 5.

As long as a vertex v is contained in less than $f+1$ clusters, v is still considered a free vertex, thus all edges incident to v are kept in the spanner. A vertex is called *saturated* if it is contained in exactly $f+1$ clusters. Note that, since a saturated vertex is contained in exactly $f+1$ clusters, the same edge could be a bridge for up to $(f+1)^2$ pairs of clusters.

Each pair of clusters X, Y will be joined by a set of bridges $B(X,Y)$. We add an edge (x,y), with $x \in X$ and $y \in Y$, to the current bridge set $B(X,Y)$ if and only if, before its insertion, all of the following conditions hold:

(a) there are at most f edges in $B(X,Y)$ incident on x;
(b) there are at most f edges in $B(X,Y)$ incident on y;
(c) $|B(X,Y)| \leq f \cdot (f+1)$.

Thus, the final degree of each endpoint in $B(X,Y)$ is at most $f+1$, and if $|B(X,Y)| = f \cdot (f+1) + 1$ then there must be at least $f+1$ distinct endpoints in X and at least $f+1$ distinct endpoints in Y.

Theorem 4. *The subgraph S defined above is an f-vertex fault-tolerant (3,2)-spanner of G.*

Proof. (Sketch) We must show that, given a path π of length ℓ from vertex v to vertex w in $G \setminus F$, with $|F| \leq f$, a path of length at most $3 \cdot \ell + 2$ from v to w exists in $S \setminus F$. We concentrate on pairs of adjacent saturated vertices in the path, since all edges incident to non-saturated vertices are maintained in the spanner. Let $\{u, v\}$ be such a pair. Vertex u is saturated, thus it is contained in exactly $f+1$ clusters, with distinct centers. Since $|F| \leq f$, at least one of those centers must be in $V \setminus F$. Let C_u be the corresponding cluster. The same holds for v, and let C_v be a cluster containing v whose center is in $V \setminus F$.

Edge (u, v) is in the original graph, and hence at some point it must have been considered by the algorithm as a candidate bridge joining clusters C_u and C_v. If edge (u, v) is in $B(C_u, C_v)$, then it is also in $S \setminus F$. Otherwise, we show now that a path of length at most 3 exists in $S \setminus F$ joining the centers of C_u and C_v. Since $(u, v) \notin B(C_u, C_v)$, at least one of the conditions (a), (b), (c) above was violated when processing edge (u, v). If condition (a) was violated, i.e., the degree of u was equal to $f+1$, then for each $F \subseteq V$, $|F| \leq f$, there must be at least an edge $(u, z) \in B(C_u, C_v)$ where $z \notin F$ and $z \in C_v$, and so the theorem holds. The same argument applies to v if condition (b) was violated. Otherwise, condition (c) was violated and $|B(C_u, C_v)| = f \cdot (f+1) + 1$. Hence, there are at least $f+1$ endpoints of $B(C_u, C_v)$ in C_u and at least $f+1$ endpoints of $B(C_u, C_v)$ in C_v, each having degree at most $f+1$. If at most f of them are faulty, at most $f \cdot (f+1)$ edges are missing in $B(C_u, C_v) \setminus F$, thus at least one edge e is in $B(C_u, C_v) \setminus F$. Hence, there is a path of length 3 joining C_u and C_v through e.

The above argument can be extended to longer sequences of saturated vertices $u_0, u_1, \ldots, u_k$ in π by considering pairs $\langle u_0, u_1 \rangle, \langle u_1, u_2 \rangle, \ldots, \langle u_{k-1}, u_k \rangle$: it is possible to show that for each F with $|F| \leq f$ there is a path of length $3 \cdot k$ in $S \setminus F$ joining the center of one cluster containing u_0 to the center of one cluster containing u_k. This gives a path of length $3 \cdot k + 2$ from u_0 to u_k in $S \setminus F$. □

Theorem 5. *The f-vertex fault-tolerant (3,2)-spanner S of graph G defined above contains $O(f^{4/3} n^{4/3})$ edges.*

Proof. (Sketch) The number of edges in the spanner strongly depends on the clustering threshold ϑ, i.e., on the lower bound for the size of a cluster. Each edge in S is contained in at least one of the following sets (recall that in our fault-tolerant spanner, the same edge could be a cluster edge, a free edge and a bridge edge at the same time):

- **cluster edges**: edges joining a clustered vertex v to the center of each cluster containing v. They can be at most $f+1$ for each vertex v, giving a total of $O(fn)$ edges;

free edges: edges having at least one free endpoint, i.e., an endpoint that belongs to at most f clusters. Given any vertex v, it can be adjacent to at most $\vartheta - 1$ free vertices, where ϑ is the clustering threshold, otherwise a new cluster centered on v would be created containing those vertices. Thus, the total number of free edges is $O(\vartheta n)$;

bridge edges: there are at most $f(f+1)+1$ bridge edges for each pair of clusters. The number of clusters is at most $(f+1)n/\vartheta$, since each vertex is contained in at most $f+1$ clusters, and each cluster contains at least ϑ vertices. The total number of bridge edges is bounded by $f(f+1)+1$ times the number of cluster pairs, that is $O(f^2 \cdot (fn/\vartheta)^2) = O(f^4 n^2/\vartheta^2)$.

The total number of edges in the spanner is thus bounded by $O(fn + \vartheta n + f^4 n^2/\vartheta^2)$. By setting the clustering threshold to $\vartheta = f^{4/3} n^{1/3}$, we have that in the worst case the size of the spanner can be at most $O(f^{4/3} n^{4/3})$. □

4.1 Fault-Tolerant Spanners: The Algorithm

The stream of edges is processed as in Section 2.2, by creating a cluster each time the algorithm finds a vertex with a sufficient number of free neighbors. If an edge has at least one free endpoint, the edge is kept in the set **CandidateFree**, even if both endpoints are already contained in some clusters, since it could participate in creating new overlapping clusters in the future. On the opposite, if both endpoints are saturated, the edge is added to the set **CandidateBridges**.

While processing a batch of edges, we mark each edge endpoint by a flag indicating whether the vertex is saturated or not. In order to decide whether a vertex is saturated or not, it is sufficient to know the number of clusters containing it. Saturated vertices are maintained with the data structure described in Section 2.3.

When an edge (x, y) joining two saturated vertices is processed, we add to the set **CandidateBridges** a quadruple $\langle X, Y, x, y\rangle$ for each of the $(f+1)^2$ pairs of clusters X, Y, respectively containing vertices x and y, except when $X = Y$. When the size of **CandidateBridges** exceeds $(f+1)^2 N_c^2$, where N_c is the current number of clusters, we discard bridges in excess as follows. We must ensure that:

- $|\{\langle X, Y, x, \cdot\rangle\}| \le f+1$ for each cluster pair X, Y, and for each $x \in X$;
- $|\{\langle X, Y, \cdot, y\rangle\}| \le f+1$ for each cluster pair X, Y, and for each $y \in Y$;
- $|\{\langle X, Y, \cdot, \cdot\rangle\}| \le f \cdot (f+1)+1$ for each cluster pair X, Y.

The set is lexicographically sorted according to the quadruple $\langle X, Y, x, y\rangle$, taking only the first $f+1$ edges having the same X, Y, x. Then we sort the sequence according to X, Y, y, x, taking only the first $f+1$ edges having the same X, Y, y. Finally, we scan again the sequence taking only the first $f(f+1)+1$ edges having the same X, Y. This process only requires sorting and sequential scans of the candidate bridge set, and produces a family of bridge sets that fulfills the conditions required.

Theorem 6. *The above algorithm computes an f-vertex fault-tolerant (3,2)-spanner in $\tilde{O}\left(f^2 \cdot \frac{m}{B}\right)$ I/O's.*

Proof. (Sketch) The main changes with respect to the algorithm for (3,2)-spanners consist in the fact that a vertex may belong to up to $f+1$ clusters and that there can be up to $f \cdot (f+1)+1$ bridges for each pair of clusters.

Set **VertexStatus** exploits the same data structure described in Section 2.3, storing for each vertex v the list of clusters containing v. The whole data structure is stored in $O(\frac{f \cdot n}{B})$ external memory blocks.

Each edge e inserted into the set **CandidateBridges** is replicated $O(f^2)$ times, one for each pair of clusters joined by e. The total number of edges inserted into the data structure is thus $O(f^2 \cdot m)$, giving a total of $\tilde{O}(f^2 \cdot \frac{m}{B})$ I/O's for deleting bridges in excess. □

References

1. Ausiello, G., Demetrescu, C., Franciosa, P.G., Italiano, G.F., Ribichini, A.: Graph spanners in the streaming model: An experimental study. Algorithmica 55(2), 346–374 (2009)
2. Ausiello, G., Franciosa, P.G., Italiano, G.F.: Small stretch spanners on dynamic graphs. Journal of Graph Algorithms and Applications 10(2), 365–385 (2006)
3. Ausiello, G., Franciosa, P.G., Italiano, G.F.: Small stretch (α, β)-spanners in the streaming model. Theoretical Computer Science 410(36), 3406–3413 (2009)
4. Baswana, S.: Dynamic algorithms for graph spanners. In: Azar, Y., Erlebach, T. (eds.) ESA 2006. LNCS, vol. 4168, pp. 76–87. Springer, Heidelberg (2006)
5. Baswana, S.: Streaming algorithm for graph spanners - single pass and constant processing time per edge. Inf. Process. Lett. 106(3), 110–114 (2008)
6. Baswana, S., Kavitha, T., Mehlhorn, K., Pettie, S.: New constructions of (α, β)-spanners and purely additive spanners. In: Proc. of the 16th Annual ACM-SIAM Symposium on Discrete Algorithms (SODA '05), pp. 672–681 (2005)
7. Baswana, S., Sarkar, S.: Fully dynamic algorithm for graph spanners with poly-logarithmic update time. In: Proc. of the 9th Annual ACM-SIAM Symposium on Discrete Algorithms (SODA '08), pp. 1125–1134 (2008)
8. Baswana, S., Sen, S.: A simple and linear time randomized algorithm for computing sparse spanners in weighted graphs. Random Struct. Algorithms 30(4), 532–563 (2007)
9. Bentley, J.L., Saxe, J.B.: Decomposable searching problems I: Static-to-dynamic transformation. J. Algorithms 1(4), 301–358 (1980)
10. Chechik, S., Langberg, M., Peleg, D., Roditty, L.: Fault-tolerant spanners for general graphs. In: Proc. of 41st Annual ACM Symposium on Theory of Computing (STOC'09), pp. 435–444 (2009)
11. Cohen, E.: Fast algorithms for constructing t-spanners and paths with stretch t. SIAM Journal on Computing 28(1), 210–236 (1998)
12. Elkin, M.: Streaming and fully dynamic centralized algorithms for constructing and maintaining sparse spanners. In: Arge, L., Cachin, C., Jurdziński, T., Tarlecki, A. (eds.) ICALP 2007. LNCS, vol. 4596, pp. 716–727. Springer, Heidelberg (2007)
13. Elkin, M., Zhang, J.: Efficient algorithms for constructing $(1+\varepsilon, \beta)$-spanners in the distributed and streaming models. Distributed Computing 18(5), 375–385 (2006)

14. Feigenbaum, J., Kannan, S., McGregor, A., Suri, S., Zhang, J.: Graph distances in the streaming model: the value of space. In: Proc. of the 16th Annual ACM-SIAM Symposium on Discrete Algorithms (SODA '05), pp. 745–754 (2005)
15. Liestman, A.L., Shermer, T.: Additive graph spanners. Networks 23, 343–364 (1993)
16. Ian Munro, J., Paterson, M.: Selection and sorting with limited storage. Theor. Comput. Sci. 12, 315–323 (1980)
17. Peleg, D., Shäffer, A.: Graph spanners. Journal of Graph Theory 13, 99–116 (1989)
18. Pettie, S.: Low distortion spanners. In: Arge, L., Cachin, C., Jurdziński, T., Tarlecki, A. (eds.) ICALP 2007. LNCS, vol. 4596, pp. 78–89. Springer, Heidelberg (2007)
19. Roditty, L., Thorup, M., Zwick, U.: Deterministic constructions of approximate distance oracles and spanners. In: Caires, L., Italiano, G.F., Monteiro, L., Palamidessi, C., Yung, M. (eds.) ICALP 2005. LNCS, vol. 3580, pp. 261–272. Springer, Heidelberg (2005)
20. Thorup, M., Zwick, U.: Spanners and emulators with sublinear distance errors. In: Proc. of 17th Annual ACM-SIAM Symposium on Discrete Algorithms (SODA '06), pp. 802–809 (2006)
21. Venkatesan, G., Rotics, U., Madanlal, M.S., Makowsky, J.A., Pandu Rangan, C.: Restrictions of minimum spanner problems. Information and Computation 136(2), 143–164 (1997)
22. Vitter, J.S.: External memory algorithms and data structures: dealing with massive data. ACM Computing Surveys 33(2), 209–271 (2001)

Factorization of Cartesian Products of Hypergraphs

Alain Bretto and Yannick Silvestre

Université de Caen, Département d'Informatique, GREYC CNRS UMR-6072, Campus II, Bd Marechal Juin BP 5186, 14032 Caen cedex, France
alain.bretto@info.unicaen.fr, yannick.silvestre@info.unicaen.fr

Abstract. In this article we present the L2-section, a tool used to represent a hypergraph in terms of an "advanced graph" and results leading to first algorithm, in $O(nm)$, for a bounded-rank, bounded-degree hypergraph H, which factorizes H in prime factors. The paper puts a premium on the characterization of the prime factors of a hypergraph, by exploiting isomorphisms between the layers in the 2-section, as returned by a standard graph factorization algorithm, such as the one designed by IMRICH and PETERIN.

1 Introduction

Cartesian products of graphs have been studied since the 1960s by VIZING and SABIDUSSI. They independently showed ([Sab60]), that for every finite connected graph there is a unique (up to isomorphism) prime decomposition of the graph into factors. This fundamental theorem was the starting point for research concerning the relations between a Cartesian product and its factors [IPv97, Bre06]. Some of the questions raised are still open, as in the case of the Vizing's conjecture[1]. These relations are of particular interest as they allow us to break down problems by transfering algorithmic complexity from the product to the factors.

Some examples of problems that can be made easier by studying factors rather than the whole product include: classical problems, like the determination of the chromatic number, as the chromatic number of a Cartesian product is the maximum of the chromatic number of each factor; or the detemination of the independence number (see [IKR08]). More original numbers or properties are also investigated, especially in coloring theory: the antimagicness of a graph [ZS09] or the game chromatic number [Pet07], to quote a few. These numbers, easily computable thanks to Cartesian product operations, are graphical invariants. Graph products offer an interesting framework as soon as these graphical invariants are involved. This is the reason why they are studied with various applications; most of networks used in the context of parallel and distributed computation are Cartesian products: the hypercube, grid graphs, etc. In this context, the problem

[1] This conjecture expressed by Vizing in 1968 states that the domination number of the Cartesian product of graphs is greater than the product of the domination numbers of its factors

M.T. Thai and S. Sahni (Eds.): COCOON 2010, LNCS 6196, pp. 173–181, 2010.

of finding a "Cartesian" embedding of an interconnection network into another is of fundamental importance and thus has gained considerable attention. Cartesian products are also used in telecommunications [Ves02].

In 2006, Imrich and Peterin [IP07] gave an algorithm able to compute the prime factorization of connected graphs in linear time and space, making the use of Cartesian product decomposition particularly attractive.

Hypergraph theory has been introduced in the 1960s as a generalization of graph theory. A lot of applications of hypergraphs have been developed since (for a survey see [Bre04]). Cartesian products of hypergraphs can be defined in a same way as graphs, and similarly it is easier to study the hypergraph factors than the product. They also support graphical invariants (see [Bre06]), as it is the case for the linearity, conformity, transversal and matching number, Helly property, and it is generally possible to extend graphical invariants discovered on graphs to them.

Summuary of the Results

In this paper we present an algorithm which gives the prime decomposition of a Cartesian product of hypergraphs. It is the first algorithm of recognition of Cartesian products of hypergraphs, up to our knowledge. This one is based on the algorithm of Imrich and Peterin [IP07] but it is easily adaptable to any algorithm which factorizes Cartesian products of graphs. Hypergraphs store more informations than graphs can (for fixed parameters); we explicit how a hypergraph can be seen as an "advanced" graph by introducing the *L2-section tool*. Some mathematical properties of the *L2-section* of a hypergraph help us then to design a recognition algorithm. By making an arrangement (called $\mathcal{R}^*$-*Cartesian joins*) on the factors returned by the recognition algorithm working on the 2-section, it finally releases the prime factors of a given hypergraph in $O(mn)$ time, when the rank and the degree of the hypergraph are constant.

Preliminaries

The cardinality of a set A is denoted by $|A|$. The set $\mathcal{P}_2(A)$ is the set of pairs $\{x, y\}$ such that $x, y \in A$ and $x \neq y$. A *hypergraph* H on a set of vertices V is a pair (V, E) where E is a set of non-empty subsets of V called *hyperedges* such that $\bigcup_{e \in E} e = V$. The set V may be written $\mathcal{V}(H)$. This implies in particular that every vertex is included in at least one hyperedge. If $\bigcup_{e \in E} e \neq V$, H is called a *pseudo-hypergraph*. A hypergraph is *simple* if no hyperedge is contained in another one.

In the sequel, we suppose that hypergraphs are simple and that no hyperedge is a *loop*, that is, the cardinality of a hyperedge is at least 2. Moreover we suppose that they are *connected*, a hypergraph being *connected* if there is a path between any pair of hyperedges. The number of hyperedges of a hypergraph H is denoted by $m(H)$ or simply m when unambiguous. It is convenient to define a *simple graph* as a particular case of simple hypergraph where every hyperedge is of size

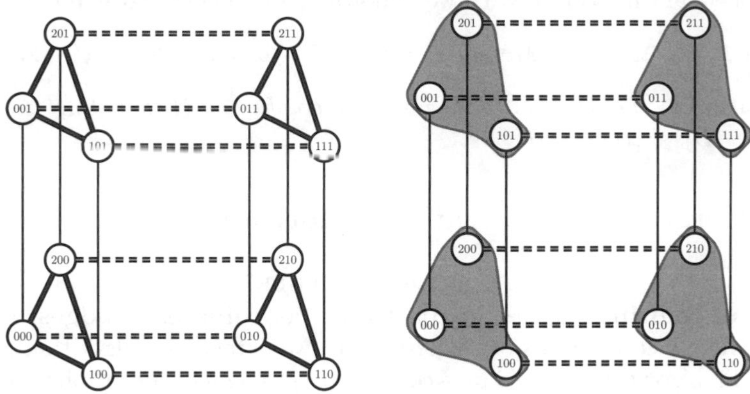

Fig. 1. To the left, the Cartesian product $T \Box K_2 \Box K_2 = T \Box K_2^2$, where T stands for the triangle with vertex set $\{0, 1, 2\}$, and K_2 the graph $(\{0, 1\}, \{\{0, 1\}\})$. Here we drew the layers with different dash patterns, and erased brackets and comas to name the nodes. For instance, T-layers are solid and thick, while K_2-layers are either dashed or doubled. We filled the nodes with their coordinates (without brackets). Notice that even if two isomorphic factors intervene in this product, their respective layers are considered as different as the coordinates of the nodes they are incident to only have one coordinate varying, but not the same. When considered isolated, "dashed" layers in the figure have their second coordinate varying while "thick" ones have their third coordinate varying. To the right the Cartesian product of hypergraphs $H = T' \Box K_2^2$ where T' stands for the hypergraph $T' = (\{0, 1, 2\}, \{\{0, 1, 2\}\})$.

2. So we suppose every graph $G = (V, E)$ to verify $\bigcup_{\{x,y\} \in E} \{x, y\} = V$. Given a graph $G = (V, E)$ and A a subset of E, we define $G(A) = (S, A)$ as a subgraph of G where $S = \{x \in a \in A\}$. For $E' \subseteq E$ the set $H' = (\bigcup_{e \in E'} e = V, E')$ is the *partial hypergraph* generated by E'. The *rank* of H is $r(H) = \max\{|e| : e \in E(H)\}$, written r when unambiguous.

The *2-section* $[H]_2$ of a hypergraph $H = (V, E)$ is the graph $G = (V, E')$ where $E' = \bigcup_{e \in E} \mathcal{P}_2(e)$, that is, two vertices are adjacent in G iff they belong to a same hyperedge. Notice that every hyperedge of H is a clique of $[H]_2$ and $\bigcup_{e \in E} e = V$ implies $\bigcup_{\{x,y\} \in E'} \{x, y\} = V$. Finally, an *isomorphism* from the hypergraph $H = (V, E)$ to the hypergraph $H' = (V', E')$ is a bijection f from V to V' such that, for every $e \subseteq V$, $e \in E$ iff $\forall x \in e$, $f(x) \in f(e)$. If H is isomorphic to H', we write $H \cong H'$. Let $H_1 = (V_1, E_1)$ and $H_2 = (V_2, E_2)$ be hypergraphs. The *Cartesian product* of H_1 and H_2 is the hypergraph $H_1 \Box H_2$ with set of vertices $V_1 \times V_2$ and set of edges:

$$E_1 \Box E_2 = \underbrace{\{\{x\} \times e : x \in V_1 \text{ and } e \in E_2\}}_{A_1} \cup \underbrace{\{e \times \{u\} : e \in E_1 \text{ and } u \in V_2\}}_{A_2}.$$

Note that up to the isomorphism the Cartesian product is commutative and associative. That allows us to denote simply by $(v_1, \ldots, v_k)$ the vertices of $V_1 \times \ldots \times V_k$. The figure 1 illustrates the notion of Cartesian product of hypergraphs.

We conclude this section with well-known facts about Cartesian products:

Lemma 1. *$A_1 \cap A_2 = \emptyset$. Moreover, $|e \cap e'| \leq 1$ for any $e \in A_1$, $e' \in A_2$.*

Proposition 1. *Let H_1 and H_2 be two hypergraphs and H their Cartesian product. Then $[H_1]_2 \Box [H_2]_2 = [H]_2$.*

2 Hypergraph Factorization Algorithm

In the sequel, we use some of the results from [IP07].
Cartesian coloring, layers, projections, coordinates, j-edges. In order to find prime factorizations of hypergraphs, we extend the algorithm given in [IP07]. This algorithm is based on a coloring of the edges of the graph G revealing the Cartesian structure of G. In the following, such a coloring will be adapted for hypergraphs and evoked as "*Cartesian coloring of the hyperedges*". Indeed, the hypergraph H to be factorized contains isomorphic copies of the factors which lay as proper subgraphs of H. These isomorphic copies are called *layers.* If $H = H_1 \Box \ldots \Box K_k$, a H_i*-layer* $(1 \leqslant i \leqslant k)$ can be defined more formally as a partial hypergraph $H_i^* = (V_i^*, E_i^*)$ such that E_i^* is a maximal set of edges where the coordinates of their endpoints are fixed except the ith one, and $V_i^* = \bigcup_{e \in E_i^*} e$. By coloring with the same color these layers, we thus reveal what factors H is made of. In the sequel, the colorings we will deal with will mainly refer to these edge colorings. From this angle of view, layers can be identified to the colors covering their edges. For all $w \in V$, $H_i^w = (V_i^w, E_i^w)$ will stand for the H_i-layer incident to w. For a product $H = H_1 \Box H_2 \Box \ldots \Box H_l$, we define the *projection on the ith coordinate* $p_i : \mathcal{V}(H) \to \mathcal{V}(H_i)$ as the mapping $p_i : (x_1, x_2, \ldots, x_l) \mapsto x_i$. Note that $p_{i|\mathcal{V}(H_i^w)}$ is a hypergraph isomorphism, for all $w \in \mathcal{V}(H)$. It can be easily shown that every hyperedge of H is contained in exactly one layer. Moreover the hyperedge sets of the layers partition the hyperedge set of H. If we decide to assign edges colors corresponding to the layer they belong to, we may call j-edges edges laying in a H_j-layer, and may assign them the colour j.

Lemma 2 (Square lemma). *[IP07] Let G be a graph. If two edges of G are adjacent edges which belong to non-isomorphic layers, then these edges lay in a unique induced square.*

A straightforward consequence of the Square Lemma is that any triangle of G is necessarily contained in a single layer. Another consequence is that opposite edges in squares have the same color. In the sequel, we will say that the *square lemma is verified* when two adjacent edges with different colors lay in exactly one induced square. From the Square lemma we easily get the following result, also given in [IP07]: every clique in a graph is contained a single layer.
The extension to hypergraphs of the algorithm of [IP07] uses L2-sections.

Definition 1 ($L2$-section). *Let $H = (V, E)$ be a hypergraph, we define the L2-section $[H]_{L2}$ of H as the triple $\Gamma = (V, E', \mathcal{L})$ where (V, E') is the 2-section of H and $\mathcal{L} : E' \to \mathcal{P}(E)$ is defined by $\mathcal{L}(\{x, y\}) = \{e : x, y \in e \in E\}$. Conversely,*

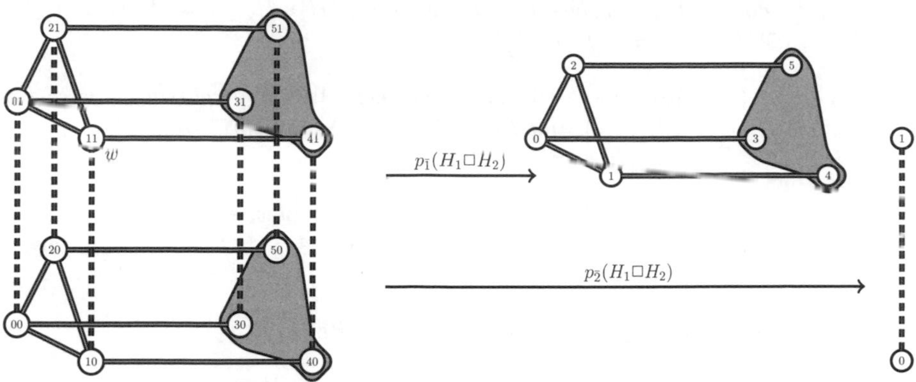

Fig. 2. We adopt the same conventions of representations of the vertices in the drawing as the ones available in figure 1. To the left, the hypergraph $H_1 \Box H_2$ where $H_1 = (\{0,1,2,3,4,5\}, \{\{0,1\},\{0,2\},\{1,2\},\{0,3\},\{1,4\},\{2,5\},\{3,4,5\}\})$ and where $H_2 = K_2$, $\mathcal{V}(K_2) = \{0,1\}$. Dashed edges emphasize K_2-layers. The given decomposition is the prime decomposition of the given hypergraph. The 2-section of the given hypergraph, with its Cartesian coloring is shown in the figure 1 at the left (up to isomorphisms). You can notice that the colors of $H_1 \Box H_2$ can be obtained by merging color classes in figure 1, left. If w stands for the vertex $(1,1)$ then H_1^w is the induced partial hypergraph with vertices $\{(0,1),(1,1),(2,1),(3,1)(4,1),(5,1)\}$. We note that, as expected, only the first coordinate vary for the vertices of this layer. The hypergraph H_2^w is simply $(\{(1,1),(1,0)\}, \{\{(1,1),(1,0)\}\})$. The hypergraph H_1 is represented to the left. Finally we observe that $H_1 = p_1(H) \cong H_1^w$ and $H_2 \cong H_2^w$, more precisely we have $p_1(H_1^w) = H_1$ and $p_2(H_2^w) = H_2$.

if $\Gamma = (V, E', \mathcal{L})$ is a triple where $\mathcal{L}$ is a function from E' to $\mathcal{P}(E)$, we define the inverse L2-section $[\Gamma]_{L2}^{-1} = (V, E)$ of Γ as the pseudo-hypergraph with $E = \bigcup_{e_i \in E'} (\mathcal{L}(e_i))$. The mapping $\mathcal{L}$ is also referred to as a labelling on edges.

It is easy to check that if Γ is a L2-section then $[\Gamma]_{L2}^{-1}$ is a hypergraph. Not surprisingly the following result comes from the definition.

Proposition 2. *For all hypergraphs H and L2-sections Γ we have $[[H]_{L2}]_{L2}^{-1} = H$ and $[[\Gamma]_{L2}^{-1}]_{L2} = \Gamma$.*

The Cartesian product operation is now extended to L2-sections.

Definition 2 (Cartesian product of $L2$-sections). *Let $\Gamma_1 = (V_1, E_1', \mathcal{L}_1)$ and $\Gamma_2 = (V_2, E_2', \mathcal{L}_2)$ be the L2-sections of $H_1 = (V_1, E_1)$ and $H_2 = (V_2, E_2)$. We define their Cartesian product $\Gamma_1 \Box \Gamma_2$ as the triple $(V, E', \mathcal{L}_1 \Box \mathcal{L}_2)$ where:*

- *(V, E') is the Cartesian product of (V_1, E_1') and (V_2, E_2').*
- *$\mathcal{L}_1 \Box \mathcal{L}_2$ is the map from $E' = E_1' \Box E_2'$ to $\mathcal{P}(E_1 \Box E_2)$ defined by:*

$$\mathcal{L}_1 \Box \mathcal{L}_2(\{(x,u),(y,v)\}) = \begin{cases} \{\, \{x\} \times e : e \in \mathcal{L}_2(\{u,v\}) \,\} \text{ if } x = y. \\ \{\, e \times \{u\} : e \in \mathcal{L}_1(\{x,y\}) \,\} \text{ if } u = v. \end{cases}$$

Lemma 3. *For all hypergraphs* H_1, H_2 *we have:* $[H_1 \Box H_2]_{L2} = [H_1]_{L2} \Box [H_2]_{L2}$ *and* $[[H_1]_{L2} \Box [H_2]_{L2}]^{-1}_{L2} = [[H_1]_{L2}]^{-1}_{L2} \Box [[H_2]_{L2}]^{-1}_{L2}$.

Definition 3 (Isomorphism of L2-sections). *An isomorphism between two L2-sections* $\Gamma_1 = (V_1, E'_1, \mathcal{L}_1)$ *and* $\Gamma_2 = (V_2, E_2', \mathcal{L}_2)$ *is a bijection* f *from* V_1 *to* V_2 *such that:*

a) $\{x, y\} \in E'_1$ *if and only if* $\{f(x), f(y)\} \in E'_2$, *for all* $x, y \in V_1$.
b) $e \in \mathcal{L}_1(\{x, y\})$ *if and only if* $\{f(z) : z \in e\} \in \mathcal{L}_2(\{f(x), f(y)\})$, *for all* $x, y \in V_1$ *and* $e \subseteq V_1$.

To express that Γ_1 *and* Γ_2 *are isomorphic, we write:* $\Gamma_1 \cong \Gamma_2$

Lemma 4. *Let* H *and* H' *be two hypergraphs. Then* $H \cong H'$ *is equivalent to* $[H]_{L2} \cong [H']_{L2}$. *If moreover* H *and* H' *are conformal then these statements are equivalent to* $[H]_2 \cong [H']_2$.

Note that the Cartesian product is commutative and associative on L2-sections. That allows us to overlook parenthesis for Cartesian products of L2-sections. Recall that, given a hypergraph H and its "Cartesian" coloring, layers in its 2-section are isomorphic, what may not be the case of the layers on its L2-section (e.g. the labelled subgraphs are not isomorphic even if the non-labelled versions of these subgraphs are). This is why layers for the 2-section in figure 1 do not coincide with the ones of the hypergraph the 2-section comes from, given in figure 2.

In the sequel, the notions of H_i-layers, H_i^w etc. are naturally extended to L2-sections with similar notations (Γ_i-layers, Γ_i^w etc.) by adding the mapping $\mathcal{L}$ to the graph as defined in definition 1.

Definition 4 (G_j-adjacent layers and Γ_j-adjacent layers). *Let* H *be a hypergraph, with 2-section* G *and L2-section* Γ. *We will say that* G_j *(resp.* Γ_j*)-*adjacent layers *are vertex-disjoint layers* G', G'' *(resp.* Γ', Γ''*) such that there exists an edge* $\{x, x'\}$ *laying in a* G_j*- (resp.* Γ_j*-) layer where* $x \in \mathcal{V}(G')$ *and* $y \in \mathcal{V}(G'')$.

The following definition is intended to specify how layers are related one to the other: some colors connect the incident layers by maintaining a mapping between them which is an L2-isomorphism, although others do not. Clearly, when one color does not "define" an isomorphism between two adjacent layers, there is not the slightest risk that the involved layers can be considered in different labelled prime layers. Then this pattern can be envisaged as an obstruction for the colors to define prime hypergraph factors. This is precisely what expresses the $\mathcal{R}^*$ relation introduced below. It will be used further, to build the Cartesian joins, which can be seen as mergings of layers so that the new colors resulting from the mergings are good candidates for the definition of the prime factors.

Definition 5 (Induction of isomorphisms of Γ_i-layers by Γ_j-layers, relations $\mathcal{A}_{i,j}$ and $\mathcal{R}_{i,j}$). *We will say that* Γ_j-layers induce an isomorphism between Γ_i-layers *$(i \neq j)$ if the following property $\mathcal{A}_{i,j}$ is verified:* "For all $w, w' \in \mathcal{V}(\Gamma)$, let Γ_i^w and $\Gamma_i^{w'}$ be two Γ_j-adjacent layers. The graph isomorphism between G_i^w and $G_i^{w'}$ which maps x to x' such that $\{x, x'\}$ is a j-edge gives rise to an isomorphism of $L2$-sections between Γ_i^w to $\Gamma_i^{w'}$." *If this relation $\mathcal{A}_{i,j}$ is not verified and $i \neq j$, we will say that i is in relation with j and we will denote it by $i\mathcal{R}j$.*

In the sequel we will consider the reflexive, symmetric and transitive closure of $\mathcal{R}$, denoted by $\mathcal{R}^*$.

Definition 6 (Set Col of colors of a Cartesian coloring, equivalence classes colors $\bar{i} \in$ Col$/\mathcal{R}^*$, $\bar{i}$-edges). *The relation $\mathcal{R}^*$ is defined as the symmetric, reflexive and transtive closure of $\mathcal{R}$. Given Col the set of the colors involved in the Cartesian coloring of a graph G, the set of these $\mathcal{R}^*$-equivalence classes of colors will be denoted Col$/\mathcal{R}^*$. The equivalence class of colors of i, $i \in \{1, \ldots, k\}$ will be denoted by $\bar{i}$.*
We will talk of $\bar{i}$-edges for edges colored in one color $\mathcal{R}^*$-equivalent to i.

Definition 7 (Cartesian joins $\Gamma_{\bar{i}}^w = (V_{\bar{i}}^w, E_{\bar{i}}^{\prime w}, \mathcal{L}_{\bar{i}}^w)$, projection $p_{\bar{i}}$). *Let H be a hypergraph, $\Gamma = (V, E', \mathcal{L})$ its L2-section, $G = G_1 \square \ldots \square G_k$ its 2-section, and Col the set of the colors involved in the Cartesian coloring of G. Given w a vertex, we define $G_{\bar{i}}^w = (V_{\bar{i}}^w, E'^w_{\bar{i}})$ (resp. $\Gamma_{\bar{i}}^w = (V_{\bar{i}}^w, E'^w_{\bar{i}}, \mathcal{L}_{\bar{i}}^w)$) as the connected component of $\bar{i}$-edges adjacent to w in G (resp. Γ). The graph $G_{\bar{i}}^w$ and the labelled graphs $\Gamma_{\bar{i}}^w$ are called the $\mathcal{R}^*$-induced Cartesian joins; they will be referred to as (Cartesian) joins in the sequel. We also redefine the* projection $p_{\bar{i}}$ *as the mapping which associates to a vertex in the Cartesian product $\prod_{\bar{i} \in Col/\mathcal{R}^*} \Gamma_{\bar{i}}$ the corresponding coordinate in the $\bar{i}$-class.*

These graphs can also be qualified as $G_{\bar{i}}$- (resp. $\Gamma_{\bar{i}}$-) layers; compared with traditional G_i- or Γ_i- layers, these ones are built on color classes, instead of single colors. One can also think of it as layers defined on colors resulting from the merging of elementary colors, the last ones being induced by the Cartesian coloring of the graph G. Notice, moreover, that Cartesian joins are L2-sections.

Proposition 3. *Let H be a hypergraph, $\Gamma = (V, E', \mathcal{L})$ its L2-section, G its 2-section and Col the set of the colors involved in the Cartesian coloring of G. For every w, for every $i \in \{1, \ldots, k\}$, we have: $\Gamma_{\bar{i}}^w \cong \Gamma_{\bar{i}}^{w'}$.*

Now that we know $\Gamma_{\bar{i}}$-layers are isomorphic, we can generalize the notion of induction of Γ_i-layers by Γ_j-layers, by simply considering classes of colors insted of single colors. It also allows us to define $\Gamma_{\bar{i}}$ as the L2-section $p_{\bar{i}}(\Gamma_{\bar{i}}^w)$, for all w.

Lemma 5. *Let $\Gamma = (V, E', \mathcal{L})$ be the Cartesianly colored L2-section of a hypergraph H, and let G_i be the layers of the 2-section G of H. Then, for all $\bar{i} \in Col/\mathcal{R}^*$, $G_{\bar{i}} \cong \prod_{k \in \bar{i}} G_k$.*

Algorithm 1. Hypergraph-prime decomposition

Require: A hypergraph $H = (V, E)$.
Return: The prime factors of H, that is $H_1, H_2, \ldots, H_l$

Compute $\Gamma = (V, E', \mathcal{L})$, the L2-section of H and $G = (V, E')$.
Run the algorithm of [IP07] on G and call $G_1 = (V_1, E'_1), \ldots, G_k = (V_k, E'_k)$ its prime factors.
Define $\mathcal{L}_i$ as the restriction of $\mathcal{L}$ to E'_i, $\Gamma_i = (V_i, E'_i, \mathcal{L}_i)$.
Define the graph $J =$ (Col=$\{1, 2, \ldots, k\}$,$E_{\text{Col}} := \emptyset$) whose connected components (CC) will "correspond to" the prime factors of H.
Define the set of the CCs as $S = \{C_1, \ldots, C_k\}$ with $C_i = \emptyset$ initially, $\forall i \in \{1, \ldots, k\}$.
Define c the investigated CC as 0, and T as an empty array connecting the colors to the CC they belong to.
Let $l = 0$ the number of the investigated CC.
For $i = 1$ to k **do**
  If $T[i]$ is defined **then**
    $c = T[i]$
  Else
    $l = l + 1$, $c = l$, $T[i] = c$
  End If
  For $j = 1$ to k, $j > i$, $j \notin C_c$ **do**
    If $i\mathcal{R}j$ OR $j\mathcal{R}i$ **then**
      $E_{Col} = E_{Col} \cup \{i, j\}$, $C_c = C_c \cup \{j\}$, $T[j] = c$
    End If
  EndFor
EndFor
Return $H_1 = [\Gamma_{C_1}]^{-1}_{L2}, \ldots, H_l = [\Gamma_{C_l}]^{-1}_{L2}$.

Theorem 1. *Let $\Gamma = (V, E', \mathcal{L})$ be the L2-section of a hypergraph H. Let G be the 2-section of H. Suppose $G = G_1 \square G_2 \square \ldots \square G_l$ where $G_i = (V_i, E'_i)$, $i \in \{1, \ldots, l\}$ are layers in G (up to isomorphisms). Define $\Gamma_i = (V_i, E'_i, \mathcal{L}_i)$, where $\mathcal{L}_i$ is the restriction of $\mathcal{L}$ to E'_i. Then*

$$H \cong \prod_{\bar{i} \in (Col/\mathcal{R}^*)} [\Gamma_{\bar{i}}]^{-1}_{L2} .$$

Theorem 2. *Let $\Gamma = (V, E', \mathcal{L})$ be the L2-section of a hypergraph H. Let G be the 2-section of H. Then $H \cong \prod_{\bar{i} \in (Col/\mathcal{R}^*)} [\Gamma_{\bar{i}}]^{-1}_{L2}$ is a prime decomposition of Γ.*

We introduce now an algorithm (algorithm 1) which gives the prime factorization of hypergraphs, derived from theorem 1. The idea is the following. From the connected hypergraph H it first builds the $L2$-section Γ of H. Then it runs the algorithm of Imrich and Peterin which colors the edges of the unlabelled underlying graph G. The color i is used for all edges of all layers that belong to the same factor G_i. When obtained the prime factorization of G, say $G_1, \ldots, G_k$ of G, we determine the $\mathcal{R}^*$-induced Cartesian joins thanks to $\mathcal{L}$. To determine the joins, we consider a graph whose vertices are the colors obtained from the prime decomposition of G (i.e. 1, ..., k), and whose edges express an "obstruction" for the prime factors of G to be considered as labelled prime factors of Γ (this obstruction exists from the moment the colors are $\mathcal{R}$-related). Then the Cartesian joins are the connected components in this new graph.

Finally, by using the inverse L2-section operation, we build back the factors of H.

Theorem 3. *Algorithm 1 is correct.*

This is deduced from the previous results.

If we rely on special data structures, the overall complexity of this algorithm is in $O((\log_2 n)^2 r^3 m \Delta^2))$, where Δ stands for the maximal degree of H, and n stands for the number of vertices of H. This leads to a polynomial algorithm in $O(nm)$ when r, Δ are fixed.

References

[Ber89] Berge, C.: Hypergraphs. North Holland, Amsterdam (1989)

[Bre04] Bretto, A.: Introduction to hypergraph theory and its use in engineering and image processing. Advances in imaging and electron physics 131 (2004)

[Bre06] Bretto, A.: Hypergraphs and the helly property. Ars Combinatoria 78, 23–32 (2006)

[IKR08] Imrich, W., Klavžar, S., Rall, D.F.: Topics in graph theory. Graphs and their Cartesian product. A K Peters, Wellesley (2008)

[IP07] Imrich, W., Peterin, I.: Recognizing cartesian products in linear time. Discrete Mathematics 307, 472–483 (2007)

[IPv97] Imrich, W., Pisanski, T., Žerovnik, J.: Recognizing cartesian graph bundles. Discrete Mathematics 167-168, 393–403 (1997)

[Pet07] Peterin, I.: Game chromatic number of cartesian product graphs. Electronic Notes in Discrete Mathematics 29, 353–357 (2007)

[Sab60] Sabidussi, G.: Graph multiplication. Mathematische Zeitschrift 72 (1960)

[Ves02] Vesel, A.: Channel assignment with separation in the cartesian product of two cycles. In: Proceedings of the 24th International Conference on Information Technology Interfaces (2002)

[Viz63] Vizing, V.G.: The cartesian product of graphs. Vycisl. Sistemy 9, 30–43 (1963)

[ZS09] Zhang, Y., Sun, X.: The antimagicness of the cartesian product of graphs. Theor. Comput. Sci. 410(8-10), 727–735 (2009)

Minimum-Segment Convex Drawings of 3-Connected Cubic Plane Graphs
(Extended Abstract)

Sudip Biswas[1], Debajyoti Mondal[1],
Rahnuma Islam Nishat[2], and Md. Saidur Rahman[1]

[1] Department of Computer Science and Engineering,
Bangladesh University of Engineering and Technology (BUET), Bangladesh
[2] Institute of Information and Communication Technology,
Bangladesh University of Engineering and Technology (BUET), Bangladesh
sudippp@yahoo.com, debajyoti_mondal_cse@yahoo.com,
nishat.buet@gmail.com, saidurrahman@cse.buet.ac.bd

Abstract. A convex drawing of a plane graph G is a plane drawing of G, where each vertex is drawn as a point, each edge is drawn as a straight line segment and each face is drawn as a convex polygon. A maximal segment is a drawing of a maximal set of edges that form a straight line segment. A minimum-segment convex drawing of G is a convex drawing of G where the number of maximal segments is the minimum among all possible convex drawings of G. In this paper, we present a linear-time algorithm to obtain a minimum-segment convex drawing Γ of a 3-connected cubic plane graph G of n vertices, where the drawing is not a grid drawing. We also give a linear-time algorithm to obtain a convex grid drawing of G on an $(\frac{n}{2}+1) \times (\frac{n}{2}+1)$ grid with at most $s_n + 1$ maximal segments, where $s_n = \frac{n}{2} + 3$ is the lower bound on the number of maximal segments in a convex drawing of G.

Keywords: Graph drawing, Convex drawing, Minimum-segment, Grid drawing, Cubic graph.

1 Introduction

A *straight-line drawing* Γ of a plane graph G is a plane drawing of G where each vertex of G is drawn as a point and each edge of G is drawn as a straight line segment. Any two clockwise consecutive edges incident to a vertex v in Γ form an *angle* θ_v at the vertex v. For $\theta_v = 180°$, we call θ_v a *straight angle*. We call a set of edges S a *maximal segment* in Γ if S is a maximal set of edges that form a straight line segment in Γ. A *minimum-segment convex drawing* of G is a convex drawing of G, where the number of maximal segments is the minimum among all possible convex drawings of G. We call a straight-line drawing of G a *straight-line grid drawing* where the vertices are drawn on integer grid points.

Convex drawings of plane graphs is one of the classical and widely studied drawing styles for plane graphs. Although not every plane graph has a convex

M.T. Thai and S. Sahni (Eds.): COCOON 2010, LNCS 6196, pp. 182–191, 2010.

drawing, every 3-connected plane graph has such a drawing [10]. Several algorithms are known for finding convex drawings of plane graphs which improve various aesthetic qualities of the drawings [1,2,3]. Dujmović *et al.* first addressed the problem of obtaining a drawing of a planar graph with few segments [4]. They have also shown that any 3-connected cubic plane graph (i.e., a plane graph where every vertex of the graph has degree three) admits a drawing with at most $n + 2$ maximal segments. Recently, Rahman *et al.* have given a linear-time algorithm for computing a minimum-segment drawing of a "series-parallel graph" with the maximum degree three [9]. These recent works motivated us to study the problem of finding minimum-segment convex drawings of plane graphs.

In this paper we give a linear-time algorithm to obtain a minimum-segment convex drawing Γ of a 3-connected cubic plane graph G with n vertices. We also give a linear-time algorithm to obtain a convex grid drawing of G on an $(\frac{n}{2}+1) \times (\frac{n}{2}+1)$ grid with at most $s_n + 1$ maximal segments, where $s_n = \frac{n}{2} + 3$ is the lower bound on the number of maximal segments in a convex drawing of G. Although several drawing styles are known for 3-connected cubic plane graphs [5,6,7], to the best of our knowledge this is the first work on near-optimal minimum-segment convex drawing where the drawing is a grid drawing.

We now present an outline of our algorithm for grid drawing. Let G be a 3-connected cubic plane graph with n vertices. We partition the input graph G into several vertex disjoint subsets by a "canonical decomposition" described in [8]. We add each subset one after another to construct a drawing Γ of G incrementally. At each addition we ensure that each vertex of degree three in the resulting drawing gets a straight angle except two of the vertices at the initial subset. We add the last subset in such a way that, at the end of the construction, there are at least $n - 4$ straight angles which are associated with $n - 4$ different vertices of G. Using this property of Γ we prove that Γ has at most $s_n + 1$ maximal segments. Figure 1 depicts a "canonical decomposition" of a 3-connected cubic plane graph G . Figures 1(a)–(g) illustrate the incremental construction of a convex grid drawing Γ of G.

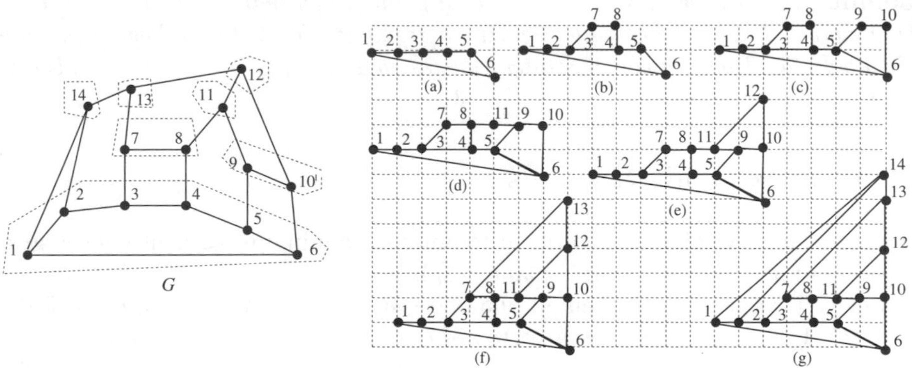

Fig. 1. Illustration of the algorithm for convex grid drawing

The rest of this paper is organized as follows. Section 2 presents some definitions and preliminary results. Section 3 gives a linear-time algorithm for obtaining a convex grid drawing Γ of a 3-connected cubic plane graph G with at most $s_n + 1$ maximal segments. Section 4 gives a linear-time algorithm to obtain a minimum-segment convex drawing Γ of G where Γ is not a grid drawing. Section 5 concludes the paper suggesting some future works.

2 Preliminaries

For the graph theoretic terminologies used in this paper see [8].

Let G be a 3-connected plane graph. An *internally convex* drawing of G is a straight-line drawing of G where all the inner faces are drawn as convex polygons. Let (v_1, v_2) be an edge on the outer face of G and let $\pi = (V_1, V_2, ..., V_m)$ be an ordered partition of V, that is $V_1 \cup V_2 \cup ... \cup V_m = V$ and $(V_i \cap V_j) = \phi$ for $i \neq j$. We denote by G_k, $1 \leq k \leq m$, the subgraph of G induced by $V_1 \cup V_2 \cup ... \cup V_k$ and by C_k the outer cycle of G_k. Let $\{v_1, v_2, \ldots, v_p\}$, $p \geq 3$, be a set of outer vertices consecutive on C_k such that degree$(v_1) \geq 3$, degree(v_2) =degree$(v_3) = \ldots =$ degree$(v_{p-1}) = 2$, and degree$(v_p) \geq 3$ in G_k. Then we call the set $\{v_2, \ldots, v_{p-1}\}$ an *outer chain* of G. We now describe the properties of a canonical decomposition π of G with the outer edge (v_1, v_2) in the following (a)-(b) [8].

(a) V_1 is the set of all vertices on the inner face containing the edge (v_1, v_2). V_m is a singleton set containing an outer vertex v such that v is a neighbor of v_1 and $v \notin \{v_1, v_2\}$.
(b) For each index k, $2 \leq k \leq m-1$, all vertices in V_k are outer vertices of G_k and the following conditions hold: (1) if $|V_k| = 1$, then the vertex in V_k has two or more neighbors in G_{k-1} and has at least one neighbor in $G - G_k$; and (2) If $|V_k| > 1$, then V_k is an outer chain $\{z_1, z_2, ..., z_l\}$ of G_k.

We now have the following lemma whose proof is omitted in this extended abstract.

Lemma 1. *Let G be a 3-connected cubic plane graph with n vertices and let Γ be a convex drawing of G with x maximal segments. Then the following (a) and (b) hold. (a) The number of straight angles in Γ is $\frac{3n}{2} - x$, (b) the number of maximal segments in Γ is at least $\frac{n}{2} + 3$.*

3 Convex Grid Drawings

In this section we give an algorithm to obtain a minimum-segment convex grid drawing of a 3-connected cubic plane graph.

Let G be a 3-connected cubic plane graph with a canonical decomposition $\pi = (V_1, V_2, ..., V_m)$. Let z be a vertex of G and $P(z)$, $x(z)$, $y(z)$ be the (x, y)-coordinates of z, the x-coordinate of z and the y-coordinate of z, respectively. We will later associate a set with each vertex z of G, which we denote by $L(z)$. By $\Gamma(G)$ we denote a drawing of G.

Let $V_1 = (z_1 = v_1, z_2, ..., z_{l-1} = v_3, z_l = v_2)$. We draw G_1 by a triangle as follows. Set $P(z_i) = (i-1, 0)$ for $1 \leq i < l$, $P(z_l) = (l-1, -1)$ and $L(z_i) = \{z_i\}$ for $1 \leq i \leq l$. Let $\Gamma(G)$ be a straight-line drawing of G. We now install $V_2, V_3, ..., V_m$ one after another to construct $\Gamma(G_2), \Gamma(G_3), \ldots, \Gamma(G_m) = \Gamma(G)$, respectively. V_k is either a singleton set or an outer chain of G_k, but in the algorithm we will treat both cases uniformly. We now explain how to install V_k to $\Gamma(G_{k-1})$. We denote by $C_k = (w_1 = v_1, w_2, ..., w_t = v_2)$ the outer cycle of G_k. Let w_p and w_q be the leftmost and rightmost neighbors of V_k on C_{k-1} where $1 < k \leq m$. For each $V_k = \{z_1, z_2, \ldots, z_l\}$, $1 < k < m$, we set $L(z_1) = \{z_1\} \cup (\bigcup_{x=p}^{t} L(w_x))$ and $L(z_i) = (\bigcup_{x=i}^{l} \{z_x\}) \cup (\bigcup_{x=q}^{t} L(w_x))$, where $2 \leq i \leq l$. For each V_k, $2 \leq k \leq m$, let $D_x = |x(w_q) - x(w_p)|$ and $D_y = |y(w_q) - y(w_p)|$. We now have Lemma 2.

Lemma 2. *Let $G_k = V_1 \cup V_2 \cup ... \cup V_k$, $1 \leq k \leq m-1$. Then G_k admits a straight-line drawing $\Gamma(G_k)$ which is internally convex and the slopes of the maximal segments in $\Gamma(G_k)$ are in $\{0, 1, \infty, \lambda_1, \lambda_2\}$ where λ_1 and λ_2 are the slopes of (v_1, v_2) and (v_2, v_3), respectively. Moreover, each vertex $v \notin \{v_1, v_2, v_3\}$ of degree three in $\Gamma(G_k)$ has a straight angle and no two vertices of degree two have the same x-coordinate in $\Gamma(G_k)$.*

Outline of the Proof. We will prove the claim by induction on k.

The case for $k = 1$ is trivial since G_1 is drawn as a triangle, where v_1, v_2 and v_3 are the corner vertices of the triangle. The slopes of the maximal segments in $\Gamma(G_1)$ are in $\{0, \lambda_1, \lambda_2\}$. We may thus assume that k is greater than one and the claim holds for $\Gamma(G_{k-1})$. We now add $V_k = (z_1, z_2, ..., z_l)$ to $\Gamma(G_{k-1})$ to obtain $\Gamma(G_k)$. As stated earlier,we denote by $C_k = (w_1 = v_1, w_2, ..., w_t = v_2)$ the outer cycle of G_k and we denote by w_p and w_q the leftmost and rightmost neighbors of V_k on C_{k-1}, $1 < k < m$. Clearly, each V_k, $2 \leq k \leq m-1$, adds one face f_k to $\Gamma(G_{k-1})$. We assume that each of the vertices v_1, v_2 and v_3 has a straight angle throughout our drawing method. This assumption helps us to avoid repetitive descriptions considering different cases for v_1, v_2 and v_3. We now have the following four cases to consider.

***Case* 1:** Both of w_p and w_q have straight angles in $\Gamma(G_{k-1})$.

Since the vertices $w_{p+1}, \ldots, w_{q-1}$ must be of degree three, each of those vertices must have a straight angle by induction hypothesis. Therefore, $y(w_p) = y(w_q)$ when $w_p \neq v_3$ and $w_q \neq v_2$. Moreover, it is trivial to observe that $y(w_p) > y(w_q)$ when $w_q = v_2$. Hence if $|V_k| = 1$, then we set $P(z_1)$=$(x(w_q), y(w_p) + D_x)$. Otherwise we shift $\bigcup_{i=q}^{t} L(w_i)$ by $|V_k| - D_x$ unit to the right when $|V_k| - D_x > 0$. Then we set $P(z_i) = (x(w_p)+i, y(w_p)+1)$, $1 \leq i < l$, and $P(z_l) = (x(w_q), y(w_p)+1)$.

***Case* 2:** Only w_q has a straight angle in $\Gamma(G_{k-1})$.

Consider first the case where the slope of (w_{p-1}, w_p) is $+1$. Then $y(w_p) \geq y(w_q)$ since each vertex $v \in \{w_{p+1}, \ldots, w_{q-1}\}$ has degree three in G_{k-1} and has a straight corner in $\Gamma(G_{k-1})$ when $v \neq v_3$ by induction hypothesis. Therefore, if $|V_k| = 1$, we set $P(z_1) = (x(w_q), y(w_p) + D_x)$. Otherwise we shift $\bigcup_{i=q}^{t} L(w_i)$ to the right by $|V_k| - D_x$ units when $|V_k| - D_x > 0$. Then we set $P(z_i)$=$(x(w_p) +$

$i, y(w_p)+1)$ for $1 \le i < l$ and $P(z_l) = (x(w_q), y(w_p)+1)$. Consider next the case where the slope of (w_{p-1}, w_p) is 0. By a similar way as shown above it can be shown that $y(w_p) > y(w_q)$. We shift $\bigcup_{i=q}^{t} L(w_i)$ to the right by $|V_k| - D_x$ units when $|V_k| - D_x > 0$. For $|V_k| = 1$, we set $P(z_1) = (x(w_q), y(w_p))$. Otherwise we set $P(z_i) = (x(w_p) + i, y(w_p))$, $1 \le i < l$, and $P(z_l) = (x(w_q), y(w_p))$.

Case 3: Only w_p has a straight angle in $\Gamma(G_{k-1})$.

We now consider the case where the slope of (w_q, w_{q+1}) is 0. Then the slope of (w_{q-1}, w_q) is +1 and by a similar way as in Case 2 it can be shown that $y(w_p) < y(w_q)$. If $|V_k| = 1$, we set $P(z_1) = (x(w_p) + D_y, y(w_q))$. Otherwise we shift $\bigcup_{i=q}^{t} L(w_i)$ to the right by $|V_k| + D_y - D_x$ units when $|V_k| + D_y - D_x > 0$. Then we set $P(z_i)$=$(x(w_p) + D_y + i - 1, y(w_q))$ for $1 \le i \le l$ as illustrated in Fig. 2(a). Consider next the case where the slope of (w_q, w_{q+1}) is ∞. If the slope of (w_{q-1}, w_q) is 0 then in a similar way as in Case 2 one can observe that $y(w_p) = y(w_q)$. Otherwise (w_{q-1}, w_q) belongs to a maximal segment with the slope +1 which implies that $y(w_p) < y(w_q)$ and $D_x > D_y$. Therefore, if $|V_k| = 1$, we set $P(z_1) = (x(w_q), y(w_p)+D_x)$ as illustrated in Fig. 2(b). Otherwise we shift $\bigcup_{i=q}^{t} L(w_i)$ to the right by $|V_k| + D_y - D_x$ units when $|V_k| + D_y - D_x > 0$ and set $P(z_i)$=$(x(w_p) + D_y + i, y(w_q) + 1)$, $1 \le i < l$, and $P(z_l)$=$(x(w_q), y(w_q) + 1)$, as illustrated in Fig. 2(c).

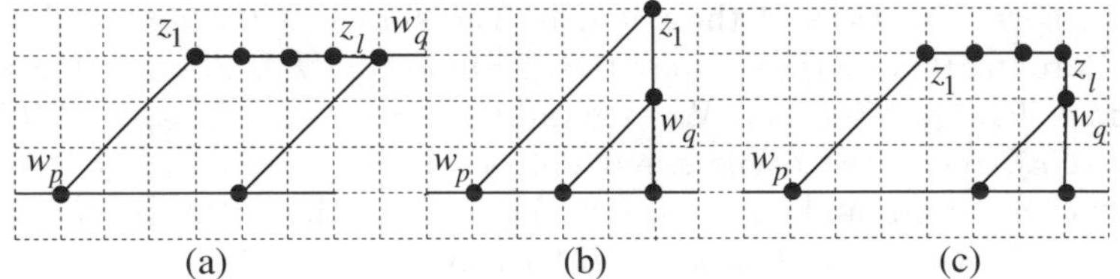

Fig. 2. Only w_p has a straight angle

Case 4: None of w_p and w_q has a straight angle in $\Gamma(G_{k-1})$.

Subcase 4a: The slope of (w_{p-1}, w_p) is 1 and the slope of (w_q, w_{q+1}) is ∞.

Consider first the case where $|V_k| = 1$. If, $y(w_p) < y(w_q)$ then (w_{q-1}, w_q) belongs to a maximal segment with the slope +1 and $D_x > D_y$. Therefore we set $P(z_1) = (x(w_q), y(w_p) + D_x)$. We next consider the case where $|V_k| > 1$ and $y(w_p) \ge y(w_q)$. We shift $\bigcup_{i=q}^{t} L(w_i)$ to the right by $|V_k| - D_x$ units when $|V_k| - D_x > 0$ and set $P(z_i) = (x(w_p) + i, y(w_p) + 1)$, $1 \le i < l$, and $P(z_l) = (x(w_q), y(w_p) + 1)$. Otherwise $|V_k| > 1$ and $y(w_p) < y(w_q)$ and we shift $\bigcup_{i=q}^{t} L(w_i)$ to the right by $|V_k|+D_y-D_x$ units when $|V_k|+D_y-D_x > 0$. Then we set $P(z_i)$=$(x(w_p)+D_y+i, y(w_q)+1)$, $1 \le i < l$, and $P(z_l) = (x(w_q), y(w_q)+1)$.

Subcase 4b: The slope of (w_{p-1}, w_p) is 1 and the slope of (w_q, w_{q+1}) is 0.

We first consider the case where $y(w_p) \ge y(w_q)$. Then we choose a vertex w_{q-i} for the smallest i, $p < q - i < q$, where w_{q-i} has one edge with slope ∞

or two edges with slope 0. Clearly, there exists such a w_{q-i}. We set $P(w_j) = (x(w_{q-i}), y(w_j))$ where $q-i < j \leq q$. For $|V_k| = 1$, we set $P(z_1) = (x(w_q), y(w_p)+D_x)$ as illustrated in Fig. 3(a). After this modification, every w_j, $q-i \leq j < q$, which had a straight angle in $\Gamma(G_{k-1})$ still has a straight angle. Otherwise we shift $\bigcup_{i=q}^{t} L(w_i)$ to the right by $|V_k| - D_x$ units when $|V_k| - D_x > 0$. Then we set $P(z_i) = (x(w_p)+i, y(w_p)+1)$, $1 \leq i < l$, and $P(z_l) = (x(w_q), y(w_p)+1)$ as illustrated in Fig. 3(b). We next consider the case where $y(w_p) < y(w_q)$. Then we shift $\bigcup_{i=q}^{l} L(w_i)$ to the right by $|V_k|+D_y-D_x$ units when $|V_k|+D_y-D_x > 0$ and set $P(z_i) = (x(w_p)+D_y+i-1, y(w_q))$, $1 \leq i \leq l$, as illustrated in Fig. 3(c).

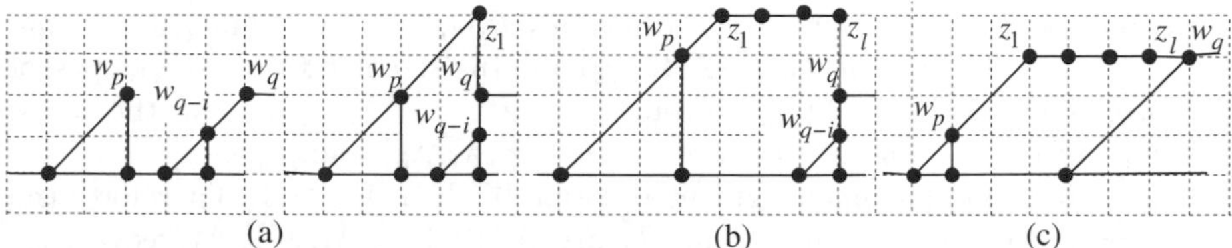

Fig. 3. None of w_p, w_q has straight angle, the slope of (w_{p-1}, w_p) is 1, the slope of (w_q, w_{q+1}) is 0

Subcase 4c: The slope of (w_{p-1}, w_p) is 0 and the slope of (w_q, w_{q+1}) is ∞.

We consider first the case where $y(w_p) > y(w_q)$. Let $|V_k| = 1$, then we set $P(z_1) = (x(w_q), y(w_p))$. Otherwise $|V_k| > 1$ and we shift $\bigcup_{i=q}^{t} L(w_i)$ to the right by $|V_k|-D_x$ units when $|V_k|-D_x > 0$ and set $P(z_i) = (x(w_p)+i, y(w_p))$, $1 \leq i < l$, and $P(z_l) = (x(w_q), y(w_p))$. We next consider the case where $y(w_p) \leq y(w_q)$. Then we choose a vertex w_{p+i} for the smallest i, $p < p+i < q$, where w_{p+i} has one edge with slope 1 or two edges with slope 0. Clearly, there exists such a w_{p+i}. We set $P(w_j) = (x(w_j)+y(w_j)-y(w_{p+i}), y(w_j))$ where $p \leq j < p+i$. After this modification, every w_j, $p < j \leq p+i$, which had a straight angle in $\Gamma(G_{k-1})$ still has a straight angle. We now shift $\bigcup_{i=q}^{t} L(w_i)$ to the right by $|V_k|+D_y-D_x$ units when $|V_k|+D_y-D_x > 0$. Let $|V_k| = 1$, then we set $P(z_1) = (x(w_q), y(w_p)+D_x)$. Otherwise $|V_k| > 1$ and we set $P(z_i) = (x(w_p)+D_y+i, y(w_q)+1)$, $1 \leq i < l$, and $P(z_l) = (x(w_q), y(w_q)+1)$.

Subcase 4d: The slope of (w_{p-1}, w_p) is 0 and the slope of (w_q, w_{q+1}) is 0.

Consider the case where $y(w_p) > y(w_q)$. Then we choose a vertex w_{q-i} for the smallest i, $p < q-i < q$, where w_{q-i} has one edge having slope ∞ or two edges having slope 0. Clearly, there exists such a w_{q-i}. We set $P(w_j) = (x(w_{q-i}), y(w_j))$ for $q-i < j \leq q$. After this modification, every w_j, $q-i \leq j < q$, which had a straight angle in $\Gamma(G_{k-1})$ still has a straight angle. Let $|V_k| = 1$, then we set $P(z_1) = (x(w_q), y(w_p))$. Otherwise $|V_k| > 1$ and we shift $\bigcup_{i=q}^{t} L(w_i)$ to the right by $|V_k|-D_x$ units when $|V_k|-D_x > 0$ and set $P(z_i) = (x(w_p)+i, y(w_p))$, $1 \leq i < l$ and $P(z_l) = (x(w_q), y(w_p))$. Consider next the case where $y(w_p) < y(w_q)$. Then we choose a vertex w_{p+i} for the smallest i, $p < p+i < q$, where w_{p+i} has one edge with slope 1 or two edges with slope 0. Clearly, there exists such a w_{p+i}. We set $P(w_j) = (x(w_j)+y(w_j)-y(w_{p+i}), y(w_j))$ where $p \leq j < p+i$. After this modification, every $w_j, p < j \leq p+i$, which had a straight angle in $\Gamma(G_{k-1})$ still has a straight angle. We now shift $\bigcup_{i=q}^{t} L(w_i)$ to the right by $|V_k|+D_y-D_x$ units when $|V_k|+D_y-D_x > 0$. We set $P(z_i) = (x(w_p)+D_y+i-1, y(w_q))$, $1 \leq i \leq l$. Finally, if $y(w_p) = y(w_q)$ we shift $\bigcup_{i=q}^{t} L(w_i)$ to the right by $|V_k|-D_x+1$ units when $|V_k|-D_x+1 > 0$. We set $P(z_i) = (x(w_p)+i, y(w_p))$, $1 \leq i \leq l$.

By induction hypothesis, $\Gamma(G_{k-1})$ is internally convex and the slopes of the maximal segments in $\Gamma(G_{k-1})$ are in $\{0, 1, \infty, \lambda_1, \lambda_2\}$. Moreover, each vertex $v \notin \{v_1, v_2, v_3\}$ of degree three in $\Gamma(G_{k-1})$ has a straight angle and no two vertices of degree two have the same x-coordinate in $\Gamma(G_{k-1})$. By Lemma 3, shifting of $\bigcup_{i=q}^{t} L(w_i)$ to the right keeps $\Gamma(G_{k-1})$ internally convex and the slopes of the maximal segments in $\Gamma(G_{k-1})$ remains in $\{0, 1, \infty, \lambda_1, \lambda_2\}$. In all the cases, the slope of (w_p, z_1) and (z_l, w_q) are in $\{0, 1, \infty, \lambda_1, \lambda_2\}$ and (z_1, z_2), (z_2, z_3),..., (z_{l-1}, z_l) form one maximal segment of slope 0 when $|V_k| > 1$. Therefore, the slopes of the maximal segments in G_k are in $\{0, 1, \infty, \lambda_1, \lambda_2\}$. Moreover, w_p and w_q are the two new vertices which become vertices of degree three in G_k. Each of w_p and w_q obtains a straight angle according to our drawing method. According to the installation of V_k, the vertices of V_k do not have any concave angle inside f_k. Moreover, the vertices of f_k which are in $G_{k-1} - \{w_p, w_q\}$ are the vertices of degree three in $\Gamma(G_{k-1})$ and each of these vertices has a straight angle by the induction hypothesis. These straight angles remain the same after the shift by Lemma 3. Therefore, f_k is a convex polygon and $\Gamma(G_k)$ is internally convex. It is observable form the properties of canonical decomposition that, all the vertices of degree two are on C_{k-1} and the vertices $w_{p+1}, \ldots, w_{q-1}$ are of degree three on C_{k-1}. According to the installation of V_k, the x-coordinates of the vertices $z_1, \ldots, z_l$ are different and for each vertex v of degree two on C_{k-1}, either $x(v) < x(z_1)$ or $x(z_l) > x(v)$. Moreover, the vertices of degree two on C_{k-1} have different x-coordinates by induction hypothesis . Therefore, no two vertex of degree two has the same x-coordinate in $\Gamma(G_k)$. □

The proof of Lemma 2 gives a method of obtaining $\Gamma(G_{m-1})$. Let $(w_p = v_1)$, v_4 and w_q be the three neighbors of V_m where $x(w_p) < x(v_4) < x(w_q)$. We now set $P(V_m) = (x(v_2), y(v_4) + x(v_2) - x(v_4))$ and add V_m to $\Gamma(G_{m-1})$ to complete the drawing $\Gamma(G_m) = \Gamma(G)$. It is obvious that the addition of V_m does not create any edge crossing. Let λ_3 be the slope of (v_1, V_m). Then clearly all the slopes of $\Gamma(G)$ is in $\{0, 1, \infty, \lambda_1, \lambda_2, \lambda_3\}$. Thus we have an algorithm for obtaining a convex grid drawing of a 3-connected cubic graph which we call Algorithm **Cubic Drawing**. We now have Lemma 3 and Lemma 4 where the proof of Lemma 3 has been omitted in this extended abstract.

Lemma 3. *Let $G_k = V_1 \cup V_2 \cup \ldots \cup V_k$, $1 \leq k \leq m-1$ where $\Gamma(G_k)$ is a drawing of G_k obtained by Algorithm **Cubic Drawing**. Let no two vertices of degree two in $\Gamma(G_k)$ have the same x-coordinate. Let $C_k = (w_1 = v_1, w_2, \ldots, w_t = v_2)$ be the outer cycle of $\Gamma(G_k)$ and let δ be any integer. If the slope of (w_{i-1}, w_i), $2 \leq i \leq t$, is not ∞ and we shift $\bigcup_{i}^{t} L(w_i)$ by δ units to the right, then $\Gamma(G_k)$*

remains internally convex and the number of slopes in $\Gamma(G_k)$ does not increase. Moreover, the slopes of all the maximal segments except (v_1, v_2) and (v_2, v_3) remain the same and no two vertices of degree two have the same x-coordinate.

Lemma 4. *Let G be a 3-connected cubic plane graph with n vertices. Then Algorithm* **Cubic Drawing** *produces a convex drawing $\Gamma(G)$ of G on atmost $(\frac{n}{2}+1) \times (\frac{n}{2}+1)$ grid.*

Proof. Let $W_{\Gamma(G)}$ and $H_{\Gamma(G)}$ be the width and height of $\Gamma(G)$, respectively. Then one can easily observe that $H_{\Gamma(G)} \leq W_{\Gamma(G)}$. We now calculate $W_{\Gamma(G)}$.

According to the reasoning presented in Cases 1–4 of the proof of Lemma 2, if the shift is δ units to the right then $W_{\Gamma(G_k)} = W_{\Gamma(G_{k-1})} + \delta$. If $\delta = |V_k| + D_y - D_x$ then $\delta < |V_k|$ since $D_x \geq D_y + 1$. If $\delta = |V_k| - D_x$ then $\delta < |V_k|$ since $D_x \geq 1$. If $\delta = |V_k| - D_x + 1$ then $\delta < |V_k|$ since $D_x \geq 2$. Finally, if there is no shift then $W_{\Gamma(G_k)} = W_{\Gamma(G_{k-1})}$. Therefore, the width in each step increases by at most $|V_k| - 1$. The installation of V_m creates two inner faces and the installation of each V_i, $1 \leq i \leq m-1$, creates one inner face. Let the number of inner faces of G be F. Then the number of partitions is $F - 1 = \frac{n}{2}$ by Euler's formula. For the installation of each V_i, $1 < i < m$, the width of the drawing increases by at most $|V_k| - 1$. Moreover, the installation of V_m does not require any shift. Therefore, $W_{\Gamma(G)}$ can be at most $|V_1| + \sum_{i=2}^{\frac{n}{2}}(|V_i| - 1) = n - \sum_{i=2}^{\frac{n}{2}} 1 = n - (\frac{n}{2} - 1) = \frac{n}{2} + 1$. Thus the drawing requires at most $(\frac{n}{2}+1) \times (\frac{n}{2}+1)$ grid. □

Theorem 1. *Let G be a 3-connected cubic plane graph. Then Algorithm* **Cubic Drawing** *gives a convex drawing of G in $O(n)$ time with at most $s_n + 1$ maximal segments where s_n is the lower bound on the number of maximal segments in a convex drawing of G.*

Proof. The case for $n = 4$ is trivial and hence we may assume that n is greater than 4. We construct $\Gamma(G_{m-1})$ by installing V_1, V_2,...,V_{m-1} one after another. Then we install V_m to obtain $\Gamma(G_m) = \Gamma(G)$. Let w_p, v_4 and w_q be the three neighbors of V_m where $x(w_p) < x(w_m) < x(w_q)$. Since all the vertices other than $(v_1 = w_p)$, v_4 and w_q are of degree three in G_{m-1}, each of those vertices of degree three except v_2 and v_3 has exactly one straight angle by Lemma 2. Therefore, there are $n-1$ vertices and at least $n-6$ straight angles in $\Gamma(G_{m-1})$ when $w_q \neq v_2$. One can easily observe that, Algorithm **Cubic Drawing** installs V_m in such a way that each of v_4 and w_q obtains a straight angle. Thus the number of straight angles in $\Gamma(G)$ is at least $n - 6 + 2 = n - 4$, in total. Similarly, if $w_q = v_2$ then there are $n-5$ straight angles in $\Gamma(G_{m-1})$ and V_m is installed in such a way that the number of straight angles in $\Gamma(G)$ becomes $n-4$, in total. Let x be the number of maximal segments in $\Gamma(G)$. Then by Lemma 1, $\Gamma(G)$ has at least $\frac{3n}{2} - x = n - 4$ straight angles and at most $x = \frac{n}{2} + 4$ maximal segments. By Lemma 1 the lower bound on the number of maximal segments s_n in a convex drawing of G is $\frac{n}{2} + 3$. Thus the number of maximal segments in $\Gamma(G)$ is at most $s_n + 1$. To obtain a linear-time implementation of the Algorithm **Cubic Drawing**, we use a method similar to the implementation in [3]. □

4 Minimum-Segment Convex Drawings

In this section we give an algorithm, which we call **Draw-Min-Segment**, to obtain a minimum-segment convex drawing $\Gamma(G)$ of a 3-connected cubic plane graph G with $n \geq 6$ vertices in linear time, where $\Gamma(G)$ is not a grid drawing.

We now describe Algorithm **Draw-Min-Segment**. We use canonical decomposition to obtain $V_1, \ldots, V_m$ using the same technique as the one in Section 3. We now draw G_1 by a triangle as follows. Set $P(v_i) = (i-1, 0)$ where $1 \leq i < l$, $P(v_l) = (l-1, -1)$. We add $V_1, \ldots, V_m$ one after another to obtain $\Gamma(G_1), \ldots, \Gamma(G_m) = \Gamma(G)$. Let w_p and w_q be the leftmost and the rightmost neighbors of V_k on $C(G_{k-1})$ where $1 < k \leq m$. We install $V_k = (z_1, \ldots, z_l)$ in such a way that $(z_1, z_2), \ldots, (z_{l-1}, z_l)$ form a segment of slope 0 and the following (a) and (b) hold for each index k, $2 \leq k < m$.

(a) If w_p has a straight angle in $\Gamma(G_{k-1})$, then slope of (w_p, z_1) is $+1$. Otherwise w_p has no straight angle in $\Gamma(G_{k-1})$, then slope of (w_p, z_1) is the same as the slope of (w_{p-1}, w_p).
(b) If $x(w_q)$ is the maximum among all the x-coordinates of the vertices of $\Gamma(G_{k-1})$, then slope of (z_l, w_q) is ∞; otherwise if w_q has a straight angle in $\Gamma(G_{k-1})$, then slope of (z_l, w_q) is -1 and if w_q has no straight angle in G_{k-1}, then slope of (z_l, w_q) is the same as the slope of (w_q, w_{q+1}).

We now have the following lemma.

Lemma 5. *Let G be a 3-connected cubic plane graph with $n \geq 6$ vertices and $\pi = (V_1, V_2, ..., V_m)$ be a canonical decomposition of the vertices of G with outer edge (v_1, v_2). Let $G_k = V_1 \cup V_2 \cup ... \cup V_k$ and $\Gamma(G_k)$ be a drawing of G_k obtained by Algorithm **Draw-Min-Segment**, where $1 \leq k < m$. Then each vertex of degree three in $\Gamma(G_k)$ has a straight angle except the vertices v_1 and v_2.*

Proof. The case for $\Gamma(G_1)$ is trivial and we may thus assume that k is greater than one. By the induction hypothesis, each vertex of degree three in $\Gamma(G_{k-1})$ has a straight angle except the vertices v_1 and v_2. Let w_p and w_q be the leftmost and the rightmost neighbors of V_k on C_{k-1}. By a case analysis similar to the one in Lemma 4, one can observe that, addition of V_k with $\Gamma(G_{k-1})$ to obtain $\Gamma(G_k)$ creates two new vertices of degree three, which are w_p and w_q. Each of w_p and w_q has a straight angle by (Co:a) and (Co:b) when $w_p \notin \{v_1, v_2\}$ and $w_q \notin \{v_1, v_2\}$. Therefore, each vertex of degree three in $\Gamma(G_k)$ has a straight angle except v_1 and v_2. □

We now describe the installation of V_m. Let w_p, v_4 and w_q be the three neighbors of V_m, where $x(w_p = v_1) < x(v_4) < x(w_q)$. We place V_m in such a way that (w_4, V_m) and (w_q, V_m) obtain the slopes $+1$ and ∞, respectively. Then we simply draw the edge (v_1, V_m).

Theorem 2. *Let G_m be a 3-connected cubic graph with $n \geq 6$ and $\Gamma(G)$ be a drawing of G obtained by Algorithm **Draw-Min-Segment**. Then $\Gamma(G)$ is a minimum-segment convex drawing of G obtained in $O(n)$ time.*

Proof. One can use a technique similar to the one presented in the proof of Theorem 1 to prove the claim. □

5 Conclusions

In this paper, we have given a linear-time algorithm to obtain a convex drawing of a 3-connected cubic plane graph. Keszegh *et al.* showed that every graph G with the maximum degree three has a straight-line drawing in the plane using edges of at most five different slopes [7]. It is interesting to observe that the drawing produced by our algorithm uses only six slopes. It is left as a future work to reduce the grid size of the drawing. It also remains as an open problem to obtain minimum-segment convex drawings of other classes of planar graphs.

Acknowledgement

This work is done in Graph Drawing & Information Visualization Laboratory of the Department of CSE, BUET established under the project "Facility Upgradation for Sustainable Research on Graph Drawing & Information Visualization" supported by the Ministry of Science and Information & Communication Technology, Government of Bangladesh. We thank BUET for providing necessary support and the anonymous referees for their useful suggestions which helped us to improve the grid size of the drawing.

References

1. Chiba, N., Onoguchi, K., Nishizeki, T.: Drawing plane graphs nicely. Acta Informatica 22(2), 187–201 (1985)
2. Chiba, N., Yamanouchi, T., Nishizeki, T.: Linear algorithms for convex drawings of planar graphs. In: Progress in Graph Theory, pp. 153–173 (1984)
3. Chrobak, M., Kant, G.: Convex grid drawings of 3-connected planar graphs. International Journal of Computational Geometry and Applications 7(3), 211–223 (1997)
4. Dujmović, V., Suderman, M., Wood, D.R.: Really straight graph drawings. In: Pach, J. (ed.) GD 2004. LNCS, vol. 3383, pp. 122–132. Springer, Heidelberg (2005)
5. Kant, G.: Hexagonal grid drawings. In: Mayr, E.W. (ed.) WG 1992. LNCS, vol. 657, pp. 263–276. Springer, Heidelberg (1993)
6. Kant, G.: Drawing planar graphs using the canonical ordering. Algorithmica 16(1), 4–32 (1996)
7. Keszegh, B., Pach, J., Pálvölgyi, D., Tóth, G.: Drawing cubic graphs with at most five slopes. Computational Geometry 40(2), 138–147 (2008)
8. Nishizeki, T., Rahman, M.S.: Planar Graph Drawing. World Scientific, Singapore (2004)
9. Samee, M.A.H., Alam, M.J., Adnan, M.A., Rahman, M.S.: Minimum segment drawings of series-parallel graphs with the maximum degree three. In: 16th International Symposium on Graph Drawing, pp. 408–419. Springer, Heidelberg (2009)
10. Tutte, W.T.: Convex representations of graphs. London Mathematical Society 10, 304–320 (1960)

On Three Parameters of Invisibility Graphs*

Josef Cibulka[1], Jan Kynčl[2], Viola Mészáros[2,3], Rudolf Stolař[1], and Pavel Valtr[2]

[1] Department of Applied Mathematics,
Charles University, Faculty of Mathematics and Physics,
Malostranské nám. 25, 118 00 Praha 1, Czech Republic
cibulka@kam.mff.cuni.cz, ruda@kam.mff.cuni.cz

[2] Department of Applied Mathematics and
Institute for Theoretical Computer Science,
Charles University, Faculty of Mathematics and Physics,
Malostranské nám. 25, 118 00 Praha 1, Czech Republic
kyncl@kam.mff.cuni.cz

[3] Bolyai Institute, University of Szeged,
Aradi vértanúk tere 1, 6720 Szeged, Hungary
viola@math.u-szeged.hu

Abstract. The invisibility graph $I(X)$ of a set $X \subseteq \mathbb{R}^d$ is a (possibly infinite) graph whose vertices are the points of X and two vertices are connected by an edge if and only if the straight-line segment connecting the two corresponding points is not fully contained in X.

We settle a conjecture of Matoušek and Valtr claiming that for invisibility graphs of planar sets, the chromatic number cannot be bounded in terms of the clique number.

1 Introduction

The *invisibility graph* $I(X)$ of a set $X \subseteq \mathbb{R}^d$ is a graph whose vertices are the points of X and two vertices are connected by an edge if and only if the straight-line segment connecting the two corresponding points is not fully contained in X. Let $\chi(G)$ be the chromatic number of a graph G and let $\omega(G)$ be its clique number. For a set $X \subseteq \mathbb{R}^d$ we define $\gamma(X)$ to be the minimum possible number of convex subsets of X that cover X. Further, let $\chi(X) := \chi(I(X))$ and $\omega(X) := \omega(I(X))$.

Observe that $\omega(X) \leq \chi(X) \leq \gamma(X)$ for any set X.

If a planar set X is closed, then $\gamma(X)$ can be bounded by a function of $\omega(X)$. This was proved by Breen and Kay [1] and the current best known upper bound is $\gamma(X) \leq O(\omega(X)^3)$ by Matoušek and Valtr [5]. From the other direction, there exist examples by Matoušek and Valtr [5] with $\gamma(X) \geq \Omega(\omega(X)^2)$.

* Work on this paper was supported by the project 1M0545 of the Ministry of Education of the Czech Republic. The third author was also supported by OTKA Grant K76099 and by the grant no. MSM0021620838 of the Ministry of Education of the Czech Republic. The first and fourth authors were also supported by the Czech Science Foundation under the contract no. 201/09/H057. The first, second and fifth authors were partially supported by project GAUK 52110.

M.T. Thai and S. Sahni (Eds.): COCOON 2010, LNCS 6196, pp. 192–198, 2010.

However, if we don't restrict ourselves to closed sets, there is no upper bound on $\gamma(X)$ even for sets with $\omega(X) = 3$. An example is the disk D_λ with λ one-point holes punctured in the vertices of a regular convex λ-gon near the boundary of D_λ, for which $\omega(D_\lambda) = 3$, but $\gamma(D_\lambda) = \lceil \lambda/2 \rceil + 1$ (see [5]).

A *one-point hole* in a set $X \subset \mathbb{R}^d$ is a point that forms a path-connected component of $\mathbb{R}^d \setminus X$. Let $\lambda(X)$ be the number of one-point holes in the set X.

The example of the set D_λ led to studying the properties of planar sets with a limited number of one-point holes by Matousek and Valtr [5]. In particular, they proved the following theorem.

Theorem 1 (Matoušek and Valtr [5]). *Let $X \subseteq \mathbb{R}^2$ be a set with $\omega(X) = \omega < \infty$ and $\lambda(X) = \lambda < \infty$. Then*

$$\gamma(X) \le O(\omega^4 + \lambda\omega^2).$$

For any $\omega \ge 3$ and $\lambda \ge 0$ they also found sets X with $\omega(X) = \omega$, $\lambda(X) = \lambda$ and $\gamma(X) \ge \Omega(\omega^3 + \omega\lambda)$.

Matoušek and Valtr [5] conjectured that for an arbitrary planar set X, the value of $\gamma(X)$ is bounded by a function of $\chi(X)$. Thus $\chi(X)$ cannot be bounded by a function of $\omega(X)$ as shows the above example with D_λ. As the main result of this paper, we settle this conjecture.

The *tower function* $T_l(k)$ is defined recursively as $T_0(k) = k$ and $T_h(k) = 2^{T_{h-1}(k)}$. Its inverse is the iterated logarithm $\log^{(l)}(n)$, that is, $\log^{(0)}(n) = n$ and $\log^{(l)}(n) = \log(\log^{(l-1)}(n))$.

Theorem 2. *Any set $X \subseteq \mathbb{R}^2$ with $\chi(X) = \chi$ satisfies*

$$\gamma(X) \le O(\chi^4 + 2^{4T_2(\chi)} \cdot \chi^3).$$

2 Proof of Theorem 2

In this and the next section we will use the following observation about one-point holes.

Observation 3. *Let q be a one-point hole in a set $X \subseteq \mathbb{R}^d$ with $\omega(x) < \infty$. For every vector $x \in \mathbb{R}^d$ there is an $\epsilon > 0$ such that the open segment between q and $q + \epsilon x$ is contained in X.*

Proof. For contradiction, suppose that there is a vector x such that the open segment $r(q, q + \epsilon x)$ is not contained in X for any $\epsilon > 0$. Now either the whole segment $r(q, q + \epsilon x)$ is contained in $\mathbb{R}^d \setminus X$ for some $\epsilon > 0$, or there is an infinite sequence of points of X on $r(q, q + x)$ such that there is at least one point of $\mathbb{R}^d \setminus X$ between every two of them. In the first case, the hole is no longer a path-connected component of $\mathbb{R}^d \setminus X$. In the second case, the sequence of points forms a clique of infinite size. Either way, we get a contradiction. □

Points $q_1, q_2, \ldots, q_n$ are in *clockwise convex position* if they form the vertices of a convex polygon in the clockwise order. In particular, they are in general position and, for every $i < j$, points $q_{j+1}, q_{j+2}, \ldots, q_n$ lie in the same half-plane of the line determined by q_i and q_j.

To derive Theorem 2 from Theorem 1, we need to show that the number of one-point holes is bounded from above by a function of the chromatic number. The following lemma shows this in the special case of one-point holes in convex position. Lemma 5 then allows to generalize it to the case of one-point holes in an arbitrary position.

Lemma 4. *Let $X \subseteq \mathbb{R}^2$ be a set with n one-point holes $q_1, q_2, \ldots, q_n$ in clockwise convex position. Then*

$$n \leq T_2(\chi(X)).$$

Proof. For a pair of indices $1 \leq i < j \leq n$ let $l(i,j)$ be the line containing the point q_i and passing in a small distance $d(i,j) > 0$ (to be specified later) from q_j in the direction from q_j such that q_j lies in the same half-plane as the points $\{q_{j+1}, \ldots q_n\}$.

For a triple of indices $1 \leq i < j < k \leq n$ let $p(i,j,k) := l(i,j) \cap l(j,k)$, which is a point near q_j.

The distances $d(i,j)$ will be set to satisfy the following conditions: The lines $l(i,j)$ are in general position. Each $l(i,j)$ contains exactly one of the holes (which is q_i). For every pair (i,j) the point q_j and all the points $p(j,k,l)$ lie in the same half-plane determined by $l(i,j)$ as the points $\{q_{j+1}, \ldots q_n\}$. Every point $p(i,j,k)$ is in X and is closer to q_j than q_i. See Figure 1.

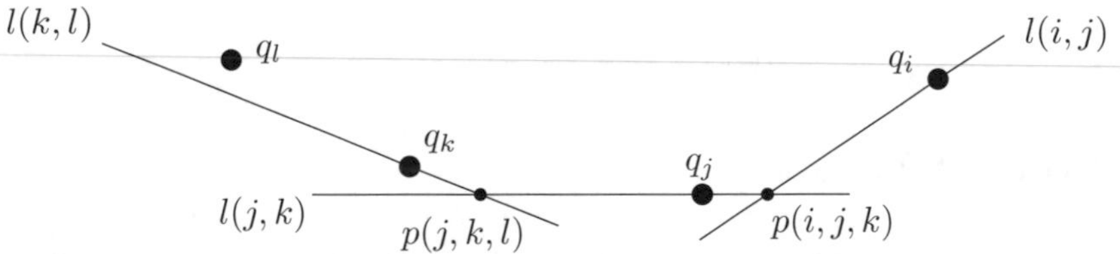

Fig. 1. An example of four one-point holes in clockwise convex position and the corresponding lines

Since the one-point holes are in clockwise convex position, there is an ϵ such that whenever all $d(i,j)$ are greater than 0 and at most ϵ, then all the conditions are satisfied except possibly the one that $p(i,j,k)$ are in X.

We start by setting the distances $d(i,n)$ to this ϵ, which allows us to place the lines $l(i,n)$. Each distance $d(i,j)$, where $j < n$, is set when all the lines $l(j,k)$ are already placed. Because $l(j,k)$ avoids q_i and by Observation 3, there is ϵ_k such that if $d(i,j) \leq \epsilon_k$, then $p(i,j,k)$ lies in X. It is now enough to take $d(i,j)$ as the minimum of ϵ and all the ϵ_k.

Consider the graph with vertex set $\binom{[n]}{3}$ and edges between vertices $\{i,j,k\}$ and $\{j,k,l\}$ for every $1 \leq i < j < k < l \leq n$. This graph is called *the shift graph* $S(n,3)$ and its chromatic number is known to be at least $\log^{(2)}(n)$ [3]. We color the vertex $\{i, j, k\}$ with the color of the point $p(i, j, k)$ in a fixed proper coloring c of X by $\chi(X)$ colors.

Assume that $n > T_2(\chi(X))$. Then there are two points $s := p(i, j, k)$ and $t := p(j, k, l)$ with $c(s) = c(t)$. Both of them lie on the line $l(j, k)$ which also contains the point q_j. To show that q_j lies between s and t we use the fact that s is the intersection of the lines $l(j, k)$ and $l(i, j)$. The points q_i and t lie in the same direction from $l(i, j)$, but q_j is closer to s. Thus s and t are connected by an edge in the invisibility graph of X, a contradiction. □

The following lemma is a slight modification of exercise 3.1.3 from [4].

Lemma 5. *Any set $P \subset \mathbb{R}^2$ of $m \cdot 2^{4n}$ points contains either m points lying on a line or $n+1$ points in convex position.*

Proof. First we show that among any mp^2 points, we can either find m points lying on a line or p points in general position. Take a set Q with no m points on a line and no p points in general position. Consider an arbitrary inclusion-wise maximal subset $S \subseteq Q$ of points in general position. Each point from $Q \setminus S$ must thus lie on a line determined by some pair of points of S. There are fewer than p^2 such lines and each contains fewer than $m - 2$ points from $P \setminus S$. Hence the total number of points of Q is less than mp^2.

From the Erdős-Szekeres theorem [2], any set of 4^n points in general position contains $n + 1$ points in convex position.

Combining these two results we obtain that any set of $m \cdot 2^{4n}$ points in $\mathbb{R}^2$ contains either m points lying on a line or 2^{2n} points in general position and hence $n + 1$ points in convex position. □

Lemma 6. *Any set $X \subseteq \mathbb{R}^2$ with $\lambda(X) = \lambda < \infty$ and $\chi(X) = \chi$ satisfies*

$$\lambda < \chi \cdot 2^{4T_2(\chi)}.$$

Proof. By Lemma 5, any set P of $\chi \cdot 2^{4T_2(\chi)}$ points in $\mathbb{R}^2$ contains either χ points on a line or $T_2(\chi)+1$ points in general position. If X contains a line with χ one-point holes then X cannot be colored with χ colors. In the second case we get the contradiction using Lemma 4. □

We now have all the ingredients to prove Theorem 2.

Proof of Theorem 2. By Lemma 6, the number of one-point holes in X satisfies $\lambda \leq \chi \cdot 2^{4R_3(4;\chi)}$. Thus by Theorem 1

$$\begin{aligned}\gamma(X) &\leq O(\omega^4 + \lambda\omega^2) \\ &\leq O(\chi^4 + \chi \cdot 2^{4R_3(4;\chi)} \cdot \chi^2).\end{aligned}$$ □

3 Higher Dimensions

It is possible that $\gamma(X)$ is bounded by a function of the chromatic number even in $\mathbb{R}^3$. However, an analogue of Theorem 1 will have to consider also holes larger than one-point holes. But the lemmas of Section 2 generalize to an arbitrary dimension. We present these generalizations, because we find them interesting on their own.

We say that a set $\{q_1, \ldots, q_n\}$ of n points in $\mathbb{R}^d$ satisfies the *same side condition* if they are in general position and for every d-tuple $q_{i_1}, q_{i_2}, \ldots, q_{i_d}$, where $i_1 < i_2 < \cdots < i_d$, the set of points $\{q_{i_d+1}, \ldots, q_n\}$ lies in one open half-space determined by the hyperplane spanned by $q_{i_1}, q_{i_2}, \ldots, q_{i_d}$.

Lemma 7. *Let $X \subseteq \mathbb{R}^d$ be a set with n one-point holes $q_1, q_2, \ldots, q_n$ satisfying the same side condition. Then*

$$n \leq T_{2d-2}(\chi(X)).$$

Proof. For a set of indices $1 \leq i_1 < \cdots < i_d \leq n$ let $h(i_1, i_2, \ldots, i_d)$ be the hyperplane containing the points $q_{i_1}, \ldots, q_{i_{d-1}}$ and a point q'_{i_d} near q_{i_d} selected so that the points $\{q_{i_d}, q_{i_d+1}, \ldots q_n\}$ are all in one half-space determined by $h(i_1, \ldots, i_d)$.

For a set of indices $1 \leq i_1 < \cdots < i_{2d-2} \leq n$ let

$$l(i_1, i_2, \ldots, i_{2d-2}) := h(i_1, \ldots, i_d) \cap h(i_2, \ldots, i_{d+1}) \cap \cdots \cap h(i_{d-1}, \ldots, i_{2d-2})$$

and for a set of indices $1 \leq i_1 < \cdots < i_{2d-1} \leq n$ let

$$p(i_1, i_2, \ldots, i_{2d-1}) := h(i_1, \ldots, i_d) \cap h(i_2, \ldots, i_{d+1}) \cap \cdots \cap h(i_d, \ldots, i_{2d-1}).$$

Observe that $l(i_1, i_2, \ldots, i_{2d-2})$ is a line containing the point $q_{i_{d-1}}$. Also observe that $p(i_1, i_2, \ldots, i_{2d-1})$ is a point in the intersection of the lines $l(i_1, \ldots, i_{2d-2})$ and $l(i_2, \ldots, i_{2d-1})$ lying near q_{i_d}. Since hyperplanes determined by sets of one-point holes $\{q_{i_1}, q_{i_2}, \cdots q_{i_d}\}, \ldots, \{q_{i_d}, q_{i_{d+1}}, \cdots q_{i_{2d-1}}\}$ are in general position, for every ϵ there is a δ such that if we move by δ one of the points determining one of the hyperplanes, then the intersection point of the hyperplanes moves by less than ϵ. The new set of hyperplanes will be in general position again so the same holds for this new set as well. By moving from points $q_{i_{2d-1}}, q_{i_{2d-2}} \cdots, q_{i_d}$ one by one we can get to points $q'_{i_{2d-1}}, q'_{i_{2d-2}} \cdots, q'_{i_d}$, respectively. In every step, we use the ϵ smaller than the minimum distance between any two holes minus the distance between the q_{i_d} and the actual intersection point of the hyperplanes in that step. Using Observation 3 we get the proper choice of $q'_{i_d}, q'_{i_{d+1}} \cdots, q'_{i_{2d-1}}$ ensuring the point $p(i_1, i_2, \ldots, i_{2d-1})$ to be inside X.

As in the planar case, we obtain the shift graph $S(n, 2d-1)$ whose chromatic number is known to be at least $\log^{(2d-2)}(n)$ [3]. We color its vertices with colors of the corresponding points $p(i_1, i_2, \ldots, i_{2d-1})$ in a fixed proper coloring c of X by $\chi(X)$ colors.

If $n > T_{2d-2}(\chi(X))$ then there are two points $s := p(i_1, \ldots, i_{2d-1})$ and $t := p(i_2, \ldots, i_{2d})$ with $c(s) = c(t)$. Both of them lie on the line $l(i_2, \ldots, i_{2d-1})$ which also contains the point q_{i_d}. To show that q_{i_d} lies between s and t we observe that s is the intersection of the line $l(i_2, \ldots, i_{2d-1})$ and the hyperplane $h(i_1, \ldots, i_d)$. The point q_{i_d} is near this hyperplane and on the same side as all the points

$q_{i_d+1}, q_{i_d+2}, \ldots, q_n$. Because t is near $q_{i_{d+1}}$, it lies in the same direction from s as q_{i_d}, but q_{i_d} is closer to s. Thus s and t are connected by an edge in the invisibility graph of X, a contradiction. □

Let $g_1(n) := n$. For any $n, d > 1$, let $g_d(n)$ be the smallest number g such that any set $P \subset \mathbb{R}^d$ of g points contains either $g_{d-1}(n)$ points lying on a hyperplane or an n-tuple $\{q_1, \ldots, q_n\} \subset P$ of points that satisfy the same side condition. In other words, any set of g points in $\mathbb{R}^d$ contains either n points on a line or there is an affine subspace of $\mathbb{R}^d$ with n points of P that satisfy the same side condition. If no such number g exists, then we say that $g_d(n)$ is infinite.

In the special case of $d = 2$, Lemma 5 proves $g_2(n) \leq n2^{4n}$.

The following lemma is a slight modification of exercise 5.4.3 from [4].

Lemma 8. *The value $g_d(n)$ is finite for all $n, d > 0$.*

Proof. The *orientation* of a $(d+1)$-tuple of points $(p_1, \ldots, p_{d+1})$ in $\mathbb{R}^d$ is the sign of the determinant of the matrix A whose columns are the d vectors $p_2 - p_1, p_3 - p_1, \ldots, p_{d+1} - p_1$. The orientation is equal to 0 if and only if the points $(p_1, \ldots, p_{d+1})$ lie on a hyperplane. Otherwise it is equal to $+1$ or -1 and determines on which side of the hyperplane spanned by $p_1, \ldots, p_d$ the point p_{d+1} lies.

Let $R_k(l_1, l_2, \ldots, l_c; c)$ be the Ramsey number denoting the smallest number r such that if the hyperedges of a complete k-uniform hypergraph on r vertices are colored with c colors, then for some color i, the hypergraph contains a complete sub-hypergraph on l_i vertices whose hyperedges are all of color i.

We will prove the lemma by showing that $g_d(n) \leq R_{d+1}(n, g_{d-1}(n), n; 3)$ for all $d, n > 1$. We take any set P of at least $R_{d+1}(n, g_{d-1}(n), n; 3)$ points. Then we color the $(d+1)$-tuples of points from P with the colors $-1, 0, +1$ determined by the orientation of the $(d+1)$-tuple.

If there are $g_{d-1}(n)$ points such that each $(d+1)$-tuple of them has orientation 0 and thus lies on a hyperplane, then all $g_{d-1}(n)$ points lie on a common hyperplane.

Otherwise there are, without loss of generality, n points $q_1, \ldots, q_n$ such that each $(d+1)$-tuple of them has orientation $+1$. Then for every d-tuple $(q_{i_1}, q_{i_2}, \ldots, q_{i_d})$, the points $q_{i_d+1}, \ldots, q_n$ lie in one half-space determined by the hyperplane spanned by $\{q_{i_1}, q_{i_2}, \ldots, q_{i_d}\}$. □

Lemma 9. *Any set $X \subseteq \mathbb{R}^d$ with $\lambda(X) = \lambda < \infty$ and $\chi(X) = \chi$ satisfies*

$$\lambda < g_d(T_{2d-2}(\chi)).$$

Proof. We proceed by induction on d.

In $\mathbb{R}^1$, any set X with λ one-point holes needs at least $\lambda + 1$ colors. Thus $\lambda < \chi(X) + 1 = g_1(T_0(\chi))$.

Suppose that X has a set P of $g_d(T_{2d-2}(2d; \chi))$ one-point holes. By Lemma 7, there is no set $P' \subseteq P$ of $T_{2d-2}(2d; \chi)$ one-point holes satisfying the same side condition. Thus by the definition of g_d, there are $g_{d-1}(T_{2d-2}(\chi)) \geq g_{d-1}(T_{2d-4}(\chi))$ one-point holes on one hyperplane and we obtain a contradiction with the induction hypothesis. □

References

1. Breen, M., Kay, D.C.: General decomposition theorems for m-convex sets in the plane. Israel Journal of Mathematics 24, 217–233 (1976)
2. Erdős, P., Szekeres, G.: A combinatorial problem in geometry. Compos. Math. 2, 463–470 (1935)
3. Hell, P., Nešetřil, J.: Graphs and Homomorphisms. Oxford University Press, Oxford (2004)
4. Matoušek, J.: Lectures on Discrete Geometry. Springer, New York (2002)
5. Matoušek, J., Valtr, P.: On visibility and covering by convex sets. Israel Journal of Mathematics 113, 341–379 (1999)
6. Perles, M., Shelah, S.: A closed $(n+1)$-convex set in $\mathbb{R}^2$ is a union of n^6 convex sets. Israel Journal of Mathematics 70, 305–312 (1990)

Imbalance Is Fixed Parameter Tractable

Daniel Lokshtanov[1], Neeldhara Misra[2], and Saket Saurabh[2]

[1] Department of Informatics, University of Bergen, Norway
daniello@ii.uib.no
[2] Institute of Mathematical Sciences, Chennai, India
{neeldhara,saket}@imsc.res.in

Abstract. In the Imbalance Minimization problem we are given a graph $G = (V, E)$ and an integer b and asked whether there is an ordering $v_1 \dots v_n$ of V such that the sum of the imbalance of all the vertices is at most b. The imbalance of a vertex v_i is the absolute value of the difference between the number of neighbors to the left and right of v_i. The problem is also known as the Balanced Vertex Ordering problem and it finds many applications in graph drawing. We show that this problem is fixed parameter tractable and provide an algorithm that runs in time $2^{O(b \log b)} \cdot n^{O(1)}$. This resolves an open problem of Kára et al. [COCOON 2005].

1 Introduction

Graph layout problems are combinatorial optimization problems where the objective is to to find a permutation of the vertex set that optimizes some function of interest. In this paper we focus on the problem of determining the *imbalance* of a given graph. Given a permutation $v_1 \dots v_n$ on the vertex set we define the left and right neighborhood of v_i to be $N_L(v_i) = N(v_i) \cap \{v_1 \dots v_{i-1}\}$ and $N_R(v_i) = N(v_i) \cap \{v_{i+1} \dots v_n\}$ respectively. The *imbalance* of v_i is $||N_L(v_i)| - |N_R(v_i)||$ and the imbalance of the graph G given the ordering $v_1 \dots v_n$ is simply the sum of the imbalances of the individual vertices. The imbalance of G is the minimum imbalance of G over all orderings of V, and an ordering yielding this imbalance is called an *optimal* ordering.

The Imbalance Minimization problem was first defined by Biedl et al. [2]. Here, we are given a graph $G = (V, E)$ on n vertices and m edges and an integer b and asked whether the imbalance of G is at most b. The problem finds a variety of applications in graph drawing [9,10,16,19,20]. The computational aspects of graph imbalance were studied by Biedl et al. [2], who show that computing the imbalance of G is NP-hard. Later Kára et al. [11,12] proved that in fact the problem remains NP-hard for planar graphs with maximum degree six and for 5-regular graphs. In the same paper Kára et al. [11,12] showed that for every fixed value of b the problem can be solved in time $O(n^b(n + m))$. They ask whether it is possible to give an algorithm for Imbalance Minimization such that the exponent of n is independent of b. More specifically, they ask for the existence of a *fixed parameter tractable* algorithm which is an algorithm with running time

M.T. Thai and S. Sahni (Eds.): COCOON 2010, LNCS 6196, pp. 199–208, 2010.

$f(b) \cdot n^{O(1)}$ for some function f depending only on b. Parameterized Complexity provides a framework for the study of such algorithms. We refer to [6,7,15] for an introduction to parameterized algorithms. Subsequently the problem was studied by Gaspers et al. [8] who prove that IMBALANCE MINIMIZATION in fact is equivalent to GRAPH CLEANING. Gaspers et al. [8] use this equivalence to obtain an algorithm to check whether the input graph G has imbalance at most b running in time $O(n^{\lfloor b/2 \rfloor}(n+m))$.

In this paper we provide an algorithm running in time $b^{O(b)} \cdot n^{O(1)}$, thereby improving the previous best algorithm by Gaspers et al. [8] running in time $O(n^{\lfloor b/2 \rfloor}(n+m))$ and resolving the question posed by Kára et al [11,12]. Whereas the previous algorithms are purely combinatorial, we exploit a connection between the imbalance, cutwidth and treewidth of the input graph and combine this with a dynamic programming approach. On the way we prove that given a graph G with treewidth at most t and maximum degree Δ an optimal ordering of G can be computed in time $t! \cdot \Delta^{O(t)} \cdot n^{O(1)}$. This shows that the IMBALANCE MINIMIZATION problem is fixed parameter tractable when parameterized by the treewidth and maximum degree of the input graph. This is in contrast to the seemingly similar CUTWIDTH MINIMIZATION problem for which only an $O(n^{t^2 \Delta})$ time algorithm is known [18].

2 Preliminaries

We use $G = (V, E)$ to refer to a graph with vertex set V (with $n := |V|$), and $T = (V_T, F)$ to denote trees on the vertex set V_T, with one distinguished vertex $r \in V_T$ called the *root* of the tree. The vertices of the tree T will be called *nodes.* For a tree node $t \in V_T$, use the notation T_t to mean the set of all vertices $t' \in T$ that are descendants of t.

A graph $G' = (V', E')$ is a *subgraph* of G if $V' \subseteq V$ and $E' \subseteq E$. The subgraph G' is called an *induced subgraph* of G if $E' = \{uv \in E \mid u, v \in V'\}$, in this case, G' is also called the subgraph *induced by* V' and denoted with $G[V']$. By $N(u)$ we denote the (open) neighborhood of u, that is the set of all vertices adjacent to u, and by $N[u] = N(u) \cup \{u\}$. By degree of a vertex u we mean $|N(u)|$ and denote it by $d(u)$.

We use $[n]$ to refer to the set $\{1, 2, \ldots, n\}$. In the interest of simplicity, we let $V = [n]$ (we will continue to use u, v and so on to denote vertices). Further, if π is a permutation of the vertex set, we define the *left* and *right* neighborhoods of v, respectively, as:

$$N_l(v, \pi) := N(v) \cap \{u \in V \mid \pi(u) < \pi(v)\}$$
$$N_r(v, \pi) := N(v) \cap \{u \in V \mid \pi(u) > \pi(v)\}$$

Definition 1. (Imbalance of G with respect to π.) *Given a graph $G = (V, E)$ and a permutation $\pi : V \to V$, the imbalance of G with respect to π, $\mathcal{I}(G, \pi)$, is given by*

$$\mathcal{I}(G, \pi) := \sum_{v \in V} ||N_l(v, \pi)| - |N_r(v, \pi)||.$$

Definition 2. (Imbalance of G) *Given a graph $G = (V, E)$ the imbalance of $\mathcal{G}$, $\mathcal{I}(G)$, is given by*

$$\mathcal{I}(G) := min_\pi\{\mathcal{I}(G, \pi)\}$$

The *treewidth* of a graph $\mathcal{G}$ is one of the most frequently used structural parameters used to examine the complexity of a problem. The definition is motivated by the desire to identify small separating sets in arbitrary graphs. We now describe the formal definition of the treewidth of $\mathcal{G}$.

Definition 3. (Treewidth) *A tree decomposition of a graph $G = (V, E)$ is a pair $(T, (B_t)_{t\in T})$, where $T = (V_T, F)$ is a tree and $(B_t)_{t\in T}$ is a family of subsets of V such that:*

1. $\cup_{t\in V_T} B_t = V$
2. *For every edge $\{v, w\} \in E$, there is a $t \in V_T$ such that $v, w \in B_t$.*
3. *For every $v \in V$, the set $B^{-1}(v) := \{t \in V_T | v \in B_t\}$ is nonempty and connected in T.*

The width of the decomposition $(T, (B_t)_{t\in T})$ is the number

$$max\{|B_t| \mid t \in V_T\} - 1.$$

The treewidth $tw(G)$ of G is the minimum of the widths of the tree decompositions of G.

If in the definitions of a tree decomposition and treewidth we restrict T to be a path, then we have the definitions of path decomposition and pathwidth. We use the notation $tw(G)$ and $pw(G)$ to denote the treewidth and the pathwidth of a graph G.

For our results, we enhance the definition of a tree decomposition $(T, (B_t)_{t\in T})$, as follows: T is a tree rooted on some node r where $B_r = \emptyset$, each of its nodes have at most two children and could be one of the following

1. **Introduce node:** A node t that has only one child t' where $B_t \supset B_{t'}$ and $|B_t| = |B_{t'}| + 1$.
2. **Forget node:** A node t that has only one child t' where $B_t \subset B_{t'}$ and $|B_t| = |B_{t'}| - 1$.
3. **Join node:** A node t with two children t_1 and t_2 such that $B_t = B_{t_1} = B_{t_2}$.
4. **Base node:** A node t that is a leaf of T, and is different from the root.

Notice that, according to the above definitions, the root r of T is either a forget or a join node. It is known that any tree decomposition can be transformed to one with the above requirements while maintaining the same width (see e.g. [5,3]). From now on, when we refer to a tree decomposition $(T, (B_t)_{t\in T})$ we presume the above requirements.

Given a tree decomposition $(T, (B_t)_{t\in T})$ and some node t of V_T, we define as $T_t = (V_t, E_t)$ the subtree of T rooted at t. Clearly, if r is the root of T, it holds that $T_r = T$. We also define $G_t = G[\cup_{s\in V_t} B_t]$.

The problem TREE-WIDTH of deciding whether a graph has treewidth k is NP-complete. It is well known that the natural parameterization of the problem is fixed-parameter tractable.

> TREE-WIDTH
> *Instance:* A graph G and $k \in \mathbb{N}$.
> *Parameter:* k.
> *Problem:* Decide if $tw(G) = k$.

For the purposes of our computation, we use an approximation algorithm that is based on a connection between tree decompositions and separators, and use a well known procedure for computing small separators of a graph. We will see that this is enough to derive our main fixed-parameter tractability result.

Proposition 1 ([1]). *There is an algorithm that, given a graph $G = (V, E)$, computes a tree decomposition of G of width at most $3k + 1$ in time*

$$\mathcal{O}(4^k \cdot k^{3.5} \cdot n^2).$$

Here k denotes the optimal treewidth, $tw(G)$, of G.

In what follows, we first show that the problem of imbalance parameterized by the treewidth and the maximum degree of the input graph is fixed-parameter tractable. Combining this with a few structural observations, we obtain the fixed-parameter tractability of the problem when parameterized by the imbalance.

3 Imbalance Parameterized by the Treewidth and the Maximum Degree

Given a graph $G = (V, E)$, recall that we are interested in checking whether G has imbalance at most b. Let $(T, (B_t)_{t \in T})$ denote a tree decomposition of the graph G of width k. We consider this problem parameterized by the treewidth and the maximum degree of of the graph, that is:

> IMBALANCE
> *Instance:* A graph G, a tree-decomposition $(T, (B_t)_{t \in T})$ of width k, and a non-negative integer b.
> *Parameter:* $\Delta(G), k$
> *Problem:* Decide if there exists a permutation $\pi : V \to V$ such that $\mathcal{I}(G, \pi) \leq b$.

A positive answer to the question can be inferred using any permutation π such that $\mathcal{I}(G, \pi)$ is at most b. If there is no such permutation, then we conclude that the answer is in the negative. Therefore, we regard a permutation π for which $\mathcal{I}(G, \pi)$ is at most b as a solution to IMBALANCE. Let $t \in V_t$, G_t be the induced subgraph corresponding to T_t and A_t be $V_t \setminus B_t$. Given a permutation

π of the vertices of B_t, we define a *partial* solution for the subgraph G_t to be a permutation on the vertex set V_t which respects the ordering on the vertices of B_t given by π and that minimizes the sum of imbalance of vertices in A_t. When we arrive at the root r of the tree, it suffices to check if (one of) the solution(s) computed at this node is a permutation π such that $\mathcal{I}(G_r, \pi) \leq b$, since $V_r = V$.

We now precisely describe what we compute at every tree node $t \in V_T$. Let X denote the set $\{0, \ldots, \Delta(G)\} \cup \{\infty\}$. Given a set $U \subseteq V$ of size q, let $\mathcal{S}_U$ denote all the bijections from $[q] \to U$ and $\mathcal{G}_U$ denote the family of functions $\{g \mid g : U \to \mathrm{X}\}$. We will use g to remember the number of left neighbors of a vertex in the current tree bag. Now consider

$$\mathcal{F}(U) = \mathcal{G}_U \times \mathcal{S}_U,$$

a family of $(\Delta(G) + 2)^q q!$ functions. Each (g, π) can be thought of as a vector of length q with entries from X paired with a bijection π (which essentially gives a permutation of the vertex set U). At node t, we compute, for every element in $\mathcal{F}(B_t)$, an element of $\mathbb{N}$ – the set of natural numbers, the latter being the partial imbalance given the restrictions encoded in (g, π). We denote this value by $f_t(g, \pi)$.

We now describe how $f_t(g, \pi)$ is computed at the node $t \in V_T$ is computed.

Base Nodes. To determine $f_t(g, \pi)$, simply check that the neighborhood conditions imposed by g are respected by the function/permutation π. For example, let $\pi(i) = v$. So the permutation π should place $g(v)$ neighbors of v to its left, and its remaining neighbors to the right. If the neighborhood constraint is indeed respected, we let $f_t(g, \pi) = 0$, otherwise set $f_t(g, \pi) = \infty$.

Introduce Nodes. Let t be the index of the bag in which the vertex u is introduced. Our goal is to compute $f_t(g, \pi)$ for all $(g, \pi) \in \mathcal{F}(B_t)$. Suppose $\pi : [|B_t| = q] \to B_t$ is a permutation of B_t. Let j denote $\pi^{-1}(u)$. Consider π' obtained from π as follows:

$$\pi'(i) = \begin{cases} \pi(i) & \text{if } i < j, \\ \pi(i+1) & \text{if } j \leq i \leq q-1 \end{cases}$$

Let the index of the child node of t be t'. Now consider $g' : B_{t'} \to \mathrm{X}$ obtained from g as follows:

$$g'(v) = \begin{cases} g(v) & \text{if } v \notin N(u) \text{ or } \pi'^{-1}(v) < j, \\ g(v) - 1 & \text{if otherwise }, \end{cases}$$

We set $f_t(g, \pi) = f_{t'}(g', \pi')$.

Forget Nodes. Let t be the node where a vertex u is forgotten and let the index of the child node of t be t'. Furthermore let $|B_t| = q$. To obtain $f_t(g, \pi)$ for all

$(g, \pi) \in \mathcal{F}(B_t)$ we do the following. Consider $\pi^{(j)} : [q+1] \rightarrow B_{t'}$, $j \in [q+1]$, obtained from π as follows:

$$\pi^{(j)}(i) = \begin{cases} \pi(i) & \text{if } i < j, \\ u & \text{if } i = j, \\ \pi(i-1) & \text{if } i > j. \end{cases}$$

Now consider $g^{(s)} : B_{t'} \rightarrow \mathrm{X}$, $s \in [d(u)]$, obtained from g as follows:

$$g^{(s)}(v) = \begin{cases} g(v) & \text{if } v \neq u, \\ s & \text{if } v = u, \end{cases}$$

We are now ready to determine $f_t(g, \pi)$. We set:

$$f_t(g, \pi) = \min \Big\{ f_{t'}(g^s, \pi^j) + |d(u) - g^s(u)| \mid j \in [q+1], s \in [d(u)] \Big\}.$$

Join Nodes. Let the index of the join node be t, and the indices of its children be t_1 and t_2. To determine $f_t(g, \pi)$, we do the following. Let $g_1 : B_{t_1} \rightarrow \mathrm{X}$ and $g_1 : B_{t_2} \rightarrow \mathrm{X}$. We let Q denote the collection of all pairs (g_1, g_2) such that $g_1 \in \mathcal{G}_{B_{t_1}}$, $g_2 \in \mathcal{G}_{B_{t_2}}$ and $g_1(v) + g_2(v) = g(v)$, for all $v \in B_t$. Note that $B_t = B_{t_1} = B_{t_2}$. Then set:

$$f_t(g, \pi) = \min \Big\{ f_{t_1}(g_1, \pi) + f_{t_2}(g_2, \pi) \mid (g_1, g_2) \in X \Big\}.$$

This completes the description of how the entries $\mathbb{T}(t)$ are computed for $t \in V_T$. The correctness of the algorithm follows from the description and the discussions preceding it. We check the unique entry corresponding to $\mathbb{T}(r)$ and check if the entry corresponding to the imbalance is at most b. If it is not so then there is indeed no permutation π such that $\mathcal{I}(G, \pi)$ is bounded by b. If the answer is yes then one can construct in polynomial time a permutation π of V such that $\mathcal{I}(G, \pi) \leq b$ using standard backtracking procedure.

Note that the time taken by the algorithm is proportional to the size of the table stored at each node, time taken to fill the table entries and the total number of nodes in the tree decomposition. The number of nodes in T is known to be linear in the number of vertices of the input Each table is of size at most $(\Delta(G) + 2)^{k+1}(k+1)!$. Each entry in the table can be filled in $(\Delta(G)^{2k+O(1)})$ (the join node could take this much amount of time). Thus the time taken by the algorithm is, therefore, $\mathcal{O}(\Delta(G)^{O(k)} k! \cdot n)$. This leads us to the following assertion:

Lemma 1. *Given a graph $G = (V, E)$ along with a tree decomposition $(\mathcal{T}, (B_t)_{t \in T})$ of G with width at most k, we can, in time $\mathcal{O}(\Delta(G)^{O(k)} k! \cdot n^{O(1)})$ check if there exists a permutation π such that $\mathcal{I}(G, \pi) \leq b$.*

4 Some Structural Observations

In this section, we establish that if the imbalance of a graph G is b, then the treewidth of the graph is at most $b/2$. More, in fact, is true: we will show that the imbalance of a graph is at least twice its *pathwidth*.

For our structural result we need an equivalent definition of pathwidth in terms of vertex separators with respect to a linear ordering of the vertices. Let G be a graph and let $\sigma = v_1 v_2 \dots v_n$ be an ordering of V. For $j \in [n]$ put $V_j = \{v_i : i \in [j]\}$ and denote by ∂V_j all vertices of V_j that have neighbors in $V \setminus V_j$. Furthermore define the set $\mathcal{E}(V_j)$ to be the set of edges with one endpoint in V_j and the other in $V \setminus V_j$. Setting $vs(G,\sigma) = \max_{i\in[n]} |\partial V_i|$, we define the *vertex separation* of G as

$$vs(G) = \min\{vs(G,\sigma) : \sigma \text{ is an ordering of } V(G)\}.$$

The following assertion is well-known. It follows directly from the results of Kirousis and Papadimitriou [14] on interval width of a graph, see also [13].

Proposition 2 ([13,14]). *For any graph G, $vs(G) = pw(G)$.*

We need one more graph parameter to establish our claims in this section. To this end, we have the following definition. Given a graph $G = (V, E)$ and an ordering $\sigma = v_1 \dots v_n$ of the vertices of G, we let $cw(G,\sigma) = \max_{i\in[n]} |\mathcal{E}(V_i)|$. We define the *cutwidth* of the graph G as

$$cw(G) = \min\{cw(G,\sigma) : \sigma \text{ is an ordering of } V(G)\}.$$

From the definitions of vertex separation and cutwidth of the graph G together with Proposition 2 we have the following observation.

Observation 1. *For a graph G, $pw(G) \le cw(G)$.*

The main result of this section is following.

Lemma 2. *Let $G = (V, E)$ be a graph then*

$$tw(G) \le pw(G) \le cw(G) \le \frac{\mathcal{I}(G)}{2}.$$

Furthermore $\Delta(G) \le \mathcal{I}(G)$ where $\Delta(G)$ is the maximum degree of G and $\mathcal{I}(G)$ is the value of minimum imbalance of G.

Proof. Let $\sigma = v_1 \dots v_n$ be an imbalance minimizing ordering of V. Define the rank of a vertex v, $r(v)$, to be $|N_r(v,\sigma)| - |N_l(v,\sigma)|$. Then it is easy to observe that $|\mathcal{E}(V_{j+1})| = |\mathcal{E}(V_j)| + r(v_{j+1})$. Expanding this one can show that for any $j \in [n-1]$, $|\mathcal{E}(V_{j+1})| = \sum_i r(v_{j+1})$.

Let π be the permutation that minimizes the imbalance of G. Now, note that:

$$\mathcal{I}(G) = \sum_{v\in V} ||N_l(v,\pi)| - |N_r(v,\pi)||.$$

Let j be the index at which the quantity $|\mathcal{E}(V_j)|$ is the maximum (with respect to π). If there are multiple indices that witness this maximum, then let j be the smallest among them. We may rewrite the sum above as:

$$\mathcal{I}(G) = \sum_{i=1}^{j} ||N_l(\pi(i)) - |N_r(\pi(i))|| + \sum_{i=j+1}^{n} ||N_l(\pi(i))| - |N_r(\pi(i))||.$$

Now we show that each of the summands above is at least the cutwidth of the graph. Observe that:

$$\sum_{i=1}^{j} ||N_l(\pi(i))| - |N_r(\pi(i))|| \geq \sum_{i=1}^{j} r_j = cw(G,\pi) \geq cw(G).$$

Let π' denote the permutation obtained by reversing π, that is, $\pi'(i) = \pi(n - i + 1)$. Then we have,

$$\begin{aligned}\sum_{i=j+1}^{n} ||N_l(\pi(i))| - |N_r(\pi(i))|| &= \sum_{i=1}^{n-j} ||N_l(\pi'(i))| - |N_r(\pi'(i))|| \\ &\geq \sum_{i=1}^{n-j} r_j = cw(G,\pi') \geq cw(G).\end{aligned}$$

This implies that $tw(G) \leq pw(G) \leq cw(G) \leq \mathcal{I}(G)/2$.

Now we show the upper bound on the maximum degree of the graph. Note that it suffices to show that the maximum degree of the graph does not exceed $2cw(G)$. Suppose not. Then there exists a vertex v such that $d(v) > 2cw(G)$. Consider the cutwidth minimizing permutation, and note that the vertex v must have more than $cw(G)$ neighbors to either it's right or left (possibly both, but this is the case at least in one direction). The cut just after or before v, depending on whether v has more than $cw(G)$ neighbors to its left or right, is more than $cw(G)$, and this is a contradiction. This completes the proof of the lemma. □

5 Imbalance as a Parameter

We now consider the following problem:

IMBALANCE
Instance: A graph G and a non-negative integer $b \in \mathbb{N}$.
Parameter: b.
Problem: Decide if there exists a permutation $\pi : V \to V$ such that $\mathcal{I}(G,\pi) \leq b$.

We use the approximation algorithm provided by Proposition 1 to check if the treewidth of the given graph is at most $1.5b + 1$. If the algorithm outputs no, we conclude, due to Lemma 2 that the imbalance of the graph is more than

b. Otherwise, we obtain a tree-decomposition of width at most $1.5b + 1$. Using Lemma 1, we can now check if the imbalance of G is at most b. Now using the bounds obtained in Lemma 2 we conclude that this will require time $b^{\mathcal{O}(b)}n^{O(1)}$, which can be simplified to $2^{\mathcal{O}(b \log b)}n^{O(1)}$.

Theorem 1. *Given a graph $G = (V, E)$ and a non-negative integer b, we can, in time $2^{\mathcal{O}(b \log b)}n^{O(1)}$ check if there exists a permutation π such that $\mathcal{I}(G, \pi)$ is not more than b.*

6 Conclusion

We have described an algorithm that checks, given a graph G and a non-negative integer b, if there exists a permutation of V for which $\mathcal{I}(G, \pi)$ is at most b. Further, whenever the answer is positive, (one of) the corresponding permutation(s) can be computed. This is achieved in $2^{\mathcal{O}(b \log b)}n^{O(1)}$ time. Whether the running time can be improved to $2^{\mathcal{O}(b)}n^{\mathcal{O}(1)}$, remains open.

It is well-known that every problem that is FPT admits a *kernel* (see [15] for definition). While graph layout problems such as TREEWIDTH, PATHWIDTH and CUTWIDTH are FPT [3,5,17] one can easily show using recently developed machinery [4] that they are unlikely to admit polynomial kernels. Giving such a lower bound for IMBALANCE seems non-trivial, while a polynomial kernel for the problem would be the first such kernel for a graph layout problem. We think that whether IMBALANCE admits a polynomial kernel is an interesting open problem.

References

1. Amir, E.: Efficient approximation for triangulation of minimum treewidth. In: UAI, pp. 7–15 (2001)
2. Biedl, T.C., Chan, T.M., Ganjali, Y., Hajiaghayi, M.T., Wood, D.R.: Balanced vertex-orderings of graphs. Discrete Applied Mathematics 148(1), 27–48 (2005)
3. Bodlaender, H.L.: A linear-time algorithm for finding tree-decompositions of small treewidth. SIAM J. Comput. 25(6), 1305–1317 (1996)
4. Bodlaender, H.L., Downey, R.G., Fellows, M.R., Hermelin, D.: On problems without polynomial kernels. J. Comput. Syst. Sci. 75(8), 423–434 (2009)
5. Bodlaender, H.L., Kloks, T.: Efficient and constructive algorithms for the pathwidth and treewidth of graphs. J. Algorithms 21(2), 358–402 (1996)
6. Downey, R.G., Fellows, M.R.: Parameterized complexity. Monographs in Computer Science. Springer, New York (1999)
7. Flum, J., Grohe, M.: Parameterized complexity theory. In: Texts in Theoretical Computer Science. An EATCS Series, Springer, Berlin (2006)
8. Gaspers, S., Messinger, M.-E., Nowakowski, R.J., Pralat, P.: Clean the graph before you draw it? Inf. Process. Lett. 109(10), 463–467 (2009)
9. Kant, G.: Drawing planar graphs using the canonical ordering. Algorithmica 16(1), 4–32 (1996)
10. Kant, G., He, X.: Regular edge labeling of 4-connected plane graphs and its applications in graph drawing problems. Theor. Comput. Sci. 172(1-2), 175–193 (1997)

11. Kára, J., Kratochvíl, J., Wood, D.R.: On the complexity of the balanced vertex ordering problem. In: Wang, L. (ed.) COCOON 2005. LNCS, vol. 3595, pp. 849–858. Springer, Heidelberg (2005)
12. Kára, J., Kratochvíl, J., Wood, D.R.: On the complexity of the balanced vertex ordering problem. Discrete Mathematics & Theoretical Computer Science 9(1) (2007)
13. Kinnersley, N.G.: The vertex separation number of a graph equals its path-width. Inf. Process. Lett. 42(6), 345–350 (1992)
14. Kirousis, L.M., Papadimitriou, C.H.: Interval graphs and searching. Discrete Mathematics 55(2), 181–184 (1985)
15. Niedermeier, R.: Invitation to fixed-parameter algorithms. Oxford Lecture Series in Mathematics and its Applications, vol. 31. Oxford University Press, Oxford (2006)
16. Papakostas, A., Tollis, I.G.: Algorithms for area-efficient orthogonal drawings. Comput. Geom. 9(1-2), 83–110 (1998)
17. Thilikos, D.M., Serna, M.J., Bodlaender, H.L.: Cutwidth i: A linear time fixed parameter algorithm. J. Algorithms 56(1), 1–24 (2005)
18. Thilikos, D.M., Serna, M.J., Bodlaender, H.L.: Cutwidth ii: Algorithms for partial w-trees of bounded degree. J. Algorithms 56(1), 25–49 (2005)
19. Wood, D.R.: Optimal three-dimensional orthogonal graph drawing in the general position model. Theor. Comput. Sci. 1-3(299), 151–178 (2003)
20. Wood, D.R.: Minimising the number of bends and volume in 3-dimensional orthogonal graph drawings with a diagonal vertex layout. Algorithmica 39(3), 235–253 (2004)

The Ramsey Number for a Linear Forest versus Two Identical Copies of Complete Graphs

I.W. Sudarsana[1], Adiwijaya[2], and S. Musdalifah[1]

[1] Combinatorial and Applied Mathematics Research Group,
Faculty of Mathematics and Natural Sciences (FMIPA),
Tadulako University (UNTAD),
Jalan Soekarno-Hatta Km. 8, Palu 94118, Indonesia
sudarsanaiwayan@yahoo.co.id, selvymusdalifah@yahoo.com
[2] Computational Science Research Group,
Faculty of Science,
Institut Teknologi Telkom,
Jl. Telekomunikasi no.1 Bandung 40257, Indonesia
adw@ittelkom.ac.id

Abstract. Let H be a graph with the chromatic number h and the chromatic surplus s. A connected graph G of order n is called H*-good* if $R(G,H) = (n-1)(h-1)+s$. We show that P_n is $2K_m$-good for $n \geq 3$. Furthermore, we obtain the Ramsey number $R(L, 2K_m)$, where L is a linear forest. In addition, we also give the Ramsey number $R(L, H_m)$ which is an extension for $R(kP_n, H_m)$ proposed by Ali et al. [1], where H_m is a cocktail party graph on $2m$ vertices.

Keywords: (G,H)-free, H-good, linear forest, Ramsey number, path.

1 Introduction

Throughout this paper we consider finite undirected simple graphs. For graphs G, H where H is a subgraph of G, we define $G - H$ as the graph obtained from G by deleting the vertices of H and all edges incident to them. The order of a graph G, $|G|$, is the number of vertices of G. The minimum (maximum) degree of G is denoted by $\delta(G)$ ($\Delta(G)$). Let A be a subset of vertices of a graph G, a graph $G[A]$ represents *the subgraph induced* by A in G. We denote a tree on n vertices by T_n, a path on n vertices by P_n and a complete graph on n vertices by K_n. *A cocktail party graph*, H_n, is a graph obtained by removing n independent edges from a complete graph of order $2n$. Two identical copies of complete graphs is denoted by $2K_n$.

Given graphs G and H, a graph F is called (G,H)*-free* if F contains no subgraph isomorphic to G and the complement of F, $\overline{F}$, contains no subgraph isomorphic to H. Any (G,H)-free graph on n vertices is denoted by (G,H,n)-free. *The Ramsey number* $R(G,H)$ is defined as the smallest natural number n such that no (G,H,n)-free graph exists.

Ramsey Theory studies the conditions when a combinatorial object contains necessarily some smaller given object. The role of Ramsey number is to quantify

M.T. Thai and S. Sahni (Eds.): COCOON 2010, LNCS 6196, pp. 209–215, 2010.

some of the general existential theorems in Ramsey theory. The Ramsey number $R(G, H)$ is called *the classical Ramsey number* if both G and H are complete graphs and in short denoted by $R(p, q)$ when $G \simeq K_p$ and $H \simeq K_q$. It is a challenging problem to find the exact values of $R(p, q)$. Until now, according to the survey in Radziszowski [10] there are only nine exact values of $R(p, q)$ which have been known, namely for $p = 3$, $q = 3, 4, 5, 6, 7, 8, 9$ and $p = 4$, $q = 4, 5$. In the relation with the theory of complexity, Burr [6] stated that for given graphs G, H and positive integer n, determining whether $R(G, H) \leq n$ holds is NP-hard. Furthermore in [11], we can find a rare natural example of a problem higher than NP-hard in the polynomial hierarchy of computational complexity theory, that is, Ramsey arrowing is $\prod_2^p$-complete.

Since it is very difficult to determine $R(p, q)$, one turns out to consider the problem of Ramsey numbers concerning the general graphs G and H, which are not necessarily complete, such as the Ramsey numbers for path versus path, tree versus complete graph, path versus cocktail party graph and so on. This makes the problem on the graphs Ramsey number become more interesting, especially for the union of graphs.

Let k be a positive integer and G_i be a connected graph with the vertex set V_i and the edge set E_i for $i = 1, 2, ..., k$. *The union of graphs*, $G \simeq \bigcup_{i=1}^k G_i$, has the vertex set $V = \bigcup_{i=1}^k V_i$ and the edge set $E = \bigcup_{i=1}^k E_i$. If $G_1 \simeq G_2 \simeq ... \simeq G_k \simeq F$, where F is an arbitrary connected graph, then we denote the union of graphs by kF. The union of graphs is called *a forest* if G_i is isomorphic to T_{n_i} for every i. In particular, if $G_i \simeq P_{n_i}$ for every i then the union of graphs is called *a linear forest*, — denoted by L.

Let H be a graph with the chromatic number h and the chromatic surplus s. The chromatic surplus of H, s, is the minimum cardinality of a color class taken over all proper h colorings of H. A connected graph G of order n is called *H-good* if $R(G, H) = (n - 1)(h - 1) + s$. In particular, for tree T_n versus complete graph K_m, Chvátal [5] showed that T_n is K_m-good with $s = 1$. Other results concerning H-good graphs with the chromatic surplus one can be found in Radziszowski [10]. Recent results on the Ramsey number for the union of graphs consisting of H-good components with $s = 1$ can be found in [2], [3], [8]. Other results concerning the Ramsey number for the union of graphs containing no H-good components can be seen in [9], [12], [13].

However, there are many graphs that have the chromatic surplus greater than one. For example, the graphs $2K_m$ and H_m have the same chromatic surplus, that is 2. Therefore, in this paper we show that P_n is $2K_m$-good for $n \geq 3$. Based on this result, we obtain the Ramsey number $R(L, 2K_m)$. In addition, we give the Ramsey number $R(L, H_m)$ which is an extension for $R(kP_n, H_m)$ proposed by Ali et al. [1].

Let us note firstly the previous theorems and lemma used in the proof of our results.

Theorem 1. *(Gerencser and Gyarfas [7]).* $R(P_n, P_m) = n + \lfloor \frac{m}{2} \rfloor - 1$, *for* $n \geq m \geq 2$.

Theorem 2. *(Ali et al. [1]). Let P_n be a path on n vertices and H_m be a cocktail party graph on $2m$ vertices. Then, $R(kP_n, H_m) = (n-1)(m-1) + (k-1)n + 2$, for $n, m \geq 3$ and $k \geq 1$.*

Lemma 1. *(Bondy [4]). Let G be a graphs of order n. If $\delta(G) \geq \frac{n}{2}$ then either G is pancyclic or n is even and $G \simeq K_{\frac{n}{2},\frac{n}{2}}$.*

2 The Ramsey Goodness for P_n versus $2K_m$

The theorem in this section deals with the Ramsey goodness for a path versus two identical copies of complete graphs. First we need to prove the following two lemmas.

Lemma 2. *Let t, n be positive integers and P_n be a path on $n \geq 2$ vertices and K_2 be a complete graph on 2 vertices. Then,*

$$R(P_n, tK_2) = \begin{cases} n + t - 1 & \text{if } t \leq \lfloor \frac{n}{2} \rfloor, \\ 2t + \lfloor \frac{n}{2} \rfloor - 1 & \text{if } t > \lfloor \frac{n}{2} \rfloor. \end{cases}$$

Proof. We separate the proof into two cases.

Case 1. $t \leq \lfloor \frac{n}{2} \rfloor$

To prove the upper bound $R(P_n, tK_2) \leq n + t - 1$ we use induction on t. For $t = 1$, the assertion is hold from the trivial Ramsey number $R(P_n, K_2) = n$. Assume that the lemma is true for $t - 1$. We shall show that the lemma is also valid for t. Let F be an arbitrary graph on $n + t - 1$ vertices containing no P_n. We will show that $\overline{F}$ contains tK_2. By induction on t, $\overline{F}$ contains $(t-1)K_2$. Let $B = \{a_1, b_1, ..., a_{t-1}, b_{t-1}\}$ be the vertex set of $(t-1)K_2$ in $\overline{F}$, where $a_i b_i$ are the independent edges in $\overline{F}$ for $i = 1, 2, ..., t-1$. By a contrary, suppose that $\overline{F}$ contains no tK_2. Let $A = V(F) \backslash B$, clearly $|A| = n - t + 1$. Then the subgraph $F[A]$ of F forms a K_{n-t+1}. Otherwise, if there exists two independent vertices in $F[A]$, say x and y, then the vertex set $\{x, y\} \cup B$ forms a tK_2 in $\overline{F}$.

Let us now consider the relation of the vertices in $F[A]$ and B. If a_i (or b_i) is not adjacent to one vertex in $F[A]$ then b_i (or a_i) must be adjacent to all other vertices in $F[A]$ since otherwise we will get two independent edges between $\{a_i, b_i\}$ and $F[A]$ in $\overline{F}$, together with B, the vertices form a tK_2 in $\overline{F}$. Without loss of generality, we may assume that each b_i is adjacent to all but at most one vertex in $F[A]$. Let us consider the subgraph $F[D]$ of F with $D = A \cup \{b_1, b_2, ..., b_{t-1}\}$. Clearly, the subgraph $F[D]$ has n vertices and $\delta(F[D]) \geq n - t$. Since $t \leq \lfloor \frac{n}{2} \rfloor$ then $\delta(F[D]) \geq \lceil \frac{n}{2} \rceil \geq \frac{n}{2}$. Lemma 1 now applies, the subgraph $F[D]$ contains a cycle of order n. This is a contradiction because there is no P_n in F. Therefore, $\overline{F}$ contains tK_2.

Next it can be verified that $K_{n-1} \cup K_{t-1}$ is a (P_n, tK_2)-free graph on $n + t - 2$ vertices and hence $R(P_n, tK_2) \geq n + t - 1$.

Case 2. $t > \lfloor \frac{n}{2} \rfloor$

In this case, Theorem 1 implies that $R(P_n, tK_2) \leq R(P_n, P_{2t}) = 2t + \lfloor \frac{n}{2} \rfloor - 1$.

Conversely, $K_{\lfloor \frac{n}{2} \rfloor -1} + \overline{K}_{2t-1}$ is a (P_n, tK_2)-free graph on $2t + \lfloor \frac{n}{2} \rfloor - 2$ vertices and therefore $R(P_n, tK_2) \geq 2t + \lfloor \frac{n}{2} \rfloor - 1$. The proof is now complete. □

Lemma 3. *Let K_m be a complete graph on m vertices and P_3 be a path on 3 vertices. Then, $R(P_3, 2K_m) = 2m$.*

Proof. We prove the upper bound $R(P_3, 2K_m) \leq 2m$ by induction on m. We have $R(P_3, 2K_2) = 4$ from Lemma 2 and therefore the assertion holds for $m = 2$. Now assume that the assertion is true for $m-1$, namely $R(P_3, 2K_{m-1}) \leq 2(m-1)$. We shall show that the lemma is also valid for m. Let F be an arbitrary graph on $2m$ vertices containing no P_3. We will show that $\overline{F}$ contains $2K_m$. By trivial Ramsey number $R(P_2, 2K_m) = 2m$, F contains P_2 or $\overline{F}$ contains $2K_m$. If $\overline{F}$ contains $2K_m$ then the proof is complete. Now consider that F contains P_2 and let u and v be the two vertices of P_2. Clearly, the subgraph $F - P_2$ of F has $2(m-1)$ vertices. By induction hypothesis on m, the complement of $F - P_2$ contains $2K_{m-1}$. Since F contains no P_3 then the vertices u and v are not adjacent to any vertices in $2K_{m-1}$ and hence of course $\{u, v\}$ and $2K_{m-1}$ form a $2K_m$ in $\overline{F}$.

It is easy to verify that $\overline{K}_{2m-1}$ is $(P_3, 2K_m)$-free graph on $2m-1$ vertices and hence of course $R(P_3, 2K_m) \geq 2m$. This concludes the proof. □

Now we are ready to prove the following theorem.

Theorem 3. *Let P_n be a path of order $n \geq 3$ and K_m be a complete graph of order $m \geq 2$. Then, $R(P_n, 2K_m) = (n-1)(m-1) + 2$.*

Proof. Let us consider a graph $G \simeq (m-1)K_{n-1} \cup K_1$. It can be verified that G contains no P_n and $\overline{G}$ contains no $2K_m$. Therefore, G is $(P_n, 2K_m)$-free graph on $(n-1)(m-1)+1$ vertices and of course $R(P_n, 2K_m) \geq (n-1)(m-1)+2$.

Next we prove the upper bound $R(P_n, 2K_m) \leq (n-1)(m-1)+2$ by induction on $n+m$. For $m = 2$ and $n = 3$, the assertion holds from Lemmas 2 and 3, respectively. Now assume that the assertion is true for $n+m-1$, namely

(1). $R(P_{n-1}, 2K_m) \leq (n-2)(m-1) + 2$ and
(2). $R(P_n, 2K_{m-1}) \leq (n-1)(m-2) + 2$.

We shall show that the theorem is also valid for $n+m$. Let F be an arbitrary graph on $(n-1)(m-1)+2$ vertices. We will show that F contains P_n or $\overline{F}$ contains $2K_m$. By induction hypothesis on n in (1), F contains P_{n-1} or $\overline{F}$ contains $2K_m$. If $\overline{F}$ contains $2K_m$ then it finishes the proof. Now consider that F contains P_{n-1} and let u and v be the two end vertices of the path P_{n-1}. It can be verified that the subgraph $F - P_{n-1}$ of F has $(n-1)(m-2)+2$ vertices. By induction hypothesis on m in (2), the subgraph $F - P_{n-1}$ contains P_n or the complement of $F - P_{n-1}$ contains $2K_{m-1}$. If the subgraph $F - P_{n-1}$ contains P_n then the proof is done. Therefore, the complement of $F - P_{n-1}$ contains $2K_{m-1}$. If u or v is adjacent to one vertex in $2K_{m-1}$ then we have P_n in F. Conversely, if u and v are not adjacent to any vertices in $2K_{m-1}$ then we obtain $2K_m$ in $\overline{F}$. So, $R(P_n, 2K_m) = (n-1)(m-1)+2$. The proof is done. □

3 The Ramsey Numbers for L versus $2K_m$ or H_m

The following two theorems deal with the Ramsey numbers for the union of graphs containing H-good components with $s = 2$. In particular, for a linear forest versus $2K_m$ or H_m. First we need to prove the following lemma.

Lemma 4. *Let K_m be a complete graph on $m \geq 2$ vertices and P_n be a path on $n \geq 3$ vertices. Then, $R(kP_n, 2K_m) = (n-1)(m-1) + (k-1)n + 2$, for $k \geq 1$.*

Proof. We prove the upper bound $R(kP_n, 2K_m) \leq (n-1)(m-1)+(k-1)n+2$ by induction on k. For $k = 1$, the assertion holds from Theorem 3. Assume that the assertion is true for $k-1$, that is $R((k-1)P_n, 2K_m) \leq (n-1)(m-1)+(k-2)n+2$. We shall show that the lemma is also valid for k. Let F be an arbitrary graph on $(n-1)(m-1)+(k-1)n+2$ vertices and suppose that $\overline{F}$ contains no $2K_m$. We will show that F contains kP_n. By inductive hypothesis, F contains $(k-1)P_n$. Thus, the subgraph $F - (k-1)P_n$ of F has $(n-1)(m-1)+2$ vertices. Now by Theorem 3, the subgraph $F - (k-1)P_n$ contains P_n and hence we obtain kP_n in F.

Next construct a graph $G \simeq (m-2)K_{n-1} \cup K_{kn-1} \cup K_1$. It is not hard to verify that G is a $(kP_n, 2K_m)$-free graph on $(n-1)(m-1)+(k-1)n+1$ vertices and therefore $R(kP_n, 2K_m) \geq (n-1)(m-1)+(k-1)n+2$. The proof is now complete. □

Now we are ready to prove the following theorem.

Theorem 4. *For integers $k \geq 1$, let $n_k \geq n_{k-1} \geq ... \geq n_1 \geq 3$ be integers. Let K_m be a complete graph on $m \geq 2$ vertices, P_{n_i} be a path on n_i vertices for $i = 1, 2, ..., k$ and $L \simeq \bigcup_{i=1}^{k} l_i P_{n_i}$. Then,*

$$R(L, 2K_m) = \max_{1 \leq i \leq k} \left\{ (n_i - 1)(m-2) + \sum_{j=i}^{k} l_j n_j + 1 \right\}, \tag{1}$$

where l_i is the number of paths of order n_i in L.

Proof. Let $t = (n_{i_0} - 1)(m-2) + t_0 + 1$ be the maximum of the right side of the Eq. (1) achieved for i_0, where $t_0 = \sum_{j=i_0}^{k} l_j n_j$. Now construct a graph $F \simeq (m-2)K_{n_{i_0}-1} \cup K_{t_0-1} \cup K_1$. Since $n_i \geq 3$ for every $i = 1, 2, ..., k$ then F does not contain at least one component P_{n_i} of L and hence F contains no L. Note that $\overline{F} \simeq K_{n_{i_0}-1,...,n_{i_0}-1,t_0-1,1}$ is a complete m-partite graph, where the smallest partite consists of one vertex. So, there is no $2K_m$ in $\overline{F}$. Thus, F is a $(L, 2K_m)$-free graph on $t-1$ vertices and therefore $R(L, 2K_m) \geq t$.

In order to show that $R(L, 2K_m) \leq t$ we argue as follows. Let U be an arbitrary graph on t vertices and suppose that $\overline{U}$ contains no $2K_m$. We will show that U contains L by induction on k. For $k = 1$, the theorem is true from Lemma 4. Let us assume that the theorem holds for $k-1$. We shall show that the theorem is also valid for k. Note that $t \geq (n_k - 1)(m-1) + (l_k - 1)n_k + 2$. Thus, by Lemma 4, U contains $l_k P_{n_k}$. By definition of t, we obtain the following fact.

$$|U - l_k P_{n_k}| = t - l_k n_k \geq \max_{1 \leq i \leq k-1} \left\{ (n_i - 1)(m-2) + \sum_{j=i}^{k-1} l_j n_j + 1 \right\}. \quad (2)$$

By induction hypothesis, the subgraph $U - l_k P_{n_k}$ of U contains $\bigcup_{i=1}^{k-1} l_i P_{n_i}$ and together with $l_k P_{n_k}$ we have $L \simeq \bigcup_{i=1}^{k} l_i P_{n_i}$ in U. Therefore, $R(L, 2K_m) = t$. This completes the proof. □

The following theorem is an extension of Theorem 2 proposed by Ali et al. in [1].

Theorem 5. *Let $k \geq 1$ and $m \geq 3$. Let $n_k \geq n_{k-1} \geq ... \geq n_1 \geq 3$ be integers. Let H_m be a cocktail party graph on $2m$ vertices, P_{n_i} be a path on n_i vertices for $i = 1, 2, ..., k$ and $L \simeq \bigcup_{i=1}^{k} l_i P_{n_i}$. Then,*

$$R(L, H_m) = \max_{1 \leq i \leq k} \left\{ (n_i - 1)(m-2) + \sum_{j=i}^{k} l_j n_j + 1 \right\}, \quad (3)$$

where l_i is the number of paths of order n_i in L.

Proof. Let $t = (n_{i_0} - 1)(m-2) + t_0 + 1$ be the maximum of the right side of the Eq. (3) achieved for i_0 with $t_0 = \sum_{j=i_0}^{k} n_j$. Now we construct a graph $G \simeq (m-2)K_{n_{i_0}-1} \cup K_{t_0 - 1} \cup K_1$. Since $n_i \geq 3$ for every $i = 1, 2, ..., k$ then G does not contain at least one component P_{n_i} of L and hence G contains no L. Note that $\overline{G} \simeq K_{n_{i_0}-1, ..., n_{i_0}-1, t_0-1, 1}$ is a complete m-partite graph which the smallest partite has one vertex. Then, $\overline{G}$ contains no H_m. So, G is a (L, H_m)-free graph on $t - 1$ vertices and hence $R(L, H_m) \geq t$.

We will prove the upper bound $R(L, H_m) \leq t$ by induction on k. Let F be an arbitrary graph of order t and suppose that $\overline{F}$ contains no H_m. We shall show that F contains L. From Theorem 2, we can see that the assertion is true for $k = 1$. Now let us assume that the theorem also holds for $k > 1$. Note that $t \geq (n_k - 1)(m-1) + 2$. So, by Theorem 2, F contains $l_k P_{n_k}$. From the definition of t, we get the following fact.

$$|F - l_k P_{n_k}| = t - l_k n_k \geq \max_{1 \leq i \leq k-1} \left\{ (n_i - 1)(m-2) + \sum_{j=i}^{k-1} l_k n_j + 1 \right\}. \quad (4)$$

By induction hypothesis, the subgraph $F - l_k P_{n_k}$ of F contains $\bigcup_{i=1}^{k-1} l_i P_{n_i}$ and hence $F \supseteq \bigcup_{i=1}^{k} l_i P_{n_i}$. This completes the proof. □

4 Conclusion

To conclude this paper let us present the following two conjectures to work on.

Conjecture 1. *Let T_n be a tree on $n \geq 3$ vertices and K_m be a complete graph on $m \geq 2$ vertices. Then, $R(T_n, 2K_m) = (n-1)(m-1) + 2$.*

Conjecture 2. *For integers $k \geq 1$, let $n_k \geq n_{k-1} \geq ... \geq n_1 \geq 3$ be integers. Let K_m be a complete graph on $m \geq 2$ vertices, T_{n_i} be a tree on n_i vertices for $i = 1, 2, ..., k$ and $F \simeq \bigcup_{i=1}^{k} l_i T_{n_i}$. Then,*

$$R(F, 2K_m) = \max_{1 \leq i \leq k} \left\{ (n_i - 1)(m-2) + \sum_{j=i}^{k} n_j + 1 \right\}, \tag{5}$$

where l_i is the number of trees of order n_i in F.

References

1. Ali, K., Baig, A.Q., Baskoro, E.T.: On the Ramsey number for a linear forest versus a cocktail party graph. J. Combin. Math. Combin. Comput. 71, 173–177 (2009)
2. Baskoro, E.T., Hasmawati, Assiyatun, H.: The Ramsey number for disjoint unions of trees. Discrete Math. 306, 3297–3301 (2006)
3. Bielak, H.: Ramsey numbers for a disjoint of some graphs. Appl. Math. Lett. 22, 475–477 (2009)
4. Bondy, J.A.: Pancyclic graph. J. Combin. Theory Ser. B 11, 80–84 (1971)
5. Chvátal, V.: Tree complete graphs Ramsey number. J. Graph Theory 1, 93 (1977)
6. Burr, S.A.: Determining generalized Ramsey numbers is NP-hard. Ars Combin. 17, 21–25 (1984)
7. Gerencser, L., Gyarfas, A.: On Ramsey-Type problems, Annales Universitatis Scientiarum Budapestinensis. Eotvos Sect. Math. 10, 167–170 (1967)
8. Hasmawati, Baskoro, E.T., Assiyatun, H.: The Ramsey number for disjoint unions of graphs. Discrete Math. 308, 2046–2049 (2008)
9. Hasmawati, Assiyatun, H., Baskoro, E.T., Salman, A.N.M.: Ramsey numbers on a union of identical stars versus a small cycle. In: Ito, H., Kano, M., Katoh, N., Uno, Y. (eds.) KyotoCGGT 2007. LNCS, vol. 4535, pp. 85–89. Springer, Heidelberg (2008)
10. Radziszowski, S.P.: Small Ramsey numbers. Electronic J. Combin. DS1 12 (August 4, 2009), `http://www.combinatorics.org`
11. Schaefer, M.: Graph Ramsey theory and the polynomial hierarchy. J. Comput. Sys. Sci. 62, 290–322 (2001)
12. Sudarsana, I.W., Baskoro, E.T., Assiyatun, H., Uttunggadewa, S.: On the Ramsey numbers of certain forest respect to small wheels. J. Combin. Math. Combin. Comput. 71, 257–264 (2009)
13. Sudarsana, I.W., Baskoro, E.T., Assiyatun, H., Uttunggadewa, S., Yulianti, L.: The Ramsey number for the union of graphs containing at least one H-good component (Under revision for publication in Utilitas Math.)

Optimal Binary Space Partitions in the Plane*

Mark de Berg and Amirali Khosravi

TU Eindhoven, P.O. Box 513, 5600 MB Eindhoven, The Netherlands

Abstract. An optimal BSP for a set S of disjoint line segments in the plane is a BSP for S that produces the minimum number of cuts. We study optimal BSPs for three classes of BSPs, which differ in the splitting lines that can be used when partitioning a set of fragments in the recursive partitioning process: *free* BSPs can use any splitting line, *restricted* BSPs can only use splitting lines through pairs of fragment endpoints, and *auto-partitions* can only use splitting lines containing a fragment. We obtain the two following results:

- It is NP-hard to decide whether a given set of segments admits an auto-partition that does not make any cuts.
- An optimal restricted BSP makes at most 2 times as many cuts as an optimal free BSP for the same set of segments.

1 Introduction

Motivation. Many problems involving objects in the plane or some higher-dimensional space can be solved more efficiently if a hierarchical partitioning of the space is given. One of the most popular hierarchical partitioning schemes is the *binary space partition*, or BSP for short [1]. In a BSP the space is recursively partitioned by hyperplanes until there is at most one object intersecting the interior of each cell in the final partitioning. Note that the splitting hyperplanes not only partition the space, they may also cut the objects into fragments.

The recursive partitioning can be modeled by a tree structure, called a BSP *tree*. Nodes in a BSP tree correspond to subspaces of the original space, with the root node corresponding to the whole space and the leaves corresponding to the cells in the final partitioning. Each internal node stores the hyperplane used to split the corresponding subspace, and each leaf stores the object fragment intersecting the corresponding cell.[1]

BSPs have been used in numerous applications. In most of these applications, the efficiency is determined by the *size* of the BSP tree, which is equal to the total number of object fragments created by the partitioning process. As a result,

* This research was supported by the Netherlands' Organisation for Scientific Research (NWO) under project no. 639.023.301.

[1] When the objects are $(d-1)$-dimensional—for example, a BSP for line segments in the plane—then it is sometimes required that the cells do not have any object in their interior. In other words, each fragment must end up being contained in a splitting plane. The fragments are then stored with the splitting hyperplanes containing them, rather than at the leaves. In particular, this is the case for so-called auto-partitions.

M.T. Thai and S. Sahni (Eds.): COCOON 2010, LNCS 6196, pp. 216–225, 2010.

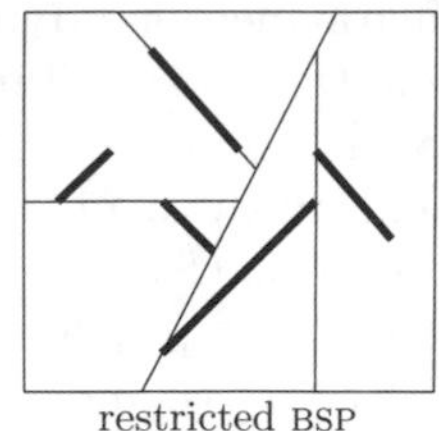

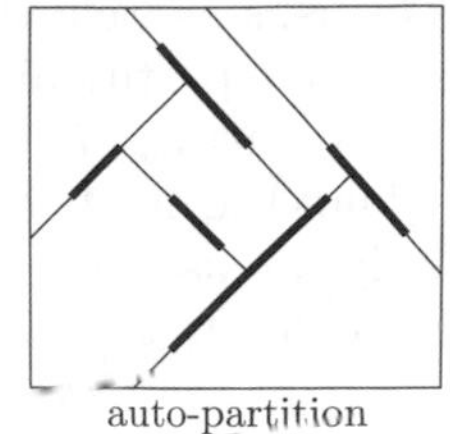

Fig. 1. The three types of BSPs, drawn inside a bounding box of the scene. Note that, as is usually done for auto-partitions, we have continued the auto-partition until the cells are empty.

many algorithms have been developed that create small BSPs; see the survey paper by Tóth [8] for an overview. In all these algorithms, bounds are proved on the *worst-case size* of the computed BSP *over all sets of* n *input objects* from the class of objects being considered. Ideally, one would like to have an algorithm that computes a BSP that is *optimal for the given input*, rather than optimal in the worst-case. In other words, given an input set S, one would like to compute a BSP that is optimal (that is, makes the minimum number of cuts) for S.

In the plane, one can compute an optimal rectilinear BSP for n axis-parallel segments in $O(n^5)$ time using dynamic programming [3]. Another result related to optimal BSPs is that for any set of (not necessarily rectilinear) disjoint segments in the plane one can compute a so-called *perfect* BSP in $O(n^2)$ time, if it exists [2]. (A perfect BSP is a BSP in which none of the objects is cut.) If a perfect BSP does not exist, then the algorithm only reports this fact; it does not produce any BSP in this case. Thus for arbitrary sets of segments in the plane it is unknown whether one can efficiently compute an optimal BSP.

Problem statement and our results. In our search for optimal BSPs, we consider three types of BSPs. These types differ in the splitting lines they are allowed to use. Let S denote the set of n disjoint segments for which we want to compute a BSP, and suppose at some point in the recursive partitioning process we have to partition a region R. Let $S(R)$ be the set of segment fragments lying in the interior of R. Then the three types of BSPs can use the following splitting lines to partition R.

- *Free* BSP*s* can use any splitting line.
- *Restricted* BSP*s* must use a splitting line containing (at least) two endpoints of fragments in $S(R)$. We call such a splitting line a *restricted splitting line.*
- *Auto-partitions* must use a splitting line that contains a segment from $S(R)$.

Fig. 1 illustrates the three types of BSPs. Note that an auto-partition is only allowed to use splitting lines containing a fragment lying in the region to be split; it is not allowed to use a splitting line that contains a fragment lying in a different region. Also note that when a splitting line contains a fragment—such splitting lines must be used by auto-partitions, but may be used by the other types of BSPs as well—then that fragment is no longer considered in the rest of the recursive partitioning process. Hence, it will not be fragmented further.

We use $\text{OPT}_{\text{free}}(S)$ to denote the minimum number of cuts in any free BSP for S. Thus the number of fragments in an optimal free BSP for S is $n+\text{OPT}_{\text{free}}(S)$. Similarly, we use $\text{OPT}_{\text{res}}(S)$ and $\text{OPT}_{\text{auto}}(S)$.

Clearly, $\text{OPT}_{\text{free}}(S) \leqslant \text{OPT}_{\text{res}}(S) \leqslant \text{OPT}_{\text{auto}}(S)$. It is well known that for some sets of segments $\text{OPT}_{\text{res}}(S) < \text{OPT}_{\text{auto}}(S)$; indeed, it is easy to come up with an example where $\text{OPT}_{\text{res}}(S) = 0$ and $\text{OPT}_{\text{auto}}(S) = n/3$. Nevertheless, auto-partitions seem to perform well in many situations. Moreover, the collection of splitting lines to choose from in an auto-partition is smaller than for restricted or free BSPs, and so it seems that computing optimal auto-partitions might be easier than computing optimal restricted or free BSPs. Unfortunately, our hope to find an efficient algorithm for computing optimal auto-partitions turned out to be idle: in Section 2 we prove that computing optimal auto-partitions is NP-hard. In fact, even deciding whether a set of segments admits a perfect auto-partition is NP-hard. This should be contrasted to the result mentioned above, that deciding whether a set of segments admits a perfect restricted BSP can be done in $O(n^2)$ time. (Notice that when it comes to perfect BSPs, there is no difference between restricted and free BSPs: if there is a perfect free BSP then there is also a perfect restricted BSP [2].) Hence, optimal auto-partitions seem more difficult to compute than optimal restricted or free BSPs. Our hardness proof is based on a new 3-SAT variant, *monotone planar* 3-SAT , which we define and prove NP-complete in Section 2. We believe this new 3-SAT variant is interesting in its own right, and may find applications in other NP-completeness proofs. Indeed, it has already been used in a recent paper [9] to prove the NP-hardness of a problem on so-called switch graphs.

We turn our attention in Section 3 to unrestricted and free BSPs. In particular, we study the relation between optimal free BSPs and optimal restricted BSPs. In general, free BSPs are more powerful than restricted BSPs: in his MSc thesis[4], Clairbois gave an example of a set of segments for which the optimal free BSP makes one cut while the optimal restricted BSP makes two cuts, and he also proved that $\text{OPT}_{\text{res}}(S) \leqslant 3 \cdot \text{OPT}_{\text{free}}(S)$ for any set S. In Section 3 we improve this result by showing that $\text{OPT}_{\text{res}}(S) \leqslant 2 \cdot \text{OPT}_{\text{free}}(S)$ for any set S.

2 Hardness of Computing Perfect Auto-Partitions

Recall that an auto-partition of a set S of disjoint line segments in the plane is a BSP in which, whenever a subspace is partitioned, the splitting line contains one of the fragments lying in that subspace. We call an auto-partition *perfect* if none of the input segments is cut, and we consider the following problem.

PERFECT AUTO-PARTITION

Input: A set S of n disjoint line segments in the plane.

Output: YES if S admits a perfect auto-partition, NO otherwise.

We will show that PERFECT AUTO-PARTITION is NP-hard. Our proof is by reduction from a special version of the satisfiability problem, which we define and prove NP-complete in the next subsection.

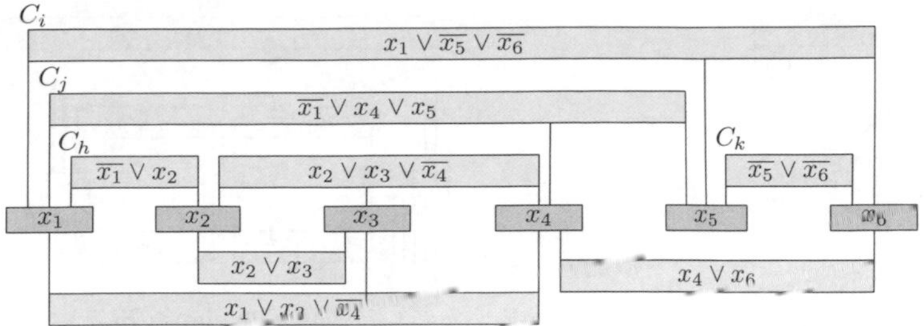

Fig. 2. A rectilinear representation of a planar 3-SAT instance

Planar monotone 3-SAT. Let $\mathcal{U} := \{x_1, \ldots, x_n\}$ be a set of n boolean variables, and let $\mathcal{C} := C_1 \wedge \cdots \wedge C_m$ be a CNF formula defined over these variables, where each clause C_i is the disjunction of at most three variables. Then 3-SAT is the problem of deciding whether such a boolean formula is satisfiable. An instance of 3-SAT is called *monotone* if each clause is monotone, that is, each clause consists only of positive variables or only of negative variables. 3-SAT is NP-complete, even when restricted to monotone instances [5].

For a given (not necessarily monotone) 3-SAT instance, consider the bipartite graph $\mathcal{G} = (\mathcal{U} \cup \mathcal{C}, \mathcal{E})$, where there is an edge $(x_i, C_j) \in \mathcal{E}$ if and only if x_i or its negation $\overline{x_i}$ is one of the variables in the clause C_j. Lichtenstein [7] has shown that 3-SAT remains NP-complete when $\mathcal{G}$ is planar. Moreover, as shown by Knuth and Raghunatan [6], one can always draw the graph $\mathcal{G}$ of a planar 3-SAT instance as in Fig. 2: the variables and clauses are drawn as rectangles with all the variable-rectangles on a horizontal line, the edges connecting the variables to the clauses are vertical segments, and the drawing is crossing-free. We call such a drawing of a planar 3-SAT instance a *rectilinear representation.* PLANAR 3-SAT remains NP-complete when a rectilinear representation is given.

Next we introduce a new version of 3-SAT, which combines the properties of monotone and planar instances. We call a clause with only positive variables a *positive clause*, a clause with only negative variables a *negative clause*, and a clause with both positive and negative variables a *mixed clause.* Thus a monotone 3-SAT instance does not have mixed clauses. Now consider a 3-SAT instance that is both planar and monotone. A *monotone rectilinear representation* of such an instance is a rectilinear representation where all positive clauses are drawn on the positive side of (above) the variables and all negative clauses are drawn on the negative side of (below) the variables. Our 3-SAT variant is defined as follows.

PLANAR MONOTONE 3-SAT

Input: A monotone rectilinear representation of a planar monotone 3-SAT instance.

Output: YES if the instance is satisfiable, NO otherwise.

PLANAR MONOTONE 3-SAT is obviously in NP. We will prove that it is NP-hard by a reduction from PLANAR 3-SAT. Let $\mathcal{C} = C_1 \wedge \cdots \wedge C_m$ be a given rectilinear representation of a planar 3-SAT instance defined over the variable set $\mathcal{U} = \{x_1, \ldots, x_n\}$. We call a variable-clause pair *inconsistent* if the variable is negative

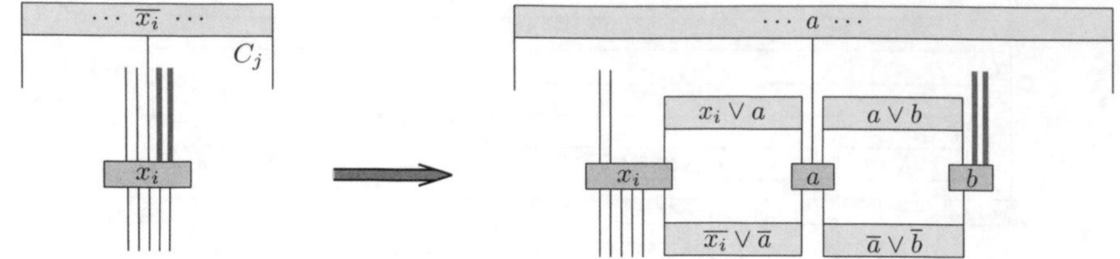

Fig. 3. Getting rid of an inconsistent variable-clause pair

(positive) in that clause while the clause is placed on the positive (negative) side of the variables. If a rectilinear representation does not have inconsistent variable-clause pairs, then it must be monotone. Indeed, any monotone clause must be placed on the correct side of the variables, and there cannot be any mixed clauses. We convert the given instance $\mathcal{C}$ step by step into an equivalent instance with a monotone planar representation, in each step reducing the number of inconsistent variable-clause pairs by one.

Let $(\overline{x_i}, C_j)$ be an inconsistent pair; inconsistent pairs involving a positive variable in a clause on the negative side can be handled similarly. We get rid of this inconsistent pair as follows. We introduce two new variables, a and b, and modify the set of clauses as follows.

- In clause C_j, replace $\overline{x_i}$ by a.
- Introduce the following four clauses: $(x_i \vee a) \wedge (\overline{x_i} \vee \overline{a}) \wedge (a \vee b) \wedge (\overline{a} \vee \overline{b})$.
- In each clause containing x_i that is placed on the positive side of the variables and that connects to x_i to the right of C_j, replace x_i by b.

In following lemma (proved in the full version), $\mathcal{C}'$ is the new set of clauses.

Lemma 1. *$\mathcal{C}$ is satisfiable if and only if $\mathcal{C}'$ is satisfiable.*

Fig. 3 shows how this modification is reflected in the rectilinear representation. (In this example, there are two clauses for which x_i is replaced by b, namely the ones whose edges to x_i are drawn fat and in grey.) We keep the rectangle for x_i at the same location. Then we shift the vertical edges that now connect to b instead of x_i a bit to the right to make room for a and b and the four new clauses. This way we keep a valid rectilinear representation.

By applying the above conversion to each of the at most $3m$ inconsistent variable-clause pairs, we obtain a new planar monotone 3-SAT instance with at most $13m$ clauses defined over a set of at most $n + 6m$ variables, which is satisfiable if and only if $\mathcal{C}$ is satisfiable.

Theorem 1. PLANAR MONOTONE 3-SAT *is* NP-*complete.*

From planar monotone 3-SAT to perfect auto-partitions. Let $\mathcal{C} = C_1 \wedge \cdots \wedge C_m$ be a planar monotone 3-SAT instance defined over a set $\mathcal{U} = \{x_1, \ldots, x_n\}$ of variables, with a monotone rectilinear representation. We show how to construct a set S of line segments in the plane that admits a perfect auto-partition if and only if $\mathcal{C}$ is satisfiable. The idea is illustrated in Fig. 4.

The variable gadget. For each variable x_i there is a gadget consisting of two segments, s_i and $\overline{s_i}$. Setting x_i = TRUE corresponds to extending s_i before $\overline{s_i}$, and setting x_i = FALSE corresponds to extending $\overline{s_i}$ before s_i.

The clause gadget. For each clause C_j there is a gadget consisting of four segments, $t_{j,0}, \ldots, t_{j,3}$. The segments in a clause form a cycle, that is, the splitting line $\ell(t_{j,k})$ cuts the segment $t_{j,(k+1) \bmod 4}$. This means that a clause gadget, when considered in isolation, would generate at least one cut. Now suppose that the gadget for C_j is crossed by the splitting line $\ell(s_i)$ in such a way that $\ell(s_i)$ separates the segments $t_{j,0}, t_{j,3}$ from $t_{j,1}, t_{j,2}$, as in Fig. 4. Then the cycle is broken by $\ell(s_i)$ and no cut is needed for the clause. But this does not work when $\ell(\overline{s_i})$ is used before $\ell(s_i)$, since then $\ell(s_i)$ is blocked by $\ell(\overline{s_i})$ before crossing C_j.

The idea is thus as follows. For each clause $(x_i \vee x_j \vee x_k)$, we want to make sure that the splitting lines $\ell(s_i)$, $\ell(s_j)$, and $\ell(s_k)$ all cross the clause gadget. Then by setting one of these variables to TRUE, the cycle is broken and no cuts are needed to create a perfect autopartition for the segments in the clause. We must be careful, though, that the splitting lines are not blocked in the wrong way—for example, it could be problematic if $\ell(\overline{s_k})$ would block $\ell(s_i)$—and also that clause gadgets are only intersected by the splitting lines corresponding to the variables in that clause. Next we show how to overcome these problems.

Detailed construction. From now on we assume that the variables are numbered according to the monotone rectilinear representation, with x_1 being the leftmost variable and x_n being the rightmost variable.

The gadget for a variable x_i will be placed inside the unit square $[2i-2, 2i-1] \times [2n-2i, 2n-2i+1]$, as illustrated in Fig. 5. The segment s_i is placed with one endpoint at $(2i-2, 2n-2i)$ and the other endpoint at $(2i-\frac{3}{2}, 2n-2i+\varepsilon_i)$ for some $0 < \varepsilon_i < \frac{1}{4}$. The segment $\overline{s_i}$ is placed with one endpoint at $(2i-1, 2n-2i+1)$ and the other endpoint at $(2i-1-\overline{\varepsilon_i}, 2n-2i+\frac{1}{2})$ for some $0 < \overline{\varepsilon_i} < \frac{1}{4}$. Next we specify the slopes of the segments, which determine the values ε_i and $\overline{\varepsilon_i}$, and the placement of the clause gadgets.

The gadgets for the positive clauses will be placed to the right of the variables, in the horizontal strip $[-\infty, \infty] \times [0, 2n-1]$; the gadgets for the negative clauses will be placed below the variables, in the vertical strip $[0, 2n-1] \times [-\infty, \infty]$. We describe how to choose the slopes of the segments s_i and to place the positive clauses; the segments $\overline{s_i}$ and the negative clauses are handled in a similar way.

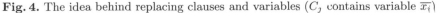

Fig. 4. The idea behind replacing clauses and variables (C_j contains variable $\overline{x_i}$)

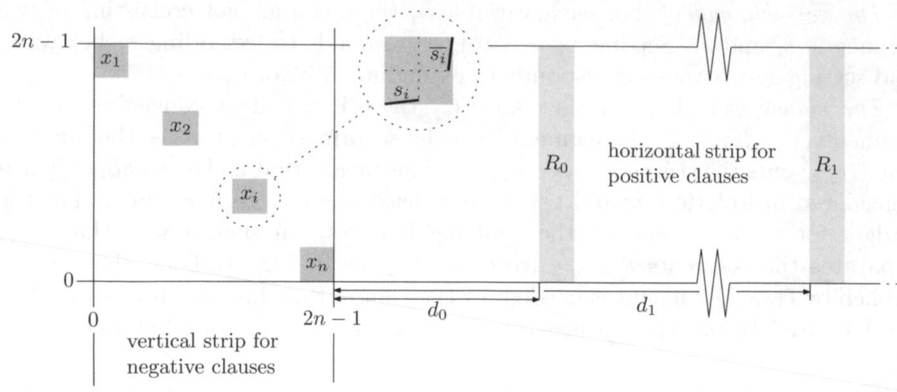

Fig. 5. Placement of the variable gadgets and the clause gadgets (not to scale)

Consider the set $\mathcal{C}^+$ of all positive clauses in our 3-SAT instance, and the way they are placed in the monotone rectilinear representation. We call the clause directly enclosing a clause C_j the *parent* of C_j. In Fig. 2 for example C_i is the parent of C_j and C_k but it is not the parent of C_h. Now let $\mathcal{G}^+ = (\mathcal{C}^+, \mathcal{E}^+)$ be the directed acyclic graph where each clause C_j has an edge to its parent (if it exists), and consider a topological order on the nodes of $\mathcal{G}^+$. We define the *rank* of a clause C_j, denoted by $\text{rank}(C_j)$, to be its rank in this topological order. Clause C_j will be placed at certain distance from the variables that depends on its rank. More precisely, if $\text{rank}(C_j) = k$ then C_j is placed in a $1 \times (2n+1)$ rectangle R_k at distance d_k from the line $x = 2n - 1$ (see Fig. 5), where $d_k := 2 \cdot (2n)^{k+1}$.

Before describing how the clause gadgets are placed inside these rectangles, we define the slopes of the segments s_i. Define $\text{rank}(x_i)$, the rank of a variable x_i (with respect to the positive clauses), as the maximum rank of any clause it participates in. Now the slope of s_i is $\frac{1}{2 \cdot d_k}$, where $k = \text{rank}(x_i)$. Recall that x_i is placed inside the unit square $[2i-2, 2i-1] \times [2n-2i, 2n-2i+1]$. The proof of the following lemma is given in the full version of the paper.

Lemma 2. *Let x_i be a variable, and $\ell(s_i)$ be the splitting line containing s_i.*
(i) For all x-coordinates in the interval $[2i-2, 2n-1+d_{\text{rank}(x_i)}+1]$, the splitting line $\ell(s_i)$ has a y-coordinate in the range $[2n-2i, 2n-2i+1]$.
(ii) The splitting line $\ell(s_i)$ intersects all rectangles R_k with $0 \leqslant k \leqslant \text{rank}(x_i)$.
(iii) The splitting line $\ell(s_i)$ does not intersect any rectangle R_k with $k > \text{rank}(x_i)$.

We can now place the clause gadgets. Consider a clause $C = (x_i \vee x_j \vee x_k) \in \mathcal{C}^+$, with $i < j < k$; the case where C contains only two variables is similar. By Lemma 2(ii), the splitting lines $\ell(x_i), \ell(x_j), \ell(x_k)$ all intersect the rectangle $R_{\text{rank}(C)}$. Moreover, by Lemma 2(i) and since we have placed the variable gadgets one unit apart, there is a 1×1 square in $R_{\text{rank}(C)}$ just above $\ell(s_i)$ that is not intersected by any splitting line. Similarly, just below $\ell(s_k)$ there is a square that is not crossed. Hence, if we place the segments forming the clause gadget as in Fig. 6, then the segments will not be intersected by any splitting line.

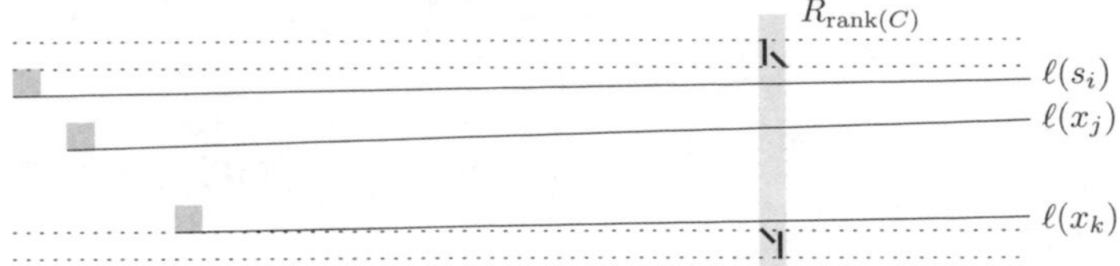

Fig. 6. Placement of the segments forming a clause gadget

Moreover, the splitting lines of segments in the clause gadget—they either have slope -1 or are vertical—will not intersect any other clause gadget. This finishes the construction.

One important property of our construction is that clause gadgets are only intersected by splitting lines of the variables in the clause. Another important property has to do with the blocking of splitting lines by other splitting lines. Recall that the rank of a variable is the maximum rank of any clause it participates in. We say that a splitting line $\ell(s_i)$ is *blocked* by a splitting line $\ell(s_j)$ if $\ell(s_j)$ intersects $\ell(s_i)$ between s_i and $R_{\text{rank}(x_i)}$. This is dangerous, since it may prevent us from using $\ell(s_i)$ to resolve the cycle in the gadget of a clause containing x_i. The next lemma (proved in the full version) states these two key properties.

Lemma 3. *The variable and clause gadgets are placed such that:*
(i) The gadget for any clause $(x_i \vee x_j \vee x_k)$ is only intersected by the splitting lines $\ell(s_i)$, $\ell(s_j)$, and $\ell(s_k)$. Similarly, the gadget for any clause $(\overline{x_i} \vee \overline{x_j} \vee \overline{x_k})$ is only intersected by the splitting lines $\ell(\overline{s_i})$, $\ell(\overline{s_j})$, and $\ell(\overline{s_k})$.
(ii) A splitting line $\ell(s_i)$ can only be blocked by a splitting line $\ell(s_j)$ or $\ell(\overline{s_j})$ when $j \geqslant i$; the same holds for $\ell(\overline{s_i})$.

Lemma 3 implies the following theorem, which is the main result of this section. The following theorem immediately implies that finding a perfect auto-partition in $3D$ is also NP-hard: just take a planar instance in the plane $z = 0$ and replace every line segment by a vertical rectangle. For the proof see the full version.

Theorem 2. PERFECT AUTO-PARTITION *is* NP-*complete.*

3 Optimal Free BSPs versus Optimal Restricted BSPs

Let S be a set of n disjoint line segments in the plane. In this section we will show that $\text{OPT}_{\text{res}}(S) \leqslant 2 \cdot \text{OPT}_{\text{free}}(S)$ for any set S. It follows from the lower bound of Clairbois[4] that this bound is tight.

Consider an optimal free BSP tree $\mathcal{T}$ for S. Let ℓ be the splitting line of the root of $\mathcal{T}$, and assume without loss of generality that ℓ is vertical. Let P_1 be the set of all segment endpoints to the left or on ℓ, and let P_2 be the set of segment endpoints to the right of ℓ. Let CH_1 and CH_2 denote the convex hulls of P_1 and P_2, respectively. We follow the same global approach as Clairbois [4]. Namely, we replace ℓ by a set L of three or four restricted splitting lines that do

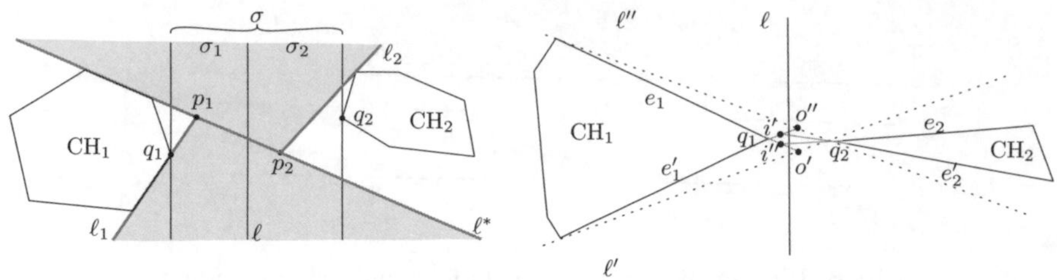

Fig. 7. Left: Illustration for Lemma 4, note that ℓ_1, ℓ_2 and ℓ^* are restricted splitting lines. Right: The case that i, i' are on the left and o' and o'' are on the right side of ℓ.

not intersect the interiors of CH_1 and CH_2, and are such that CH_1 and CH_2 lie in different regions of the partition induced by L. In Fig. 7, for instance, we would replace ℓ by the lines ℓ^*, ℓ_1, ℓ_2. The regions not containing CH_1 and CH_2—the grey regions in Fig. 7—do not contain endpoints, so inside them we can simply make splits along any segments intersecting the regions. After that, we recursively convert the BSPs corresponding to the two subtrees of the root to restricted BSPs. The challenge in this approach is to find a suitable set L, and this is where we will follow a different strategy than Clairbois.

Observe that the segments that used to be cut by ℓ will now be cut by one or more of the lines in L. Another potential cause for extra cuts is that existing splitting lines that used to end on ℓ may now extend further and create new cuts. This can only happen, however, when ℓ crosses the region containing CH_1 and/or the region containing CH_2 in the partition induced by L (the white regions in Fig. 7); if ℓ is separated from these regions by the lines in L, then the existing splitting lines will actually be shortened and thus not create extra cuts. Hence, to prove our result, we will ensure the following properties:

(I) the total number of cuts made by the lines in L is at most twice the number of cuts made by ℓ.
(II) in the partitioning induced by L, the regions containing CH_1 and CH_2 are not crossed by ℓ.

The lines in L are of three types. They are either *inner tangents* of CH_1 and CH_2, or *extensions* of edges of CH_1 or CH_2, or they pass through a vertex of CH_1 (or CH_2) and the intersection of another line in L with a segment in S.

We denote the vertex of CH_1 closest to ℓ by q_1 and we denote the vertex of CH_2 closest to ℓ by q_2 (with ties broken arbitrarily). Let σ be the strip enclosed by the lines through q_1 and q_2 parallel to ℓ, and for $i \in \{1,2\}$ let σ_i denote the part of σ lying on the same side of ℓ as CH_i—see also Fig. 7. (When q_1 lies on ℓ, then σ_1 will just be a line; this does not invalidate the coming arguments.). The proof of the following lemma is given in the full version.

Lemma 4. *Let ℓ^* be a restricted splitting line separating* CH_1 *from* CH_2. *Suppose there are points $p_1 \in \ell^* \cap \sigma_1$ and $p_2 \in \ell^* \cap \sigma_2$ such that, for $i \in \{1,2\}$, the line ℓ_i through p_i and tangent to* CH_i *that separates* CH_i *from ℓ is a restricted* *splitting line (after the addition of ℓ^*). Then we can find a set L of three partition lines satisfying conditions (I) and (II) above.*

To show we can always find a set L satisfying conditions (I) and (II), we distinguish six cases. To this end we consider the two inner tangents ℓ' and ℓ'' of CH_1 and CH_2, and look at which of the points q_1 and q_2 lie on which of these lines. Cases (a)–(e) are handled by applying Lemma 4, case (f) needs a different approach. We discuss case (a), which is representative for the first five cases and all other cases, including case (f), are discussed in the full version.

Case (a): Both ℓ' and ℓ'' do not contain any of q_1, q_2. Let e_1 and e_2 be the edges of CH_1 and CH_2 incident to and below q_1 and q_2 respectively. Let $\ell(e_1)$ and $\ell(e_2)$ be the lines through these edges, and assume without loss of generality that $\ell(e_1) \cap \ell(e_2) \in \sigma_1$. We can now apply Lemma 4 with $\ell^* = \ell(e_2)$, and $p_1 = \ell(e_1) \cap \ell(e_2)$, and $p_2 = q_2$. That we can always replace ℓ with a set of segments such that both conditions (I) and (II) holds. As shown in the full version the same holds for cases (b)-(f) which leads to the following theorem.

Theorem 3. *For any set S of disjoint segments in the plane,* $\text{OPT}_{\text{res}}(S) \leqslant 2 \cdot \text{OPT}_{\text{free}}(S)$.

References

1. de Berg, M., Cheong, O., van Kreveld, M., Overmars, M.: Computational Geometry: Algorithms and Applications, 3rd edn. Springer, Heidelberg (2008)
2. de Berg, M., de Groot, M.M., Overmars, M.H.: Perfect binary space partitions. Comput. Geom. Theory Appl. 7, 81–91 (1997)
3. de Berg, M., Mumford, E., Speckmann, B.: Optimal BSPs and rectilinear cartograms. In: Proc. 14th Int. Symp. Adv. Geographic Inf. Syst. (ACM-GIS), pp. 19–26 (2006)
4. Clairbois, X.: On Optimal Binary Space Partitions. MSc thesis, TU Eindhoven (2006)
5. Garey, M.R., Johnson, D.S.: Computers and Intractability: A Guide to the Theory of NP-Completeness. W.H. Freeman and Co., New York (1979)
6. Knuth, D.E., Raghunathan, A.: The problem of compatible representatives. Discr. Comput. Math. 5, 422–427 (1992)
7. Lichtenstein, D.: Planar formulae and their uses. SIAM J. Comput. 11, 329–343 (1982)
8. Tóth, C.D.: Binary space partitions: recent developments. In: Goodman, J.E., Pach, J., Welzl, E. (eds.) Combinatorial and Computational Geometry. MSRI Publications, vol. 52, pp. 525–552. Cambridge University Press, Cambridge (2005)
9. Katz, B., Rutter, I., Woeginger, G.: An algorithmic study of switch graphs. In: Paul, C. (ed.) WG 2009. LNCS, vol. 5911, pp. 226–237. Springer, Heidelberg (2009)

Exact and Approximation Algorithms for Geometric and Capacitated Set Cover Problems

Piotr Berman[1], Marek Karpinski[2,⋆], and Andrzej Lingas[3,⋆⋆]

[1] Department of Computer Science and Engineering, Pennsylvania State University
berman@cse.psu.edu

[2] Department of Computer Science, University of Bonn
marek@cs.uni-bonn.de

[3] Department of Computer Science, Lund University
Andrzej.Lingas@cs.lth.se

Abstract. First, we study geometric variants of the standard set cover motivated by assignment of directional antenna and shipping with deadlines, providing the first known polynomial-time exact solutions.

Next, we consider the following general (non-necessarily geometric) capacitated set cover problem. There is given a set of elements with real weights and a family of sets of the elements. One can use a set if it is a subset of one of the sets in the family and the sum of the weights of its elements is at most one. The goal is to cover all the elements with the allowed sets.

We show that any polynomial-time algorithm that approximates the uncapacitated version of the set cover problem with ratio r can be converted to an approximation algorithm for the capacitated version with ratio $r + 1.357$.

The composition of these two results yields a polynomial-time approximation algorithm for the problem of covering a set of customers represented by a weighted n-point set with a minimum number of antennas of variable angular range and fixed capacity with ratio 2.357.

1 Introduction

In this paper, we study special geometric set cover problems and capacitated set cover problems.

In particular, the shapes of geometric sets we consider correspond to those of potential directional antenna ranges. Several geometric covering problems where a planar point set is to be covered with a minimum number of objects of a given shape have been studied in the literature, e.g., in [5,6,9,12].

On the other hand, a capacitated set cover problem can be seen as a generalization of the classical bin packing problem (e.g., see [7]) to include several types of bins. Thus, we are given a set of elements $\{1, \ldots, n\}$, each with a demand d_i,

⋆ Research supported in part by DFG grants and the Hausdorff Center research grant EXC59-1.

⋆⋆ Research supported in part by VR grant 621-2008-4649.

M.T. Thai and S. Sahni (Eds.): COCOON 2010, LNCS 6196, pp. 226–234, 2010.

a set of subsets of $\{1,\ldots,n\}$ (equivalently, types of bins), and an upper bound d on set capacity. The objective is to partition the elements into a minimum number of copies of the subsets (bins) so the total demand of elements assigned to each set copy does not exceed d.

Capacitated set cover problems are useful abstraction in studying the problems of minimizing the number of directional antennas. The use of directional antennas in cellular and wireless communication networks steadily grows [2,17,19,18]. Although such antennas can only transmit along a narrow beam in a particular direction they have a number of advantages over the standard ones. Thus, they allow for an additional independent communication between the nodes in parallel [18], they also attain higher throughput, lower interference, and better energy-efficiency [2,17,19].

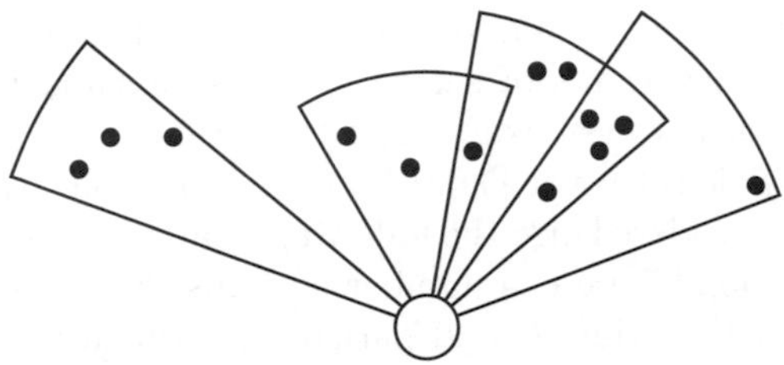

Fig. 1. The sectors correspond to the reaches of directional antennas

We consider the following problem of optimal placement of directional antennas in wireless networks.

There is a base station coupled with a network infrastructure. The station transfers information to and from a number of customers within the range of directional antennas placed at this station. Each customer has fixed position and demand on the transmission capacity. The demands are unsplittable, thus a customer can be assigned only to a single antenna. One can choose the orientation and the angular range of an antenna. When the angular range is narrower an antenna can reach further so the area covered by any antenna is always the same. There is a common limit on the total bandwidth demand that can be assigned to an antenna. The objective is to minimize the number of antennas.

Berman et al. [3] termed this problem as MinAntVar and provided an approximation polynomial-time algorithm with ratio 3. They also observed in [3] that even when the angular range of antennas is fixed, MinAntVar cannot be approximated in polynomial time with ratio smaller than 1.5 by a straightforward reduction from Partition (see [11]).

We provide a substantially better polynomial-time approximation algorithm for MinAntVar achieving the ratio of 2.357. Our algorithm is based on two new results which are of independent interest in their own rights.

The first of these results states that a cover of the set of customers with the minimum number of antennas without the demand constraint can be found in polynomial time. Previously, only a polynomial-time approximation with ratio 2 as well as an integrality gap with set cover ILP were established for this problem in [3].

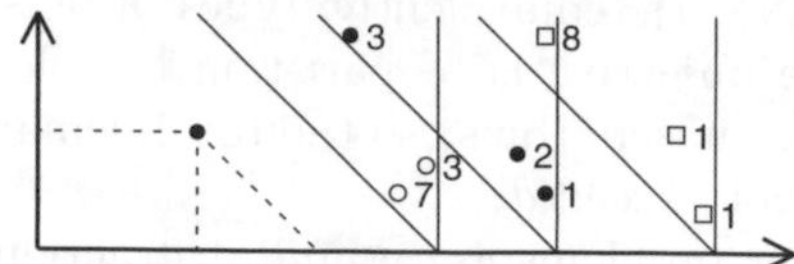

Fig. 2. The X coordinate of an item i encodes t_i and the Y coordinate encodes p_i. Shipment has capacity 10. The numbers indicate the weights. Items which are to be shipped together must be enclosed by an angle.

The second result shows that generally, given an approximate solution with ratio r to an instance of (uncapacitated) set cover, one can find a solution to a corresponding instance of the capacitated set cover, where each set has the same capacity, within $r + 1.357$ of the optimum. This result is especially useful when applied to variants of set cover whose uncapacitated versions admit close approximations or even exact algorithms running in polynomial time, e.g., variants of geometric set cover [5,6,9,12] or in particular MinAntVar.

Berman et al. considered also the following related problem which they termed as BinSchedule [3]. There is a number of items to be delivered. The i-th item has a weight d_i, arrival time t_i and patience p_i, which means that it has to be shipped at latest by $t_i + p_i$. Given a capacity of a single shipment, the objective is minimize the number of shipments.

Similarly as Berman et al. could adopt their approximation for MinAntVar to obtain an approximation with ratio 3 for BinSchedule [3], we can adopt our approximation for MinAntVar to obtain a polynomial-time approximation algorithm with ratio 2.357 for BinSchedule.

Our third main result is a PTAS for a dual problem to capacitated set cover where the number of sets (e.g., antennas) to use is fixed and the task is to minimize the maximum set load, in case the sets correspond to line intervals or arcs. In the application to directional antennas, the aforementioned correspondence comes from fixing the radius and hence also the angular range of the antennas and the problem has been termed as MinAntLoad in [3]. The task is to minimize the maximum load of an antenna. In [3], there has been solely presented a polynomial-time approximation with ratio 1.5 for MinAntLoad.

Marginally, we also discuss the approximability of the generalization of MinAntVar to include several bas stations for antennas, and in particular show its APX-hardness already in the uncapacitated case.

Organization: In Section 2 we present problem definitions and notations. In Section 3, we derive our polynomial-time dynamic programming method for the uncapacitated variant of MinAntVar. In Section 4, we show our general method of the approximate reduction of the capacitated vertex cover to the corresponding uncapacitated one. By combing it with the method of Section 3, we obtain the 2.357 approximation for MinAntVar. Next, in Section 5, we present the PTAS for MinAntLoad, or more generally, for minimizing the maximum load in capacitated set cover of bounded cardinality, in case the sets correspond to intervals or arcs. In the final section, we briefly discuss the approximability of the multi base-station generalization of MINANTVAR.

2 Preliminaries

This section presents terminology and notation used throughout this paper.

We use U to denote $\{1, 2, \ldots, n\}$. If $x_i \in \mathbb{R}$ are defined for $i \in U$ (e.g., d_i) and $A \subset U$ then $x(A) = \sum_{i \in A} x_i$ (e.g., $d(A) \sum_{i \in A} d_i$).

An instance of the set cover problem is given by a family $\mathcal{S}$ of subsets of $U - \{1, \ldots, n\}$. A cover is $\mathcal{C} \subset \mathcal{S}$ such that $\bigcup_{A \in \mathcal{C}} A = U$. The objective is to minimize $|\mathcal{C}|$. An instance of capacitated set cover also specifies d_i for $i \in U$. A capacitated cover is a family of sets $\mathcal{C}$ satisfying (i) for each $A \in \mathcal{C}$ there exists $B \in \mathcal{S}$ s.t. $A \subset B$, while $d(A) \leq 1$, and (ii) $\bigcup_{A \in \mathcal{C}} A = U$. Again, the objective is to minimize $|\mathcal{C}|$.

For each $j \in U$, we denote its radial coordinates by (r_j, θ_j), where r_j stands for the radius and θ_j for the angle. We define an angle sector with radius bound δ as

$$\mathcal{R}(r, \alpha, \delta) = \{j \in U : \ r_j \leq r \text{ and } \theta_j = \alpha + \beta \text{ with } 0 \leq \beta \leq \delta\}.$$

In MINANTVAR as well as in its uncapacitated variant, U is the set of customers with radial coordinates defined in respect to the position of the base station. This is a variant of capacitated (or uncapacitated) set cover where $\mathcal{S}$ consists of sets of customers that can be within range of a single antenna, *i.e.* of the form $\mathcal{R}(r, \alpha, \rho(r))$, where $\rho(r)$ is the angular width of an antenna with radial reach r.

The trade-off function ρ is decreasing; to simplify the proofs, we assume that $\rho(r) = 1/r$. We can change the r-coordinates to obtain exactly the same family of antenna sets as for arbitrary ρ.

3 Uncapacitated Cover by Antenna Sets

To simplify proofs, we will ignore the fact that the radial coordinate has a "wrap-around". We also renumber the customers so $\theta_i < \theta_{i+1}$ for $1 \leq i < n$. Observe that if $\theta_i = \theta_j$ and $r_i \geq r_j$ then every antenna set that contains i also contains j, so we can remove j from the input.

It suffices to consider only $n(n+1)/2$ different antenna sets. For such an antenna set A, let $i = \min A$, $j = \max A$. If $i = j$, we denote A as $A[i,i] = \{i\}$, and if $i < j$, we set $r(i,j) = (\theta_j - \theta_i)^{-1}$ and define $A[i,j] = \mathcal{R}(r(i,j), \theta_i, 1/r(i,j))$. (This definition is more complicated when the "wrap-around" is allowed.) Because $A \subseteq A[i,j]$ we can use $A[i,j]$ in our set cover instead of A.

We say that points i and j are compatible, denoted $i \heartsuit j$, if $i \leq j$ and there exists an antenna set that contains $\{i,j\}$. If $i = j$ then $i \heartsuit j$ is obvious; if $i < j$ then $i \heartsuit j$ is equivalent to $\{i,j\} \subseteq A[i,j]$ which in turn is equivalent to $r_i, r_j \leq r(i,j)$. If $i \heartsuit j$, we set $S[i,j] = \{k : \ i \leq k \leq j\} \setminus A[i,j]$.

We solve our minimum cover problem by dynamic programming. Our recursive subproblem is specified by a compatible pair i,j and its objective is to compute the size $C[i,j]$ of minimum cover of $S[i,j]$ with antenna sets. If we modify the input by adding the points 0 and $n+1$ with coordinates $(\varepsilon, \theta_1 - 1)$ and $(\varepsilon, \theta_n + 1)$ then our original problem reduces to computing $C[0, n+1]$.

If $S[i,j] = \varnothing$ then $C[i,j] = 0$. Otherwise, $S[i,j] = \{a_0, \ldots, a_{m-1}\}$, where $a_k < a_{k+1}$ for $k = 0, \ldots, m-2$.

We define a weighted graph $G_{i,j} = (V_{i,j}, E_{i,j}, c)$, where $V_{i,j} = \{0, \ldots, m\}$, $(k, \ell+1) \in E_{i,j}$ iff $a_k \heartsuit a_\ell$ and for an edge $(k, \ell+1)$, we define the cost $c(k, \ell+1) = 1 + C[a_k, a_\ell]$.

Note that $G_{i,j}$ is acyclic. Therefore, we can find a shortest (i.e., of minimum total cost) path from 0 to m in time $O(|E_{i,j}|) = O(n^2)$ [8]. Let d be the length of this path. We will argue that $C[i,j] = d$.

First, we show a cover of $S[i,j]$ with d antenna sets. A path from 0 to m in $G_{i,j}$ is an increasing sequence, and a path edge (u, v) with cost c corresponds to a cover of $\{a_u, a_{u+1}, \ldots, a_{v-1}\}$ with $A[a_u, a_{v-1}]$ and $c - 1$ antenna sets that cover $S[a_u, a_{v-1}]$.

Conversely, given a cover $\mathcal{C}$ of $S[i,j]$, we can obtain a path with cost $|\mathcal{C}|$ in $G_{i,j}$ that connects 0 with m.

For $A[k,\ell] \in \mathcal{C}$, we say that $\ell - k$ is its *width*. To make a conversion from a cover $\mathcal{C}$ of $S[i,j]$ to a path in $G_{i,j}$, we request that $\mathcal{C}$ has the minimum sum of widths among the minimum covers of $S[i,j]$.

This property of $\mathcal{C}$ implies that if $A[k,\ell] \in \mathcal{C}$ then:

$k, \ell \in S[i,j]$,
k and ℓ are not covered by $\mathcal{C} - \{A[k,\ell]\}$ (otherwise we eliminate $A[k,\ell]$ from $\mathcal{C}$ or replace it with a set that has a smaller width).

Note that for each pair of sets $A[k,\ell], A[k',\ell'] \in \mathcal{C}$, where $k < k'$, one of two following cases applies:

1. $\ell < k'$, i.e., $A[k,\ell]$ precedes $A[k',\ell']$;
2. $\ell' < \ell$, i.e., $A[k',\ell']$ is nested in $A[k,\ell]$.

Let $\mathcal{D}$ be the family of those sets in $\mathcal{C}$ that are not nested in others. Clearly $\mathcal{D}$ can be ordered by the leftmost elements in the sets. Note that if $A[k,\ell] \in \mathcal{D}$ then for some f, g, c, we have

$a_f = k \in S[i,j]$,
$a_g = \ell \in S[i,j]$,
$c - 1$ sets of $\mathcal{C}$ are nested in $A[k,\ell]$ and they cover $S[i,j]$,
$(f, g+1)$ is an edge in $G_{i,j}$ with cost c,
$g + 1 = m$ or $A[a_{g+1}, \ell'] \in \mathcal{D}$ for some ℓ'.

These $(f, g+1)$ edges form a path that connects 0 with m with cost $|\mathcal{C}|$.

Our dynamic programming algorithm solves the $n(n+1)/2$ subproblems specified by compatible pairs i, j in a non-decreasing order of the differences $j - i$. In the reduction of a subproblem to already solved subproblems the most expensive is the construction of the graph $G_{i,j}$ and finding the shortest path in it, both take quadratic time. Hence, we obtain our main result in this section.

Theorem 1. *The uncapacitated version of the problem of minimum covering with antenna sets n points, i.e., the restriction of* MinAntVar *to the case where all point demands are zero, is solvable in time* $O(n^4)$ *and space* $O(n^2)$ [1].

Previously, only a polynomial-time approximation algorithm with ratio two was known for the uncapacitated version of MinAntVar [3].

4 From Set Cover to Capacitated Set Cover

By the discussion in the previous section, it is sufficient to consider only $O(n^2)$ antenna sets in an instance of MinAntVar on n points. Hence, MinAntVar is a special case of minimum capacitated set cover.

Since we can determine a minimum uncapacitated set cover of an instance of MinAntVar by ignoring the demands and running the dynamic programming method given in the previous section, we shall consider the following more general situation.

We are given an instance of the general problem of minimum capacitated set cover and an approximation with ratio r for minimum set cover of the corresponding instance of minimum set cover obtained by removing the demands. The objective is to find a good approximation of a minimum capacitated set cover of the input instance.

We can obtain an approximation with ratio $r+1.692$ for minimum capacitated set cover on the base of an approximation with ratio r for minimum uncapacitated set cover $\mathcal{U}^*$ by running a simple greedy FFD algorithm (see Fig. 3). Our analysis of this algorithm in part resembles that of the first-fit heuristic for bin-packing [7,10], but the underlying problems are different. It yields the approximation ratio $r+1.692$. By refining the algorithm and its analysis, we can improve the factor substantially to $r+1.357$. Because of space considerations for the proof the reader is referred to the full version.

Theorem 41. *Let an instance of capacitated set cover be specified by a universe set* $P = \{1, ..., n\}$, *demands* $d_i \geq 0$ *for each* $i \in P$, *and a family* $\mathcal{S}$ *of subsets of* P. *If an approximation with ratio r for minimum set cover of the uncapacitated version of the instance (i.e., where the demands are removed) is given then a capacitated set cover of the input instance of size at most* $r+1.357$ *times larger than the optimum can be determined in polynomial time.*

Corollary 42. *There exists a polynomial-time approximation algorithm for the problem of* MinAntVar *with ratio* 2.357.

By the reduction of BinSchedule to MinAntVar given in [3], we also obtain the following corollary.

Corollary 43. *There exists a polynomial-time approximation algorithm for the problem of* BinSchedule *with ratio* 2.357.

---

[1] Very recently, M. Patrascu found a tricky way of improving the time complexity of an equivalent problem to a cubic one [16].

```
𝒬 ← ∅
for (U ∈ 𝒰*)
    while (U ≠ ∅)
        Q ← ∅
        for (i ∈ U, with d_i non-decreasing)
            if (d(Q) + d_i ≤ 1)
                insert i to Q
                remove i from U and P
        insert Q to 𝒬
```

Fig. 3. FFD, First Fit Decreasing algorithm for converting a cover into a capacitated cover

5 PTAS for MinAntLoad

In MinAntLoad problem, the radius of antennas is fixed and the number m of antennas that may be used is specified. The task is to minimize the maximum load of an antenna. A polynomial-time approximation for this problem achieving ratio 1.5 is presented in [3].

In the dual problem MinAnt, the maximum load is fixed and the task is to minimize the number of antennas. Recall that achieving an approximation ratio better than 1.5 for the latter problem requires solving the following problem equivalent to Partition.

Suppose that all demands can be covered with a single set, the load threshold is D and the sum of all demands is to $2D$. Decide whether or not two antennas are sufficient (which holds if and only if one can split the demands into two equal parts).

However, in case of the corresponding instance of MinAntLoad, we can apply FPTAS for the SubsetSum problem [14] in order to obtain a good approximation for the minimization of the larger of the two loads.

If all demands can be covered by a single antenna set (and the sum of demands is arbitrary) then MinAntLoad problem is equivalent to that of minimizing the makespan while scheduling jobs on m identical machines. Hochbaum and Shmoys showed a PTAS for this case in [13].

Interestingly enough, the PTAS of Hochbaum and Shmoys can be modified for MinAntLoad, while it does not seem to be the case with their practical algorithms that have approximation ratios of 6/5 and 7/6 [13].

Theorem 2. *MinAntLoad for n points admits an approximation with ratio $1+\varepsilon$ in time $n^{\frac{1}{\varepsilon}\ln\frac{1}{\varepsilon}+O(1)}$.*

Proof. See the full version. ❑

Note that the only geometric property of antennas with fixed radius that we used to design the PTAS for MinAntLoad is their correspondence to intervals or arcs. Hence, we obtain the following generalization of Theorem 2.

Theorem 3. *The problem of minimizing the maximum load in a capacitated set cover where the sets correspond to intervals or arcs admits a PTAS.*

6 Extentions to Multi-base MinAntVar

Our general method of approximating with ratio $r+1.357$ minimum capacitated set cover on the base of an approximate solution with ratio r to the corresponding minimum (uncapacitated) set cover can be also used to approximate optimal solutions to the natural extension of MinAntVar to include several base stations. It is sufficient to combine it with known approximation algorithms for geometric set cover, e.g., [5,6,12]. In this way, we can obtain an approximation with ratio $O(\log OPT)$ for the multi-base variant of MinAntVar, where OPT is the size of minimum uncapacitated set cover with antennas, see the full version. We can also prove the APX-hardness of the multi-base uncapacitated variant of MinAntVar by a reduction from a Minimum Line Covering Problem [4], see the full version.

Acknowledgments

The authors are grateful to Martin Wahlen for discussions on MinAntLoad, to David Ilcinkas, Jurek Czyzowicz and Leszek Gasieniec for preliminary discussions on MinAntVar, and to Mihai Patrascu for informing on his recent time improvement.

References

1. Aronov, B., Ezra, E., Sharir, M.: Small-size ϵ-Nets for Axis-Parallel Rectangles and Boxes. In: Proc. STOC 2009 (2009)
2. Bao, L., Garcia-Luna-Aceves, J.: Transmission scheduling in ad hoc networks with directional antennas. In: Proc. ACM MOBICOM 2002, pp. 48–58 (2002)
3. Berman, P., Kasiviswanathan, S.P., Urgaonkar, B.: Packing to Angles and Sectors. In: Proc. SPAA 2007 (2007)
4. Broden, B., Hammar, M., Nilsson, B.J.: Guarding Lines and 2-Link Polygons is APX-Hard. In: Proc. CCCG 2001 (2001)
5. Brönnimann, H., Goodrich, M.T.: Almost optimal set covers in finite VC-dimension. Discrete and Computational Geometry 14(4), 463–479 (1995)
6. Clarkson, K.L., Varadarajan, K.R.: Improved approximation algorithms for vertex cover. In: Proc. ACM SoCG 2005, pp. 135–141 (2005)
7. Coffman, E.G., Garey, M.R., Johnson, D.S.: Approximation algorithms for bin packing: a survey. In: Hochbaum, D.S. (ed.) Approximation Algorithms for NP-hard problems, pp. 46–93. PWS Publishing (1997)
8. Cormen, T.H., Leiserson, C.E., Rivest, R.L.: Introduction to Algorithms. The MIT Press, Cambridge (1990)
9. Erlabach, T., Jansen, K., Seidel, E.: Polynomial-Time Approximation Schemes for Geometric Intersection Graphs. SIAM J. Comput. 34(6), 1302–1323
10. Garey, M.R., Graham, R.L., Johnson, D.S., Yao, A.C.: Resource constrained scheduling as generalized bin-packing. J. Comb. Th. Ser. A 21, 257–298 (1976); cited from Coffman, E.G., Garey, M.R., Johnson, D.S.: Approximation algorithms for bin packing: a survey. In: Hochbaum, D.S. (ed.) Approximation Algorithms for NP-hard Problems, p. 50

11. Garey, M.R., Johnson, D.S.: Computers and Intractability. A Guide to the Theory of NP-completeness. W.H. Freeman and Company, New York (2003)
12. Hochbaum, D.S., Maass, W.: Approximation schemes for covering and packing in image processing and VLSI. Journal of the ACM 32(1), 130–136 (1985)
13. Hochbaum, D.S., Shmoys, D.B.: Using dual approximation algorithms for scheduling problems: theoretical and practical results. Journal of the ACM 34(1), 144–162 (1987)
14. Kellerer, H., Pfershy, U., Speranza, M.: An Efficient Approximation Scheme for the Subset Sum Problem. JCSS 66(2), 349–370 (2003)
15. Kellerer, H., Pferschy, U.: Improved Dynamic Programming in Connection with an FPTAS for the Knapsack Problem. J. Comb. Optim. 8(1), 5–11 (2004)
16. M. Patrascu. Personnal Communication (August. 2009)
17. Peraki, C., Servetto, S.: On the maximum stable throughput problem in random networks with directional antennas. In: Proc. ACM MobiHoc 2003, pp. 76–87 (2003)
18. Spyropoulos, A., Raghavendra, C.S.: Energy efficient communication in ad hoc networks using directional antennas. In: Proc. IEEE INFOCOM 2002 (2002)
19. Yi, S., Pei, Y., Kalyanaraman, S.: On the capacity improvement of ad hoc wireless networks using directional antennas. In: Proc. ACM MobiHoc 2003, pp. 108–116 (2003)

Effect of Corner Information in Simultaneous Placement of K Rectangles and Tableaux

Shinya Anzai[1], Jinhee Chun[1], Ryosei Kasai[1],
Matias Korman[2], and Takeshi Tokuyama[1]

[1] Graduate School of Information Sciences, Tohoku University, Japan
{anzai,jinhee,ryosei,tokuyama}@dais.is.tohoku.ac.jp
[2] Computer Science Department, Université Libre de Bruxelles, Belgium
mkormanc@ulb.ac.be

Abstract. We consider the optimization problem of finding k nonintersecting rectangles and tableaux in $n \times n$ pixel plane where each pixel has a real valued weight. We discuss existence of efficient algorithms if a corner point of each rectangle/tableau is specified.

1 Introduction

In his Programming Pearls column in Communications of ACM, Jon Bentley introduced the maximum subarray problem [5] motivated from a practical problem in pattern matching. The original problem was computation of the maximum sum subarray in a two dimensional $n \times n$ array of real numbers, which we call the *maximum-weight rectangle problem* in this paper. This problem was used to design a maximum likelihood estimator of two digital pictures. Bentley demonstrated the importance of elegant algorithm design techniques in the solution for its one-dimensional version (the maximum subarray problem), and challenged readers with a research problem to give the complexity of maximum-weight rectangle problem. Many readers (including some famous names such as E. W. Dijkstra and S. Mahaney) answered $O(n^3)$ time solutions appeared in the column of the next issue. It is not easy to improve the $O(n^3)$ time bound, and only improvement by sublogarithmic factors have been attained so far [16,3], while only the trivial $\Omega(n^2)$ lower bound is known.

In general, let $\mathbf{P}$ be an $n \times n$ pixel plane, and consider a family $\mathcal{F} \subset 2^{\mathbf{P}}$ of *pixel regions*. A pixel of $\mathbf{P}$ is the unit square $p(i,j) = [i-1,i] \times [j-1,j]$ where $1 \le i \le n$ and $1 \le j \le n$. A corner point of a pixel is referred to a grid point. The pixel $p = p(i,j)$ has a real value $W(p) = W(i,j)$ called the weight of p. Thus, we have a two dimensional array $W = (W(i,j))(1 \le i,j \le n)$ and consider the following *maximum-weight region* problem:

Find a region $R \in \mathcal{F}$ maximizing $W(R) = \sum_{p \in R} W(p)$.

The maximum-weight region problem (MWRP) has several applications such as image processing [1], data mining [11,17], surface approximation [6,7], and radiation therapy [7], as well as the pattern matching application given in [5].

The difficulty of the problem depends on the family $\mathcal{F}$. If $\mathcal{F} = 2^{\mathbf{P}}$, the problem is trivial, since R is obtained as the set of all pixels with positive weights. On the other hand, the problem is NP-hard if $\mathcal{F}$ is the set of all connected regions in $\mathbf{P}$ in the usual

M.T. Thai and S. Sahni (Eds.): COCOON 2010, LNCS 6196, pp. 235–243, 2010.

4-neighbor topology [1]. The following is a list of known families for which MWRP can be solved efficiently: rectangles [5,16,2], based monotone regions, x-monotone regions[1], rectilinear convex regions [17], staircase convex regions centered at a pixel r (stabbed unions of rectangles) [6], and digital star-shaped regions [9].

Recently, Chun et al. [8] gave an efficient algorithm to compute the maximum weight region decomposable into k based-monotone regions if we are given k axis-aligned base-lines, where a based monotone region for a horizontal (resp. vertical) line ℓ is a union of a set of column segments (resp. row segments) intersecting ℓ. This paper also deals with MWRP problems for disjoint unions of k fundamental shapes, namely, rectangles and staircase tableaux.

1.1 Our Problems and Results

The maximum-weight rectangle problem was generalized by Bae and Takaoka [2,3] so that the k largest weight rectangles are computed. However, those k maximum weight rectangles may overlap each other; therefore, we instead want to find placement of k nonoverlapping rectangles maximizing the total weight or maximizing the minimum of the weights of k rectangles. We call the former problem Max-Sum kRP (maximum weight sum k rectangles placement problem) and the latter Max-Min kRP. See Figure 1 for a solution of Max-Sum kRP.

Bae and Takaoka [4] also studied *k-disjoint maximum subarrays problem* in a two-dimensional array. The problem is defined in a greedy fashion such that we find the maximum weight rectangle, then find the maximum weight rectangle in the remaining part, and so on. Although their algorithm attains a very fast $O(n^3 + kn^2 \log n)$ time complexity, the output family of k rectangles does not always attain the maximum sum of weights.

A very naive solution of the kRP problems is to find all $O(n^4)$ possible locations for each rectangle and consider all possible combinations of k of them. This is done in $O(n^{4k})$ time, and hence the problem is in class P if k is a constant. It is not difficult to reduce the running time to $O(n^{2k+1})$: indeed, any column intersects at most k rectangles, and we can sweep the pixel plane to run a dynamic programming keeping candidate combinations of at most $2k$ edges of intersecting rectangles. One drawback of this solution is its high ($O(n^{2k+1})$) space complexity. On the other hand, the Max-Min kRP problem is NP-hard if k is considered as an input variable.

Thus, a natural question is whether the problem is fixed parameter tractable (FPT) with respect to the parameter k. In other words, we want to find an $O(f(k)n^c)$ time algorithm where f is some function and c is a constant independent of k. Unfortunately, we do not know how can the above $O(n^{2k+1})$ time complexity be improved in the general case (indeed, we suspect that the problem is $W(1)$-hard). Thus, in order to analyze the structure of complexity of the problem, we consider an easier situation where additional information is given. In this paper, as the additional information, we are given a set S of k grid points as a part of input, and find the optimal set of k nonintersecting rectangles such that each point p of S defines a left side corner of a rectangle (we also give indication that p is a upper-left or a lower-left corner). There are $O(n^{2k})$ possible sets of corner points, and $O(2^{2k})$ choices of types (upper or lower) of corners. Therefore, this may well refine the $O(n^{2k+1})$ complexity of the above mentioned algorithm.

2	3	-4	-2	2	3	-4	3
-1	0	-2	-6	1	0	-2	-6
-2	5	1	-2	-2	5	1	-2
4	-2	-1	3	4	-2	-1	3
2	3	-4	10	2	3	-4	-5
1	0	-2	-6	1	0	-2	-6
-2	-5	1	-2	-2	5	1	2
4	-2	-1	-3	4	-2	-1	3

Fig. 1. Max-Sum placement of 4 rectangles

2	3	-4	-2	2	3	-4	3
-1	0	-2	-6	1	0	-2	-6
-2	5	1	-2	-2	5	1	-2
4	-2	-1	3	4	-2	-1	3
2	3	-4	10	2	3	-4	-5
1	0	-2	-6	1	0	-2	-6
-2	-5	1	-2	-2	5	1	-2
4	-2	-1	-3	4	-2	-1	3

Fig. 2. Max-Sum placement of 4 tableaux

We give an algorithm for the Max-Min kCRP problem (Max-Min k cornered rectangles placement) with $O(kn^2 + n^2 \log n)$ time and $O(kn + n^2)$ space complexities (note that the input size is n^2). We also show that the Max-Sum kCRP problem has $O((2k)^{k+1}n^2)$ time and $O(n^2)$ space FPT algorithm.

Even if corner points are not given, our result implies that the k rectangle placement problem can be solved in $O(n^{2k+2})$ time using linear space if k is a constant (since we can exhaustively try all possible combinations of locations of corner points). This is a space efficient method, and allows a lot of parallelism compared with the above mentioned methods.

We further consider the *staircase tableau*[1] (we call *tableau* in short), which is a union of rectangles sharing a given grid point p as their fixed (say, lower-left) corner. A tableau is obtained by chopping a rectangle with a monotone rectilinear path, thus it is considered as a "right triangle" in the rectilinear world. We consider the problem of finding k disjoint staircase tableaux maximizing the total weight (as shown in Figure 2). We show that this problem has an $O(2^{2k^2}n^3)$ time FPT algorithm if we are given the corner position of each tableau and if each tableau has its corner in either its lower-left or upper-left position.

1.2 Related Problems

Map labeling. In a map labeling problem, we have a set of k pinning points, n (typically rectilinear) obstacles in the plane, and a set $L(p)$ of rectangular labels for each p. We want to place exactly one label of $L(p)$ for each p such that each label does not overlap obstacles nor other labels. The map labeling problem can be related to our problem: We simulate the plane as a pixel plane, and represent obstacles as a set of pixels with $-\infty$ weights. We give a positive (typically unit) weight to other pixels. Then, this particular Max-Min kCRP problem is a discrete version of the *elastic label placement problem* [12], where we are allowed to use labels (with an area constraint) that have p as one of its left corners. We can control weights to give different types of label sets, and methodologies in map labeling are utilized in our kRP problems.

[1] This object is called Young diagram (or Young tableau if each pixel has some data such as weight) in combinatorics.

Dissection of rectilinear polygon and a cutting stock problem. Given a rectilinear polygon Q with holes, we can dissecting it into minimum number of rectangles in polynomial time [15]. Suppose that we further find a dissection of Q into rectangles maximizing the total area of k maximum rectangles. We can divide the plane by the horizontal and vertical lines going through vertices of Q to have a grid that is mapped to the pixel plane such that $W(p) = -\infty$ if $p \notin Q$ and $W(p)$ is the original area of p if $p \in Q$. Thus, the problem becomes a Max-Sum kRP problem. If Q is a plate of wood, the problem is a cutting stock problem where we would like to cut out k axis-parallel rectangular pieces from it. We may also optimize quality caused from scars and bad textures in the plate by refining the pixels and giving suitable weights to measure the quality. Our algorithms for kCRP imply a semi-automatic system to give the optimal solution if an expert gives the positions of the corners.

Image segmentation. Separating an object from its background in an image is a key problem in image processing. This operation is commonly called *image segmentation*. In a pixel plane $\mathbf{P}$ representing a picture, each pixel p has a real value $f(p)$ representing the brightness of the pixel. The segmented image should be a pixel region with a nice geometric property. The quality of the segmentation depends on the separation of brightness in the image and background. Asano et al. [1] proposed an *optimization-based* image segmentation method that gives a robust solution with theoretical guarantee. It defines a family $\mathcal{F}$ of grid regions, and finds the region $R \in \mathcal{F}$ optimizing a convex objective function. Their framework gives a method to compute the optimal segmentation provided that the MWRP is solved efficiently for $\mathcal{F}$. Our new algorithms enable us to segment optimal disjoint k rectangles with given corner positions (on the left side), and k tableaux.

2 Placing Rectangles

Given a set S of k grid points $p_1, p_2, .., p_k$ and a sequence of bits $\mathbf{b} = b_1, b_2, .., b_k$, we say that k non-overlapping rectangles $R_1, R_2, .., R_k$ are a *feasible assignment* if p_i is the lower-left (resp. upper-left) corner of R_i if $b_i = 0$ (resp. $b_i = 1$) for each i. Let $W(R_i)$ be the sum of weights of pixels in the rectangle R_i.

2.1 Max-Min Problem

In the Max-Min CRP, we find a feasible assignment such that the smallest weight of R_i is maximized. The following theorem follows from the idea of *Left side ordered map labeling* algorithm given in [13]:

Theorem 1. *Max-Min CRP can be solved in $O(\min\{kn^2 + n^2 \log n, n^2 \log \Gamma\})$ time, where Γ is the maximum of absolute values of weights of all possible rectangles.*

Proof. First we consider the decision problem in which we find a feasible set such that $W(R_i) \geq \theta$ for a given threshold value. We remark a data structure of size $O(n^2)$ can answer the weight of any rectangle in $O(1)$ time since it is represented as a linear combination of weights of at most five rectangles containing either $p(1, 1)$ or $p(1, n)$. We sort p_i in a increasing lexicographic order from left to right, where the x-coordinate value is the main key and the y-coordinate value is the sub key. A rectangle is called *rich* if its weight is not less than θ. Now, we do the plane sweep from right to left , visiting points in reverse order. Suppose that we have already placed the rectangles for p_k through p_{j+1}. Then, we examine rectangles cornered by p_j that do not overlap any placed rectangles so far from shortest to tallest (and narrowest to widest among the same height ones) by scanning the antipodal corner pixel to p_j until we find a rich rectangle. Since a shorter rectangle less affects to the placement of rectangles at remaining points $p_1, p_2, \ldots, p_{j-1}$ on the left of p_j, we can easily prove that this is always the best possible choice. The algorithm fails when no such rich rectangle exists.

In the whole process, each pixel is scanned at most once. Thus, the decision version is solved in $O(n^2)$ time. We now can do binary search to solve Max-Min CRP using the decision problem as a subroutine. If Γ is huge, we search among the (at most) kn^2 possible rectangle weights by applying median-finding. □

Note: We need $O(kn^2)$ space if we store all the threshold values. However, the required space can be reduced to $O(n(n+k))$ by using a standard randomized selection method.

Remark. If the bit sequence $\mathbf{b}$ is not given and k is considered as a variable in complexity, the problem is NP-hard. The hardness can be shown analogously to hardness results of map labeling (e.g. [10,12]). Of course, the number of combinations of $\mathbf{b}$ is 2^k, and hence the problem has an FPT algorithm when $\mathbf{b}$ is not known. The NP-hardness of the general Max-Min kRP can be also obtained by modifying an NP-hardness proof of disjoint rectangle covering of a point set [14]. We omit hardness proofs in this version.

2.2 Max-Sum Problem

We next consider the Max-Sum kCRP problem of finding the feasible assignment maximizing the total sum $\sum_{i=1}^{k} W(R_i)$.

In order to understand the problem, we start with the case in which $\mathbf{b} = \mathbf{0}$ (i.e., each R_i has p_i as its lower-left corner). We subdivide the pixel plane into $(k+1)^2$ small rectangular cells by vertical and horizontal lines going through each point in S. Consider a cell C, and suppose that two rectangles R_i and R_j in a feasible assignment intersect C. Since the rectangles grow from lower-left corners, both of R_i and R_j must contain the lower-left corner of C. In particular, they overlap and we have a contradiction. Hence, at most one rectangle can intersect C.

We guess which rectangle intersects C for each C; in other words, we give label i ($i = 1, 2, .., k$) to C if R_i intersects C (0 if it intersects no R_i). There are $(k+1)^{(k+1)^2}$ such label assignments, and if the assignment is feasible, the union of regions with a label i should form a rectangle $Rect(i)$ that has p_i as its lower-left corner. We can reduce the search space by just guessing the opposite cell of the rectangle $Rect(i)$ for each i. This way, the total number of possible assignments is reduced to k^{2k}. Once we have a right guess, R_i is obtained as the maximum weight rectangle in $Rect(i)$ cornered by p_i, and we can compute R_i for all i in linear time. Therefore, we can compute the Max-Sum CRP in $O(k^{2k}n^2)$ time if $\mathbf{b} = \mathbf{0}$.

For any general **b**, we consider the same subdivision, and guess the rectangles intersecting C. In this case, up to two rectangles can intersect C; one contains the lower-left corner and the other contains the upper-left one. Thus, C may have two labels i and j such that $b_i + b_j = 1$ (see an example in Figure 3).

We sweep the plane from right to left and run a dynamic programming algorithm. The dynamic programming maintains the intersections of rectangles and the vertical lines. If we naively do this, it costs $O(n^{2k+1})$ time as mentioned in the introduction. Let us consider vertical lines through points of S that divides the plane into $k+1$ slabs. Let us consider the i-th slab (from the left) B_i, which consists of $k+1$ cells. We next decompose B_i into rectangles called components. If there is a cell with two labels a and b, we define the component $Z(a, b)$ as the union of cells with either a or b as their labels. If a label c does not share any cell, we define the component $Z(c)$ as the union of cells with the label c. We ignore empty cells and decompose B_i into components.

Let V_i be the set of generated components in slab B_i, and let $V = \cup_{i=1}^{k+1} V_i$. We define the graph $G = (V, E)$ on V, where the directed edge (v, u) from v towards u in G is in E if u is a component in B_i, v is a component in B_{i+1} and u and v share a label.

Lemma 1. *Suppose that G corresponds to a feasible placement of k rectangles. Then, for any node v of G, both the indegree and outdegree of v is at most one. In other words, G is a union of directed paths.*

Proof. Assume that there exists a component $u = Z(a, b)$ with labels a and b in slab B_i adjacent to two components $v = Z(a)$ and $v' = Z(b)$ in slab B_{i+1}. By definition, there exists a cell C in B_i with both labels a and b. Since both a and b appear in slab B_{i+1}, the cell C' located to the right of C must also have both labels a and b. In particular, B_i must have the component $Z(a, b)$ instead of two separate components $Z(a)$ and $Z(b)$. Thus, we have contradiction and the indegree of u is at most one. The discussion for the outdegree is similar. □

Now, we can run our dynamic programming on each directed path of G separately. Indeed, a component is only affected by components with labels in common, and hence we can independently sweep on each directed path to obtain the optimal solution within each path; in other words, in the union of components on the path.

Theorem 2. *Max-Sum CRP problem can be solved in $O(k^{2k+1}n^2)$ time.*

Proof. At most two rectangles intersect each component v. Moreover, the height of one of the horizontal edges of each rectangle is known. Thus, the intersection of the rectangles at the left side edge of a component v is determined by the position of at most two horizontal edges. At most n_v^2 combinations are possible, where n_v is the height of the component v (i.e., the sum of heights of cells in the component). Each rectangle may penetrate or start within v. To each rectangle in v we add the previous weight-sum that is inherited from the parent node of v. So, it is easy to show that the update of the DP table can be done in $O(n_v^2)$ time. In total, the DP algorithm runs in $O(kn^2)$ time, since the sum of n_v^2 over all $v \in V$ is $O(kn^2)$ (note that we have only $k+1$ slabs). Since at most k^{2k} labels are possible, the theorem is shown. □

3 Placement of k Tableaux

In contrast to the case of rectangles, there are exponential number of different tableaux in $\mathbf{P}$. If k is a constant, we can compute the Max-Sum (or Max-Min) k disjoint tableaux in polynomial time by doing a plane-sweep and maintaining (at most) k intersecting segments with the current column. We can then use a DP table to keep the $O(n^{2k})$ possibilities of such segments. Naively, this can be done in $O(n^{4k+1})$ time.

Our question is how we can improve it if we are give the set of k corner points as a part of input. Note that we allow the case in which the same point is the origin of more than one tableaux of different orientations. We show that the problem has an FPT algorithm if each tableau has its corner in either its lower-left or upper-left position: We call them up-tableau and down-tableau, respectively. The time complexity if $O(2^{2k^2}n^3)$, which implies that the naive $O(n^{4k+1})$ bound can be improved to $O(n^{2k+3})$ even if the the corner positions are unknown, since the number of combinations of k corner points is $O(n^{2k})$. We focus on Max-Sum problem, since Max-Min problem can be solved similarly.

Analogous to the rectangle case, we subdivide the pixel plane by the vertical and horizontal lines going through each points in S into rectangular cells. We give labels of tableaux intersecting each cell as shown in Figure 4. As before, a cell can have at most one up-tableaux and one down-tableau label.

Let Q_i be the union of cells with label i. The optimal region R_i must be contained in Q_i for $i = 1, 2, .., k$. Moreover, Q_i must be a staircase tableau. For each i, there are at most ${}_{2k}C_k \approx 2^{2k}/\sqrt{k} < 2^{2k}$ possible candidate regions of Q_i, and hence we need to consider at most $(2^{2k})^k = 2^{2k^2}$ combinations of a family of k such regions. Let us assume that we have correct regions $Q_1, Q_2, \ldots, Q_k$ such that $R_i \subseteq Q_i$ for each $i = 1, 2, \ldots, k$.

Similarly to the case of rectangles, we consider vertical slabs B_1, B_2,. . .,B_{k+1}, and decompose each slab into components such that each component is a union of cells corresponding to either a fixed pair of labels or a label. We define the set V_i of components in B_i, $V = \cup_{i=1}^{k+1} V_i$, and the graph $G = (V, E)$ connecting components sharing the same label in adjacent slabs with a directed edge towards left. Figure 5 gives the graph for the labeling given in Figure 4.

Lemma 2. *If G corresponds to a feasible placement of k talbeaux, G is a directed forest such that each edge is directed towards the root of a tree in the forest.*

Proof. Assume that there exists a component $u = Z(a, b)$ with labels a and b in slab B_i that is adjacent to components $Z(a)$ and $Z(b)$ in B_{i-1}. By definition, there exists a cell C in B_i with both labels a and b. As in Lemma 1, the cell C' located to the left of C in B_{i-1} must also have both labels a and b. Thus, B_{i-1} must have the component $Z(a, b)$ instead of the two separate components $Z(a)$ and $Z(b)$. Thus, we have contradiction and the outdegree of u is at most one, thus G is a directed forest. We remark that the indegree may become two since a tableau shrinks to the right. □

We run a dynamic programming on each tree of the forest G from the leaves to the root. At a vertex u associated with a component of width m_u and height n_u, we inherit from its children (if any) the information of the best tableaux to the right u. The information is given for each possible intersection patterns of tableaux at the right edge of the component. The number of patterns is bounded by $O(n_u^2)$ since at most two tableaux intersect the component. We now find the best solution for each possible intersection pattern at its left edge in $O(n_u^2 m_u)$ time by using dynamic programming within the component. Since we sweep leftwards, the tableaux cannot decrease in height. We find the optimal placement in each tree, and merge them to find the global solution. The total time complexity is $O(\sum_{u \in V} n_u^2 m_u) = O(n^3)$ and we have the following:

Theorem 3. *The maximum weight placement of k up or down tableaux with given k corner positions is computed in $O(2^{2k^2} n^3)$ time.*

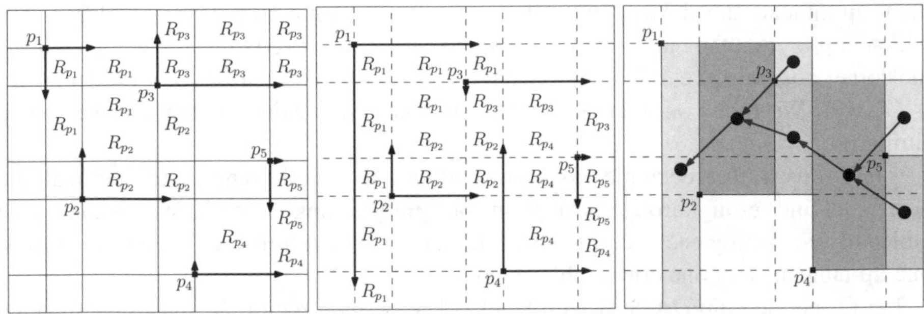

Fig. 3. Labels of cells for Max-Sum kCRP

Fig. 4. Labeling for tableaux placement

Fig. 5. Graph G for tableaux placement

4 Concluding Remarks

The following is a list of open problems: (1) kCRP allowing four different positions of rectangles. (2) kRP where we prelocate a point lying on the left edge of each rectangle. (3) k tableaux placement allowing all four types of tableaux. (4) Max-Sum k square placement with given center points of squares. (Max-min problem is easy). (5) Placement of k digital star regions (see [9] for definition).

Acknowledgement. The authors gratefully acknowledge to Sang Won Bae for helpful discussions. This work is partially supported by MEXT grant on basic research (B) 22300001 and young researcher grant (B) 21700004.

References

1. Asano, T., Chen, D.Z., Katoh, N., Tokuyama, T.: Efficient Algorithms for Optimization-Based Image Segmentation. Int. J. Comput. Geometry Appl. 11(2), 145–166 (2001)
2. Bae, S.E., Takaoka, T.: Improved Algorithms for the K-Maximum Subarray Problem. The Computer Journal 49(3), 358–374 (2006)
3. Bae, S.E., Takaoka, T.: A Sub-Cubic Time Algorithm for the K-Maximum Subarray Problem. In: Tokuyama, T. (ed.) ISAAC 2007. LNCS, vol. 4835, pp. 751–762. Springer, Heidelberg (2007)

4. Bae, S.E., Takaoka, T.: Algorithms for K-Disjoint Maximum Subarrays. Int. J. Found. Comput. Sci. 18(2), 319–339 (2007)
5. Bentley, J.: Algorithm Design Techniques. ACM Commu. 27(9), 865–871 (1984); Also found in Bentley, J.: Programming Pearls, 2nd edn. ACM Press, New York (2000)
6. Chen, D.Z., Chun, J., Katoh, N., Tokuyama, T.: Efficient algorithms for approximating a multi-dimensional voxel terrain by a unimodal terrain. In: Chwa, K.-Y., Munro, J.I.J. (eds.) COCOON 2004. LNCS, vol. 3106, pp. 238–248. Springer, Heidelberg (2004)
7. Chen, D.Z., Hu, X.S., Luan, S., Wu, X., Yu, C.X.: Optimal Terrain Construction Problems and Applications in Intensity-Modulated Radiation Therapy. Algorithmica 42(3-4), 265–288 (2005)
8. Chun, J., Kasai, R., Korman, M., Tokuyama, T.: Algorithms for Computing the Maximum Weight Region Decomposable into Elementary Shapes. In: Dong, Y., Du, D.-Z., Ibarra, O. (eds.) ISAAC 2009. LNCS, vol. 5878, pp. 1166–1174. Springer, Heidelberg (2009)
9. Chun, J., Korman, M., Nöllenburg, M., Tokuyama, T.: Consistent Digital Rays. In: Proc. 24th ACM SoCG, pp.355–364 (2008)
10. Formann, M., Wanger, F.: A Packing Probelm with Applications to Lettering of Maps. In: Proc. 7th ACM SoCG, pp. 281–290 (1991)
11. Fukuda, T., Morimoto, Y., Morishita, S., Tokuyama, T.: Data Mining with optimized two-dimensional association rules. ACM Trans. Database Syst. 26(2), 179–213 (2001)
12. Iturriaga, C.: Map Labeling Problems, Ph. D Thesis, Waterloo University (1999)
13. Koike, A., Nakano, S.-I., Nishizeki, T., Tokuyama, T., Watanabe, S.: Labeling Points with Rectangles of Various Shapes. Int. J. Comput. Geometry Appl. 12(6), 511–528 (2002)
14. Korman, M.: Theory and Applications of Geometric Optimization Problems in Rectilinear Metric Spaces, Ph. D Thesis, Tohoku University (2009)
15. Soltan, V., Gorpinevich, A.: Minimum Dissection of a Rectilinear Polygon with Arbitrary Holes into Rectangles. Disc. Comput. Geom. 9, 57–79 (1993)
16. Tamaki, H., Tokuyama, T.: Algorithms for the Maximum Subarray Problem Based on Matrix Multiplication. In: Proc. 9th SODA, pp. 446–452 (1998)
17. Yoda, K., Fukuda, T., Morimoto, Y., Morishita, S., Tokuyama, T.: Computing Optimized Rectilinear Regions for Association Rules. In: Proc. KDD 1997, pp. 96–103 (1997)

Detecting Areas Visited Regularly

Bojan Djordjevic[1,2] and Joachim Gudmundsson[2]

[1] School of Information Technology, Sydney University, Sydney, Australia
[2] NICTA* Sydney, Locked Bag 9013, Alexandria NSW 1435, Australia
{bojan.djordjevic,joachim.gudmundsson}@nicta.com.au

Abstract. We are given a trajectory $\mathcal{T}$ and an area $\mathcal{A}$. $\mathcal{T}$ might intersect $\mathcal{A}$ several times, and our aim is to detect whether $\mathcal{T}$ visits $\mathcal{A}$ with some regularity, e.g. what is the longest time span that a GPS-GSM equipped elephant visited a specific lake on a daily (weekly or yearly) basis, where the elephant has to visit the lake *most* of the days (weeks or years), but not necessarily on *every* day (week or year). We call this a *regular pattern* with *period* of one day (week or year, respectively).

We consider the most general version of the problem defined in [8], the case where we are not given the period length of the regular pattern but have to find the longest regular pattern over all possible period lengths. We give an exact algorithm with $\mathcal{O}(n^{3.5}\log^3 n)$ running time and an approximate algorithm with $\mathcal{O}(\frac{1}{\varepsilon}n^3\log^2 n)$ running time.

We also consider the problem of finding a region that is visited regularly if one is not given. We give exact and approximate algorithms for this problem when the period length is fixed.

1 Introduction

Technological advances of location-aware devices and mobile phone networks have made it possible to easily get access to large amounts of tracking data. As a result many different areas including geography, market research, database research, animal behaviour research, surveillance, security and transport analysis involve to some extent the study of movement patterns of entities [2,9,13]. This has triggered an increasing amount of research into developing algorithms and tools to support the analysis of trajectory data [10]. Examples are detection of flock movements [3,5,11], leadership patterns [4], commuting patterns [14,15,17], regular visit patterns [8] and identification of popular places [12].

In this paper we consider the problem of finding regular visit patterns introduced in [8]. Consider a trajectory obtained by tracking an elephant [1]; it is easy to detect which areas are important for the elephant, i.e. where it spends a certain amount of its time. However, ideally we would like to be able to detect if this area is visited with some regularity which might indicate that it could be used as a grazing or mating area during certain times of year. Another example occurs when tracking a person. Again, it is easy to detect which areas

* NICTA is funded by the Australian Government's Backing Australia's Ability initiative, in part through the Australian Research Council.

M.T. Thai and S. Sahni (Eds.): COCOON 2010, LNCS 6196, pp. 244–253, 2010.

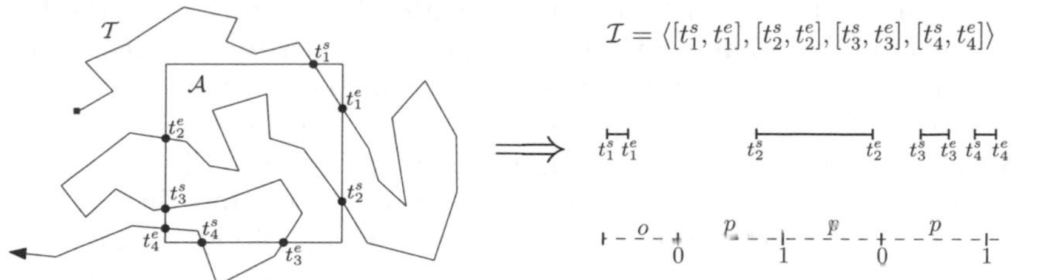

Fig. 1. A trajectory $\mathcal{T}$ and an area $\mathcal{A}$ is shown. From this, we derive the sequence of intervals $\mathcal{I}$ from which we obtain the sequence of regular time points and also $s(o,p)$.

are important for her, such as home, work, local shopping areas and the cricket ground. But it would be interesting to find out if she goes to the cricket ground with some regularity, for example batting practice every Wednesday night. Note however, that the visits may be regular even though the cricket ground is not visited *every* Wednesday evening. It might be a regular event even though it only takes place 50% of all Wednesday evenings.

The above examples give rise to the following problem, as illustrated in Fig. 1. Given an area $\mathcal{A}$ in some space and a trajectory $\mathcal{T}$, i.e. a time-stamped sequence of points in the space, one can generate a sequence of n time intervals $\mathcal{I} = \langle I_1 = [t_1^s, t_1^e], \ldots, I_n = [t_n^s, t_n^e] \rangle$ for which the entity generating the trajectory is within $\mathcal{T}$ intersects the area $\mathcal{A}$. Some of these intervals might be important for our application, while others are not (e.g. a person's visit to the cricket ground on a Sunday to watch a match is not interesting for detecting regular practice that occurs on Wednesdays). Hence, we look at whether $\mathcal{T}$ intersects $\mathcal{A}$ for a sequence of regular time points.

For modelling regularity among the sequence of time points, we introduce two important notions: the *period length* p and the *offset* o. For a fixed period length p and offset o (with $o < t_1^s + p$), we have a sequence of time points between t_1^s and t_n^e uniquely determined in the following way: all time points are equidistant with distance p and the first time point is at time o (e.g. if o is chosen to be a 'Wednesday at 19:30', and p equals 7 days, then all these time points will correspond to all 'Wednesday at 19:30' for all weeks). Having the entire sequence of regular time points, the problem is to find the longest subsequence of consecutive time points such that $\mathcal{T}$ intersects $\mathcal{A}$ with high density. We model the density as a value $c \in [0,1]$ and require that $\mathcal{T}$ intersects $\mathcal{A}$ for at least $(100 \cdot c)$% of the times. To each time point we associate a value of 1 if $\mathcal{T}$ intersects $\mathcal{A}$ and 0 otherwise. We call the collection of these bits the *bitstring* $s(o,p)$ for a particular p and o and we call the longest pattern the *longest dense substring*, or LDS.

In [8] we consider a number of variations of this problem. For the most restricted case, where both the period length and the offset are given, we give an optimal linear time algorithm to find the longest dense pattern.

And for the least restricted case where neither the period length nor the offset are given an $\mathcal{O}(\frac{n^5}{c})$ time algorithm was given in [7], which was later improved to $\mathcal{O}(\frac{n^4}{c})$ in [8]. This algorithm finds the period length that gives the longest dense pattern. Note that we restrict the period length to be longer than the longest interval in $\mathcal{I}$. Without this restriction we could get arbitrarily long regular patterns by making p much smaller than the interval lengths.

In this paper we give a $\mathcal{O}(n^{3.5}\log^3 n)$ time exact and a $\mathcal{O}(\frac{1}{\varepsilon}n^3\log^2 n)$ time approximation algorithm for the most general problem, improving on the previous $\mathcal{O}(n^4)$ time algorithm. We also solve the problem of finding a region that is visited regularly when one is not given. The paper is organised as follows. In Section 2 we give algorithms for finding regular patterns in bitstrings in a dynamic setting. In Section 3 we define the *configuration space* of the problem and in Section 3.1 we prove a number of results about the complexity of the configuration space. In Section 3.2 we solve the regular pattern problem by sweeping the configuration space. In Section 3.3 we show that the previous results, which assumed that universe size $U = t_n^e - t_1^s$ is small, carry over to the case where the universe size is polynomially bounded. In Section 4 we define the problem of finding regions from trajectory data, and in Sections 4.1 and 4.2 we solve the continuous and discrete versions of the problem.

2 Working with Bitstrings

Fixing the period length p and an offset o gives a bitstring $s(o,p)$. By changing the period or the offset by a small amount we get a different bitstring $s(o+\delta o, p+\delta p)$ which differs from the previous bitstring in only one bit. In Section 3.2 we will repeatedly change a single bit in the bitstring and look for the longest patern. We would like to avoid recalculating the longest dense substring from scratch in situations like this, hence we are interested in a dynamic setting where we allow single bit updates.

We assume we are given a density $c \in [0,1]$ and a bitstring $B = \langle b_1, \ldots, b_n \rangle$ of length n, and would like to find the longest dense substring. Theorem 1 in [8] can be used to solve this problem in $O(n)$ time.

We are given a list of updates $\mathcal{U} = \langle u_1, \ldots, u_m \rangle$ where u_i is the index of the bit that is flipped during update step i. After each update step we would like to report the longest dense substring S_i. In this paper we are not interested in computing the longest pattern after *each* update, but rather the *overall* longest dense substring, i.e. instead of finding the longest pattern S_i after every update i, we only need the S_i with the maximum length. We call this the *overall longest pattern.*

Lemma 1. *Given ε, c, a bitstring $B = \langle b_1, \ldots, b_n \rangle$ of length n, and m single bit update steps $\mathcal{U} = \langle u_1, \ldots, u_m \rangle$, there is an algorithm to compute an ε-approximate solution of the overall longest pattern in $\mathcal{O}(\frac{1}{\varepsilon}(n + m\log n))$ time using $\mathcal{O}(n+m)$ space.*

Proof. This is a simple consequence of Theorem 2, Remark 2 in [8]. The length of the bitstring is n so the tree data structure in [8] can be built in $\mathcal{O}(\frac{n}{\varepsilon})$ time, the event queue, i.e. the sorted list of update steps $\mathcal{U}$, is already given, and the number of updates required is $\mathcal{O}(m)$. □

In the exact case we can prove the following.

Lemma 2. *Given c, a bitstring $B = \langle b_1, \ldots, b_n \rangle$ of length n, and m single bit update steps $\mathcal{U} = \langle u_1, \ldots, u_m \rangle$, there is an algorithm to compute the overall longest pattern in $\mathcal{O}(\sqrt{n}(n+m)\log^2(n+m))$ time using $\mathcal{O}((n+m)\log(n+m))$ space.*

Proof. We transform this into a geometric problem in the following way. Consider an $m \times n$ grid with the y axis representing the index in the bitstring and the x axis the update steps. For each index $i \in \{1, \ldots, n\}$ we will build horizontal line segments that specify during which update steps the bit b_i was active.

If we want to know the state of the bitstring after update u_i we simply look at the intersection of the vertical line $x = u_i$ with the segments. The bit b_i is 1 if and only if the vertical line intersects a segment with y coordinate i. We therefore want to select the vertical line that maximises the length of the longest dense substring of the bitstring induced by that vertical line. This problem is solved in Section 5 in [8] in time $\mathcal{O}(n^{\frac{3}{2}} \log^2 n)$ for the case where $m = \mathcal{O}(n)$. The same technique can be used for arbitrary m and gives the required running time. □

The previous theorems will be used in Section 3.2, where we show that the longest pattern over all period lengths and offsets can be reduced to this type of dynamic problem.

3 Configuration Space

In [8] we showed that it is sufficient to consider only a finite number of period-offset pairs. In this section we present a graphical representation of these pairs as cells in a line arrangement. In this representation adjacent cells represent two period-offset pairs whose bitstrings differ by only one bit. This is the crucial property that allows us to improve on the results in [8].

Without loss of generality we assume that $0 < t_1^s < U$, where $U = t_n^e - t_1^s$. The time intervals can always be translated to have this property. This immediately gives $t_n^e < 2U$. For convenience we also let o be in the range $[0, p)$ rather than $[t_1^s, t_1^s + p)$. This does not affect the results because it can at most double the number of bits in each bitstring.

We define a *configuration* to be an assignment of p and o. There are only a finite number of configurations that produce different bitstrings. In this section we show how to efficiently enumerate all interesting configurations and the bitstrings they produce.

We consider the 2D *configuration space* of all configurations. This space is divided into regions that produce the same bitstring. If we fix a period length p then we effectively partition the time line into pieces of length p. Consider an event at time t. We define the *relative position* of the event $r = \frac{t}{p}$, or equivalently $\frac{1}{p} = \frac{1}{t}r$. Hence, we define the configuration space with the axes r and $\frac{1}{p}$. Consider the $\frac{1}{p}$ versus r space in Fig. 2(a) and an event at time $t > 0$. This event is mapped to the line $\frac{1}{p} = \frac{1}{t}r$. This is a result of the fact that an event can be represented by infinitely many (p, r) pairs.

Now consider an interval $[t_1, t_2]$ on the timeline. Events at t_1 and t_2 correspond to the two lines in Fig. 2(b) and events inside the interval to points within the wedge defined by the two lines. This wedge represents all (p, r) pairs that correspond to the time interval $[t_1, t_2]$.

Finally consider all time intervals in the sequence $\mathcal{I}$. Each interval maps to a wedge so $\mathcal{I}$ will map to a set of wedges $\mathcal{W}_0$ in Fig. 2(c).

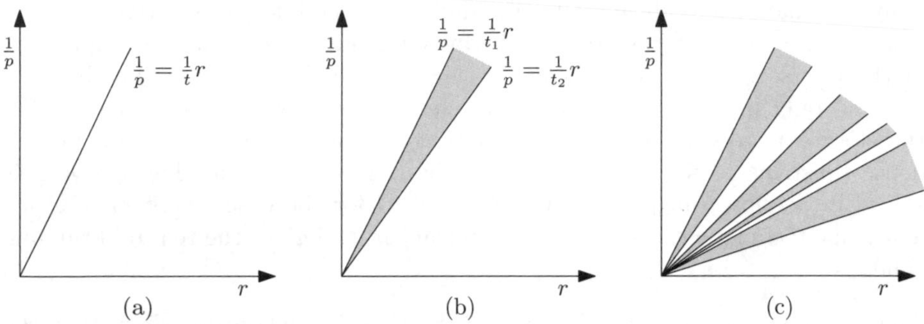

Fig. 2. (a) A single point, (b) an interval, and (c) a set of intervals in the configuration space

Let us fix a period length $p = p_0$. In the configuration space this means that we are only interested in the horizontal line $\frac{1}{p} = \frac{1}{p_0}$. The intersection of this line with a wedge W gives the relative positions of the events in the corresponding interval $I(W)$. If we also fix an offset $o \in [0, p_0)$ then the bitstring will be composed of the events at $t \in \{o, o + p_0, o + 2p_0, \ldots, o + kp_0\}$, where $k = \mathcal{O}(\frac{U}{p_{min}})$. This corresponds to relative positions at $r \in \{\frac{o}{p_0}, \frac{o}{p_0} + 1, \ldots, \frac{o}{p_0} + k\}$, see Fig. 3(a). The bit at each of those points will be a 1 if the point is inside a wedge and a 0 otherwise.

We have shown that fixing a period and an offset gives a set of k points on a horizontal line, at unit distance from each other. The bitstring is created by assigning each of those points a 1 or a 0 depending on whether it is inside or outside a wedge in $\mathcal{W}_0$.

An equivalent way of modelling this is to create k translated copies $\mathcal{W}_1, \ldots \mathcal{W}_k$ of $\mathcal{W}_0$ and translate them so that the apex of $\mathcal{W}_i$ is at $(-i, 0)$, see Fig. 3(b). Consider a point at $(\frac{o}{p_0}, \frac{1}{p_0})$ in this new diagram. This point will stab the same wedge in $\mathcal{W}_i$ as the point at $(\frac{o}{p_0} + i, \frac{1}{p_0})$ did in $\mathcal{W}_0$ in the previous diagram. Consequently the set of wedges stabbed by this point will be exactly the same as the set of wedges stabbed by the k points in the previous diagram. We can therefore find the bitstring corresponding to a point by finding all wedges it stabs. Furthermore the bitstrings of two configurations, given as two points, will be the same if the points stab the same set of wedges.

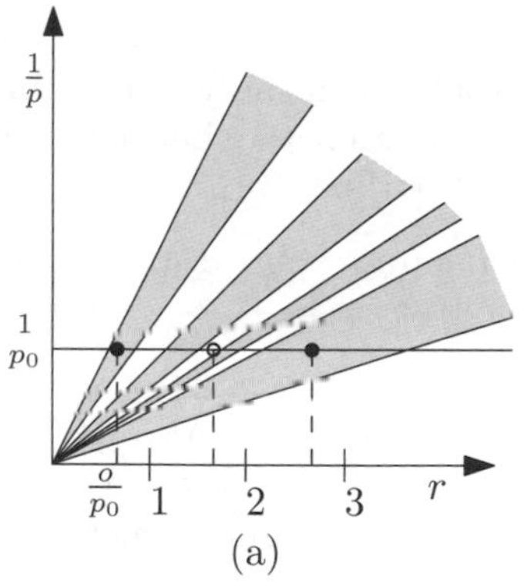

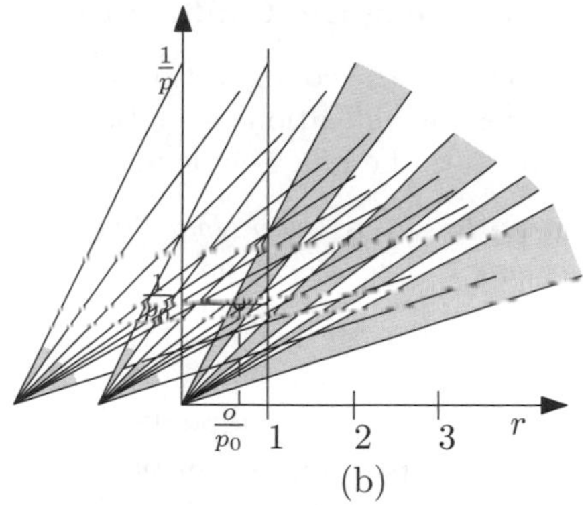

Fig. 3. (a) Fixing p and o gives points on a horizontal line. (b) Equivalently the single point stabs exactly the same wedges when they are translated.

We call the region $[0, 1) \times [0, \frac{1}{p_{min}}]$ the configuration space of the problem. The wedges in $\bigcup_{i=0}^{k} \mathcal{W}_i$ partition it into regions such that all configurations inside a region produce the same bitstring.

3.1 Properties of the Configuration Space

Let $\mathcal{A}$ be the arrangement in the configuration space defined by the wedges in $\bigcup_{i=0}^{k} \mathcal{W}_i$.

Observation 1. *The bitstrings of two adjacent cells in $\mathcal{A}$ differ by exactly one bit.*

This property suggests the following algorithm: we walk through the arrangement and apply single bit updates at every step. The algorithm is described in detail in Section 3.2. First we bound the complexity $\mathcal{A}$. Consider a line l in $\mathcal{W}_0$ and the k translated copies of l. A horizontal segment of length 1 can intersect at most two copies so we have the following result.

Lemma 3. *A horizontal segment $[0, 1] \times \{y\}$ intersects $\mathcal{O}(n)$ lines in the configuration space.*

Sharir [16] has shown that an arrangement of n segments has complexity $\mathcal{O}(nk)$ if any horizontal line intersects at most k segments. Lemma 3 allows us to apply this bound to $\mathcal{A}$ which gives:

Lemma 4. *The complexity of $\mathcal{A}$ is $\mathcal{O}(n^2 k)$.*

Proof. There are $\mathcal{O}(nk)$ segments in $\bigcup_{i=0}^{k} \mathcal{W}_i$ and a horizontal line in the configuration space (which is only the region $[0, 1) \times [0, \frac{1}{p_{min}}]$) intersects only $\mathcal{O}(n)$ of them by Lemma 3. □

3.2 Sweeping the Configuration Space

We can sweep the configuration space with a horizontal segment from bottom to top using a modification of the Bentley-Ottman intersection searching algorithm [6]. We state the following lemma with only a proof outline.

Lemma 5. *There is an algorithm to sweep the configuration space $\mathcal{A}$ using $\mathcal{O}(n)$ space and $\mathcal{O}(n^2 k \log n)$ time.*

Proof (outline). By Lemma 3 the horizontal sweepline intersects $\mathcal{O}(n)$ segments so the Bentley-Ottman data structure only requires $\mathcal{O}(n)$ space. Also when the sweepline hits the top endpoint of a segment it will also hit the bottom endpoint of a segment parallel to that one. This makes it possible to find the next event point without storing a sorted list of all $\mathcal{O}(nk)$ endpoints.

Theorem 1. *Given ε, c and a set of intervals $\mathcal{I} = \langle I_1 = [t_1^s, t_1^e], \ldots, I_n = [t_n^s, t_n^e]\rangle$, there is an algorithm to compute an ε-approximate longest dense pattern in $\mathcal{O}(\frac{1}{\varepsilon} n^2 k \log(n+k))$ time using $\mathcal{O}(n+k)$ space.*

Theorem 2. *Given c and a set of intervals $\mathcal{I} = \langle I_1, \ldots, I_n\rangle$, there is an algorithm to compute the longest dense pattern in $\mathcal{O}(n^2 k (n+k)^{\frac{1}{2}} \log^2(n+k))$ time using $\mathcal{O}((n+k)\log(n+k))$ space.*

Proof (outline). By Observation 1 only one bit changes from a region to an adjacent one. We can walk through the arrangement by walking only from a region to an adjacent one in such a way that we visit each region at least once and on average at most a constant number of times. Recording the update at each step allows us to use the algorithms in Section 2.

3.3 Removing Dependence on *k*

Recall that $k = \mathcal{O}(\frac{U}{p_{min}})$. We will assume that $\frac{U}{p_{min}}$ is polynomially bounded, i.e. that there exists a constant g such that $\frac{U}{p_{min}} = \mathcal{O}(n^g)$, so that $\log k = \log \frac{U}{p_{min}} = \mathcal{O}(\log n)$. This is a common assumption in computational geometry. In this case we can improve show the following results.

Theorem 3. *Given ε, c and a set of intervals $\mathcal{I} = \langle I_1 = [t_1^s, t_1^e], \ldots, I_n = [t_n^s, t_n^e]\rangle$, there is an algorithm to compute an ε-approximate longest dense pattern in $\mathcal{O}(\frac{1}{\varepsilon c} n^3 \log \frac{n}{c} \log n)$ time using $\mathcal{O}(\frac{n}{c})$ space.*

Theorem 4. *Given c and a set of intervals $\mathcal{I} = \langle I_1, \ldots, I_n\rangle$, there is an algorithm to compute the longest dense pattern in $\mathcal{O}(c^{-\frac{3}{2}} n^{\frac{7}{2}} \log^2 \frac{n}{c} \log n)$ time using $\mathcal{O}(\frac{n}{c} \log \frac{n}{c})$ space.*

4 Unknown Location

Here we consider a more general problem. We are given only the trajectory $\mathcal{T}$ without the region of interest $\mathcal{A}$ and we want to find the regions that are visited regularly.

The trajectory $\mathcal{T}$ is given as a sequence of points in the plane sampled at regular time intervals. W.l.o.g. we let the unit of time be the time between samples so that the trajectory is sampled at times $0, 1, 2, \ldots, n$. Formally, $\mathcal{T} = \langle (x_1, y_1), (x_2, y_2), \ldots, (x_n, y_n) \rangle$, if the entity was at (x_i, y_i) at time i. Consider the points that are specified in $\mathcal{T}$ and assume that the sampling rate is high enough so that these points accurately represent the path.

In addition to the trajectory $\mathcal{T}$ we are given a positive real number r, a real number $c \in [0, 1]$ and a period length p. We want to find an axis aligned square of side length r, called an r-region, that has the longest regular pattern with density at least c. We can define two classes of problems. In the *continuous position* case we allow the square to be positioned anywhere in the plane, while in the *discrete position* case the center of the square has to be on a $\delta \cdot r$ grid for a given value of δ. This is essentially a $(1+\delta)$ approximation of the continuous position problem.

We give exact and approximation algorithms for these problems. The running time $T(n)$ and space $S(n)$ of the algorithms is summarised in this table:

	Approximate	Exact
Continuous position	$T(n) = \mathcal{O}(\frac{1}{\varepsilon} \cdot n^2 \log n)$ $S(n) = \mathcal{O}(n)$	$T(n) = \mathcal{O}(n^{\frac{5}{2}} \log^2 n)$ $S(n) = \mathcal{O}(n \log n)$
Discrete position	$T(n) = \mathcal{O}(\frac{1}{\varepsilon\delta^2} \cdot n \log n)$ $S(n) = \mathcal{O}(\frac{1}{\delta^2} \cdot n)$	$T(n) = \mathcal{O}(\frac{1}{\delta^2} \cdot n^{\frac{3}{2}} \log^2 \frac{n}{\delta})$ $S(n) = \mathcal{O}(\frac{1}{\delta^2} \cdot n \log \frac{n}{\delta})$

4.1 Continuous Position

For each point $p_i(x_i, y_i)$ on the trajectory $\mathcal{T}$ we define S_i to be the axis aligned square centered at p_i with sides r. The point p_i is included in an r-region if and only if the center of the r-region is inside S_i. The set of all squares $S_1, S_2, \ldots, S_n$ partitions the plane into $\mathcal{O}(n^2)$ cells, see Figure 4(a). Two r-regions centered at different points within the same cell will cover exactly the same set of points. Furthermore two r-regions centered in adjacent cells will differ in exactly one point, i.e. one of them will cover an extra point.

To simplify the analysis we assume that any horizontal or vertical line contains an edge of at most one square S_i.

Consider the dual graph of this planar partition, which is also a planar graph, shown in Figure 4(b). We apply the same technique as in Section 3.2 to walk this dual graph and apply single bit updates at each node. This gives us the following theorems, which we state without proof.

Theorem 5. *Given a trajectory $\mathcal{T} = \langle (x_1, y_1), \ldots, (x_n, y_n) \rangle$, positive real numbers ε, c and r, and a positive integer p there is an algorithm to find the axis aligned square of side length r with the ε-approximate longest pattern of period length p and density at least c in $\mathcal{O}(\frac{1}{\varepsilon} \cdot n^2 \log n)$ time and $\mathcal{O}(n)$ space.*

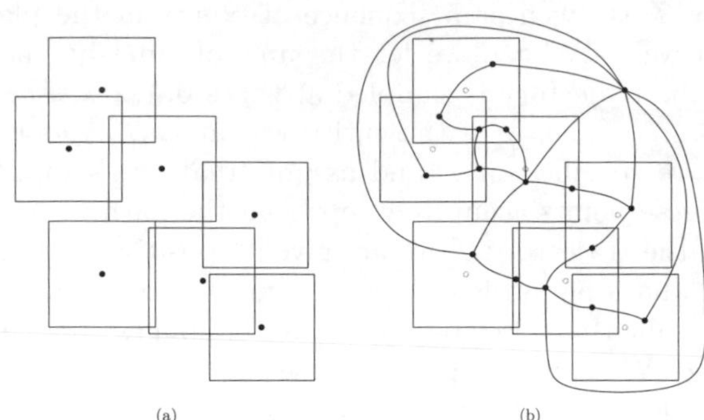

Fig. 4. (a) A partition of the plane by the squares S_i. (b) The dual graph of the partition.

Theorem 6. *Given a trajectory $\mathcal{T} = \langle (x_1, y_1), \ldots, (x_n, y_n) \rangle$, positive real numbers c and r, and a positive integer p there is an algorithm to find the axis aligned square of side length r with the longest pattern of period length p and density at least c in $\mathcal{O}(n^{\frac{5}{2}} \log^2 n)$ time and $\mathcal{O}(n \log n)$ space.*

4.2 Discrete Position

Again consider the squares $S_1, S_2, \ldots, S_n$ centered at the points in the trajectory and let $\mathcal{S}$ be the union of these squares. It is sufficient to consider only the points on a $(r \cdot \delta) \times (r \cdot \delta)$ grid that are contained in $\mathcal{S}$ as candidates for the center of the region. There are $\mathcal{O}(\frac{n}{\delta^2})$ such points. Testing each point independently would give an $\Omega(n^2)$ time algorithm but we can again exploit locality properties to get the following algorithms.

Theorem 7. *Given a trajectory $\mathcal{T} = \langle (x_1, y_1), \ldots, (x_n, y_n) \rangle$, positive real numbers ε, δ, c and r, and a positive integer p there is an algorithm to find the $r \cdot \delta$ grid centered axis aligned square of side length r with the ε-approximate longest pattern of period length p and density at least c in $\mathcal{O}(\frac{1}{\varepsilon\delta^2} \cdot n \log n)$ time and $\mathcal{O}(\frac{1}{\delta^2} \cdot n)$ space.*

Theorem 8. *Given a trajectory $\mathcal{T} = \langle (x_1, y_1), \ldots, (x_n, y_n) \rangle$, positive real numbers δ, c and r, and a positive integer p there is an algorithm to find the δ-approximate axis aligned square of side length r with the longest pattern of period length p and density at least c in $\mathcal{O}(\frac{1}{\delta^2} \cdot n^{\frac{3}{2}} \log^2 \frac{n}{\delta})$ time and $\mathcal{O}(\frac{1}{\delta^2} \cdot n \log \frac{n}{\delta})$ space.*

References

1. Save the elephants, `http://www.save-the-elephants.org`
2. Wildlife tracking projects with GPS GSM collars (2006), `http://www.environmental-studies.de/projects/projects.html`
3. Al-Naymat, G., Chawla, S., Gudmundsson, J.: Dimensionality reduction for long duration and complex spatio-temporal queries. In: Proceedings of the 22nd ACM Symposium on Applied Computing, pp. 393–397. ACM, New York (2007)

4. Andersson, M., Gudmundsson, J., Laube, P., Wolle, T.: Reporting leadership patterns among trajectories. In: Proceedings of the 22nd ACM Symposium on Applied Computing, pp. 3–7. ACM, New York (2007)
5. Benkert, M., Gudmundsson, J., Hübner, F., Wolle, T.: Reporting flock patterns. Computational Geometry—Theory and Applications (2007)
6. Bentley, J.L., Ottmann, T.A.: Algorithms for reporting and counting geometric intersections. IEEE Transactions on Computers C-28, 643–647 (1979)
7. Djordjevic, B., Gudmundsson, J., Pham, A., Wolle, T.: Detecting regular visit patterns. In: 16th European Symposium on Algorithms, pp. 344–355 (2008)
8. Djordjevic, B., Gudmundsson, J., Pham, A., Wolle, T.: Detecting regular visit patterns. Algorithmica (to appear)
9. Frank, A.U.: Socio-Economic Units: Their Life and Motion. In: Frank, A.U., Raper, J., Cheylan, J.P. (eds.) Life and motion of socio-economic units. GISDATA, vol. 8, pp. 21–34. Taylor & Francis, London (2001)
10. Gudmundsson, J., Laube, P., Wolle, T.: Movement Patterns in Spatio-Temporal Data. In: Encyclopedia of GIS. Springer, Heidelberg (2008)
11. Gudmundsson, J., van Kreveld, M.: Computing longest duration flocks in trajectory data. In: Proceedings of the 14th ACM Symposium on Advances in GIS, pp. 35–42 (2006)
12. Gudmundsson, J., van Kreveld, M., Speckmann, B.: Efficient detection of motion patterns in spatio-temporal sets. GeoInformatica 11(2), 195–215 (2007)
13. Güting, R.H., Schneider, M.: Moving Objects Databases. Morgan Kaufmann Publishers, San Francisco (2005)
14. Lee, J.-G., Han, J., Whang, K.-Y.: Trajectory clustering: a partition-and-group framework. In: SIGMOD '07: Proceedings of the 2007 ACM SIGMOD international conference on Management of data, pp. 593–604. ACM Press, New York (2007)
15. Mamoulis, N., Cao, H., Kollios, G., Hadjieleftheriou, M., Tao, Y., Cheung, D.: Mining, indexing, and querying historical spatiotemporal data. In: Proceedings of the 10th ACM SIGKDD International Conference On Knowledge Discovery and Data Mining, pp. 236–245. ACM, New York (2004)
16. Sharir, M.: On k-sets in arrangement of curves and surfaces. Discrete & Computational Geometry 6, 593–613 (1991)
17. Vlachos, M., Kollios, G., Gunopulos, D.: Discovering similar multidimensional trajectories. In: Proceedings of the 18th International Conference on Data Engineering (ICDE '02), pp. 673–684 (2002)

Tile-Packing Tomography Is $\mathbb{NP}$-hard

Marek Chrobak[1], Christoph Dürr[2], Flavio Guíñez[3], Antoni Lozano[4], and Nguyen Kim Thang[5]

[1] Department of Computer Science, University of California, Riverside, USA Research partially supported by National Science Foundation, grant CCF-0729071
[2] CNRS and Lab. of Computer Science of the Ecole Polytechnique, France
[3] DIM, Universidad de Chile, Chile
[4] Technical University of Catalonia, Barcelona, Spain Research partially supported by CICYT projects TIN2007-68005-C04-03 and TIN2008-06582-C03-01
[5] Department of Computer Science, Aarhus University, Denmark

Abstract. Discrete tomography deals with reconstructing finite spatial objects from their projections. The objects we study in this paper are called tilings or tile-packings, and they consist of a number of disjoint copies of a fixed tile, where a tile is defined as a connected set of grid points. A row projection specifies how many grid points are covered by tiles in a given row; column projections are defined analogously. For a fixed tile, is it possible to reconstruct its tilings from their projections in polynomial time? It is known that the answer to this question is affirmative if the tile is a bar (its width or height is 1), while for some other types of tiles $\mathbb{NP}$-hardness results have been shown in the literature. In this paper we present a complete solution to this question by showing that the problem remains $\mathbb{NP}$-hard for *all* tiles other than bars.

1 Introduction

Discrete tomography deals with reconstructing finite spatial objects from their low-dimensional projections. Inverse problems of this nature arise naturally in medical computerized tomography, electron tomography, non-destructive quality control, timetable design and a number of other areas. This wide range of applications inspired significant theoretical interest in this topic and led to studies of computational complexity of various discrete tomography problems. For an extensive and detailed coverage of practical and theoretical aspects of this area, we refer readers to the book by Kuba and Herman, see [7,6].

In this paper we consider the problem of reconstructing a tile packing from its row and column projections. Formally, consider the integer grid of dimension $m \times n$, consisting of all cells $(i, j) \in [0, m) \times [0, n)$. In the paper, we will often use the matrix notation and terminology, using terms "row" and "column", with rows numbered top-down and columns numbered from left to right, so that the-upper left cell is $(0, 0)$.

We define a *tile* to be any finite connected set T of grid cells. By "connected" we mean that for any two cells of T there is a path inside T between these

M.T. Thai and S. Sahni (Eds.): COCOON 2010, LNCS 6196, pp. 254–263, 2010.

cells and any two consecutive cells on this path are adjacent. The *width* and *height* of T are defined in the obvious manner, as the dimensions of the smallest $h \times w$ rectangle containing T. If $w = 1$ or $h = 1$, then T is called a *bar*. By $T + (i, j) = \{(x + i, y + j) : (x, y) \in T\}$ we denote the translation of T by the vector (i, j). We also refer to $T + (i, j)$ as a *translated copy* (or just *copy*) of T.

A *tile packing of the $m \times n$ grid using T* — or *T-packing*, in short, if m and n are understood from context — is a disjoint partial covering of the grid with translated copies of T. Formally, a T-packing is defined by a set D of translation vectors such that all translated copies $T + (i, j)$, for all $(i, j) \in D$, are contained in the $m \times n$ grid and are pairwise disjoint. We stress here that we do not require the tiles to completely cover the grid — such packings, in the literature, are sometimes called partial tilings. Without loss of generality, throughout the paper, we will be assuming that the tile T used in packing is in a *canonical position* in the upper-left corner of the grid, that is $\min\{x : (x, y) \in T\} = \min\{y : (x, y) \in T\} = 0$.

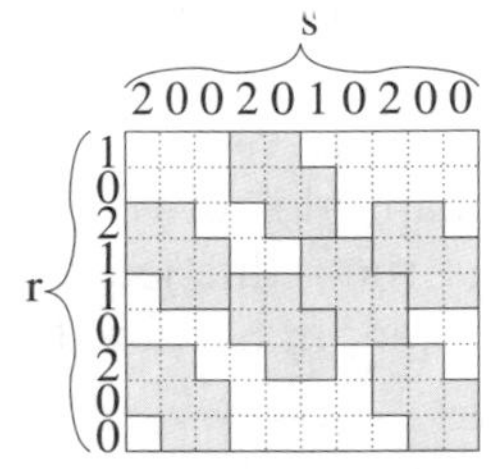

Fig. 1. A tile packing of the 9×10 grid and its projections

To simplify notation, instead of couting how many grid points are covered by tiles in a given row (or column), we count how many tiles *start* in a given row (column), which is equivalent up to some base-change. So the *row* and *column projections* of a packing D are defined as a pair $r \in \mathbb{N}^m$ and $s \in \mathbb{N}^n$ of vectors such that

$$r_i = |\{j : (i, j) \in D\}| \quad \text{and} \quad s_j = |\{i : (i, j) \in D\}|.$$

For example, consider tile $T = \{(0, 0), (0, 1), (1, 0), (1, 1), (1, 2), (2, 1), (2, 2)\}$. Figure 1 shows an example of a T-packing. This packing is $D = \{(0, 3), (2, 0), (2, 7), (3, 5), (4, 3), (6, 0), (6, 7)\}$.

We study the problem of reconstructing tile packings from its horizontal and vertical projections. More formally, for any fixed tile T, the problem is defined as follows:

Tile Packing Tomography Problem ($\mathsf{TPTP}(T)$)**.** The instance of $\mathsf{TPTP}(T)$ consists of vectors $r \in \mathbb{N}^m$, $s \in \mathbb{N}^n$. The objective is to decide if there is a T-packing D whose projections are r and s.

This problem has been introduced in [2] and shown to be NP-hard for some particular tiles. In [4], the proof technique has been adapted to show NP-hardness for any rectangular tile, i.e. a tile that consists of all cells $(i, j) \in [0, h) \times [0, w)$ for some dimensions $w, h \geq 2$.

On the positive side, when T is a single cell, classical work of Ryser [10] on projections of 0-1 matrices provides a characterization of vectors that correspond to projections of T-packings and provides a simple polynomial-time algorithm for that case. The ideas from [10] were extended in [3,8] to the case when T is a bar. In [1], polynomial-time algorithms were given for restricted special cases. The complexity status was unknown for all other tiles, and the current paper completes the picture by proving the following theorem.

Theorem 1. *Problem* TPTP(T) *is* $\mathbb{NP}$*-complete for any tile* T *that is not a bar.*

The general structure of our proof resembles those introduced in [2] and [4], although the reductions we present are substantially more difficult, since the generality of our result means that we cannot take advantage of a specific shape of the tile, and that we need to base the construction on some generic properties shared by infinitely many tiles. Our techniques take advantage of Ryser's structure results for 0-1 matrices, in combination with some arguments based on linear algebra.

After reviewing some background information in Section 2, we introduce the main idea of the reduction in Section 3, by formulating the overall framework of the reduction and conditions on T required for this reduction to be correct. Then, in Section 4 we show that all non-bar tiles satisfy these conditions.

2 Main Tools

In this section we briefly review two concepts that will play crucial role in our proofs: affine independence and Ryser's theorem.

Affine independence. Vectors $v_1, v_2, ..., v_k \in \mathbb{R}^n$ are called *affinely independent* if the unique solution of equations $\sum_{i=1}^{k} \alpha_i = 0$ and $\sum_{i=1}^{k} \alpha_i v_i = 0$ is $\alpha_1 = \alpha_2 = ... = \alpha_k = 0$. It is easy to show that the following three conditions are equivalent:

(ai1) $v_1, v_2, ..., v_k$ are affinely independent,
(ai2) $v_2 - v_1, v_3 - v_1, ..., v_k - v_1$ are linearly independent,
(ai3) $(v_1, 1), (v_2, 1), ..., (v_k, 1)$ are linearly independent.

We will refer to vectors $v_i - v_1$, $i = 2, 3, ..., k$, in (ai2), as *difference vectors.* Condition (ai2) is useful in verifying affine independence. For example, $(1, 1)$, $(3, 4), (5, 5)$ are affinely independent because, the difference vectors $(3, 4)-(1, 1) = (2, 3)$ and $(5, 5) - (1, 1) = (4, 4)$ are linearly independent.

Condition (ai3) implies that if $v_1, v_2, ..., v_k$ are affinely independent then for any vector v and constant β the equations $\sum_{i=1}^{k} \alpha_i v_i = v$ and $\sum_{i=1}^{k} \alpha_i = \beta$ have a unique solution $\alpha_1, \alpha_2, ..., \alpha_k$.

For any vector $v \in \mathbb{R}^n$ and any set of indices $i_1, i_2, ..., i_b \in [0, n)$, define the $(i_1, i_2, ..., i_b)$*-restriction of* v to be the vector $v' \in \mathbb{R}^b$ that consists only of the coordinates i_t, $t = 1, ..., b$, of v. For example, the $(0, 3, 4)$-restriction of $v = (4, 3, 1, 0, 7, 9, 5)$ is $v' = (4, 0, 7)$. For any set of vectors $v_1, v_2, ..., v_k \in \mathbb{R}^n$, to show that they are affinely independent it is sufficient to show that their $(i_1, i_2, ..., i_b)$-restrictions are affinely independent, for some set of indices $i_1, i_2, ..., i_b$.

Ryser's theorem. Ryser [9] studied structure of 0-1 matrices with given projections. We adapt his characterization of these matrices and express it in terms of tile packings.

Fix a tile T and let $I \subseteq [0, m)$ be a set of rows and $J \subseteq [0, n)$ a set of columns. We say that a tile copy $T + (i, j)$ *belongs to* $I \times J$ if $i \in I, j \in J$. Note that here we do not require inclusion of $T + (i, j)$ in $I \times J$.

Define $\xi_{I,J} = \max_D |D \cap (I \times J)|$, where the maximum is taken over all T-packings D of the $m \times n$ grid. Thus $\xi_{I,J}$ is the maximum number of copies of T that can belong to $I \times J$ in a T-packing without overlapping (and without any restriction on their projections).

For a set I or rows, denote $r(I) = \sum_{i \in I} r_i$. Analogously, $s(J) = \sum_{j \in J} s_j$, for a set J of columns. By $\bar{I} = [0, m) - I$ and $\bar{J} = [0, n) - J$ we denote the complements of these two sets.

Consider a T-packing D with projections r, s. Then we have

$$r(I) - s(\bar{J}) = |D \cap (I \times J)| - |D \cap (\bar{I} \times \bar{J})|.$$

By definition, $|D \cap (I \times J)| \leq \xi_{I,J}$. Therefore we obtain the following lemma (inspired by [9]).

Lemma 1. *Let I be a set of rows and J a set of columns. If $r(I) - s(\bar{J}) = \xi_{I,J}$ then every T-packing D with projections r, s satisfies $|D \cap (I \times J)| = \xi_{I,J}$ and $|D \cap (\bar{I} \times \bar{J})| = 0$.*

3 General Proof Structure

For each non-bar tile T, we show a polynomial-time reduction from the *3-Color Tomography Problem* introduced in [5] and shown to be NP-hard in [4]. In that problem, an object to be reconstructed is a set of "atoms" (in our terminology, single cells) colored red (R), green (G) or blue (B). The instance contains separate projections for each color. The formal definition is this:

3-Color Tomography Problem (3CTP). The instance consists of six vectors $r^R, r^G, r^B \in \mathbb{N}^m$, $s^R, s^G, s^B \in \mathbb{N}^n$. The objective is to decide whether there is a $m \times n$ matrix M with values from $\{R, G, B\}$ such that, for each color $c \in \{R, G, B\}$, $r^c_x = |\{y : M_{xy} = c\}|$ for each x and $s^c_y = |\{x : M_{xy} = c\}|$ for each y.

From now on, assume that T is some non-bar fixed tile of width w and height h. Let $\mathcal{I}$ be an instance of 3CTP for some $m \times n$ matrix specified by six projections $r^R, r^G, r^B, s^R, s^G, s^B$. We will map $\mathcal{I}$ into an instance $\mathcal{J}$ of TPTP(T) for an $m' \times n'$ grid with projections r, s, such that $\mathcal{I}$ has a matrix M with projections $r^R, r^G, r^B, s^R, s^G, s^B$ if and only if $\mathcal{J}$ has a T-packing with projections r, s.

Without loss of generality we assume that, for every color c, we have $\sum_x r^c_x = \sum_y s^c_y$, for every row x we have $\sum_c r^c_x = m$ and for every column y we have $\sum_c s^c_y = n$. Otherwise $\mathcal{I}$ is of course unfeasible, so we could take $\mathcal{J}$ to be any fixed unfeasible instance of TPTP(T).

We now describe $\mathcal{J}$. We will choose a grid of size $m' \times n'$ for $m' = mk$ and $n' = n\ell$, where k and ℓ are positive integer constants to be specified later. We will use the term *block* for a $k \times \ell$ grid. We can partition our $m' \times n'$ grid into mn rectangles of dimension $k \times \ell$, and we can think of each such rectangle as a translated block. The rectangle $[xk, (x+1)k) \times [y\ell, (y+1)\ell)$ will be referred to as the block (x, y).

Next, we need to specify the projections r and s. We will describe these projections in a somewhat unusual way, by fixing three packings of a block denoted D^R, D^G, and D^B (obviously, corresponding to the three colors), and then expressing r and s as linear combinations of these packings. More specifically, denoting by $\bar{r}^c$ and $\bar{s}^c$ the horizontal and vertical projections of packing D^c, for each $c \in \{R, G, B\}$, we define

$$r_{xk+i} = \sum_c r_x^c \cdot \bar{r}_i^c \quad \text{and} \quad s_{y\ell+j} = \sum_c s_y^c \cdot \bar{s}_j^c. \tag{1}$$

for every $i \in [0, k)$, $j \in [0, \ell)$, $x \in [0, m)$, and $y \in [0, n)$. The idea is that replacing each cell in a solution to the 3CTP instance r, s by the color-corresponding block, gives a solution to the TPTP(T) instance $\bar{r}, \bar{s}$.

To complete the description of the reduction, it still remains to define the three packings D^R, D^G, and D^B. This will be done in the next section. In the remainder of this section we establish conditions that will guarantee correctness of our reduction.

Our three packings will be designed to satisfy the following two requirements:

Requirement 1: Vectors $\bar{r}^R, \bar{r}^G, \bar{r}^B$ are affinely independent and vectors $\bar{s}^R, \bar{s}^G, \bar{s}^B$ are affinely independent. Note that, by property (ai3), this implies that for any vector v there is at most one possible way to represent it in a form $v = n_R\bar{r}^R + n_G\bar{r}^G + n_B\bar{r}^B$, where $n_R + n_G + n_B = n$.

Requirement 2: In any packing D of $\mathcal{J}$ with projections r, s, the restriction of D to each block of the grid has projections equal to $\bar{r}^c, \bar{s}^c$, for some $c \in \{R, G, B\}$.

Lemma 2. *Assume that the three packings D^R, D^G, D^B satisfy Requirements 1 and 2. Then $\mathcal{I}$ has a solution if and only if $\mathcal{J}$ has a solution.*

Proof: ($\Rightarrow$) Let $M \in \{R, G, B\}^{m \times n}$ be a solution to $\mathcal{I}$. We transform M into the following packing D for the $m' \times n'$ grid:

$$D = \bigcup_{x \in [0,m)} \bigcup_{y \in [0,n)} (D^{M_{xy}} + (xk, y\ell)).$$

In other words, if $M_{xy} = c$ then block (x, y) of the $m' \times n'$ grid contains a copy of D^c. By simple inspection, the projections of D are indeed equal to the vectors r and s in (1).

($\Leftarrow$) For the converse, suppose that there is a packing D with projections r, s. By Requirement 2, every block of the $m' \times n'$ grid has projections $\bar{r}^c$ and $\bar{s}^c$, for some $c \in \{R, G, B\}$. We then *associate* this block with color c. We can thus define a matrix $M \in \{R, G, B\}^{m \times n}$ such that $M_{xy} = c$ if block (x, y) of D is associated with color c.

We now need to show that M is a solution for $\mathcal{I}$. To this end, fix some arbitrary $0 \le x < m$ and consider vector $v = (r_{xk}, r_{xk+1}, ..., r_{(x+1)k-1})$, which is the projection of the "row" of all blocks (x, y), for all y. By the construction, v can be written as $v = n_R\bar{r}^R + n_G\bar{r}^G + n_B\bar{r}^B$, where $n_R = r_x^R$, $n_G = r_x^G$, and $n_B = r_x^B$. Now, using Requirement 1, we obtain that this representation is unique under the assumption that $n_R + n_G + n_B = n$. We can thus conclude that the projection of row x of M is correct, that is $|\{y : M_{xy} = c\}| = r_x^c$ for all c. By the same argument, column projections of M are correct as well, completing the proof. □

In summary, to complete the proof for the given tile T, we need to do this: (i) define a rectangular $k \times \ell$ block with three packings D^R, D^G, D^B, (ii) show that the row projections of D^R, D^G, D^B and the corresponding column projections are affinely independent (Requirement 1), and (iii) show that in any solution to $\mathcal{J}$, each block (x, y) has projections equal to those of one of D^c, for some c (Requirement 2). We show the construction of such block packings in the next section.

4 Construction of Block Packings

As in the previous section, T is a fixed (but arbitrary) non-bar tile. We call (i, j) a *conflicting vector* if T and $T + (i, j)$ overlap, that is $T \cap (T + (i, j)) \neq \emptyset$. Since T is not a bar, it has a conflicting vector (i, j) with $i, j \neq 0$. For example one of $(1, 1)$ or $(-1, 1)$ is conflicting, since T is connected.

For the construction of the proof, fix $(-p, q)$, with $p, q \neq 0$, to be a conflicting translation vector of T that maximizes the L_1 norm under the constraint that none of the coordinates is 0. So any vector (i, j) with $i, j \neq 0$ and $|i| + |j| > |p| + |q|$ is not conflicting. Without loss of generality, we will be assuming that $p, q > 0$, for otherwise we can flip T horizontally or vertically and give the proof for the resulting tile.

Let a be the smallest positive integer such that $(ap, 0)$ is not a conflicting vector. Similarly let b be the smallest positive integer such that $(0, bq)$ is not a conflicting vector. Without loss of generality we assume that $a \leqslant b$, since otherwise we can exchange the roles of columns and rows in the proof.

We now divide the proof into four cases, and for each of them we show that Requirements 1 and 2 are satisfied.

4.1 Case $a = 1$ and $b = 1$

In this case, we use the following three packings:

$$D^R = \{(p, 0), (p, q)\}, \quad D^G = \{(0, q), (p, q)\}, \quad \text{and } D^B = \{(p, q)\}.$$

The values of k and ℓ are chosen to be the smallest integers for which these three packings are contained in the $k \times \ell$ grid.

The packings are depicted on Figure 2. The squares represent possible positions for tiles. Two positions are connected with a solid line, if the difference of the positions is a conflicting vector. That means that no packing can contain simultaneously a tile in both positions. Dashed lines indicate non-conflicting

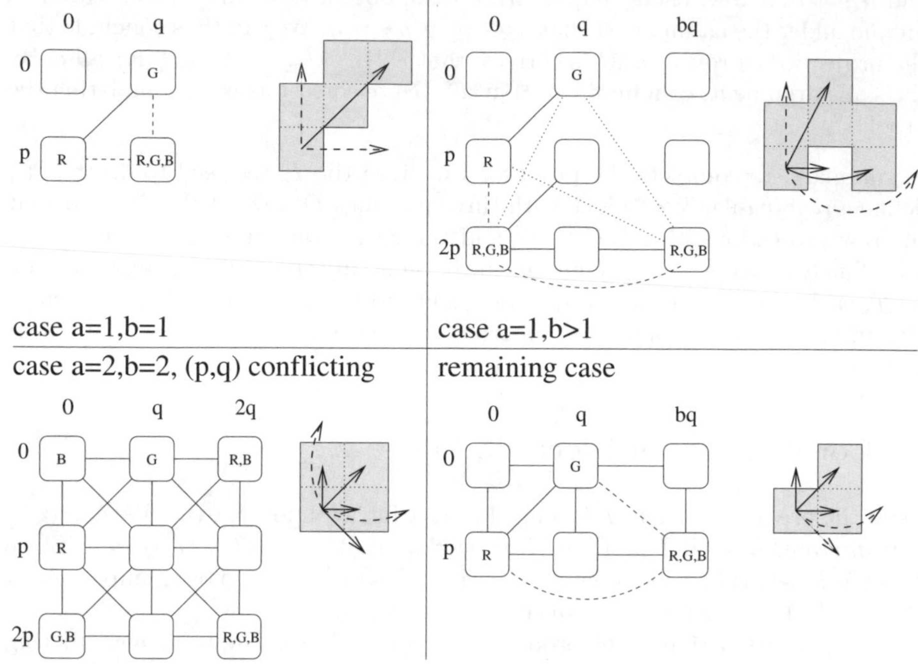

Fig. 2. For each case of the proof, the three packings (left-hand side) and an example of a tile (right-hand side) satisfying this case. Dotted vectors are non-conflicting by maximality of $(-p, q)$. In the figure of the third case, all vectors that are not depicted (for readability) are non-conflicting.

vectors, i.e. they connect pairs of compatible positions. We show lines only between position pairs relevant to the proof. In the figure we mark with letter c the tile positions of D^c, for $c \in \{R, G, B\}$. On the right-hand side of the figure we show a tile satisfying the case conditions for illustration. Again, solid vectors show conflicting translations and dashed vectors non-conflicting translations.

We first verify Requirement 1. The $(0,p)$-restrictions of $\bar{r}^R, \bar{r}^G, \bar{r}^B$ are, respectively, $(0,2)$, $(1,1)$ and $(0,1)$, and the $(0,q)$-restrictions of $\bar{s}^R, \bar{s}^G, \bar{s}^B$ are, respectively, $(1,1)$, $(0,2)$ and $(0,1)$. For both the row and column projections, routine calculations show that their restrictions are affinely independent.

We now focus on Requirement 2. Let r, s be the projections obtained by the reduction, and consider a packing D with these projections. We use Lemma 1, with I being the set of all row indices that are p modulo k, and J being the set of all column indices that are q modulo ℓ. By inspecting the definition of the projections we have $r(I) - s(\bar{J}) = mn$, which is $|I \times J|$, so Lemma 1 applies. Therefore every block in D contains a tile at position (p, q) and none at position $(0, 0)$. The remaining possible positions for tiles are $(p, 0), (0, q)$, but both cannot be occupied in a same block. This shows that every block of D is one of the packings D^R, D^G, D^B.

4.2 Case $a = 1$ and $b \geq 2$

In this case, we use the following three packings:

$$D^B = \{(2p, 0), (2p, bq)\} \qquad D^R = D^B \cup \{(p, 0)\} \qquad D^G = D^B \cup \{(0, q)\}.$$

The values of k and ℓ are chosen to be the smallest integers for which these three packings are contained in the $k \times \ell$ grid.

The $(0, p, 2p)$-restrictions of $\bar{r}^R, \bar{r}^G, \bar{r}^B$ are linearly independent vectors $(2, 0, 1)$, $(1, 1, 1)$, $(1, 0, 1)$; therefore $\bar{r}^R, \bar{r}^G, \bar{r}^B$ are affine independent. By a similar argument, we obtain that the corresponding column projection vectors $\bar{s}^R, \bar{s}^G, \bar{s}^B$ are affine independent as well. Thus Requirement 1 holds.

Now we verify Requirement 2. The row $2p$ of a block can contain at most 2 tiles. Let I be the set of rows i with $i \bmod k = 2p$. By inspecting the definition of the projections we have $r(I) = 2mn$, so every block in a solution D must contain exactly 2 tiles in row $2p$, and they are at positions $(2p, 0), (2p, bq)$. Now let J be the set of all columns j with $j \bmod \ell = bq$. We only have $s(J) = mn$, so in every block of D, the positions $(0, bq), (p, bq)$ are empty. The tile at $(2p, 0)$ forces position (p, q) to be empty. By the case assumption $a = 1$, there is no conflict between $(2p, 0)$ and $(p, 0)$. By maximality of $(-p, q)$ there is no conflict between positions $(2p, 0)$ and $(0, q)$ or between $(2p, bq)$ and $(p, 0)$ or $(2p, bq)$ and $(0, q)$. That leaves only 3 positions where the block packings can differ. Let d be the number of blocks in D with a tile in $(0, 0)$. Similarly let e be the numbers of blocks in D with a tile in $(0, q)$. Now we use the fact that in the original instance we had $\sum r_i^G = \sum s_j^G$, let n_G denote this quantity. This time let I be the set of rows i with $i \bmod k = 0$ and J the set of columns j with $j \bmod \ell = q$. Then $r(I) = d + e$ and $r(I) = n_G$. But also $r(J) = e$ and $r(J) = n_G$ as well, which shows $d = 0$. Therefore every block packing in D is one of D^R, D^G, D^B.

4.3 Case $a = 2$, $b = 2$ and Vector (p, q) Conflicting

In this case we assume $a = 2$, $b = 2$ and that the vector (p, q) is conflicting. Since $(-p, q)$ is conflicting as well, this makes the construction very symmetric. The three packings used in this case are:

$$\begin{aligned} D^R &= \{(0, 2q), (p, 0), (2p, 2q)\}, \\ D^G &= \{(0, q), (2p, 0), (2p, 2q)\}, \\ D^B &= \{(0, 0), (0, 2q), (2p, 0), (2p, 2q)\}. \end{aligned}$$

Again, the values of k and ℓ are chosen to be the smallest integers for which these three packings are contained in the $k \times \ell$ grid. The construction is illustrated in Figure 2. The idea behind this construction is similar to the reduction used in [2] to show NP-hardness of the packing problem for the 2×2 square tile.

The $(0, p, 2p)$-restrictions of $\bar{r}^R, \bar{r}^G, \bar{r}^B$ are $(1, 1, 1)$, $(1, 0, 2)$, $(2, 0, 2)$, therefore $\bar{r}^R, \bar{r}^G, \bar{r}^B$ are affinely independent. By symmetry the same holds for $\bar{s}^R, \bar{s}^G, \bar{s}^B$.

Now we verify Requirement 2. Let D be a packing with projections r, s. By the conflicting vectors there can be at most 2 tiles in the rows $p, 2p$ of a block.

Let $I = \{i : i \bmod k \in \{p, 2p\}\}$. Since $r(I) = 2mn$, and there are mn blocks in D, every block of D must contain exactly 2 tiles in rows $p, 2p$. By symmetry the same holds for columns $q, 2q$. There are only 4 packings satisfying these properties, D^R, D^G, D^B and some packing $D^A = \{(0, 2q), (2p, 0), (2p, 2q)\}$. Let n_R, n_G, n_B, n_A be the respective numbers of these different block packings in D. Since D has row projection r, we have

$$3\sum r_x^R + 3\sum r_x^G + 4\sum r_x^B = 3n_R + 3n_G + 4n_B + 3n_A,$$

and by the assumption $\sum_c \sum_x r_x^c = mn$ we have also

$$\sum r_x^R + \sum r_x^G + \sum r_x^B = mn = n_R + n_G + n_B + n_A.$$

Now let $I = \{i : i \bmod k = p\}$. Then $r(I) = \sum_x r_x^R$, but also $r(I) = n_R$. For $J = \{j : j \bmod \ell = q\}$ and the assumption $\sum r_x^G = \sum s_y^G$ we obtain $s(J) = \sum_y s_y^G = \sum_x r_x^G = n_G$. So we are left with the equalities

$$\sum r_x^B = n_B + n_A, \qquad 4\sum r_x^B = 4n_B + 3n_A,$$

from which we conclude $n_A = 0$. This verifies Requirement 2.

4.4 The Remaining Case

Assume now that none of the previous cases holds. This means that either

(a) $a \geq 2$ and $b \geq 3$, or
(b) $a = 2$, $b = 2$ and vector (p, q) is not conflicting.

We claim that the translation vector $(p, (b-1)q)$ is not conflicting. Indeed, in case (b) above it follows by case assumption and in case (a) it follows from the maximality of $(-p, q)$. Therefore in any block of D both positions $(0, q)$ and (p, bp) could contain a tile.

We use the following three packings (see Figure 2):

$$D^R = \{(p, 0), (p, bq)\}, \quad D^G = \{(0, q), (p, bq)\}, \quad \text{and } D^B = \{(p, bq)\}.$$

Again, the values of k and ℓ are chosen to be the smallest integers for which these three packings are contained in the $k \times \ell$ grid.

The $(0, p)$-restrictions of $\bar{r}^R, \bar{r}^G, \bar{r}^B$ are $(0, 2)$, $(1, 1)$, $(0, 1)$, and their difference vectors $(0, 1)$, $(1, 0)$ are linearly independent. Therefore $\bar{r}^R, \bar{r}^G, \bar{r}^B$ are affine independent. The $(0, q, 2q)$-restrictions of $\bar{s}^R, \bar{s}^G, \bar{s}^B$ are linearly independent vectors $(1, 0, 1)$, $(0, 1, 1)$, $(0, 0, 1)$; therefore $\bar{s}^R, \bar{s}^G, \bar{s}^B$ are affine independent. Thus Requirement 1 holds.

Now we verify Requirement 2. The proof is slightly more involved than in the previous cases. Fix some T-packing D with projections r, s. First we observe that every block of D must contain exactly one tile in column bq, by using the same arguments as in the previous case. That leaves us with six possible block packings, and we introduce notation for their cardinalities:

1. n_G is the number of packings $\{(0, q), (p, bq)\}$ or $\{(p, q), (0, bq)\}$,
2. n_R is the number of packings $\{(p, 0), (p, bq)\}$,
3. n_W is the number of packings $\{(p, bq\})$,
4. n_A is the number of packings $\{(0, 0), (p, bq)\}$ or $\{(p, 0), (0, bq)\}$,
5. n_B is the number of packings $\{(0, 0), (0, bq)\}$,
6. n_C is the number of packings $\{(0, 0), (p, q), (0, bq)\}$,
7. n_D is the number of packings $\{(0, bq)\}$.

Let I be the set of all rows i with $i \bmod k = 0$ and J the set of all columns j with $j \bmod \ell = q$. Then by inspecting the projection definitions, we have $r(I) = \sum_x r_x^G$ and $s(J) = \sum_y s_y^G = r(I)$. Since r and s are the projections of D, we also have $r(I) = n_G + n_A + 2n_B + 2n_C + n_D$ and $s(J) = n_G + n_C$. This shows $n_A = n_B = n_C = n_D = 0$, completing the analysis of this case and the proof of NP-hardness.

References

1. Brunetti, S., Costa, M.C., Frosini, A., Jarray, F., Picouleau, C.: Reconstruction of binary matrices under adjacency constraints, pp. 125–150. Birkhauser, Boston (2007)
2. Chrobak, M., Couperus, P., Dürr, C., Woeginger, G.: On tiling under tomographic constraints. Theoretical Computer Science 290(3), 2125–2136 (2003)
3. Dürr, C., Goles, E., Rapaport, I., Rémila, E.: Tiling with bars under tomographic constraints. Theoretical Computer Science 290(3), 1317–1329 (2003)
4. Dürr, C., Guíñez, F., Matamala, M.: Reconstructing 3-colored grids from horizontal and vertical projections is NP-hard. In: Proc. 17th Annual European Symposium on Algorithms, pp. 776–787 (2009)
5. Gardner, R., Gritzmann, P., Prangenberg, D.: On the computational complexity of determining polyatomic structures by X-rays. Theoretical Computer Science 233, 91–106 (2000)
6. Herman, G., Kuba, A.: Advances in Discrete Tomography and Its Applications. Birkhäuser, Basel (2007)
7. Kuba, A., Herman, G.T.: Discrete tomography: Foundations, Algorithms and Applications. Birkhäuser, Basel (1999)
8. Picouleau, C.: Reconstruction of domino tiling from its two orthogonal projections. Theoretical Computer Science 255, 437–447 (2001)
9. Ryser, H.J.: Matrices of zeros and ones. Bulletin of the American Mathematical Society 66, 442–464 (1960)
10. Ryser, H.J.: Combinatorial Mathematics. Mathematical Association of America/Quinn & Boden, Rahway/New Jersey (1963)

The Rectilinear k-Bends TSP

Vladimir Estivill-Castro, Apichat Heednacram, and Francis Suraweera

IIS, Griffith University, Nathan, QLD, 4111, Australia
{v.estivill-castro,f.suraweera}@griffith.edu.au,
apichat.heednacram@student.griffith.edu.au

Abstract. We study a hard geometric problem. Given n points in the plane and a positive integer k, the Rectilinear k-Bends Traveling Salesman Problem asks if there is a piecewise linear tour through the n points with at most k bends where every line-segment in the path is either horizontal or vertical. The problem has applications in VLSI design. We prove that this problem belongs to the class FPT (fixed-parameter tractable). We give an algorithm that runs in $O(kn^2 + k^{4k}n)$ time by kernelization. We present two variations on the main result. These variations are derived from the distinction between line-segments and lines. Note that a rectilinear tour with k bends is a cover with k line-segments, and therefore a cover by lines. We derive FPT-algorithms using bounded-search-tree techniques and improve the time complexity for these variants.

1 Introduction

The Minimum Bends Traveling Salesman Problem seeks a tour through a set of n points in the plane, consisting of the least number of straight lines, so that the number of bends in the tour is minimized. Minimizing the number of bends in the tour is desirable in applications such as the movement of heavy machinery because the turns are considered very costly. Both, general and rectilinear, versions of this problem are studied in the literature. In the general version, the lines could be in any configuration whereas in the rectilinear version, the line-segments[1] are either horizontal or vertical. The general version of the problem is NP-complete [2]. The hardness of the rectilinear version remains open, however, Bereg et al. [3] suspect that it is NP-complete because the Rectilinear Line Cover in 3 dimensions (or higher) is NP-complete [8]. The rectilinear version of the problems received considerable attention during 1990's [5,9,10] and recently [1,3,4,13], since much of the interest in the rectilinear setting have been motivated by applications in VLSI. In the context of VLSI design, the number of bends on a path affects the resistance and hence the accuracy of expected timing and voltage in chips [10]. Stein and Wagner [12] solved approximately the rectilinear version of the Minimum Bends Traveling Salesman Problem. They gave a 2-approximation algorithm that runs in

[1] A line is unbounded whereas a line-segment is bounded.

M.T. Thai and S. Sahni (Eds.): COCOON 2010, LNCS 6196, pp. 264–277, 2010.

$O(n^{1.5})$ time. However, no polynomial-time exact algorithm is known for this rectilinear tour problem despite the motivating applications in VLSI.

In classical complexity theory, NP-completeness is essentially a tag for intractability to finding exact solutions to optimization problems. However, parameterized complexity theory [6,7,11] offers FPT (fixed-parameter tractable) algorithms, which require polynomial time in the size n of the input to find these exact answers, although exponential time may be required on a parameter k.

We reformulate the Rectilinear Minimum Bends Traveling Salesman Problem as a parameterized problem and we call it the Rectilinear k-Bends Traveling Salesman Problem. From the parameterized complexity perspective, we show that the problem in general belongs to the class FPT by kernelization. As such, it can be solved exactly and in polynomial time for small values of the parameter. The requirement that line-segments of the tour are hosted exclusively by a line leads to a different variant of the problem. Another variant also emerges if we require that the same line-segment orientation cover points on the same line. We provide FPT-algorithms with improved complexity for these two variants of the rectilinear tour problem. Our algorithms for these variants are based on bounded-search-tree techniques.

2 Rectilinear Tours

We define the Rectilinear k-Bends Traveling Salesman Problem formally as follows. Given a set S of n points in the plane, and a positive integer k, we are asked if there is a piecewise linear tour (which may self-intersect) through the n points in S with at most k bends where every line-segment in the path is either horizontal or vertical (the tour must return to its starting point). An instance of the Rectilinear k-Bends Traveling Salesman Problem is encoded as the pair (S, k), and we call the solution a *rectilinear tour*. In this rectilinear version, the standard convention restricts the tour to 90° turns. A 180° turn is considered two 90° turns with a zero-length line-segment in between. If $n \geq 3$, it is always possible to transform a tour with a 180° turns into a tour with only proper 90° turns and line-segments of positive length. With these conventions, every 90° turn consists of one horizontal line-segment and one vertical line-segment, both of positive length. Thus, we assume $n \geq 3$, we also accept that there are no tours with an odd number of bends and that the required number k of bends is even. A rectilinear tour must have at least 4 bends.

Lemma 1. *If there exists a rectilinear tour with at most k bends, the number of horizontal line-segments is at most $k/2$ and the number of vertical line-segments is at most $k/2$.*

Proof. If there exists a tour with at most k bends, there are at most k line-segments. In a rectilinear tour, the number of horizontal line-segments is equal to the number of vertical line-segments. There cannot be more than $k/2$ horizontal line-segments and no more than $k/2$ vertical line-segments. □

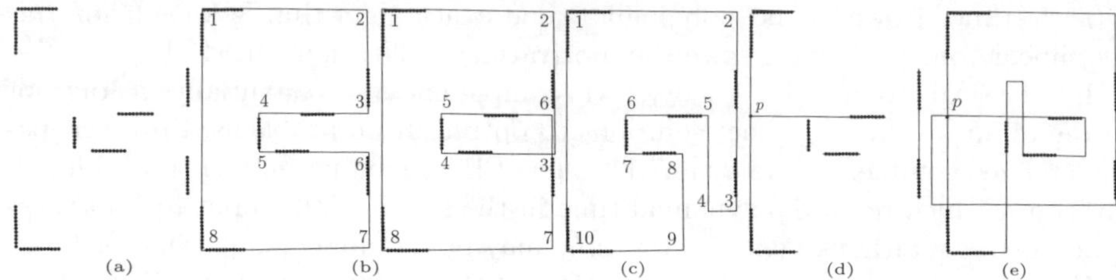

Fig. 1. Three types of the RECTILINEAR k-BENDS TRAVELING SALESMAN PROBLEM emerge as we considered the legal line-segments that are part of the tour

We distinguish 3 types of rectilinear tours that derive from the distinction between line-segment and line. Fig. 1 illustrates this. In the first case, we require that if l is the line containing a line-segment s of the tour (i.e. $l \cap s = s$), then the line-segment s covers all the points in $S \cap l$. In the second case, if a point p is on a line l used by the tour, then there must be a segment of the tour with the same orientation as l covering p. The third type does not have any of the above constraints. For illustration, consider the set of points in Fig. 1 (a). Each vertical cluster or horizontal cluster of points is numerous enough to force being covered by at least one line-segment of the tour with minimum bends. Without any constraint, the two tours with 8 bends in Fig. 1 (b) are optimal; however, in both there are two vertical line-segments that share a common line (in the second one, the two segments $\overline{2,3}$ and $\overline{6,7}$ are not drawn exactly co-linear so the reader can appreciate that the tour self intersects). The first constraint will require that all the points under these two line-segments be covered by only one line segment. In Fig. 1 (c) we see that this constraint forces the tour to travel over the large line-segment $\overline{2,3}$ and the minimal tour has now 10 bends. The second constraint can be illustrated if we add a point p to S as per Fig. 1 (d). This new point lies on lines used by two types of line-segments, one horizontal and one vertical and both of these types of line-segments must cover the point (that is, the point p cannot be already covered by the vertical line segment, because it belongs to a line where there is a horizontal line-segment of the tour). Note that this constraint is satisfied by points that are located at a bend. That is, if a bend is placed at a data point q, this constraint is automatically satisfied for q because the horizontal line-segment at q plus all other horizontal line-segments on the same horizontal line will cover q (and symmetrically for the vertical line-segment). In Fig. 1 (d) all the line-segments drawn must be contained in a line-segment of a minimum bends tour satisfying the second constraint (which now has 12 turns; see Fig. 1 (e)).

We first show that the problem in general (without any constraints) is FPT. The first variant requires that one line-segment covers all the points on the same line while the second variant requires the same orientation be represented by a line-segment that covers the points. We prove that the first variant and the second variant are also FPT but the time complexity of the FPT-algorithms here is smaller than in the general setting.

2.1 Tours without Constraints

We now proceed with the kernelization approach by presenting some reduction rules. Kernelization is central to parameterized complexity theory because a decision problem is in FPT if and only if it is kernelizable [11]. Intuitively, kernelization self-reduces the problem efficiently to a smaller problem using reduction rules. The rules are applied repeatedly until none of the rules applies. If the result is a problem of size no longer dependent on n, but only on k, then the problem is kernelizable because the kernel, the hard part, can be solved by exhaustive search in time that depends only on the parameter (even if it is exponential time).

Reduction Rule 1. *If $k \geq 4$ and all points in S lie on only one rectilinear line, then the instance (S, k) of the* RECTILINEAR k-BENDS TRAVELING SALESMAN PROBLEM *is a YES-instance.*

The next rule is derived from Lemma 1.

Reduction Rule 2. *If the minimum number of rectilinear line-segments needed to cover a set S of n points in the plane is greater than k, then the instance (S, k) of the* RECTILINEAR k-BENDS TRAVELING SALESMAN PROBLEM *is a NO-instance.*

We refer to a set of k rectilinear lines that cover the points in S as a k-cover. If (S, k) is a YES-instance of the RECTILINEAR k-BENDS TRAVELING SALESMAN PROBLEM, then the tour induces a k-cover. In fact, we can discover lines that host line-segments of any tour.

Lemma 2. *Let (S, k) be a YES-instance of the* RECTILINEAR k-BENDS TRAVELING SALESMAN PROBLEM. *Let l be a rectilinear line through $1 + k/2$ or more co-linear points. Then the line l must host a line-segment of any tour T with k or fewer bends.*

Proof. Without loss of generality, assume l is a vertical line. In contradiction to the lemma, assume there is no vertical segment on l for a tour T that covers with k bends. Then, the $1 + k/2$ points in $S \cap l$ would be covered by horizontal lines in T. According to Lemma 1, this contradicts T has k or fewer bends. □

The rectilinear line through S' in the proof above may be represented by separate line-segments of a witness tour. We now describe how to compute a k-cover if one exists. Consider a preprocessing of an input instance that consists of repeatedly finding $1 + k/2$ or more co-linear points and on a rectilinear line (that is, they are on a vertical or horizontal line). This process can be repeated until $k + 1$ rectilinear lines, each covering $1 + k/2$ or more different points are found, or no more rectilinear lines covering $1 + k/2$ points are found. In the first case, we have $k + 1$ disjoint subsets of points each co-linear and each with $1 + k/2$ or more points. When this happens, we halt indicating a NO-instance. By Lemma 2, even if each of the $k + 1$ lines hosts only one line-segment in the tour, we would still have more than k line-segments. In the second case, once we discover that we cannot find a line covering $1 + k/2$ points and not exceeded k repetitions, we have a problem kernel.

Lemma 3. *Any instance (S, k) of the of the* Rectilinear k-Bends Traveling Salesman Problem *can be reduced to a kernel S' of size at most $k^2/2$.*

Proof. Let S' be the set of points after we cannot repeat the removal of points covered by a rectilinear line covering $1 + k/2$ or more points. Recall that if we repeated the removal more than k times, we know it is a NO-instance. If we repeated no more than k times and it is a YES-instance, a witness tour T would have matched hosting lines with the lines removed. Also T is a k-cover of S'. So the lines in T are rectilinear and each covers no more than $k/2$ points. This means $|S'| \leq k^2/2$. □

From the above lemma, if we have a kernel of size larger than $k^2/2$, then it is a NO-instance. Algorithmically, we can either determine that we have a NO-instance in polynomial time in k and in n, or we have a kernel where we still have to determine if it is a YES or NO-instance. What follows resolves this issue. With the next lemma we prove that each best tour is always equivalent to a tour with the same number of bends but where line-segments on the same line are not disjoint (as the two tours with 8 bends in Fig. 1 (b)).

Lemma 4. *Every optimal rectilinear tour T that has two disjoint line-segments hosted by the same line can be converted into a tour T' with the same number of bends and where the line segments have no gap.*

Proof. Consider first the case the two disjoint line segments s_1 and s_2 are traversed by T in the same direction. Fig. 2 (a) and (b) shows the transformation of the two line segments s_1 and s_2 in the same hosting line l_p into a new tour by enlarging both s_1 and s_2 and flipping the direction of two bends. Clearly, there are no more bends and although the direction of the path between these two bends is reversed, we still have a well formed tour. Note that Fig. 2 (a) deals with the case when the bends share a label[2] while (b) is the case the two bends share no label. Once this case is clear, the case where two disjoint line segments s_1 and s_2 are traveled by T in opposite directions is also clear, although now 4 bends are involved. Fig. 2 (c) illustrates this. □

Moreover, the transformation in the proof above always increases the length of the tour. So if we apply it again, it will not undo the work done by its previous application. Thus we can apply it repeatedly until there are never two disjoint line-segments in an optimal tour. In particular, we can assume optimal tours have no gap between co-linear line-segments like in Fig. 2 (a).

Let L_k be the set of rectilinear lines found by kernelization having at least $1 + k/2$ co-linear points. We know $|L_k| \leq k$ and all these lines have segments that are part of the tour. Given a vertical line $l \in L_k$, $cover(l) = S \cap l$. we let h_{max} be the horizontal line through the point $p_{max} \in cover(l)$ with the largest y coordinate, while h_{min} is the horizontal line through the point

[2] We say the turns have a common label if the turns are NE at p and NW at q (with N in common) or the turns are SE at p and SW at q (with S in common) where N, S, W, and E stand for North, South, West and East respectively.

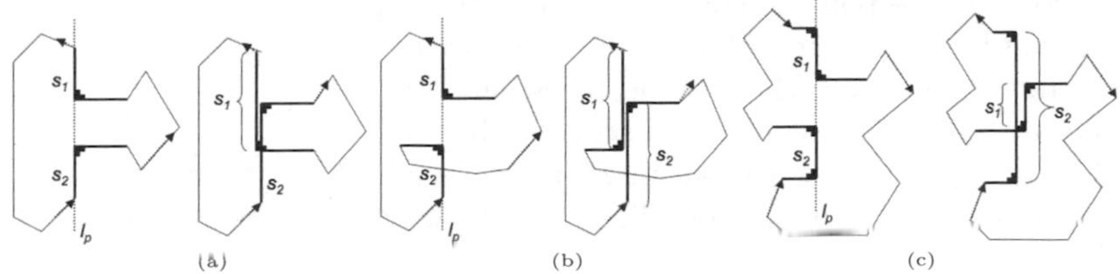

Fig. 2. The tour T is changed preserving the number of bends. Thick lines correspond to segments of the tour, while thin lines indicate the Jordan curve of the tour somewhere in the plane.

$p_{min} \in cover(l)$ with the smallest y coordinate. The line $h_{(max-i)}$ is the horizontal line through the i-th point in $cover(l)$ below p_{max} while $h_{(min+i)}$ is the horizontal line through the i-th point in $cover(l)$ above p_{min}, for $1 \leq i \leq (k/2-1)$. We expand the set L_k as follows. For every vertical line $l \in L_k$, we add k horizontal lines $h_{max}, h_{(max-1)}, \ldots, h_{(max-k/2+1)}$ and $h_{min}, h_{(min+1)}, \ldots, h_{(min+k/2-1)}$ to L_k. Symmetrically, for every horizontal line $l \in L_k$, we add also k vertical lines $v_{max}, v_{(max-1)}, \ldots, v_{(max-k/2+1)}$ and $v_{min}, v_{(min+1)}, \ldots, v_{(min+k/2-1)}$ to L_k, where v_{max} passes through $p_{max} \in cover(l)$ with the largest x-coordinate and v_{min} passes through the point $p_{min} \in cover(l)$ with the smallest x-coordinate. The lines $v_{(max-i)}$ and $v_{(min+i)}$ are defined in a similar way to that of $h_{(max-i)}$ and $h_{(min+i)}$. Note that, if l covers less than k different points, we add a line that is orthogonal to l on every point in $cover(l)$. We call L_k with all these additional lines the set L'_k and $|L'_k| \leq k^2$. Let H be all the horizontal lines through a point in the kernel S' and let V be all the vertical lines through a point in S'. Thus, $|H| \leq k^2/2$ and $|V| \leq k^2/2$. Now, we add to L'_k all the lines in V and all the lines in H. The new set $R = L'_k \cup H \cup V$ has quadratic size in k. Moreover the set I of all intersections of two lines in R has also polynomial size in k (that is, $O(k^4)$). We will argue that a rectilinear tour of k bends exists for any given instance if and only if a tour with k bends exists with bends placed on I.

Lemma 5. *If the instance has a rectilinear tour T with k or fewer bends, then a tour can be built with k or fewer bends and all the bends are at I, the set of all intersections of lines in R where $R = L'_k \cup H \cup V$.*

Proof. We will show that for every YES-instance, we can transform the witness tour T to a tour T' with the same number of bends where every line-segment in T' is hosted by a line in R. From this, it follows that the set I hosts the possible positions for the bends in T'.

Let p be any point in S. If this is a YES-instance, there is a witness tour that has a line-segment l_p covering the point p. Moreover, because of Lemma 4, we can assume that if $s_1, \ldots, s_i$ are line-segments hosted by a rectilinear line l_p, there are no gaps; that is, $\cup_{j=1}^{i} s_j$ is not disjoint. If the point p was from the kernel, we are done because the line-segment is hosted on $H \subseteq R$ or on $V \subseteq R$. Otherwise, the point p was from a line discovered in kernelization. If the point p is covered by some s_j in the same orientation as the line discovered by kernelization, we are done. If the point is covered in the k-bends witness tour by a segment orthogonal to the line discovered in kernelization, the case becomes delicate. Assume p is on a vertical line l_p discovered by the kernelization, but in the witness tour T', p is covered in a horizontal line-segment over a line h_p. If h_p was discovered by kernelization, we are done, the same if it was a line in H. Therefore, line h_p is either above or below $\cup_{j=1}^{i} s_j$ because we are in a case where p is not covered by $\cup_{j=1}^{i} s_j$ (see Fig. 3, for example). Moreover, there cannot be more than $k/2$ points in the same situation as p. Otherwise, the witness structure would have more than $k/2$ horizontal line-segments (Lemma 1) at those positions. Therefore, the rank of p from either end of l_p must be no more than $k/2$. That is, p is in one of the lines $h_{max}, h_{(max-1)}, \ldots, h_{(max-k/2+1)}$ or $h_{min}, h_{(min+1)}, \ldots, h_{(min+k/2-1)}$ that we have in L'_k. Since $h_p \subseteq L'_k$, we have $h_p \subseteq R$. □

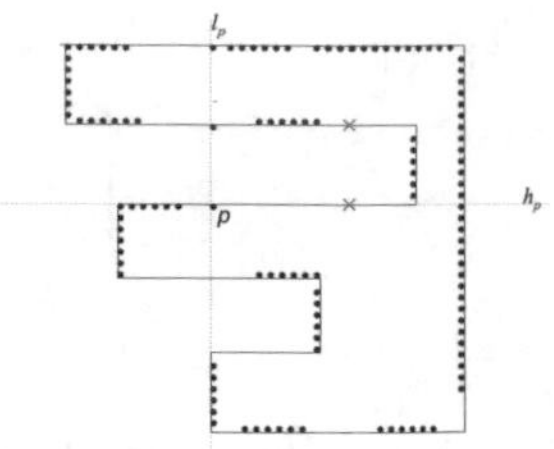

Fig. 3. The point p is covered by a horizontal line that is not found in the set H or the set L_k

Theorem 1. *The* Rectilinear k-Bends Traveling Salesman Problem *is FPT.*

Proof. The algorithm computes R and searches exhaustively all tours over the lines in R. For example, a naive algorithm tests all subsets of size k of I to decide whether these k candidate-intersections can be completed into a tour with at most k bends and all the n points lie on the line-segments that make up the tour. Note that the k candidate-intersections of lines essentially host the bends in the tour. Since the number $|I|$ of intersections by lines in R is $O(k^4)$, the number of subsets of size k is bounded by $\binom{k^4}{k} = O(k^{4k}/k!)$. Testing all permutations to cover all tours result in time bounded by $O((k!)(n)(k^{4k}/k!)) = O(k^{4k}n)$. Kernelization can be performed in $O(kn^2)$ time. Thus, we can decide the Rectilinear k-Bends Traveling Salesman Problem in $O(kn^2 + k^{4k}n)$ time. □

2.2 Tours with Constraints

Now, we decide the problem for the rectilinear tour with at most k bends for the two variants introduced earlier. In fact, for both variants, the first phase computes several k-covers (candidates set L of k lines that cover all the n points in S). However, instead of kernelization, in order to identify hosting lines, the technique here will be a bounded-search-tree. The second phase checks if each candidate k-cover can constitute a tour based on such k-cover. To find these k-covers, we consider a search tree as illustrated by Fig. 4. In every node of the tree, the set L is a partial cover of S, and in some leaves it will be a k-cover. In the root of the tree, the set L is empty. In each internal node of the tree, we choose a point $p \in S \setminus cover(L)$ and we explore the two possibilities of enlarging L by analyzing the situation at p. Note that if the given instance were a YES-instance, every point in S is covered by a horizontal line (H-line), or a vertical line (V-line) (or both, if a bend or crossing is placed at the point). The point p is marked as covered with an H-line or V-line, depending on the branch of the tree. Also, the chosen line is added to L. The points that fall into the same class with p are also marked as covered so that we do not consider these points again in the next recursive call.

We keep track of how many vertical and horizontal assignments have been made and we emphasize that we do not assign more than $k/2$ horizontal lines and also no more than $k/2$ vertical lines. Each branch of the tree stops when the upper bound is reached or when every point is marked as covered. At the leaves of the tree, we have at most k lines that cover the set of n points, or exactly k lines that do not cover; in this case, we know the candidate set of lines cannot be completed into a covering tour. Therefore, the depth of the tree is at most k. This matches the pattern of a bounded-search-tree. The depth of the tree is bounded by the parameter and the fan-out is a constant (in this case the constant is 2).

Let L be the set of lines that cover the set S at a leaf of the tree where $|L| \leq k$. We only consider tours where every line-segment covers at least one point because for every tour T that covers S, there is an equivalent tour T' where every line-segment of the tour covers at least one point[3]. If T is a tour, we let $lines(T)$ be the set of lines used by the line-segments in T. Each line in $lines(T)$ covers at least one point and $lines(T)$ is a cover with k or fewer orthogonal lines. These observations allow us to state the following lemma.

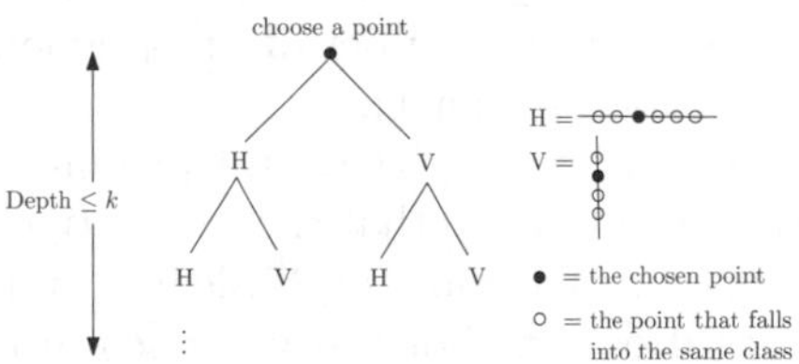

Fig. 4. Computing candidate sets of k lines

Lemma 6. *If the instance has a tour T with k or fewer bends, then there is a leaf of the tree at the end of phase one where the cover made by the lines in L is consistent with the cover by $lines(T)$ (i.e. $L \subseteq lines(T)$).*

Proof. The algorithm of the first phase picks up a point p at each node. This point must be covered by $lines(T)$ with a horizontal or a vertical line. Therefore, there is a child that agrees with $lines(T)$ for this node. Traveling the path from the root of the tree down the child that agrees with $lines(T)$, we reach a leaf where each line in L must be in $lines(T)$. □

In the second phase, we investigate if those leaves of the first phase that cover the n points can result in a tour with the constraints of the variants.

[3] If a line-segment in the tour covers no points, it can be translated in parallel until it is placed over one point.

Tours where one line-segment covers all the points on the same line: Here, we require that every line-segment s of a tour T must cover all the points in S on the line that includes s. In this special case of the RECTILINEAR k-BENDS TRAVELING SALESMAN PROBLEM, each candidate set L of lines at a leaf of the tree (recall Fig. 4) results in a candidate set of line-segments since we can simply connect the extreme points in $cover(l)$ for each line $l \in L$ to get the candidate line-segments. We can explore exhaustively if these candidate line-segments result in a tour. In the worst case, we have k line-segments from the first phase. These k line-segments are organized into $k!$ orders in a possible tour. In each of these orders, we can connect the line-segments (in the order chosen) one by one to form a tour. There are at most 4 ways to connect the line-segment l_i to the consecutive line-segment $l_{(i \bmod k)+1}$ for $i \in \{1, 2, \ldots, k\}$. Let a and b be the extreme points of l_i, while c and d are the extreme points of $l_{(i \bmod k)+1}$. We can connect these two line-segments as ac, ad, bc or bd (see Fig. 5). This means a total of $(k!)(4^k)$ tests. In some cases when we connect two line-segments together, the extension of these two line-segments may be enough to make a turn, therefore no additional line-segment is required. In some cases, it requires two additional line-segments as shown in Fig. 5. These extra line-segments cover no points, but they can be added in constant time when constructing the tour. Note that if the total number of line-segments in the final tour exceeds k, we simply answer no (Lemma 1).

We now analyze the total running time of our algorithm. It is obvious that the search tree for the first phase in Fig. 4 has at most $O(2^k)$ leaves and $O(2^{k-1})$ internal nodes. However, not every branch in the tree has to be explored. We explore only the branches that have the equal number of H-lines and V-lines. This is equivalent to choosing (among the k levels) a subset of size $k/2$ where to place the H-lines (or V-lines). Another way to recognize this is that each path from the root to a leaf in Fig. 4 is a word of length at most k with exactly $k/2$ symbols H and $k/2$ symbols V. Based on this analysis we reduce the size of the tree to $\binom{k}{\frac{k}{2}}$ which is simplified to $O(2^k/\sqrt{k})$ using Stirling's approximation. The work at each internal node is to choose a point, record the line and mark the associated points that are covered by that line. The dominant work is the computation at the leaves of the tree. Here we perform the tests to cover all tours that require time bounded by $O(\frac{2^k}{\sqrt{k}}(k!)(4^k)n)$ which is simplified to $O((2.95k)^k n)$. The time complexity is exponential in the parameter but linear in the size of the input. This gives the following result.

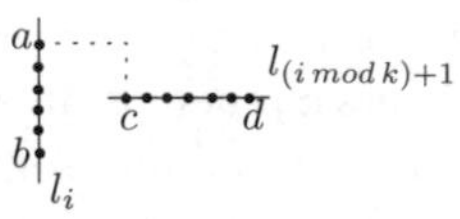

Fig. 5. One of four possible ways of joining two consecutive lines

Theorem 2. *The* RECTILINEAR k-BENDS TRAVELING SALESMAN PROBLEM *where one line-segment covers all the points on the same line is FPT.*

Tours where the same line-segment orientation cover points on the same line: In this special case, a point that lies on a line hosting a horizontal-segment of the tour must be covered by a horizontal line-segment of the tour (possibly another horizontal-line segment, and possibly also a vertical line-segment). The trick is that it cannot be covered only by a vertical line-segment. The symmetric condition holds for points in a line hosting a vertical line-segment. We call this the *no distinct type of line-segment condition.*

In the first phase, a leaf that may hold a YES-instance has a candidate set L of no more than k lines and $L \subseteq lines(T)$. We expand this set of candidate lines as follows. For every vertical line $l \in L$ we add two horizontal lines h_{max} and h_{min}. The line h_{max} is the horizontal line through the point $p \in cover(l)$ with the largest y coordinate, while h_{min} is the horizontal line through the point $p \in cover(l)$ with the smallest y coordinate. Symmetrically, for every horizontal line $l \in L$, we add two lines v_{max} and v_{min} where v_{max} passes through $p \in l$ with the largest x-coordinate and v_{min} passes through the point $p \in l$ with the smallest x-coordinate. Note that L with all these additional lines has size linear in k. In what follows, we call the set L at a leaf with these additional lines, the set C of lines. Our aim is the next result.

Lemma 7. *If the instance has a tour T with k or fewer bends (and meeting the no distinct type of line-segment condition), then there is a leaf of the tree at the end of phase one where a tour can be built with k or fewer bends and all the bends are at intersections of lines in C.*

The proof shows that if we have a YES-instance, we can transform the witness tour T to a tour T' with the same numbers of bends, where every line-segment covers at least one point and $lines(T') \subseteq C$. From this, it follows that the intersections of all lines in C hosts the possible positions for the bends in T'.

The argument shows that every time we have a line-segment $\overline{pq}$ in a tour with its hosting line in $lines(T) \setminus C$, we can find a covering tour T' with the same bends and leading to the same leaf in phase one, the line-segment $\overline{pq}$ is not used and more line-segments in T' have their hosting lines in C. Consider a line-segment $\overline{pq}$ in a tour T that is a witness that the leaf is a YES-instance, but the line l hosting $\overline{pq}$ in the tour is such that $l \notin C$ (i.e. $l \in lines(T) \setminus C$). Without loss of generality, assume $\overline{pq}$ is horizontal (if $\overline{pq}$ were vertical, we rotate S and the entire discussion by 90°). Also, we can assume that $\overline{pq}$ covers at least one point in S and T has minimal number of bends. Let l_1 be the line-segment in the tour before $\overline{pq}$ and l_2 the line-segment in the tour after $\overline{pq}$.

Claim 1. *For all $p' \in S \cap \overline{pq} \setminus \{p,q\}$, there is a vertical line $l \in L$ (and thus $l \in C$) such that $p' \in cover(l)$.*

Proof. Let $p' \in S \cap \overline{pq} \setminus \{p,q\}$. Then p' is covered by a vertical line in L, because L covers S, and if p' was covered by a horizontal line, then the line hosting $\overline{pq}$ would be in L and $L \subseteq C$. This contradicts that the line hosting $\overline{pq}$ is not in C. If $\emptyset = S \cap \overline{pq} \setminus \{p,q\}$ the claim is vacuously true. □

Points in $S \cap \overline{pq} \setminus \{p,q\}$ are covered in T by vertical line-segments (if any point p' was covered only by $\overline{pq}$ and not a vertical segment in T through the vertical line at p', then T' would not satisfy the no distinct type of line-segment condition).

We will now distinguish two cases (refer to Fig. 6). In the first case, the tour T makes a U-return shape reversing direction, while in the second case, the tour makes a zig-zag shape and continues in the same direction. The bends at p and q of the line-segment $\overline{pq}$ make a U-return if they have one common label. The bends at p and q make a zig-zag of the line-segment $\overline{pq}$ if they have no common label. In this case the turns are NE at p and SW at q (with no letter label in common) or SE at p and NW at q (with no letter label in common). Without loss of generality, note that also a horizontal reflection along $\overline{pq}$ can be made so the other subcases can be ignored and we can assume the cases are as the two drawings of Fig. 6.

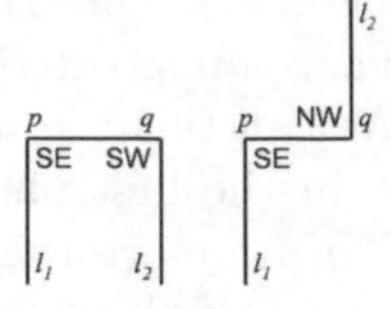

Fig. 6. The line $l \notin L$ hosts $\overline{pq}$ from T

Case 1: In this case, we obtain the corresponding equivalent tour by shifting $\overline{pq}$ vertically. This is possible because all points in $S \cap \overline{pq} \setminus \{p, q\}$ are already covered by other vertical lines of T. In fact, if there is any point in $S \cap \{(x, y)|p_x \leq x \leq q_x \wedge y \geq p_y\}^4$, by shifting $\overline{pq}$ vertically (up) we can (without increasing the number of bends) overlaps with h_{max} for some vertical line in L. In fact, the set $S \cap \{(x, y)|p_x \leq x \leq q_x \wedge y \geq p_y\}$ is not empty because $\overline{pq}$ has at least one point covered by a vertical line in L (L is a cover and we assumed the horizontal line hosting $\overline{pq}$ is not in L). It is important to note that our tour T' may self-intersect, but that is not a constraint for the problem.

Case 2: This setting has the following subcases. First we show that if $l_2 \notin L$, we can also change to a tour T' where now the set $lines(T') \setminus C$ is smaller. The symmetric argument shows that we can do this also if $l_1 \notin L$. Finally, the case left is when both $l_1, l_2 \in L$.

Subcase 2.a: If $l_2 \notin L$, then $q \notin S$. Because the line-segment $\overline{qq'}$ hosted by l_2 must cover one point $q' \in S$ and L is a cover, the point q' is covered by a horizontal line $l_3 \in L$. Fig. 7 (a) shows that the tour cannot have a SE turn at q', because then we can make a tour with 2 fewer bends contradicting the minimality of the witness tour (neither $\overline{qq'}$ nor $\overline{pq}$ are needed).

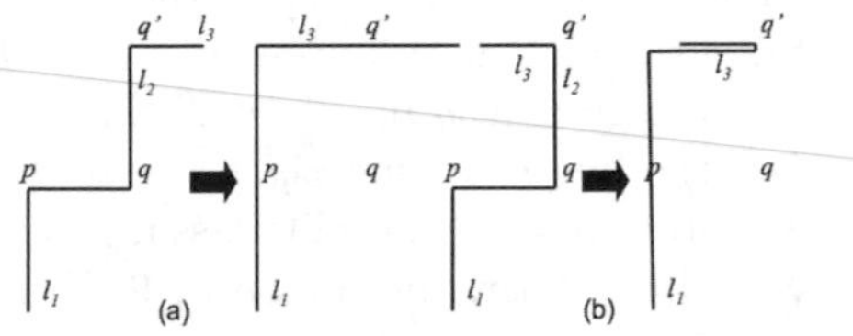

Fig. 7. The subcases if $l_2 \notin L$ can be converted and eliminate l_2

Thus, the turn at q' must be a SW turn. Fig. 7 (b) shows that a tour with a 180°-bend is equivalent and l_2 and the line hosting $\overline{pq}$ are not need. This makes $lines(T') \setminus C$ smaller by two lines.

Subcase 2.b: If $l_1 \notin L$, then $p \notin S$. The arguments is analogous to the previous case.

---

4 Here p_x is the x-coordinate of point x, thus $S \cap \{(x, y)|p_x \leq x \leq q_x \wedge y \geq p_y\}$ is all points in S above or on $\overline{pq}$.

Subcase 2.c: Now we must have $l_1, l_2 \in L$. In this case, we make a cut and join operation to obtain a new tour that now is as in Case 1 (that is, we have a U-turn and not a zigzag).

First note that in this case there must be points in S covered by l_2 below or including q. Otherwise, the line h_{min} from l_2 can be made to coincide with $\overline{pq}$ and we are done. Similarly, there must be points in S covered by l_1 above p, otherwise $\overline{pq}$ can be made to coincide with h_{max} for l_1. Note also that there must be points in S above q covered by l_2 otherwise shifting $\overline{pq}$ can be made to coincide with h_{max} for l_2. Also, by an analogous argument, there must be points in S below p covered by l_1. The tour T must use a vertical line-segment to cover the points in l_2 below q,[5] because the tour complies with the no distinct type of line-segment condition. Assume that the tour T is traveled in the direction l_1, then $\overline{pq}$, then l_2 and let $\overline{p'q'}$ be a segment of this tour under q hosted by l_2, with $p'_y > q'_y$. In the chosen traversal of the tour T, $\overline{p'q'}$ may be traveled from p' to q' or from q' to p' (refer to Fig. 8).

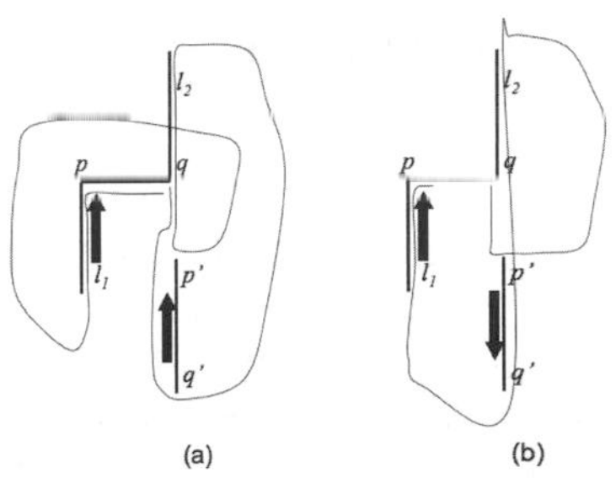

Fig. 8. The tour T is changed so there is no zig-zag

If T travels from p' to q', we build a new tour that travels from q to p' and continues along T but in reverse, which ensures we will meet l_2 and then q again, but we now continue until q'. From here we continue along T in the same direction, which ensures we reach p. This new tour now makes a U-turn at $\overline{pq}$ and has the same set L as T and the same number of bends. For the case that T travels from q' to p', the new tour travels from $\overline{pq}$ to q'. Then along T in reverse order which guarantees arriving at q from l_2. We continue until p' and then in reverse order in T until we reach p again. Once again, this conversion reduces this case to Case 1 before.

We can now carry out the following procedure at each leaf. The number of intersections between any pair of lines in C is bounded by $O(k^2)$. We enumerate all words with repetitions of length k over the alphabet of intersections. We require repetitions since we need to consider tours with 180°-bends. A word like this encodes where the bends of a tour will be. For each of these words, we generate each of the possible placements given by the 4 types of 90°-bends (NE, NW, SE, SW) at each intersection. A placement of bends can be tested (in polynomial time) if it can be in fact completed into a tour and whether such tour covers S. The running time is bounded by $O(\frac{2^k}{\sqrt{k}} \frac{(2k^2)^k}{k!}(4^k)kn)$ which is simplified to $O((43.5k)^k n)$. This has clearly exponential complexity in k, but the overall algorithm has polynomial complexity in n and demonstrates the following result.

---

[5] An analogous argument happens if the tour T uses a vertical line-segment to cover points above p covered by l_1.

Theorem 3. *Under the constraint that a tour satisfies the no distinct type of line-segment condition, the* Rectilinear k-Bends Traveling Salesman Problem *is FPT.*

3 Conclusions

We have presented three FPT algorithms for different variants of the Rectilinear k-Bends Traveling Salesman Problem. We summarize these results in Table 1. It is apparent that the complexity of the algorithms is slightly better as more constraints are placed on the solution. Also, as we argued with an example, a solution for the general case may not be a solution for either of the constrained cases, and a solution of the constrained cases may require more bends although it is a solution for the general case. Therefore, the fact that one variant is FPT does not imply the other variant is FPT. While we have discussed the decision-version of the problem, it is not hard to adapt to an FPT-algorithm for the optimization version. For example, we can use an optimization-version FPT-algorithm to find a k'-cover where k' is the number of covering lines and $O(k')$ is a bound for k where k is the minimum number of bends. In fact, it has been shown [12] that $k \leq 2k' + 2$. That is, we can approximate the minimum number of bends in FPT-time with respect to the sought minimum (or use a polynomial-time approximation algorithm [12] to obtain the approximate minimum). Then the decision problem can be used to find the exact value of the minimum in FPT-time (using binary search, for example).

Table 1. FPT algorithms for finding a rectilinear tour with at most k bends

Types of Rectilinear Tour	Time Complexities
1) general case (without constraints)	$O(kn^2 + k^{4k}n)$
2) same line-segment orientation cover points on the same line	$O((43.5k)^k n)$
3) one line-segment covers all the points on the same line	$O((2.95k)^k n)$

References

1. Arkin, E., Bender, M., Demaine, E., Fekete, S., Mitchell, J., Sethia, S.: Optimal covering tours with turn costs. SIAM Journal of Computing 35(3), 531–566 (2005)
2. Arkin, E., Mitchell, J., Piatko, C.: Minimum-link watchman tours. Information Processing Letters 86(4), 203–207 (2003)
3. Bereg, S., Bose, P., Dumitrescu, A., Hurtado, F., Valtr, P.: Traversing a set of points with a minimum number of turns. Discrete and Computational Geometry 41, 513–532 (2009)
4. Collins, M.: Covering a set of points with a minimum number of turns. Int. J. Comput. Geometry Appl. 14(1-2), 105–114 (2004)
5. de Berg, M., van Kreveld, M., Nilsson, B., Overmars, M.: Shortest path queries in rectilinear worlds. Int. J. Comput. Geometry Appl. 2(3), 287–309 (1992)
6. Downey, R., Fellows, M.: Parameterized Complexity. Monographs in Computer Science. Springer, New York (1999)

7. Flum, J., Grohe, M.: Parameterized Complexity Theory. Texts in Theoretical Computer Science. Springer, Berlin (2006)
8. Hassin, R., Megiddo, N.: Approximation algorithms for hitting objects with straight lines. Discrete Applied Mathematics 30(1), 29–42 (1991)
9. Lee, D., Chen, T., Yang, C.: Shortest rectilinear paths among weighted obstacles. In: 6th ACM Symp. on Comput. Geometry (SCG '90), pp. 301–310. ACM, NY (1990)
10. Lee, D., Yang, C., Wong, C.: Rectilinear paths among rectilinear obstacles. Discrete Applied Mathematics 70(3), 185–215 (1996)
11. Niedermeier, R.: Invitation to Fixed-Parameter Algorithms. Oxford Lecture Series in Mathematics and its Applications, vol. 31. Oxford University Press, New York (2006)
12. Stein, C., Wagner, D.: Approximation algorithms for the minimum bends traveling salesman problem. In: Aardal, K., Gerards, B. (eds.) IPCO 2001. LNCS, vol. 2081, pp. 406–422. Springer, Heidelberg (2001)
13. Wagner, D., Drysdale, R., Stein, C.: An $O(n^{5/2} \log n)$ algorithm for the rectilinear minimum link-distance problem in three dimensions. Comput. Geom. Theory Appl. 42(5), 376–387 (2009)

Tracking a Generator by Persistence*,**

Oleksiy Busaryev, Tamal K. Dey, and Yusu Wang

Department of Computer Science and Engineering,
The Ohio State University, Columbus OH 43210, USA
{busaryev,tamaldey,yusu}@cse.ohio-state.edu

Abstract. The persistent homology provides a mathematical tool to describe "features" in a principled manner. The persistence algorithm proposed by Edelsbrunner et al. [5] can compute not only the persistent homology for a filtered simplicial complex, but also representative generating cycles for persistent homology groups. However, if there are dynamic changes either in the filtration or in the underlying simplicial complex, the representative generating cycle can change wildly. In this paper, we consider the problem of tracking generating cycles with temporal coherence. Specifically, our goal is to track a chosen essential generating cycle so that the changes in it are "local". This requires reordering simplices in the filtration. To handle reordering operations, we build upon the matrix framework proposed by Cohen-Steiner et al. [3] to swap two consecutive simplices, so that we can process a reordering directly. We present an application showing how our algorithm can track an essential cycle in a complex constructed out of a point cloud data.

1 Introduction

The notion of prominent feature in a geometric shape plays important role in various scientific studies and engineering applications. The persistent homology introduced by Edelsbrunner et al. [5] provides a mathematical tool along with a combinatorial algorithm that aids capturing these features in a principled manner. Various aspects of persistent homology have been studied including its stability against noise [1,2] and maintenance under changes in the filtration [3]. In this paper, we elaborate upon the last aspect, that is, maintaining persistence and a persistent generator when the simplicial complex and its filtration changes.

Let $\mathcal{K}$ be a simplicial complex. A sequence of its simplices in non-decreasing order of their dimensions induces a *filtration* of $\mathcal{K}$. A persistent pairing of simplices can be obtained from a filtration following the persistence algorithm [5]. A simplex is a *creator* if it creates a new homology class and it is a *destroyer* if it kills a homology class. A destroyer pairs with a creator if it kills a class created by the creator. Each destroyer pairs with a unique creator. The unpaired creators get associated with homology classes that survive and represent essential classes in the homology of $\mathcal{K}$. Given a filtration, the persistence algorithm computes the

* The full version of this paper is available at authors' webpages.

** Authors acknowledge the support of NSF grants CCF-0830467, CCF-0915996, CCF-0747082 and DBI-0750891.

M.T. Thai and S. Sahni (Eds.): COCOON 2010, LNCS 6196, pp. 278–287, 2010.

pairing and also the representative cycles of the classes that survive. If the filtration changes, the persistent pairing changes - so do the essential cycles. Similar phenomena happen when $\mathcal{K}$ changes with insertion and deletion of simplices.

Our main goal is to track a chosen essential generator under insertions and deletions of simplices. This also requires reordering of simplices in the filtration. Such operations have been considered before [1,3]. Specifically, Cohen-Steiner et al. [3] gave an efficient algorithm to swap two consecutive simplices in a filtration and maintain persistent pairing using a matrix view of the persistence. Insertions and deletions can be implemented using such swaps. We observe that reordering (and thus insertions and deletions) can be implemented more directly avoiding repeated swaps. Although this does not improve the worst-case time analysis, it indeed saves computing time in practice as our empirical results confirm.

The essential cycle that we track is kept associated with the first creator in the filtration that remains unpaired. With each update necessary changes need to be made in this cycle. However, if the persistence algorithm is executed, this cycle may change wildly. We show how to maintain all invariants necessary to detect persistent pairs and still make only "local" changes to the cycle being tracked.

2 Background on Persistence

Simplicial homology. Let $\mathcal{K}$ be a simplicial complex. Under $\mathbb{Z}_2$ coefficients, a *p-chain* is a subset of p-simplices of $\mathcal{K}$. The set of p-chains, together with modulo 2 addition, forms a free abelian group C_p called the *p-th chain group* of $\mathcal{K}$. The *boundary* $\partial_p(\sigma)$ of a k-simplex σ is the set of its $(p-1)$-faces, and that of a p-chain is the modulo 2 addition of the boundaries of its simplices. The boundary operator defines a homomorphism $\partial_p : \mathsf{C}_p \to \mathsf{C}_{p-1}$.

Now define a *p-cycle* to be a p-chain with empty boundary and a *p-boundary* to be a p-chain in the image of ∂_{p+1}. The spaces of p-cycles and p-boundaries are called the *p-th cycle group* $\mathsf{Z}_p = \ker \partial_p$ and the *p-th boundary group* $\mathsf{B}_p = \operatorname{im} \partial_{p+1}$, respectively, and they are both subgroups of C_p. The *p-th homology group* is the quotient group $\mathsf{H}_p(\mathcal{K}) = \mathsf{Z}_p/\mathsf{B}_p$. Elements of H_p are the *homology classes* $c+\mathsf{B}_p = \{c+b \mid b \in \mathsf{B}_p\}$ for p-cycles c; c is also referred to as a *generating cycle* of the homology class $[c] = c+\mathsf{B}_p$. Two p-cycles c and d are *homologous* if $[c]=[d]$, that is, $c+d \in \mathsf{B}_p$ is the boundary of some $(p+1)$-chain.

The rank of H_p is called the *p-th Betti number* of $\mathcal{K}$, denoted by $\beta_p = \operatorname{rank} \mathsf{H}_p$. A *basis* of H_p is a minimal set of homology classes that generates H_p. A set of p-cycles $C = \{c_1, \ldots, c_l\}$ *generates* H_p, if $\{[c_i]\}$ forms a basis for H_p; $|C| = \beta_p$.

Persistent homology. A *filtration* of the simplicial complex $\mathcal{K}$ is a sequence of nested subcomplexes $\mathcal{K}_0 \subset \mathcal{K}_1 \subset \cdots \subset \mathcal{K}_m = \mathcal{K}$. Consider the homomorphism $h_p^{i,j} : \mathsf{H}_p(\mathcal{K}_i) \hookrightarrow \mathsf{H}_p(\mathcal{K}_j)$ induced by the inclusion $\mathcal{K}_i \subset \mathcal{K}_j$. We say that a homology class $h \in \mathsf{H}_p(\mathcal{K}_r)$ was *created* at time $i \leq r$ if it does not have a preimage in $\mathsf{H}_p(\mathcal{K}_{i-1})$ under $h_p^{i-1,r}$, but has a preimage in $\mathsf{H}_p(\mathcal{K}_i)$ under $h_p^{i,r}$. A homology class $h \in \mathsf{H}_p(\mathcal{K}_r)$ is *destroyed* at time $j > r$ if its image in $\mathsf{H}_p(\mathcal{K}_{j-1})$ is non-trivial, but its image in $\mathsf{H}_p(\mathcal{K}_j)$ is trivial. If h is created at i and destroyed at j, then it has *persistence* $j-i$ and produces a *persistence pairing* (i,j).

An ordering $(\sigma_1, \ldots, \sigma_m)$ of simplices in $\mathcal{K}$ induces a filtration Π such that $\mathcal{K}_i = \mathcal{K}_{i-1} \cup \{\sigma_i\}$. We also write $\Pi = (\sigma_1, \sigma_2, \ldots, \sigma_m)$ for simplicity. A valid filtration requires that each simplex has a larger index than any of its faces.

Matrix view. Consider a simplicial complex $\mathcal{K}$ and a filtration $\Pi = (\sigma_1, \sigma_2, \ldots, \sigma_m)$. Let D denote the boundary matrix for $\mathcal{K}$; i.e., $D[i][j] = 1$ if $\sigma_i \in \partial\sigma_j$, and $D[i][j] = 0$ otherwise. Persistent pairing can be computed by the matrix reduction approach [3,5]. Since we use $\mathbb{Z}_2$ coefficients, matrices we consider have entries only 0 or 1, and we use modulo 2 arithmetic during matrix multiplications.

Given any matrix M, let $\mathrm{row}_M[i]$ and $\mathrm{col}_M[j]$ denote the i-th row and j-th column of M, respectively. We abuse the notation slightly to let $\mathrm{col}_M[j]$ denote also the chain $\{\sigma_i \mid M[i][j] = 1\}$. Let $\mathrm{low}_M[j]$ denote the row index of the last 1 in the j-th column of M, which we call the *low-row index* of the column j. It is undefined for empty columns. The matrix M is *reduced* (or is in *reduced form*) if $\mathrm{low}_M[j] \neq \mathrm{low}_M[j']$ for any $j \neq j'$; that is, no two columns share the same low-row indices. We define a matrix M to be *upper-triangular* if all of its diagonal elements are 1, and there is no entry $M[i][j] = 1$ with $i > j$.

Proposition 1 ([3]). *Let $R = DV$, where R is reduced and V is upper-triangular. Then the simplices σ_i and σ_j form a persistent pair if and only if $\mathrm{low}_R[j] = i$.*

Furthermore, let c_j and c'_j be the p- and $(p-1)$-chains corresponding to the columns $\mathrm{col}_V[j]$ and $\mathrm{col}_R[j]$ respectively where $R = DV$. Then, $c'_j = \partial_p(c_j)$.

Note that there are many R and V for a fixed D forming the decomposition as described above. Proposition 1 implies that the persistent pairing is independent of the contents of R and V.

3 Efficient Updates

We now describe how to perform reordering of simplices. Given a filtration $\Pi = (\sigma_1, \sigma_2, \ldots, \sigma_m)$, we consider operations MOVERIGHT(i, j) and MOVELEFT(i, j), which move σ_i to the j-th position, with $i < j$ and $i > j$, respectively. Insertion and deletion can be implemented using these operations.

To handle these movements, we could perform a sequence of the *transposition* operations as defined in [3] till σ_i is moved to its new position. Here, a transposition is simply MOVERIGHT$(i, i+1)$ or MOVELEFT$(i, i-1)$. Our approach will use the same framework as [3], but handle the movement directly to remove certain overhead. Specifically, instead of scanning the elements of a certain row $j-i$ times (once per transposition), we only scan it once for the entire operation. We present experimental results at the end of this section showing the practical improvement.

In the high level, we will maintain matrices R and V with the following invariants as the filtration changes:

I1. $R = DV$ where D is the boundary matrix for $\mathcal{K}$ with respect to filtration.
I2. R is in reduced form, and V is upper-triangular.

If I1 and I2 are maintained, we can read the persistent pairs by Proposition 1. Moreover, for σ_i that is a creator, $\mathrm{col}_V[i]$ represents a cycle created by σ_i. If σ_i is not paired, then $\mathrm{col}_V[i]$ represents a non-trivial homology class of $\mathcal{K}$.

Implementation of matrix updates involves column additions. To maintain invariant I1, whenever we add two columns in R, we add them in V. To maintain V upper-triangular, we always add a column to a column on its right.

3.1 Moving Right

This operation moves a simplex σ_i to the j-th position with $j > i$. To reflect this change, we should move the i-th column and row of D, R, and V, to the j-th position. Let $\widehat{D}, \widehat{R}$, and $\widehat{V}$ be the resulting matrices. They still satisfy $\widehat{R} = \widehat{D}\widehat{V}$; however, $\widehat{V}$ may not be upper-triangular and $\widehat{R}$ may not be reduced.

First, moving a column in V to the right does not destroy its upper-triangular property; however, moving a row down may destroy it. Specifically, $\mathrm{row}_{\widehat{V}}[j]$ may now contain 1's for indices in $[i, j-1]$. Similarly, moving down a row in R may destroy its reduced form. In both cases we can perform a similar procedure which we call RESTORE to restore the desired properties.

Restoring upper-triangular property. Consider the original matrices $R = DV$. Let $I_1 = i$ and let $\{I_2, \ldots, I_s\}$ be the set of indices within $[i+1, j]$ that have entry 1 in $\mathrm{row}_V[i]$. To keep V upper-triangular after we move $\mathrm{row}_V[i]$ to the j-th position, we need to cancel these 1's. To do this, we can add $\mathrm{col}_V[i]$ to each of $\mathrm{col}_V[I_l]$ for $1 < \ell \leq s$. Adding columns in V induces the same additions in R to maintain $R = DV$, which may destroy its reduced form. Fortunately, it turns out that we can maintain R reduced for all columns except possibly $\mathrm{col}_R[i]$.

Let $\rho_l = \mathrm{low}_R[I_l]$. Consider adding column I_1 to I_2 in both V and R to obtain V' and R'. Depending on whether $\rho_1 > \rho_2$ or not, one of them becomes the low-row index of $\mathrm{col}_{R'}[I_2]$ after addition, and the other one becomes a *free low-row index* which can potentially become the low-row index of some other column later. We call the column from R whose low-index now becomes free the *donor column* and its index the *donor index* i_d. Now, to cancel $V'[i][I_3]$, instead of adding $\mathrm{col}_V[i]$ to $\mathrm{col}_{V'}[I_3]$, we add the column $\mathrm{col}_V[i_d]$ in V to $\mathrm{col}_{V'}[I_3]$. This addition can also remove the entry 1 from the i-th element in $\mathrm{col}_{V'}[I_3]$ while not creating any entry 1 in $\mathrm{col}_{V'}[I_3]$ for row indices larger than I_3. Hence the column $\mathrm{col}_{V'}[I_3]$ now satisfies the upper-triangular property. The advantage of adding the donor column is that the matrix R' stays reduced except possibly for the i-th column. We repeat this procedure to cancel each entry 1 in the row $\mathrm{row}_V[i]$ for indices in $\{I_2, \ldots, I_s\}$, maintaining the donor index at each iteration. The entire algorithm is summarized below. It returns new matrices R and V, and also the last donor columns in both R and V which can potentially contribute a free low row index.

Algorithm 1. $\textsc{Restore}(R, V, \mathbb{I} = \{I_1, \ldots, I_s\})$

$d\_id \leftarrow I_1$; $d\_low \leftarrow \mathrm{low}_R[d\_id]$; $d\_col_R \leftarrow \mathrm{col}_R[d\_id]$; $d\_colV \leftarrow \mathrm{col}_V[d\_id]$
for $k \leftarrow 2$ to s **do**
  $new\_d\_low \leftarrow \mathrm{low}_R[I_k]$; $new\_d\_colR \leftarrow \mathrm{col}_R[I_k]$; $new\_d\_colV \leftarrow \mathrm{col}_V[I_k]$
  $\mathrm{col}_V[I_k] \leftarrow \mathrm{col}_V[I_k] + d\_colV$; $\mathrm{col}_R[I_k] \leftarrow \mathrm{col}_R[I_k] + d\_colR$
  if $d\_low > new\_d\_low$ **then**
    $d\_id \leftarrow I_k$; $d\_low \leftarrow new\_d\_low$; $d\_colR \leftarrow new\_d\_colR$; $d\_colV \leftarrow new\_d\_colV$
  end if
end for
return (R, V, d_colR, d_colV)

Restoring reduced form. Let R_1 and V_1 be the matrices returned by procedure $\textsc{Restore}(R, V, \mathbb{I})$. If we now move the i-th column and row of matrices R_1, D, V_1 to the j-th position, the new decomposition $\widehat{R} = \widehat{D}\widehat{V}$ guarantees that $\widehat{V}$ is upper-triangular. But $\widehat{R}$ may not be reduced. In particular, (1) the new column $\mathrm{col}_{\widehat{R}}[j]$ may not be reduced since we did not reduce $\mathrm{col}_R[i]$ in $\textsc{Restore}$; and (2) the row $\mathrm{row}_{\widehat{R}}[j]$ corresponding to the simplex σ that we moved may render several columns in $\widehat{R}$ having the same low-row index j. For (1), it turns out that we can simply replace $\mathrm{col}_{\widehat{R}}[j]$ with $\mathrm{col}_R[i_d]$, where i_d is the last donor index after performing the procedure $\textsc{Restore}$ above. This will give a unique low-row index to $\mathrm{col}_{\widehat{R}}[j]$ (by the definition of the donor column), and $\mathrm{col}_{\widehat{V}}[j] = \mathrm{col}_V[i_d]$ is a valid column in V. For (2), observe that the indices of those columns having low-row index j in $\widehat{R}$ are $\mathbb{J} = \{J_1, \ldots, J_t\}$ such that for each $l \in [1, t]$, the original low-row index $\mathrm{low}_R[J_l]$ falls inside the range $[i, j]$ and $\mathrm{row}_R[i][J_l]$ (i.e., $R[i][J_l]$) is 1. To resolve these conflicts, we call exactly the same algorithm $\textsc{Restore}$, for matrices R_1, V_1 before moving the i-th columns and rows, but with the set $\mathbb{J}$ substituting the role of $\mathbb{I}$. The entire algorithm for handling a right-movement operation is summarized below.

Algorithm 2. $\textsc{MoveRight}(R, V, i, j)$

compute $\mathbb{I} = \{I_1, \ldots, I_s\}$
compute $\mathbb{J} = \{J_1, \ldots, J_t\}$
compute $(R_1, V_1, d\_colR, d\_colV) \leftarrow \textsc{Restore}(R, V, \mathbb{I})$
compute $(R_2, V_2, d\_colR', d\_colV') \leftarrow \textsc{Restore}(R_1, V_1, \mathbb{J})$
Let $\hat{R}$ and $\hat{V}$ denote matrices after moving the i-th column and row of R_2 and V_2 to the j-th column and row, respectively
$\mathrm{col}_{\widehat{R}}[j] \leftarrow d\_colR$; $\mathrm{col}_{\widehat{V}}[j] \leftarrow d\_colV$

Proposition 2. *Suppose we move a p-simplex from the i-th position to the j-th position in a filtration of a simplicial complex $\mathcal{K}$ that has n simplices of dimensions $p-1$, p, and $p+1$. Let $R = DV$ where invariants I1 and I2 hold under the given filtration, and $\widehat{R}$ and $\widehat{V}$ be the matrices obtained by* $\textsc{MoveRight}(R, V, i, j)$. *Then, I1 and I2 hold for the new matrices $\widehat{R} = \widehat{D}\widehat{V}$. The algorithm runs in $O(kn)$ time, where $k = (s+t) \leq 2(j-i)$ and s, t are as described above.*

3.2 Other Operations

MoveLeft(i, j) can be handled by a procedure somewhat dual to MoveRight in same time complexity, and we omit the description here. The following property of MoveRight and MoveLeft is used in justifying our tracking algorithm.

Proposition 3. *Operation* MoveRight(i, j) *cannot change a destroyer to a creator other than the simplex being moved. Operation* MoveLeft(i, j) *cannot change a creator to a destroyer other than the simplex being moved. None of the operation changes the status of a simplex whose index is not between i and j.*

We can use MoveLeft and MoveRight to handle insertions and deletions. The routine InsertAt(σ, i) inserts a simplex σ at location i, and can be implemented by first inserting it at the end of the filtration, and then moving it to the i-th location. The routine DeleteAt(i) deletes the i-th simplex, and can be achieved by first moving it to the end of the filtration, and then removing it.

3.3 Experiments on Reordering

As in [3], we use a sparse matrix representation based on unsorted lists to represent matrices R and V. The advantage is that the size of the data structure is proportional to the number of 1's in each matrix, instead of m^2 (m is the number of simplices in the input complex). The disadvantage is, as each row is not stored in a linear order, searching for an element in it requires scanning this row. The main advantage of our MoveRight(i, j) and MoveLeft(i, j) procedures is that this cost is needed only once, instead of $j - i$ times when using the transposition algorithm from [3]. For example, in MoveRight(i, j), although we move σ_i over $j - i$ columns, we spend linear time only for those columns that will potentially be affected (which are the columns indexed by $\mathbb{I}$ and $\mathbb{J}$ in the procedure MoveRight), and constant time for other columns. Handling the movement using $j - i$ transpositions, however, will require linear time for every transposition to check whether current swapping will affect any column.

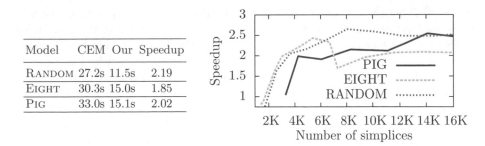

Model	CEM	Our	Speedup
Random	27.2s	11.5s	2.19
Eight	30.3s	15.0s	1.85
Pig	33.0s	15.1s	2.02

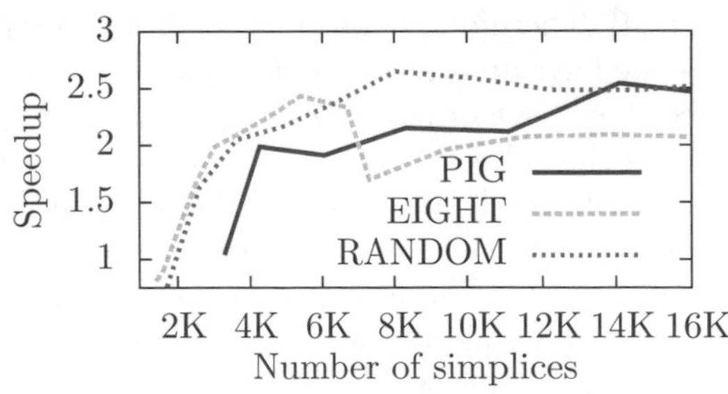

We tested our algorithm on three point cloud models, including a randomly generated set of 1000 points. For each model we produced 12 simplicial complexes of variable sizes; for each complex, 10 scenarios of 10000 random moves were generated. In the table above we show the running time for each model as well as the speedup relative to the algorithm from [3] denoted with acronym CEM. Running times and speedup values were separately averaged over 120 test cases.

4 Tracking a Generator

We now consider the problem of tracking an essential generator in a simplicial complex under insertions and deletions of simplices. A p-cycle G is *essential* if $[G]$ is not zero in $\mathsf{H}_p(\mathcal{K})$.

The class $[G]$ may become trivial when inserting a $(p+1)$-simplex. Essentiality of G can be detected using matrices R and V. Defining whether the class $[G]$ disappears or not is more subtle. Of course, $[G]$ may disappear only with deletions of simplices. If a simplex which is not in G is deleted, $[G]$ is not affected. Now consider the case when a simplex $\alpha \in G$ is being deleted. We cannot delete α if it has co-faces; hence α can have no co-faces at the time of its deletion. But in that case $[G]$ is always destroyed. Therefore, any deletion of a simplex from G always destroys its class. To overcome this technical difficulty, we define existence of a class under deletion differently: instead of deleting a simplex, we delete a simplex and all its co-faces and check whether $[G]$ survives or not.

In particular, let $\widehat{\mathcal{K}}$ be the complex obtained by deleting α and its co-faces in $\mathcal{K}$ where $\alpha \in G$. Consider the map $i_p : \mathsf{H}_p(\widehat{\mathcal{K}}) \hookrightarrow \mathsf{H}_p(\mathcal{K})$ induced by the inclusion $\widehat{\mathcal{K}} \subset \mathcal{K}$. The homology class $[G]$ in $\mathsf{H}_p(\mathcal{K})$ may have a pre-image in $\mathsf{H}_p(\widehat{\mathcal{K}})$ under i_p, in which case we say that the class $[G]$ *exists* in $\mathsf{H}_p(\widehat{\mathcal{K}})$. $[G]$ does not exist in $\mathsf{H}_p(\widehat{\mathcal{K}})$ if and only if α did not have any $(p+1)$-dimensional co-face in $\mathcal{K}$. In this case, $[G]$ is destroyed and the tracking of G ends.

A motivating application. In many applications the input shape is represented by a discrete points sample. Inference of geometry and topology from such data is a topic of active research. For topological inferences, often a data structure called *Rips complex* is built on top of these points. Essential generators in the Rips complex approximate the essential generators in the sampled space [4]. Therefore, if we track such a generator, say under motion of the vertices, we have an idea of the deformation of the true generator in the underlying space. For a set of points $P = \{p_1, p_2, \ldots, p_n\}$, the Rips complex $\mathcal{R}^\varepsilon(P)$ is defined as the collection of all simplices whose vertices have pairwise distances at most ε. Obviously, the topological change in $\mathcal{R}^\varepsilon(P)$ occurs only at *critical events* when the distance between two vertices p_i, p_j changes from larger to smaller than ε or vice versa. The former corresponds to an *insertion* event that inserts the edge $e = p_i p_j$, as well as all the higher-dimensional simplices it induces. The latter corresponds to a *deletion* event that deletes the edge $e = p_i p_j$ together with all its co-faces. This is why we consider the tracking of a cycle with respect to insertion and deletion of a simplex together with its co-faces.

Suppose e has k co-faces either before its deletion or after its insertion. Then using the reordering operations described above, we can maintain a set of generating cycles of any dimension p in $O(km^2)$ time, where $m = |\mathcal{K}|$.

However, during the updates, generating cycles can change wildly. See the right figure for an example. The initial generating cycle for a cylinder is C_1. After we remove the edge e and its incident triangle t, the creator of the new cycle C_1' may change to say e_1, which may come later than all edges in the cycle C_2. In this case, the generating cycle will suddenly change to C_2.

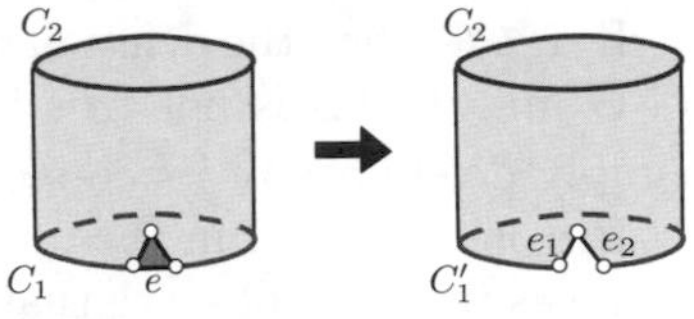

We wish to track a given input generating cycle G so that there is good temporal coherence in the cycle we produce (the meaning of this is made more precise later). We also keep the generator G *simple.* A p-cycle C is simple if one cannot decompose it into two independent p-cycles C_1 and C_2 with $C = C_1 + C_2$. Simplicity of cycles is often required in graphics and visualization applications.

Tracking invariant. The basic operation of the dynamic updates is as follows: given a simplicial complex $\mathcal{K}$, we either insert a p-simplex α and all of its faces if that are not already in $\mathcal{K}$, or remove α together with all its co-faces. To track a p-cycle, we maintain a decomposition $R = DV$ for the current simplicial complex, and the tracked cycle we return is always the chain $G = \mathrm{col}_V[\theta]$, where θ is the first p-dimensional creator in the filtration. We call this cycle the *first p-cycle.* We guarantee that G is essential and simple throughout the updates, and its change after insertions or deletions is "local" in some sense. If G becomes non-essential, its tracking ends. The following observation provides some intuition behind the choice of using the first p-cycle as the essential cycle we are tracking.

Claim. Let $G = \mathrm{col}_V[\theta]$ be the cycle created by the first p-dimensional creator θ in Π. G must be simple, and $[G]$ is essential if and only if θ is unpaired.

Insertions. A simplex α is always inserted at the end of the filtration. This adds one rightmost column and one bottom row to the matrix D. The new rightmost column can be reduced in $O(m^2)$ time, and no other columns/rows in R are affected. Hence θ is still the first p-creator, and $\mathrm{col}_V[\theta]$ (i.e. G) remains the same. However, if the dimension of α is $p + 1$, α may be paired with θ. In this case, $[G]$ is trivial in the new complex, and the tracking of G ends.

Claim. We can update the reduced matrix decomposition after inserting a simplex α and its k co-faces in $O(km^2)$ time. Either $[G]$ is trivial and the tracking terminates, or we still return G as the new generator.

Deletions. Tracking G after deletion is much more involved. We delete simplices in decreasing order of dimensions. After deletion of each simplex, we update the matrices R and V using DeleteAt operation. The status of the p-cycle G and θ can potentially be affected only when deleting p-simplices. However, whether $[G]$ is destroyed or not depends also on the $(p+1)$-simplices deleted. Hence from now on, we discuss only the basic case of deleting a p-simplex α and its $(p + 1)$-dimensional co-faces. For deleting simplices of dimension other than p and $p+1$, there is no additional operations required other than the standard DeleteAt operations. Now let $\mathcal{K}$ and $\widehat{\mathcal{K}}$ denote the simplicial complex before and after deleting the p-simplex α and its $(p + 1)$-dimensional co-faces, respectively.

If $\alpha \notin G$, then the deletion of it changes neither the status nor the composition of G and θ (this is not true in general, but holds for the first p-creator θ). We simply delete α and its co-faces, and update R and V for $\widehat{\mathcal{K}}$ appropriately.

If α is contained in the cycle G, and α did not have any $(p+1)$-dimensional co-faces in $\mathcal{K}$, we observe that $[G]$ is destroyed and the tracking of G ends.

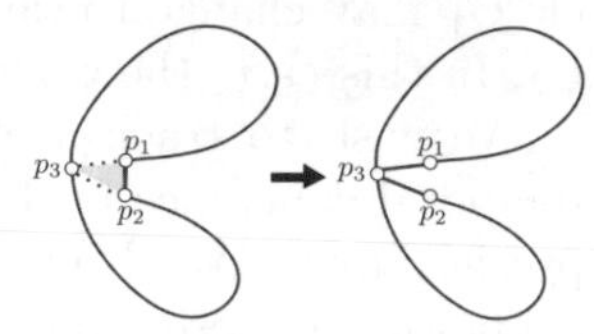

The remaining case is that there exists at least one $(p+1)$-simplex τ incident to α in $\mathcal{K}$, and thus $[G]$ exists in $\mathsf{H}_p(\widehat{\mathcal{K}})$. Consider the cycle $G' = G + \partial\tau$ which is homologous to G in $\mathcal{K}$: G' must exists in $\widehat{\mathcal{K}}$, $[G']$ is necessarily non-trivial in $\mathsf{H}_p(\widehat{\mathcal{K}})$ as the pre-image of $[G]$ under the map i_p, and the change between G and G' is local (differs by at most p simplices). Our goal is thus to return G' as the new generator $\widehat{G}$ as long as it is simple. If, however, G' is not simple anymore, then we aim to identify a simple sub-cycle $\widehat{G}$ of G' such that $[\widehat{G}]$ is still non-trivial. Intuitively, this generating cycle is still close to G. Unfortunately, a cycle $\widehat{G}$ that is both simple and homologous to G' may not exist. See the top-right figure for an example where we are to delete the edge p_1p_2 and its only co-face is the triangle $p_1p_2p_3$. After the deletion, the only representative cycle of the class $[G]$ has two sub-cycles and is thus not simple. We now describe how these goals are achieved.

First, delete all co-faces of α, including τ, using the algorithm from Sec. 3. To handle the deletion of α, let $\sigma_0 = \alpha, \ldots, \sigma_r$ be the set of p-faces of τ that are contained in G, and $\beta_1, \ldots, \beta_u$ the remaining set of p-faces of τ. We first perform the following: (1) move $\beta_1, \ldots, \beta_u$ immediately after θ (to the right of θ); (2) delete $\sigma_0 (= \alpha)$ and move $\sigma_1, \ldots, \sigma_r$ to the right of β_u. This can be completed in $O((r+u)m^2) = O(pm^2)$ time using operations from Sec. 3. Let $R_1 = \widehat{D}V_1$ be the resulting decomposition of the new filtration Π_1. Obviously, β_u is necessarily a creator, as it creates the cycle G' $(= G + \partial\tau)$ that cannot exist before β_u is introduced. However, note that G' is not necessarily the cycle stored at $\mathrm{col}_{V_1}[\beta_u]$ at this point. Specifically, β_u may not be the first p-dimensional creator. One or more β_is for $1 \le i < u$ may become creators (by Proposition 3, all other simplices older than θ necessarily remain as destroyers). Suppose β_j is the first creator in Π_1 and consider the cycle $\mathrm{col}_{V_1}[\beta_j]$ that it creates.

case i. If there is a simplex $\gamma \in \mathrm{col}_{V_1}[\beta_j]$ that is not in G', then we move γ to the right of β_u. This renders γ a creator (as it creates $\mathrm{col}_{V_1}[\beta_j]$) and β_j a destroyer, as otherwise, β_j could not have been the first p-creator in Π_1.

case ii. If all simplices in $\mathrm{col}_{V_1}[\beta_j]$ are contained in G', then G' is no longer simple (as it contains the cycle $\mathrm{col}_{V_1}[\beta_j]$ in it). It is necessary to break G'.

(ii.a) If β_j is unpaired, then our algorithm terminates with β_j being the new first p-creator and $\widehat{G} = \mathrm{col}_{V_1}[\beta_j]$ which is necessarily a subset of G'.

(ii.b) If β_j is paired, then we modify $G' \leftarrow G' + \mathrm{col}_{V_1}[\beta_j]$. From now on we target to track the new G', which is a sub-cycle of the previous G'. We next move β_j to the right of β_u.

In the two cases above, either our algorithm returns with a new first p-creator β_j with $j \in [1, u)$ and the update for deletion is over (case ii.a), or the first

p-creator is getting closer to β_u (cases i and ii.b). If it is the latter, we repeat the above procedure till either the algorithm returns in case ii.a or β_u becomes the first p-creator, in which case, the update terminates with $\widehat{G} = \mathrm{col}_{V_1}[\beta_u]$. This procedure can be called at most $u \leq p$ number of time, leading to an $O(pm^2)$ update time for deleting a p-simplex σ.

Claim. The above algorithm deletes a p-simplex α and its k co-faces in $O((p+k)m^2)$ time. It can detect whether $[G]$ is destroyed after the deletion or not, and terminates the tracking if and only if $[G]$ is destroyed. If $[G]$ remains, it returns a simple essential cycle $\widehat{G}$ such that either $[\widehat{G}] = [G]$ and $\widehat{G}$ differs from G by at most p simplices, or $\widehat{G}$ is a subset of $G + \partial\tau$ for some $(p+1)$-coface τ of α.

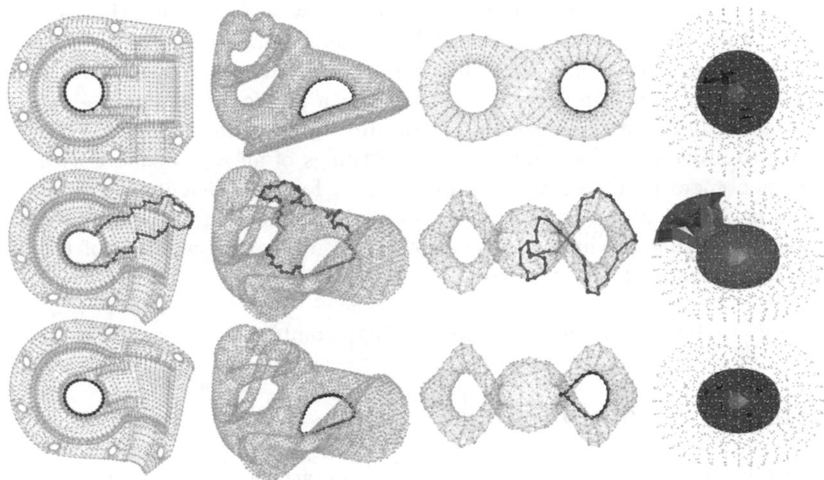

Fig. 1. Tracking a generator in a Rips complex as the points move: the original generator (top row), the generator after several updates if tracking is not executed (middle row), and if tracking is enabled (bottom row). The rightmost column shows tracking of a 2-cycle.

References

1. Carlsson, G., de Silva, V., Morozov, D.: Zigzag persistent homology and real-valued functions. In: Proc. 25th Annu. Sympos. Comput. Geom., pp. 247–256 (2009)
2. Cohen-Steiner, D., Edelsbrunner, H., Harer, J.: Stability of persistence diagrams. Discr. & Comput. Geom. 37, 103–120 (2007)
3. Cohen-Steiner, D., Edelsbrunner, H., Morozov, D.: Vines and vineyards by updating persistence in linear time. In: Proc. 22nd Annu. Sympos. Comput. Geom., pp. 119–134 (2006)
4. Dey, T.K., Sun, J., Wang, Y.: Approximating loops in a shortest homology basis from point data. In: 26th Annu. Sympos. Comput. Geom. (to appear, 2010)
5. Edelsbrunner, H., Letscher, D., Zomorodian, A.: Topological persistence and simplification. Discr. & Comput. Geom. 28, 511–533 (2002)

Auspicious Tatami Mat Arrangements

Alejandro Erickson[1], Frank Ruskey[1], Mark Schurch[2], and Jennifer Woodcock[1]

[1] Department of Computer Science, University of Victoria, Canada
[2] Department of Mathematics and Statistics, University of Victoria, Canada
{ate,ruskey,mschurch,jwoodcoc}@uvic.ca

Abstract. We introduce tatami tilings, and present some of the many interesting questions that arise when studying them. Roughly speaking, we are considering tilings of rectilinear regions with 1×2 dimer tiles and 1×1 monomer tiles, with the constraint that no four corners of the tiles meet. Typical problems are to minimize the number of monomers in a tiling, or to count the number of tilings in a particular shape. We determine the underlying structure of tatami tilings of rectangles and use this to prove that the number of tatami tilings of an $n \times n$ square with n monomers is $n2^{n-1}$. We also prove that, for fixed-height, the number of tatami tilings of a rectangle is a rational function and outline an algorithm that produces the coefficients of the two polynomials of the numerator and the denominator.

Keywords: dimer, monomer, tatami, tilings, combinatorics, enumeration.

1 What Is a Tatami Tiling?

Traditionally, a tatami mat is made from a rice straw core, with a covering of woven soft rush straw. Originally intended for nobility in Japan, they are now available in mass-market stores. The typical tatami mat occurs in a 1×2 aspect ratio and various configurations of them are used to cover floors in houses and temples. By parity considerations it may be necessary to introduce mats with a 1×1 aspect ratio in order to cover the floor of a room. Such a covering is said to be "auspicious" if no four corners of mats meet at a point. Hereafter, we only consider such auspicious arrangements, since without this constraint the problem is the classical and well-studied dimer tiling problem [1], [2]. Following Knuth [3], we will call the auspicious tatami arrangements, *tatami tilings*. The enumeration of tatami tilings that use only dimers (no monomers) was solved in [4].

Perhaps the most commonly occurring instance of tatami tilings is in paving stone layouts of driveways and sidewalks, where the most frequently used paver has a rectangular shape with a 1×2 aspect ratio. Two of the most common patterns, the "herringbone" and the "running bond", have the tatami property (see Fig. 1). Consider a driveway of the shape in Fig. 1. How can it be tatami tiled with the least possible number of monomers? The answer to this question could be interesting both because of aesthetic appeal, and because it could save work, since to make a monomer a worker typically cuts a 1×2 paver in half.

M.T. Thai and S. Sahni (Eds.): COCOON 2010, LNCS 6196, pp. 288–297, 2010.

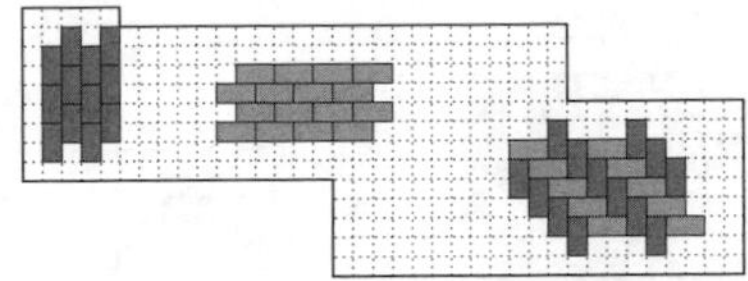

Fig. 1. The dimers shown illustrate, from left-to-right, the vertical running bond pattern, the horizontal running bond pattern, and the herringbone pattern. Ignoring those patterns, what is the least number of monomers among all tatami tilings of this region? The answer is provided at the end of the paper (Fig. 6).

Before attempting to study tatami tilings in general orthogonal regions it is crucial to understand them in rectangles, and our results are about tatami tilings of rectangles. Here is an outline of the paper. In Section 2 we determine the structure of tatami tilings in a rectangle. This structural characterization has important algorithmic implications, for example, it reduces the size of the description of a tiling from $O(rc)$ to $O(\max\{r, c\})$ and may be used to generate tilings quickly. In Section 3 we provide some counting results for tatami tilings in a rectangle: The number of tilings of an $n \times n$ square with n monomers is $n2^{n-1}$ and for a fixed number of rows r, the ordinary generating function of the number of tilings of an $r \times n$ rectangle is a rational function. In Section 4 we return to the question of tatami tiling general orthogonal regions and introduce the "magnetic water strider problem". Additional conjectures and open problems are also introduced in this section.

2 The Structure of Tatami Tilings: T-Diagrams

We show that all tatami tilings have an underlying structure which partitions the grid into blocks, where each block is filled with either the vertical or horizontal running bond. We describe this structure precisely and prove some results for rectangular regions.

Notice that in Fig. 2(a), wherever a horizontal and vertical dimer share an edge, the placement of another dimer is forced to preserve the tatami condition. These herringbone formations, called *rays*, propagate themselves to the boundary of the grid and do not intersect one another. Between the rays, there are only vertical or horizontal running bonds. Wherever a running bond meets the grid boundary, either a *smooth* or *jagged* edge is formed. Jagged describes edges in which dimers alternate with monomers and otherwise an edge is smooth.

In a rectangular grid, a ray may be forced by one of four possible *sources*. To discover what they are we consider the tiles that can cover the grid squares at the beginning of a ray. Referring to the inline diagrams, the innermost square may not be covered by a vertical dimer because this would extend the ray. If it is covered by a horizontal dimer, this is called a *hamburger*. Otherwise it is covered by a monomer in which case we consider the grid square beside

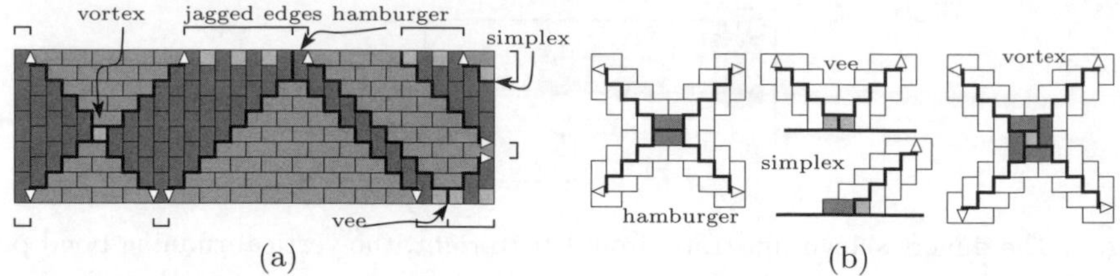

Fig. 2. (a) A tatami tiling whose **T**-diagram contains all four types of features, showing jagged edges. (b) The four types of features with their local tilings.

it . If it is covered by a monomer the source is called a *vee* , by a vertical dimer the source is called a *vortex* , and if it is covered by a horizontal dimer it is called a *simplex* . Each of these four sources forces at least one ray in the tiling and all rays begin at either a hamburger, vee, vortex or simplex. These source-ray combinations are depicted in Fig. 2(b).

The set of bold curves in each of the four source-ray drawings in Fig. 2(b), is called a *feature*. Just as rays cannot intersect, neither can features in a tatami tiling, so a *feature-diagram* refers to a set of non-intersecting features drawn in a grid. A feature-diagram admits a tatami tiling if the grid can be tiled so that the tiles propagating from different features agree, as depicted in Fig. 2(b). In this case it is called a **T**-*diagram*. A feature diagram is a **T**-diagram if and only if rays bounding the same block admit compatible bond patterns and the distance between them has the correct parity. Some incompatible configurations and distance parities are shown in Figures 3(c) and 3(d).

This characterization guarantees a verification that a given feature-diagram is a **T**-diagram, which is linear in the perimeter of the grid. Furthermore, a Tatami tiling can be validated using only the information at its boundaries (see Figures 3(b) and 3(a)).

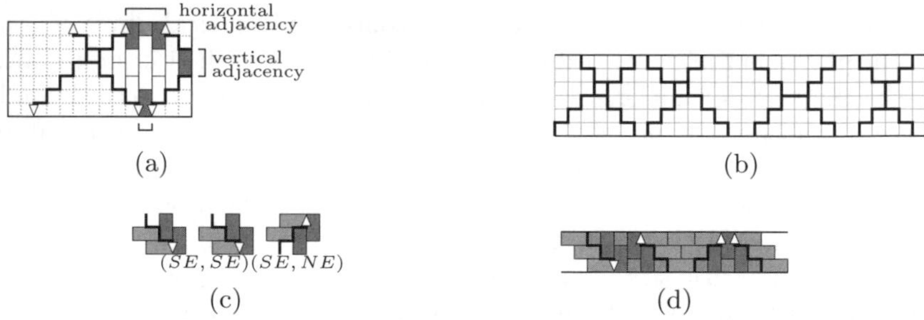

Fig. 3. (a) Fill a contiguous block of a **T**-diagram with the bond pattern by laying along the boundaries first and then filling from the outside inward. (b) Hamburgers and vortices can be differentiated by their endpoints. (c) Incompatible "adjacent" rays. (d) Distance parities between rays generalize inductively from these.

3 Counting Results

Let $T(r, c, m)$ be the number of tatami tilings of a rectangular grid with r rows, c columns, and m monomers. Also, $T(r, c)$ will denote the sum

$$T(r, c) = \sum_{m \geq 0} T(r, c, m).$$

We begin by giving necessary conditions for $T(r, c, m)$ to be non-zero.

Theorem 1. *If $T(r, c, m) > 0$, then m has the same parity as rc and $m \leq \max(r+1, c+1)$.*

Proof. Let r, c and m be such that $T(r, c, m) > 0$ and let d be the number of grid squares covered by dimers in an $r \times c$ tatami tiling so that $m = rc - d$. Since d is even, m must have the same parity as rc.

We may assume that $r \leq c$, and prove that $m \leq c + 1$. The first step is to prove that a monomer on a side boundary of any tiling can be mapped to the top or bottom. We then observe that for any pair of consecutive monomers on the top or bottom boundary, there is a horizontal dimer on the opposite boundary, and for any vortex there are horizontal dimers on each of the top and bottom boundaries. With this it is possible to show the bound on the number of monomers.

Let T be a tatami tiling of the $r \times c$ grid with a monomer on the West boundary, touching neither the South nor the North boundary. Such a monomer is on a jagged edge of a block of horizontal dimers. Define a *diagonal slice* to be a set of dimers in this block which form a stairway from this monomer to either the top or bottom of the grid, as in Fig. 4. If such a diagonal slice exists it can be flipped diagonally, which changes the orientation of its dimers and maps the monomer to the other end of the slice. Suppose neither slice exists, then they must each be impeded by a distinct ray. Such rays have this horizontal block to the left so one of them is directed SE and the other NE and they meet the right side before intersecting. Referring to Fig. 4, $\alpha + \beta + 3 = \gamma + \delta + 1 \leq r$ while $c \leq \alpha + \gamma = \beta + \delta$ and $r \leq c$. Thus $\alpha + \beta + 3 \leq \alpha + \gamma$ implying that $\beta < \gamma$. But at the same time, $\gamma + \delta + 1 \leq \beta + \delta$ implies that $\gamma < \beta$, which is a contradiction. Therefore one of the diagonal slices exists and the monomer can be moved.

Assume that no monomers are strictly on the sides of the tiling, so that all monomers are either on the top or bottom boundaries or in vortices. Let α be the number of vortices.

Consider a vee on the North boundary and the block of horizontal dimers directly South. This block must reach the South boundary, otherwise, by an argument similar to the above, we would have $c < r$. Both above and below any vortex, there are blocks of horizontal dimers which reach the top and bottom boundaries. More precisely, there is an injective function from the set of vees and vortices[2] to horizontal monomers on the North or South boundaries.

Define two binary sequences, one for the top row of grid squares and the other for the bottom row, where there is a 1 for each square covered by a monomer

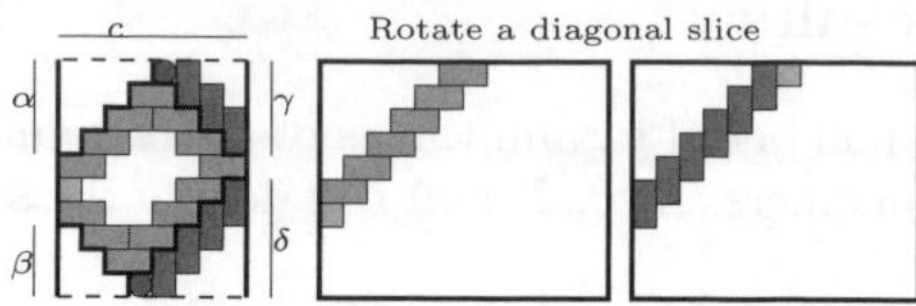

Fig. 4. If both diagonal slices are impeded then there cannot be as many columns as rows. Note that vees are implicitly impossible.

and 0s elsewhere. There are α distinct pairs of 0s in each sequence, one for each vortex, so we remove a 0 from each pair. Now we have two sequences, each of length $c - \alpha$ and there is an injective function from consecutive pairs of 1s to consecutive pairs of 0s. For each pair of 1s, remove one of the corresponding 0s and squeeze it between the pair of 1s. There are no two adjacent 1s in the new sequences, whose lengths shall be called s and t, and we have $s+t = 2c-2\alpha$. The total number of 1s is at most $\lceil \frac{s}{2} \rceil + \lceil \frac{t}{2} \rceil$, which is at most $c-\alpha+1$ implying that there are at most $c+1$ monomers in total, adding back in the α monomers in vortices. This is all that was required because we proved earlier that the number also has the same parity as rc.

The maximum is achieved in a bond pattern. □

The converse of Theorem 1 is false. For example $T(9, 13, 1) = 0$.

3.1 Square Tatami Tilings

In this Section, we count $T(n, n, n)$, the number of tilings of an $n \times n$ square with n monomers. The number of these for each n is a surprisingly nice sequence. We will frequently refer to flipping diagonal slices of a tiling, which is as described in Theorem 1, or, equivalently, to flipping a monomer in a particular direction. We give the following Lemmas without proof since they are intuitively true.

Lemma 1. *A tiling of a rectangle has the maximum number of monomers for its dimensions if and only if it has no hamburgers and no vortices. For an $n \times n$ square, the maximum number of monomers is n.*

Lemma 2. *Every tiling of an $n \times n$ square with n monomers can be generated from a single trivial tiling (up to rotational symmetry) by flipping diagonal slices. Further, every such tiling has exactly two monomers in corners and they are in adjacent corners.*

Theorem 2. *The number of $n \times n$ tilings with n monomers, $T(n, n, n)$, is $n2^{n-1}$.*

Proof. We will count the $n \times n$ tilings with n monomers up to rotational symmetry by fixing two monomers in the northern corners. Thus, let $S(n) = T(n, n, n)/4$. We will give a combinatorial proof that $S(n)$ satisfies the following recurrence:

$$S(n) = 2^{n-2} + 4S(n-2) = n2^{n-3} \text{ where } S(1) = S(2) = 1.$$

We count 2^{n-2} of the tilings directly and then find a one-to-one correspondence between the remaining tilings and one quarter of the $n-2 \times n-2$ tilings with

$n-2$ monomers. The even and odd cases have very similar proofs and as such we prove only the even case here.

The trivial case for even n, shown in Fig. 5(a) for $n = 8$, is a horizontal running bond tiling with monomers in the northern corners and $n/2$ monomers on both the West and East boundaries. We break the recurrence into two cases based on what happens to the southern-most (yellow) monomer on each of these boundaries. We will call these monomers w and e respectively.

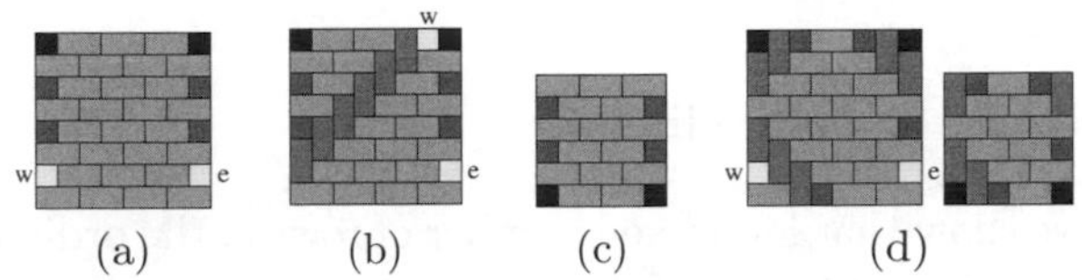

Fig. 5. (a) Trivial case for an 8×8 square with 8 monomers. (b) Flipping a southern-most (yellow) monomer northward. (c) Trivial case for a 6×6 square with 6 monomers. (d) An 8×8 tiling with its associated 6×6 tiling.

The diagonal slices corresponding to w and e can be flipped in a northerly or southerly direction, or not flipped. If w (e) is flipped in a northerly direction, no orthogonal diagonal slices can be flipped (including that corresponding to e (w)). Further, each of the remaining $n-3$ non-fixed monomers can be independently flipped in the same orientation (though not necessarily the same direction) as w (e), as shown in Fig. 5(b). This gives 2^{n-3} possibilities for each of w and e, resulting in a total of 2^{n-2} tilings.

If w or e are flipped in a southerly direction, they can be flipped independently of each other and of other non-fixed monomers. Thus for each tiling which has w and e fixed, there are four corresponding tilings. As such, we now fix w and e in the trivial tiling as well as the already-fixed northern corner (black) monomers. To count the remaining tilings, we take an inductive approach by finding a one-to-one correspondence between the remaining $n \times n$ tilings and one quarter of the $n-2 \times n-2$ tilings with $n-2$ monomers.

There are $n-4$ (red) monomers on the East and West edges of the trivial case that we have not fixed. Consider the 180 degree rotation of the trivial case for the $n-2 \times n-2$ tilings with $n-2$ monomers, which has fixed (black) monomers in the two southern corners, as shown in Fig. 5(c) for $n-2 = 6$. Associate the $n-4$ non-fixed monomers of this tiling with the $n-4$ non-fixed monomers of the $n \times n$ trivial case in a natural way: with those in the same relative position paired.

There are $S(n-2)$ ways of flipping the monomers of the $n-2 \times n-2$ trivial case, and thus $S(n-2)$ ways of flipping the corresponding monomers of the $n \times n$ trivial case. Intuitively it is easy to see that this works. To be rigorous, however, we need to show that if two diagonal slices cannot both be flipped in the larger square, the same is true for the corresponding slices in the smaller square, and vice versa. To do so, observe that the conflict between two flips depends entirely on the distance of the monomers from the horizontal centerline of the trivial tiling.

For any pair of monomers on opposite edges, let d_W and d_E be the distances from the horizontal centerline, with negative distance below the line and positive distance above. If $d_W + d_E > 0$ ($d_W + d_E < 0$) then the two monomers cannot both be flipped southward (northward). This distance from the centerline is preserved between the associated monomers in the larger and smaller squares and thus the compatibilities between flips of monomers are also preserved.

This completes the recurrence defined at the beginning of the proof and establishes the theorem for even n. □

3.2 Fixed Height Tatami Tilings

In this section we show that for a fixed number of rows r, the ordinary generating function of the number of tilings of an $r \times c$ rectangle is a rational function. We show that, for each value of r, the number of fixed-height tilings satisfies a system of linear recurrences with constant coefficients. We will derive the recurrences for small values of r and then discuss an algorithm which can be used for larger values of r.

Let $T_r(z)$ denote the ordinary generating function $T_r(z) = \sum_{c\geq 0} T(r,c)z^c$.

Theorem 3. *For $h = 1, 2, 3$ then generating functions $T_h(z)$ are*

$$T_1(z) = \frac{1+z}{1-z-z^2}, \qquad T_2(z) = \frac{1+2z^2-z^3}{1-2z-2z^3+z^4}, \qquad \textit{and}$$

$$T_3(z) = \frac{1+2z+8z^2+3z^3-6z^4-3z^5-4z^6+2z^7+z^8}{1-z-2z^2-2z^4+z^5+z^6}.$$

Proof. We only provide a proof outline for height 3. In this proof we focus on explaining the system of recurrences and leave out the technical details of determining the generating function which can be carried out using Maple. We begin by considering the possible unique left-hand side starts of a tatami tiling in a $3 \times c$ rectangle.

Case 1: The tiling begins with a dimer placed vertically in the first column. There are actually two ways to do this since the vertical tile could be placed in the first two rows or the second two rows. We will just count it as one for now and double the result from this case in the end. Let $A(c)$ be the number of $3 \times c$ tilings associated with each of these cases.

Case 2: The tiling begins with a monomer tile placed in the center (row 2). Let $B(c)$ be the number of tilings associated with this case.

Case 3: The tiling begins with a dimer tile placed horizontally in the center (row 2). Let $C(c)$ be the number of tilings associated with this case.
The following diagram illustrates the three cases.

Case 1: $A(c)$ Case 1: $A(c)$ Case 2: $B(c)$ Case 3: $C(c)$

These cases are mutually exclusive and exhaustive, thus $T(3, c) = 2A(c) + B(c) + C(c) - 3$. We subtract 3 since for each case we define the number of empty tilings to be one so this is tiling is counted 4 times. Now we determine recurrences for each of $A(c), B(c)$ and $C(c)$.

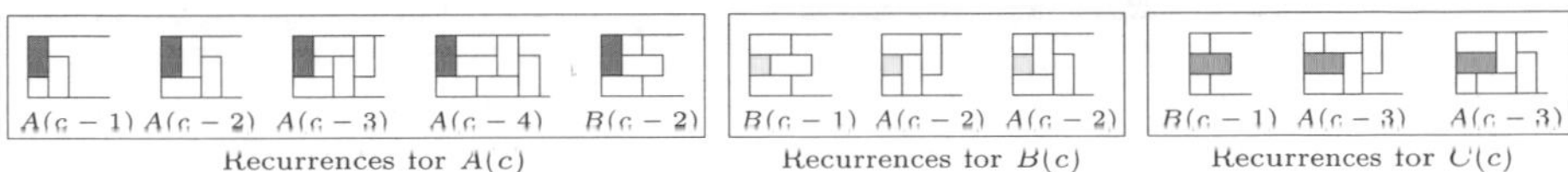

Recurrences for $A(c)$ Recurrences for $B(c)$ Recurrences for $C(c)$

By consideration of the various cases that can occur, as illustrated in the digram above, we are led to the following three recurrence relations:

$$A(c) = A(c-1) + A(c-2) + A(c-3) + A(c-4) + B(c-2)$$
$$B(c) = B(c-2) + 2A(c-2) \quad \text{and} \quad C(c) = B(c-1) + 2A(c-3).$$

Let $\mathcal{A}(z)$, $\mathcal{B}(z)$, and $\mathcal{C}(z)$ denote the generating functions for $A(c), B(c)$ and $C(c)$ respectively. Using the initial conditions $A(0) = 1, A(1) = 1, A(2) = 3, A(3) = 7$, $B(0) = 1, B(1) = 1, B(2) = 1$, $C(0) = 1, C(1) = 0, C(2) = 4$ we can solve for $\mathcal{A}(z)$, $\mathcal{B}(z)$, and $\mathcal{C}(z)$, eventually obtaining the expression for $T_3(z)$ given in the statement of the Theorem. □

For larger values of r the number of cases increases rapidly so we develop an algorithm to aid in determining the recurrences. For the $r = 3$ case discussed above we wanted to have the least number of cases possible to help simplify the calculations.

We begin by considering all options of tiling the first column with the added allowance that horizontal tiles may lie partially in the second column. We store the placement of the tiles for each of the tilings of the first column. Each of these tilings will correspond to a different case in our system of linear recurrences. In the next step we attempt to tile the second column for each way that we tiled the first, again allowing horizontal tiles into the next column. The word "attempt" is used because not all of the first column tile configurations permit a valid tatami tiling of the second column. For each second column tiling the algorithm then runs a check against the stored cases to define the recurrence.

In essence, the first column tilings represent a set of boundary configurations, say S, and for each element in S we set up a recurrence by matching each of its second column boundary configurations with an element in S. For initial conditions we determine the number of ways to complete an r by 2 rectangle for each first column tiling. The system of recurrences is then solved using Maple. Note that the algorithm produces a system of linear recurrences with constant coefficients which implies the generating functions for the system will be rational functions. Hence, for a fixed number of rows r, the generating function for the number of tatami tilings of an $r \times c$ rectangle is a rational function. The output of our algorithm for $r = 10$ has degree 65 in the numerator and degree 56 in the denominator.

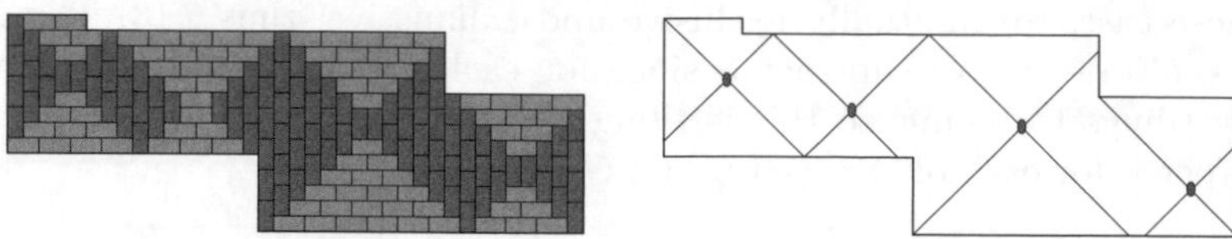

Fig. 6. On the left: The solution to the question posed in Figure 1; no monomers are required to tatami tile the region. On the right: A legal configuration of six magnetic water striders in an orthogonal "pond". Note that no further striders may be added.

4 Conjectures and Further Research

In this section we list some open problems and conjectures.

- It would be interesting to extend the structural analysis to orthogonal regions. We believe that the main structural components are the same as they were for rectangles, but there are a few subtleties to be clarified at inside corners.
- What is the computational complexity of determining the least number of monomers that can be used to tile an orthogonal region given the segments that form the boundary of the region and the unit size of each dimer/monomer?
- Interpreted as a matching problem on a grid graph G, a tatami tiling is a matching M with the property that $G - M$ contains no 4-cycles. Note that there is always such a matching (e.g., take the "running bond" layout on the infinite grid graph and then restrict it to G). However, if we insist on a perfect matching, then the problem is equivalent to our "perfect" driveway paving problem from the introduction. Thus there are a variety of natural problems about tatami tilings of an arbitrary graph.
- It turns out that the problem of minimizing the number of monomers in a tiling is related to what we call the "magnetic water strider problem". This time the orthogonal region is a pond populated by water striders. A water strider is an insect that rides atop water in ponds by using surface tension. Its 4 longest legs jut out at 45 degrees from its body. In the fancifully named magnetic water strider problem, we require the body to be aligned north-south. Furthermore its legs support it, not by resting on the water, but by extending to the boundary of the pond. Naturally, the legs of the striders are not allowed to intersect. A legal configuration of magnetic water striders in an orthogonal pond is shown on the right in Fig. 6. Here the problem is a maximization problem, namely what is the largest number of magnetic waters striders that a pond can support?
- Various games would also be of interest. Given an orthogonal region players take turns placing dimers (or dimers and monomers); each placement must satisfy the tatami constraint and the last player who can move wins. A similar game could be played with magnetic water striders. The URL, `http://www.theory.cs.uvic.ca/~cos/oku/` is reserved for such games.

– Given $r + c$ triples of numbers (h, v, m), one for each row and one for each column, is there a tatami tiling which has h horizontal dimers, v vertical dimers, and m monomers in the respective row or column? Similar questions are important in tomography [5].

Acknowledgements

Thanks to Donald Knuth for his comments on an earlier draft of this paper.

References

1. Kenyon, R., Okounkov, A.: What is ... a dimer? Notices of the American Mathematical Society 52, 342–343 (2005)
2. Stanley, R.P.: On dimer coverings of rectangles of fixed width. Discrete Applied Mathematics 12(1), 81–87 (1985)
3. Knuth, D.E.: The Art of Computer Programming, vol. 4, fascicle 1B. Addison-Wesley, Reading (2009)
4. Ruskey, F., Woodcock, J.: Counting fixed-height tatami tilings. Electronic Journal of Combinatorics 16, 20 (2009)
5. Dürr, C., Guiñez, F., Matamala, M.: Reconstructing 3-colored grids from horizontal and vertical projections is NP-hard. In: Fiat, A., Sanders, P. (eds.) ESA 2009. LNCS, vol. 5757, pp. 776–787. Springer, Heidelberg (2009)
6. Anzalone, N., Baldwin, J., Bronshtein, I., Petersen, T.K.: A reciprocity theorem for monomer-dimer coverings. In: Morvan, M., Rémila, É. (eds.) Discrete Models for Complex Systems, DMCS '03. DMTCS Proceedings, Discrete Mathematics and Theoretical Computer Science, vol. AB, pp. 179–194 (2003)
7. Pachter, L.: Combinatorial approaches and conjectures for 2-divisibility problems concerning domino tilings of polyominoes. Electronic Journal of Combinatorics 4, 2–9 (1997)

Faster Generation of Shorthand Universal Cycles for Permutations

Alexander Holroyd[1], Frank Ruskey[2,⋆], and Aaron Williams[2]

[1] Dept. of Computer Science, University of Victoria, Canada
[2] Microsoft Research, Redmond, WA, USA

Abstract. A universal cycle for the k-permutations of $\langle n \rangle = \{1, 2, ..., n\}$ is a circular string of length $(n)_k$ that contains each k-permutation exactly once as a substring. Jackson (Discrete Mathematics, 149 (1996) 123–129) proved their existence for all $k \leq n-1$. Knuth (*The Art of Computer Programming, Volume 4*, Fascicle 2, Addison-Wesley, 2005) pointed out the importance of the $k = n-1$ case, where each $(n-1)$-permutation is "shorthand" for exactly one permutation of $\langle n \rangle$. Ruskey-Williams (ACM Transactions on Algorithms, in press) answered Knuth's request for an explicit construction of a shorthand universal cycle for permutations, and gave an algorithm that creates successive symbols in worst-case $O(1)$-time. This paper provides two new algorithmic constructions that create successive blocks of n symbols in $O(1)$ amortized time within an array of length n. The constructions are based on: (a) an approach known to bell-ringers for over 300 years, and (b) the recent shift Gray code by Williams (SODA, (2009) 987-996). For (a), we show that the majority of changes between successive permutations are full rotations; asymptotically, the proportion of them is $(n-2)/n$.

1 Introduction

A *universal cycle* (or *Ucycle*) [1] is a circular string containing every object of a particular type exactly once as substring. For example, consider the circular string in Figure 1(a). Starting from 12 o'clock, and proceeding clockwise, its substrings of length three are

$$432, 321, 214, 142, \ldots, 413, 132, 324, 243. \tag{1}$$

In total there are 24 substrings, and these substrings include every 3-permutation of $\langle 4 \rangle$ exactly once. For this reason, Figure 1(a) is a *Ucycle for the* 3*-permutations of* $\langle 4 \rangle$. Let $\Pi_k(n)$ denote the set of all k-permutations of $\langle n \rangle$. Notice that $|\Pi_k(n)| = (n)_k := n(n-1)\cdots(n-k+1)$ (the *falling factorial*). In the special case $k = n$, we use $\Pi(n)$ to represent the permutations of $\langle n \rangle$. Jackson proved that Ucycles of $\Pi_k(n)$ exist whenever $k < n$ [4]. On the other hand, the reader may easily verify that Ucycles of $\Pi(n)$ do not exist when $n \geq 3$.

This section describes four additional interpretations for Ucycles of $\Pi_{n-1}(n)$, then discusses applications, relevant history, and outlines our new results.

⋆ Research supported in part by NSERC.

M.T. Thai and S. Sahni (Eds.): COCOON 2010, LNCS 6196, pp. 298–307, 2010.

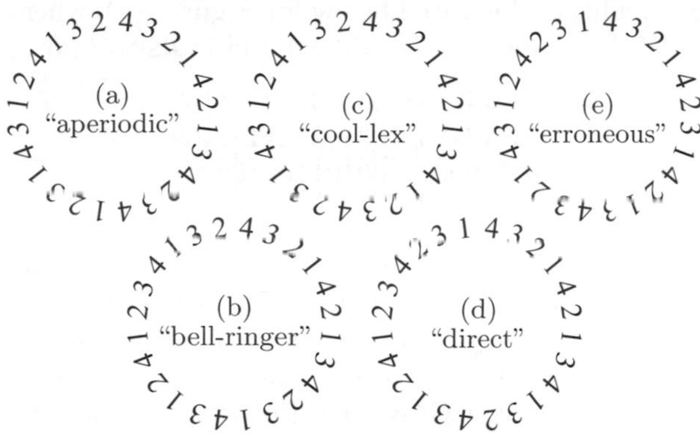

Fig. 1. (a)-(d) are Ucycles for $\Pi_3(4)$, or equivalently shorthand Ucycles for $\Pi(4)$. The symbol 4 is periodic in (b)-(e) since it is in every nth position. (e) is not a Ucycle for $\Pi_3(4)$ due to two types of errors: an erroneous substring 242, and an extra copy of 142.

1.1 Interpretations and Applications

For the first interpretation, notice that $|\Pi_{n-1}(n)| = n_{n-1} = n! = |\Pi(n)|$. Each $(n-1)$-permutation of $\langle n \rangle$ extends to a unique permutation of $\langle n \rangle$ by appending its *missing symbol* from $\langle n \rangle$. For example, the first string in (1) is 432, and it is missing the symbol 1. Thus, the substring 432 is *shorthand* for 4321. Similarly, the substrings in (1) are shorthand for the following permutations

$$4321, 3214, 2143, 1423, 4213, \ldots, 2413, 4132, 1324, 3241, 2431. \tag{2}$$

For this reason, Ucycles for $\Pi_{n-1}(n)$ can be interpreted as Ucycles for permutations called *shorthand Ucycles for* $\Pi(n)$. The substrings comprising $\Pi_{n-1}(n)$ are the Ucycle's *substrings*, and the extended strings comprising $\Pi(n)$ are the Ucycles's *permutations*. This leads to the remaining interpretations.

The second interpretation is the *binary representation*. Given a length $n-1$ substring in a shorthand Ucycle for $\Pi(n)$, the *next symbol* is the symbol that follows this substring in the shorthand Ucycle, and the *next substring* is the length $n-1$ substring that ends with this next symbol. That is, if $s_1 s_2 \cdots s_{n-1}$ is a substring in a shorthand Ucycle for $\Pi(n)$, then the next symbol is some $x \in \langle n \rangle$, and the next substring is $s_2 s_3 \cdots s_{n-1} x$. Since $s_2 s_3 \cdots s_{n-1} x$ is in $\Pi_{n-1}(n)$, then there are only two choices for x. Either x is s_1, or x is the missing symbol from $s_1 s_2 \cdots s_{n-1}$. This dichotomy gives rise to the binary representation. The ith bit in the binary representation is 1 if the substring starting at position i has its next symbol equal to its first symbol; otherwise the next symbol equals its missing symbol and the ith bit in the binary representation is 0. The binary representation can be visualized by placing two copies of the shorthand Ucycle for $\langle n \rangle$ above itself, with the second copy left-shifted $(n-1)$ positions. This comparison vertically aligns the first symbol of each length n substring with its next symbol. Accordingly, 1s are recorded precisely when the vertically aligned symbols are equal. This is illustrated below for Figure 1(a), where $\underline{4}$ denotes the first symbol in the shorthand Ucycle (above) and in its rotation (below).

$$\begin{array}{c} \underline{4}32142134234123143124132 \\ 142134234123143124132\underline{4}32 \\ \hline 001100011000101100100011 \end{array}$$

To check the binary string, notice that its first bit is 0. This is because the first substring is 432, and its first symbol 4 does not equal its next symbol 1. On the other hand, the third bit in the binary string is 1. This is because the third substring is 214, and its first symbol 2 equals its next symbol 2. (As above, the shorthand Ucycle for $\Pi(n)$ is assumed to "start" with n $(n-1)$ $\cdots$ 2.)

The third interpretation is a Gray code for permutations. If $p_1p_2\cdots p_n$ is a permutation in a shorthand Ucycle for $\Pi(n)$, then the *next permutation* begins with $p_2p_3\cdots p_{n-1}$. Therefore, the next permutation is either $p_2p_3\cdots p_{n-1}p_np_1$ or $p_2p_3\cdots p_{n-1}p_1p_n$. These cases are obtained by rotating the first symbol of $p_1p_2\cdots p_n$ into one of the last two positions. More precisely, if the ith bit in the binary representation is 0 or 1, then the ith permutation is transformed into the $(i+1)$st permutation by the rotation $\sigma_n = (1\ 2\ \cdots\ n)$ or $\sigma_{n-1} = (1\ 2\ \cdots\ n-1)$, respectively. Thus, permutations in a shorthand Ucycle for $\Pi(n)$ are in a *circular Gray code* using σ_n and σ_{n-1}. For example, the first bit in the binary string representation for (2) is 0, so σ_4 transforms the first permutation in (2), 4321, into the second permutation in (2), 3214. Conversely, every circular Gray code of $\Pi(n)$ using σ_n and σ_{n-1} provides a shorthand Ucycle for $\Pi(n)$ by appending the first symbols of each permutation.

The fourth interpretation is an equivalence to Hamiltonian cycles in the directed Cayley graph on $\Pi(n)$ with generators σ_n and σ_{n-1} (see [7]).

The binary representation provides a natural application for shorthand Ucycles of $\Pi(n)$: $n!$ permutations are encoded in $n!$ bits. The Gray code interpretation also provides applications. The rotations σ_n and σ_{n-1} can also be described respectively as *prefix shifts* of length n and $n-1$. Prefix shifts of length n and $n-1$ can be performed as basic operations within linked lists or circular arrays. In particular, a prefix shift of length n simply increments the starting position within a circular array, whereas a prefix shift of length $n-1$ increments the starting position and then performs an adjacent-transposition of the last two symbols. This is illustrated below with permutation 123456 (below left) being followed by 234561 or 234516 (below right).

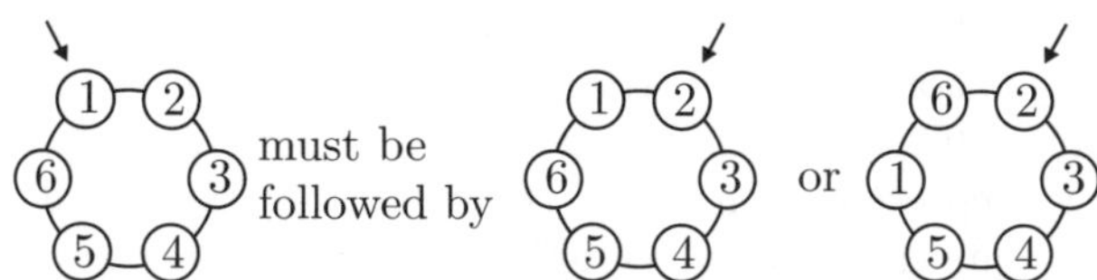

1.2 History

Ucycles for combinatorial objects were introduced by Chung, Diaconis, and Graham [1] as natural generalizations of *de Bruijn cycles* [2], which are circular strings of length 2^n that contain every binary string of length n exactly once as substring. They pointed out that Ucycles of $\Pi(n)$ do not exist, and suggested instead using order-isomorphism to represent the permutations. A string is *order-isomorphic* to a permutation of $\langle n \rangle$ if the string contains n distinct integers whose relative orders are the same as in the permutation.

Knuth suggested the use of what we call *shorthand-isomorphism*, and asked for an explicit construction [5]. An explicit construction was discovered by Ruskey-Williams [7], who also provided an algorithm that creates each successive symbol in the Ucycle (or bit in the binary representation) in worst-case O(1)-time. The construction has the property that the symbol n appears in every nth position in the Ucycle. For this reason, it is said to have a *periodic symbol.* See Figure 1(d) for the construction when $n = 4$, and notice that 4 appears in every 4th position. (Figure 1(a) illustrates that shorthand Ucycles of $\Pi(n)$ do not necessarily contains a periodic symbol, and it is also obvious that a shorthand Ucycles of $\Pi(n)$ can contain at most one periodic symbol when $n \geq 3$.)

When a shorthand Ucycle for $\Pi(n)$ has n as its periodic symbol, then its remaining symbols are divided into blocks of length $n-1$. Furthermore, each block must be a distinct permutation of $\langle n-1 \rangle$. Given this situation, the permutations of $\langle n-1 \rangle$ are called the *sub-permutations.* Figure 2 summarizes the substrings, permutations, binary representation, and sub-permutations of Figure 1(d).

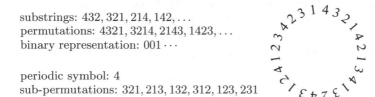

substrings: 432, 321, 214, 142, ...
permutations: 4321, 3214, 2143, 1423, ...
binary representation: 001 ⋯

periodic symbol: 4
sub-permutations: 321, 213, 132, 312, 123, 231

Fig. 2. The direct construction for $n = 4$

Given this terminology, there is a succinct description of the construction in [7]: The permutations for n are the sub-permutations for $n+1$. For example, the Ucycle for $n = 3$ is 321312, and so the permutations are 321, 213, 132, 312, 123, 231. Notice that the permutations are identical to the sub-permutations found in Figure 2. For this reason, the construction in [7] can be described as the *direct construction.* The binary representation of the direct construction also has a nice description. If $x_1x_2 \cdots x_{n!}$ is the binary string representation for n, then

$$001^{n-2}\,\overline{x}_1 001^{n-2}\,\overline{x}_2 \cdots 001^{n-2}\,\overline{x}_{n!} \tag{3}$$

is the binary string representation for $n+1$, where $\overline{x_i} = 1 - x_i$, and 1^{n-2} represents $n-2$ copies of 1. For example, the binary representation for $n = 2$ is 00, and so the binary representation for $n = 3$ and $n = 4$ are $00\,\bar{0}\,00\,\bar{0} = 001001$ and

$$001\,\bar{0}\,001\,\bar{0}\,001\,\bar{1}\,001\,\bar{0}\,001\,\bar{0}\,001\,\bar{1} = 001100110010001100110010.$$

1.3 Our New Approach

This paper expands upon [7] by providing two additional constructions. Our first construction is based on a technique from change-ringing (English-style church bell ringing). The general method (involving "half-hunts", in change-ringing terminology) is more than 300 years old (see [3,8]), but we believe that our application of it to the Ucycle problem is novel. See Figure 1(b).

The second construction uses the recently discovered shift Gray code of Williams [9], which generalizes *cool-lex order*. Aside from the intrinsic interest of having additional answers to Knuth's query, both of these constructions have two distinct advantages over the previous construction found in [7].

First, the constructions have fewer 1s in the binary representations. The bell-ringer algorithm is particularly advantageous in this regard since the vast majority of bits are 0 (Theorem 3); asymptotically, the proportion of them is $(n-2)/n$.

Second, the resulting algorithms are faster in the following scenario: Suppose an application wants a shorthand Ucycle for the permutations of $\{1, 2, \ldots, n\}$. Since $n!$ is prohibitively large, the shorthand universal is given to the application n symbols at a time within an array of length n. Our new algorithm updates this array in average-case $O(1)$-time. In other words, it provides successive blocks of n symbols in *constant amortized time*. This is an improvement over the previous algorithm [7], which would have required $\Omega(n)$-time to obtain the next n symbols.

Sections 2 and 3 describe these new constructions. In both cases the strategy is to generate a list of $\Pi(n)$ that will become the sub-permutations for a shorthand Ucycle for $\Pi(n+1)$. When proving that the result is in fact a shorthand Ucycle for $\Pi(n+1)$, we use the following definition and notation.

Definition 1. *Let $P = p_1, p_2, \ldots, p_{n!}$ be a list of permutations. By $\mathsf{R}(P)$ we denote the cyclic string $(n{+}1)p_1(n{+}1)p_2 \cdots (n{+}1)p_{n!}$. The list P is* recyclable *if $\mathsf{R}(P)$ is a shorthand universal cycle for the permutations of $\langle n{+}1\rangle$.*

Figure 1(e) illustrates that not every order of $\Pi(n)$ is recyclable. Theorem 1 proves that its two types of "errors" are the only obstacles to being recyclable.

Theorem 1. *A circular list of permutations $P = p_1, p_2, \ldots, p_{n!}$ is recyclable if and only if the following two conditions are satisfied.*

- *If α and β are successive permutations, then $\alpha_i^{-1} - \beta_i^{-1} \leq 1$ for all $i \in \langle n\rangle$.*
- *If α and β are two successive permutations and α' and β' are also two successive permutations, then whenever there is a $j \in \langle n\rangle$ such that*

$$\alpha_{j+1} \cdots \alpha_n \beta_1 \cdots \beta_{j-1} = \alpha'_{j+1} \cdots \alpha'_n \beta'_1 \cdots \beta'_{j-1},$$

then $\alpha = \alpha'$ (and hence $\beta = \beta'$).

Proof. Let $X = \mathsf{R}(P) = (n{+}1)p_1(n{+}1)p_2 \cdots (n{+}1)p_{n!}$. The first condition guarantees that any n successive symbols from X are distinct; that is, they are all n-permutations of $\langle n{+}1\rangle$. The condition says that no symbol in the successor moves left more than one position (it could move right).

The second condition will guarantee that all of the length n substrings of X are distinct. Since the length of X is obviously $(n+1)!$, this will finish the proof. Let ω be a substring of X of length n. Clearly, if ω does not contain $n+1$ then it is distinct, since in that case we must have $\omega = p_i$ for some i. If ω contains $n+1$, then $\omega = \alpha_{j+1} \cdots \alpha_n (n+1) \beta_1 \cdots \beta_{j-1}$ for some value of $j \in \langle n \rangle$ and successive permutations α and β. If ω is not distinct, then there would be some other two permutations α' and β' such that $\omega' = \alpha'_{j+1} \cdots \alpha'_n (n+1) \beta'_1 \cdots \beta'_{j-1}$. □

2 The Bell-Ringer Construction and 7-Order

In order to describe the bell-ringing-inspired Ucycle, we introduce an ordering of the permutations of $\langle n \rangle$ that we call "seven order" and denote 7-order. It is a recursive method in which every permutation π of $\langle n-1 \rangle$ is expanded into n permutations of $\langle n \rangle$ by inserting n in every possible way into π. These insertions are done in a peculiar order that is reminiscent the way the number seven is normally written, and is what inspires us to call it 7-order. That is, the first permutation is $n\pi$, the second is πn, and the remaining permutations are obtained by moving the n one position to the left until it is in the second position. We use the notation 7_n to denoted the 7-order of $\langle n \rangle$. Thus $7_2 = 21, 12$ and $7_3 = 321, 213, 231, 312, 123, 132$ and the first 4 permutations of 7_4 are $4321, 3214, 3241, 3421$.

Theorem 2. *The list 7_n is recyclable.*

Proof. We use Theorem 1. The first condition is clearly met by 7_n. To verify the second condition, our strategy is to show that every symbol in α is determined by $\beta_1 \cdots \beta_{j-1}$ and $\alpha_{j+1} \cdots \alpha_n$. First note that either n is in $\alpha_{j+1} \cdots \alpha_n$ or it is in $\beta_1 \cdots \beta_{j-1}$. If n is in $\alpha_{j+1} \cdots \alpha_n$, then $\alpha = \beta_1 \cdots \beta_{j-1} x \alpha_{j+1} \cdots \alpha_n$, where $x = \langle n \rangle \setminus \{\beta_1, \ldots, \beta_{j-1}, \alpha_{j+1}, \ldots, \alpha_n\}$. If $n = \beta_k$ is in $\beta_1 \cdots \beta_{j-1}$, but $n \neq \beta_1$, then $\alpha = \beta_1 \cdots \beta_{k-1} \beta_{k+1} n \beta_{k+2} \cdots \beta_{j-1} x \alpha_{j+1} \cdots \alpha_n$. If $\beta_1 = n$ then the result follows by induction. □

We define the *bell-ringer order* to be the order of permutations obtained by recycling 7_n.

2.1 The Binary Interpretation of Bell-Ringer Order

In this subsection we infer the recursive 0/1 structure of $\mathsf{R}(7_n)$. Since the ns are n apart, every nth and $(n+1)$st bit is 0. (This is because $n s_1 s_2 \cdots s_{n-1} \in \Pi(n)$ and $s_1 s_2 \cdots s_{n-1} n \in \Pi(n)$ when $s_1 s_2 \cdots s_{n-1}$ is a sub-permutation.) We thus may think of the 0/1 string as having the form

$$\Upsilon(n) \;=\; 00\; B(n)_1\; 00\; B(n)_2\; 00\; \cdots\; 00\; B(n)_{(n-1)!},$$

where $|B(n)_j| = n-2$. The initial 0 represents the σ_n that takes $n(n-1)\cdots 21$ to $(n-1)\cdots 21n$. We will now describe how to get $\Upsilon(n+1)$ from $\Upsilon(n)$.

We use several strings which are defined below. We omit n from the first two notations since it will be clear from context. Let

$$A = 1^{n-2}, \quad Z_k = 0^{n-k-2}10^{k-1}, \text{ and } \quad X_{n-1} = A00Z_100Z_200\cdots00Z_{n-3}.$$

Note that $|A| = |Z_k| = n-2$ and $|X_n| = (n-1)(n-2)$. Given $\Upsilon(n)$, the 0/1 string for $n+1$ is

$$\Upsilon(n+1) \;=\; 00\ X_n\ 00\ 1B(n)_1\ 00\ X_n\ 00\ 1B(n)_2\ 00\ \cdots\ 00\ X_n\ 00\ 1B(n)_{(n-1)!}.$$

Here are the bitstrings for $\Upsilon(4)$ and $\Upsilon(5)$, where the initial case is $\Upsilon(3) =$ 00 1 00 1 $=$ 00 $B(3)_1$ 00 $B(3)_2$. First, $\Upsilon(4) =$ 00 11 00 01 00 1$\underline{1}$ 00 11 00 01 00 1$\underline{1}$, which can also be written as $\Upsilon(4) =$ 00 A 00 Z_1 00 $1B(3)_1$ 00 A 00 Z_1 00 $1B(3)_2 =$ 00 X_2 $1B(3)_1$ 00 $X_2$00 $1B(3)_2$. And $\Upsilon(5)$ is

$$
\begin{aligned}
&00\ 111\ 00\ 001\ 00\ 010\ 00\ 1\underline{11}\ \ (= 00\ A\ 00\ Z_1\ 00\ Z_2\ 00\ 1B(4)_1 = 00X_3001B(4)_1)\\
&00\ 111\ 00\ 001\ 00\ 010\ 00\ 1\underline{01}\ \ (= 00\ A\ 00\ Z_1\ 00\ Z_2\ 00\ 1B(4)_2 = 00X_3001B(4)_2)\\
&00\ 111\ 00\ 001\ 00\ 010\ 00\ 1\underline{11}\ \ (= 00\ A\ 00\ Z_1\ 00\ Z_2\ 00\ 1B(4)_3 = 00X_3001B(4)_3)\\
&00\ 111\ 00\ 001\ 00\ 010\ 00\ 1\underline{11}\ \ (= 00\ A\ 00\ Z_1\ 00\ Z_2\ 00\ 1B(4)_4 = 00X_3001B(4)_4)\\
&00\ 111\ 00\ 001\ 00\ 010\ 00\ 1\underline{01}\ \ (= 00\ A\ 00\ Z_1\ 00\ Z_2\ 00\ 1B(4)_5 = 00X_3001B(4)_5)\\
&00\ 111\ 00\ 001\ 00\ 010\ 00\ 1\underline{11}\ \ (=00\ A\ 00\ Z_1\ 00\ Z_2\ 00\ 1B(4)_6 = 00X_3001B(4)_6).
\end{aligned}
$$

Theorem 3. *The number of 1s in $\Upsilon(n)$ is $2((n-1)!-1)$.*

Proof. If c_n is the number of 1s then our recursive construction implies that $c_{n+1} = c_n + 2(n-2)(n-2)!$ with $c_3 = 2$. The solution of this recurrence relation is $2((n-1)!-1)$. □

Asymptotically, this means that the relative frequency of σ_{n-1} transitions is about $2/n$ and the relative frequency of σ_n transitions is asymptotically $(n-2)/n$. This answers an open question listed at the end of [7]; it asks whether there is a Ucycle whose binary representation uses more 1s than 0s. The bell-ringer listing clearly does so.

2.2 Iterative Rules

Now let us describe a rule for transforming one permutation of $\langle n-1\rangle$ in 7-order into the next. This is useful for efficiently generating 7-order and the bell-ringer shorthand Ucycle, as well as proving that it is indeed a Ucycle.

Let $\boldsymbol{s} = s_1s_2\cdots s_{n-1} \in \Pi(n-1)$. Let h be the index such that $s_h = n-1$. If $h = 2$, then let i be the maximum value such that

$$s_1\ s_2\ \cdots\ s_i = s_1\ (n{-}1)\ (n{-}2)\ \cdots\ (n{-}i{+}1)$$

and let j be chosen to maximize the value of s_j such that $i+1 \le j \le n-1$. (Notice that j is undefined when $i = n-1$, and otherwise $j > i+1$.) The next permutation in 7-order denoted $7(\boldsymbol{s})$ and is obtained from the following formula

$$\begin{cases} s_1 s_2 \cdots s_{h-2} s_h s_{h-1} s_{h+1} s_{h+2} \cdots s_{n-1} & \text{if } h > 2 \quad \text{(4a)} \\ s_2 s_3 \cdots s_i s_1 s_{i+2} s_{i+2} \cdots s_{j-2} s_j s_{j-1} s_{j+1} s_{j+2} \cdots s_{n-1} & \text{if } h{=}2 \;\&\; s_1{<}n{-}i \quad \text{(4b)} \\ s_2 s_3 \cdots s_{n-1} s_1 & \text{otherwise.} \quad \text{(4c)} \end{cases}$$

To see why (4) generates 7-order, one can simply compare each of the cases to the recursive definition of seven order:

- (4c) is performed when the largest symbol is in the first position of $\boldsymbol{s}$, and the result is the first symbol of $\boldsymbol{s}$ is moved into the last position;
- (4a) is performed when the largest symbol is in neither of the first two positions of $\boldsymbol{s}$, and the result is the largest symbol moves one position left;
- (4b) is performed when the largest symbol is in the second position, and the result is that symbols $s_2 s_3 \cdots s_i$ move one position to the left by recursion, and then the jth symbol moves one position to the left.

Algorithmically, successive iterations of (4) can be generated by a constant amortized time (CAT) algorithm when $\boldsymbol{s}$ is stored in array. That is, the 7-order for $\Pi(n-1)$ can be generated in $O((n-1)!)$-time. To do this, one needs to simply keep track of the position of the largest symbol in a variable h. More precisely, given the value of h, (4c) is performed $1 \cdot (n-2)!$ times, and involves $O(n-1)$ work each time. Similarly, (4a) is performed $(n-2) \cdot (n-2)!$ times, and involves $O(1)$ work. Finally, (4b) is performed $1 \cdot (n-2)!$ times, and involves $O(n-1)$ work each time. Therefore, the overall implementation $O((n-1)!)$-time since

$$n \cdot (n-2)! + n \cdot (n-2)! + (n-2) \cdot (n-2)! = 3n - 2 \cdot (n-2)! \leq 4 \cdot (n-1)!.$$

This proves the following theorem.

Theorem 4. *7-order for $\Pi(n-1)$ can be generated in $O((n-1)!)$-time when the permutations are stored in an array. Using the same algorithm, the bell-ringer shorthand Ucycle for $\Pi(n)$ can be generated in $O((n-1)!)$-time when successive blocks of length n are stored in an array (and the first element of the array is fixed at value n).*

The iterative rule in (4) also allows us to state a simple iterative rule for directly generating the permutations from the bell-ringer Ucycle.

Theorem 5. *Let $\boldsymbol{s} = s_1 s_2 \cdots s_n \in \Pi(n)$ be a permutation in the bell-ringer shorthand Ucycle for $\Pi(n)$, where m is the maximum value of s_1 and s_n, and k is the maximum value such that n $(n-1)$ $\cdots$ k appears in the permutation as a circular substring. If $k-1 \leq m \leq n-1$, then the next permutation is $s_2 s_3 \cdots s_{n-1} s_1 s_n$. Otherwise, (if $m = n$ or $k-1 < m$) the next permutation is $s_2 s_3 \cdots s_n s_1$.*

Proof. Omitted. □

3 Cool-Lex Construction

This section discusses the *cool-lex order* for $\Pi(n)$ [9]. Given a permutation, a *prefix left-shift* moves the symbol at a given position into the first position of the resulting permutation. This operation is denoted by $\lhd$ as follows

$$\lhd(s_1 s_2 \cdots s_n,\ j) = s_j s_1 s_2 \cdots s_{j-1} s_{j+1} s_{j+2} \cdots s_n.$$

A *prefix right-shift* is the inverse operation and involves moving the symbol in the first position into a given position. This operation is denoted by $\rhd$ as follows

$$\rhd(s_1 s_2 \cdots s_n,\ j) = s_2 s_3 \cdots s_{j-1} s_1 s_{j+1} s_{j+2} \cdots s_n.$$

Cool-lex order is generated by a single operation that is repeated over and over again. The operation was originally stated in terms of prefix left-shifts, but for the purposes of this document it will be useful to restate the definition in terms of prefix right-shifts. Both the *cool left-shift* and *cool right-shift* involve the notion of a *non-increasing prefix* which is defined below.

Definition 2 (⅂). *If $\boldsymbol{s} = s_1 s_2 \ldots s_n$ is a string, then the* non-increasing prefix *of $\boldsymbol{s}$ is*

$$⅂(\boldsymbol{s}) = s_1 s_2 \cdots s_j$$

where j is the maximum value such that $s_{j-1} \geq s_j$ for all $2 \leq j \leq ⅂(\boldsymbol{s})$.

For example, ⅂(55432413) = 55432 and ⅂(33415312) = 33.

Given a list of strings, the *reflected* list begins with the last string in the original list and ends with the first string in the original list. In particular, the reflected cool-lex order for $\Pi(n)$ is generated by repeated applications of the cool right-shift which is defined below.

Definition 3. *Let $\boldsymbol{s} = s_1 \cdots s_n$ and $\boldsymbol{s}' = s_2 s_3 \cdots s_n$ and $k' = |⅂(\boldsymbol{s}')|$. Then,*

$$\overrightarrow{\mathsf{cool}}(\boldsymbol{s}) = \begin{cases} \rhd(\boldsymbol{s},\ k'+1) & \text{if } k' \leq n-2 \text{ and } s_1 > s_{k'+1} \quad (5a) \\ \rhd(\boldsymbol{s},\ k'+2) & \text{if } k' \leq n-2 \text{ and } s_1 < s_{k'+1} \quad (5b) \\ \rhd(\boldsymbol{s},\ n) & \text{otherwise (if } k' \geq n-1). \quad (5c) \end{cases}$$

For example, $\overrightarrow{\mathsf{cool}}$ circularly generates the following list of $\Pi(3)$. These lists for $\Pi(n)$ are denoted by $\overrightarrow{\mathcal{C}}(n)$, as in $\overrightarrow{\mathcal{C}}(3) = 321\ 213\ 123\ 231\ 312\ 132$.

We now prove that $\overrightarrow{\mathcal{C}}(n)$ becomes a universal cycle for $\Pi_n(n+1)$ after prefixing $n+1$ as a periodic symbol. For example, when $n = 3$

$$\overrightarrow{\mathcal{C}}(3) = \ 321\ \ 213\ \ 123\ \ 231\ \ 312\ \ 132$$
$$\mathsf{R}(\overrightarrow{\mathcal{C}}(3)) = 432142134123423143124132$$

and $\mathsf{R}(\overrightarrow{\mathcal{C}}(3))$ is a universal cycle for the 3-permutations of $\langle 4 \rangle$. In general, if $\mathcal{L}$ is a list of $\Pi(n)$ then $\mathsf{R}(\mathcal{L})$ denotes the result of prefixing $n+1$ to every permutation in $\mathcal{L}$ and then concatenating the resulting strings together. Using this notation, the main result can be stated as follows.

Theorem 6. *The string* $\mathsf{R}(\overrightarrow{\mathcal{C}}(n))$ *is a universal cycle for* $\Pi_n(n+1)$.

Proof. This result follows from a judicious application of Lemma 1, below, that is stated informally as follows: For every value of j satisfying $0 \le j \le n-1$, there are consecutive strings in $\overrightarrow{\mathcal{C}}(n)$ that contain the last j symbols of $\boldsymbol{p}$ as a prefix of one string, and the first $(n-1)-j$ symbols as a suffix of the previous string. The rest of the proof is omitted due to space limitations. □

Lemma 1. *If* $\boldsymbol{p} \in \Pi_{n-1}(n)$ *and* j *satisfies* $0 \le j \le n-1$, *then there exists* $\boldsymbol{s} = s_1 s_2 \cdots s_n \in \Pi(n)$ *followed by* $\boldsymbol{t} = t_1 t_2 \cdots t_n \in \Pi(n)$ *in* $\overrightarrow{\mathcal{C}}(\mathcal{L})$ *such that*

$$s_{j+2} s_{j+3} \cdots s_n t_1 t_2 \cdots t_j = \boldsymbol{p}. \tag{6}$$

In other words, there exist consecutive strings in $\overrightarrow{\mathcal{C}}(n)$ *whose concatenation has* $\boldsymbol{p}$ *as a substring. Moreover, the substring uses* j *symbols from the second string.*

Proof. The proof of this technical lemma is omitted here. □

References

1. Chung, F., Diaconis, P., Graham, R.: Universal cycles for combinatorial structures. Discrete Mathematics 110, 43–59 (1992)
2. de Bruijn, N.G.: A Combinatorial Problem. Koninkl. Nederl. Acad. Wetensch. Proc. Ser A 49, 758–764 (1946)
3. Duckworth, R., Stedman, F.: Tintinnalogia (1668)
4. Jackson, B.: Universal cycles of k-subsets and k-permutations. Discrete Mathematics 149, 123–129 (1996)
5. Knuth, D.E.: The Art of Computer Programming. Generating All Tuples and Permutations, Fascicle 2, vol. 4. Addison-Wesley, Reading (2005)
6. Ruskey, F., Williams, A.: The coolest way to generate combinations. Discrete Mathematics 309, 5305–5320 (2009)
7. Ruskey, F., Williams, A.: An explicit universal cycle for the $(n-1)$-permutations of an n-set. ACM Transactions on Algorithms (in press)
8. White, A.T.: Fabian Stedman: The First Group Theorist? The American Mathematical Monthly 103, 771–778 (1996)
9. Williams, A.: Loopless Generation of Multiset Permutations Using a Constant Number of Variables by Prefix Shifts. In: Proceedings of the Twentieth Annual ACM-SIAM Symposium on Discrete Algorithms, SODA 2009, New York, NY, USA, January 4-6, pp. 987–996.

The Complexity of Word Circuits

Xue Chen[1], Guangda Hu[1], and Xiaoming Sun[2,⋆]

[1] Department of Computer Science and Technology, Tsinghua University
[2] Institute for Theoretical Computer Science and Center for Advanced Study, Tsinghua University

Abstract. A *word circuit* [1] is a directed acyclic graph in which each edge holds a w-bit word (i.e. some $x \in \{0,1\}^w$) and each node is a gate computing some binary function $g : \{0,1\}^w \times \{0,1\}^w \to \{0,1\}^w$. The following problem was studied in [1]: How many binary gates are needed to compute a ternary function $f : (\{0,1\}^w)^3 \to \{0,1\}^w$. They proved that $(2+o(1))2^w$ binary gates are enough for any ternary function, and there exists a ternary function which requires word circuits of size $(1-o(1))2^w$. One of the open problems in [1] is to get these bounds tight within a low order term. In this paper we solved this problem by constructing new word circuits for ternary functions of size $(1+o(1))2^w$. We investigate the problem in a general setting: How many k-input word gates are needed for computing an n-input word function $f : (\{0,1\}^w)^n \to \{0,1\}^w$ (here $n \geq k$). We show that for any fixed n, $(1-o(1))2^{(n-k)w}$ basic gates are necessary and $(1+o(1))2^{(n-k)w}$ gates are sufficient (assume w is sufficiently large). Since word circuit is a natural generalization of boolean circuit, we also consider the case when w is a constant and the number of inputs n is sufficiently large. We show that $(1 \pm o(1))\frac{2^{wn}}{(k-1)n}$ basic gates are necessary and sufficient in this case.

1 Introduction

Word circuit, defined in [1], is an acyclic graph where each edge holds a w-bit word and each node computes some binary word function $g : \{0,1\}^w \times \{0,1\}^w \to \{0,1\}^w$. In this paper we extends this definition so that each node computes a k-input word function $g : (\{0,1\}^w)^k \to \{0,1\}^w$, where k is a parameter. We call this k-input word function a *basic gate*, or simply gate if no confusion. For a word circuit C, the size of it is defined as the number of basic gates used in C. It is a natural question to ask: *How many basic gates are needed for computing an n-input word function $f : (\{0,1\}^w)^n \to \{0,1\}^w$?* Here the number of input $n \geq k$. We use symbols $x_1, x_2, \ldots, x_n$ to denote the input of the word function (or the word circuit), and symbols $b_1, b_2, \ldots, b_{nw}$ to denote the input bits, i.e. $x_i = b_{(i-1)w+1} b_{(i-1)w+2} \cdots b_{iw}$ ($i = 1, \ldots, n$).

⋆ Corresponding author. xiaomings@tsinghua.edu.cn. Supported in part by the National Natural Science Foundation of China Grant 60553001, 60603005, 60621062, the National Basic Research Program of China Grant 2007CB807900,2007CB807901 and Tsinghua University Initiative Scientific Research Program 2009THZ02120.

M.T. Thai and S. Sahni (Eds.): COCOON 2010, LNCS 6196, pp. 308–317, 2010.

This problem was first considered by Hansen et al. [1] for binary word gates ($k = 2$) and ternary word functions ($n = 3$). They proved that $(1-o(1))2^w$ gates are necessary and $(2 + o(1))2^w$ gates are sufficient. One of the open problem remains is to get the bound tight within a lower order term.

We answered this question in this work. We give a tight bound for the generalized problem: For every fixed n, there exist an n-input word function which needs $(1 - o(1))2^{(n-k)w}$ basic gates to compute, and $(1 + o(1))2^{(n-k)w}$ basic gates are always sufficient to build a circuit computing any n-input word function. Here w is sufficient large (the $o(\cdot)$ notation approximates as growing of w). These are proved in section 2. The lower bound is proved by Shannon's counting arguments [2]. The upper bound is much more sophisticated. In [1] they build a formula for ternary word functions. Here we balanced the role of the three input words, and build a "grid" circuit. More efforts are needed for the construction of the general case.

We also consider the problem on another aspect. That is, the word length w is fixed, while the desired number of inputs n is sufficient large. Boolean circuits can be considered as the special case when $w = 1$ and $k = 2$. A well known result for this case is that $(1 \pm o(1))\frac{2^n}{n}$ gates are necessary and sufficient (see [2–4]). As a generalization we show that for any fixed w and k, $(1 \pm o(1))\frac{2^{wn}}{(k-1)n}$ basic gates are necessary and sufficient. This result is given in section 3. The lower bound is proved by a similar counting argument as above. The upper bound is a modification of Lupanov's construction for boolean circuits.

2 Fixed Number of Input, Large Word Length

In this section, we assume that k, n are constants and w is sufficiently large.

2.1 Lower Bound

First, we have the following counting lemma:

Lemma 1. *A word circuit of size s can compute at most $(s+n-1)^{sk}2^{sw2^{kw}}/s!$ different n-input functions $f : (\{0,1\}^w)^n \to \{0,1\}^w$.*

Proof. For each gate there are at most $(s + n - 1)^k$ ways to choose its k inputs, namely the output of the other $(s-1)$ gates and $x_1, x_2, \ldots, x_n$. There are $(2^w)^{2^{kw}}$ different types of a gate. Finally each circuit is counted $s!$ times for the gates can be numbered in $s!$ different ways. Thus the lemma is proved. □

By this lemma, the lower bound is shown as following:

Theorem 1. *For any constant $n > k \geq 2$, there exists a word function f : $(\{0,1\}^w)^n \to \{0,1\}^w$ so that no word circuit computing f consists of less than $(1 - o(1))2^{(n-k)w}$ basic gates.*

Proof. The number of different functions $f : (\{0,1\}^w)^n \to \{0,1\}^w$ is $2^{w2^{nw}}$. By Lemma 1 we have

$$(s+n-1)^{sk} 2^{sw2^{kw}} \geq 2^{w2^{nw}} s! > 2^{w2^{nw}}.$$

Take the logarithm of both sides, we have

$$sk \log_2(s+n-1) + sw2^{kw} > w2^{nw}.$$

An easy calculation gives $s \geq (1 - O(\frac{w}{2^{kw}}))2^{(n-k)w}$. □

2.2 Upper Bound for $k = 2, n = 3$

Here we give a matched upper bound for the special case $k = 2, n = 3$, which was first considered in [1].

Theorem 2. *Every ternary function $f : (\{0,1\}^w)^3 \to \{0,1\}^w$ can be computed by a word circuit that consists of $(1 + o(1))2^w$ binary gates.*

Proof. We use x_1, x_2, x_3 to denote the 3 input words as described previously. Let $x_{2,1}$ be the first $\lfloor w/2 \rfloor$ bits of x_2, and $x_{2,2}$ be the last $\lceil w/2 \rceil$ bits.

Partition the $2^{w+\lfloor w/2 \rfloor}$ possibilities of input $(x_1, x_{2,1})$ into $r_1 = \lceil \frac{2^{w+\lfloor w/2 \rfloor}}{2^w - 1} \rceil = 2^{\lfloor w/2 \rfloor} + 1$ sets $Q_{1,1}, Q_{1,2}, \ldots, Q_{1,r_1}$, so that $|Q_{1,j}| \leq 2^w - 1 (1 \leq j \leq r_1)$. (The way to divide is arbitrary.) For each $Q_{1,j}$, we build a gate $h_{1,j}$ taking x_1, x_2 as inputs.[1] If $(x_1, x_{2,1}) \in Q_{1,j}$, $h_{1,j}$ outputs the index of $(x_1, x_{2,1})$ in $Q_{1,j}$ (a unique w-bit integer from 1 to $|Q_{1,j}|$), otherwise the output is 0^w.

Similarly, we partition the $2^{\lceil w/2 \rceil + w}$ possibilities of input $(x_{2,2}, x_3)$ into $r_2 = \lceil \frac{2^{\lceil w/2 \rceil + w}}{2^w - 1} \rceil = 2^{\lceil w/2 \rceil} + 1$ sets $Q_{2,1}, Q_{2,2}, \ldots, Q_{2,r_2}$, whose size is at most $2^w - 1$. We build a gate $h_{2,j}$ corresponding to $Q_{2,j}$ $(1 \leq j \leq r_2)$, namely the input of $h_{2,j}$ is x_2, x_3, and the output is 0^w if $(x_{2,2}, x_3) \notin Q_{2,j}$, or the index of $(x_{2,2}, x_3)$ in $Q_{2,j}$ otherwise.

For any input x_1, x_2, x_3, only one of outputs of gates $h_{1,1}, h_{1,2}, \ldots, h_{1,r_1}$ is not 0^w. This is because these gates correspond to sets $Q_{1,1}, Q_{1,2}, \cdots, Q_{1,r_1}$, which partition all the possibilities of $(x_1, x_{2,1})$. Similarly, only one of outputs of $h_{2,1}, h_{2,2}, \cdots, h_{2,r_2}$ is not 0^w. If we know two gates h_{1,j_0} and h_{2,l_0} do not equal to 0^w for some $j_0 \in \{1 \ldots, r_1\}$, $l_0 \in \{1 \ldots, r_2\}$, we can identify the values of x_1, x_2, x_3 (and thus $f(x_1, x_2, x_3)$).

For each $h_{1,j}$ $(1 \leq j \leq r_1)$, we construct r_2 gates $g_{j,1}, g_{j,2}, \ldots, g_{j,r_2}$ like a chain (we call these gates the *j-th chain*, see Figure 1) in the following way:

$$g_{j,1}(h_{1,j}, h_{2,1}) = \begin{cases} h_{1,j} & \text{if } h_{2,1} = 0^w; \\ 0^w & \text{if } h_{1,j} = 0^w; \\ f(x_1, x_2, x_3) & \text{otherwise, } x_1, x_2, x_3 \text{ are known.} \end{cases}$$

[1] Here we use a k-input gate as a 2-input one. In later proof we will use a k-input gate as a 2-input or 1-input one without claim.

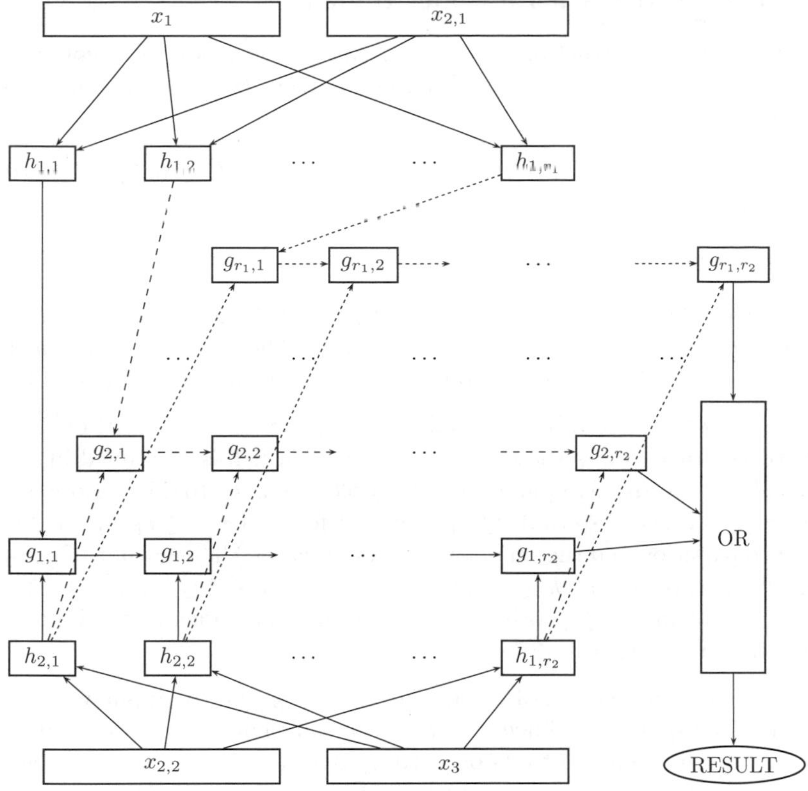

Fig. 1. Full construction for $k = 2, n = 3$

For $2 \leq l \leq r_2$,

$$g_{j,l}(g_{j,l-1}, h_{2,l}) = \begin{cases} g_{j,l-1} & \text{if } h_{2,l} = 0^w; \\ 0^w & \text{if } g_{j,l-1} = 0^w; \\ f(x_1, x_2, x_3) & \text{otherwise, } x_1, x_2, x_3 \text{ are known.} \end{cases}$$

It is easy to see that if $h_{1,j} = 0^w$, the end of the j-th chain g_{j,r_2} must output 0^w. Let's consider the case $h_{1,j} \neq 0^w$. As claimed above, there is only one of $h_{2,1}, h_{2,2}, \ldots, h_{2,r_2}$ nonzero, say $h_{2,l}$. By the construction, all gates in the j-th chain other than $g_{j,l}$ outputs its first input word. Thus the first input of $g_{j,l}$ must be $h_{1,j}$, which is nonzero. Therefore x_1, x_2, x_3 can be identified at $g_{j,l}$ and the gate is well defined. The output of gates $g_{j,l}, g_{j,l+1}, \ldots, g_{j,r_2}$ are all $f(x_1, x_2, x_3)$.

At last, we compute the disjunction of all gates g_{j,r_2} $(1 \leq j \leq r_1)$. This takes $(r_1 - 1)$ 2-input OR gates. The final output is the desired function value. (See Figure 1 for the full construction) The total number of gates is $r_1 + r_2 + r_1 \times r_2 + r_1 - 1 = 2 \times (2^{\lfloor w/2 \rfloor} + 1) + 2^{\lceil w/2 \rceil} + (2^{\lceil w/2 \rceil} + 1) \times (2^{\lfloor w/2 \rfloor} + 1) = 2^w + O(2^{w/2}) = (1 + o(1))2^w$. □

2.3 Proofs for the General Upper Bound

Now we give the construction for general cases. The following theorem shows an upper bound for any fixed k and n that matches the lower bound stated in Theorem 1.

Theorem 3. *For any fixed $n > k \geq 2$, every function $f : (\{0,1\}^w)^n \to \{0,1\}^w$ can be computed by a word circuit that consists of $(1+o(1))2^{(n-k)w}$ basic gates.*

Proof. The construction contains two steps:

1. We divide the whole input $(b_1, b_2, \ldots, b_{nw})$ into k parts $S_1, S_2, \ldots, S_k$, each contains $\lfloor nw/k \rfloor$ or $\lceil nw/k \rceil$ consecutive bits. The first step is for each part S_i $(1 \leq i \leq k)$, build a word circuit C_i satisfying the following conditions:

   For every $i \in [k]$, let a_i be the number of bits in S_i ($a_i = \lfloor nw/k \rfloor$ or $\lceil nw/k \rceil$). Word circuit C_i takes these a_i bits as input (which is contained in at most $\lceil a_i/w \rceil + 1$ words). We partition the whole input set $\{0,1\}^{a_i}$ into r_i subsets $Q_{i,1}, Q_{i,2}, \ldots, Q_{i,r_i}$ so that $|Q_{i,j}| \leq 2^w - 1$ for all $1 \leq j \leq r_i$. (We define the way to partition and the number r_i precisely later.) The output of circuit C_i contains r_i words $h_{i,1}, h_{i,2}, \ldots, h_{i,r_i}$. For every $1 \leq j \leq r_i$, $h_{i,j} = 0^w$ if the input is not in set $Q_{i,j}$, otherwise $h_{i,j}$ returns the index of the input in set $Q_{i,j}$ (a w-bit integer in $[2^w - 1]$).

   Here is the construction of C_i: Let $y_1, y_2, \ldots, y_{n_i}$ be the input words of C_i, then $n_i \leq \lceil a_i/w \rceil + 1$. Since $n > k$ and w sufficiently large, we have $n_i \geq 2$. Let u_j be the number of bits of word y_j that are in set S_i (the *useful* input bits in word y_j of circuit C_i), we have $\sum_{j=1}^{n_i} u_j = a_i$. Since the partition of S_i are consecutive, there are at most two of $\{u_1, u_2, \ldots, u_{n_i}\}$ small than w. W.l.o.g. let's assume they are u_{n_i-1} and u_{n_i}, and assume $u_{n_i-1} \geq u_{n_i}$. (If there is only one less than w, assume it is u_{n_i}). Now we partition the $2^{u_1+u_2}$ possibilities of (y_1, y_2) into $m = \lceil \frac{2^{u_1+u_2}}{2^w-1} \rceil$ sets $T_1^1, T_2^1, \ldots, T_m^1$, each contains at most $2^w - 1$ elements. For each set T_j^1 $(1 \leq j \leq m)$, we build a gate z_j^1 taking y_1, y_2 as input. The output of z_j^1 is 0^w if $(y_1, y_2) \notin T_j^1$, otherwise it is the index of (y_1, y_2) in set T_j^1. These m gates are the first layer.
   Next we build the second layer (if $n_i > 2$) on the outputs of the gates $z_1^1, z_2^1, \ldots, z_m^1$ and input y_3. Partition the $2^{u_1+u_2+u_3}$ possibilities of (y_1, y_2, y_3) into $m \times 2^{u_3}$ sets: $T_{j,l}^2 = \{(y_1, y_2, y_3) \mid (y_1, y_2) \in T_j^1, y_3 = l\}$ $(1 \leq j \leq m, 1 \leq l \leq 2^{u_3})$. For every j, we build 2^{u_3} gates taking the output of z_j^1 and the input y_3 as input, say $z_{j,1}^2, z_{j,2}^2, \ldots, z_{j,2^{u_3}}^2$. The gate $z_{j,l}^2$ $(1 \leq l \leq 2^{u_3})$ outputs z_j^1 if $y_3 = l$, or 0^w otherwise. (See Figure 2)
   One can easily check $z_{j,l}^2$ outputs 0^w if $(y_1, y_2, y_3) \notin T_{j,l}^2$, and it is the index (a unique w-bit integer for each (y_1, y_2, y_3)) otherwise.
   The third layer (for $n_i > 3$) is build on the outputs of the second layer and y_4. For each output of the second layer, say z_{j,l_1}^2 $(1 \leq j \leq m, 1 \leq l_1 \leq 2^{u_3})$. We build 2^{u_4} gates z_{j,l_1,l_2}^3 $(1 \leq l_2 \leq 2^{u_4})$ on z_{j,l_1}^2 and y_4 in the same way as above: The gate z_{j,l_1,l_2}^3 outputs z_{j,l_1}^2 if $y_4 = l$, and it is 0^w otherwise.

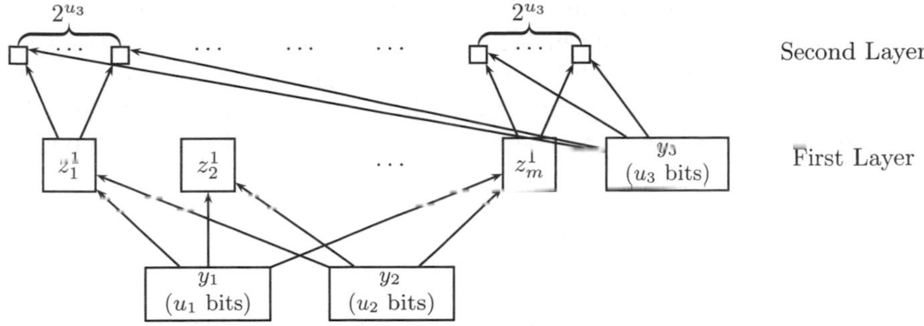

Fig. 2. First two layers of C_i

Continue this to build $n_i - 1$ layers in total. One can see that the output of the last layer satisfies our requirement of C_i and the corresponding sets $(1 \leq j \leq m, 1 \leq l_1 \leq 2^{u_3}, 1 \leq l_2 \leq 2^{u_4}, \ldots, 1 \leq l_{n_i-2} \leq 2^{u_n})$

$$T^{n_i-1}_{j,l_1,l_2,\ldots,l_{n_i-2}} = \{(y_1, y_2, \ldots, y_n) \mid (y_1, y_2) \in T^1_j, y_3 = l_1, y_4 = l_2, \ldots, y_n = l_{n_i-2}\}$$

can be put into a one to one mapping with the $Q_{i,1}, Q_{i,2}, \ldots, Q_{i,r_i}$. Thus

$$r_i = m2^{u_3+u_4+\cdots+u_{n_i}} = \left\lceil \frac{2^{u_1+u_2}}{2^w - 1} \right\rceil 2^{u_3+u_4+\cdots+u_{n_i}} < 2^{a_i-w} + O(1)2^{u_3+u_4+\cdots+u_{n_i}}.$$

By $n > k$ and $a_i \geq \lfloor nw/k \rfloor$, one can see $u_1 + u_2 - w \to \infty$ as $w \to \infty$. It follows that $2^{u_3+u_4+\cdots+u_{n_i}}/2^{a_i-w} = 2^{w-u_1-u_2} = o(1)$. Therefore $r_i = (1 + o(1))2^{a_i-w}$, and the number of gates used in C_i is $O(r_i) = O(2^{a_i-w}) = O(2^{(n/k-1)w})$.

2. This step is to build circuit on the outputs of $C_1, C_2, \ldots, C_k$. Say the output of C_1: $h_{1,1}, h_{1,2}, \ldots, h_{1,r_1}$ are the *master line*, the outputs of $C_2, C_3, \ldots, C_k$ are the *slave lines*. We list the Cartesian production of slave lines as (the order is unimportant) $p_1 = (h_{2,1}, h_{3,1}, \ldots, h_{k,1})$, $p_2 = (h_{2,2}, h_{3,1}, \ldots, h_{k,1})$, $\ldots$, $p_R = (h_{2,r_2}, h_{3,r_3}, \ldots, h_{k,r_k})$, where $R = r_2 r_3 \cdots r_k$.
For each $h_{1,j}$ in the master line, we construct a list of gates $g_{j,1}, g_{j,2}, \ldots, g_{j,R}$ linking like a chain (we call them the j-*th* chain). $g_{j,1}$ takes the output of $h_{1,j}$ and gates in p_1 as input. For $2 \leq l \leq R$, $g_{j,l}$ takes the output of $g_{j,l-1}$ and gates in p_l as input.
The gate $g_{j,l}$ functions as following:
(1) Outputs the first input word if one of the other $k - 1$ input words is 0^w.
(2) Otherwise if the first input word is 0^w, the gate outputs 0^w.

(3) If the first input is also nonzero, we claim that $x_1, x_2, \ldots, x_n$ can be determined by the input. As $h_{i,j}$ is nonzero only when the bits in S_i are in $Q_{i,j}$, we see the bits in $S_2, S_3, \ldots, S_k$ can all be determined. Also we see all gates in the j-th chain other than $g_{j,l}$ must be the first case, i.e. output its first input word. Thus the first input of $g_{j,l}$ must equal to $h_{1,j}$, and the bits in S_1 can also be determined. The gate $g_{j,l}$ outputs $f(x_1, x_2, \ldots, x_n)$ in this case.

Formally, these gates are defined as following:

$$g_{j,1}(h_{1,j}, p_1) = \begin{cases} h_{1,j} & \text{if one of gates in } p_1 \text{ is } 0^w; \\ 0^w & \text{if } h_{1,j} = 0^w; \\ f(x_1, \ldots, x_n) & \text{otherwise, all } x_1, \ldots, x_n \text{ are known.} \end{cases}$$

For $2 \leq l \leq R$,

$$g_{j,l}(g_{j,l-1}, p_l) = \begin{cases} g_{j,l-1} & \text{if one of gates in } p_l \text{ is } 0^w; \\ 0^w & \text{if } g_{j,l-1} = 0^w; \\ f(x_1, \ldots, x_n) & \text{otherwise, all } x_1, \ldots, x_n \text{ are known.} \end{cases}$$

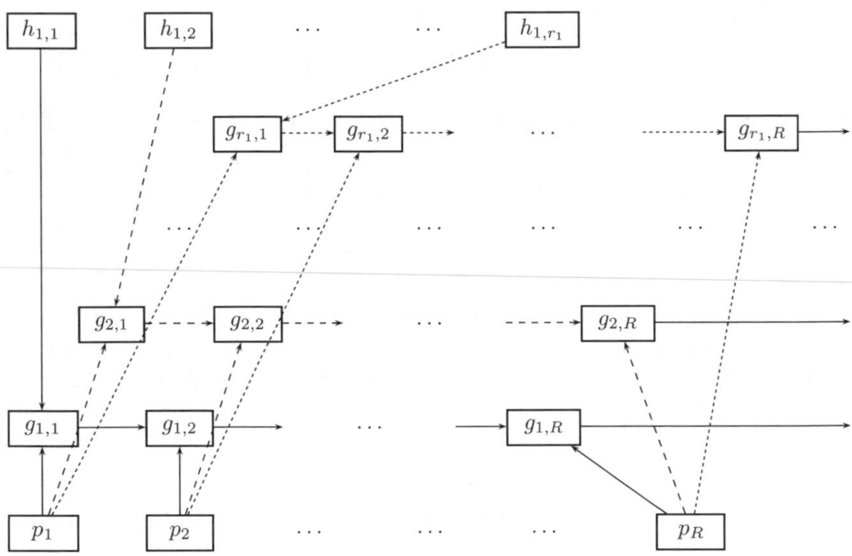

Fig. 3. Construction of $g_{j,l}$ $(1 \leq j \leq r_1, 1 \leq l \leq R)$

As we have shown that at most one gate in the j-th chain does not transfer its first input forward, one can see the last gate

$$g_{j,R} = \begin{cases} 0^w & \text{if bits in } S_1 \text{ are not in } Q_{1,j}; \\ f(x_1, x_2, \ldots, x_n) & \text{otherwise.} \end{cases}$$

Then we use $(r_1 - 1)$ gates to compute the disjunction of $g_{1,R}, g_{2,R}, \ldots, g_{r_1,R}$. This part of circuit can be any tree consisting of 2-input OR gates that takes

the outputs of $g_{1,R}, g_{2,R}, \ldots, g_{r_1,R}$ as leaves. The root of this tree outputs the value of $f(x_1, x_2, \ldots, x_n)$.
In this step, $r_1 \times R + r_1 - 1 = r_1 r_2 \cdots r_k + r_1 - 1 = (1 + o(1))2^{(n-k)w}$ gates are used in total.

The total number of gates is

$$k \cdot O(2^{(n/k-1)w}) + (1 + o(1))2^{(n-k)w} = (1 + o(1))2^{(n-k)w}.$$

Thus the theorem is proved. □

3 Fixed Word Length, Large Number of Input

In this section, we assume that w, k are fixed and n is sufficiently large. First we give the lower bound using similar counting argument as the previous section.

Theorem 4. *For any fixed $w \geq 1, k \geq 2$ and sufficiently large number n, there exists a function $f : (\{0,1\}^w)^n \to \{0,1\}^w$ so that no circuit consists of less than $(1 - o(1))\frac{2^{nw}}{(k-1)n}$ k-input gates computes f.*

Proof. Similarly as in Theorem 1, by Lemma 1 we have

$$(s + n - 1)^{sk} 2^{sw2^{kw}} / s! \geq 2^{w2^{nw}}.$$

Take the logarithm of both sides, by Stirling's Formula, we have

$$sk \log_2(s + n - 1) + sw2^{kw} - s \log_2 s + s \log_2 e - \frac{1}{2} \log_2 s + O(1) \geq w2^{nw},$$

For sufficiently large n, it implies $s \geq (1 - o(1))\frac{2^{nw}}{(k-1)n}$. □

To show that $(1 + o(1))\frac{2^{nw}}{(k-1)n}$ gates are sufficient to compute every function, we consider a function $f : (\{0,1\}^w)^n \to \{0,1\}^w$ as a combination of w one-bit output functions $(\{0,1\}^w)^n \to \{0,1\}$. We first construct word circuits that computes these one-bit functions.[2]

Lemma 2. *For any fixed $w \geq 1, k \geq 2$, every function $F : (\{0,1\}^w)^n \to \{0,1\}$ can be computed by a word circuit of size at most $(1 + o(1))\frac{2^{nw}}{(k-1)nw}$.*

Proof. The following proof is based on Lupanov's construction for boolean circuits ([3, 4]).

1. First use nw gates to break the input words into bits: For each input word x_i $(1 \leq i \leq n)$, we build w gates taking x_i as input. The j-th gate $(1 \leq j \leq w)$ outputs the j-th bit of x_i. We simply use $b_1, b_2, \ldots, b_{nw}$ to denote the output.

[2] Here the highest bit of the output word is the result, all lower bits are not used. In later proof we will use a w-bit word as a single bit without claim.

2. Let $t = \lceil 3\log_2(nw) \rceil$, we use binary AND/OR gates to compute all minterms on $\{b_1, b_2, \ldots, b_t\}$ and $\{b_{t+1}, b_{t+2}, \ldots, b_{nw}\}$. In this step, $O(2^t + 2^{nw-t})$ gates are enough.
3. Divide the 2^t possibilities of $(b_1, b_2, \ldots, b_t)$ into p sets $Q_1, Q_2, \ldots, Q_p$, so that each contains at most $s = nw - \lceil 5\log_2(nw) \rceil$ elements and $p = \lceil 2^t/s \rceil$.
For $1 \le i \le p$ and a 0-1 string v of length $|Q_i|$, we build gates to compute the following value:

$$F^1_{i,v} = \begin{cases} j\text{-th bit of } v & \text{if } (b_1, b_2, \ldots, b_t) \text{ is the } j\text{-th element of } Q_i; \\ 0 & \text{otherwise.} \end{cases}$$

For given i and v, $F^1_{i,v}$ equals to the disjunction of some minterms on $\{b_1, b_2, \ldots, b_t\}$. Since the length of v is at most s, we see each minterm is used as most 2^s times. Thus this step takes at most $2^s 2^t$ gates.
4. Every possibility of $(b_1, b_2, \ldots, b_t)$ is corresponding to a tuple (i, j), where $(b_1, b_2, \ldots, b_t)$ is the j-th element of Q_i. Thus we may write $F(b_1, b_2, \ldots, b_{nw})$ as $F(i, j, b_{t+1}, b_{t+2}, \ldots, b_{nw})$. Define $G(i, b_{t+1}, b_{t+2}, \ldots, b_{nw})$ to be the 0-1 string of length $|Q_i|$, that the j-th bit equals to $F(i, j, b_{t+1}, b_{t+2}, \ldots, b_{nw})$ $(1 \le i \le p, 1 \le j \le |Q_i|)$.
For $1 \le i \le p$ and a 0-1 string v of length $|Q_i|$, we build gates to compute the following value:

$$F^2_{i,v} = \begin{cases} 1 & \text{if } G(i, b_{t+1}, b_{t+2}, \ldots, b_{nw}) = v; \\ 0 & \text{otherwise.} \end{cases}$$

Similarly, $F^2_{i,v}$ equals to the disjunction of some minterms on the last $nw - t$ bits $\{b_{t+1}, b_{t+2}, \ldots, b_{nw}\}$, say $e_1, e_2, \ldots, e_d$. If $d \le k$, we can compute $F^2_{i,v}$ by one gate. Otherwise we can use $\lceil (d-1)/(k-1) \rceil$ k-input OR gates (only the highest bit of every word is used) to compute it: The first gate takes k minterms as input, all other gate takes at most $k-1$ minterms and the output of the previous gate as input. The output of the last gate is the disjunction. (See Figure 4)

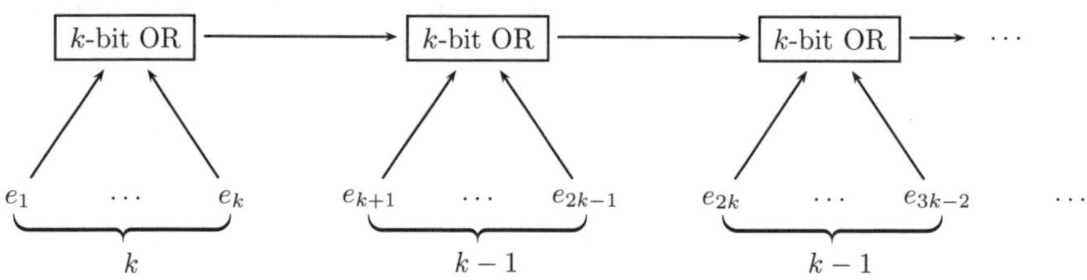

Fig. 4. Disjunction of $e_1, e_2, \ldots, e_d$

Every minterm on $\{b_{t+1}, b_{t+2}, \ldots, b_{nw}\}$ is used at most p times. Among the OR gates there are only $O(p2^s)$ (the number of all possible (i, v)) taking less than $k-1$ minterms as input. Thus this step takes at most $p2^{nw-t}/(k-1) + O(p2^s)$ gates.

5. It is not difficult to see $F = \bigvee_i \bigvee_v (F^1_{i,v} \bigwedge F^2_{i,v})$ (see [3, 4] for details). Thus we can compute F by the result of the previous steps. This step takes $O(p2^s)$ gates.

The gates used in total is

$$O(2^t + 2^{nw-t}) + 2^s 2^t + p2^{nw-t}/(k-1) + O(p2^s) = (1 + o(1))\frac{2^{nw}}{(k-1)nw}.$$

Thus the lemma is proved. □

By Lemma 2 we get an upper bound matching the lower bound in Theorem 4 immediately.

Theorem 5. *For any fixed $w \geq 1, k \geq 2$, every function $f : (\{0,1\}^w)^n \to \{0,1\}^w$ can be computed by a circuit that consists of $(1 + o(1))\frac{2^{nw}}{(k-1)n}$ k-input gates.*

Proof. The function f can be considered as a combination of w one-bit output functions. We construct circuits for each of these functions, then use at most $w - 1 = O(1)$ gates to combine the w outputs into one word. The whole circuit takes $(1 + o(1))\frac{2^{nw}}{(k-1)n}$ gates. □

Acknowledgments

The third author would like to thank Kristoffer Hansen and Peter Miltersen for helpful discussions.

References

1. Hansen, K.A., Lachish, O., Miltersen, P.B.: Hilbert's thirteenthv problem and circuit complexity. In: Dong, Y., Du, D.-Z., Ibarra, O. (eds.) ISAAC 2009. LNCS, vol. 5878, pp. 153–162. Springer, Heidelberg (2009)
2. Lupanov, O.B.: A method of circuit synthesis. In: Izvestiya VUZ, Radio_zika, pp. 120–140 (1959)
3. Shannon, C.E.: The synthesis of two-terminal switching circuits. Bell System Technical Journal 28, 59–98 (1949)
4. Wegener, I.: The Complexity of Boolean Functions, pp. 87–92. John Wiley & Sons Ltd./ B. G. Teubner (1987)

On the Density of Regular and Context-Free Languages

Michael Hartwig

Faculty of Information Technology, Multimedia University
Cyberjaya, Malaysia
michael.jua.hartwig@gmail.com

Abstract. The density of a language is defined as the function $d_L(n) = |L \cap \Sigma^n|$ and counts the number of words of a certain length accepted by L. The study of the density of regular and context-free languages has attracted some attention culminating in the fact that such languages are either sparse, when the density can be bounded by a polynomial, or dense otherwise. We show that for all nonambiguous context-free languages the number of accepted words of a given length n can also be computed recursively using a finite combination of the number of accepted words smaller than n, or $d_L = \sum_{j=1}^{k} u_j d_L(n-j)$. This extends an old result by Chomsky and provides us with a more expressive description and new insights into possible applications of the density function for such languages as well as possible characterizations of the density of higher languages.

Keywords: context-free languages, regular languages, density.

1 Introduction

> "Given a regular language, it is often useful to know how many words of a certain length are in the language." [1]

This study of the density of regular and context-free languages has already a longer history [3]. Flajolet could show that regular languages are either sparse or dense [5], a fact that was recently generalized to context-free languages [6]. A language is hereby called sparse, if its density function $d_L(n)$ can be bounded from above by a polynomial (i.e., there exists a polynomial p such that $d_L(n) \leq p(n)$). On the other hand, if there exists a real number $h > 1$ such that $d_L(n) \geq h^n$ for infinitely many $n \geq 0$ then L is called dense [8]. Notice that the language a^*b^* is a sparse language, while the language that includes all words over a binary alphabet that start with the letter a (i.e., $a(a+b)^*$) is dense. Given the fact that only sparse regular languages have the power to restrict NP complete problems such that they become polynomially solvable [8], such languages have been studied in more detail and characterised as SLRE [1,3] or bounded regular languages [9]. (Nevertheless, it is not difficult to see that the majority of all regular languages are dense.) Eisman suggested then that the density function

M.T. Thai and S. Sahni (Eds.): COCOON 2010, LNCS 6196, pp. 318–327, 2010.

could also be used in some application areas such as streaming algorithms, where "rapid computation must be performed (often in a single pass)" [7].

In this paper we'd like to extend a characterization of the density of regular languages given already by Chomsky stating that every rational function satisfies a linear recurrence relation [2]. In other words, for each regular language L there exists an initial length n_0 such that $\forall n \geq n_0 \cdot d_L(n) = \sum_{j=1}^{k} u_j d_L(n-j)$. A fact that must not have been known to Flajolet and his colleagues, as above mentioned dense/sparse characterization is a direct consequence of this result. As shown here, Chomsky's result holds also for all nonambiguous context-free languages while there there exist some ambiguous context-free languages with a different density.

Examples

Let $e = (ab + aab)^*$, then $d_{L_e}(n) = d_{L_e}(n-2) + d_{L_e}(n-3)$.
Let $e = (ab + aab)^* b^*$, then $d_{L_e}(n) = d_{L_e}(n-1) + d_{L_e}(n-2) - d_{L_e}(n-4)$.
Let $L = a^n b^n$, then $d_L(n) = d_L(n-2)$.

It is also often interesting to study the acceptance probability $Acc(L, n) = |L \cap \Sigma^n| \,/\, |\Sigma^n|$ of a given language rather than its density, e.g., the ratio between the number of accepted words and all possible words of a given length. As mentioned above, $a(a + b)^*$ has exponential density but it has only stable acceptance probability as $Acc(a(a + b)^*, n) = 0.5$, which in this case describes the quantity of accepted words more appropriately.

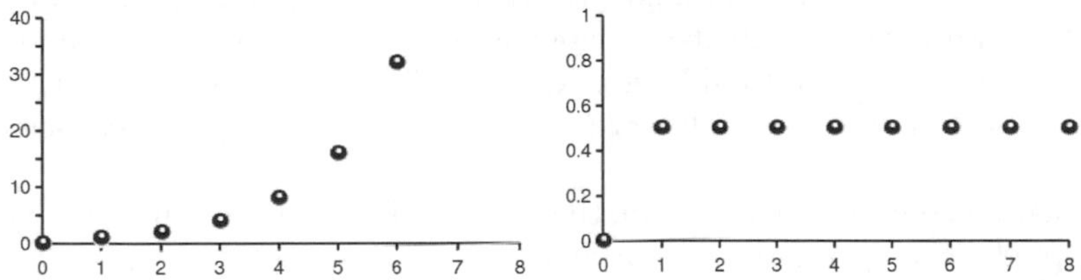

Fig. 1. $d_{L(e)}(n)$ and $Acc(L(e), n)$ for $e = a(a + b)^*$

In [11] we could show that for all starheight-one regular languages we can find a step width d decomposing the acceptance probability graph into d chains. By this we mean that, $Acc(L, n_0 + d), Acc(L, n_0 + 2d), Acc(L, n_0 + 3d), \ldots$ is either an increasing, decreasing or stable function. Generalising now those thoughts to all nonambiguous context-free languages it can be said that such languages are not only sparse or dense but can rather be split into d sparse or dense languages. This provides us with a much more expressive characterization of the density of nonambiguous context-free languages than the formerly known dense/sparse distinction giving hope that density properties will play a more vital role in complexity theory than now. The description presented here offers especially possibilities to describe the density of many more classes in the Chomsky hierarchy in a simple, uniform and coherent way for which first incentives are provided.

The paper is then divided as follows. The following chapter introduces necessary definitions. Chapter 3 shows that the density of all nonambiguous languages can be described recursively extending Chomsky's result. Chapter 4 discusses some properties of the acceptance probability of mentioned languages. Chapter 5 provides an ambiguous context-free language with a different density function before the work is summarized and incentives for further research are given.

2 Preliminaries

The paper follows standard formal language notations. The alphabet for all strings is $\Sigma = \{a, b\}$, the length of a string w is given by $|w|$. All sets $L_1, L_2, ..$ are considered subsets of Σ^*. A regular expression e over Σ is built from all symbols in Σ, the symbols λ and $\emptyset$, the binary operators $+$, $\cdot$ and the unary operator *. The language specified by a regular expression is denoted by $L(e)$ or L_e and is referred to as a regular language. $REG(c)$ specifies all regular languages having a star height of c or less.

A context-free grammar (CFG) is given with the structure $G = (N, T, S, P)$, where N is a finite set of nonterminals, $T = \{a, b\}$ a finite set of terminals, $S \in N$ the start symbol and $P \subseteq V \times (V + \Sigma)^*$ a finite set of production rules. We denote with $\Rightarrow_G$ a derivation step, with $\Rightarrow_G^*$ its reflexive, transitive closure and with $L(G) = \{w|S \Rightarrow_G^* w\}$ the language generated by G. If G has at least one word with two different derivations, G is said to be ambiguous. It is well known that there exist context-free languages with only ambiguous grammars, which we will call ambiguous context-free languages. We furthermore call an ordered, possibly infinite tree T a total language tree if T contains nodes from $(N+T)^*$ and shows the derivation of all words in the language using a fixed derivation strategy[4].

As mentioned in the introduction, the density of a language counts the number of accepted words per given length and is defined as $d_L(n) = |L \cap \Sigma^n|$, while the acceptance probability of a language is defined as the ratio between the number of accepted words $d_L(n)$ and the number of all words of a given length, $Acc(L, n) = |L \cap \Sigma^n| \; / \; |\Sigma^n|$. For reading purposes we often write $d_L(n)$ instead of $d_{L(G)}(n)$ in the rest of the paper.

3 The Density of Regular and Nonambiguous Context-Free Languages

The main purpose of the paper is to extend the recursive description provided by Chomsky for the density of all regular languages to nonambiguous languages. In order to prove the main result, some lemmata are provided.

Lemma
Let L_1, L_2 be languages for which we can give initial length's n_1, n_2 and integers $k_1, u_{1,1}, .., u_{1,k_1}$, $k_2, u_{2,1}, .., u_{2,k_2}$ such that $\forall n \geq n_1, n_2$:

$$d_{L_1}(n) = \sum_{j_1=1,\dots,k_1} u_{1,j_1} d_{L_1}(n - j_1),$$

$$d_{L_2}(n) = \sum_{j_2=1,\dots,k_2} u_{2,j_2} d_{L_2}(n - j_2),$$

then we can also describe the density of $L = L_1 \cup L_2$ *with constants* $n', k', u'_1, .., u'_{k'}$ *such that* $\forall n \geq n'$:

$$d_L(n) = \sum_{j'=1,\dots,k'} u'_{j'} d_L(n - j').$$

Proof
Having $d_L(n) = d_{L_1}(n) + d_{L_2}(n)$, $d_L(n-1) = d_{L_1}(n-1) + d_{L_2}(n-1)$, $\dots$, *we can can replace each occurrence of* $d_{L_2}(n-i)$ *with* $d_L(n-i) - d_{L_1}(n-i)$. *This will give us a recursive characterisation of the density of* $L(n)$ *as follows.*

$$d_L(n) = \sum_{j_1=1,\dots,k_1} u_{1,j_1} d_{L_1}(n - j_1) + \sum_{j_2=1,\dots,k_2} [u_{2,j_2} d_L(n - j_2) - u_{2,j_2} d_{L_1}(n - j_2)]$$

Or,

$$d_L(n) = \sum_{j_1=1,\dots,max(k_1,k_2)} (u_{1,j_1} - u_{2,j_1}) d_{L_1}(n - j_1) + \sum_{j_2=1,\dots,k_2} u_{2,j_2} d_L(n - j_2)$$

We can now use above formula to calculate $d_L(n+1), d_L(n+2), \dots$ *and constantly replace every use of* $d_{L_1}(n+j)$ *with terms from lower levels using the information from its recursive description,* $d_{L_1}(n) = \sum_{j_1=1,\dots,k_1} u_{1,j} d_{L_1}(n-j_1)$. *This will give us an equational system that allows us to eliminate all terms referring to* L_1. □

Take note: Above lemma holds especially in the case where one of the languages is finite.

Lemma
Let L *be a language for which we can give an initial length* n_0 *and integers* $k, u_1, .., u_k$ *such that* $\forall n \geq n_0$:

$$d_L(n) = \sum_{j=1,\dots,k} u_j d_L(n - j) + c$$

then there exist an initial length n'_0 *and integers* $k', u'_1, .., u'_k$ *such that* $\forall n \geq n'_0$:

$$d_L(n) = \sum_{j'=1,\dots,k'} u'_{j'} d_L(n - j')$$

Proof
We can give the new description. It computes the value of c *out of the difference between* $d_L(n-1)$ *and a recursive computation for it leaving out* c.

$$d_L(n) = \sum_{j=1,\dots,k} u_j d_L(n - j) + d_L(n-1) - \sum_{j=1,\dots,k} u_j d_L(n - 1 - j)$$

□

Lemma

Assume L to be a nonambiguous and infinite context-free language. For every long enough word $w = w_1w_3w_2 \in L$ we can find the following paths in every total language tree for L.

$$w_1N_1w_2 \Rightarrow^* w_1w_3w_2,$$
$$w_1N_1w_2 \Rightarrow^* w_1w_4N_1w_5w_2 \Rightarrow^* w_1w_4w_3w_5w_2,$$
$$\ldots$$
$$w_1N_1w_2 \Rightarrow^* w_1w_4^*N_1w_5^*w_2 \Rightarrow^* w_1w_4^*w_3w_5^*w_2.$$

Proof

The argument follows similar pumping lemmata proofs. If L has an infinite number of words then for all words long enough there must exist a derivation from a nonterminal to itself in the total language tree. In other words, we can give a nonterminal N_1 that was in the derivation of w with $N_1 \Rightarrow^* w_4N_1w_5$. Additionally, N_1 produces also a derivation to a word, say w_3. Those derivations can then be repeated to produce all other words mentioned above. □

In the following we will then call the subtree of the total language tree that includes above mentioned paths a *collar*. The existence of those collars allows us to prove the desired property.

Lemma

Let L be a nonambiguous context-free language. Then there exists an initial length n_0 and integers $k, u_1, .., u_k$ such that $\forall n \geq n_0$:

$$d_L(n) = \sum_{j=1}^{k} u_j d_L(n-j)$$

Proof

Let L be a nonambiguous context-free language. Then there exist a nonambiguous context-free grammar G and total language tree T for it. In the following we will cover T inductively with collars and combinations thereof. Having L nonambiguous we can simply add the density of subtrees to obtain the density of the complete tree. As already shown, every word long enough must be part of one or more collars. The number of leafs produced by such a collar substituting hereby $t_1 = |w_1w_3w_2|$, and $t_2 = |w_4w_5|$ in the previous lemma can then be easily given with $d(t_1) = 1, d(n) = d(n - t_2)$

The pumping lemma tells us then that there is a length n_0 after which w_1 in $w = w_1N_1w_2$ was itself developed making use of a cycle from one nonterminal to itself. This means that there is only a finite number of collars having no cycle in w_1. All other collars can therefore only be derived based on this finite number of collars using the rules in the grammar in any combination of the following ways.

1. *(UNION) Two subtrees T_1, T_2 are combined. (A typical grammar would be $S \Rightarrow A|B, A \Rightarrow \ldots, B \Rightarrow \ldots$). If there exists a density description of required*

form for T_1 and T_2 then there exist also such a description for the union of them as demonstrated earlier.

2. *(REPETITION) A subtree T with density $d_T(n) = \sum_{j=1,\ldots,k} u_j d_T(n-j)$ is infinitely often repeated. (A typical grammar could be given with $S \Rightarrow AB, A \Rightarrow a|aA, B \Rightarrow b|bB$ producing a highest collar for ab^n and successively starting a new collar at every new length with $a^2b^n, a^3b^n, \ldots$) The new subtree covering all those infinitely many repetitions of T_1 has then a density of $d_{T'}(n) = \sum_{j'=1,\ldots,k'} u'_j d_{T'}(n-j') + c$, where c counts the number of repetitions at each new level.*
3. *(RECURSION) A subtree T is recursively called and added. This specifies the case that always a complete copy of what has been already developed is newly added (and could therefore be also called a second order repetition). The simplest case is given with T being a collar. Then the next subtree has an infinite number of repetitions of this T. The third subtree again infinitely often calls the second subtree, etc. A typical grammar would be $S \Rightarrow SaN|a, N \Rightarrow bN|bb$. Its total language tree could be split into the following collars leaving out the leaf a.*

$$\begin{aligned}
\textit{first subtree} &: a^2b^n\\
\textit{second subtree} &: a^2b^2ab^n, a^2b^3ab^n, \ldots\\
\textit{third subtree} &: a^2b^2a^2b^2ab^n, a^2b^2a^2b^3ab^n, \ldots\\
& a^2b^3a^2b^2ab^n, a^2b^3a^2b^3ab^n, \ldots\\
& a^2b^4a^2b^2ab^n, a^2b^4a^2b^3ab^n, \ldots\\
& \ldots
\end{aligned}$$

The number of leafs produced by such a recursion is, fortunately, easily computed. If T has density $d_T(n) = \sum_{j=1,\ldots,k} u_j d_T(n-j)$ and a new recursion is added at level c deeper then the density of the whole subtree covering all recursive calls can be given with $d_{T'}(n) = \sum_{j=1,\ldots,k} u_j d_{T'}(n-j) + d_{T'}(n-c)$.

As there is only a finite number of rules in the grammar giving rise to only a finite number of application of above cases the lemma holds. □

Example

Let L be again $S \Rightarrow SaN|a, N \Rightarrow bN|bb$ as in the last part of above proof. The leftmost subtree of its total language tree, call it t_1, produces only the word a having a density of $d_{t_1}(|a|) = d_{t_1}(1) = 1$. The second leftmost subtree t_2 produces the words a^2b^n in a collar and its density can be given with $d_{t_2}(a^2b^2) = d_{t_2}(4) = 1, d_{t_2}(n) = d_{t_2}(n-1)$. This collar t_2 is then recursively called at every third level producing all other collars and leaves. (The rule $S \Rightarrow SaN$ causing the recursion adds hereby the letter a and the rule $N \Rightarrow Nb|bb$ produces a minimal word bb giving a distance of $c = 3$ to each new recursion.) The density of L can therefore be calculated with

$$\begin{aligned}
d_L(1) = d_L(4) = d_L(5) = 1,&\\
d_L(n) &= d_L(n-1) + d_L(n - (|a| + |bb|)),\\
&= d_L(n-1) + d_L(n-3).
\end{aligned}$$

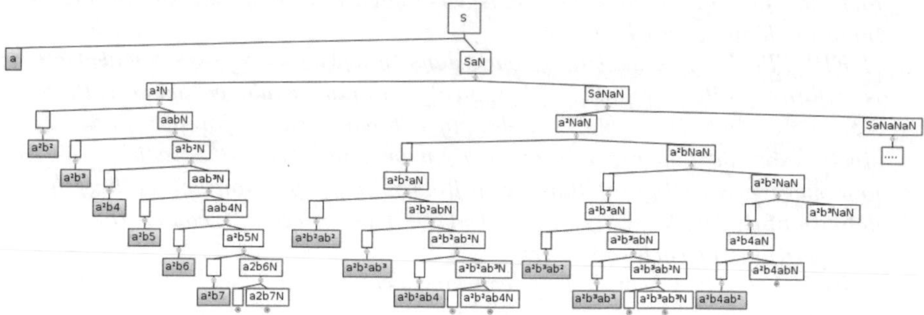

Fig. 2. Total language tree for $S \Rightarrow SaN|a$, $N \Rightarrow bN|bb$ with a leftmost subtree deriving a, a collar t_2 deriving a^2b^n and a recursive call of t_2

4 Properties of Nonambiguous Context-Free Acceptance Probability

The following lemma shows that the acceptance probability of nonambiguous context-free languages can be characterized by a finite number of chains growing towards a limit. A chain of step width d contains hereby the elements $Acc(n_0), Acc(n_0 + d), Acc(n_0 + 2d), \ldots$ for every n_0.

Lemma
Let L be a nonambiguous context-free language. Then there exist an initial length n_0 and step width d such that for all $n > n_0$:

$$\lim_{i \to \infty} \frac{Acc(L, n + (i + 1)d)}{Acc(L, n + (i)d)} = c_n$$

Proof
See the proof in [11] which now applies also to nonambiguous context-free languages.

It is obvious that the growth rate c of such a chain decides whether a chain is sparse, dense, increasing, decreasing or stable. We claim then that chains of nonambiguous context-free and regular languages are not only growing towards fixed limits but also monotone and the difference between two neighbouring elements in a chain is decreasing. In other words, $|Acc(L, n_0 + (i+2)d) - Acc(L, n_0 + (i+1)d)| \leq |Acc(L, n_0 + (i + 1)d) - Acc(L, n_0 + id)|$. The existence of such chains can be used to separate above languages from languages of higher classes in a new and interesting way. It is well known that the density of higher functions (although we do not yet have a full description) exhibits some differences to the characterisations presented so far. The most notable is hereby certainly the so-called phase transition of NP problems. This demonstrates that the density of a class of languages can indeed indicate complexity properties and justifies further research.

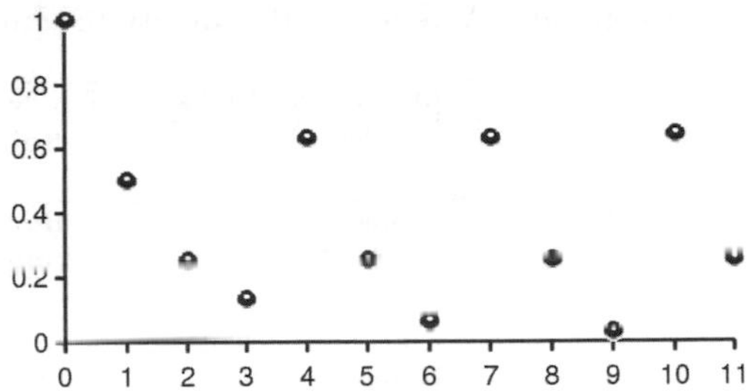

Fig. 3. Acceptance probability graph of a typical nonambiguous context-free language ($L = a^n b^{2n} + ab(\Sigma\Sigma\Sigma)^* + (bbb)^* a(\Sigma\Sigma\Sigma)^*$ has three chains: one decreasing, increasing and stable chain)

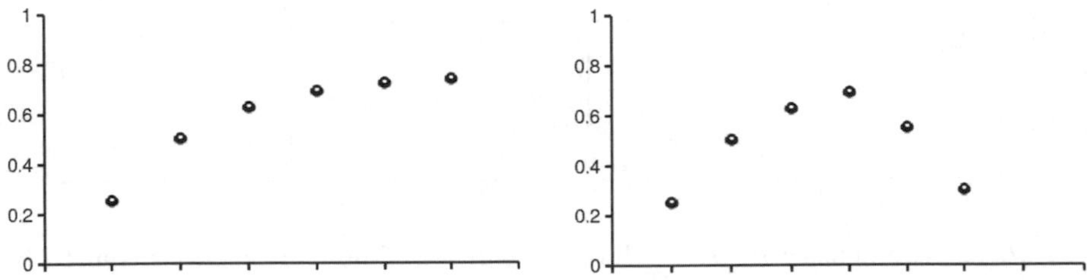

Fig. 4. Nonambiguous context-free versus higher acceptance probability graphs

5 On the Density of Ambiguous Context-Free Languages

Surprisingly there exist also some ambiguous context-free languages with such a recursive density description (For instance, let $L = \{a^i b^j a^k | i = j \text{ or } i = k\}$ then $d_L(n) = d_L(n-1) + d_L(n-6) - d_L(n-7)$). However, this can not be generalized.

Lemma
There exist an ambiguous language L for which we can ***not*** *give a length n_0 and integers $k, u_1, .., u_k$ such that $\forall n \geq n_0$:*

$$d_L(n) = \sum_{j=1,\ldots,k} u_j d_L(n-j)$$

Proof
Let $L = \Sigma^* - \{baba^2ba^3b \ldots ba^m b : m = 1, 2, 3, \ldots\}$. This language has density $d_L(n) = 2^n$ except for $n = 1, 3, 6, 10, \ldots (= \sum_{m=1,2,\ldots} m)$ with density $d_L(n) = 2^n - 1$. Obviously the density of L does not allow a recursive description in above sense because of the constantly growing distances between the elements having a density of $2^n - 1$. □

However, the density of above language can be computed by making reference to only one smaller term although being not from a fixed smaller length. In

Table 1. Density and Acceptance Probability (Summary)

Density	Acceptance Probability	Classes & Examples
$d_L(n) = \sum_{j=1}^{0} d_L(n-j) = 0$	stable (0)	**finite** $\lambda + aba + bb$
$d_L(n) = \sum_{j=1}^{1} 2d_L(n-j) = 2d_L(n-1)$	stable	**finite time**, **one free** Σ^*, $a\Sigma^* + \Sigma^* a$
$d_L(n) = \sum_{j=1}^{1} u_j d_L(n-j) = u_1 d_L(n-1)$	constantly decreasing, stable	**one star over symbol**, $ab^*b + \Sigma^* a$
$d_L(n) = \sum_{j=k}^{k} u_j d_L(n-j) = u_k d_L(n-k)$	one constantly decreasing or stable chain	**one star at heigh** 1, $a(bb)^*b$
$d_L(n) = \sum_{j=1}^{k} u_j d_L(n-j)$	finite number of limit-growing chains	**REG(1), REG(n), Nonambiguous CF**, $a^n b^n$
$d_L(n) = \sum_{j=1}^{n} u(n,j) d_L(n-j)$ *(Claim)*	*infinite number of finitely different limit-growing chains (Claim)*	**Ambiguous CF**, $\Sigma^* - \{ba^1ba^2b \dots ba^mb\}$

fact, L has the density $d_L(n) = d_L(n - \lceil \sqrt{2n}\, \rceil)$. This can be re-written to be more similar to previously used density functions yielding a density that could be understood as a function having an infinite number of limit-growing chains with a constantly increasing step width. Such a characterisation would still be different from known densities exhibited by higher problems. A complete description of the density of all context-free languages (besides the sparse/dense criterion) is not known to the author and subject to further research.

$$d_L(n) = \sum_{j=1,\dots,k} u(n,j) \cdot d_L(n-j), where$$

$$with\ u(n,j) = \begin{cases} 1, & if\ j = \lceil \sqrt{2n}\, \rceil \\ 0, & otherwise \end{cases}$$

6 Conclusions

The research into the density of regular and context-free languages has flourished and will continue to offer new insights. This paper extended an old lemma by Chomsky stating that the density of all regular languages can be described recursively as the sum of a finite number of the density of smaller terms to all nonambiguous context-free languages. This allows us to divide the acceptance probability graph of such languages into a finite number of limit-growing chains. This differs from the density of ambiguous context-free languages who seem to have an infinite number of finitely different limit-growing chains and should be studied next. Adding some simple results from [10], the work presented can be summarized in the following table.

References

1. Szilard, A., Yu, S., Zhang, K., Shallit, J.: Characterizing Regular Languages with Polynomial Densities. In: Havel, I.M., Koubek, V. (eds.) MFCS 1992. LNCS, vol. 629, pp. 494–503. Springer, Heidelberg (1992)
2. Chomsky, N., Miller, G.A.: Finite-state lanuages. Information and Control 1, 91–112 (1958)
3. Rozenberg, G., Salomaa, A.: Handbook of Formal Languages. In: Regular Languages, ch. 2, Springer, Heidelberg (1997)
4. Cohen, D.I.A.: Introduction to Computer Theory. Wiley, Chichester (1996)
5. Flajolet, P.: Analytic Models and Ambiguity of Context-Free Languages. Theor. Comput. Sci. 49, 283–309 (1987)
6. Incitti, R.: The Growth Function of Context-Free Languages. Theor. Comput. Sci. 255(1-2), 601–605 (2001)
7. Eisman, G., Ravikumar, B.: Approximate Recognition of Non-regular Languages by Finite Automata. In: ACSC, pp. 219–228 (2005)
8. Demaine, E., López-Ortiz, A., Munro, J.: On Universally Easy Classes for NP-Complete Problems. Theor. Comput. Sci. 304(1-3), 471–476 (2003)
9. Habermehl, P.: A Note on SLRE (2000), `http://citeseer.ist.psu.edu/375870.html`
10. Hartwig, M.: Acceptance Probability of Lower Regular Languages and Problems using Little Resources. In: MMU International Symposium on Information and Communications Technologies (M2USIC 2006), Petaling Jaya, Malaysia, Nov 16 - 17 (2006)
11. Hartwig, M., Phon-Amnuaisuk, S.: Regular Languages up to Star Height 1 and the Difference Shrinking Acceptance Probability. In: Proceedings of TMFCS-08 (2008)

Extensions of the Minimum Cost Homomorphism Problem

Rustem Takhanov

Department of Computer and Information Science, Linköping University,
SE-581 83 Linköping, Sweden
g-rusta@ida.liu.se, takhanov@mail.ru

Abstract. Assume D is a finite set and R is a finite set of functions from D to the natural numbers. An instance of the minimum R-cost homomorphism problem ($MinHom_R$) is a set of variables V subject to specified constraints together with a positive weight c_{vr} for each combination of $v \in V$ and $r \in R$. The aim is to find a function $f : V \to D$ such that f satisfies all constraints and $\sum_{v \in V} \sum_{r \in R} c_{vr} r(f(v))$ is maximized.

This problem unifies well-known optimization problems such as the minimum cost homomorphism problem and the maximum solution problem, and this makes it a computationally interesting fragment of the valued CSP framework for optimization problems. We parameterize $MinHom_R$ by *constraint languages*, i.e. sets Γ of relations that are allowed in constraints. A constraint language is called *conservative* if every unary relation is a member of it; such constraint languages play an important role in understanding the structure of constraint problems. The dichotomy conjecture for $MinHom_R$ is the following statement: if Γ is a constraint language, then $MinHom_R$ is either polynomial-time solvable or NP-complete. For $MinHom$ the dichotomy result has been recently obtained [Takhanov, STACS, 2010] and the goal of this paper is to expand this result to the case of $MinHom_R$ with conservative constraint language. For arbitrary R this problem is still open, but assuming certain restrictions on R we prove a dichotomy. As a consequence of this result we obtain a dichotomy for the conservative maximum solution problem.

1 Introduction

Constraint satisfaction problems (CSP) and valued constraint satisfaction problems ($VCSP$) are natural ways of formalizing a large number of computational problems arising in combinatorial optimization, artificial intelligence, and database theory. CSP has the following two equivalent formulations: (1) to find an assignment of values to a given set of variables, subject to constraints on the values that can be assigned simultaneously to specified subsets of variables, and (2) to find a homomorphism between two finite relational structures A and B. $VCSP$ is a "soft" version of CSP where constraint relations are replaced by functions from set of tuples to some totally ordered set with addition operation (for example, rational numbers). A solution is defined as an assignment to variables that maximize a functional which is equal to a sum of constraint

M.T. Thai and S. Sahni (Eds.): COCOON 2010, LNCS 6196, pp. 328–337, 2010.

functions applied to corresponding variables. Applications of CSPs and $VCSP$s arise in the propositional logic, database and graph theory, scheduling, biology and many other areas. CSP and its subproblems has been intensively studied by computer scientists and mathematicians since the 70s, and recently attention has been paid to its modifications such as $VCSP$. Considerable attention has been given to the case where the constraints are restricted to a given finite set of relations Γ, called a constraint language [2,5,12,18]. For example, when Γ is a constraint language over the boolean set $\{0, 1\}$ with four ternary predicates $x \vee y \vee z$, $\overline{x} \vee y \vee z$, $\overline{x} \vee \overline{y} \vee z$, $\overline{x} \vee \overline{y} \vee \overline{z}$ we obtain 3-SAT. For every constraint language Γ, it has been conjectured that $CSP(\Gamma)$ is either in P or NP-complete [5]. An analogous situation appears in $VCSP$ where the constraint language is defined as a set of "soft" predicates.

We believe that problems like minimum cost homomorphism problem ($MinHom$) has an intermediate position between CSPs and $VCSP$s which makes their structure important for understanding the relationship between "hard" and "soft" constraints in optimization. In the minimum cost homomorphism problem, we are given variables subject to constraints and, additionally, costs on variable/value pairs. Now, the task is not just to find any satisfying assignment to the variables, but one that minimizes the total cost. In the context of $VCSP$ this is equivalent to addition of "soft" constraints equal to characteristic functions of one element sets. We will consider a weighted version of this problem.

Definition 1. *Suppose we are given a finite domain set A, a finite constraint language $\Gamma \subseteq \bigcup_{k=1}^{\infty} 2^{A^k}$ and a finite set of functions $R \subseteq \{r : A \to \mathbb{N}\}$. Denote by $MinHom_R(\Gamma)$ the following minimization task:*
Instance: *A triple (V, C, W) where*

- *V is a set of variables;*
- *C is a set of constraints, where each constraint $C \in \mathsf{C}$ is a pair (s, ρ), such that*
  - *$s = (v_1, \ldots, v_m)$ is a tuple of variables of length m, called the constraint scope;*
  - *ρ is an element of Γ with arity m, called the constraint relation.*
- *Weights $w_{vr} \in \mathbb{N}, v \in V, r \in R$.*

Solution: *A function f from V to A, such that, for each variable $v \in V$, $f(v) \in A$, and for each constraint $(s, \rho) \in \mathsf{C}$, with $s = (v_1, \ldots, v_m)$, the tuple $(f(v_1), \ldots, f(v_m))$ belongs to ρ.*
Measure: $\sum_{v \in V} \sum_{r \in R} w_{vr} r(f(v))$.

Definition 2. *For $R^* = \{e_i | i \in A\}$, $MinHom_{R^*}(\Gamma)$ is called* minimum cost homomorphism problem *where $e_i : A \to \mathbb{N}$ denotes a characteristic function of $\{i\} \subseteq A$.*

We will write $MinHom$ instead of $MinHom_{R^*}$ for short. $MinHom$ has applications in defence logistics [10] and machine learning [4]. Complete classification of constraint languages Γ for which $MinHom(\Gamma)$ is polynomial-time solvable has recently been obtained in [19]. The question for which directed graphs H the problem $MinHom(\{H\})$ is polynomial-time solvable was considered in [7,8,9,10,11]. Maximum Solution Problem ($MaxSol$), which is defined analogously to $MinHom$, but with $A \subseteq \mathbb{N}$ and a functional of the form $\sum_{v \in V} w_v f(v)$ to maximize, was investigated in a series of papers [13,14,15]. It is easy to see that if $n = \max_{s \in A} s + 1$ and $R = \{n - x\}$, then $MinHom_R(\Gamma) = MaxSol(\Gamma)$. In this paper, we will assume that a constraint language Γ contains all unary predicates over a domain set A and approach the problem of characterizing the complexity of $MinHom_R(\Gamma)$ in its most general form by algebraic methods. When R satisfies certain conditions, we obtain a dichotomy for $MinHom_R(\Gamma)$, i.e., if $MinHom_R(\Gamma)$ is not polynomial-time solvable, then it is NP-hard. As a consequence, we obtain a dichotomy for conservative $MaxSol$.

In Section 2, we present some preliminaries together with results connecting the complexity of $MinHom_R$ with conservative algebras. The main dichotomy theorem is stated in Section 3 and its proof is divided into several parts which can be found in Sections 4-6. Finally, in Section 7 we present directions for future research.

1.1 Algebraic Structure of Tractable Constraint Languages

Recall that an optimization problem A is called NP-hard if some NP-complete language can be recognized in polynomial time with the aid of an oracle for A. We assume that $P \neq NP$.

Definition 3. *Suppose we are given a finite set A and a constraint language $\Gamma \subseteq \bigcup_{k=1}^{\infty} 2^{A^k}$. The language Γ is said to be R-*tractable *if, for every finite subset $\Gamma' \subseteq \Gamma$, the task $MinHom_R(\Gamma')$ is polynomial-time solvable, and Γ is called R-*NP-hard *if there is a finite subset $\Gamma' \subseteq \Gamma$, such that the task $MinHom_R(\Gamma')$ is NP-hard.*

First, we will state some standard definitions from universal algebra.

Definition 4. *Let $\rho \subseteq A^m$ and $f : A^n \to A$. We say that the function (operation) f* preserves *the predicate ρ if, for every $(x_1^i, \ldots, x_m^i) \in \rho, 1 \leq i \leq n$, we have that $(f(x_1^1, \ldots, x_1^n), \ldots, f(x_m^1, \ldots, x_m^n)) \in \rho$.*

For a constraint language Γ, let $Pol(\Gamma)$ denote the set of operations preserving all predicates in Γ. Throughout the paper, we let A denote a finite domain and Γ a constraint language over A. We assume the domain A to be finite.

Definition 5. *A constraint language Γ is called* a relational clone *if it contains every predicate expressible by a first-order formula involving only*

a) predicates from $\Gamma \cup \{=^A\}$*;*
b) conjunction; and
c) existential quantification.

First-order formulas involving only conjunction and existential quantification are often called *primitive positive (pp) formulas*. For a given constraint language Γ, the set of all predicates that can be described by pp-formulas over Γ is called the *closure* of Γ and is denoted by $\langle \Gamma \rangle$.

For a set of operations F on A, let $Inv\,(F)$ denote the set of predicates preserved under the operations of F. Obviously, $Inv\,(F)$ is a relational clone. The next result is well-known [1,6].

Theorem 6. *For a constraint language* Γ *over a finite set* A*,* $\langle \Gamma \rangle = Inv\,(Pol\,(\Gamma))$*.*

Theorem 6 tells us that the Galois closure of a constraint language Γ is equal to the set of all predicates that can be obtained via pp-formulas from the predicates in Γ.

Theorem 7. *For any finite constraint language* Γ*, finite* $\Gamma' \subseteq \langle \Gamma \rangle$ *and finite* $R \subseteq \{r : A \rightarrow \mathbb{N}\}$ *there is a polynomial time reduction from* $MinHom_R\,(\Gamma')$ *to* $MinHom_R\,(\Gamma)$*.*

The previous theorem tells us that the complexity of $MinHom_R\,(\Gamma)$ is basically determined by $Inv\,(Pol\,(\Gamma))$, i.e., by $Pol\,(\Gamma)$. That is why we will be concerned with the classification of sets of operations F for which $Inv\,(F)$ is a tractable constraint language.

Definition 8. *An* algebra *is an ordered pair* $\mathcal{A} = (A, F)$ *such that* A *is a nonempty set (called a universe) and* F *is a family of finitary operations on* A*. An algebra with a finite universe is referred to as a finite algebra.*

Definition 9. *An algebra* $\mathcal{A} = (A, F)$ *is called* R-tractable *if* $Inv(F)$ *is a* R*-tractable constraint language and* $\mathcal{A}$ *is called* R-NP-hard *if* $Inv(F)$ *is an* R*-NP-hard constraint language.*

For $B \subseteq A$, define $R_B = \{f|_B | f \in R\}$, where $f|_B$ is a restriction of f on a set B. We will use the term MinHom-tractable (NP-hard) instead of $\{e_i | i \in A\}$-tractable (NP-hard) and, in case $A = \{0, 1\}$, the term min-tractable (NP-hard) instead of $\{x\}$-tractable (NP-hard).

We only need to consider a very special type of algebras, so called *conservative* algebras.

Definition 10. *An algebra* $\mathcal{A} = (A, F)$ *is called* conservative *if for every operation* $f \in F$ *we have that* $f\,(x_1, \ldots, x_n) \in \{x_1, \ldots, x_n\}$*.*

Since we assume that Γ is a constraint language with all unary relations over the domain set A, then $\mathcal{A} = (A, Pol\,(\Gamma))$ is conservative. Besides conservativeness of constraint languages we will make some additional restrictions on function sets R.

Definition 11. *Suppose we are given a finite set of functions* $R \subseteq \{r : A \to \mathbb{N}\}$. *Denote by* $G(R) = (A, E(R))$ *a directed graph with a set of vertices* A *and an edge set* $E(R) = \{(a,b) | \exists r \in R\ r(a) > r(b)\}$. *The* $UG(R)$ *is the graph* $G(R)$ *with all edges considered as undirected. We will call* G a preference graph *and* UG an undirected preference graph.

In the sequel, we will assume that a graph UG is complete. It is easy to see that $UG(\{e_i | i \in A\})$, $UG(\{n - x\})$ are complete and our results can be applied to $MinHom$ and $MaxSol$.

2 Boolean Case and the Necessary Local Conditions

The first step to understand the structure of R-tractable algebras is to understand the boolean case. Well-known structure of boolean clones [16] helps us to prove the following theorem.

Theorem 12. *A boolean clone* H *is MinHom-tractable if either* $\{x \wedge y, x \vee y\} \subseteq H$ *or* $\{(x \wedge \overline{y}) \vee (\overline{y} \wedge z) \vee (x \wedge z)\} \subseteq H$, *where* $\overline{x}, x \wedge y, x \vee y$ *denote negation, conjunction and disjunction. Otherwise,* H *is MinHom-NP-hard. A conservative boolean clone* H *is min-tractable if either* $\{x \wedge y\} \subseteq H$ *or* $\{(x \wedge \overline{y}) \vee (\overline{y} \wedge z) \vee (x \wedge z)\} \subseteq H$. *Otherwise,* H *is min-NP-hard.*

Definition 13. *We call a pair of vertices* $\{a, b\}$ *in* $G(R)$ a MinHom-pair *if they have arcs in both directions and we call* $\{a, b\}$ a min-pair *if they have an arc in one direction only.*

Suppose $f \in F$. By $\overset{a}{\underset{b}{\downarrow}} f$, we mean $a \neq b$ and $f(a,b) = f(b,a) = b$. Guided by Theorem 12 we obtain the following definition.

Definition 14. *Let* F *be a conservative functional clone over* A *and* $R \subseteq \{r : A \to \mathbb{N}\}$. *We say that* F *satisfies the* necessary local conditions for R *if and only if*

(1) for every MinHom-pair $\{a, b\}$ *of* $G(R)$, *either*
- *(1.a) there exists* $f_1, f_2 \in F$ *s.t.* $\overset{a}{\underset{b}{\downarrow}} f_1$ *and* $\overset{a}{\underset{b}{\uparrow}} f_2$; *or*
- *(1.b) there exists* $f \in F$ *s.t.* $f|_{\{a,b\}}(x,x,y) = f|_{\{a,b\}}(y,x,x) = f|_{\{a,b\}}(y,x,y) = y$,

(2) for every min-pair $\{a, b\}$ *of* $G(R)$ *such that* $(a,b) \in E(R)$, *either*
- *(2.a) there exists* $f \in F$ *s.t.* $\overset{a}{\underset{b}{\downarrow}} f$; *or*
- *(2.b) there exists* $f \in F$ *s.t.* $f|_{\{a,b\}}(x,x,y) = f|_{\{a,b\}}(y,x,x) = f|_{\{a,b\}}(y,x,y) = y$.

Theorem 15. *Suppose* F *is a conservative functional clone. If* F *is* R*-tractable and* $UG(R)$ *is complete, then it satisfies the necessary local conditions for* R. *If* F *does not satisfy the necessary local conditions for* R, *then it is* R*-NP-hard.*

As in case of $MinHom$, the necessary local conditions are not sufficient for R-tractability of a conservative clone. Let $M = \{B|B \subseteq A, |B| = 2, F|_B$ contains different binary commutative functions$\}$, $M^o = \{(a,b) \mid \{a,b\} \in M\}$ and $\overline{M} = \{B|B \subseteq A, |B| = 2\} \setminus M$.

Introduce an undirected graph without loops $T_F^R = (M^o \cap E(R), P)$ where $P = \left\{ \langle (a,b), (c,d) \rangle \mid (a,b), (c,d) \in M^o \cap E(R), \text{ there is no } f \in F : \begin{smallmatrix} a & c \\ \downarrow & \downarrow \\ b & d \end{smallmatrix} f \right\}$.

Theorem 16. *Suppose F satisfy the necessary local conditions for R and $UG(R)$ is complete. If the graph $T_F^R = (M^o \cap E(R), P)$ is bipartite, then F is R-tractable. Otherwise, F is R-NP-hard.*

A proof for NP-hard case of Theorem 16 will be omitted since it is basically the same as in case of $MinHom$ [19].

3 Multi-sorted MinHom and Its Tractable Case

Definition 17. *For any collection of sets $\mathsf{A} = \{A_i | i \in I\}$, and any list of indices $i_1, \ldots, i_m \in I$, a subset ρ of $A_{i_1} \times \ldots \times A_{i_m}$, together with the list $(i_1, \ldots, i_m)$, will be called* a multi-sorted relation over A with arity m and signature $(i_1, \ldots, i_m)$. *For any such relation ρ, the signature of ρ will be denoted $\sigma(\rho)$.*

Definition 18. *Let Γ be a set of multi-sorted relations over a collection of sets $\mathsf{A} = \{A_i | i \in I\}$. The multi-sorted $MinHom$ problem over Γ, denoted $MMinHom(\Gamma)$, is defined to be the minimization problem with*
Instance: *A quadruple $(V, \delta, \mathsf{C}, W)$ where*

- *V is a set of variables;*
- *δ is a mapping from V to I, called the domain function;*
- *C is a set of constraints, where each constraint $C \in \mathsf{C}$ is a pair (s, ρ), such that*
  - *$s = (v_1, \ldots, v_m)$ is a tuple of variables of length m, called the constraint scope;*
  - *ρ is an element of Γ with arity m and signature $(\delta(v_1), \ldots, \delta(v_m))$, called the constraint relation.*
- *Weights $w_{va} \in \mathbb{N}, v \in V, a \in A_{\delta(v)}$.*

Solution: *A function f from V to $\bigcup_{i \in I} A_i$, such that, for each variable $v \in V$, $f(v) \in A_{\delta(v)}$, and for each constraint $(s, \rho) \in \mathsf{C}$, with $s = (v_1, \ldots, v_m)$, the tuple $(f(v_1), \ldots, f(v_m))$ belongs to ρ.*
Measure: $\sum_{v \in V} w_{vf(v)}$.

We can consider any multi-sorted relation ρ over $\mathsf{A} = \{A_i | i \in I\}$ as an ordinary relation ρ^{A} over a set $\bigcup_{i \in I} A_i$ where $A_i, i \in I$ are considered to be disjoint. If Γ is a set of multi-sorted relations over $\mathsf{A} = \{A_i | i \in I\}$, then Γ^{A} denotes a set of relations of Γ considered as relations over $\bigcup_{i \in I} A_i$. It is easy to see that $MMinHom(\Gamma)$ is equivalent to $MinHom(\Gamma^{\mathsf{A}})$.

Definition 19. *A set of multi-sorted relations over* A, Γ, *is said to be* MinHom-tractable, *if* Γ^{A} *is MinHom-tractable. A set of multi-sorted relations over* A, Γ, *is said to be* MinHom-NP-complete, *if* Γ^{A} *is MinHom-NP-complete.*

Definition 20. *Let* A *be a collection of sets.* An n-ary multi-sorted operation t on A *is defined by a collection of interpretations* $\left\{t^A | A \in \mathsf{A}\right\}$, *where each* t^A *is an* n*-ary operation on the corresponding set* A*. The multi-sorted operation* t *on* A *is said to be* a polymorphism *of a multi-sorted relation* ρ *over* A *with signature* $(\delta(1), \ldots, \delta(m))$ *if, for any* $(a_{11}, \ldots, a_{m1}), \ldots, (a_{1n}, \ldots, a_{mn}) \in \rho$, *we have*

$$t\begin{pmatrix} a_{11} & \cdots & a_{1n} \\ \vdots & & \vdots \\ a_{m1} & \cdots & a_{mn} \end{pmatrix} = \begin{pmatrix} t^{\delta(1)}\left(a_{11}, \ldots, a_{1n}\right) \\ \vdots \\ t^{\delta(m)}\left(a_{m1}, \ldots, a_{mn}\right) \end{pmatrix} \in \rho$$

For any given set of multi-sorted relations Γ, $MPol\left(\Gamma\right)$ denotes the set of multi-sorted operations which are polymorphisms of every relation in Γ.

Definition 21. *Suppose a set of operations* H *over* D *is conservative and* $B \subseteq \left\{\{x,y\} | x, y \in D, x \neq y\right\}$*. A pair of binary operations* $\phi, \psi \in H$ *is called a* tournament pair *on* B*, if* $\forall \{x,y\} \in B\ \phi\left(x,y\right) = \phi\left(y,x\right), \psi\left(x,y\right) = \psi\left(y,x\right), \phi\left(x,y\right) \neq \psi\left(x,y\right)$ *and for arbitrary* $\{x,y\} \in \overline{B}$, $\phi\left(x,y\right) = x, \psi\left(x,y\right) = x$*. An operation* $m \in H$ *is called* arithmetical *on* B*, if* $\forall \{x,y\} \in B\ m\left(x,x,y\right) = m\left(y,x,x\right) = m\left(y,x,y\right) = y$*.*

The following theorem is a simple consequence of the main result of [19].

Theorem 22. *Let* Γ *be a constraint language over* A *containing all unary relations and* $B \subseteq \left\{\{a,b\} | a, b \in A, a \neq b\right\}$*. If* $Pol\left(\Gamma\right)$ *contains operations* ϕ, ψ, m *such that*

- ϕ, ψ *is a tournament pair on* B*,*
- m *is arithmetical on* $\overline{B}$*,*

then Γ *is MinHom-tractable.*

The following theorem is a generalization of the previous one.

Theorem 23. *Let* Γ *be a set of multi-sorted relations over a collection of finite sets* $\mathsf{A} = \{A_1, \ldots, A_n\}$ *containing all unary multi-sorted relations. Assume that* $B_i \subseteq \left\{\{a,b\} | a, b \in A_i, a \neq b\right\}$*. If* $MPol\left(\Gamma\right)$ *contains a multi-sorted operations* ϕ, ψ, m *such that*

- ϕ^{A_i}, ψ^{A_i} *is a tournament pair on* B_i*,*
- m^{A_i} *is arithmetical on* $\overline{B_i}$*,*

then Γ^{A} *is MinHom-tractable.*

4 Structure of R-Tractable Algebras

Definition 24. *Suppose a set of operations H over D is conservative and $O \subseteq \{(x,y)\,|x,y \in D, x \neq y\}$. A pair of binary operations $\phi, \psi \in H$ is called a* weak tournament pair *on O, if*

(1) $\forall\, a,b$ such that $(a,b),(b,a) \in O$: $\overset{a}{\underset{b}{\downarrow}}\phi, \overset{a}{\underset{b}{\uparrow}}\psi$ or $\overset{a}{\underset{b}{\uparrow}}\phi, \overset{a}{\underset{b}{\downarrow}}\psi$;

(2) $\forall\, a,b$ such that $(a,b) \in O, (b,a) \notin O$: $\overset{a}{\underset{b}{\downarrow}}\phi, \overset{a}{\underset{b}{\uparrow}}\psi$ or $\overset{a}{\underset{b}{\uparrow}}\phi, \overset{a}{\underset{b}{\downarrow}}\psi$ or $\overset{a}{\underset{b}{\downarrow}}\phi, \overset{a}{\underset{b}{\downarrow}}\psi$;

(3) $\forall\, a,b$ such that $(a,b) \notin O, (b,a) \notin O$: $\phi|_{\{a,b\}}(x,y) = x, \psi|_{\{a,b\}}(x,y) = y$.

For a binary operation $f \in F$ define $Com(f) = \{\{a,b\} : f|_{\{a,b\}}$ is commutative$\}$. Consider any binary operation $f_{max} \in F$ which has maximal set $Com(f_{max})$, i.e. there is no $f \in F$ such that $Com(f_{max}) \subset Com(f)$. Since for any binary operations a, b, $Com(a(x,y)) \cup Com(b(x,y)) = Com(a(b(x,y), b(y,x)))$, we conclude that $Com(f_{max}) = \bigcup_{f \in F} Com(f)$. Define $Com^o(f_{max}) = \{(a,b)\,|\,\{a,b\} \in Com(f_{max})\}$.

Theorem 25. *If F satisfies the necessary local conditions for R, $UG(R)$ is complete and $T_F^R = (M^o \cap E(R), P)$ is bipartite, then there is a pair $\phi, \psi \in F$ which a weak tournament pair on $Com^o(f_{max}) \cap E(R)$.*

Theorem 26. *If F satisfies the necessary local conditions for R and $\overline{Com(f_{max})} \neq \emptyset$, then F contains an arithmetical operation on $\overline{Com(f_{max})}$.*

5 Proof of Theorem 16

Suppose we have some instance (V, C, W) of $MinHom_R(Inv(F))$. Let $Sol(V, \mathsf{C}, W)$ denote the set of its solutions. By Theorem 25, there is a pair $\phi, \psi \in F$ which is a weak tournament pair on $Com^o(f_{max}) \cap E(R)$.

Until the end of this section, the pair ϕ, ψ will be fixed. For any $f, g : V \to A$, define $\phi(f,g), \psi(f,g) : V \to A$ such that $\phi(f,g)(v) = \phi(f(v), g(v))$, $\psi(f,g)(v) = \psi(f(v), g(v))$. Since any constraint relation of C is preserved by ϕ, ψ, we have $\phi(f,g), \psi(f,g) \in Sol(V, \mathsf{C}, W)$ for $f, g \in Sol(V, \mathsf{C}, W)$.

Lemma 27. *For any measure $M(f) = \sum_{v \in V} \sum_{r \in R} w_{vr} r(f(v))$ we have*

$$M(\phi(f,g)) + M(\psi(f,g)) \leq M(f) + M(g)$$

Lemma 28. *Suppose $A_v = \{f(v)|f \in Sol(V, \mathsf{C}, W)\}$ and there is $a, b \in A_v$ such that $\overset{a}{\underset{b}{\downarrow}}\psi$, $\overset{a}{\underset{b}{\downarrow}}\phi$. Then we have*

$$\min_{f \in Sol(V,\mathsf{C},W), f(v)=b} M(f) \leq \min_{f \in Sol(V,\mathsf{C},W), f(v)=a} M(f)$$

Proof. Suppose that $f^* \in Sol(V, \mathsf{C}, W), f^*(v) = a$ and $f^{**} \in Sol(V, \mathsf{C}, W), f^{**}(v) = b$ such that

$$M(f^*) = \min_{f \in Sol(V,\mathsf{C},W), f(v)=a} M(f), M(f^{**}) = \min_{f \in Sol(V,\mathsf{C},W), f(v)=b} M(f)$$

Since $\phi(f^*, f^{**})(v) = b$, $\psi(f^*, f^{**})(v) = b$, then by Lemma 27

$$2M(f^{**}) \leq M(\phi(f^*, f^{**})) + M(\psi(f^*, f^{**})) \leq M(f^*) + M(f^{**})$$ □

Now we are ready to describe polynomial-time algorithm for $MinHom(Inv(F))$, which conceptually follows the proof of Theorem 8.3 in [3]. First we compute sets $A_v = \{f(v) | f \in Sol(V, \mathsf{C}, W)\}$ for every $v \in V$. This could be done polynomial-time, because, by result of [2], $CSP(Inv(F))$ is polynomial-time solvable by 3-consistency algorithm: adding a constraint $(v, \{a\})$ to C we can find whether $a \in A_v$ or not. After this operation, we iteratively delete all elements a from every A_v such that there is $b \in A_v : \underset{b}{\overset{a}{\downarrow}} \psi, \underset{b}{\overset{a}{\downarrow}} \phi$. This operation will not increase a minimum of optimized functional due to Lemma 28. Afterwards, we update every constraint pair $((v_1, \ldots, v_m), \rho) \in \mathsf{C}$ by defining $\rho = \rho \cap A_{v_1} \times \ldots \times A_{v_m}$.

Consider a collection of sets $\mathsf{A} = \{A_v | v \in V\}$. Obviously, $|\mathsf{A}| < 2^{|A|}$. For a pair $((v_1, \ldots, v_m), \rho) \in \mathsf{C}$, a predicate ρ can be considered as multi-sorted predicate over A with signature $(A_{v_1}, \ldots, A_{v_m})$. Moreover, for any $p \in F$ this multi-sorted predicate preserves a polymorphism $\{p^B | B \in \mathsf{A}\}$ where $p^B = p|_B$.

By Theorem 26, F contains an operation m which is arithmetical on $\overline{Com(f_{max})}$. Consider multi-sorted polymorphisms $\{\phi^B | B \in \mathsf{A}\}, \{\psi^B | B \in \mathsf{A}\}, \{m^B | B \in \mathsf{A}\}$. It is easy to see that this polymorphisms satisfy conditions of Theorem 23 and the set of multi-sorted predicates $\{\rho | (s, \rho) \in \mathsf{C}\}$ is tractable. Therefore, we can polynomially solve our problem.

6 Directions for Further Work

As it was said in the introduction, $MinHom_R$ fits the framework of Valued CSP ($VCSP$) [3]. The following subproblems of $VCSP$ generalize $MinHom$ and are not investigated yet.

1. What happens with $MinHom_R$ when $UG(R)$ is not complete. We believe that Theorem 16 is true for this case too.

2. Suppose we have a finite valued constraint language Γ, i.e. a set of valued predicates over some finite domain set. If Γ contains all unary valued predicates, we call $VCSP(\Gamma)$ a conservative $VCSP$. This name is motivated by the fact that in this case the multimorphisms (which is a generalization of polymorphisms for valued constraint languages [3]) of Γ must consist of conservative functions. Since there is a well-known dichotomy for conservative CSPs [2], we suspect that there is a dichotomy for conservative $VCSPs$.

References

1. Bodnarcuk, V.G., Kalužnin, L.A., Kotov, N.N., Romov, B.A.: Galois theory for Post algebras. Kibernetika 3(1-10), 5 (1–9) (in Russian)
2. Bulatov, A.A.: Tractable conservative Constraint Satisfaction Problems. In: 18th Annual IEEE Symposium on Logic in Computer Science, pp. 321–330 (2003)

3. Cohen, D., Cooper, M., Jeavons, P.: Generalising submodularity and horn clauses: Tractable optimization problems defined by tournament pair multimorphisms. Theor. Comput. Sci. 401, 36–51 (2008)
4. Daniels, H., Velikova, M.: Derivation of monotone decision models from non-monotone data. Tilburg University, Center Internal Report 30 (2003)
5. Feder, T., Vardi, M.Y.: The computational structure of monotone monadic SNP and constraint satisfaction: A study through datalog and group theory. SIAM Journal on Computing 28, 57–104 (1999)
6. Geiger, D.: Closed Systems of Functions and Predicates. Pacific Journal of Mathematics 27, 95–100 (1968)
7. Gupta, A., Hell, P., Karimi, M., Rafiey, A.: Minimum cost homomorphisms to reflexive digraphs. In: Laber, E.S., Bornstein, C., Nogueira, L.T., Faria, L. (eds.) LATIN 2008. LNCS, vol. 4957, pp. 182–193. Springer, Heidelberg (2008)
8. Gutin, G., Hell, P., Rafiey, A., Yeo, A.: A dichotomy for minimum cost graph homomorphisms. European Journal of Combinatorics 29, 900–911 (2008)
9. Gutin, G., Hell, P., Rafiey, A., Yeo, A.: Minimum cost and list homomorphisms to semicomplete digraphs. Discrete Appl. Math. 154, 890–897 (2006)
10. Gutin, G., Rafiey, A., Yeo, A., Tso, M.: Level of repair analysis and minimum cost homomorphisms of graphs. Discrete Applied Mathematics 154, 881–889 (2006)
11. Gutin, G., Rafiey, A., Yeo, A.: Minimum Cost Homomorphism Dichotomy for Oriented Cycles. In: Fleischer, R., Xu, J. (eds.) AAIM 2008. LNCS, vol. 5034, pp. 224–234. Springer, Heidelberg (2008)
12. Jeavons, P.: On the Algebraic Structure of Combinatorial Problems. Theoretical Computer Science 200, 185–204 (1998)
13. Jonsson, P., Kuivinen, F., Nordh, G.: Max Ones generalised to larger domains. SIAM Journal on Computing 38, 329–365 (2008)
14. Jonsson, P., Nordh, G.: Generalised integer programming based on logically defined relations. In: Královič, R., Urzyczyn, P. (eds.) MFCS 2006. LNCS, vol. 4162, pp. 549–560. Springer, Heidelberg (2006)
15. Jonsson, P., Nordh, G., Thapper, J.: The maximum solution problem on graphs. In: Kučera, L., Kučera, A. (eds.) MFCS 2007. LNCS, vol. 4708, pp. 228–239. Springer, Heidelberg (2007)
16. Post, E.: The two-valued iterative systems of mathematical logic. Annals of Mathematical Studies 5 (1941)
17. Rafiey, A., Hell, P.: Duality for Min-Max Orderings and Dichotomy for Min Cost Homomorphisms, http://arxiv.org/abs/0907.3016v1
18. Schaefer, T.J.: The complexity of satisfiability problems. In: 10th ACM Symposium on Theory of Computing, pp. 216–226 (1978)
19. Takhanov, R.: A dichotomy theorem for the general minimum cost homomorphism problem. In: STACS (to appear, 2010)

The Longest Almost-Increasing Subsequence

Amr Elmasry*

Max-Planck Institut für Informatik, Saarbrücken, Germany
elmasry@mpi-inf.mpg.de

Abstract. Given a sequence of n elements, we introduce the notion of an almost-increasing subsequence in two contexts. The first notion is the longest subsequence that can be converted to an increasing subsequence by possibly adding a value, that is at most a fixed constant c, to each of the elements. We show how to optimally construct such subsequence in $O(n \log k)$ time, where k is the length of the output subsequence. As an exercise, we show how to produce in $O(n^2 \log k)$ time a special type of subsequences, that we call subsequences obeying the triangle inequality, by using as a subroutine our algorithm for the above case. The second notion is the longest subsequence where every element is at least the value of a monotonically non-decreasing function in terms of the r preceding elements (or even with respect to every r elements among those preceding it). We show how to construct such subsequence in $O(n^r \log k)$ time.

1 Introduction

The longest increasing subsequence (LIS) is a subsequence of maximum length where every element is greater than the previous element. The longest increasing subsequence problem refers to either producing the subsequence or just finding its length. The problem was first tackled by Robinson [15] seventy years ago. The classical dynamic-programming algorithm for the problem, which appears in many algorithmic textbooks [8], is due to Schensted [16]. This algorithm runs in $O(n \log k)$, where n is the length of the input sequence, and k is the length of the longest increasing subsequence. Knuth [13] gave generalizations to the problem with relations to Young tableaux. Fredman [11] showed that $O(n \log n)$ comparisons are both necessary and sufficient, to find the length or produce the subsequence, in the comparison-tree model. The same lower bound was also proven for the algebraic decision-tree model [14]. If the input sequence is a permutation of the integers 1 to n, some algorithms were introduced that construct the longest increasing subsequence in $O(n \log \log n)$ time [7,12].

The problem is important in practice. Several other problems involve a LIS construction (see, for example, [5]). It has lately gained even more practical importance as it is used in the MUMmer system [9] for aligning genomes.

A related problem is the longest common subsequence (LCS) problem, which considers two sequences and locates a series of entries that appear in the same order in both sequences. Note that we can apply the LCS algorithms to a sequence and its sorted outcome to get a longest increasing subsequence.

* Supported by Alexander von Humboldt Fellowship.

M.T. Thai and S. Sahni (Eds.): COCOON 2010, LNCS 6196, pp. 338–347, 2010.

Several variants of the LIS problem have been introduced. The longest increasing subsequence of a circular list (LICS) assumes the input sequence to be circular. A randomized algorithm for the LICS that runs in expected $O(n^{3/2} \log n)$ time is given in [3]. The best worst-case bound known is $O(n^2)$, and can be achieved using techniques from [4]. Another variant is to find the longest increasing subsequences that lie in all the sliding windows with a specified width. An algorithm that runs in time proportional to the size of the output subsequences plus an additive bound for constructing one LIS is given in [4]. A generalization of the LIS problem is discussed in [2], where a fixed set of permutations is given and the task is to compute, for a given input sequence, the longest subsequence that is order isomorphic to one of the given permutations. The LIS problem applies when these given permutations are the identity permutations.

The combinatorics of the problem are of no less importance. Starting with the work of Erdös and Szekeres, the length of a LIS in a random permutation was investigated. The complete limiting distribution of the length of the LIS of a permutation of length n chosen uniformly at random is given in [6]. The expected length of a LIS is shown to be close to $2\sqrt{n}$.

Suppose one is considering the process of monitoring the performance of an activity. We say that the activity is well performing once it is well performing in comparison with a large number of accredited historical snapshots where it was as well performing when deploying the same criteria. Picking the largest number of points when the activity is strictly performing better among such previously selected points is too restricted and unfair. The notion need to be relaxed to reflect a good progress without necessarily being the best selected so far. We need a relaxed version of the LIS problem.

In this paper, we introduce a variant of the LIS problem that we call the longest almost-increasing subsequence (LaIS) problem. We define such notion in two contexts. In Section 2, we allow a drop of at most a constant value from the maximum element that appeared so far. In Section 3, we only restrict an element to be more than a defined monotonically non-decreasing function (say the median or the average) of some of its predecessors. Using dynamic programming, we show how to efficiently solve both versions. For the first case, we give an algorithm that runs in $O(n \log k)$ time, which is asymptotically optimal (our algorithm can be used to find the LIS by setting the constant to zero). For the second case, we give an algorithm that runs in $O(n^r \log k)$ time, where r is the number of predecessors considered. Practically, we should therefore consider a constant number of such elements. It is still to be investigated whether the case when r grows with n has polynomial-time solutions or not. In Section 4, we consider the problem of finding the longest subsequence $\langle y_1, y_2, \ldots, y_k \rangle$ that satisfies : $\forall_{j1<j2<j3}\ y_{j3} \geq y_{j2} - y_{j1}$. We call such condition the *triangle inequality.* By applying the algorithm of Section 2 as a subroutine, we give an $O(n^2 \log k)$ algorithm to produce the longest subsequence that obeys the triangle inequality. Note that the function in the case of the triangle inequality is a difference function (i.e. not monotonically increasing), and hence the algorithm in Section 3 cannot be applied.

2 Problem I

Given a sequence $\langle x_1, x_2, \ldots x_n \rangle$ and a constant c, our goal is to construct the longest subsequence $\langle y_1, y_2, \ldots, y_k \rangle$ such that $\forall_i \; y_i \geq \max_{j=1}^{i-1} y_j - c$.

Recursive Formulation

Let $LaIS(h, i)$ denote the longest almost-increasing subsequence among the elements of the subsequence $\langle x_1, \ldots, x_i \rangle$, such that the largest element included is x_h. We show that these two parameters fully characterize any $LaIS$, and recursively express the $LaIS$ in terms of solutions to smaller problems. The key observation is that $LaIS(h, i)$ can be split into two independent subsequences (except for the value of h), following the relation:

$$LaIS(h, i) = LaIS(h, h) \cdot T(h, i),$$

where "$\cdot$" is the concatenation operation, and $T(h, i)$ is the subsequence including every element x_j among $\langle x_{h+1}, \ldots, x_i \rangle$ satisfying $x_h - c \leq x_j < x_h$.

The second observation is that $LaIS(h, h)$ can be expressed as:

$$LaIS(h, h) = LaIS(i', h - 1) \cdot \langle x_h \rangle,$$

where $length(LaIS(i', h - 1))$ is the maximum among all i' satisfying $x_{i'} \leq x_h$.

The Basic Algorithm

The algorithm proceeds in n iterations. After the i-th iteration, the algorithm maintains for each element x_j, among the first i elements, the longest almost-increasing subsequence whose largest element is x_j. For each such subsequence, it is enough to keep track of its length l_j, and the index p_j of the largest element among the elements preceding x_j. During the i-th iteration, two tasks are performed: The first task is to find the longest almost-increasing subsequence whose largest element is x_i. This subsequence is constructed by appending x_i to the longest subsequence found so far whose largest element is at most x_i. More formally, we look for an index $i' < i$ such that $l_{i'} \geq l_j$ among all indexes $j < i$ having $x_j \leq x_i$. The length l_i is then set to $l_{i'} + 1$, and the index p_i is set to i'. The second task is to append x_i to every subsequence found so far whose largest element is larger than x_i and at most $x_i + c$. More formally, for all $j < i$, set l_j to $l_j + 1$ if $x_i < x_j \leq x_i + c$. After the n-th iteration, the length of the longest almost-increasing subsequence is the maximum length l_m among the l_i's stored by the algorithm. To construct the longest almost-increasing subsequence, we make use of the p_i's to produce the subsequence in reverse order. Using the element x_m corresponding to the maximum length l_m, scan every element x_i, from $i = n$ to $m + 1$, and output the elements satisfying $x_m - c \leq x_i < x_m$ followed by x_m. Let x_t be the last element of the subsequence output in the reverse order, scan every element x_i from $i = t-1$ to p_t+1, and output the elements satisfying $x_{p_t} - c \leq x_i < x_{p_t}$ followed by x_{p_t}. The previous step is repeated until we get a value of p_t that indicates the first element of the subsequence (when, for example, $p_t = t$).

A straightforward implementation of the previous algorithm would use two linked lists (or two arrays) each of size n; one for the l_i's and another for the p_i's. This implementation runs in $O(n^2)$ time.

The Improved Algorithm

To achieve the claimed $O(n \log k)$ bound, the inner loop must be executed in $O(\log k)$ time. Instead of storing one l_i corresponding to each x_i, we store *at most* one element for every length value. Namely, after the i-th iteration, we store $y_l \leftarrow x_h$ corresponding to length l, where $length(LaIS(h,i)) = l$ and $x_h \leq x_{h'}$ for all h' satisfying $length(LaIS(h',i)) = l' \geq l$. It follows that $y_l \leq y_{l'}$ when $l < l'$. To show that these elements are enough to construct all the longest subsequences, consider two elements $x_h > x_{h'}$ where $length(LaIS(h',i)) \geq length(LaIS(h,i))$. Consider any almost-increasing sequence that contains $LaIS(h,i)$ as a prefix subsequence, we can replace $LaIS(h,i)$ with $LaIS(h',i)$ and get a valid (almost-increasing) sequence with at least the same length.

In the sequel, we call the second task of the i-th iteration of the basic algorithm, which involves incrementing the length of every subsequence whose largest element is larger than x_i and at most $x_i + c$, a *length-shift*. Apart from the length-shift, the algorithm is quite similar to that of constructing the longest increasing subsequence [16]. If we use an array to store the sequence $\langle y_1, y_2, \ldots \rangle$, following the algorithm in [16], we will be able to use binary search to locate the predecessor of x_i and $x_i + c$ in $O(\log k)$ time. But, a length-shift, which is now performed by sequentially overwriting every y_j in a specified range using $y_{j+1} \leftarrow y_j$, would require $O(k)$ time. This results in an implementation that runs in $O(nk)$ time; still not the claimed bound.

To efficiently implement a length-shift, we store the y_j's as a linked list, where a node holding y_j points to a node holding its successor element $y_{j'}$. Note that, as a result of the length-shifts, j' need not be equal to $j + 1$. With every node, we also store an integer d that is the difference between the length this node represents and the length represented by the predecessor node. Once a range of nodes is determined, a length-shift can be efficiently performed on this range by incrementing d for the first node in the range and decrementing d for the successor of the last node in the range. If the value of d for this successor node becomes 0 following the decrement, the node is deleted from the list (it is overwritten). During each iteration, a search is performed for the predecessor of x_i. A node that contains x_i and a value of d equals 1 is then inserted in this position. To construct the subsequence, we still maintain an array for the p_i's. This array is modified once per iteration, and later used to produce the output.

To summarize, at each of the n iterations the algorithm performs: two predecessor searches, one successor finding, an insertion, a possible deletion, plus a length-shift that is done in constant time. See the pseudo-code for Algorithm 1.

Algorithm 1. pseudo-code for Problem I.

```
for i = 1 to n do
    v ← new_node()
    v.value ← x_i
    v.d ← 1
    v.index ← i
    pred ← search(x_i)
    if (pred ≠ null) then
        p_i ← pred.index
    else
        p_i ← i
    end if
    insert(v)
    s ← successor(search(x_i + c))
    if (s ≠ null) then
        s.d ← s.d − 1
        if (s.d == 0) then
            delete(s)
        end if
    end if
end for
m ← tail_node().index
for i = n down to m + 1 do
    if (x_m − c ≤ x_i < x_m) then
        print x_i
    end if
end for
print x_m
t ← m
while (p_t ≠ t) do
    for i = t − 1 down to p_t + 1 do
        if (x_{p_t} − c ≤ x_i < x_{p_t}) then
            print x_i
        end if
    end for
    print x_{p_t}
    t ← p_t
end while
```

Each of these operations can be executed in $O(\log k)$ time when building a balanced search structure over the nodes of the linked list. We may use, for example, an AVL tree [1] to achieve $O(\log k)$ cost per operation, or a splay tree [17] for an amortized $O(\log k)$ cost per operation. For practical purposes, it would be better to keep a linked list of pointers to find the successor of a given node in constant time, accounting for at least three pointers per node for the mentioned structures. The best search structure for this application, from the practical point of view, is the jumplist [10]. Using a jumplist: a search, an insertion and a deletion require $O(\log k)$ amortized time, finding the successor requires constant time, and we only use and maintain two pointers per node.

3 Problem II

Given a sequence $\langle x_1, x_2, \ldots x_n \rangle$, our goal is to construct the longest subsequence $\langle y_1, y_2, \ldots, y_k \rangle$ such that $y_j \geq f(y_{j-1}, y_{j-2})$, where the function $f(.,.)$ is a monotonically non-decreasing function in terms of its two parameters.

The Basic Algorithm

We use a two-dimensional upper-triangular array. An entry $l_{a,b}$ represents the length of the longest subsequence that ends with x_b preceded by x_a. To be able to efficiently construct the subsequence, we also maintain a two-dimensional array of indexes p. During the i-th iteration, every column $j < i$ is sequentially scanned. We find the index i' such that $l_{i',j} \geq l_{a,j}$ among all indexes a having $x_i \geq f(x_a, x_j)$. The length $l_{j,i}$ is then set to $l_{i',j}+1$, and the index $p_{j,i}$ is set to i'. After the n-th iterations, the length of the longest almost-increasing subsequence is the value of the maximum entry in the array, say $l_{m,m'}$. The longest almost-increasing subsequence in reverse order will be: $x_{m'}, x_m, x_{p_{m,m'}}, \ldots$, until we get a value of $p_{t,t'}$ that indicates the first element of the subsequence (when, for example, $p_{t,t'} = t$). See the pseudo-code for Algorithm 2.

A straightforward implementation of this algorithm runs in $O(n^3)$ time.

The Improved Algorithm

To achieve the $O(n^2 \log k)$ bound, the same idea that is used in the solution for the longest increasing subsequence [16] is again deployed. We store a two-dimensional array, this time, having k rows and n columns. Every row represents a length from 1 to k, and every column i represents element x_i. An entry indexed by (l, i) in this two-dimensional array represents a valid subsequence with length l that ends with x_i, moreover it should have the smallest element x_j preceding x_i among those with the same length l. Again, the smallest possible element x_j is the best that we can use for constructing longer subsequences. When considering x_i to be appended to the current solutions, instead of sequentially scanning a column by the innermost loop, we use binary search to get a sequence with maximum length that ends with an element smaller than x_i. Of course we again use the fact that the elements stored in one column are monotonically increasing.

Extensions

The previous algorithm can be extended to construct the longest subsequence that has every element bounded below by a value of a function that is monotonically non-decreasing in terms of its r immediate predecessors, instead of only two. An r-dimensional array is used instead of the two-dimensional array, for a total running time of $O(n^r \log k)$.

Algorithm 2. pseudo-code for Problem II.

```
for i = 1 to n do
   for j = i to n do
      l_{i,j} ← 2
   end for
end for
for i = 2 to n do
   for j = 1 to i − 1 do
      i' ← j
      for a = 1 to j − 1 do
         if (x_i ≤ f(x_a, x_j)  &&  l_{a,j} > l_{i',j}) then
            i' ← a
         end if
      end for
      if (i' ≠ j) then
         l_{j,i} ← l_{i',j} + 1
      end if
      p_{j,i} ← i'
   end for
end for
m ← 1
m' ← 2
for i = 1 to n do
   for j = i + 1 to n do
      if (l_{i,j} > l_{m,m'}) then
         m ← i
         m' ← j
      end if
   end for
end for
print x_{m'}, x_m
t ← m
t' ← m'
while (p_{t,t'} ≠ t) do
   tt ← p_{t,t'}
   print x_{tt}
   t' ← t
   t ← tt
end while
```

The case where every element is at least the value a monotonically non-decreasing function of any two preceding elements can be solved in the same asymptotic bound. Since the function is non-decreasing, the validity of appending an element to a subsequence solely depends on the value of the largest two elements in this subsequence. It follows that if we store the partial solutions in the form of a length value together with the largest two elements among those forming the subsequence, this would be enough to extend the subsequences and recalculate these parameters by appending a new element. Once more, using the trick of storing the value of the first of these two elements and indexing these by the length values, we achieve the $O(n^2 \log k)$ bound. Naturally, the ideas are extended to include r elements, instead of two, for an $O(n^r \log k)$ time bound.

It may perhaps be surprising that the task of producing the longest subsequence where every element depends on all its predecessors may be easier. This is possible if the underlying function has the property that, when extending a subsequence by appending a new element, the new value of the function can be efficiently recomputed using the old value, the length, and the new element. (This is, for example, the case with the average.) The problem can then be solved with a slight modification to the original algorithm for finding the longest increasing subsequence, just in $O(n \log k)$ time. The reason is that, in this case, any valid subsequence can be only characterized in terms of the value of the function and the length of the subsequence.

4 Subsequences Satisfying the Triangle Inequality

In this section, we consider the problem of computing the longest subsequence satisfying the triangle inequality. More precisely, given a sequence $\langle x_1, x_2, \ldots, x_n \rangle$, our goal is to compute the longest subsequence $\langle y_1, y_2, \ldots, y_k \rangle$ such that

$$y_{j2} \leq y_{j1} + y_{j3}, \; \forall \; j1 < j2 < j3 \tag{1}$$

We show how this subsequence can be produced by solving $2n$ instances of the longest almost-increasing subsequence problem as defined in Section 2. For a given sequence, let $\langle y_1, y_2, \ldots, y_k \rangle$ be a subsequence satisfying (1). The following lemma gives a necessary and sufficient condition for a feasible subsequence, and hence leads to a natural algorithm.

Lemma 1. *Let $\langle y_1, y_2 \ldots, y_k \rangle$ be a sequence, with $y_s = \min_{i=1}^{k} y_i$, then the following conditions are necessary and sufficient for the sequence to satisfy the triangle inequality.*

1. $y_s + y_i \geq y_j \; \forall \; i < j \leq s$.
2. $y_s + y_j \geq y_i \; \forall \; s \leq i < j$.

Proof. The conditions are clearly necessary. To show sufficiency, we need to show that if the conditions hold, then the triangle inequality holds for all triples. Consider any three elements y_{j1}, y_{j2}, and y_{j3}, where $j1 < j2 < j3$, and assume that the conditions hold. We need to show that $y_{j1} + y_{j3} \geq y_{j2}$. There are four cases to consider, and we show that the triangle inequality holds in all cases.

1. $j1 < j2 < j3 \leq s$: Since $y_s + y_{j1} \geq y_{j2}$, and $y_{j3} \geq y_s$, the result follows.
2. $s \leq j1 < j2 < j3$: Since $y_s + y_{j3} \geq y_{j2}$, and $y_{j1} \geq y_s$, the result follows.
3. $j1 < j2 \leq s \leq j3$: Since $y_s + y_{j1} \geq y_{j2}$, and $y_{j3} \geq y_s$, the result follows.
4. $j1 \leq s \leq j2 < j3$: Since $y_s + y_{j3} \geq y_{j2}$, and $y_{j1} \geq y_s$, the result follows.

□

An algorithm directly follows from the previous lemma. The algorithm runs for n iterations. At the i-th iteration, we impose that x_i is the smallest element of the resulting subsequence, i.e. $x_i = y_s$. We then execute the following steps: Remove all elements smaller than x_i. Compute the longest almost-increasing subsequence L_1 of the sequence $\langle x_i, x_{i-1}, \ldots, x_1 \rangle$ (the *reverse* of $\langle x_1, x_2 \ldots, x_i \rangle$), with $c = x_i$. Compute the longest almost-increasing subsequence L_2 of the sequence $\langle x_{i+1}, \ldots, x_n \rangle$, again with $c = x_i$. The concatenation of the *reverse* of L_1 with L_2 gives the longest subsequence satisfying the triangle inequality with x_i as the minimum element. Repeating for all x_i and returning the longest subsequence found give us the desired solution. The algorithm therefore performs $2n$ calls to the *LaIS* algorithm of Section 2, for a total of $O(n^2 \log k)$ time.

5 Conclusion

We have introduced the longest almost-increasing subsequence problem as an extension to the well-studied longest increasing subsequence problem, and obtained an optimal $O(n \log k)$ algorithm for this problem. We then considered the problem of computing the longest subsequence where every element is at least as large as a function of the previous r elements, and presented an $O(n^r \log k)$ algorithm assuming the function is monotonic. The intimidating question is whether this algorithm can be improved to work in polynomial or quasi-polynomial time in terms of both n and r, even by producing an approximate outcome. It would be interesting to investigate if these problems have other interesting applications. We also showed how the longest almost-increasing subsequence problem can be used to obtain a solution, that runs in $O(n^2 \log k)$ time, to the problem of computing the longest subsequence with the triangle inequality. The question we leave unanswered is the possibility of improving the running time to $o(n^2 \log k)$.

Acknowledgment

I would like to thank Rajiv Raman and Saurabh Ray for introducing the problem and for several useful discussions.

References

1. Adelson-Velskii, G., Landis, E.: On an information organization algorithm. Doklady Akademia Nauk SSSR 146, 263–266 (1962)
2. Albert, M., Aldred, R., Atkinson, M., Ditmarsch, H., Handley, B., Handley, C., Opatrny, J.: Longest subsequences in permutations. Australian J. Combinatorics 28, 225–238 (2003)
3. Albert, M., Atkinson, M., Nussbaum, D., Sack, J., Santoro, N.: On the longest increasing subsequence of a circular list. Information Processing Letters 101, 55–59 (2007)
4. Albert, M., Golnski, A., Hamel, A., Lopez-Ortiz, A., Rao, S., Safari, M.: Longest increasing subsequences in sliding windows. Theoretical Computer Science 321, 405–414 (2004)

5. Apostolico, A., Atallah, M., Hambrusch, S.: New clique and independent set algorithms for circle graphs. Discrete Appl. Math. 36, 1–24 (1992)
6. Baik, J., Deift, P., Johannsson, K.: On the distribution of the length of the longest increasing subsequence of random permutations. J. American Math. Society 12, 1119–1178 (1999)
7. Chang, M., Wang, F.: Efficient algorithms for the maximum weight clique and maximum weight independent set problems on permutation graphs. Information Processing Letters 76, 7–11 (2000)
8. Cormen, T., Leiserson, C., Rivest, R., Stein, C.: Introduction to Algorithms, 2nd edn. The MIT Press, Cambridge (2001)
9. Delcher, A., Kasif, S., Fleischmann, R., Paterson, J., White, O., Salzberg, S.: Alignment of whole genomes. Nucl. Acis. Res. 27, 2369–2376 (1999)
10. Elmasry, A.: Deterministic jumplists. Nordic J. Comp. 12(1), 27–39 (2005)
11. Fredman, M.: On computing the length of longest increasing subsequence. Disc. Math. 11, 29–35 (1975)
12. Hunt, J., Szymanski, T.: A fast algorithm for computing longest common subsequences. ACM Comm. 20, 350–353 (1977)
13. Knuth, D.: Permutations, matrices, and generalized Young tableaux. Pacific J. Math. 34, 709–727 (1970)
14. Ramanan, P.: Tight $\Omega(n \log n)$ lower bound for finding a longest increasing subsequence. International J. Comp. Math. 65(3-4), 161–164 (1997)
15. Robinson, G.: On representations of the symmetric group. American J. Math. 60, 745–760 (1938)
16. Schensted, C.: Longest increasing and decreasing subsequences. Canadian J. Math. 13, 179–191 (1961)
17. Sleator, D., Tarjan, R.E.: Self-adjusting binary search trees. J. ACM 32(3), 652–686 (1985)

Universal Test Sets for Reversible Circuits
(Extended Abstract)

Satoshi Tayu, Shota Fukuyama, and Shuichi Ueno

Department of Communications and Integrated Systems
Tokyo Institute of Technology, Tokyo 152-8550-S3-57, Japan
{tayu,ueno}@lab.ss.titech.ac.jp

Abstract. A set of test vectors is complete for a reversible circuit if it covers all stuck-at faults on the wires of the circuit. It has been known that any reversible circuit has a surprisingly small complete test set, while it is NP-hard to generate a minimum complete test set for a reversible circuit. A test set is universal for a family of reversible circuits if it is complete for any circuit in the family. We show minimum universal test sets for some families of CNOT circuits.

1 Introduction

The power consumption and heat dissipation are major issues for VLSI circuits today. Landauer [3] showed that conventional irreversible circuits necessarily dissipate heat due to the erasure of information. Bennett [1] showed, however, that heat dissipation can be avoided if computation is carried out without losing any information. This motivates the study of reversible circuits. Furthermore, reversible circuits have potential applications in nanocomputing [4], digital signal processing [8], and quantum computing [5].

In order to ensure the functionality and durability of reversible circuits, testing and failure analysis are extremely important during and after the design and manufacturing. It has been known that testing of reversible circuits is relatively easier than conventional irreversible circuits in the sense that few test vectors are needed to cover all stuck-at faults, while it is NP-hard to generate a minimum complete test set for a reversible circuit. This paper considers universal test sets for some families of reversible circuits.

1.1 Reversible Circuits

A function $f{:}\{0,1\}^n \rightarrow \{0,1\}^n$ is called a *permutation* if it is bijective. A permutation f is *linear* if $f(\boldsymbol{x} \oplus \boldsymbol{y}) = f(\boldsymbol{x}) \oplus f(\boldsymbol{y})$ for any $\boldsymbol{x}, \boldsymbol{y} \in \{0,1\}^n$, where $\oplus$ is the bitwise XOR operation.

A gate is *reversible* if the Boolean function it computes is a permutation. If a reversible gate has k input and output wires, it is called a $k \times k$ *gate*. A circuit is *reversible* if all gates are reversible and are interconnected without fanout or feedback. If a reversible circuit has n input and output wires, it is called an $n \times n$ *reversible circuit*.

M.T. Thai and S. Sahni (Eds.): COCOON 2010, LNCS 6196, pp. 348–357, 2010.

An $n \times n$ reversible circuit is constructed by starting with n wires, forming the basic circuit, and iteratively concatenating a reversible gate to some subset of the output wires of the previous circuit. Thus, a reversible circuit C can be represented as a sequence of reversible gates G_i: $C = G_1 G_2 \ldots G_m$.

1.2 Fault Testing

We focus our attention on detecting stuck-at faults in a reversible circuit C which fix the values of wires to either 0 or 1. Let $W(C)$ be the set of all wires of C. $W(C)$ consists of all output wires of C and input wires to the gates in C. $W(C)$ is the set of all possible fault locations in C. For an $n \times n$ reversible circuit C, a *test* is an input vector in $\{0,1\}^n$. A test set T is said to be *complete* for C if for every possible single or multiple stuck-at fault on $W(C)$, there exists a test $\boldsymbol{t} \in T$ which detects the fault. It is known that there exists a complete test set for any reversible circuit [6]. Let $\tau(C)$ be the minimum size of a complete test set for a reversible circuit C. It is also known that computing $\tau(C)$ is NP-hard [9], while $\tau(C) = O(\log |W(C)|)$ for any reversible circuit C [6], and there exists a reversible circuit C such that $\tau(C) = \Omega(\log \log |W(C)|)$ [9].

1.3 Universal Test Sets

Let $\mathcal{D}_n$ be the set of all $n \times n$ reversible circuits. A test set $T \subseteq \{0,1\}^n$ is said to be *universal* for $\mathcal{D} \subseteq \mathcal{D}_n$ if T is complete for any circuit in $\mathcal{D}$. It is known that there exists a universal test set for any $\mathcal{D} \subseteq \mathcal{D}_n$[6]. Let $\sigma(\mathcal{D})$ be the minimum size of a universal test set for $\mathcal{D}$. Since the value of a wire of an $n \times n$ reversible circuit is set to 0 by exactly 2^{n-1} input vectors, and to 1 by the remaining 2^{n-1} input vectors, we have the following. (See Theorem IV in Section 2)

Theorem I. [6] *$\sigma(\mathcal{D}) \leq 2^{n-1} + 1$ for any $\mathcal{D} \subseteq \mathcal{D}_n$.* □

1.4 CNOT Circuits

A *k-CNOT gate* is a $(k+1) \times (k+1)$ reversible gate. It passes some k inputs, referred to as control bits, to the outputs unchanged, and inverts the remaining input, referred to as target bit, if the control bits are all 1. The 0-CNOT gate is just an ordinary NOT gate. A *CNOT gate* is a k-CNOT gate for some k. Some CNOT gates are shown in Fig. 1, where a control bit and target bit are denoted by a black dot and ring-sum, respectively. A *CNOT circuit* is a reversible circuit consisting of only CNOT gates. Any Boolean function can be implemented by a CNOT circuit, since the 2-CNOT gate can implement the NAND function. Any permutation can also be implemented by a CNOT circuit. A k-CNOT circuit is a CNOT circuit consisting of only k-CNOT gates. A 1-CNOT circuit is called a *linear reversible circuit.* A linear reversible circuit computes a linear permutation, and any linear permutation can be implemented by a linear reversible circuit [7]. The linear reversible circuits are an important class of reversible circuits with applications to quantum computation [7].

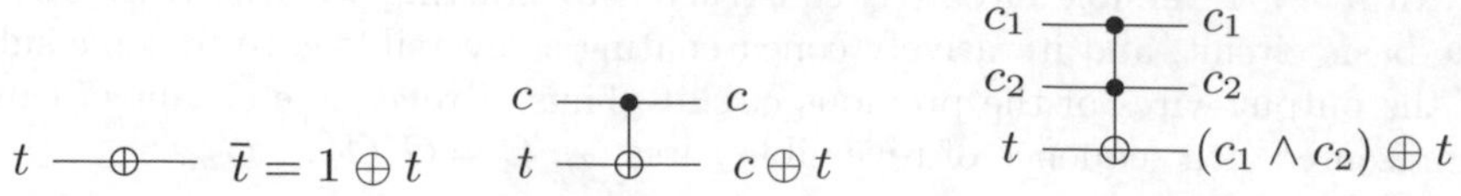

t —⊕— $\bar{t} = 1 \oplus t$

(a) 0-CNOT gate.

c —●— c
t —⊕— $c \oplus t$

(b) 1-CNOT gate.

c_1 —●— c_1
c_2 —●— c_2
t —⊕— $(c_1 \wedge c_2) \oplus t$

(c) 2-CNOT gate.

Fig. 1. CNOT gates

Let $\mathcal{C}_n$ be the set of all $n \times n$ CNOT circuits. Let $\mathcal{C}_{n,k \geq i}$ be the set of all $n \times n$ CNOT circuits consisting of k-CNOT gates for $k \geq i$, $\mathcal{C}_{n,k \leq i}$ be the set of all $n \times n$ CNOT circuits consisting of k-CNOT gates for $k \leq i$, and $\mathcal{C}_{n,k=i} = \mathcal{C}_{n,k \geq i} \cap \mathcal{C}_{n,k \leq i}$. Notice that $\mathcal{C}_n = \mathcal{C}_{n,k \geq 0}$. The following is immediate from Theorem I.

Theorem II. [6] $\sigma(\mathcal{C}_{n,k \leq 2}) \leq \sigma(\mathcal{C}_n) \leq 2^{n-1} + 1$. □

Since the set of n unit vectors is universal for $\mathcal{C}_{n,k \geq 2}$, we have the following.

Theorem III. [2] $\sigma(\mathcal{C}_{n,k \geq 2}) \leq n$. □

1.5 Our Results

We show the following theorems complementing the results in the previous section.

Theorem 1. $\sigma(\mathcal{C}_{n,k=1}) = n + 1$. □

Theorem 2. $\sigma(\mathcal{C}_{n,k \leq 1}) = n + 1$. □

Theorem 3. $\sigma(\mathcal{C}_{n,k \leq 2}) = 2^{n-1} + 1$. □

Corollary 1. $\sigma(\mathcal{C}_n) = 2^{n-1} + 1$. □

Theorem 4. $\sigma(\mathcal{C}_{n,k=b}) = \lceil n/(b-1) \rceil$ *for any* $n \geq 3$ *and* $2 \leq b \leq n-1$. □

Theorem 5. $\sigma(\mathcal{C}_{n,k \geq b}) = \lceil n/(b-1) \rceil$ *for any* $n \geq 3$ *and* $2 \leq b \leq n-1$. □

Corollary 2. $\sigma(\mathcal{C}_{n,k \geq 2}) = n$ *for any* $n \geq 2$. □

Theorem 6. $\sigma(\mathcal{C}_{n,k \geq 1}) = 2^{n-1} + 1$. □

Theorem 1 shows that the linear reversible circuits have a surprisingly small universal test set. Theorem 2 is a generalization of Theorem 1. Theorems 1 and 2 are proved in Sections 3 and 4, respectively.

Corollary 1 is immediate from Theorem 3. Theorem 3 and Corollary 1 show that the equalities hold in the inequalities of Theorem II. They also show that the size of a minimum universal test set for the general CNOT circuits is exponentially large, as suspected. Theorem 3 is proved in Section 5.

Corollary 2 is a special case of Theorem 5 when $b = 2$, which shows that the equality holds in the inequality of Theorem III. Theorems 4 and 5 are proved in Section 6. (An outline of the proof is presented in the extended abstract, due to space limitations.)

Theorem 6 shows that $b \geq 2$ is essential for Theorem 5. Theorem 6 is proved in Section 7.

It is an interesting open question to find a polynomial time algorithm generating a complete test set of size $O(\log |W(C)|)$ for a reversible circuit C.

2 Preliminaries

A wire w of a reversible circuit C is said to be *i-controllable* by a test set T if the value of w can be set to i by a vector of T $(i = 0, 1)$. A wire w is said to be *controllable* by T if w is both 0- and 1-controllable by T. A reversible circuit C is said to be *i-controllable* by T if each wire of C is i-controllable by T $(i = 0, 1)$. A reversible circuit C is said to be *controllable* by T if C is both 0- and 1-controllable by T. The following characterization for complete test sets is shown in [6].

Theorem IV. [6] *A test set T for a reversible circuit C is complete if and only if C is controllable by T.* □

A test set $T \subseteq \{0, 1\}^n$ is said to be *i-universal* for $\mathcal{D} \subseteq \mathcal{D}_n$ if every circuit of $\mathcal{D}$ is i-controllable by T $(i = 0, 1)$. It should be noticed that T is a universal test set for $\mathcal{D}$ if and only if T is both 0- and 1-universal for $\mathcal{D}$.

A family $\mathcal{D}$ of reversible circuits is said to be *hereditary* if $\mathcal{D}$ satisfies the following property: if $C = G_1G_2 \ldots G_m$ is in $\mathcal{D}$ then a subcircuit $C' = G_1G_2 \ldots G_i$ is also in $\mathcal{D}$ for any $i \leq m$. Notice that $\mathcal{C}_{n,k \geq i}$, $\mathcal{C}_{n,k \leq i}$, and $\mathcal{C}_{n,k=i}$ are hereditary for any $n \geq 1$ and $i \geq 0$. Let $C = G_1G_2 \ldots G_m$ be a circuit in a hereditary family $\mathcal{D}$ of reversible circuits, and w be any wire of C. If w is an output wire of gate G_j then w is an output wire of a subcircuit $C' = G_1G_2 \ldots G_j$. Since C' is also in $\mathcal{D}$, we have the following.

Lemma 1. *A test set T is i-universal for a hereditary family $\mathcal{D}$ of reversible circuits if and only if every output wire of any reversible circuit in $\mathcal{D}$ is i-controllable by T $(i = 0, 1)$.* □

3 Proof of Theorem 1

We consider that $\{0, 1\}^n$ is an n-dimensional vector space of column vectors over GF(2). The action of an $n \times n$ linear reversible circuit C can be represented by a linear transformation over $\{0, 1\}^n$, and represented as multiplication by a non-singular $n \times n$ matrix $A(C)$ over $\{0, 1\}$:

$$A(C)\boldsymbol{x} = \boldsymbol{y},$$

where $\boldsymbol{x}$ [$\boldsymbol{y}$] is a column vector in $\{0,1\}^n$ whose i-th entry contains the value of the i-th bit of the input [output] of C. A 1-CNOT gate G is denoted by $G[c,t]$, where c and t are the control bit and target bit of G, respectively. The action of a 1-CNOT gate corresponds to multiplication by an elementary matrix, which is the identity matrix with one off-diagonal entry set to one. The elementary matrix $A(G)$ for a 1-CNOT gate $G[c,t]$ is the identity matrix with (t,c)-entry set to one. It should be noted that the multiplication by $A(G)$ performs a row operation, the addition of the c-th row of a matrix to the t-th row. Applying a series of 1-CNOT gates corresponds to multiplying an input vector by a series of elementary matrices, or equivalently to performing a series of row operations on the input vector. Let $C = G_1G_2 \dots G_m$ be an $n \times n$ linear reversible circuit represented as a sequence of 1-CNOT gates G_i, and A_i be an $n \times n$ elementary matrix for G_i. Then the action of C is represented by

$$A_m A_{m-1} \cdots A_1 \boldsymbol{x} = \boldsymbol{y},$$

where $\boldsymbol{x}$ and $\boldsymbol{y}$ are input and output vectors in $\{0,1\}^n$. In other words,

$$A(C) = A_m A_{m-1} \cdots A_1.$$

Let $T = \{\boldsymbol{t}_1, \boldsymbol{t}_2, \dots, \boldsymbol{t}_s\} \subseteq \{0,1\}^n$, and $\boldsymbol{t}_i = (t_{1,i}, t_{2,i}, \cdots, t_{n,i})'$, where $\boldsymbol{v}'$ is the transpose of $\boldsymbol{v}$. Define a matrix

$$M(T) = \begin{bmatrix} t_{1,1} & t_{1,2} & \dots & t_{1,s} \\ t_{2,1} & t_{2,2} & \dots & t_{2,s} \\ \vdots & \vdots & \ddots & \vdots \\ t_{n,1} & t_{n,2} & \dots & t_{n,s} \end{bmatrix}$$

consisting of column vectors $\boldsymbol{t}_1, \boldsymbol{t}_2, \dots, \boldsymbol{t}_s$.

We prove the theorem by a series of lemmas.

Lemma 2. *A set of tests $T \subseteq \{0,1\}^n$ is 1-universal for $\mathcal{C}_{n,k=1}$ if and only if T contains a basis for $\{0,1\}^n$.*

Proof. Suppose that T contains a basis for $\{0,1\}^n$. Then $r(M(T)) = n$, where $r(B)$ is the rank of a matrix B. Let C be a linear reversible circuit. Then the i-th column of a matrix $A(C)M(T)$ is the output vector for $\boldsymbol{t}_i$. Since $A(C)$ is non-singular, $r(A(C)M(T)) = n$, i.e., every row of $A(C)M(T)$ contains a one element. Therefore, every output wire of C is 1-controllable by T. Thus from Lemma 1, T is 1-universal for $\mathcal{C}_{n,k=1}$.

Suppose contrary that T does not contain a basis for $\{0,1\}^n$. Then, $r(M(T)) < n$, and there exists a sequence of elementary row operations of $M(T)$ resulting in a zero row vector $\mathbf{0} = (0,0,\dots,0)$. Let $A_1, A_2, \dots, A_m$ be elementary matrices corresponding to the sequence of such elementary row operations, and G_i be a 1-CNOT gate corresponding to A_i ($1 \le i \le m$). Then for a linear reversible circuit $C = G_1G_2 \dots G_m$, the i-th column of a matrix $A_m A_{m-1} \cdots A_1 M(T)$ is the output for $\boldsymbol{t}_i$. Since a row of $A_m A_{m-1} \cdots A_1 M(T)$ is a zero row vector, the output wire of C corresponding to the zero row vector is not 1-controllable by T. Thus from Lemma 1, T is not 1-universal for $\mathcal{C}_{n,k=1}$. □

Lemma 3. *No basis for* $\{0,1\}^n$ *is* 0*-universal for* $\mathcal{C}_{n,k=1}$.

Proof. Let T be a basis for $\{0,1\}^n$. Since $M(T)$ is non-singular, there exists a sequence of elementary row operations of $M(T)$ resulting in a sum row vector $\mathbf{1} = (1, 1, \ldots, 1)$. Let $A_1, A_2, \ldots, A_m$ be elementary matrices corresponding to the sequence of such elementary row operations, and G_i be a 1-CNOT gate corresponding to A_i $(1 \leq i \leq m)$. Then, for a linear reversible circuit $C = G_1 G_2 \ldots G_m$, the i-th column of a matrix $A_m A_{m-1} \ldots A_1 M(T)$ is the output for $\boldsymbol{t}_i$. Since a row of $A_m A_{m-1} \ldots A_1 M(T)$ is a sum row vector, the output wire of C corresponding to the sum row vector is not 0-controllable by T. Thus from Lemma 1, T is not 0-universal for $\mathcal{C}_{n,k=1}$. □

Since a set consisting of just a zero (column) vector is 0-universal for $\mathcal{C}_{n,k=1}$ as easily seen, we have the following.

Lemma 4. *A set of the zero vector and the basis vectors of a basis for* $\{0,1\}^n$ *is a minimum universal test set for* $\mathcal{C}_{n,k=1}$. □

This completes the proof of Theorem 1.

4 Proof of Theorem 2

Since $\sigma(\mathcal{C}_{n,k\leq 1}) \geq \sigma(\mathcal{C}_{n,k=1})$ by definition, $\sigma(\mathcal{C}_{n,k\leq 1}) \geq n+1$ by Theorem 1. Thus, it suffices to show that a set of the zero vector and the basis vectors of a basis for $\{0,1\}^n$ is also a universal test set for $\mathcal{C}_{n,k\leq 1}$.

Let C be a circuit in $\mathcal{C}_{n,k\leq 1}$. We denote a 0-CNOT gate G on bit t by $G[t]$. Let C^+ be a circuit in $\mathcal{C}_{n+1,k=1}$ obtained from C by adding an additional bit $n+1$ and replacing each 0-CNOT gate $G[t]$ of C by a 1-CNOT gate $G[n+1, t]$.

Let $T = \{\boldsymbol{t}_0, \boldsymbol{t}_1, \boldsymbol{t}_2, \ldots, \boldsymbol{t}_n\} \subseteq \{0,1\}^n$, where $\boldsymbol{t}_0 = \mathbf{0}'$, $\{\boldsymbol{t}_1, \boldsymbol{t}_2, \ldots, \boldsymbol{t}_n\}$ is a basis for $\{0,1\}^n$, and $\boldsymbol{t}_i = (t_{1,i}, t_{2,i}, \ldots, t_{n,i})'$. Let $\boldsymbol{t}_i^+ = (t_{1,i}, t_{2,i}, \ldots, t_{n,i}, 1)' \in \{0,1\}^{n+1}$, and $T^+ = \{\boldsymbol{t}_0^+, \boldsymbol{t}_1^+, \boldsymbol{t}_2^+, \ldots, \boldsymbol{t}_n^+\} \subseteq \{0,1\}^{n+1}$. Then

$$M(T^+) = \begin{bmatrix} 0 & t_{1,1} & t_{1,2} & \ldots & t_{1,n} \\ 0 & t_{2,1} & t_{2,2} & \ldots & t_{2,n} \\ \vdots & \vdots & \vdots & \ddots & \vdots \\ 0 & t_{n,1} & t_{n,2} & \ldots & t_{n,n} \\ 1 & 1 & 1 & \ldots & 1 \end{bmatrix}.$$

Notice that T^+ is a basis for $\{0,1\}^{n+1}$, since $\{\boldsymbol{t}_1, \boldsymbol{t}_2, \ldots, \boldsymbol{t}_n\}$ is a basis for $\{0,1\}^n$. Notice also that $\boldsymbol{y}_i = (y_{1,i}, y_{2,i}, \ldots, y_{n,i})'$ is the output vector of C for input vector $\boldsymbol{t}_i$ if and only if $\boldsymbol{y}_i^+ = (y_{1,i}, y_{2,i}, \ldots, y_{n,i}, 1)'$ is the output vector of C^+ for input vector $\boldsymbol{t}_i^+$ by the definition of C^+. Thus we have that

$$A(C^+)M(T^+) = \begin{bmatrix} y_{1,0} & y_{1,1} & y_{1,2} & \cdots & y_{1,n} \\ y_{2,0} & y_{2,1} & y_{2,2} & \cdots & y_{2,n} \\ \vdots & \vdots & \vdots & \ddots & \vdots \\ y_{n,0} & y_{n,1} & y_{n,2} & \cdots & y_{n,n} \\ 1 & 1 & 1 & \ldots & 1 \end{bmatrix}.$$

It should be noted that $A(C^+)M(T^+)$ is non-singular since both $A(C^+)$ and $M(T^+)$ are non-singular. It follows that each row of $A(C^+)M(T^+)$ except the last row has a zero element, for otherwise $A(C^+)M(T^+)$ has a sum row vector other than the last row. Thus, every output wire of C is 0-controllable by T, and T is 0-universal for $\mathcal{C}_{n,k\leq 1}$ by Lemma 1. It also follows that each row of $A(C^+)M(T^+)$ has a one element, for otherwise $A(C^+)M(T^+)$ has a zero row vector. Thus, every output wire of C is 1-controllable by T, and T is 1-universal for $\mathcal{C}_{n,k\leq 1}$ by Lemma 1.

Thus, we conclude that T is a universal test set for $\mathcal{C}_{n,k\leq 1}$, which completes the proof of Theorem 2.

5 Proof of Theorem 3

Since $\sigma(\mathcal{C}_{n,k\leq 2}) \leq 2^{n-1}+1$ by Theorem II, it suffices to show the following.

Lemma 5. $\sigma(\mathcal{C}_{n,k\leq 2}) \geq 2^{n-1}+1$.

Proof. Since $\mathcal{C}_{1,k\leq 2} = \mathcal{C}_{1,k=0}$, it is easy to see that $\sigma(\mathcal{C}_{1,k\leq 2}) = 2 = 2^{1-1}+1$. Since $\mathcal{C}_{2,k\leq 2} = \mathcal{C}_{2,k\leq 1}$, we have by Theorem 2 that $\sigma(\mathcal{C}_{2,k\leq 2}) = 2+1 = 2^{2-1}+1$.

We now assume that $n \geq 3$. Let T be a subset of $\{0,1\}^n$ with $|T| \leq 2^{n-1}$. A permutation $f : \{0,1\}^n \to \{0,1\}^n$ is said to be *even* if f can be represented as the product of an even number of transpositions. It is shown in [8] that any even permutation can be implemented by a circuit in $\mathcal{C}_{n,k\leq 2}$. Let $V \subseteq \{0,1\}^n$ be the set of all vectors having one at the n-th entry. Since $|T| \leq 2^{n-1}$ and $|V| = 2^{n-1}$, it is easy to see that there exists an even permutation $f : \{0,1\}^n \to \{0,1\}^n$ such that $f(T) \subseteq V$. Let C be a circuit in $\mathcal{C}_{n,k\leq 2}$ that implements f. Then the n-th bit output wire of C is not 0-controllable by T, and we conclude that T is not a universal test set for $\mathcal{C}_{n,k\leq 2}$ by Lemma 1. Thus, $\sigma(\mathcal{C}_{n,k\leq 2}) \geq 2^{n-1}+1$. □

6 Proof of Theorems 4 and 5

Since $\sigma(\mathcal{C}_{n,k=b}) \leq \sigma(\mathcal{C}_{n,k\geq b})$, it suffices to show the following.

Lemma 6. $\sigma(\mathcal{C}_{n,k\geq b}) \leq \lceil n/(b-1) \rceil$ *for* $n \geq 3$ *and* $2 \leq b \leq n-1$. □

Lemma 7. $\sigma(\mathcal{C}_{n,k=b}) \geq \lceil n/(b-1) \rceil$ *for* $n \geq 3$ *and* $2 \leq b \leq n-1$. □

6.1 Proof of Lemma 6

For $1 \leq j \leq \lceil n/(b-1) \rceil$, define $\boldsymbol{t}_j = (t_{1,j}, t_{2,j}, \ldots, t_{n,j})' \in \{0,1\}^n$ as

$$t_{i,j} = \begin{cases} 1 \text{ if } (j-1)(b-1)+1 \leq i \leq \min\{j(b-1), n\} \\ 0 \text{ otherwise,} \end{cases}$$

and let $T = \{\boldsymbol{t}_j \mid 1 \leq j \leq \lceil n/(b-1) \rceil\} \subseteq \{0,1\}^n$. For any circuit $C \in \mathcal{C}_{n,k\geq b}$, the output vector of C for an input vector $\boldsymbol{t}_j$ is also $\boldsymbol{t}_j$, since any $\boldsymbol{t}_j$ has at most

$b-1$ one elements ($1 \le j \le \lceil n/(b-1) \rceil$). By definition, the i-th entry of $\boldsymbol{t}_j$ is one if and only if $j = \lceil i/(b-1) \rceil$. Therefore, for any $1 \le i \le n$, there exist $\boldsymbol{t}_j$ and $\boldsymbol{t}_k \in T$ such that $\boldsymbol{t}_j$ and $\boldsymbol{t}_k$ have one and zero at the i-th entry, respectively. Thus the output wires of C are controllable by T, and we conclude that T is a universal test set for $\mathcal{C}_{n,k \ge b}$ by Lemma 1. Since $|T| = \lceil n/(b-1) \rceil$, we have the lemma. □

6.2 Proof of Lemma 7(Sketch)

6.2.1 Proof for the Case of $n = 3$

We have $b = 2$, and will show that $\sigma(\mathcal{C}_{3,k\ge 2}) \ge 3 = \lceil 3/(2-1) \rceil$. We denote by $\overline{\boldsymbol{v}}$ the complement of a vector $\boldsymbol{v}$ such that $\boldsymbol{v} \oplus \overline{\boldsymbol{v}} = \mathbf{1}$. For a gate G of a circuit C, $G(\boldsymbol{v})$ is the output vector of G generated by an input vector $\boldsymbol{v}$ of C. The following lemma is shown in [9].

Lemma I. *A test set $T = \{\boldsymbol{v}_1, \boldsymbol{v}_2\}$ for a circuit $C \in \mathcal{C}_{n,k\ge 1}$ is complete if and only if T satisfies the following conditions:*

(i) $\boldsymbol{v}_2 = \overline{\boldsymbol{v}_1}$, and
(ii) $G(\boldsymbol{v}_i) = \boldsymbol{v}_i$ ($i = 1, 2$) for every gate G of C. □

We now prove that $\tau(C) \ge 3$ for circuit C shown in Fig. 2. For any test set $T = \{\boldsymbol{v}, \overline{\boldsymbol{v}}\}$, $\boldsymbol{v}$ or $\overline{\boldsymbol{v}}$ has two one elements. Thus, T does not satisfy condition (ii) of Lemma I, and so T is not complete for C.

Since $C \in \mathcal{C}_{3,k\ge 2}$, we have $\sigma(\mathcal{C}_{3,k\ge 2}) \ge \tau(C) \ge 3$. □

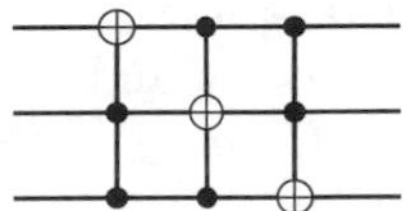

Fig. 2. Reversible circuit C in $\mathcal{C}_{3,k=2}$

6.2.2 Technical Lemmas

Before proving the case of $n \ge 4$, we need some preliminaries. Let $[n] = \{1, 2, \ldots, n\}$, and $X \subsetneq [n]$. For an integer $\tau \in [n] - X$, let $G[X;\tau]$ be the $|X|$-CNOT gate with control bits $c \in X$ and the target bit τ. Such a gate is called an *X-CNOT gate.* For $X_i \subset [n]$, let $\mathcal{C}_n[X_1, X_2, \ldots, X_m]$ be the class of CNOT circuits consisting of only X_i-CNOT gates ($1 \le i \le m$). For $\boldsymbol{v} \in \{0,1\}^n$, let $\chi(\boldsymbol{v})$ be the set of integers such that $\boldsymbol{v}$ has one at the i-th entry if and only if $i \in \chi(\boldsymbol{v})$. For a CNOT circuit C, we denote by $C(\boldsymbol{v})$ the output vector of C for input $\boldsymbol{v}$. Since any X-CNOT gate does not change $\boldsymbol{v}$ if $X \not\subseteq \chi(\boldsymbol{v})$, we have the following.

Lemma 8. *Let $X_i \subseteq [n]$ for $1 \le i \le m$, and $\boldsymbol{v} \in \{0,1\}^n$ such that $X_i \not\subseteq \chi(\boldsymbol{v})$ for all i with $1 \le i \le m$. Then, $C(\boldsymbol{v}) = \boldsymbol{v}$ for $C \in \mathcal{C}_n[X_1, X_2, \ldots, X_m]$.* □

For two vectors $\boldsymbol{u} = (u_1, u_2, \ldots, u_n)'$ and $\boldsymbol{v} = (v_1, v_2, \ldots, v_n)'$, let $\boldsymbol{u} \vee \boldsymbol{v} = (u_1 \vee v_1, u_2 \vee v_2, \ldots, u_n \vee v_n)'$. By definition, $\chi(\boldsymbol{u} \vee \boldsymbol{v}) = \chi(\boldsymbol{u}) \cup \chi(\boldsymbol{v})$.

Lemma 9. *Let $\boldsymbol{u}, \boldsymbol{v} \in T$, $X_1 \subseteq \chi(\boldsymbol{u})$, and $X_2 \subseteq \chi(\boldsymbol{v})$. Then, there exists a circuit $C \in \mathcal{C}_n[X_1, X_2]$ such that $C(\boldsymbol{u}) = \boldsymbol{v}$.*

Proof Sketch. There exist circuits $C_1 \in \mathcal{C}_n[X_1]$ and $C_2 \in \mathcal{C}_n[X_2]$ such that $C_1(\boldsymbol{u}) = \boldsymbol{u} \vee \boldsymbol{v}$ and $C_2(\boldsymbol{u} \vee \boldsymbol{v}) = \boldsymbol{v}$. Thus by concatenating C_1 and C_2, we obtain a desired circuit C. □

For $1 \leq i \leq n - b + 1$, define $\boldsymbol{\beta}_i^{\langle b \rangle} = (\beta_{1,i}, \beta_{2,i}, \ldots, \beta_{n,i})'$ as

$$\beta_{j,i} = \begin{cases} 1 & \text{if } i \leq j \leq i + b - 1, \\ 0 & \text{otherwise.} \end{cases}$$

By definition, $\chi(\boldsymbol{\beta}_i^{\langle b \rangle}) = \{i, i+1, \ldots, i+b-1\}$. For $m \leq n - b + 1$, define that $B_m = \{\boldsymbol{\beta}_i^{\langle b \rangle} \mid 1 \leq i \leq m\}$. Let $\mathcal{T}_{n,b} \subseteq \{0,1\}^n$ be the set of vectors which have at least b one elements.

Lemma 10. *Let $m \leq n - b + 1$, and $\boldsymbol{v}_i \in \{0,1\}^n$ be vectors such that $\boldsymbol{v}_i \in \mathcal{T}_{n,b}$ for $1 \leq i \leq m$. Then, there exists a b-CNOT circuit C_m such that $C_m(\boldsymbol{v}_i) = \boldsymbol{\beta}_i^{\langle b \rangle}$ for $1 \leq i \leq m$.*

Proof Sketch. The proof is done by induction on m. Initially, a circuit C_1 is obtained from Lemma 9. Assume C_j is given. From Lemma 9, there is a $[\chi(\boldsymbol{\beta}_{j+1}^{\langle b \rangle}), \chi(C_j(\boldsymbol{v}_{j+1}))]$-circuit C with $C(C_j(\boldsymbol{v}_{j+1})) = \boldsymbol{\beta}_{j+1}^{\langle b \rangle}$. Since any gate in C does not change $\boldsymbol{\beta}_i^{\langle b \rangle}$ for $i \leq j$, C_{j+1} is obtained by concatenating C_j and C. □

6.2.3 Proof for the Case of $n \geq 4$

We omit the proof for the case of $b = 2$, and assume $b \geq 3$. Let $T \subset \{0,1\}^n$ be any test set consisting of m tests for $m \leq \lceil n/(b-1) \rceil - 1$. Let $T_1 = T \cap \mathcal{T}_{n,b}$ and $T_2 = T - T_1$. Since $|T_1| \leq \lceil n/(b-1) \rceil - 1$, all the vectors in T_1 have zero at the same entry. We assume without loss of generality that all the vector in T_1 have zero at the n-th entry. We can prove that $|T_2| \leq n - b$. Thus by Lemma 9, there is a b-CNOT circuit C with the n-th entry of $C(\boldsymbol{t})$ is zero for all $\boldsymbol{t} \in T_2$. Since any b-CNOT gate does not change $\boldsymbol{t} \in T_1$, the n-th output wire of C is not 1-controllable by $T = T_1 \cup T_2$.

7 Proof of Theorem 6

To prove Theorem 6, it suffices to show $\sigma(\mathcal{C}_{n,k\geq 1}) \geq 2^{n-1} + 1$. Let $\text{val}(\boldsymbol{v}) = \sum_{i=1}^{n} 2^{i-1} v_i$, and $\boldsymbol{r}_m \in \{0,1\}^n$ be the reverse of the n-bit binary representation of integer m. Then, $\text{val}(\boldsymbol{r}_m) = m$. For $\boldsymbol{v}, \boldsymbol{w} \in \{0,1\}^n$, if $\text{val}(\boldsymbol{v}) < \text{val}(\boldsymbol{w})$ then $\chi(\boldsymbol{w}) \not\subseteq \chi(\boldsymbol{v})$. Thus by Lemmas 8 and 9, we have the following.

Lemma 11. *Let $\boldsymbol{u}, \boldsymbol{v} \in \{0,1\}^n$ be non-zero vectors. Then, there exists a circuit C in $\mathcal{C}_n[\chi(\boldsymbol{u}), \chi(\boldsymbol{v})] \subset \mathcal{C}_{n,k\geq 1}$ such that $C(\boldsymbol{u}) = \boldsymbol{v}$, and that $C(\boldsymbol{w}) = \boldsymbol{w}$ for all vectors $\boldsymbol{w}$ with* $\text{val}(\boldsymbol{w}) \leq \min\{\text{val}(\boldsymbol{u}), \text{val}(\boldsymbol{v})\} - 1$. □

Lemma 12. *For $l \leq 2^n - 1$, let $T = \{\boldsymbol{t}_0, \boldsymbol{t}_1, \ldots, \boldsymbol{t}_l\} \subseteq \{0,1\}^n$ with $\boldsymbol{t}_0 = \boldsymbol{0}$. Then, there exists a circuit C in $\mathcal{C}_{n,k\geq 1}$ such that $C(\boldsymbol{t}_i) = \boldsymbol{r}_i$ for $0 \leq i \leq l$.*

Proof. We show the lemma by induction on l. Initial case is clear since $\boldsymbol{t}_0 = \boldsymbol{r}_0$.
Assume that the lemma holds for $l = l'$ with $l' \leq 2^n - 2$, and we show that the lemma also holds for $l = l' + 1$. By induction hypothesis, there is a circuit $C_{l'}$ satisfying $C_{l'}(\boldsymbol{t}_i) = \boldsymbol{r}_i$ for $0 \leq i \leq l'$. Since $C_{l'}$ implements a permutation, $C_{l'}(\boldsymbol{t}_{l'+1}) \neq \boldsymbol{r}_i$ for $0 \leq i \leq l'$. Thus by Lemma 11, there exists a circuit C in $\mathcal{C}_{n,k\geq 1}$ such that $C(C_{l'}(\boldsymbol{t}_{l'+1})) = \boldsymbol{r}_{l'+1}$ and $C(\boldsymbol{r}_i) = \boldsymbol{r}_i$ for $i \leq l'$. Thus by concatenating $C_{l'}$ and C, we obtain a circuit $C_{l'+1}$ in $\mathcal{C}_{n,k\geq 1}$ satisfying $C_{l'+1}(\boldsymbol{t}_i) = \boldsymbol{r}_i$ for $i \leq l' + 1$. □

Since $\boldsymbol{r}_i$ has 0 at the n-th entry for $i \leq 2^{n-1} - 1$, we have the following by Lemmas 1 and 12.

Corollary 3. *A test set $T \subseteq \{0,1\}^n$ with $\boldsymbol{0} \in T$ is not 1-universal for $\mathcal{C}_{n,k\geq 1}$ if $|T| \leq 2^{n-1}$.* □

Similarly, we can prove the following by considering $T' = \{\boldsymbol{t}_0, \boldsymbol{t}_1, \ldots, \boldsymbol{t}_{l'}\} = \{0,1\}^n - T$ and a circuit $C \in \mathcal{C}_{n,k\geq 1}$ satisfying $C(\boldsymbol{t}_i) = \boldsymbol{r}_i$ for $i \leq l'$, where $l' = 2^n - 1 - |T|$ and $\boldsymbol{t}_0 = \boldsymbol{0}$.

Corollary 4. *A test set $T \subseteq \{0,1\}^n$ with $\boldsymbol{0} \notin T$ is not 0-universal for $\mathcal{C}_{n,k\geq 1}$ if $|T| \leq 2^{n-1}$.* □

From Corollaries 3 and 4, we conclude that $T \subseteq \{0,1\}^n$ is not universal for $\mathcal{C}_{n,k\geq 1}$ if $|T| \leq 2^{n-1}$, and we have Theorem 6.

References

1. Bennett, C.: Logical reversiblity of computation. IBM J. Res. Dev. 17(6), 525–532 (1973)
2. Chakraborty, A.: Synthesis of reversible circuits for testing with universal test set and c-testability of reversible iterative logic array. In: Proc. of the 18th International Conference on VLSI Design, pp.249–254 (2005)
3. Landauer, R.: Irreversibility and heat generation in the computing process. IBM J. Res. Dev. 5(3), 183–191 (1961)
4. Merkle, R.: Two types of mechanical reversible logic. Nanotechnology 4(2), 114–131 (1993)
5. Nielsen, M., Chuang, I.: Quantum Computation and Quantum Information. Cambridge University Press, Cambridge (2000)
6. Patel, K., Hayes, J., Markov, I.: Fault testing for reversible circuits. IEEE Trans. Computer-Aided Design 23(8), 1220–1230 (2004)
7. Patel, K., Markov, I., Hayes, J.: Optimal synthesis of linear reverisible circuits. Quantum Information and Computation 8(3&4), 282–294 (2008)
8. Shende, V., Prasad, A., Markov, I., Hayes, J.: Synthesis of reversible logic circuits. IEEE Trans. Computer-Aided Design 22(6), 710–722 (2003)
9. Tayu, S., Ito, S., Ueno, S.: On fault testing for reversible circuits. IEICE Trans. Inf. & Syst. E91-D(12), 2770–2775 (2008)

Approximate Counting with a Floating-Point Counter

Miklós Csűrös

Department of Computer Science and Operations Research, University of Montréal
csuros@iro.umontreal.ca

Abstract. When many objects are counted simultaneously in large data streams, as in the course of network traffic monitoring, or Webgraph and molecular sequence analyses, memory becomes a limiting factor. Robert Morris [*Communications of the ACM*, 21:840–842, 1978] proposed a probabilistic technique for approximate counting that is extremely economical. The basic idea is to increment a counter containing the value X with probability 2^{-X}. As a result, the counter contains an approximation of $\lg n$ after n probabilistic updates, stored in $\lg \lg n$ bits. Here we revisit the original idea of Morris. We introduce a binary floating-point counter that combines a d-bit significand with a binary exponent, stored together on $d + \lg \lg n$ bits. The counter yields a simple formula for an unbiased estimation of n with a standard deviation of about $0.6 \cdot n2^{-d/2}$.

We analyze the floating-point counter's performance in a general framework that applies to any probabilistic counter. In that framework, we provide practical formulas to construct unbiased estimates, and to assess the asymptotic accuracy of any counter.

1 Introduction

Motivation. An elementary information-theoretic argument shows that $\lceil \lg(n+1) \rceil$ bits are necessary to represent integers between 0 and n (lg denotes binary logarithm throughout the paper). Counting thus takes logarithmic space. Certain applications need to be more economical because they need to maintain many counters simultaneously while, say, tracking patterns in large data streams. Such is the case typically in network processors and embedded systems, where time and space constraints impose strict requirements on computational solutions. For instance, when measuring network traffic, per-packet updates may occur more often than allowed by Dynamic RAM access times. Faster, Static RAMs are prohibitively expensive for storing exact counters [1,2].

Memory may become a limiting factor even on mainstream desktop computers in demanding applications such as Webgraph analyses [3]. Numerous bioinformatics studies also require space-efficient solutions when searching for recurrent motifs in protein and DNA sequences. These frequent sequence motifs are associated with mobile, structural, regulatory or other functional elements, and have been studied since the first molecular sequences became available [4]. Some recent studies have concentrated on patterns involving long oligonucleotides, i.e.,

M.T. Thai and S. Sahni (Eds.): COCOON 2010, LNCS 6196, pp. 358–367, 2010.

"words" of length 16–40 over the 4-letter DNA alphabet, revealing potentially novel regulatory features [5,6], and general characteristics of copying processes in genome evolution [7,8]. Hashtable-based indexing techniques [9] used in homology search and genome assembly procedures also rely on counting in order to identify repeating sequence patterns. In these applications, billions of counters need to be handled, making implementations difficult in mainstream computing environments. The need for many counters is aggravated by the fact that the counted features often have heavy-tailed frequency distributions [7,3,8], and there is thus no "typical" size for individual counters that could guide the memory allocation at the outset. As a numerical example, consider a study [7] of the 16-mer distribution in the human genome sequence, which has a length surpassing three billion. More than four billion (4^{16}) different words need to be counted, and the counter values span more than sixteen binary magnitudes even though the average 16-mer occurs only once or twice.

The idea of approximate counting. One way to greatly reduce memory usage is to relax the requirement of exact counts. Namely, approximate counting to n is possible using $\lg\lg n + O(1)$ bits with probabilistic techniques [10,11]. The idea of probabilistic counting was introduced by Morris [11]. In the simplest case, a counter is initialized as $X = 0$. The counter is incremented by one at the occurrence of an event with probability 2^{-X}. The counter is meant to track the magnitude of the true number of events. More precisely, after n events, the expected value of 2^X is exactly $(n + 1)$.

A generalization of the binary Morris counter is the so-called *q-ary counter* with some $r \geq 1$ and $q = 2^{1/r}$. In such a setup, the counter is incremented with probability q^{-X}. The actual event count is estimated as $f(X)$, using the transformation $f(x) = \frac{q^x-1}{q-1} = \frac{2^{x/r}-1}{2^{1/r}-1}$. The function f yields an unbiased estimate, as $\mathbb{E}f(X) = n$ after n probabilistic updates. The accuracy of a probabilistic counting method is characterized by the variance of the estimated count. For the q-ary counter,

$$\operatorname{Var} f(X) = (q-1)\frac{n(n+1)}{2}, \tag{1}$$

which is approximately $\frac{\ln 2}{2r}n^2$ for large n and r. The parameter r governs the tradeoff between memory usage and accuracy. The counter stores X (with $n = f(X)$) using $\lg r + \lg\lg n$ bits; larger r thus increases the accuracy at the expense of higher storage costs.

Approximate counting is perhaps the simplest of many related problems where probabilistic techniques can lead to efficient solutions (e.g., estimating the number of distinct elements in a multiset of size n using $\lg\lg n + O(1)$ bits [12]). Flajolet [10] gave a precise analysis of the q-ary counter. Kirschenhoffer and Prodinger [13] performed the same analysis using Rice's method instead of the Mellin transform. Kruskal and Greenberg [14] analyzed approximate counting in a general framework (see Section 2) from a Bayesian viewpoint, assuming a known distribution for the true count.

Approximate counting revisited. The main goal of this study is to introduce a novel algorithm for approximate counting. Our *floating-point counter* is defined with the aid of a design parameter $M = 2^d$, where d is a nonnegative integer. As we discuss later, M determines the tradeoff between memory usage and accuracy, analogously to parameter r of the q-ary counter. The procedure relies on a uniform random bit generator RandomBit(). Algorithm FP-Increment below shows the incrementation procedure for a floating-point counter, initialized with $X = 0$. Notice that the first M updates are deterministic.

```
FP-Increment(X)                        // returns new value of X
set t ← ⌊X/M⌋                          // bitwise right shift by d positions
while t > 0 do
    if RandomBit() = 1 then return X
    set t ← t − 1
return X + 1
```

The counter value $X = 2^d \cdot t + u$, where u denotes the lower d bits, is used to estimate the actual count $f(X) = (M + u) \cdot 2^t - M$. The estimate reaches n with $d + \left\lceil \lg\lg \frac{n+M}{M-1/2} \right\rceil$ bits. The estimate's standard deviation is $\frac{c}{\sqrt{M}} n$ where c fluctuates between about 0.58 and 0.61 asymptotically (see Corollary 3 for a precise characterization). The random updates in the floating-point counter occur with exact integer powers 2^{-i}, and such random values can be generated using an average of 2 random bits. Specifically, the FP-Increment procedure uses an expected number of $\left(2 - \frac{t}{2^t - 1}\right)$ calls to RandomBit().

Notice that a q-ary counter with $r = M$ has asymptotically the same memory usage, and a standard deviation of about $\frac{0.59}{\sqrt{r}} n$ (see Eq. (1)). Our algorithm thus has similar memory usage and accuracy as q-ary counting. The floating-point counter is more advantageous in two aspects. First, the updates at the beginning are deterministic, i.e., small values are exactly represented with convenience. But more importantly, our counter can be implemented with a few elementary integer and bitwise operations, whereas a q-ary counter needs floating-point arithmetic (for random increments with irrational probabilities and the decoding by f) which may not be available on a specialized processor.

The rest of the paper is organized as follows. In order to quantify the performance of floating-point counters, we found it fruitful to develop a general analysis of probabilistic counting, which is of independent mathematical interest. Section 2 presents the general results. First, Theorem 1 shows that every probabilistic counting method has a unique unbiased estimator f with $\mathbb{E} f(X) = n$ after n probabilistic updates. Second, Theorem 2 shows that the accuracy of any such method is computable directly from the counter value. Finally, Theorem 3 gives relatively simple upper and lower bounds on the asymptotic accuracy of the unbiased estimator. Section 3 presents floating-point counters in detail, and mathematically characterizes their utility by relying on the results of Section 2. Section 3 further illustrates the theoretical analyses with simulation experiments. The proofs of the theorems are omitted due to space constraints.

2 Probabilistic Counting

For a formal discussion of probabilistic counting, consider the Markov chain formed by the successive counter values.

Definition 1. *A* counting chain *is a Markov chain* $(X_n : n = 0, 1, \ldots)$ *with*

$$X_0 = 0; \tag{2a}$$

$$\mathbb{P}\Big\{X_{n+1} = k+1 \;\Big|\; X_n = k\Big\} = q_k \tag{2b}$$

$$\mathbb{P}\Big\{X_{n+1} = k \;\Big|\; X_n = k\Big\} = 1 - q_k, \tag{2c}$$

where $0 < q_k \leq 1$ *are the transition probabilities defining the counter.*

It is a classic result associated with probabilities in pure-birth processes [15] that the *n-step probabilities* $p_n(k) = \mathbb{P}\{X_n = k\}$ are computable by a simple recurrence. In case of probabilistic counting, we want to infer n from the value of X_n alone through a computable function f. A given probabilistic counting method is defined by the transition probabilities and the function f. As we will see later (Theorem 1), the transition probabilities determine a unique function f that gives an unbiased estimate of the update count n.

Definition 2. *A function* $f : \mathbb{N} \mapsto \mathbb{N}$ *is an* unbiased count estimator *for a given counting chain if and only if* $\mathbb{E}f(X_n) = n$ *holds for all* $n = 0, 1, \ldots$.

In the upcoming discussions, we assume that the probabilistic counting method uses an unbiased count estimator f. The merit of a given method is gauged by its dispersion, as defined below.

Definition 3. *The* dispersion *of the counter is the coefficient of variation* $A_n = \frac{\sqrt{\operatorname{Var} f(X_n)}}{\mathbb{E}f(X_n)}$.

The theorems below provide an analytical framework for evaluating probabilistic counters. Theorem 1 shows that the unbiased estimator is uniquely defined by a relatively simple expression involving the transition probabilities. Theorem 2 shows that the uncertainty of the estimate can be determined directly from the counter value. Theorem 3 gives a practical bound on the asymptotic dispersion of the counter.

Theorem 1. *The function*

$$f(0) = 0 \tag{3a}$$

$$f(k) = \frac{1}{q_0} + \frac{1}{q_1} + \ldots + \frac{1}{q_{k-1}}. \qquad \{k > 0\} \tag{3b}$$

uniquely defines the unbiased count estimator f *for any given set of transition probabilities* $(q_k : k = 0, 1, \ldots)$. *Hence, for any given counting chain, we can determine efficiently an unbiased estimator. Conversely, we can compute the counting chain for any given* $f : \mathbb{N} \mapsto [0, \infty)$ *provided that* $f(0) = 0$ *and* $\forall k : f(k+1) \geq f(k) + 1$.

Definition 4. *The* variance function *for a given counting chain is defined by*

$$g(0) = 0 \tag{4a}$$

$$g(k) = \frac{1-q_0}{q_0^2} + \frac{1-q_1}{q_1^2} + \cdots + \frac{1-q_{k-1}}{q_{k-1}^2} \qquad \{k > 0\} \tag{4b}$$

Theorem 2 below shows that the dispersion is computable directly from the counter value for any counting chain. The statement has a practical relevance: g quantifies the uncertainty of the estimate f, The variance function is used later to evaluate the asymptotic dispersion (see Theorem 3).

Theorem 2. *The variance function g of Definition 4 provides an unbiased estimate for the variance of f from Theorem 1. Specifically,*

$$\operatorname{Var} f(X_n) = \mathbb{E} g(X_n) \tag{5}$$

holds for all $n \geq 0$, where the moments refer to the space of n-step probabilities.

Theorem 3 is the last main result of this section. The statement relates the asymptotics of the variance function, the unbiased count estimator, and the counting chain's dispersion.

Theorem 3. *Let A_n be the dispersion of Definition 3, and let*

$$B_k = \frac{\sqrt{g(k)}}{f(k)}. \tag{6}$$

Let $\liminf_{k\to\infty} B_k = \mu$. *Suppose that* $\limsup_{k\to\infty} B_k = \lambda < 1$ *(and, thus, $\mu < 1$). Then*

$$\frac{\mu}{\sqrt{1-\mu^2}} \leq \liminf_{n\to\infty} A_n; \qquad \limsup_{n\to\infty} A_n \leq \frac{\lambda}{\sqrt{1-\lambda^2}}. \tag{7}$$

Example. Consider the case of a q-ary counter, where $q_i = q^{-i}$ with some $q > 1$. Theorem 1 automatically gives the unbiased count estimator $f(k) = \sum_{i=0}^{k-1} q_i^{-1} = \frac{q^k-1}{q-1}$. Theorem 2 yields the variance function $g(k) = \sum_{i=0}^{k-1}\big(q_i^{-2} - q_i^{-1}\big) = \frac{q^{2k}-1}{q^2-1} - \frac{q^k-1}{q-1}$. In order to use Theorem 3, observe that $\mu = \lambda = \lim_{k\to\infty}\sqrt{\frac{g(k)}{f^2(k)}} = \sqrt{\frac{q-1}{q+1}} < 1$. Therefore, we obtain the known result [10] that $\lim_{n\to\infty} A_n = \sqrt{\frac{\lambda^2}{1-\lambda^2}} = \sqrt{\frac{q-1}{2}}$.

Counting in $(1+c)\lg\lg n + O(1)$ ***space.*** Theorem 1 confirms the intuition that the transition probabilities must be exponentially decreasing in order to achieve storage on $\lg\lg n + O(1)$ bits. Otherwise, with subexponential $q_k^{-1} = 2^{o(k)}$, one would have $f(k) = 2^{o(k)}$, leading to $\lg n = o(k)$. In other words, $\lg\lg n + O(1)$ bits would not suffice to store the counter value k. It is possible, however, to devise other exotic counting schemes with $O(\log\log n)$ memory usage by "reverse engineering" the memory requirements. Consider the *subexponential counter*

$\mathsf{SubExp}(\beta)$ for $0 < \beta < 1$ defined by the transition probabilities $q_k = 1/\big(\exp((k+1)^\beta) - \exp(k^\beta)\big)$, yielding the unbiased estimator $f(k) = \exp(k^\beta) - 1$, and a memory requirement of $\beta^{-1} \lg \ln n + O(1)$. Using continuous approximations, $q_k^{-1} \approx \beta k^{\beta-1} \exp(k^\beta)$, and $g(k) \approx \int_{x=1}^{k} \beta^2 x^{2\beta-2} \exp(2x^\beta)\, dx$. Now, $B_k = \sqrt{g(k)}/f(k) \sim k^{(\beta-1)/2} \sqrt{\beta/2}$, since $\lim_{k\to\infty} \frac{B_k^2}{\frac{\beta}{2} k^{\beta-1}} \approx \lim_{k\to\infty} \frac{\int_{x=1}^{k} \beta^2 x^{2\beta-2} e^{2x^\beta}\, dx}{(e^{k^\beta} - 1)^2 \frac{\beta}{2} k^{\beta-1}} = 1$ by l'Hospital's rule. Acoodingly (plug in k with $f(k) = n$, i.e., $k = \ln^{1/\beta}(n+1)$), we conjecture that the $(1+c) \lg \ln n + O(1)$ memory usage of the $\mathsf{SubExp}((c+1)^{-1})$ counter entails that A_n goes to 0 at a slow speed of $O\big(\sqrt{1/\log^c n}\big)$.

Bayesian count estimation. The same general framework for approximate counting is used by Kruskal and Greenberg [14] in a Bayesian setting. In particular, they are interested in the case when the true count N is a random variable. The actual count is estimated as $\phi(k) = \mathbb{E}\big[N \mid X_N = k\big]$. They further consider the case of geometrically distributed N, as well as an improper prior $\mathbb{P}\{N = n\} = \epsilon$. For both distributions, $\phi(k) = \mathbb{E}\big[N \mid X_N = k\big] = f(k+1) - 1$ and $\mathrm{Var}\big[N \mid X_N = k\big] = g(k+1)$ where f and g are defined in Theorems 1 and 2 above. Kruskal and Greenberg motivate their results by suggesting that the unbiased count estimator may be difficult to find since in order to verify the condition $\mathbb{E} f(X_n) = n$, it may be complicated to calculate the expectation for an arbitrary nonlinear function f. Theorem 1 shows that this is not so: the unbiased estimator has a fairly simple form for any counting chain.

3 Floating-Point Counters

The counting chain for a floating-point counter is defined using a design parameter $M = 2^d$ with some nonnegative integer d:

$$\mathbb{P}\Big\{X_{n+1} = k+1 \,\Big|\, X_n = k\Big\} = 2^{-\lfloor k/M \rfloor}; \tag{8a}$$

$$\mathbb{P}\Big\{X_{n+1} = k \,\Big|\, X_n = k\Big\} = 1 - 2^{-\lfloor k/M \rfloor}. \tag{8b}$$

Figure 1 illustrates the states of the floating-point counter. The counter's designation becomes apparent from examining the binary representation of the counter value k. Write $k = Mt + u$ with $t = \lfloor k/M \rfloor$ and $u = k \bmod M$; i.e., u corresponds to the lower d bits of k, and t corresponds to the remaining upper bits. The pair (t, u) is essentially a floating-point representation of the true count n, where t is the exponent, and u is a d-bit significand without the hidden bit for the leading '1.' More precisely, Theorem 1 applies with $q_k = 2^{-\lfloor k/M \rfloor}$, and leads to the following corollary.

Corollary 1. *The unbiased estimator for $k = Mt + u$ is*

$$f(k) = f(t, u) = (M + u)2^t - M. \tag{9}$$

Theorem 2 yields the following variance function.

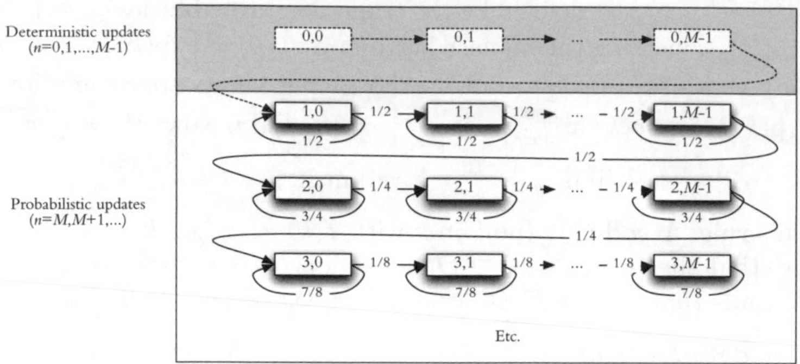

Fig. 1. States of the counting Markov chain. Each state is labeled with a pair (t, u), where $(u + M)$ are the most significant digits and t is the number of trailing zeros for the true count.

Corollary 2. *The variance function for the floating-point counter is*

$$g(k) = g(t, u) = \left(\frac{M}{3} + u\right)4^t - (M + u)2^t + \frac{2}{3}M. \tag{10}$$

Combining Corollaries 1 and 2, we get the following bounds on the dispersion.

Corollary 3. *The dispersion of the floating-point counter is asymptotically bounded as*

$$\sqrt{\frac{1}{3M-1}} \leq \liminf_{n\to\infty} A_n; \qquad \limsup_{n\to\infty} A_n \leq \sqrt{\frac{3}{8M-3}}.$$

Proof. By Equations (9) and (10), we have $\lim_{t\to\infty} \frac{g(t,u)}{f^2(t,u)} = \frac{\frac{M}{3}+u}{(M+u)^2}$. Considering the extreme values at $u = 0$ and $u = \lfloor M/3 \rfloor$ or $u = \lceil M/3 \rceil$, respectively:

$$\mu^2 = \liminf_{k\to\infty} \frac{g(k)}{f^2(k)} = \frac{1}{3M}; \quad \lambda^2 = \limsup_{k\to\infty} \frac{g(k)}{f^2(k)} \leq \lim_{t\to\infty} \frac{g(t, \frac{M}{3})}{f^2(t, \frac{M}{3})} = \frac{3}{8M}. \tag{11}$$

Plugging these limits into Theorem 3 leads to the Corollary. □

For large $M = 2^d$, the bounds of Corollary 3 become $\limsup_{n\to\infty} A_n \lessapprox 2^{-d/2}\sqrt{3/8} \approx 0.612 \cdot 2^{-d/2}$ and $\liminf_{n\to\infty} A_n \gtrapprox 2^{-d/2}\sqrt{1/3} \approx 0.577 \cdot 2^{-d/2}$.

The dispersion is thus comparable to the dispersion of a q-ary counter with $q = 2^{2^{-d}}$, which is approximately $2^{-d/2}\sqrt{0.5 \cdot \ln 2} \approx 0.589 \cdot 2^{-d/2}$. The memory requirements of the two counters are equivalent: in order to count up to $n = f(k)$, $d + \lg\lg n$ bits are necessary.

Scaled floating-point counters. A similar floating-point technique of approximate counting is described by Stanojević [2]. In his solution (for measuring network traffic with per-packet counter updates), a so-called "small active counter" splits a fixed number of ℓ bits into a d-bit significand u and an $(\ell - d)$-bit exponent t, in successive scale regimes $\sigma = 1, 2, \ldots$. The scaling factor σ is stored

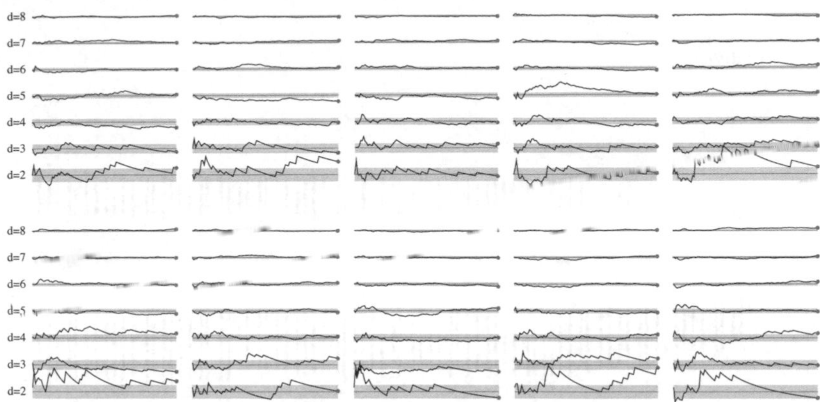

Fig. 2. Error trajectories for floating-point counters (**top**) and q-ary counters (**bottom**). Each trajectory follows the appropriate counting chain in a random simulated run. The lines trace the relative error $(f(X_n) - n)/n$ for floating-point counters with d-bit mantissa, and comparable q-ary counters with $q = 2^{1/r}$ where $r = 2^d$. The shaded areas indicate a relative error of $\pm 0.59 \cdot 2^{-d/2}$. The dots at the end of the trajectories denote the final value for $n = 100000$.

separately. (The same scaling factor is used for a whole set of counters, but the analysis is performed for a single counter only.) The counter is incremented with a probability of $2^{-\sigma\lfloor X/M \rfloor}$ at scaling σ, along with suitable adjustements to X at scale changes. Stanojević infers the appropriate unbiased estimator, and calculates the coefficient of variation (B_k of Theorem 3) for the hitting time $f(k) = \min\{n \colon X_n = k\}$ at a fixed scaling σ. Consider the scaled version of the floating-point counter where $q_k = q^{-\lfloor k/M \rfloor}$ with some $q > 1$ ($q = 2$ in the basic version and $q = 2^\sigma$ with scaling σ). By Theorems 1 and 2,

$$f(k) = f(t, u) = \Big(\frac{M}{q-1} + u\Big)q^t - \frac{M}{q-1}$$

$$g(k) = g(t, u) = \Big(\frac{M}{q^2-1} + u\Big)q^{2t} - \Big(\frac{M}{q-1} + u\Big)q^t + M\frac{q}{q^2-1}$$

with $t = \lfloor k/M \rfloor$ and $u = M\{k/M\} = k - tM$. Consequently, $\lg M + \lg \log_q n$ bits are needed to reach n. The extremes for B_k are attained with $u = 0$, and with $u = \lfloor M/(q+1) \rfloor$ or $u = \lceil M/(q+1) \rceil$:

$$\mu = \liminf_{k\to\infty} B_k = \lim_{t\to\infty} \frac{\sqrt{g(t,0)}}{f(t,0)} = M^{-1/2} \cdot \sqrt{\frac{q-1}{q+1}}$$

$$\lambda = \limsup_{k\to\infty} B_k \le \lim_{t\to\infty} \frac{\sqrt{g(t, M/(q+1))}}{f(t, M/(q+1))} = M^{-1/2} \cdot \sqrt{\frac{q^2-1}{4q}}$$

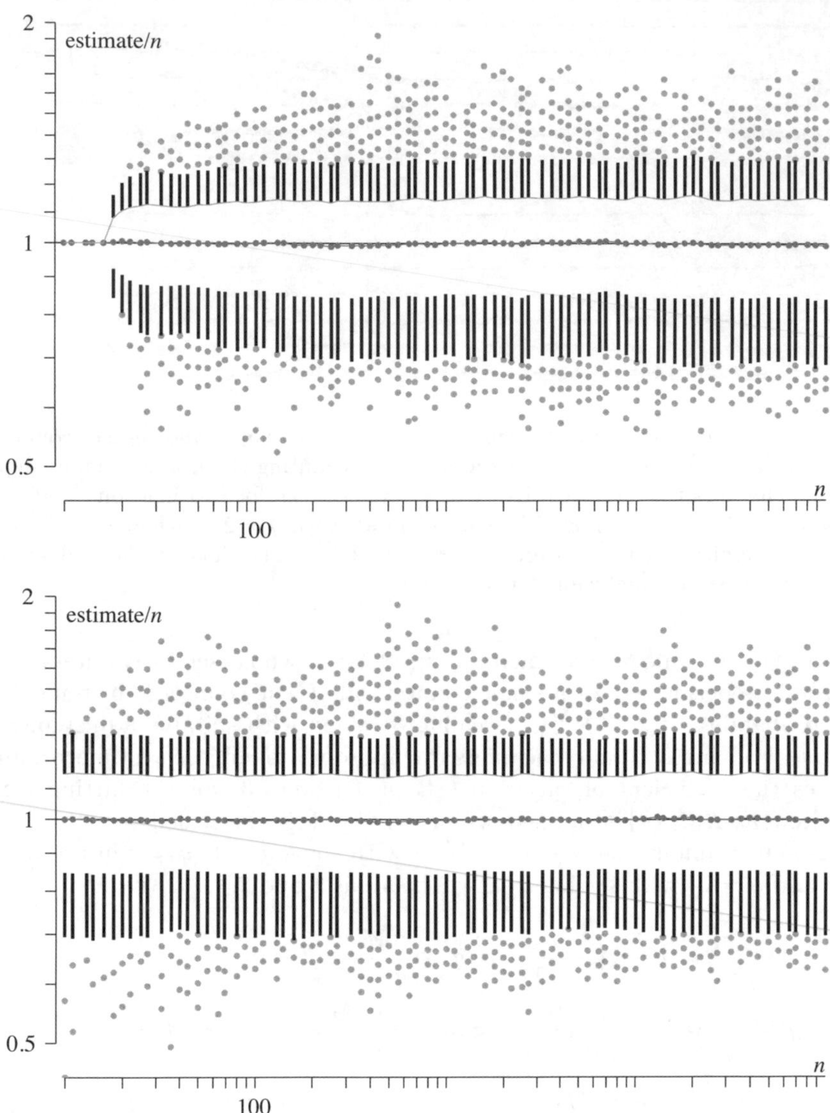

Fig. 3. Distribution of the estimates for a floating-point counter (**top**) and a comparable q-ary counter (**bottom**). Each plot depicts the result of 1000 experiments, in which a floating-point counter with $d = 4$-bit mantissa, and a q-ary counter with $q = 2^{1/16}$ were run until $n = 100,000$. The dots in the middle follow the averages; the black segments depict the standard deviations (for each σ, they are of length σ spaced at σ from the average), and grey dots show outliers that differ by more than $\pm 2\sigma$ from the average. The shading highlights the asymptotic dispersion of the q-ary counter ($\approx 0.59 \cdot 2^{-d/2}$).

By Theorem 3, the dispersion is bounded as

$$\sqrt{\frac{q-1}{(q+1)M-(q-1)}} \le \liminf_{n\to\infty} A_n; \qquad \limsup_{n\to\infty} A_n \le \sqrt{\frac{q^2-1}{4qM-(q^2-1)}}.$$

Simulations. Figures 2 and 3 compare the performance of the floating-point counters with equivalent base-q counters in simulation experiments. The equivalence is manifest on Figure 2 that illustrates the trajectories of the estimates by the different counters. Figure 3 plots statistics about the estimates across multiple experiments: the estimators are clearly unbiased, and the two counters display the same accuracy.

Acknowledgments

I am very grateful to Philippe Flajolet for valuable suggestions on improving previous versions of the manuscript. This work has been supported by a grant from the Natural Sciences and Engineering Research Council of Canada.

References

1. Estan, C., Varghese, G.: New directions in traffic measurement and accounting: Focusing on the elephants, ignoring the mice. ACM TOCS 21, 270–313 (2003)
2. Stanojević, R.: Small active counters. In: Proceedings INFOCOM, pp. 2153–2161 (2007)
3. Donato, D., Laura, L., Leonardi, S., Millozzi, S.: Large-scale properties of the Webgraph. Eur. Phys. J. B 38, 239–243 (2004)
4. Karlin, S.: Statistical signals in bioinformatics. PNAS 102, 13355–13362 (2005)
5. Jones, N.C., Pevzner, P.A.: Comparative genomics reveals unusually long motifs in mammalian genomes. Bioinformatics 22, e236–e242 (2006)
6. Rigoutsos, I., Huynh, T., Miranda, K., Tsirigos, A., McHardy, A., Platt, D.: Short blocks from the noncoding parts of the human genome have instances within nearly all known genes and relate to biological processes. PNAS 103, 6605–6610 (2006)
7. Csűrös, M., Noé, L., Kucherov, G.: Reconsidering the significance of genomic word frequencies. Trends Genet. 23, 543–546 (2007)
8. Sindi, S.S., Hunt, B.R., Yorke, J.A.: Duplication count distributions in DNA sequences. Phys. Rev. E 78, 61912 (2008)
9. Ning, Z., Cox, A.J., Mullikin, J.C.: SSAHA: A fast search method for large DNA databases. Genome. Res. 11(10), 1725–1729 (2001)
10. Flajolet, P.: Approximate counting: A detailed analysis. BIT 25, 113–134 (1985)
11. Morris, R.: Counting large number of events in small registers. CACM 21(10), 840–842 (1978)
12. Flajolet, P., Fusy, É., Gandouet, O., Meunier, F.: HyperLogLog: the analysis of a near-optimal cardinality estimation algorithm. In: Proceedings AofA, DMTCS Proceedings, pp. 127–146 (2007)
13. Kirschenhoffer, P., Prodinger, H.: Approximate counting: An alternative approach. RAIRO ITA 25, 43–48 (1991)
14. Kruskal, J.B., Greenberg, A.G.: A flexible way of counting large numbers approximately in small registers. Algorithmica 6, 590–596 (1991)
15. Karlin, S., Taylor, H.M.: A First Course in Stochastic Processes, 2nd edn. Academic Press, San Diego (1975)

Broadcasting in Heterogeneous Tree Networks*

Yu-Hsuan Su[1], Ching-Chi Lin[2], and D.T. Lee[3]

[1] Graduate Institute of Electronic Engineering, National Taiwan University, Taiwan
d98943034@ntu.edu.tw

[2] Department of Computer Science and Engineering,
National Taiwan Ocean University, Taiwan
lincc@mail.ntou.edu.tw

[3] Institute of Information Science, Academia Sinica, Taiwan
dtlee@iis.sinica.edu.tw

Abstract. We consider the broadcasting problem in heterogeneous tree networks. A heterogeneous tree network is represented by a weighted tree $T = (V, E)$ such that the weight of each edge denotes the communication time between the two end vertices. The broadcasting problem is to find a broadcast center such that the maximum communication time from the broadcast center to all vertices is minimized. In this paper, we propose a linear time algorithm for the broadcasting problem in a heterogeneous tree network following the postal model. As a byproduct of the algorithm, we can compute in linear time the broadcasting time of any vertex in the tree, i.e., the maximum time required to transmit messages from the vertex to every other vertex in the tree. Furthermore, an optimal sequence by which the broadcast center broadcasts its messages to all vertices in T can also be determined in linear time.

Keywords: algorithm, broadcast center, heterogeneous network, weighted tree, postal model.

1 Introduction

A heterogeneous network is a network connecting workstations with different operating systems and communication protocols. Thus, the times to communicate between any pair of workstations may be different. A heterogeneous network is represented by a weighted graph $G = (V, E)$, in which $V(G)$ represents a set of workstations and each edge $\overline{u,v} \in E(G)$ represents a connection between two adjacent workstations. The weight of each edge represents the transmission time β required to transmit messages between two adjacent workstations. In heterogeneous networks, each communication link may have different message transmission time, and each workstation may need connection time α to set up connection between two adjacent workstations to complete the message transmission. In this paper, we consider the broadcasting problem in a heterogeneous tree

* This work is partially supported by the National Science Council under the Grants No. NSC98-2221-E-001-007-MY3, NSC98-2221-E-001-008-MY3, and NSC-98-2221-E-019-029.

M.T. Thai and S. Sahni (Eds.): COCOON 2010, LNCS 6196, pp. 368–377, 2010.

network $T = (V, E)$, where the weight $w(u, v)$ of each edge $\overline{u,v} \in E(G)$, denotes the transmission time required. The broadcasting problem has been extensively studied for several decades [1, 6–9, 11–13] due to the increasing demands of heterogeneous network of workstations [7, 9]. The broadcasting problem is to find a broadcast center such that the broadcasting time from the broadcast center to all vertices in T is minimized. We consider the problem following the *postal model* [1–4, 9] as described below.

The postal model makes distinction between *connection time* α and *transmission time* β, where $\alpha > 0$ is assumed to be a constant and $\beta \geq 0$ varies from edge to edge. More specifically, the postal model assumes that the sender requires α time to set up a connection. After the sender sets up the connection, the sender is allowed to set up another connection to the next receiver while the sender is still transmitting the messages to the current receiver. For example, the sender v first sets up the connection with the receiver u_1 in α time, and then v can set up another connection to the next receiver u_2 while still transmitting the messages to u_1. The message transmission between two vertices is referred as a "call". A sender can only set up one connection to make a call to a receiver per α time and a sender can only transmit the messages to adjacent vertices. A call is said to be completed only after the sender completes transmitting the messages to the receiver. That is, if the transmission time β between the sender v and receiver u is $w(u, v)$, and the receiver u is the ith vertex called by the sender, then the call from v to u will be completed after $\alpha i + w(u, v)$ time. The receiver u cannot forward the messages until the sender completes the transmission.

In contrast to the postal model is the *telephone model*, in which the sender is allowed to set up another connection to the next receiver only after completing the transmission of messages to the current receiver. That is, only after the sender completes the transmission, can the sender set up another connection to the next receiver. For example, the sender v first sets up a connection with the receiver u_1 after α time, then the sender v can only set up another connection to the next receiver u_2 after $\alpha + w(u_1, v)$ time. For both the postal and telephone models, the receiver cannot forward the messages until the receiver finishes the receipt of the messages from the sender. Moreover, we assume that the message transmission between two adjacent vertices is *full-duplex*, i.e., the sender and receiver can exchange the messages between them simultaneously.

Some notations are introduced in order to give a formal definition of the broadcasting problem. Given a weighted tree $T = (V, E)$, the *broadcasting time* of v, denoted as $b(v, T)$, is the minimum time required to broadcast a message from v to all vertices in T. The *broadcasting time* of T, denoted as $b(T)$, is the minimum broadcasting time of any vertex $v \in V(T)$, i.e., $b(T) = \min\{b(v, T) | \ v \in V(T)\}$. The *broadcast center* $BC(T)$ of T is the set of vertices with the minimum broadcasting time, i.e., $BC(T) = \{v \mid v \in V(T), \ b(v, T) = b(T)\}$.

Problem Definition: Given a weighted tree $T = (V, E)$ in which the weight $w(u, v) \geq 0$ of an edge $\overline{u,v}$ represents the transmission time between them, the broadcasting problem is to compute the set of broadcast centers $BC(T)$

and determine the broadcasting time $b(T)$, following the postal model with a constant connection time $\alpha > 0$. We use n to denote the size of the tree T.

1.1 Previous Work

The telephone model and the postal model are the two most popular communication models in the literature. In 1978, Slater *et al.* [13] showed that computing the broadcasting time of a given vertex in an arbitrary unweighted graph is NP-complete following the telephone model. Furthermore, they also provided an $O(n)$-time algorithm to compute the set of broadcast centers and determine the broadcasting time for the unweighted trees following the telephone model. By adopting Slater *et al.*'s algorithm, Koh and Tcha [10] extended the results to provide an algorithm for the weighted trees following the telephone model. The algorithm runs in $O(n \log n)$ time due to the use of a sorting procedure and a priority queue.

For the telephone model in an unweighted graph G, Farley [5] determined the lower and upper bounds of the time required to broadcast m messages from a given vertex v to all vertices in G. For the postal model in an unweighted complete graph G, Bar-Noy and Kipnis [2] presented an algorithm to determine the minimum broadcasting time from a given broadcast center to all vertices in G. Further, they showed that the algorithm is optimal which runs in $\Theta(\lambda \log n / \log(\lambda + 1))$ time, where λ is the communication latency, i.e., the time needed when the sender starts sending messages until the receiver completes the transmission.

Given a graph G and two nonempty subsets $A \subseteq V(G)$ and $B \subseteq V(G)$ of vertices in G, the multicasting problem is to determine the minimum time to broadcast messages from all vertices in A to all vertices in B. For the telephone model in an unweighted complete graph, Khuller *et al.* [8] presented polynomial-time approximation algorithms for the single-source multicasting, i.e., $|A| = 1$, multi-source broadcasting problems, i.e., $B = V$, and multi-source multicasting problems.

In this paper, we consider the broadcasting problem in heterogeneous tree networks following the postal model. Note that our problem becomes the broadcasting problem for unweighted trees following the telephone model when $\alpha = 1$ and $\beta = 0$ for all edges.

1.2 Contributions

In this paper, we propose an $O(n)$-time algorithm for the broadcasting problem in heterogeneous tree networks following the postal model. Similar to the algorithm by Koh and Tcha [10], our algorithm is based on the concept of Slater *et al.*'s algorithm. But unlike their algorithm which uses a priority queue and a sorting procedure, resulting in an $O(n \log n)$-time algorithm, we develop an $O(n)$-time algorithm by using a new observation and a non-sorting labeling method. The two major refinements lead to a time complexity improvement from $O(n \log n)$ to $O(n)$. We further show that an optimal sequence of calls can also be obtained in $O(n)$ time.

For determining the broadcast centers $BC(T)$ in T, Slater *et al.* [13] use a bottom-up approach to iteratively update the labels of vertices and remove leaf nodes with the smallest label. Since the edges are unweighted, the value of labels are integers in $\{0, \ldots, n-1\}$. They exploit this fact to create a useful data structure to avoid the use of a priority queue and a sorting procedure. Since the labels are real numbers in our problem, it is not clear how to avoid the use of priority queue, which takes $O(\log n)$ time per operation, if one needs to select a vertex with the smallest label. However, we prove in Lemma 4 that it is sufficient to select between any two labeled vertices the one with the smaller label. With this new observation, we can perform this task in $O(n)$ time. As for update of labels of vertices in T, we propose a brand-new non-sorting procedure, and show in Theorem 1 its details.
We summarize our contributions as follows.

- The set of broadcast centers $BC(T)$ can be computed and the broadcasting time $b(T)$ can be determined in $O(n)$ time.
- An optimal sequence of calls by which the broadcast center broadcasts to all vertices in T can be determined in $O(n)$ time.
- Given a vertex $v \in V(T)$, the broadcasting time $b(v,T)$ and an optimal sequence of calls from v to all vertices in T can be determined in $O(n)$ time.

1.3 Organization

The rest of the paper is organized as follows. Section 2 describes the algorithm BROADCAST which computes the set of broadcast centers $BC(T)$ and determines the broadcasting time $b(T)$ following the postal model. In Section 3, we provide the correctness proof and timing analysis of the algorithm. Finally, we give concluding remarks and suggest some direction for future work in Section 4.

2 Algorithm BROADCAST

The algorithm adopts a greedy strategy to process the vertices in T in a bottom-up manner. For an edge $\overline{u,v}$ in T, the removal of this edge will result in two subtrees, each of which contains v and u respectively. The subtree of T containing v is denoted as $T(v,u)$, and the subtree of T containing u is denoted as $T(u,v)$. Recall that $b(v,T)$ denotes the minimum time required to broadcast a message from v to all vertices in T. Suppose that $u_1, u_2, \ldots, u_k$ are the neighbors of v in T such that $b(u_1, T(u_1,v)) + w(u_1,v) \geq b(u_2, T(u_2,v)) + w(u_2,v) \geq \ldots \geq b(u_k, T(u_k,v)) + w(u_k,v)$. If we want to broadcast messages from v to all vertices in T, it is not difficult to see that an optimal sequence of calls from v to its neighbors would be ordered as $u_1, u_2, ..., u_k$. Hence, the broadcasting time from v to all vertices in T is $b(v,T) = \max\{b(u_i, T(u_i,v)) + w(u_i,v) + \alpha i \mid 1 \leq i \leq k\}$.

Based on the above observation, the concept of our algorithm BROADCAST is as below. The algorithm initially assigns a label $t(u) = 0$ to each leaf node u. It then removes a leaf u in the current tree T iteratively and assigns a label

$t(v)$ to the vertex v, which becomes a leaf in the tree $T - \{u\}$. We always keep the leaf vertex with the largest label when removing a leaf node. It is one of the most significant differences of our algorithm from the algorithm by Slater *et al.* [13]. We will show that the tree $T - \{u\}$ still contains a broadcast center in T after removing the vertex u. Each vertex in T would be labeled exactly once except the last remaining vertex, which gets labeled twice. Let κ denote the last remaining vertex in the rest of this paper. For a vertex $v \neq \kappa$ in T, let v' be the neighbor of v such that v' is on the path from v to κ. We will show that $t(v) = b(v, T(v, v'))$, which is the minimum broadcasting time from v to all vertices in $T(v, v')$. The algorithm is detailed below.

Algorithm 1. Algorithm BROADCAST

Input: A weighted tree $T = (V, E)$ with weight $w(u, v) \geq 0$ for $\overline{u, v}$ in $E(T)$.
Output: The broadcasting time $b(T)$ and the set of broadcast centers $BC(T)$.

for each leaf $\ell \in T$ **do** $t(\ell) \leftarrow 0$;
let $BC(T) \leftarrow \emptyset, W \leftarrow \emptyset$ and $T' \leftarrow T$;
while $|V(T')| \geq 2$ **do**
  select two leaves u_x and u_y in T' arbitrarily;
  let $t(u) \leftarrow \min\{t(u_x), t(u_y)\}$, $W \leftarrow W \cup \{u\}$ and $T' \leftarrow T' - \{u\}$;
  let v be the vertex adjacent to u in T';
  if v is a leaf in T' **then**
    suppose that v is adjacent to labeled vertices $u_1, u_2, \ldots, u_k$ in W such that $t(u_1) + w(u_1, v) \geq t(u_2) + w(u_2, v) \geq \ldots \geq t(u_k) + w(u_k, v)$;
    let $t(v) \leftarrow \max\{t(u_i) + w(u_i, v) + \alpha i |\ 1 \leq i \leq k\}$;
  end if
end while
let κ be the only vertex left in T', $b(T) \leftarrow t(\kappa)$ and $BC(T) \leftarrow \{\kappa\}$;
let the neighbors of κ in T be $u_1, u_2, \ldots, u_k$ such that $t(u_1) + w(u_1, \kappa) \geq t(u_2) + w(u_2, \kappa) \geq \ldots \geq t(u_k) + w(u_k, \kappa)$;
let h be the smallest integer such that $t(u_h) + w(u_h, \kappa) + \alpha h + \alpha > b(T)$;
$BC(T) \leftarrow BC(T) \cup \{u_i |\ w(u_i, \kappa) = 0$ and $i \leq h\}$.

In the beginning of the algorithm, we set $t(\ell) = 0$ for each leaf ℓ in T. Next, in each iteration of the while loop, to keep the vertex with the largest label, we arbitrarily select two leaves u_x and u_y in the current tree T. If $t(u) = \min\{t(u_x), t(u_y)\}$, then we remove u from the current tree T. Suppose that v is the vertex adjacent to u in the current tree T. If v becomes a leaf in $T - \{u\}$, we set $t(v) = \max\{t(u_i) + w(u_i, v) + \alpha i |\ 1 \leq i \leq k\}$, where $u_1, u_2, \ldots, u_k \in W$ are the neighbors of v with $t(u_i) + w(u_i, v) \geq t(u_{i+1}) + w(u_{i+1}, v)$ for $1 \leq i \leq k - 1$. After the execution of the while loop, we have only one vertex left, denoted as κ, which is one of the broadcast centers in T. Moreover, the neighbors of κ are the only candidates for being a broadcast center of T. Suppose that h is the smallest integer such that $t(u_h) + w(u_h, \kappa) + \alpha h + \alpha > b(T)$. Let $N(\kappa)$ be the neighbors of κ in T. We will prove that for each vertex $u_i \in N(\kappa)$, u_i is a broadcast center in T if and only if $w(u_i, \kappa) = 0$ and $i \leq h$.

3 Correctness and Complexity Analysis

Let v be an arbitrary vertex in T and $u_1, ..., u_k$ be the neighbors of v. Let $\bar{T} = \bigcup_{1 \le i \le k}\{(v, u_i) + T(u_i, v)\}$ be a rooted tree of T with v as the root, and the neighbors of v are ordered so that $b(u_i, T(u_i, v)) + w(u_i, v) \ge b(u_{i+1}, T(u_{i+1}, v)) + w(u_{i+1}, v)$ for $1 \le i \le k-1$.

Intuitively, in order to shorten the broadcasting time from v to all vertices in $\bar{T}$, we will transmit the message to u_1 first, then u_2, u_3, and so on. The following lemma shows that $u_1, ..., u_k$ is in fact an optimal sequence of calls to broadcast messages from v to its neighboring vertices in $\bar{T}$. Then, it follows that the minimum time required to broadcast from v to all vertices in $\bar{T}$ is $b(v, \bar{T}) = \max\{b(u_i, T(u_i, v)) + w(u_i, v) + \alpha i \mid 1 \le i \le k\}$.

Lemma 1. *Let v be an arbitrary vertex in T and $u_1, ..., u_k$ be the neighbors of v. Let $\bar{T} = \bigcup_{1 \le i \le k}\{(v, u_i) + T(u_i, v)\}$ be a rooted tree of T with v as the root, and the neighbors of v are ordered so that $b(u_i, T(u_i, v)) + w(u_i, v) \ge b(u_{i+1}, T(u_{i+1}, v)) + w(u_{i+1}, v)$ for $1 \le i \le k-1$. Then, $u_1, u_2, \ldots, u_k$ is an optimal sequence of calls to broadcast messages from v to neighboring vertices in $\bar{T}$. Consequently, $b(v, \bar{T}) = \max\{b(u_i, T(u_i, v)) + w(u_i, v) + \alpha i \mid 1 \le i \le k\}$.*

Then we show an important result, on which the correctness proof of our algorithm is based, and also some useful properties for finding a broadcast center.

Lemma 2. *For each vertex v in T we have $t(v) = b(v, T(v, v'))$.*

Proof. Let $v_1, v_2, \ldots, v_n$ be the elimination order of vertices in T such that v_i is removed from T' before v_j if $i < j$. We prove the statement by induction on the number of vertices. Let vertex v'_1 be the neighbor of v_1 on the path from v_1 to $\kappa = v_n$. Clearly, v_1 is a leaf in T and so we have $t(v_1) = b(v_1, T(v_1, v'_1)) = 0$. Suppose that the statement holds for $i = k$. We consider $i = k+1$ below.

We first consider the case when v_{k+1} is a leaf in T. Clearly, $t(v_{k+1}) = b(v_{k+1}, T(v_{k+1}, v'_{k+1})) = 0$, where v'_{k+1} is the neighbor of v_{k+1} on the path from v_{k+1} to κ. Next, we consider the case when v_{k+1} is an internal vertex in T. Suppose that $v'_{k+1}, u_1, ..., u_\ell$ are the neighbors of v_{k+1} in T, and without loss of generality, let we assume that $t(u_i) + w(u_i, v) \ge t(u_{i+1}) + w(u_{i+1}, v)$, for $i = 1, ..., \ell - 1$. Notice that $u_i \in \{v_1, ..., v_k\}$ for $1 \le i \le \ell$. By induction hypothesis, we have $t(u_i) = b(u_i, T(u_i, v_{k+1}))$ for $1 \le i \le \ell$. Therefore, by Lemma 1, it holds that $b(v_{k+1}, T(v_{k+1}, v'_{k+1})) = \max\{t(u_i) + w(u_i, v_{k+1}) + \alpha i \mid 1 \le i \le \ell\}$. Meanwhile, according to the execution of Step 7 of the algorithm BROADCAST, we have $t(v_{k+1}) = \max\{t(u_i) + w(u_i, v_{k+1}) + \alpha i \mid 1 \le i \le \ell\}$. It follows that $t(v_{k+1}) = b(v_{k+1}, T(v_{k+1}, v'_{k+1}))$.

Lemma 3. *If $b(x_1, T(x_1, x_2)) \le b(x_2, T(x_2, x_1))$ with $\overline{x_1, x_2} \in E(T)$, then the following two statements hold :*

1. *$b(x_1, T) = \alpha + w(x_1, x_2) + b(x_2, T(x_2, x_1))$; and*
2. *$b(x_2, T) \le b(x_1, T)$.*

In the following, we show that after removing a leaf u in the current tree T, $T - \{u\}$ still contains a broadcast center by keeping the vertex with the largest label. It leads to the fact that the last remaining vertex κ is a broadcast center.

Lemma 4. *Suppose that a leaf u is deleted in the current tree T in the ith iteration of the while loop. Then we have the following results:*

1. *$BC(T) \cap V(T') \neq \emptyset$, where $T' = T - \{u\}$; and*
2. *the last remaining vertex $\kappa \in BC(T)$ and $b(\kappa, T) = b(T)$.*

Proof. Suppose that the leaf u is deleted in the ith iteration and v is the vertex adjacent to u in the current tree T. To prove $BC(T) \cap V(T') \neq \emptyset$, it suffices to show that $b(v, T) \leq b(u, T)$. We first consider the case when the current tree T contains exactly two vertices u and v. Note that by the choice of u and Lemma 2, $b(v, T(v, u)) = t(v) \geq t(u) = b(u, T(u, v))$. According to Lemma 3, since $b(v, T(v, u)) \geq b(u, T(u, v))$ with $\overline{u, v} \in E(T)$, we have $b(v, T) \leq b(u, T)$.

Next, we consider the case when the current tree T contains at least three vertices u, v, and y. Suppose that u and y are the two leaves selected in the ith iteration of the while loop and $t(u) \leq t(y)$. Let y' be the neighbor of y on the path from y to v. Similarly, it suffices to show that $b(v, T) \leq b(u, T)$. Once again, we prove that by showing $b(v, T(v, u)) \geq b(u, T(u, v))$ according to Lemma 3. Suppose to the contrary that $b(v, T(v, u)) < b(u, T(u, v))$. Since $T(y, y')$ is a subtree of $T(v, u)$, we have $b(y, T(y, y')) < b(v, T(v, u))$. By $t(u) \leq t(y)$, we have $b(u, T(u, v)) = t(u) \leq t(y) = b(y, T(y, y'))$. This implies that $b(y, T(y, y')) < b(v, T(v, u)) < b(u, T(u, v)) \leq b(y, T(y, y'))$, a contradiction. Therefore, it is in fact that $b(v, T(v, u)) \geq b(u, T(u, v))$. Hence according to Lemma 3, since $b(v, T(v, u)) \geq b(u, T(u, v))$ with $\overline{u, v} \in E(T)$, we have $b(v, T) \leq b(u, T)$.

Intuitively, a tree may contain more than one broadcast center. Below we show that the only candidates for being a broadcast center of T are the neighbors of κ. Moreover, for each vertex $u_i \in N(\kappa)$, u_i is a broadcast center of T if and only if $w(u_i, \kappa) = 0$ and $i \leq h$.

Lemma 5. *If v is the broadcast center in T, then $v \in N(\kappa) \cup \{\kappa\}$.*

Lemma 6. *For each vertex $u_i \in N(\kappa)$, if $w(u_i, \kappa) = 0$ and $i \leq h$, then we have $u_i \in BC(T)$.*

Lemma 7. *For each vertex $u_i \in N(\kappa)$, if $w(u_i, \kappa) > 0$ or $i > h$, then we have $u_i \notin BC(T)$.*

Since the only candidates for being a broadcast center of T are the neighbors of κ, the set of broadcast centers $BC(T)$ is a star. Then we have the following corollary.

Corollary 1. *The set of broadcast centers $BC(T)$ is a star.*

Below we provide the correctness proof and the timing analysis of the algorithm.

Theorem 1. *Given a weighted tree $T(V, E)$, the algorithm BROADCAST computes the set of broadcast centers and determines the broadcasting time $b(T)$ in $O(n)$ time.*

Proof. Combining Lemmas 4 to 7, we obtain the correctness proof of the algorithm. In the following, we will show that the algorithm can be implemented in $O(n)$ time. We first observe that Steps 1 and 2 take $O(n)$ time. For the while loop, there is an intuitive implementation which takes $O(k \log k)$ time to sort the values $t(u_1) + w(u_1, v), t(u_2) + w(u_2, v), \ldots, t(u_k) + w(u_k, v)$ for each vertex v in T. However, we can see that one vertex is removed from T in each iteration of the while loop. Therefore, if Step 7 takes $O(k)$ time to assign the label $t(v)$ for each vertex v in T, then the while loop can also be completed in $O(n)$ time. In the following, we introduce a useful data structure which enables us to compute labels for n vertices in $O(n)$ time without sorting.

Suppose that $u_1, u_2, \ldots, u_k$ are the neighbors of v in T such that $t(u_1) + w(u_1, v) \geq t(u_2) + w(u_2, v) \geq \ldots \geq t(u_k) + w(u_k, v)$. Note that $k = |N(v)|$. If a vertex u_p satisfies $t(u_1) + w(u_1, v) \geq t(u_p) + w(u_p, v) + \alpha k$ with $k \geq p$, then $t(u_1) + w(u_1, v) + \alpha > t(u_p) + w(u_p, v) + \alpha p$. Note that the vertex u_p would have no influence on the label of v. Therefore, in order to determine $t(v)$, we only need to consider the neighbor u_i of v such that $t(u_i) + w(u_i, v) > t(u_1) + w(u_1, v) - \alpha k$. Based on the above observation, we construct the following k linked lists.

For each list$[i]$, with $0 \leq i \leq k - 1$, the list$[i]$ contains the vertices u_j such that $\alpha i \leq (t(u_1) + w(u_1, v)) - (t(u_j) + w(u_j, v)) < \alpha(i + 1)$. Note that the vertex u_1 with $t(u_1) + w(u_1, v) = \max\{t(u_i) + w(u_i, v)\}$ can be determined by simply visiting the neighbors of v once. For each neighbor u_i of v in $N(v) - \{u_1\}$, if $t(u_i) + w(u_i, v) \leq t(u_1) + w(u_1, v) - \alpha k$, then we discard it. Otherwise, we insert the vertex into the front of its corresponding linked list. Therefore, it takes $O(k)$ time to construct these k linked lists.

We use num$[i]$ to denote the number of vertices in the list$[i]$ and let acc$[i]$ = $\sum_{j=0}^{i}$ num$[j]$ for $0 \leq i \leq k - 1$. Further, let u_{i^*} be the vertex in the list$[i]$ such that $t(u_{i^*}) + w(u_{i^*}, v) = \min\{t(u_j) + w(u_j, v) \mid u_j$ belongs to the list$[i]\}$. Clearly, for $0 \leq i \leq k - 1$, the values num$[i]$, acc$[i]$, and the vertex u_{i^*} can be determined in $O(k)$ time. For any given vertices u_x and u_y belonging to the same linked list with $x < y$, since $y - x \geq 1$ and $(t(u_x) + w(u_x, v)) - (t(u_y) + w(u_y, v)) < \alpha$, we have $t(u_y) + w(u_y, v) + \alpha y > t(u_x) + w(u_x, v) + \alpha x$. Therefore,

$$\begin{aligned} t(v) &= \max\{t(u_i) + w(u_i, v) + \alpha i \mid 1 \leq i \leq k\} \\ &= \max\{t(u_{i^*}) + w(u_{i^*}, v) + \alpha \mathrm{acc}[i] \mid 0 \leq i \leq k - 1\}. \end{aligned}$$

So the label of v, $t(v)$, can be determined in $O(k)$ time. By the above labeling method, the while loop can be completed in $O(n)$ time for n vertices.

Next, we show that the smallest integer h such that $t(u_h) + w(u_h, \kappa) + \alpha h + \alpha > b(T)$ can also be determined in $O(n)$ time in Step 12. Let q be the smallest integer such that $t(u_{q^*}) + w(u_{q^*}, \kappa) + \alpha \mathrm{acc}[q] + \alpha > b(T)$. Clearly, the list$[q]$ contains the vertex u_h. We will show that either $h = \mathrm{acc}[q]$ or $h = \mathrm{acc}[q] - 1$. For any given vertices u_x and u_y belonging to the same linked list with $y \geq x + 2$, since

$(t(u_x)+w(u_x,v))-(t(u_y)+w(u_y,v)) < \alpha$, we have $(t(u_y)+w(u_y,\kappa)+y)-(t(u_x)+w(u_x,\kappa)+x) > \alpha$. Note that $t(u_{q^*})+w(u_{q^*},\kappa)+\alpha\text{acc}[q] \leq b(T)$. Therefore, we have $h = \text{acc}[q]$ or $h = \text{acc}[q]-1$. Hence, h can be determined in $O(n)$ time. These prove the correctness and the time complexity of the algorithm.

In the following, we will show that an optimal sequence of calls by which the broadcast center broadcasts its messages to all vertices in T can be determined in $O(n)$ time.

Theorem 2. *An optimal sequence of calls by which the broadcast center broadcasts its messages to all vertices in T can be determined in $O(n)$ time.*

Proof. Suppose that $u_1, u_2, \ldots, u_k$ are the neighbors of v in T such that $t(u_1)+w(u_1,v) \geq t(u_2)+w(u_2,v) \geq \ldots \geq t(u_k)+w(u_k,v)$. For each list$[i]$, with $0 \leq i \leq k-1$, the list$[i]$ contains the vertices u_j such that $\alpha i \leq (t(u_1)+w(u_1,v))-(t(u_j)+w(u_j,v)) < \alpha(i+1)$. Suppose further that the list$[k]$ contains the vertices u_j such that $t(u_1)+w(u_1,v) \geq t(u_j)+w(u_j,v)+\alpha k$. Let num$[i]$ denote the number of vertices in the list$[i]$ and $\text{acc}[i] = \sum_{j=0}^{i} \text{num}[j]$ for $0 \leq i \leq k-1$. Further, let u_{i^*} be the vertex in the list$[i]$ such that $t(u_{i^*})+w(u_{i^*},v) = \min\{t(u_j)+w(u_j,v) \mid u_j$ belongs to the list$[i]\}$. We place the vertex u_{i^*} at the end of the list$[i]$ for $0 \leq i \leq k$.

Then, we assume that $u_{\pi(1)}, u_{\pi(2)}, \ldots, u_{\pi(k)}$ is a traversal ordering of the lists such that the list$[p]$ is traversed before the list$[q]$ if $p < q$. When traversing a list, we traverse the list from the beginning to the end of the list sequentially. Clearly, the ordering $u_{\pi(1)}, u_{\pi(2)}, \ldots, u_{\pi(k)}$ can be determined in $O(k)$ time. Since

$$
\begin{aligned}
t(v) &= \max\{t(u_i)+w(u_i,v)+\alpha i \mid 1 \leq i \leq k\} \\
&= \max\{t(u_{i^*})+w(u_{i^*},v)+\alpha\text{acc}[i] \mid 0 \leq i \leq k-1\} \\
&= \max\{t(u_{\pi(i)})+w(u_{\pi(i)},v)+\alpha\pi(i) \mid 1 \leq i \leq k\},
\end{aligned}
$$

$u_{\pi(1)}, u_{\pi(2)}, \ldots, u_{\pi(k)}$ is an optimal sequence of calls by which v broadcasts messages to its neighboring vertices in T, which completes the proof.

Using a similar method of the above arguments, one can show that the following result is true.

Theorem 3. *Given a vertex $v \in V(T)$, the broadcasting time $b(v,T)$ and an optimal sequence of calls from v to all vertices in T can be determined in $O(n)$ time.*

4 Conclusion and Future Work

We have proposed a non-sorting linear time algorithm for the broadcasting problem in a weighted tree following the postal model. The algorithm BROADCAST computes the set of broadcast centers, determines the broadcasting time $b(T)$, and an optimal sequence of calls from the broadcast center to all vertices in T. Also, given a vertex $v \in V(T)$, the broadcasting time $b(v,T)$ and an optimal sequence of calls from v to all vertices in T can be determined in linear time.

Below we present some open problems related to the broadcasting problem in heterogeneous networks following the postal model, i.e., the broadcasting p-center and the broadcasting p-median problems.

The broadcasting p-center problem is a generalization of the broadcasting problem. In the broadcasting p-center problem, we want to locate p centers on a network and partition the set of n vertices in p subsets, such that the maximum communication time from the centers to the associated subsets of vertices is minimized. On the other hand, in the broadcasting p-median problem, instead of minimizing the maximum communication time, we minimize the sum of communication times from the centers to the associated subsets of vertices.

References

1. Bar-Noy, A., Guha, S., Naor, J., Schieber, B.: Message multicasting in heterogeneous networks. SIAM J. Comput. 30(2), 347–358 (2000)
2. Bar-Noy, A., Kipnis, S.: Designing broadcasting algorithms in the postal model for message-passing systems. Math. Systems Theory 27(5), 431–452 (1994)
3. Bar-Noy, A., Kipnis, S.: Multiple message broadcasting in the postal model. Networks 29(1), 1–10 (1997)
4. Elkin, M., Kortsarz, G.: An approximation algorithm for the directed telephone multicast problem. Algorithmica 45(4), 569–583 (2006)
5. Farley, A.M.: Broadcast time in communication networks. SIAM J. Appl. Math. 39(2), 385–390 (1980)
6. Hedetniemi, S.M., Hedetniemi, S.T., Liestman, A.L.: A survey of gossiping and broadcasting in communication networks. Networks 18(4), 319–349 (1988)
7. Khuller, S., Kim, Y.-A.: Broadcasting in heterogeneous networks. Algorithmica 48(1), 1–21 (2007)
8. Khuller, S., Kim, Y.-A., Wan, Y.-C.: On generalized gossiping and broadcasting. J. Algorithms 59(2), 81–106 (2006)
9. Khuller, S., Kim, Y.-A., Wan, Y.-C.: Broadcasting on networks of workstations. Algorithmica 52(3), 1–21 (2008)
10. Koh, J.-M., Tcha, D.-W.: Information dissemination in trees with nonuniform edge transmission times. IEEE Trans. Comput. 40(10), 1174–1177 (1991)
11. Lee, H.-M., Chang, G.J.: Set to set broadcasting in communication networks. Discrete Appl. Math. 40(3), 411–421 (1992)
12. Richards, D., Liestman, A.L.: Generalizations of broadcasting and gossiping. Networks 18(2), 125–138 (1988)
13. Slater, P.J., Cockayne, E.J., Hedetniemi, S.T.: Information dissemination in trees. SIAM J. Comput. 10(4), 692–701 (1981)

Contention Resolution in Multiple-Access Channels: k-Selection in Radio Networks*

Antonio Fernández Anta[1] and Miguel A. Mosteiro[1,2]

[1] LADyR, GSyC, Universidad Rey Juan Carlos, 28933 Móstoles, Madrid, Spain
anto@gsyc.es

[2] Department of Computer Science, Rutgers University, Piscataway, NJ 08854, USA
mosteiro@cs.rutgers.edu

Abstract. In this paper, contention resolution among k contenders on a multiple-access channel is explored. The problem studied has been modeled as a k-Selection in Radio Networks, in which every contender has to have exclusive access at least once to a shared communication channel. The randomized adaptive protocol presented shows that, for a probability of error 2ε, all the contenders get access to the channel in time $(e+1+\xi)k+O(\log^2(1/\varepsilon))$, where $\varepsilon \leq 1/(n+1)$, $\xi > 0$ is any constant arbitrarily close to 0, and n is the total number of potential contenders. The above time complexity is asymptotically optimal for any significant ε. The protocol works even if the number of contenders k is unknown and collisions can not be detected.

1 Introduction

A recurrent question, in settings where a resource must be shared among many contenders, is how to make that resource available to all of them. The problem is particularly challenging if not even the number of contenders is known. The broad spectrum of settings where answers to such a question are useful makes its study a fundamental task. An example of such contention is the problem of broadcasting information in a multiple-access channel. A multiple-access channel is a synchronous system that allows a message to be delivered to many recipients at the same time using a channel of communication but, due to the shared nature of the channel, the simultaneous introduction of messages from multiple sources produce a conflict that precludes any message from being delivered to any recipient. In Radio Networks,[1] one of the instances of such a question is the problem known in the literature [3] as *Selection*. In its general version, the k-*Selection* problem [11], also known as *all-broadcast*, is solved when an unknown size-k subset of n network nodes have been able to access a unique shared channel of communication, each of them at least once. The k-Selection problem in Radio Networks and related problems have been well-studied for settings where a tight upper

* This research was supported in part by Comunidad de Madrid grant S-0505/TIC/0285; Spanish MEC grant TIN2005-09198-C02-01, TIN2008–06735-C02-01; NSF grant CCF 0632838; and EU Marie Curie International Reintegration Grant IRG 210021.

[1] As pointed out in [3], the historical developments justify the use of Radio Network to refer to any communication network where the channel is contended, even if radio communication is not actually used.

M.T. Thai and S. Sahni (Eds.): COCOON 2010, LNCS 6196, pp. 378–388, 2010.

bound on k is known. In this paper, a randomized adaptive protocol for k-Selection in Radio Networks is presented, assuming that such a knowledge is not available, the arrival of messages is batched, and conflicts to access the channel cannot be detected by all nodes. To our knowledge, this is the first k-Selection protocol in the Radio Networks literature that works in such conditions and it is asymptotically optimal for any sensible error-probability bound (up to inverse exponential in k). This protocol improves over previous work in adversarial packet contention-resolution thanks to the adaptive nature of the protocol and the knowledge of n. Given that the error probability is parametrized, this protocol can be also applied to solve k-Selection in multiple neighborhoods of a multi-hop Radio Network.

Notation and Model. Most of the following assumptions and notation are folklore in the Radio Networks literature. For details and motivation, see the survey of Chlebus [3]. We study the k-Selection problem in a Radio Network comprised of n labeled stations called *nodes*. Each node is assumed to be potentially reachable from any other node in one communication step, hence, the network is characterized as *single-hop* or *one-hop* indistinctively. Before running the protocol, nodes have no information besides n and their own label, which is assumed to be unique but arbitrary.[2] Time is supposed to be slotted in *communication steps*. Assuming that the computation time-cost is negligible in comparison with the communication time-cost, time efficiency is studied in terms of communication steps only. The piece of information assigned to a node in order to deliver it to other nodes is called a *message*. The assignment of a message is due to an external agent and such an event is called a *message arrival*. Communication among nodes is carried out by means of radio broadcast on a shared channel. If exactly one node transmits at a communication step, such a transmission is called *successful* or *non-colliding*, we say that the message was *delivered*, and all other nodes *receive* such a message. If more than one message is transmitted at the same time, a *collision* occurs, the messages are garbled, and nodes only receive *interference noise*. If no message is transmitted in a communication step, nodes receive only *background noise*. In this work, nodes can not distinguish between interference noise and background noise, thus, the channel is called *without collision detection*. Each node is in one of two states, *active* if it holds a message to deliver, or *idle* otherwise. In contrast with *oblivious* protocols, where the sequence of transmissions of a node does not depend on the transmissions received, the *adaptive* protocol presented in this paper exploits the information implicit on the occurrence of a successful transmission. In the randomized protocol presented here all active nodes use the same probability in the same communication step, a class of protocols usually called *fair*. Therefore, it is also a *uniform* protocol, i.e., all active nodes use the same protocol. As in for instance [1, 11, 7], we assume that all the k messages arrive in a *batch*, i.e. in the same communication step, a problem usually called *static* k-Selection,[3] and that each node becomes idle upon delivering its message.

Problem Definition. Given a Radio Network where a subset K of the set of n network nodes, such that $|K| = k$, are activated by message arrivals, the k-Selection problem

[2] Notice that our protocol does not make any use of the identity of a message originator. Thus, it can be used even in settings where nodes are not labeled or labels are not unique.

[3] A *dynamic* counterpart where messages arrive at different times was also studied [11].

is solved when each node in K has delivered its message. The definition given pertains to the general version of the problem where messages may arrive at different times, although in this paper we study only simultaneous, or *batched*, arrivals.

Related Work. Regarding deterministic solutions, the k-Selection problem was shown to be in $O(k \log(n/k))$ already in the 70's by giving adaptive protocols that make use of collision detection [2,8,14]. In all these results the algorithmic technique, known as *tree algorithms*, relies on modeling the protocol as a complete binary tree where the messages are placed at the leaves. Later, Greenberg and Winograd [6] showed a lower bound for that class of protocols of $\Omega(k \log_k n)$. Regarding oblivious algorithms, Komlòs and Greenberg [10] showed the existence of $O(k \log(n/k))$ solutions even without collision detection but requiring knowledge of k and n. More recently, Clementi, Monti, and Silvestri [4] showed a lower bound of $\Omega(k \log(n/k))$, which also holds for adaptive algorithms if collision detection is not available. In [11], Kowalski presented the construction of an oblivious deterministic protocol that, using the explicit selectors of Indyk [9], gives a $O(k \operatorname{polylog} n)$ upper bound without collision detection.

In the following results, availability of collision detection is assumed. Martel presented in [13] a randomized adaptive protocol for k-Selection that works in $O(k+\log n)$ time in expectation[4]. As argued by Kowalski in [11], this protocol can be improved to $O(k + \log\log n)$ in expectation using Willard's expected $O(\log\log n)$ selection protocol of [17]. In the same paper, Willard shows that, for any given protocol, there exists a choice of $k \leq n$ such that selection takes $\Omega(\log\log n)$ expected time for the class of fair selection protocols. For the case in which n is not known, in the same paper a $O(\log\log k)$ expected time selection protocol is described, again, making use of collision detection. If collision detection is not available, using the techniques of Kushilevitz and Mansour in [12], it can be shown that, for any given protocol, there exists a choice of $k \leq n$ such that $\Omega(\log n)$ is a lower bound in the expected time to get even the first message delivered.

A frequent challenging difficulty to overcome in resolving collisions is to determine which is the best probability of transmission to be used by the contenders when their number is unknown. The method of choice is then to increase or decrease such probability based on the success or failure of successive trials. When the probability of transmission is increased it is said that a *back-on* strategy is used, whereas *back-off* is the term used when such probability is decreased. A combination of both strategies is usually called *back-on/back-off*. Monotonic back-off strategies for contention resolution of batched arrivals of k packets on simple multiple access channels, a problem that can be seen as k-Selection, have been analyzed in [1]. The best strategy shown is the so-called *loglog-iterated back-off* with a makespan in $\Theta(k \log\log k/ \log\log\log k)$ with probability at least $1 - 1/k^c, c > 0$, which does not use any knowledge of k or n.

Regarding related problems, extending previous work on tree algorithms, Greenberg and Leiserson [7] presented randomized routing strategies in fat-trees for bounded number of messages. Choosing appropriate constant capacities for the edges of the fat-tree, the problem could be seen as k-Selection. However, that choice implies a logarithmic congestion parameter which yields an overall $O(k \operatorname{polylog} n)$ time. In [5],

[4] Througout this paper, log means $\log_2$ unless otherwise stated.

Gerèb-Graus and Tsantilas presented an algorithm that solves the problem of realizing arbitrary h-relations in an n-node network, with probability at least $1 - 1/n^c, c > 0$, in $\Theta(h + \log n \log\log n)$ steps. In an h-relation, each processor is the source as well as the destination of h messages. Making $h = k$ this protocol can be used to solve k-Selection. However, it requires that nodes know h.

Results and Outline. In this paper, a randomized adaptive protocol for k-Selection, in a one-hop Radio Networkwithout collision detection, that does not require knowledge of the number of contenders k, is presented. Assuming that $\varepsilon \leq 1/(n+1)$, the protocol is shown to solve the problem in $(e+1+\xi)k+O(\log^2(1/\varepsilon))$ communication steps, where $\xi > 0$ is any constant arbitrarily close to 0 with probability at least $1-2\varepsilon$. Given that the error probability is parametric, this protocol can be applied to multiple neighborhoods of a multi-hop Radio Network, adjusting the error probability in each one-hop neighborhood appropriately. To our knowledge, $O(k \log\log k/\log\log\log k)$ [1] is the best upper bound in the literature for a protocol suitable to solve k-Selection in Radio Networks (although they propose it for packet contention resolution), that works without knowledge of k, under batched arrivals, and without collision detection. By exploiting back-on/back-off and the knowledge of n, our protocol improves their time complexity. Given that k is a lower bound for this problem, the protocol is optimal (modulo a small constant factor) if $\varepsilon \in \Omega(2^{-\sqrt{k}})$. In Section 2 the details of the protocol are presented and they are analyzed in Section 3.

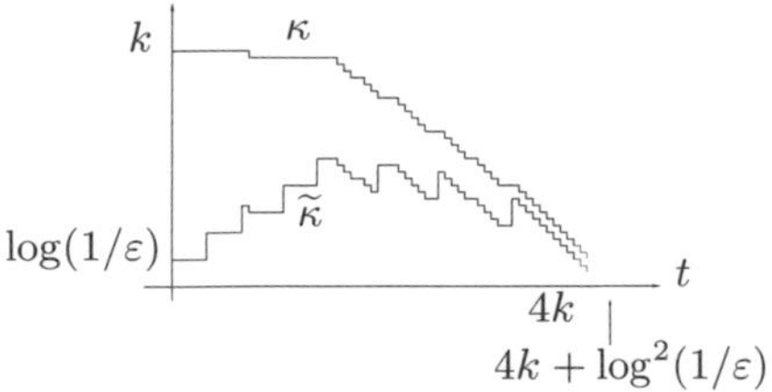

Fig. 1. Illustration of estimate progress

2 Protocol

The protocol comprises two different algorithms. Each of them is particularly suited for one of two scenarios, depending on the number of messages left to deliver. The algorithm called *BT* solves the problem for the case when that number is below a threshold (that will be defined later). The algorithm called *AT* is suited to reduce that number from the initial k to a value below that threshold. The BT algorithm uses the well-known technique of repeating transmissions with the same appropriately-suited probability until the problem is solved. The AT algorithm on the other hand is adaptive by repeatedly increasing an estimation of the messages left and decreasing such an estimation by roughly one each time a message is delivered. (Even if that successful transmission is due to the BT algorithm.) An illustration of the estimation progress is depicted in Figure 1. Further details can be seen in Algorithm 1. Both algorithms are executed interleaving their communication steps (see Task 1 in Algorithm 1). For clarity, each communication step is referred to by using the name of the algorithm executed at that step. The following notation used in the algorithm is defined for clarity: $\beta \triangleq e + \xi_\beta$, $\delta \triangleq 1 + \xi_\delta$, $\tau \triangleq 300\beta \ln(1/\varepsilon)$, $\varepsilon \triangleq$ error probability, $0 < \xi_\delta < 1$, $0 < \xi_\beta < 0.27$ and $0 < \xi_t \leq 1/2$ are constants arbitrarily close to 0, and $1/\xi_t \in \mathbb{N}$.

Algorithm 1. Pseudocode for node x.

```
1  upon message arrival do
2      t ← τ
3      κ̃ ← τ
4      start tasks 1, 2 and 3
5  Task 1
6      foreach communication-step = 1, 2, ... do
7          if communication-step ≡ 1 (mod 1/ξ_t) then          // BT-step
8              transmit ⟨x, message⟩ with prob 1/τ
9          else                                                // AT-step
10             transmit ⟨x, message⟩ with prob 1/κ̃
11             t ← t − 1
12             if t ≤ 0 then
13                 t ← τ
14                 κ̃ ← κ̃ + τ
15 Task 2
16     upon reception from other node do
17         κ̃ ← max{κ̃ − δ, τ}
18         t ← t + β
19 Task 3
20     upon message delivery stop
```

3 Analysis

For clarity, each of the algorithms comprising the protocol are first analyzed separately and later put together in the main theorem. Consider first the AT algorithm. (Refer to Algorithm 1.) Let $\widetilde{\kappa}$ be called the *density estimator*. Let a *round* be the sequence of AT-steps between increasings of the density estimator (Line 1). Let the rounds be numbered as $r \in \{1, 2, \ldots\}$ and the AT-steps within a round as $t \in \{1, 2, \ldots\}$. (E.g., round 1 is the sequence of AT-steps from initialization until Line 1 of the algorithm is executed for the first time.) Let $\kappa_{r,t}$, called the *density*, be the number of messages not delivered yet (i.e., the number of active nodes) at the beginning of AT-step t of round r. Let $\widetilde{\kappa}_{r,t}$ be the density estimator used at the AT-step t of round r. Let $X_{r,t}$ be an indicator random variable such that, $X_{r,t} = 1$ if a message is delivered at the AT-step t of round r, and $X_{r,t} = 0$ otherwise. Then, $Pr(X_{r,t} = 1) = (\kappa_{r,t}/\widetilde{\kappa}_{r,t})(1 - 1/\widetilde{\kappa}_{r,t})^{\kappa_{r,t}-1}$. Also, for a round r, let the number of messages delivered in the interval of AT-steps $[1, t)$ of r, including those delivered in BT steps, be $\sigma_{r,t}$. The following intermediate results will be useful. First, we state the following useful fact.

Fact 1. *[15, §2.68]*

$$e^{x/(1+x)} \leq 1 + x \leq e^x, 0 < |x| < 1.$$

Lemma 1. *For any round r where $\widetilde{\kappa}_{r,1} \leq \kappa_{r,1} - \gamma$, $\gamma \geq \delta(2-\delta)/(\delta-1) \geq 0$, $Pr(X_{r,t}=1)$ is monotonically non-increasing with respect to t for $\delta+1 < \widetilde{\kappa}_{r,t} \leq \kappa_{r,t}$, and $\delta < (\kappa_{r,t}-\gamma)(\kappa_{r,t}-\gamma-1)/(\kappa_{r,t}-\gamma+1)$.*

Proof. We want to show conditions such that for any t in round r, $Pr(X_{r,t}=1) \geq Pr(X_{r,t+1}=1)$. If $\kappa_{r,t} = \kappa_{r,t+1}$ the claim holds trivially. Then, let us assume instead that $\kappa_{r,t} > \kappa_{r,t+1}$. We want to show that

$$\frac{\kappa_{r,t}}{\widetilde{\kappa}_{r,t}}\left(1-\frac{1}{\widetilde{\kappa}_{r,t}}\right)^{\kappa_{r,t}-1} \geq \frac{\kappa_{r,t+1}}{\widetilde{\kappa}_{r,t+1}}\left(1-\frac{1}{\widetilde{\kappa}_{r,t+1}}\right)^{\kappa_{r,t+1}-1}.$$

Due to the BT-step between two consecutive AT-steps, at most two messages are delivered in the interval $[t, t+1)$ of r. Thus, replacing appropriately, we want to show that the following inequalities hold.

$$\frac{\kappa_{r,t}}{\widetilde{\kappa}_{r,t}}\left(1-\frac{1}{\widetilde{\kappa}_{r,t}}\right)^{\kappa_{r,t}-1} \geq \frac{\kappa_{r,t}-1}{\widetilde{\kappa}_{r,t}-\delta}\left(1-\frac{1}{\widetilde{\kappa}_{r,t}-\delta}\right)^{\kappa_{r,t}-2}, \quad (1)$$

$$\frac{\kappa_{r,t}}{\widetilde{\kappa}_{r,t}}\left(1-\frac{1}{\widetilde{\kappa}_{r,t}}\right)^{\kappa_{r,t}-1} \geq \frac{\kappa_{r,t}-2}{\widetilde{\kappa}_{r,t}-2\delta}\left(1-\frac{1}{\widetilde{\kappa}_{r,t}-2\delta}\right)^{\kappa_{r,t}-3}. \quad (2)$$

Reordering 1,

$$\frac{\widetilde{\kappa}_{r,t}-\delta-1}{\widetilde{\kappa}_{r,t}}\left(\frac{\widetilde{\kappa}_{r,t}-1}{\widetilde{\kappa}_{r,t}}\frac{\widetilde{\kappa}_{r,t}-\delta}{\widetilde{\kappa}_{r,t}-\delta-1}\right)^{\kappa_{r,t}-1} \geq \frac{\kappa_{r,t}-1}{\kappa_{r,t}}. \quad (3)$$

Using calculus, it can be seen that the left-hand side of 3 is monotonically non-increasing for $\delta+1 < \widetilde{\kappa}_{r,t} \leq \kappa_{r,t}$. The details are omitted for brevity. Then, given that $\widetilde{\kappa}_{r,t} = \widetilde{\kappa}_{r,1} - \sigma_{r,t} \leq \kappa_{r,1} - \sigma_{r,t} - \gamma \leq \kappa_{r,t} - \gamma$, it is enough to show

$$\frac{\kappa_{r,t}}{\kappa_{r,t}-1} \cdot \frac{\kappa_{r,t}-\gamma-\delta-1}{\kappa_{r,t}-\gamma} \cdot \left(\frac{\kappa_{r,t}-\gamma-1}{\kappa_{r,t}-\gamma}\frac{\kappa_{r,t}-\gamma-\delta}{\kappa_{r,t}-\gamma-\delta-1}\right)^{\kappa_{r,t}-1} \geq 1. \quad (4)$$

Again using calculus, it can be seen that the left-hand side of Inequality 4 is monotonically non-increasing on $\kappa_{r,t}$ for $\gamma \geq \delta(2-\delta)/(\delta-1)$ and $\delta < (\kappa_{r,t}-\gamma)(\kappa_{r,t}-\gamma-1)/(\kappa_{r,t}-\gamma+1)$. The details are omitted for brevity. Then, it is enough to show that, in the limit, the left-hand side of Inequality 4 tends to 1. Which can be verified using standard calculus techniques. The details are omitted for brevity. Using the same techniques, Inequality 2 can be shown to hold.

Lemma 2. *For any round r where $\kappa_{r,1} - \gamma - \tau \leq \widetilde{\kappa}_{r,1} < \kappa_{r,1} - \gamma$, $\gamma \geq 0$ and for any AT-step t in r such that*

$$\sigma_{r,t} \leq \kappa_{r,1}\frac{\ln\beta - 1}{\delta\ln\beta - 1} - \frac{(\gamma+\tau+1)\ln\beta - 1}{\delta\ln\beta - 1},$$

the probability of a successful transmission is at least $Pr(X_{r,t}=1) \geq 1/\beta$.

Proof. We want to show $(\kappa_{r,t}/\widetilde{\kappa}_{r,t})(1-1/\widetilde{\kappa}_{r,t})^{\kappa_{r,t}-1} \geq 1/\beta$. Given that nodes are active until their message is delivered, it is enough to show

$$\frac{\kappa_{r,1}-\sigma_{r,t}}{\widetilde{\kappa}_{r,1}-\delta\sigma_{r,t}}\left(1-\frac{1}{\widetilde{\kappa}_{r,1}-\delta\sigma_{r,t}}\right)^{\kappa_{r,1}-1-\sigma_{r,t}} \geq 1/\beta. \quad (5)$$

Using calculus, it can be seen that the left hand side of Inequality 5 is monotonically non-decreasing with restpect to $\widetilde{\kappa}_{r,1}$ under the conditions of the Lemma. The details are omitted for brevity. Then, it is enough to prove Inequality 5 for $\widetilde{\kappa}_{r,1} = \kappa_{r,1} - \gamma - \tau$.

$$\frac{\kappa_{r,1}-\sigma_{r,t}}{\kappa_{r,1}-\gamma-\tau-\delta\sigma_{r,t}}\cdot\left(1-\frac{1}{\kappa_{r,1}-\gamma-\tau-\delta\sigma_{r,t}}\right)^{\kappa_{r,1}-1-\sigma_{r,t}} \geq 1/\beta$$

$$\left(1-\frac{1}{\kappa_{r,1}-\gamma-\tau-\delta\sigma_{r,t}}\right)^{\kappa_{r,1}-1-\sigma_{r,t}} \geq 1/\beta.$$

Given that $\sigma_{r,t} \leq \kappa_{r,1}\frac{\ln\beta-1}{\delta\ln\beta-1} - \frac{(\gamma+\tau+1)\ln\beta-1}{\delta\ln\beta-1} < (\widetilde{\kappa}_{r,1}-(\gamma+\tau+1))/\delta$, using Fact 1, we want

$$\exp\left(\frac{\kappa_{r,1}-\sigma_{r,t}-1}{\kappa_{r,1}-\gamma-\tau-\delta\sigma_{r,t}-1}\right) \leq \beta$$

$$\frac{\kappa_{r,1}-\sigma_{r,t}-1}{\kappa_{r,1}-\gamma-\tau-\delta\sigma_{r,t}-1} \leq \ln\beta.$$

Manipulating the last expression, it can be seen that the lemma holds.

The following lemma, shows the efficiency and correctness of the AT-algorithm.

Lemma 3. *If the number of messages to deliver is more than*

$$M = 2\frac{\delta\ln\beta-1}{\ln\beta-1}\left(\sum_{j=1}^{5}(5/6)^{j-1}\tau\right) + \frac{((\delta(2-\delta)/(\delta-1))+\tau+1)\ln\beta-1}{\ln\beta-1} \in O(\log(1/\varepsilon)),$$

after running the AT-algorithm for $(e+\xi_\beta+1+\xi_\delta)k-\tau$ steps, where ξ_β and ξ_δ are constants arbitrarily close to 0, the number of messages left to deliver is reduced to at most M with probability at least $1-\varepsilon$, for $\varepsilon \leq 1/(n+1)$.

Proof. Consider the first round r such that

$$\kappa_{r,1}-\gamma-\tau \leq \widetilde{\kappa}_{r,1} < \kappa_{r,1}-\gamma, \gamma = \delta(2-\delta)/(\delta-1). \quad (6)$$

By definition of the AT algorithm, unless the number of messages left to deliver is reduced to at most M before, such a round exists. To see why, notice in Algorithm 1 that the density estimator is either increased by τ in Line 1, or decreased by δ in Line 1, or assigned τ in Line 1 or 1. After the first assignment, we have $\widetilde{\kappa}_{1,1} = \tau < \kappa_{1,1}-\gamma-\tau$, because $\kappa_{1,1} > M > 2\tau+\gamma$. We show now that condition 6 of r can not be satisfied right after decreasing the density estimator in Line 1. Consider two consecutive steps $t', t'+1$ of some round r' such that still $\widetilde{\kappa}_{r',t'} < \kappa_{r',t'}-\gamma-\tau$. If, upon a success at step

t' of r', $\widetilde{\kappa}_{r',t'+1} = \tau$ by the assignment in Line 1, and $\kappa_{r',t'+1} - \gamma - \tau \leq \widetilde{\kappa}_{r',t'+1}$, then $\kappa_{r',t'+1} \leq \tau + \gamma + \tau < M$ and we are done. If on the other hand $\widetilde{\kappa}_{r',t'+1} = \widetilde{\kappa}_{r',t'} - \delta$ by the assignment in Line 1, then $\widetilde{\kappa}_{r',t'+1} = \widetilde{\kappa}_{r',t'} - \delta < \kappa_{r',t'} - \gamma - \tau - \delta < \kappa_{r',t'+1} - \gamma - \tau$. Thus, the only way in which the density estimator gets inside the aforementioned range is by the increase in Line 1 and therefore round r exists.

We show now that, before leaving round r, at least τ messages are delivered with high probability so that in some future round $r'' > r$ the condition $\kappa_{r'',1} - \gamma - \tau \leq \kappa_{r'',1} < \kappa_{r'',1} - \gamma$ holds again. In order to do that, we divide round r in consecutive sub-rounds of size $\tau, 5/6\tau, (5/6)^2\tau, \ldots$ (The fact that a number of steps is an integer is omitted throughout for clarity.) More specifically, the sub-round S_1 is the set of AT-steps in the interval $(0, \tau]$ and, for $i \geq 2$, the sub-round S_i is the set of steps in the interval $((5/6)^{i-2}\tau, (5/6)^{i-1}\tau]$. Thus, denoting $|S_i| = \tau_i$ for all $i \geq 1$, it is $\tau_1 = \tau$ and $\tau_i = (5/6)\tau_{i-1}$ for $i \geq 2$. For each $i \geq 1$, let Y_i be a random variable such that $Y_i = \sum_{t \in S_i} X_{r,t}$. Even if no message is delivered, round r still has at least the sub-round S_1 by definition of the algorithm. Given that, according with Algorithm 1, each message delivered delays the end of round r in $\beta = e + \xi_\beta$ AT-steps, for $i \geq 2$, the existence of sub-round S_i is conditioned on $Y_{i-1} \geq 5\tau_{i-1}/(6\beta)$. We show now that with big enough probability round r has 5 sub-rounds and at least τ messages are delivered. Even if messages are delivered in every step of the 5 sub-rounds (including messages delivered in BT-steps), given that $\kappa_{r,1} > M$, the total number of messages delivered is less than $\kappa_{r,1}\frac{\ln\beta - 1}{\delta\ln\beta - 1} - \frac{(\gamma+\tau+1)\ln\beta - 1}{\delta\ln\beta - 1}$ because $\gamma = \delta(2-\delta)/(\delta-1)$. Thus, Lemma 2 can be applied and the expected number of messages delivered in S_i is $E[Y_i] \geq \tau_i/\beta$. In order to use Lemma 1, we verify first its preconditions. If, at any step t, $\kappa_{r,t} \leq M$, we are done. Otherwise, we know that $\kappa_{r,t} \geq \widetilde{\kappa}_{r,t} > \delta + 1$ and $(\kappa_{r,t} - \gamma)(\kappa_{r,t} - \gamma - 1)/(\kappa_{r,t} - \gamma + 1) > \delta$. Given that $\gamma = \delta(2-\delta)/(\delta-1)$, by Lemma 1, the random variables $X_{r,i}$ are not positively correlated, therefore, in order to bound from below the number of successful transmissions we can use the following Chernoff-Hoeffding bound [16]. For $0 < \varphi < 1$,

$$\begin{cases} Pr(Y_1 \leq (1-\varphi)\tau_1/\beta) \leq e^{-\varphi^2\tau_1/(2\beta)} \\ Pr(Y_i \leq (1-\varphi)\tau_i/\beta | Y_{i-1} \geq 5\tau_{i-1}/(6\beta)) \leq e^{-\varphi^2\tau_i/(2\beta)}, \forall i : 2 \leq i \leq 5. \end{cases}$$

Taking $\varphi = 1/6$,

$$\begin{cases} Pr(Y_1 \leq 5\tau_1/(6\beta)) \leq e^{-\varphi^2 300\ln(1/\varepsilon)/2} \\ Pr(Y_i \leq 5\tau_i/(6\beta) | Y_{i-1} \geq 5\tau_{i-1}/(6\beta)) \leq e^{-\varphi^2(5/6)^{i-1}300\ln(1/\varepsilon)/2}, \forall i : 2 \leq i \leq 5. \end{cases}$$

$$\begin{cases} Pr(Y_1 \leq 5\tau_1/(6\beta)) < e^{-2\ln(1/\varepsilon)} \\ Pr(Y_i \leq 5\tau_i/(6\beta) | Y_{i-1} \geq 5\tau_{i-1}/(6\beta)) < e^{-2\ln(1/\varepsilon)}, \forall i : 2 \leq i \leq 5. \end{cases}$$

Given that $\varepsilon \leq 1/(n+1)$ and $k \leq n$, then it holds that $\varepsilon^2 + k\varepsilon \leq 1$ which implies that $\ln(1/\varepsilon) \geq \ln(\varepsilon + k)$, therefore $e^{-2\ln(1/\varepsilon)} \leq e^{-\ln(\varepsilon+k) - \ln(1/\varepsilon)} = \varepsilon/(\varepsilon + k)$. So, more than $(5/(6(e + \xi_\beta)))\tau_i$ messages are delivered in any sub-round S_i with probability at least $1 - \varepsilon/(\varepsilon + k)$. Given that each success delays the end of round r in $\beta = e + \xi_\beta$ AT-steps, we know that, for $1 \leq i \leq 4$, sub-round S_{i+1} exists with probability at least

$1-\varepsilon/(\varepsilon+k)$. If, after any sub-round, the number of messages left to deliver is at most M, we are done. Otherwise, conditioned on these events, the total number of messages delivered over the 5 sub-rounds is at least $\sum_{j=1}^{5} Y_j > \sum_{j=1}^{5}(5/(6(e+\xi_\beta)))^j(e+\xi_\beta)^{j-1}\tau = (\tau/(e+\xi_\beta))\sum_{j=1}^{5}(5/6)^j > \tau$ because $\xi_\beta < 0.27$.

Thus, the same analysis can be repeated over the next round r'' such that $\kappa_{r'',1}-\gamma-\tau \leq \widetilde{\kappa}_{r'',1} < \kappa_{r'',1}-\gamma$. Unless the number of messages left to deliver is reduced to at most M before, such a round r'' exists by the same argument used to prove the existence of round r. The same analysis is repeated over various rounds until all messages have been delivered or the number of messages left is at most M. Then, using conditional probability, the overall probability of success is at least $(1-\varepsilon/(\varepsilon+k))^k$. Using Fact 1 twice, that probability is at least $1-\varepsilon$.

It remains to be shown the time complexity of the AT algorithm. The difference between the number of messages to deliver and the density estimator right after initialization is at most $k-\tau$. This difference is increased with each message delivered by at most $\delta-1$ and reduced at the end of each round by τ. Therefore, the total number of rounds is at most $(k-\tau+(\delta-1)k)/\tau = \delta k/\tau - 1$. Each message delivered adds only a constant factor β to the total time, whereas the other steps in each round add up to τ. Therefore, the total time is at most $(\beta+\delta)k-\tau = (e+\xi_\beta+1+\xi_\delta)k-\tau$.

The time efficiency and correctness of the BT algorithm is established in the following lemma. The proof, omitted for brevity, is a straightforward computation of the probability of some message not being delivered.

Lemma 4. *If the number of messages left to deliver is at most*

$$M = 2\frac{\delta\ln\beta-1}{\ln\beta-1}\left(\sum_{j=1}^{5}(5/6)^{j-1}\tau\right) + \frac{((\delta(2-\delta)/(\delta-1))+\tau+1)\ln\beta-1}{\ln\beta-1},$$

there exists a constant $c>0$ such that, after running the BT-algorithm for $c\log^2(1/\varepsilon)$ steps, all messages are delivered with probability at least $1-\varepsilon$.

The following theorem establishes the main result.

Theorem 2. *For any one-hop Radio Network, under the model detailed in Section 1, Algorithm 1 solves the k-selection problem within $(e+1+\xi)k+O(\log^2(1/\varepsilon))$ communication steps, where $\xi>0$ is any constant arbitrarily close to 0, with probability at least $1-2\varepsilon$ for $\varepsilon \leq 1/(n+1)$.*

Proof. From Lemmas 3 and 4, and the definition of the algorithm, the total time is $(e+1+\xi_\delta+\xi_\beta)k/(1-\xi_t)+O(\log^2(1/\varepsilon))$. Given that ξ_β, ξ_δ, and ξ_t are positive constants arbitrarily close to 0, the claim follows.

4 Conclusions and Open Problems

The general problem of enabling an unknown number of contenders the access to a shared resource was studied in this paper. The results obtained pertain to a problem of broadcasting information in a multiple-access radio-channel, but they may be straightforwardly applied to any setting that supports the same model. The specific problem studied here, k-Selection in Radio Networks, was previously studied in the literature, but assuming that a tight upper bound on the number of contenders is known. Thus, a crucial contribution of this paper was the removal of such assumption, consequently widening the scope of application of the protocol presented. Furthermore, we have assumed that messages are assigned to all nodes at the same time, increasing the potential contention for the channel with respect to scenarios where messages might arrive sparsely. To avoid collisions resulting from that contention it would be useful to have a mechanism to detect them at each node. However, we studied a more challenging scenario where only the transmitter of a message knows if it was the only one to access the channel in a time slot or not. Nonetheless, even under all these challenging conditions, the bound shown is asymptotically optimal for any sensible error-probability bound. To the best of our knowledge, the k-Selection protocol presented in this paper is the first in the Radio Networks literature that works in such conditions and is optimal.

A number of possible extensions of this work arise as natural questions that are left for future work. First, different patterns of message arrivals comprising specific application scenarios, such as Poisson arrivals and others, may also yield optimal bounds. Also, the protocol presented here improves over previous work in adversarial packet contention-resolution thanks to the adaptive nature of the protocol and the knowledge of n. Therefore, the question of how to solve the problem optimally in settings where nodes don't even know n or the feasibility of a non-adaptive optimal protocol are also important. Finally, the experimental evaluation of the protocol presented here, or others resulting from the above mentioned future work, would be useful for comparison with heuristics currently in use.

References

1. Bender, M.A., Farach-Colton, M., He, S., Kuszmaul, B.C., Leiserson, C.E.: Adversarial contention resolution for simple channels. In: 17th Ann. ACM Symp. on Parallel Algorithms and Architectures, pp. 325–332 (2005)
2. Capetanakis, J.: Tree algorithms for packet broadcast channels. IEEE Trans. Inf. Theory IT-25(5), 505–515 (1979)
3. Chlebus, B.S.: Randomized communication in radio networks. In: Pardalos, P.M., Rajasekaran, S., Reif, J.H., Rolim, J.D.P. (eds.) Handbook on Randomized Computing, vol. 1, pp. 401–456. Kluwer Academic Publishers, Dordrecht (2001)
4. Clementi, A., Monti, A., Silvestri, R.: Selective families, superimposed codes, and broadcasting on unknown radio networks. In: Proc. of the 12th Ann. ACM-SIAM Symp. on Discrete Algorithms, pp. 709–718 (2001)
5. Gerèb-Graus, M., Tsantilas, T.: Efficient optical communication in parallel computers. In: 4th Ann. ACM Symp. on Parallel Algorithms and Architectures, pp. 41–48 (1992)
6. Greenberg, A., Winograd, S.: A lower bound on the time needed in the worst case to resolve conflicts deterministically in multiple access channels. Journal of the ACM 32, 589–596 (1985)
7. Greenberg, R.I., Leiserson, C.E.: Randomized routing on fat-trees. Advances in Computing Research 5, 345–374 (1989)
8. Hayes, J.F.: An adaptive technique for local distribution. IEEE Trans. Comm. COM-26, 1178–1186 (1978)

9. Indyk, P.: Explicit constructions of selectors and related combinatorial structures, with applications. In: Proc. of the 13th Ann. ACM-SIAM Symp. on Discrete Algorithms, pp. 697–704 (2002)
10. Komlòs, J., Greenberg, A.: An asymptotically nonadaptive algorithm for conflict resolution in multiple-access channels. IEEE Trans. Inf. Theory 31, 303–306 (1985)
11. Kowalski, D.R.: On selection problem in radio networks. In: Proc. 24th Ann. ACM Symp. on Principles of Distributed Computing, pp. 158–166 (2005)
12. Kushilevitz, E., Mansour, Y.: An $\Omega(D \log(N/D))$ lower bound for broadcast in radio networks. SIAM Journal on Computing 27(3), 702–712 (1998)
13. Martel, C.U.: Maximum finding on a multiple access broadcast network. Inf. Process. Lett. 52, 7–13 (1994)
14. Mikhailov, V., Tsybakov, B.S.: Free synchronous packet access in a broadcast channel with feedback. Problemy Peredachi Inform 14(4), 32–59 (1978)
15. Mitrinović, D.S.: Elementary Inequalities. P. Noordhoff Ltd., Groningen (1964)
16. Mitzenmacher, M., Upfal, E.: Probability and Computing: Randomized Algorithms and Probabilistic Analysis. Cambridge University Press, Cambridge (2005)
17. Willard, D.E.: Log-logarithmic selection resolution protocols in a multiple access channel. SIAM Journal on Computing 15, 468–477 (1986)

Online Preemptive Scheduling with Immediate Decision or Notification and Penalties

Stanley P.Y. Fung*

Department of Computer Science, University of Leicester, Leicester, United Kingdom
pyfung@mcs.le.ac.uk

Abstract. We consider online preemptive scheduling problems where jobs have deadlines and the objective is to maximize the total value of jobs completed before their deadlines. In the first problem, preemptions are not free but incur a penalty. In the second problem, a job has to be accepted or rejected immediately upon arrival, and possibly allocated a fixed scheduling interval as well; if these accepted jobs are eventually not completed they incur a penalty (on top of not getting the value of the job). We give an algorithm with the optimal competitive ratio for the first problem, and new and improved algorithms for the second problem.

1 Introduction

Penalties. In most work on preemptive scheduling, preemption is assumed to be free. However, preemptions may actually be quite costly or have otherwise negative impact, e.g., customer dissatisfaction, so that we should aim at minimizing preemption if possible. One way to model this is to introduce *preemption penalties*, i.e., a penalty has to be paid for each preemption. Recently, several papers [11,5,13,9] studied algorithms for preemptive scheduling with penalties when the objective is to maximize the total value of completed jobs.

In this paper we introduce a related notion, perhaps more suitable from a customer's perspective, which we call *non-completion penalties*. Here preemption by itself does not cause penalty, as all the customer cares is that the job is completed on time. If a committed job is not delivered at the end, however, not just the value of the job is lost, but the algorithm needs to pay a penalty. This is akin to, for example, a customer giving a job to a company with a specific deadline, which the company promises will be completed on time, only to find out later that it cannot be completed – clearly this creates dissatisfaction to the customer. In cases where jobs are "nonpreemptable" (a job's preemption means the job is lost) then this is equivalent to preemption penalties.

With non-completion penalties, it is necessary to give algorithms the ability to refuse jobs on arrival. At the moment a job arrives, the algorithm can decide whether to accept or reject the job. If it is rejected, no value is obtained (the customer goes away) but no penalty is imposed either. But if the job is accepted

* Part of this work was done while the author was on study leave granted by University of Leicester.

M.T. Thai and S. Sahni (Eds.): COCOON 2010, LNCS 6196, pp. 389–398, 2010.

but cannot be finished by its deadline (e.g. there are other "more important" jobs arriving later), not just the value is lost but a penalty has to be paid.

We can further distinguish between *immediate notification* and *immediate decision* models: in immediate notification, the algorithm informs the customer whether the job is accepted, but not the time interval of which will it be processed, and the algorithm may change any such assigned timeslots or preempt partially executed jobs. Immediate decision is a stronger notion where additionally the algorithm has to fix (tell the customer) during which time interval the job will be scheduled (for example a company may want to advise its customers on when a delivery will take place, at the time the order is accepted).

To be more precise, here are the three models of penalties that we consider. In all cases, the penalty is proportional to the value of the job.

(Preemption Penalty). Jobs in execution can be preempted, and a penalty has to be paid for the preemption. The job can be restarted or resumed again later, depending on the model (this will be addressed shortly). No commitment need to be made at a job's arrival.

(Immediate Notification with Non-completion Penalty). At a job's arrival it must be decided immediately whether to accept or reject a job, but not when to schedule it. A rejected job will not be penalised (but also gives no value). A penalty is incurred if the job is not completed, but not for each individual preemption.

(Immediate Decision with Non-completion Penalty). At a job's arrival it must be decided immediately whether to accept or reject a job and, if it is accepted, to allocate a time interval (with length equal to the length of the job) for execution for this job. Once a job is allocated the interval, it cannot be rescheduled, although the job can be later preempted, however the preempted job cannot be run again later (hence the job is forever lost). A penalty is incurred for such a non-completed job. Equivalently, we can say a penalty is imposed on the preemption of the job. This is the exact model in [9].

Definitions and Notations. A *job* j is specified by its release time $r(j)$, deadline $d(j)$, processing time (length) $p(j)$ and value $v(j)$. The expiry time of a job is defined as $d(j) - p(j)$. For a job that is released but not completed, if it has not past its expiry time, it is called *pending*, otherwise it has *expired*. The laxity $\alpha(j)$ of a job j is $(d(j) - r(j))/p(j)$. A job is *tight* if it has laxity 1, which means it must be scheduled immediately (or rejected) on arrival. Each job of value $v(j)$ gives rise to a penalty of $\rho v(j)$ if it is preempted or not completed (depending on the model), for some constant $\rho \geq 0$.

Jobs arrive online, meaning that they are not known before they arrive. Jobs have to be completed before its deadline to receive its value. The value of a schedule is the total value of all completed jobs, minus all penalties paid. Our objective is to maximize the value of the schedule.

We will always use $\mathcal{OPT}$ to denote the optimal offline algorithm, and $\mathcal{ONL}$ to denote the online algorithm in question. When it is clear from context, they also denote the schedules produced by these algorithms. $\mathcal{OPT}$ never needs to pay any penalties for the non-completion models since it would know in advance whether a job can be completed before accepting it.

To measure the quality of online algorithms we use the standard definition of *competitive ratio*: an algorithm $\mathcal{ONL}$ is c-competitive if the value returned by $\mathcal{ONL}$ on any instance I is at least $1/c$ that of the value returned by the optimal offline algorithm $\mathcal{OPT}$ on the same instance.

Value models. The general model defined above does not admit constant competitive algorithms, even without any penalties or immediate decision/notification. Hence we focus on the following two special cases regarding the values of jobs, commonly studied in the literature: (1) The *proportional value* model, where the value of a job is equal to its length, i.e. $v(j) = p(j)$ for all j. (2) The *unit length* model, where jobs can have arbitrary values, but all jobs have the same length. Without loss of generality this length can be assumed to be 1. Thus $p(j) = 1$ for all j and $v(j)$ is independent of other parameters of the job.

Preemption models. Two different preemption models have been considered in the literature:
(1) *Preemption restart* (e.g. [12,7]): a job can be preempted, but it can only restart from the beginning if executed again later.
(2) *Preemption resume* (e.g. [1,8]): a job can be preempted, and it can resume execution at the point where it is preempted.

Note that for the case of immediate decision, there is no issue on restart or resume because the jobs are forever lost once preempted.

Previous work. The problem studied in [9] is closest to the models we study here. It gave an algorithm for the immediate decision model with proportional value, although it did not explicitly mention immediate decision. We are not aware of other results that address immediate decision/notification and penalties together in the preemptive setting, but there are other related work; we classify them based on the preemption models below.

(1) *Preemption restart:* When there are no penalties, there is a 4.56-competitive algorithm [12] and a lower bound of 4 [10] for the unit length case. For tight jobs an optimal upper bound of 4 is known [10]. (The work in [12] concern a slightly different *broadcast scheduling* problem but the results apply here.) Zheng et al [11] incorporated preemption penalties, and gave a $(2\rho + 3 + 2\sqrt{\rho^2 + 3\rho + 1})$-competitive algorithm. A lower bound of $4(1 + \rho)$ was proved in [5].

(2) *Preemption resume:* For the case without penalties, a 4-competitive optimal bound was given in [1] for the proportional value case. For the case with penalties, [5] gave a lower bound $4(1 + \rho)$.

(3) *Nonpreemptive:* Results in (1) and (2) above do not consider immediate decision or notification. The idea of immediate notification and immediate decision was proposed and studied in earlier papers [3,6,2,4], but all these papers were concerned with nonpreemptive scheduling. The proportional value case was

considered in [6], while [3,2] considered the unit length case but where jobs are unweighted. [6] also gave results when the laxity of jobs is a parameter.

Our results. In this paper we give new algorithms for the preemption or non-completion penalty models defined above. We focus on the restart model but will indicate when the results apply to the resume model as well.

- For the preemption penalty model we give a $4(1+\rho)$-competitive algorithm for the proportional value case. The algorithm and the analysis works for both restart or resume models. This matches the lower bound in [5].
- For immediate decision or notification with the proportional value model, we give an algorithm with competitive ratio $2\rho + 3 + 2\sqrt{\rho^2 + 3\rho + 2}$. This bound is better than the one in [9]. For example when $\rho = 0, 0.2, 1$ this gives $r = 5.828, 6.650, 9.899$ respectively whereas [9] gives $r = 13.32, 14.63, 20.14$ respectively. (Although [9] considers multiple processors, our algorithm can also work for the multiple processors case, with the same competitive ratio and a straightforward extension of the analysis, which we omit here.)
- For immediate decision or notification with the unit length model, we give an upper bound of $4\rho + 5 + 4\sqrt{\rho^2 + 5\rho/2 + 3/2}$ on the competitive ratio.
- We also show a lower bound for immediate decision even if jobs have arbitrarily large laxity.

Some proofs are omitted in this conference version due to space constraints.

2 Preemption Penalty

Algorithm 1. Let $\beta = 2(1+\rho)$. At any time the algorithm maintains a pool of all pending jobs. Newly released jobs are added to the pool. If the algorithm is idle, start the pending job with the earliest expiry time. If while executing a job j, another pending job k reaches its expiry time, then preempt j and start k if (i) $v(k) > \beta v(j)$ and j has preempted other jobs before, or (ii) $v(k) > \alpha v(j)$ and j has not preempted other jobs before, where $\alpha = 2(\beta - 1) \geq \beta$. In these cases j is never started again. Otherwise, k will not be started (and will therefore expire).

The algorithm borrows ideas from the one in [1], and essentially follows a simple strategy of preempting a job if another expiring one is β times more valuable than it. The exception is the first preemption after a job completion (or at the very beginning), in which case the preemption ratio is higher.

Note that the algorithm never reschedules a job after it is preempted. Thus the algorithm can be used for both resume and restart models. In the following we analyse its competitive ratio for the resume model (which implies the same ratio for the restart model, because $\mathcal{ONL}$ is the same while $\mathcal{OPT}$ is more powerful in the resume model). In the resume model, $\mathcal{OPT}$ may run a job in several parts in different time intervals. We refer to each of these as a *job fragment*.

Basic subschedules. We first define the concept of *basic subschedules*, which will be used throughout the paper. The schedule produced by $\mathcal{ONL}$ can be viewed as a sequence of preempted or completed jobs. The maximal sequence of jobs $j_1, j_2, \ldots, j_k$, such that j_i is preempted by j_{i+1} for all $i < k$ and only j_k is completed, is called a basic subschedule. We denote by $B = (v_1, \ldots, v_k)$ the values of the jobs in the basic subschedule.

Lemma 1. *For any basic subschedule $B = (v_1, \ldots, v_k)$,*
(i) If $k \geq 2$ then $v_i \leq v_k/\beta^{k-i}$ for $i \geq 2$ and $v_1 \leq v_k/(\alpha\beta^{k-2})$,
(ii) If a job j expires (and is not executed by $\mathcal{ONL}$) while $\mathcal{ONL}$ is running some job in B, then $v(j) \leq \beta v_k$ if $k \geq 2$, and $v(j) \leq \alpha v_1$ if $k = 1$.

Theorem 1. Algorithm 1 *is $4(1+\rho)$-competitive.*

Proof. In the following we ignore any penalties $\mathcal{OPT}$ may need to pay, because such penalties can only improve the competitive ratio. The total execution time of $\mathcal{OPT}$, and hence its value, can be divided into two parts: those while $\mathcal{ONL}$ is busy and those while $\mathcal{ONL}$ is idle. For the (part of) job fragments in $\mathcal{OPT}$ that are contained in times while $\mathcal{ONL}$ is busy, we simply associate them to the basic subschedule running in $\mathcal{ONL}$ at the same time. In the following we consider the (part of) job fragments in $\mathcal{OPT}$ that are run while $\mathcal{ONL}$ is idle.

Consider a period of idle time between two basic subschedules, and let f be the finishing time of the earlier basic subschedule. Let $j_1, \ldots, j_m$ be the (part of) job fragments executed by $\mathcal{OPT}$ in this idle period. For each such job fragment j_i, the reason that $\mathcal{ONL}$ is not executing the corresponding job (and instead stays idle) is either:

(1) $\mathcal{ONL}$ started the job earlier, and has either completed or preempted it. If its expiry time is on or after f, that means $\mathcal{ONL}$ started the job before its expiry time, meaning that it must be the first job in some basic subschedule B_0 (recall that the algorithm only preempts jobs if the preempting job is about to expire). In this case we associate j_i with B_0. Otherwise (its expiry time is before f), we follow (2) below.

(2) It has expired. (Note that it can have expired in $\mathcal{ONL}$ but $\mathcal{OPT}$ still runs it, because $\mathcal{OPT}$ may have completed portions of the job earlier on.) Consider all such job fragments. Their expiry times are all before f, and hence the corresponding jobs were all released before f. We can assume without loss of generality that $\mathcal{OPT}$ schedules its jobs in the Earliest Deadline First order. Thus these j_i's must come from different jobs.

Let j be the last job (fragment) executed by $\mathcal{OPT}$ in this idle period. Suppose at j's expiry time, $\mathcal{ONL}$ is running some job in some basic subschedule $B = (v_1, \ldots, v_k)$. (Note that this does not necessarily have to be the same basic subschedule that end in f; but if it is a different one, then there cannot be idle time between these two basic subschedules as j is pending.) By Lemma 1, $v(j) \leq \beta v_k$ if $k \geq 2$ and $v(j) \leq \alpha v_1$ if $k = 1$.

The expiry time of j is before f; hence its deadline is before $f + p(j)$, i.e. this execution of j in $\mathcal{OPT}$ must finish before $f + p(j)$ and hence the total execution time ℓ in this idle period is at most $p(j)$, i.e. $\ell \leq p(j) = v(j)$. This shows that the total length of job fragments done by $\mathcal{OPT}$ while $\mathcal{ONL}$ is idle, excluding those of type (1) that comes after j (they are dealt with differently as discussed above), are at most βv_k (if $k \geq 2$) or αv_1 (if $k = 1$). We associate these (parts of) job fragments of $\mathcal{OPT}$ to this basic subschedule B.

From the above discussion, all job fragments started by $\mathcal{OPT}$ while $\mathcal{ONL}$ is idle is associated with some basic subschedules. We now consider the job fragments associated with each basic subschedule. Consider a basic subschedule $B = (v_1, v_2, \ldots, v_k)$, where $k \geq 2$. (Parts of) job fragments of total length at most $v_1 + \cdots + v_k$ are started and ended within B and are therefore associated with B. In addition, there can be one job of length v_1, and job fragments of total length at most βv_k associated. (Only job fragments from one idle period can be associated with B.) Thus the total length of $\mathcal{OPT}$ jobs associated with B is at most $(v_1 + \cdots + v_k) + \beta v_k + v_1$. Hence for $k \geq 2$, $|\mathcal{OPT}| \leq \left(\frac{1}{\alpha\beta^{k-2}} + \frac{1}{\beta^{k-2}} + \cdots + 1\right) v_k + \beta v_k + \frac{1}{\alpha\beta^{k-2}} v_k = \left(\frac{2}{(2\beta-2)\beta^{k-2}} + \frac{1}{\beta^{k-2}} + \cdots + 1 + \beta\right) v_k = \left(\frac{1}{(\beta-1)\beta^{k-2}} + \frac{1-1/\beta^{k-1}}{1-1/\beta} + \beta\right) v_k = \left(\frac{1}{(\beta-1)\beta^{k-2}} - \frac{\beta}{(\beta-1)\beta^{k-1}} + \frac{\beta}{\beta-1} + \beta\right) v_k = \frac{\beta^2}{\beta-1} v_k$.

$\mathcal{ONL}$ gets a value of v_k in B, but pays penalties of $\rho(v_1 + \cdots + v_{k-1}) \leq \rho(1/\alpha\beta^{k-2} + 1/\beta^{k-2} + \cdots + 1/\beta)v_k \leq \rho(1/\beta^{k-1} + 1/\beta^{k-2} + \cdots + 1/\beta)v_k < \rho v_k/(\beta - 1)$. Thus the competitive ratio r is at most $r \leq \frac{\beta^2/(\beta-1)}{1-\rho/(\beta-1)} = \frac{\beta^2}{\beta-1-\rho}$. This is minimum when $\beta = 2(1 + \rho)$, in which case $r = 4(1 + \rho)$.

If $k = 1$, then similar analysis can show that $\mathcal{OPT}$ gets a value at most $(1 + \alpha + 1)v_1 = 2\beta v_1$ while $\mathcal{ONL}$ gets v_1, so $r = 2\beta = 4(1 + \rho)$ again. □

3 Immediate Decision

3.1 Proportional Value

Before we state the algorithm, we first define a few terms. At any point during $\mathcal{ONL}$'s execution, the job that is running is called the *current* job. A job that has been accepted on arrival (and allocated a time to run) is called *planned.* A planned job is *aborted* if it was started but preempted. A planned job is *discarded* if it was preempted before it can start. By *final schedule* we refer to the schedule produced by $\mathcal{ONL}$ containing the aborted and completed jobs (but not the discarded jobs) and when they are run. The *provisional schedule at time* t, denoted by S_t, refers to the planned schedule of $\mathcal{ONL}$ which consists of the current job at time t following by the planned jobs. For $u \geq t$, $S_t(u)$ denotes the job planned to run (or running) at time instant u of S_t. If nothing is scheduled at u (it is idle) then $S_t(u) = \emptyset$.

Algorithm 2. Let $\beta > 1$ be a parameter that depends on ρ. Its value will be specified later. When a new job j arrives, identify a set of current and/or planned jobs in the provisional schedule (this could be an empty set) such that their total value is less than $v(j)/\beta$ and their preemption would create a single time slot long enough to schedule j. (If there are more than one such job sets, choose any one.) If no such set of jobs exist then reject j. Otherwise, preempt those jobs, and allocate j to start as early as possible in the newly-created empty timeslot (leaving no idle time between the preceding job).

We can show that if a preemption happens, then it must preempt all jobs starting from some job all the way to the end of the provisional schedule, i.e. *the preempted part is always a suffix* of the provisional schedule. This allows for efficient implementation as there is only a linear number of candidate set of jobs to be considered.

We first show several key properties of the resulting schedule.

Lemma 2. *At any time t, if $S_t(u) = \emptyset$ then $S_t(u') = \emptyset$ for all $u' > u$. That is, there is no idle time in-between in the provisional schedule.*

Lemma 3. *Consider a time u. If at some time $t < u$, $S_t(u) \neq \emptyset$, then $S_{t'}(u) \neq \emptyset$ for all $t < t' \leq u$. That is, if some job is allocated to run at time u in the provisional schedule, then u will never become idle again.*

We now analyze the competitive ratio by mapping profits obtained in $\mathcal{OPT}$ to those obtained in $\mathcal{ONL}$. We partition the set of jobs completed by $\mathcal{OPT}$ into four sets: C (Completed), those completed by both $\mathcal{ONL}$ and $\mathcal{OPT}$; R (Rejected), those rejected immediately by $\mathcal{ONL}$ upon arrival; D (Discarded), those accepted by $\mathcal{ONL}$ but are subsequently preempted before they can start; A (Aborted), those accepted and started by $\mathcal{ONL}$ but preempted during execution. Let $v(X)$ denote the total value of jobs in a set X, and let $|\mathcal{OPT}|$ and $|\mathcal{ONL}|$ denote the values of the schedules $\mathcal{OPT}$ and $\mathcal{ONL}$.

We consider the final schedule of $\mathcal{ONL}$, and partition it into *busy periods* with idle time in-between.

Lemma 4. *No job in R is started by $\mathcal{OPT}$ during any idle time in the final schedule of $\mathcal{ONL}$.*

Consider a busy period which consists of k basic subschedules, i.e. with k completed jobs $j_1, \ldots, j_k$. Each completed job may be preceded by some aborted jobs, which are (transitively) aborted by the associated completed job. The following lemma argues that although a job in $\mathcal{OPT}$ can overlap several busy periods, it is sufficient to associate it with just one busy period (the one during which it is started by $\mathcal{OPT}$).

Lemma 5. *If $\mathcal{OPT}$ starts a job j in R within a busy period with completed jobs $j_1, \ldots, j_k$, then $v(j) < \beta \sum_{i=1}^{k} v(j_i)$.*

Proof. Let f be the time when the busy period ends. Since the final schedule is idle at f, by Lemma 3, the provisional schedule when j arrives is also idle at f, i.e. $S_{r(j)}(f) = \emptyset$. Hence by Lemma 2, $S_{r(j)}(t) = \emptyset$ for all times $t \in [f, \infty)$. Thus j is rejected due to jobs in this busy period only. (It cannot be due to jobs in earlier busy periods because j has not been released back then, or else $\mathcal{ONL}$ would not idle between the busy periods.) Let $q_1, \ldots, q_m$ be the jobs that are in the provisional schedule at time $r(j)$ 'blocking' j's position in $\mathcal{OPT}$. Then $v(j) < \beta \sum_{i=1}^{m} v(q_i)$. Each of these q_i's may be preempted later by other jobs, but the preempting jobs must appear in the same busy period (since there is no gap in the provisional schedule). Thus this must eventually leads to one or more of the j_i's. Each of these preempting jobs must has larger value than the total value of its preempted jobs, thus if $v(j) < \beta \sum_{i=1}^{m} v(q_i)$ then $v(j) < \beta \sum_{i=1}^{k} v(j_i)$ as well. □

Theorem 2. Algorithm 2 *has a competitive ratio of* $2\rho + 3 + 2\sqrt{\rho^2 + 3\rho + 2}$ *with* $\beta = 1 + \rho + \sqrt{\rho^2 + 3\rho + 2}$.

Proof. From the above two lemmas, it is sufficient to analyse each busy period separately: each job in R is associated with the busy period during which it is started, and each job in C, A or D is naturally associated with a busy period. From now on, we consider one busy period with k completed jobs $j_1, \ldots, j_k$, and C, R, D, A refer to jobs associated with this busy period only. Let $V = \sum_{i=1}^{k} v(j_i)$. Clearly $v(C) \leq V$. A completed job j_i preempts jobs of total value at most $v(j_i)/\beta$ by the definition of the algorithm. Each of these preempted jobs may in turn have preempted other jobs of total value at most $1/\beta$ of their own value, and so on. Thus the sum of values of aborted and discarded jobs associated with j_i is at most $v(j_i)/\beta + v(j_i)/\beta^2 + \cdots < v(j_i)/(\beta - 1)$. Therefore, $v(A) + v(D) < \sum_{i=1}^{k} \frac{1}{\beta-1} v(j_i) = \frac{1}{\beta-1} V$.

Let L be the length of this busy period. In the worst case, all aborted jobs appear in this busy period in their (almost) entirety. The previous formula can be used to give an upper bound on L due to aborted jobs. Together with completed jobs, this gives $L < \sum_{i=1}^{k} \left(\frac{v(j_i)}{\beta-1} + v(j_i) \right) = \sum_{i=1}^{k} \frac{\beta}{\beta-1} v(j_i) = \frac{\beta}{\beta-1} V$.

The total length, or value, of jobs in R associated with this busy period is at most $L + \beta V$. This is because they must be non-overlapping, and must start within this busy period; hence for those that finish before this busy period ends, their total length is at most L, while Lemma 5 guarantees the length of the (at most one) job that finishes after this busy period ends is at most βV. Therefore $|\mathcal{OPT}| = v(C) + v(A) + v(D) + v(R) \leq V + \frac{1}{\beta-1} V + (L + \beta V) \leq \left(\frac{2\beta}{\beta-1} + \beta \right) V$.

$\mathcal{ONL}$ gets a value of V for completing the jobs, but pays a penalty for all aborted and discarded jobs, thus $|\mathcal{ONL}| \geq V - \rho(\frac{1}{\beta-1} V)$. Therefore the competitive ratio r is at most $r \leq \frac{\frac{2\beta}{\beta-1} + \beta}{1 - \frac{\rho}{\beta-1}} = \frac{2\beta + \beta(\beta-1)}{\beta - 1 - \rho} = \frac{\beta^2 + \beta}{\beta - 1 - \rho}$. It can be verified by differentiation that the minimum is achieved when we set $\beta = 1 + \rho + \sqrt{\rho^2 + 3\rho + 2}$, and then $r = 2\rho + 3 + 2\sqrt{\rho^2 + 3\rho + 2}$. □

3.2 Unit Length

Algorithm 2 can also be applied to the unit length case. One difference is that, since all jobs are of the same length, preempting a single planned job (but not the current job, which only has a portion of its length left) will create enough space to fit in the new job. Thus a preemption will only preempt one job, unless the deadline of the new job is before the finishing time of the current job plus 1, in which case the algorithm may preempt the current job plus the next one. This allows for efficient implementation as the checking for preemption only need to be made against each individual job in the provisional schedule before the new job's deadline, or the current job plus the job following it.

Theorem 3. ALGORITHM 2 *has a competitive ratio of* $4\rho+5+4\sqrt{\rho^2+5\rho/2+3/2}$ *for the unit length case.*

Proof. We can follow a similar method to the proof of Theorem 2. We define C, A, D and R as before, but for the whole schedule. Similar to Theorem 2 we have $v(C) \leq V$ and $v(D) + v(A) \leq \frac{1}{\beta-1}V$, where $V = \sum v_i$ is the sum of value of all jobs completed by $\mathcal{ONL}$. However we bound R differently.

Consider the time t when a job j in R is scheduled by $\mathcal{OPT}$. Since it is rejected, the time interval $[t, t+1)$ is occupied by some jobs in the provisional schedule such that their total value is at least $1/\beta$ that of j. In fact, since jobs are of the same length, there are at most two such jobs. We associate j with these jobs; each such job carries a charge of at most β times their own value representing the fact that they 'blocked' j. The two blocking jobs will then either be completed or be aborted/discarded; thus they belong to C, A or D. Again because of the unit-length property, each job in $C/A/D$ blocks at most two jobs in R and hence receives at most two such charges. Thus we have $v(R) < 2\beta(v(C)+v(A)+v(D))$. So $|\mathcal{OPT}| = v(C) + v(A) + v(D) + v(R) \leq (2\beta+1)(v(C) + v(A) + v(D)) \leq (2\beta+1)\left(\frac{\beta}{\beta-1}\right)V$. Again $|\mathcal{ONL}| \geq V - \rho(\frac{1}{\beta-1}V)$. Thus the competitive ratio is at most $r \leq \frac{(2\beta+1)\beta/(\beta-1)}{1-\rho/(\beta-1)} = \frac{(2\beta+1)\beta}{\beta-1-\rho}$. This is minimized by choosing $\beta = 1+\rho+\sqrt{\rho^2+5\rho/2+3/2}$, in which case $r = 4\rho+5+4\sqrt{\rho^2+5\rho/2+3/2}$. □

3.3 Lower Bounds

The lower bound of $4(1+\rho)$ in [5] (either proportional value or unit length) for the preemption penalty model can be applied in the immediate decision or notification model. This is because the proof used only tight jobs. In such cases acceptance or rejection must be decided on arrival, there is no choice of when to schedule the job, and a preempted job cannot be started again later and will therefore not complete. Thus a preemption penalty is the same as a non-completion penalty.

It may be reasonable to assume in the immediate notification or decision models that all jobs come with certain laxity > 1; for example, in a home delivery situation, it is common that there are several days of lead time between order and delivery. Often, larger laxity allows for smaller competitive ratios (see e.g. [6]). However we can show:

Theorem 4. *For immediate decision with proportional value, no algorithm can be better than 2-competitive even if all jobs have arbitrarily large laxity.*

4 Remarks on Immediate Notification

Our discussion above is on immediate decision. For immediate notification, if $\rho = 0$, one may just use the algorithms for the preemption penalty model, accepting every job, since there is no penalty on later rejecting them. However, with any other positive ρ, however small, it is easy to see that such an algorithm is not competitive, or may not even give a positive value at all. Any algorithm with immediate decision (including Algorithm 2) can be applied to the immediate notification model, because it simply does not use the extra ability to restart or resume jobs. In fact, for the restart model, it gives the same competitive ratio, because $\mathcal{OPT}$ would not use preemption, and can always announce in advance when it will schedule a job. Hence both $\mathcal{OPT}$ and $\mathcal{ONL}$ give the same schedule with the same value as in the immediate decision model.

References

1. Baruah, S., Koren, G., Mao, D., Mishra, B., Raghunathan, A., Rosier, L., Shasha, D., Wang, F.: On the competitiveness of on-line real-time task scheduling. Real-Time Systems 4, 125–144 (1992)
2. Ding, J., Ebenlendr, T., Sgall, J., Zhang, G.: Online scheduling of equal-length jobs on parallel machines. In: Proc. of 15th ESA, pp. 427–438 (2007)
3. Ding, J., Zhang, G.: Online scheduling with hard deadlines on parallel machines. In: Proc. of 2nd AAIM, pp. 32–42 (2006)
4. Ebenlendr, T., Sgall, J.: A lower bound for scheduling of unit jobs with immediate decision on parallel machines. In: Proc. of 6th WAOA, pp. 43–52 (2008)
5. Fung, S.P.Y.: Lower bounds on online deadline scheduling with preemption penalties. Information Processing Letters 108(4), 214–218 (2008)
6. Goldwasser, M.H., Kerbikov, B.: Admission control with immediate notification. J. of Scheduling 6, 269–285 (2003)
7. Hoogeveen, H., Potts, C.N., Woeginger, G.J.: On-line scheduling on a single machine: Maximizing the number of early jobs. Operations Research Letters 27(5), 193–197 (2000)
8. Koren, G., Shasha, D.: D^{over}: An optimal on-line scheduling algorithm for overloaded uniprocessor real-time systems. SIAM J. on Computing 24, 318–339 (1995)
9. Thibault, N., Laforest, C.: Online time constrained scheduling with penalties. In: Proc. of 23rd IEEE Int. Symposium on Parallel and Distributed Processing (2009)
10. Woeginger, G.J.: On-line scheduling of jobs with fixed start and end times. Theoretical Computer Science 130(1), 5–16 (1994)
11. Zheng, F., Dai, W., Xiao, P., Zhao, Y.: Competitive strategies for on-line production order disposal problem. In: Megiddo, N., Xu, Y., Zhu, B. (eds.) AAIM 2005. LNCS, vol. 3521, pp. 46–54. Springer, Heidelberg (2005)
12. Zheng, F., Fung, S.P.Y., Chan, W.-T., Chin, F.Y.L., Poon, C.K., Wong, P.W.H.: Improved on-line broadcast scheduling with deadlines. In: Chen, D.Z., Lee, D.T. (eds.) COCOON 2006. LNCS, vol. 4112, pp. 320–329. Springer, Heidelberg (2006)
13. Zheng, F., Xu, Y., Poon, C.K.: On job scheduling with preemption penalties. In: Goldberg, A.V., Zhou, Y. (eds.) AAIM 2009. LNCS, vol. 5564, pp. 315–325. Springer, Heidelberg (2009)

Discovering Pairwise Compatibility Graphs

Muhammad Nur Yanhaona, Md. Shamsuzzoha Bayzid, and Md. Saidur Rahman

Department of Computer Science and Engineering
Bangladesh University of Engineering and Technology
nur.yanhaona@gmail.com, shams.bayzid@gmail.com,
saidurrahman@cse.buet.ac.bd

Abstract. Let T be an edge weighted tree, let $d_T(u, v)$ be the sum of the weights of the edges on the path from u to v in T, and let d_{min} and d_{max} be two non-negative real numbers such that $d_{min} \leq d_{max}$. Then a pairwise compatibility graph of T for d_{min} and d_{max} is a graph $G = (V, E)$, where each vertex $u' \in V$ corresponds to a leaf u of T and there is an edge $(u', v') \in E$ if and only if $d_{min} \leq d_T(u, v) \leq d_{max}$. A graph G is called a pairwise compatibility graph (PCG) if there exists an edge weighted tree T and two non-negative real numbers d_{min} and d_{max} such that G is a pairwise compatibility graph of T for d_{min} and d_{max}. Kearney *et al.* conjectured that every graph is a PCG [3]. In this paper, we refute the conjecture by showing that not all graphs are PCGs. We also show that the well known tree power graphs and some of their extensions are PCGs.

1 Introduction

Let T be an edge weighted tree and let d_{min} and d_{max} be two non-negative real numbers such that $d_{min} \leq d_{max}$. A *pairwise compatibility graph of* T *for* d_{min} *and* d_{max} is a graph $G = (V, E)$, where each vertex $u' \in V$ represents a leaf u of T and there is an edge $(u', v') \in E$ if and only if the distance between u and v in T lies within the range from d_{min} to d_{max}. T is called the *pairwise compatibility tree* of G. We denote a pairwise compatibility graph of T for d_{min} and d_{max} by $PCG(T, d_{min}, d_{max})$. A graph G is a *pairwise compatibility graph* (PCG) if there exists an edge weighted tree T and two non-negative real numbers d_{min} and d_{max} such that $G = PCG(T, d_{min}, d_{max})$. Figure 1(a) depicts an edge weighted tree T and Fig. 1(b) depicts a pairwise compatibility graph G of T for $d_{min} = 4$ and $d_{max} = 7$; there is an edge between a' and b' in G since in T the distance between a and b is six, but G does not contain the edge (a', c') since in T the distance between a and c is eight, which is larger than seven. It is quite apparent that a single edge weighted tree may have many pairwise compatibility graphs for different values of d_{min} and d_{max}. Likewise, a single pairwise compatibility graph may have many trees of different topologies as its pairwise compatibility trees. For example, the graph in Fig. 1(b) is a PCG of the tree in Fig. 1(a) for $d_{min} = 4$ and $d_{max} = 7$, and it is also a PCG of the tree in Fig. 1(c) for $d_{min} = 5$ and $d_{max} = 6$.

In the realm of pairwise compatibility graphs, two fundamental problems are the tree construction problem and the pairwise compatibility graph recognition problem. Given a PCG G, the tree construction problem asks to construct an edge weighted tree T, such that G is a pairwise compatibility graph of T for suitable d_{min} and d_{max}. The

M.T. Thai and S. Sahni (Eds.): COCOON 2010, LNCS 6196, pp. 399–408, 2010.

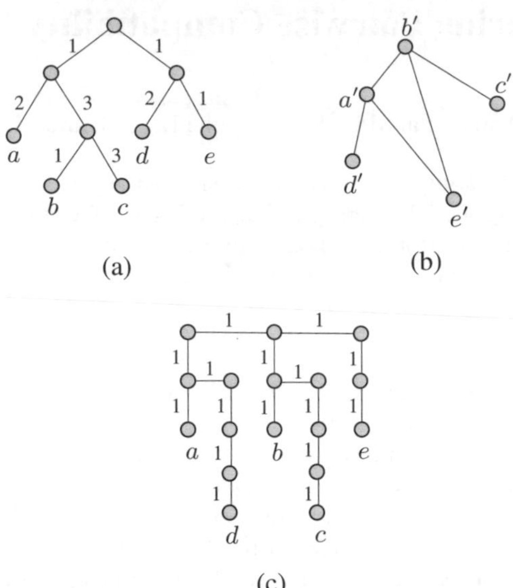

Fig. 1. (a) An edge weighted tree T_1, (b) a pairwise compatibility graph G, and (c) an edge weighted tree T_2

pairwise compatibility graph recognition problem seeks the answer whether or not a given graph is a PCG.

Pairwise compatibility graphs have their origin in *Phylogenetics*, which is a branch of computational biology that concerns with reconstructing evolutionary relationships among organisms [2,4]. Phylogenetic relationships are commonly represented as trees known as the phylogenetic trees. From a problem of collecting leaf samples from large phylogenetic trees, Kearney *et al.* introduced the concept of pairwise compatibility graphs [3]. As their origin suggests, these graphs can be used in reconstruction of evolutionary relationships. However, their most intriguing potential lies in solving the "Clique Problem." A *clique* in a graph G is a set of pairwise adjacent vertices. The *clique problem* asks to determine whether a graph contains a clique of at least a given size k. It is a well known NP-complete problem. The corresponding optimization problem, the *maximum clique problem*, asks to find the largest clique in a graph [1]. Kearney *et al.* have shown that for a pairwise compatibility graph G, the clique problem is equivalent to a "leaf sampling problem" – which is solvable in polynomial time in any pairwise compatibility tree T of G [3].

Since their inception, pairwise compatibility graphs have raised several interesting problems, and hitherto most of these problems have remained unsolved. Among the others, identifying different graph classes as pairwise compatibility graphs is an important concern. Although overlapping of pairwise compatibility graphs with many well known graph classes like chordal graphs and complete graphs is quite apparent; slight progresses have been made on establishing concrete relationships between pairwise compatibility graphs and other known graph classes. Phillips has shown that every graph of five vertices or less is a PCG [6] and Yanhaona *et al.* have shown that all cycles, cycles with a single chord, and cactus graphs are PCGs [7]. Seeing the exponentially increasing number of possible tree topologies for large graphs, the proponents of PCGs conceived that all undirected graphs are PCGs [3]. In this paper, we refute the conjecture by showing that not all graphs are PCGs.

Pairwise compatibility graphs have striking similarity, in their underlying concept, with the well studied graph roots and powers. A graph $G' = (V', E')$ is a *k-root* of a graph $G = (V, E)$ if $V' = V$ and there is an edge $(u, v) \in E$ if and only if the length of the shortest path from u to v in G' is at most k. G is called the k-*power* of G' [5]. A special case of graph power is the tree power, which requires G' to be a tree. Tree power graphs and their extensions (*Steiner* k-*power* graphs, *phylogenetic* k-*power* graphs, etc.) are by definition similar to pairwise compatibility graphs. However, the exact relationship of these graph classes with pairwise compatibility graphs was unknown. In this paper, we investigate the possibility of the existence of such a relationship, and show that tree power graphs and some of their extensions are in fact pairwise compatibility graphs. Such a relationship may serve the purpose of not only unifying related graph classes but also utilizing the method of tree constructions for one graph class in another.

The rest of the paper is organized as follows. Section 2 describes some of the definitions we have used in our paper, Section 3 shows that not all graphs are pairwise compatibility graphs. Section 4 establishes a relationship of tree power graphs and their extensions with pairwise compatibility graphs. Finally, Section 5 concludes our paper with discussions.

2 Preliminaries

In this section we define some terms that we have used in this paper.

Let $G = (V, E)$ be a simple graph with vertex set V and edge set E. The sets of vertices and edges of G are denoted by $V(G)$ and $E(G)$, respectively. An edge between two vertices u and v of G is denoted by (u, v). Two vertices u and v are *adjacent* and called *neighbors* if $(u, v) \in E$; the edge (u, v) is then said to be *incident* to vertices u and v. The *degree* of a vertex v in G is the number of edges incident to it. A *subgraph* of a graph $G = (V, E)$ is a graph $G' = (V', E')$ such that $V' \subseteq V$ and $E' \subseteq E$; we then write $G' \subseteq G$. If G' contains all the edges of G that join two vertices in V' then G' is said to be the *subgraph induced by* V'. A *path* $P_{uv} = w_0, w_1, \cdots, w_n$ is a sequence of distinct vertices in V such that $u = w_0, v = w_n$ and $(w_{i-1}, w_i) \in E$ for every $1 \leq i \leq n$. A *subpath* of P_{uv} is a subsequence $P_{w_j w_k} = w_j, w_{j+1}, ..., w_k$ for some $0 \leq j < k \leq n$. A vertex x on P_{uv} is called an *internal node* of P_{uv} if $x \neq u, v$. G is *connected* if each pair of vertices of G belongs to a path, otherwise G is *disconnected*. A set S of vertices in G is called an *independent set* of G if the vertices in S are pairwise non-adjacent. A graph $G = (V, E)$ is a *bipartite graph* if V can be expressed as the union of two independent sets; each independent set is called a *partite set*. A *complete bipartite graph* is a bipartite graph where two vertices are adjacent if and only if they are in different partite sets. A *cycle* of G is a sequence of distinct vertices starting and ending at the same vertex such that two vertices are adjacent if they appear consecutively in the list.

A *tree* T is a connected graph with no cycle. Vertices of degree one in T are called *leaves*, and the rests are called *internal nodes*. A tree T is *weighted* if each edge is assigned a number as the weight of the edge. A *subtree induced by a set of leaves* of

T is the minimal subtree of T which contains those leaves. Figure 2 illustrates a tree T with six leaves u, v, w, x, y and z, where the edges of the subtree of T induced by u, v and w is drawn by thick lines. We denote by T_{uvw} the subtree of a tree induced by three leaves u, v and w. One can observe that T_{uvw} has exactly one vertex of degree 3. We call the vertex of degree 3 in T_{uvw} the *core* of T_{uvw}. The vertex o is the core of T_{uvw} in Fig. 2. The *distance between two vertices u and v* in T, denoted by $d_T(u, v)$, is the sum of the weights of the edges on P_{uv}. In this paper we have considered only weighted trees. We use the convention that if an edge of a tree has no number assigned to it then its default weight is one. A *star* is a tree where every leaf has a common neighbor which we call the *base* of the star.

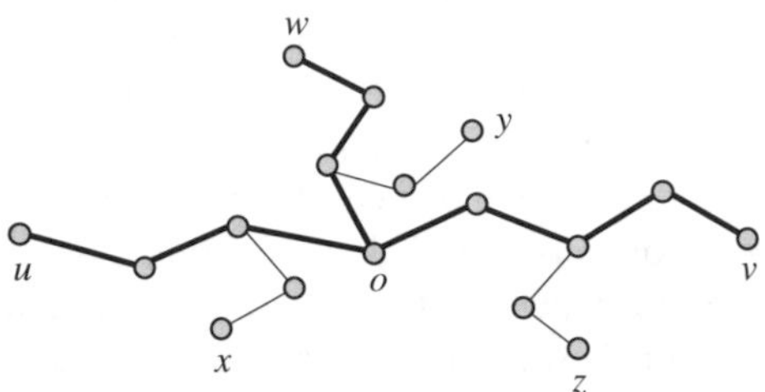

Fig. 2. Illustration for a leaf induced subtree

A graph $G = (V, E)$ is called a *phylogenetic k-power graph* if there exists a tree T such that each leaf of T corresponds to a vertex of G and an edge $(u, v) \in E$ if and only if $d_T(u, v) \leq k$, where k is a given proximity threshold. *Steiner k-power graphs* extend the notion of phylogenetic k-power. For a Steiner k-power graph the corresponding tree may have some internal nodes as well as the leaves that correspond to the vertices of the graph. Both Steiner k-power graphs and phylogenetic k-power graphs belong to the widely known family of graph powers. Another special case of graph powers is the tree power graph. A graph $G = (V, E)$ is said to have a tree power for a certain proximity threshold k if a tree T can be constructed on V such that $(u, v) \in E$ if and only if $d_T(u, v) \leq k$.

3 Not All Graphs Are PCGs

In this section we show that not all graphs are pairwise compatibility graphs, as in the following theorem.

Theorem 1. *Not all graphs are pairwise compatibility graphs.*

To prove the claim of Theorem 1 we need the following lemmas.

Lemma 1. *Let T be an edge weighted tree, and u, v and w be three leaves of T such that P_{uv} is the largest path in T_{uvw}. Let x be a leaf of T other than u, v and w. Then, $d_T(w, x) \leq d_T(u, x)$ or $d_T(w, x) \leq d_T(v, x)$.*

Proof. Let o be the core of T_{uvw}. Then each of the paths P_{uv}, P_{uw} and P_{wv} is composed of two of the three subpaths P_{uo}, P_{ow} and P_{ov}. Since $d_T(u, v)$ is the largest path in T_{uvw}, $d_T(u,v) \geq d_T(u,w)$. This implies that $d_T(u,o) + d_T(o,v) \geq d_T(u,o) + d_T(o,w)$. Hence $d_T(o,v) \geq d_T(o,w)$. Similarly, $d_T(u,o) \geq d_T(o,w)$ since $d_T(u,v) \geq d_T(w,v)$. Since T is a tree, there is a path from x to o. Let o_x be the first vertex in $V(T_{uvw}) \cap V(P_{xo})$ along the path P_{xo} from x. Then clearly o_x is on P_{uo}, P_{vo} or P_{wo}. We first assume that o_x is on P_{uo}, as illustrated in Fig. 3(a). Then $d_T(v,x) \geq d_T(w,x)$ since $d_T(w,x) = d_T(x,o) + d_T(w,o)$, $d_T(v,x) = d_T(x,o) + d_T(v,o)$ and $d_T(v,o) \geq d_T(w,o)$. We now assume that o_x is on P_{vo}, as illustrated in Fig. 3(c). Then $d_T(u,x) \geq d_T(w,x)$ since $d_T(w,x) = d_T(x,o) + d_T(w,o)$, $d_T(u,x) = d_T(x,o) + d_T(o,u)$ and $d_T(u,o) \geq d_T(w,o)$. We finally assume that o_x is on P_{wo}, as illustrated in Fig. 3(b). Then $d_T(u,x) = d_T(u,o) + d_T(o,o_x) + d_T(o_x,x)$ and $d_T(w,x) = d_T(w,o_x) + d_T(o_x,x)$. As $d_T(w,o_x) \leq d_T(w,o)$ and $d_T(u,o) \geq d_T(w,o)$, $d_T(u,x) \geq d_T(w,x)$. Likewise, $d_T(v,x) \geq d_T(w,x)$. Thus, in each case, at least one of u and v is at a distance from x that is either larger than or equals to the distance between w and x. □

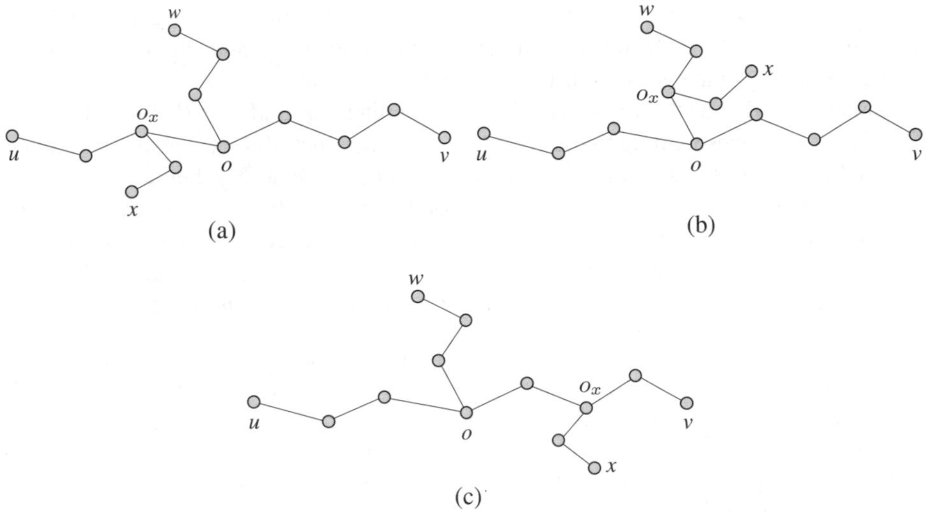

(a)

(b)

(c)

Fig. 3. Different positions of x

Lemma 2. *Let $G = (V,E)$ be a $PCG(T, d_{min}, d_{max})$. Let a,b,c,d and e be five leaves of T and let a',b',c',d' and e' be five vertices of G corresponding to the five leaves a,b,c,d and e of T, respectively. Let P_{ae} be the largest path in the subtree of T induced by the leaves a,b,c,d and e, and P_{bd} be the largest path in T_{bcd}. Then G has no vertex x' such that x' is adjacent to a', c' and e' but not adjacent to b' and d'.*

Proof. Assume for a contradiction that G has a vertex x' such that x' is a neighbor of a',c' and e' but not of b' and d'. Let x be the leaves of T corresponding to the vertex x' of G. Since P_{ae} is the largest path in T among all the paths that connect a pair of leaves from the set $\{a,b,c,d,e\}$, $\max_{y\in\{a,e\}} d_T(x,y) \geq \max_{z\in\{b,c,d\}} d_T(x,z)$ by Lemma 1. Since both a and e are adjacent to x in G, $\max_{y\in\{a,e\}} d_T(x,y) \leq d_{max}$. This implies that $d_T(x,y) \leq d_{max}$, $y \in \{a,b,c,d,e\}$. Since P_{bd} is the largest path in T_{bcd}, $\max_{y\in\{b,d\}} d_T(x,y) \geq d_T(x,c)$ by Lemma 1. Without loss of generality assume that $d_T(x,b) \geq d_T(x,c)$. Since b' and x' are not adjacent in G and $d_T(x,b) \leq d_{max}$, $d_T(x,b) < d_{min}$. Then $d_T(x,c) < d_{min}$ since $d_T(x,b) \geq d_T(x,c)$. Since $d_T(x,c) < d_{min}$, c' cannot be adjacent to x' in G, a contradiction. □

Using Lemma 2 we now present a graph which is not a PCG as in the following Lemma.

Lemma 3. *Let $G = (V,E)$ be a graph of 15 vertices, and let $\{V_1, V_2\}$ be a partition of the set V such that $|V_1| = 5$ and $|V_2| = 10$. Assume that each vertex in V_2 has exactly three neighbors in V_1 and no two vertices in V_2 has the same three neighbors in V_1. Then G is not a pairwise compatibility graph.*

Proof. Assume for a contradiction that G is a pairwise compatibility graph, i.e., $G = PCG(T, d_{min}, d_{max})$ for some T, d_{min} and d_{max}. Let P_{uv} be the longest path in the subtree of T induced by the leaves of T representing the vertices in V_1. Clearly u and v are leaves of T. Let u' and v' be the vertices in V_1 corresponding to the leaves u and v of T, respectively. Let P_{wx} be the longest path in the subtree of T induced by the leaves of T corresponding to the vertices in $V_1 - \{u', v'\}$. Clearly w and x are also the leaves of T, and let w' and x' be the vertices in V_1 corresponding to w and x of T. Since $|V_1| = 5$, T has a leaf y corresponding to the vertex $y' \in V_1$ such that $y' \notin \{u', v', w', x'\}$. Since G is a PCG of T, G cannot have a vertex adjacent to u', v' and y' but not adjacent to w' and x' by Lemma 2. However, for every combination of three vertices in V_1, V_2 has a vertex which is adjacent to only those three vertices of the combination. Thus there is indeed a vertex in V_2 which is adjacent to u', v' and y' but not to w' and x'. Hence G can not be a pairwise compatibility graph of T by Lemma 2, a contradiction. □

Lemma 3 immediately proves Theorem 1. Figure 4 shows an example of a bipartite graph which is not a PCG. Quite interestingly, however, every complete bipartite graph is a PCG. It can be shown as follows. Let $K_{m,n}$ be a complete bipartite graph with two partite sets $X = \{x_1, x_2, x_3, \cdots, x_m\}$, and $Y = \{y_1, y_2, y_3, \cdots, y_n\}$. We construct a star for each partite set such that each leaf corresponds to a vertex of the respective partite set. Then we connect the bases of the stars through an edge as illustrated in Fig. 5. Finally, we assign one as the weight of each edge. Let T be the resulting tree. Now one can easily verify that $K_{m,n} = PCG(T, 3, 3)$.

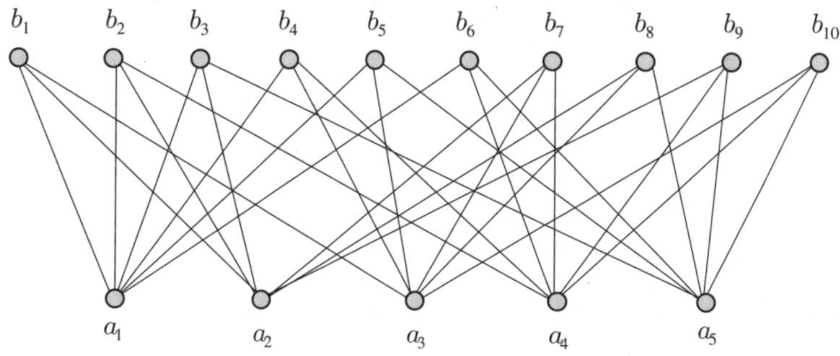

Fig. 4. Example of a graph which is not a PCG

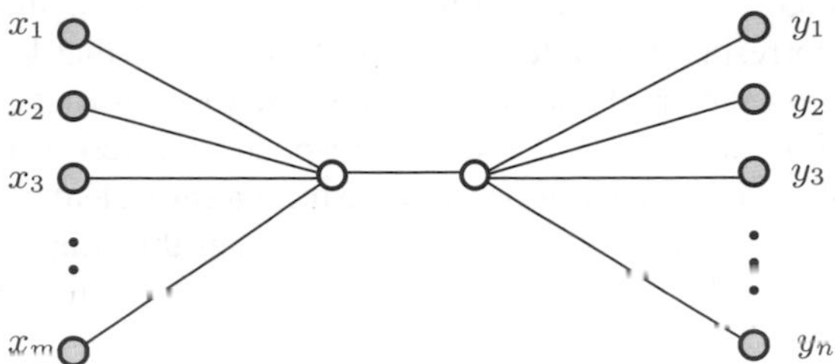

Fig. 5. Pairwise compatibility tree T of a complete bipartite graph $K_{m,n}$

Taking the graph described in Lemma 3 as a subgraph of a larger graph, we can show a larger class of graphs which is not PCG, as described in the following lemma.

Lemma 4. *Let $G = (V, E)$ be a graph, and let V_1 and V_2 be two disjoint subsets of vertices such that $|V_1| = 5$ and $|V_2| = 10$. Assume that each vertex in V_2 has exactly three neighbors in V_1 and no two vertices in V_2 has the same three neighbors in V_1. Then G is not a pairwise compatibility graph.*

Proof. Assume for a contradiction that G is PCG, i.e., $G = PCG(T, d_{min}, d_{max})$ for some T, d_{min} and d_{max}. Let H be the subgraph of G induced by $V_1 \cup V_2$. Now, let T_H be the subtree of T induced by the leaves representing the vertices in $V_1 \cup V_2$. According to the definition of leaf induced subtree, for any pair of leaves u, v in T_H, $d_{T_H}(u, v) = d_T(u, v)$. Then $H = PCG(T_H, d_{min}, d_{max})$ since $G = PCG(T, d_{min}, d_{max})$. However, H is not a PCG by Lemma 3, a contradiction. □

4 Variants of Tree Power Graphs and PCGs

In this section we will show that tree power graphs and two of their extensions are PCGs.

Tree power graphs and their extensions (Steiner k-power and phylogenetic k-power graphs) have striking resemblance, in their underlying concept, with PCGs. But does this similarity signify any real relationship? It does indeed: we find that tree power graphs and these two extensions are essentially PCGs. To establish this relationship of aforementioned three graph classes with pairwise compatibility graphs, we introduce a generalized graph class which we call "tree compatible graphs." A graph $G = (V, E)$ is a *tree compatible graph* if there exists a tree T such that all leaves and a subset of internal nodes of T correspond to the vertex set V of G, and for any two vertices $u, v \in V$; $(u, v) \in E$ if and only if $k_{min} \leq d_T(u, v) \leq k_{max}$. Here k_{min} and k_{max} are real numbers. We call G the *tree compatible graph* of T for k_{min} and k_{max}. It is quite evident from this definition that tree compatible graph comprises tree power graphs, Steiner k-power graphs, and phylogenetic k-power graphs. We now have the following theorem.

Theorem 2. *Every tree compatible graph is a pairwise compatibility graph.*

Proof. Let G be a tree compatible graph of a tree T for non-negative real numbers k_{min} and k_{max}. Then to prove the claim, it is sufficient to construct a tree T' and find two non-negative real numbers d_{min} and d_{max} such that $G = PCG(T', d_{min}, d_{max})$.

Clearly $G = PCG(T', d_{min}, d_{max})$ for $T' = T$, $d_{min} = k_{min}$ and $d_{max} = k_{max}$ if every vertex in V corresponds to a leaf in T. We thus assume that V contains a vertex which corresponds to an internal node of T. In this case we construct a tree T' from T as follows. For every internal node u of T that corresponds to a vertex in V, we introduce a surrogate internal node u'. In addition, we transform u into a leaf node by connecting u through an edge of weight λ with u'. Figure 6 illustrates this transformation. Here, in addition to the leaves of T, two internal nodes d and e correspond to the vertices in V. T' is the modified tree after transforming d and e into leaf nodes by replacing them by d' and e', respectively.

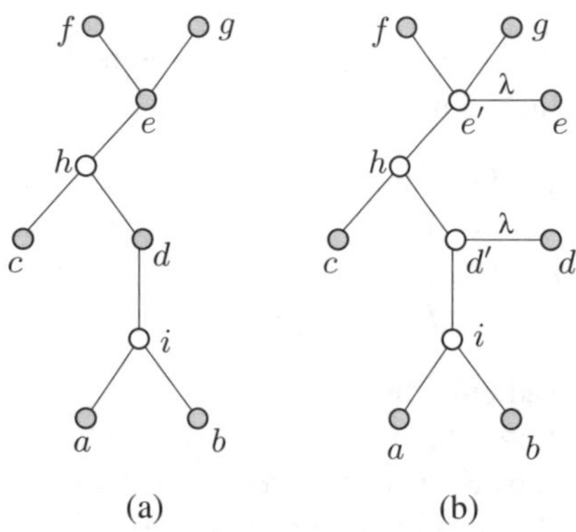

Fig. 6. (a) T and (b) T'

The aforementioned transformation transmutes the subset of internal nodes of T that participates in V into a subset of leaves in T'. Let u and v be two arbitrary nodes in T. If u and v are both leaves in T then $d_{T'}(u, v) = d_T(u, v)$. If both u and v are internal nodes of T that are contributing to V then in T' they are two leaf nodes, and $d_{T'}(u, v) = d_T(u, v) + 2\lambda$. Finally, if only one of u and v is transformed to leaf then $d_{T'}(u, v) = d_T(u, v) + \lambda$. We next define $d_{min} = k_{min}$ and $d_{max} = k_{max} + 2\lambda$. Since every vertex $u \in V$ is represented as a leaf in T', T' may be a pairwise compatibility tree of G. We will prove that T' is indeed a pairwise compatibility tree by showing that $G = PCG(T', d_{min}, d_{max})$ for an appropriate value of λ. Note that we cannot simply assign $\lambda = 0$ because, in the context of root finding as well as phylogenetics, an edge of zero weight is not meaningful. For example, if an evolutionary tree contains zero weighted edges then we may find a path of length zero between two different organisms, which is clearly unacceptable. Therefore, we have to choose a value for λ more intelligently.

According to the definition of tree compatible graphs, for every pair of vertices $u, v \in V$, $(u, v) \in E$ if and only if $k_{min} \leq d_T(u, v) \leq k_{max}$. Meanwhile, we have derived T' from T in such a way that either the distance between u and v in T' remains the same as in T, or increased by at most 2λ. Therefore, if we can prove that $d_{min} \leq d_{T'}(u, v) \leq d_{max}$ if and only if $k_{min} \leq d_T(u, v) \leq k_{max}$ then it will imply that $G = PCG(T', d_{min}, d_{max})$. Depending on the nature of the change in the distance between u and v from T to T', we have to consider three different cases.

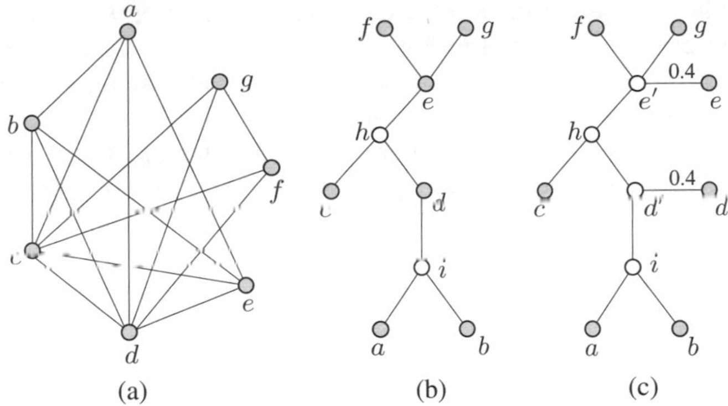

Fig. 7. (a) A tree compatible graph, (b) the corresponding tree T, and (c) the corresponding pairwise compatibility tree T'

Case 1: $d_{T'}(u, v) = d_T(u, v)$.
In this case, three possible relationships can exist among $d_T(u, v)$, k_{min} and k_{max}. First, if $d_T(u, v) < k_{min}$ then $d_{T'}(u, v) < d_{min}$ since $d_{min} = k_{min}$. Next, if $k_{min} \leq d_T(u, v) \leq k_{max}$ then $k_{min} \leq d_T(u, v) \leq k_{max}+2\lambda$. That implies, $d_{min} \leq d_T(u, v) \leq d_{max}$. Finally, let $d_T(u, v) > k_{max}$. Suppose p is the minimum difference between k_{max} and the length of a path in T that is longer than k_{max}, that is $p= \min_{u,v \in V} \{d_T(u, v) - k_{max}\}$. Then $d_T(u, v) - k_{max} \geq p$. By subtracting 2λ from both side of the inequality we get, $d_T(u, v) - k_{max} - 2\lambda \geq p - 2\lambda$. Which implies $d_{T'}(u, v) - d_{max} \geq p - 2\lambda$. Therefore, if we can ensure that $p > 2\lambda$ then $d_{T'}(u, v)$ will be larger than d_{max}.

Case 2: $d_{T'}(u, v) = d_T(u, v) + 2\lambda$.
In this case, we have to consider three scenarios as we have in case 1. First, if $k_{min} \leq d_T(u, v) \leq k_{max}$ then $k_{min} \leq d_T(u, v) + 2\lambda \leq k_{max} + 2\lambda$. Which implies $d_{min} \leq d_T(u, v) + 2\lambda \leq d_{max}$. Hence $d_{min} \leq d_{T'}(u, v) \leq d_{max}$. Next, if $d_T(u, v) > k_{max}$ then adding 2λ in both sides we get $d_T(u, v) + 2\lambda > k_{max} + 2\lambda$. That implies $d_{T'}(u, v) > d_{max}$. Finally, let assume that $d_T(u, v) < k_{min}$. Suppose q is the minimum difference between k_{min} and the length of a path in T that is smaller than k_{min}; that is $q = \min_{u,v \in V} \{k_{min} - d_T(u, v)\}$. Then $k_{min} - d_T(u, v) \geq q$. Subtracting 2λ from both sides of the inequality we get $k_{min} - d_T(u, v) - 2\lambda \geq q - 2\lambda$. Which implies $d_{min} - d_{T'}(u, v) \geq q - 2\lambda$. Therefore, if we can ensure that $q > 2\lambda$ then $d_{T'}(u, v) < d_{min}$.

Case 3: $d_{T'}(u, v) = d_T(u, v) + \lambda$.
This case is similar to case 2. By following the same reasoning as in case 2, we can show that $d_{min} \leq d_{T'}(u, v) \leq d_{max}$ if and only if $k_{min} \leq d_T(u, v) \leq k_{max}$, provided $q \geq \lambda$. If we can satisfy the inequality derived from case 2 ($q > 2\lambda$) then the inequality $q > \lambda$ will be immediately satisfied.

From our analysis of the three cases above, it is evident that if we can satisfy the two inequalities $p > 2\lambda$ and $q > 2\lambda$ simultaneously then $G = PCG(T', d_{min}, d_{max})$. We can do this by assigning λ any value smaller than $min(p, q)/2$. Thus T' is a pairwise compatibility tree of G, and hence G is a PCG. □

Figure 7(a) illustrates an example of a tree compatible graph $G = (V, E)$ and the corresponding tree T is depicted in Fig. 7(b). Here $k_{min} = 2$, $k_{max} = 4$, and the weight of every edge is one . Two internal nodes d and e along with the leaves of T correspond to the vertices in V of G. We now transfer T into T' according to the procedure described in Theorem 2. Figure 7(c) illustrates this transformation. Here, $p = q = 1$ and hence we can chose any positive value less than 0.5 for λ. Let $\lambda = 0.4$ and then, $d_{min} = k_{min} = 2$ and $d_{max} = k_{max} + 2\lambda = 4.8$. One can now easily verify that $G = PCG(T', 2, 4.8)$.

5 Conclusion

In this paper, we have proved that all graphs are not PCGs. Additionally, we have proved that tree power graphs and two of their extensions are PCGs. Our first proof establishes a necessary condition over the adjacency relationships that a graph must satisfy to be a PCG. However, a complete characterization of PCGs is not known. We left it as a future work. It would be quite challenging and significant to develop efficient algorithms for solving pairwise tree construction problem for other classes of graphs. Such algorithms may come handy in both clique finding and evolutionary relationships modeling contexts.

Acknowledgement

This work is a part of an ongoing Master's thesis which is done in Graph Drawing & Information Visualization Laboratory of the Department of CSE, BUET established under the project "Facility Upgradation for Sustainable Research on Graph Drawing & Information Visualization" supported by the Ministry of Science and Information & Communication Technology, Government of Bangladesh. We thank BUET for providing necessary supports.

References

1. Cormen, T.H., Leiserson, C.E., Rivest, R.L., Stein, C.: Introduction to Algorithms, 2nd edn. The MIT Press, Cambridge (2001)
2. Jones, N.C., Pevzner, P.A.: An Introduction to Bioinformatics Algorithms. The MIT Press, Cambridge (2004)
3. Kearney, P., Munro, I., Phillips, D.: Efficient Generation of Uniform Samples from Phylogenetic Trees. In: Benson, G., Page, R.D.M. (eds.) WABI 2003. LNCS (LNBI), vol. 2812, pp. 177–189. Springer, Heidelberg (2003)
4. Lesk, A.M.: Introduction to Bioinformatics. Oxford University Press, Oxford (2002)
5. Lin, G.H., Jiang, T., Kearney, P.E.: Phylogenetic k-root and steiner k-root. In: Lee, D.T., Teng, S.-H. (eds.) ISAAC 2000. LNCS, vol. 1969, pp. 539–551. Springer, Heidelberg (2000)
6. Phillips, D.: Uniform Sampling From Phylogenetics Trees. Master's thesis, University of Waterloo (August 2002)
7. Yanhaona, M.N., Hossain, K.S.M.T., Rahman, M.S.: Pairwise compatibility graphs. Journal of Applied Mathematics and Computing 30, 479–503 (2009)

Near Optimal Solutions for Maximum Quasi-bicliques

Lusheng Wang

Department of Computer Science, City University of Hong Kong, Hong Kong
cswangl@cityu.edu.hk

Abstract. The maximum quasi-biclique problem has been proposed for finding interacting protein group pairs from large protein-protein interaction (PPI) networks. The problem is defined as follows:

THE MAXIMUM QUASI-BICLIQUE PROBLEM: Given a bipartite graph $G = (X \cup Y, E)$ and a number $0 < \delta \leq 0.5$, find a subset X_{opt} of X and a subset Y_{opt} of Y such that any vertex $x \in X_{opt}$ is incident to at least $(1-\delta)|Y_{opt}|$ vertices in Y_{opt}, any vertex $y \in Y_{opt}$ is incident to at least $(1-\delta)|X_{opt}|$ vertices in X_{opt} and $|X_{opt}| + |Y_{opt}|$ is maximized.

The problem was proved to be NP-hard [2]. We design a polynomial time approximation scheme to give a quasi-biclique (X_a, Y_a) for $X_a \subseteq X$ and $Y_a \subseteq Y$ with $|X_a| \geq (1-\epsilon)|X_{opt}|$ and $|Y_a| \geq (1-\epsilon)|Y_{opt}|$ such that any vertex $x \in X_a$ is incident to at least $(1-\delta-\epsilon)|Y_a|$ vertices in Y_a and any vertex $y \in Y_a$ is incident to at least $(1-\delta-\epsilon)|X_A|$ vertices in X_a for any $\epsilon > 0$, where X_{opt} and Y_{opt} form the optimal solution.

1 Introduction

Proteins with interactions carry out most biological functions within living cells such as gene expression, enzymatic reactions, signal transduction, inter-cellular communications and immunoreactions. Protein-protein interaction (PPI) networks are one of the major post-genomic data sources available to biologists. They provide a comprehensive view of the global interaction structure of an organisms proteome, as well as detailed information on specific interactions. Bicliques in protein-protein interaction networks play an important role in biological research. Andreopoulos *et al.* proposed to find bicliques from large-scale protein interaction networks to identify locally significant proteins that mediate the function of modules [1]. Thomas *et al.* introduced complementary domains in [3], and showed that the complementary domains can form near bicliques in PPI networks. Bu *et al.* showed that the graph topological structures consist of biologically relevant functional groups [4]. Thus, they proposed a new method to predict the function of uncharacterized proteins based on the classification of known proteins within topological structures such as quasi-clique and quasi-biclique. Other methods for functional prediction of open reading frames coded in the genome and protein-protein interaction models can be found in [5,6].

To identify motif pairs at protein interaction sites, Li *et al.* introduced a method in which the core idea is to find bicliques (complete bipartite subgraphs) from PPI networks [7]. The first step in their algorithm is to find large

M.T. Thai and S. Sahni (Eds.): COCOON 2010, LNCS 6196, pp. 409–418, 2010.

subnetworks with all-versus-all interactions (complete bipartite subgraphs) between a pair of protein groups. They then compute conserved motifs (possible interaction sites) by multiple sequence alignments within each protein group. After that, those conserved motifs can be paired with motifs identified from other protein groups for modeling protein interaction sites. In practice, due to various reasons, some edges in a biclique may be missing and a biclique becomes a *quasi-biclique*, where every vertex in one side is incident to most (not necessarily all) of the vertices in the opposite side. Liu *et al.* showed that using quasi-bicliques instead of bicliques can give better results in practice [2,8]. Thus, finding quasi-bicliques is more important in some cases.

Let $\mathcal{G} = (\mathcal{V}, \mathcal{E})$ be an undirected graph (PPI network), where each vertex represents a protein and there is an edge connecting two vertices if the two proteins have an interaction. Since $\mathcal{G}$ is an undirected graph, any edge $(u, v) \in \mathcal{E}$ implies $(v, u) \in \mathcal{E}$. For a selected edge (u, v) in $\mathcal{G}$, in order to find the two groups of proteins having the similar pairs of binding sites, we translate the graph $\mathcal{G} = (\mathcal{V}, \mathcal{E})$ into a bipartite graph $G = (X \cup Y, E)$, where $X = \{x|(x, v) \in \mathcal{E}\}$, $Y = \{y|(u, y) \in \mathcal{E}\&y \notin X\} \cup \{y'|(u, y) \in \mathcal{E}\&y \in X\}$, and $E = \{(x, y)|(x, y) \in \mathcal{E}\&x \in X\&y \in Y\} \cup \{(x, y')|(x, y) \in \mathcal{E}\&x \in X\&y' \in Y\}$. If their exists a vertex w such that w is incident to both u and v in $\mathcal{G}$, then w is in X and a new virtual vertex w' (corresponding to the real vertex w) is in Y. In this way, we have a bipartite graph $G = (X \cup Y, E)$. A biclique in G corresponds to two subsets of vertices, say, subset A and subset B, in $\mathcal{G}$. In $\mathcal{G}$, every vertex in A is adjacent to all the vertices in B, and every vertex in B is adjacent to all the vertices in A. Moreover, $A \cap B$ may not be empty. In this case, for any vertex $w \in A \cap B$, $(w, w) \in \mathcal{E}$. This is the case, where the protein has a self-loop. Self-loops are very common in practice. When a self-loop appears, one protein molecular interacts with other identical protein molecular. For example, two identical subunit proteins can assemble together to form a homodimeric protein.

For a vertex $x \in X$ and a vertex set $Y' \subseteq Y$, the degree of x in Y' denoted as $d(x, Y')$ is the number of vertices in Y' that are adjacent to x. In other words, $d(x, Y') = |\{y|y \in Y' \text{ and } (x, y) \in E\}|$. Similarly, for a vertex $y \in Y$ and $X' \subseteq X$, we use $d(y, X') = |\{x|x \in X' \quad (x, y) \in E\}|$ to denote the degree of y in X'. For a bipartite graph $G = (X \cup Y, E)$ and a parameter $0 < \delta \leq \frac{1}{2}$, G is called a *δ-quasi-biclique* if for each $x \in X$, $d(x, Y) \geq (1 - \delta)|Y|$ and for each $y \in Y$, $d(y, X) \geq (1 - \delta)|X|$.

The maximum vertex quasi-biclique problem is defined as follows.

Definition: Given a bipartite graph $G = (X \cup Y, E)$ and $0 < \delta \leq \frac{1}{2}$, the maximum vertex δ-quasi-biclique problem is to find $X' \subseteq X$ and $Y' \subseteq Y$ such that the $X' \cup Y'$ induced subgraph is a δ-quasi-biclique and $|X'| + |Y'|$ is maximized.

When $\delta = 0$, the problem becomes the Maximum Vertex Biclique problem and can be solved in polynomial time [9]. When $0 < \delta \leq \frac{1}{2}$, the problem was proved to be NP-hard [8]. In this paper, we design a polynomial time approximation scheme to give a quasi-biclique $X' \subseteq X$ and $Y' \subseteq Y$ with $|X'| \geq (1 - \epsilon)|X_{opt}|$ and $|Y'| \geq (1 - \epsilon -)|Y_{opt}|$ such that any vertex $x \in X'$ is incident to at least

$(1-\delta-\epsilon)|Y'|$ vertices in Y' and any vertex $y \in Y'$ is incident to at least $(1-\delta-\epsilon)|X'|$ vertices in X' for any $\epsilon > 0$, where X_{opt} and Y_{opt} form the optimal solution.

2 Computing the Maximum Quasi-biclique Problem

The following lemma is well known and will be repeatedly used in our proofs See [21] for details.

Lemma 1. *Let $X_1, X_2, \ldots, X_n$ be n independent random 0-1 variables, where X_i takes 1 with probability p_i, $0 < p_i < 1$. Let $X = \sum_{i=1}^{n} X_i$, and $\mu = E[X]$. Then for any $0 < \epsilon \leq 1$,*

$$\mathbf{Pr}(X > \mu + \epsilon n) < exp(-\frac{1}{3}n\epsilon^2),$$

$$\mathbf{Pr}(X < \mu - \epsilon n) \leq exp(-\frac{1}{2}n\epsilon^2).$$

Let $G = (X \cup Y, E)$ be the input bipartite graph. Let $X_{opt} \subseteq X$ and $Y_{opt} \subseteq Y$ be the optimal biclique for the maximum quasi-biclique problem. Without loss of generality, we can assume that

Assumption 1: $|Y_{opt}| \geq |X_{opt}|$.
The basic idea of our algorithm is to (1) formulate the problem into a quadratic program problem and (2) use a random sampling approach to approximately solve the problem. In order to make the random sampling approach work, we have to make sure that

$$|X_{opt}| = \Omega(|X|) \tag{1}$$

and

$$|Y_{opt}| = \Omega(|Y|). \tag{2}$$

However, for any input bipartite graph $G = (X \cup Y, E)$, there is no guarantee that (1) and (2) hold. Here we propose a method to find a subset X' of X and Y' of Y such that for any $t > 0$, $|X_{opt}| = \Omega(|X'|)$, $|X_{opt} \cap X'| \geq \frac{t-1}{t}|X_{opt}|$, $|Y_{opt}| = \Omega(|Y'|)$, and $|Y_{opt} \cap Y'| \geq \frac{t-1}{t}|Y_{opt}|$. If we can obtain this kind of X' and Y', then we can work on the induced bipartite graph $G' = (X' \cup Y', E')$, where $E' = \{(u,v) | u \in X', v \in Y' \text{ and } (u,v) \in E\}$. Obviously, any good approximate solution of G' is also a good approximate solution of G.

Let x_i be a vertex in the bipartite graph $G = (X \cup Y, E)$. Define $D(x_i, Y)$ to be the set of vertices in Y that are incident to x_i. The following lemma tells us how to obtain X' and Y'.

Lemma 2. *For any $t > 0$, there exist k vertices x_1, x_2, ..., x_k in X for $k = \lceil \delta t \rceil$ such that $|\bigcup_{i=1}^{k} D(x_i, Y)| \leq k(|Y_{opt}| + |X_{opt}|)$ and $|Y_{opt} \cap \bigcup_{i=1}^{k} D(x_i, Y)| \geq \frac{t-1}{t}|Y_{opt}|$. Similarly, there exists k vertices y_1, y_2, ..., y_k in Y for $k = \lceil \delta t - 1 \rceil$ such that $|\bigcup_{i=1}^{k} D(y_i, X)| \leq k(|Y_{opt}| + |X_{opt}|)$ and $|X_{opt} \cap \bigcup_{i=1}^{k} D(y_i, X)| \geq \frac{t-1}{t}|X_{opt}|$.*

Though we do not know which k vertices in X we should choose, we can try all possible size k subset of X in $O(|X|^k)$ time. Thus, from now on, we assume that the k vertices $x_1, x_2, \ldots, x_k$ are known. Let $X' = \bigcup_{i=1}^k D(y_i, X)$ and $Y' = \bigcup_{i=1}^k D(x_i, Y)$. We will focus on finding a quasi-biclique in the sub-graph $G' = (X' \cup Y', E')$ of G induced by X' and Y'.

Let $X'_{opt} \subseteq X'$ and $Y'_{opt} \subseteq Y'$ be a quasi-$\delta + \frac{1}{t}$-biclique with maximum number of vertices in G'. From Lemma 2, $|X'_{opt}| + |Y'_{opt}| \geq (1 - \frac{1}{t})(|X_{opt}| + |Y_{opt}|)$ since $X' \cap X_{opt}$ and $Y' \cap Y_{opt}$ also form a quasi-$\delta + \frac{1}{t}$-biclique of size $(1 - \frac{1}{t})(|X_{opt}| + |Y_{opt}|)$. From now on, we will try to find a good approximation solution for X'_{opt} and Y'_{opt}.

If $|X'_{opt}|$ and $|Y'_{opt}|$ are approximately the same, then we have $|X'_{opt}| = \Omega(|X'|)$ and $|Y'_{opt}| = \Omega(|Y'|)$. That is, (1) and (2) hold for graph G'. Therefore, we can use quadratic programming approach to solve the problem. Nevertheless, there is no guarantee that $|X'_{opt}|$ and $|Y'_{opt}|$ are approximately the same. For any $\epsilon > 0$, we consider two cases.

Case 1: $|X'_{opt}| < \epsilon |Y'_{opt}|$. In this case, the number of vertices in Y'_{opt} will dominates the size of the whole quasi-biclique. If we select a vertex $x \in X'_{opt}$, then x and $D(x, Y')$ form a biclique of size at least $1 + (1-\delta)|d(x, Y')| \geq 1 + (1-\delta)|Y'_{opt}|$. When the value of δ is big with respect to ϵ, we do not have the desired quasi-biclique. If we try to add more vertices from Y', we have to guarantee that for every selected vertex y in Y', y is incident to at least $(1 - \delta)\%$ selected vertices in X'. This is impossible if x is the only selected vertex from X'. Therefore, we have to consider to add more vertices from both X' and Y'. It is clear that the task here is non-trivial.

In the following lemma, we will show that there exists a subset of r vertices (for some constant r) $X_r \subseteq X'$ and a subset $Y''_{opt} \subseteq Y'_{opt}$ such that X_r and Y''_{opt} form a quasi-$(\delta + \epsilon'')$-biclique with $|Y''_{opt}| \geq (1 - \epsilon'')|Y_{opt}|$ for some $\epsilon'' > 0$. Here r and ϵ'' are closely related.

Lemma 3. *Let $\frac{1}{t} = \epsilon'$. There exist a subset X'_r of X'_{opt} containing $r = \frac{2}{\epsilon'^2} \log(\frac{1}{\epsilon'})$ elements and a subset Y''_{opt} of Y'_{opt} with $|Y''_{opt}| \geq (1 - \frac{r(r-1)}{2|X_{opt}|} - 2\epsilon')|Y'_{opt}|$ such that X'_r and Y''_{opt} form a quasi-$(\delta + \frac{r(r-1)}{2|X'_{opt}|} + 2\epsilon')$-biclique.*

Based on Lemma 3, we can design an algorithm that finds a quasi-$(\delta + 4\epsilon')$-biclique with size at least $(1 - 4\epsilon' - \epsilon)(|X'_{opt}| + |Y'_{opt}|)$. Let $G' = (X' \cup Y', E')$ be the sub-graph obtained from Lemma 2. For any $\epsilon' > 0$, define $r = \frac{2}{\epsilon'^2} \log(\frac{1}{\epsilon'})$.

Case 1.1. $|X'_{opt}| \geq \frac{r(r-1)}{\epsilon'}$: When $|X'_{opt}| \geq \frac{(r(r-1)}{\epsilon'}$, $\frac{r(r-1)}{2|X'_{opt}|} \leq \epsilon'$. Thus, there exist a quasi-$\delta + 3\epsilon'$-biclique $X_r \subset X'$ and Y''_{opt} as described in Lemma 3.

We select r vertices from X'. For each subset $X_r \subseteq X'$ of r vertices $\{v_1, v_2, \ldots, v_r\}$, we define the following integer linear programming. Let $c_{i,j}$ be a constant, where $c_{i,j} = 1$ if $(v_i, u_j) \in E'$; and $c_{i,j} = 0$ if $(v_i, u_j) \notin E'$. Let y_i be a 0/1 variable, where $y_i = 1$ indicates that the vertex u_i in Y' is selected in the quasi-biclique and $y_i = 0$ otherwise.

$$y_i(\sum_{j=1}^{r} c_{i,j}) \geq (1-\delta-\frac{1}{t}-\epsilon')r \quad (3)$$

$$\sum_{i=1}^{|Y'|} y_i c_{i,j} \geq (1-\delta-3\epsilon')|Y'_{opt}| \text{ for } j=1,2,\ldots,r, \quad (4)$$

Here we do not know $|Y'_{opt}|$. However, we can guess the value of $|Y'_{opt}|$ by trying $|Y'_{opt}| = 1, 2, \ldots, |Y'|$. The integer programming problem is hard to solve. However, we can obtain a fractional solution $\bar{y}_i$ for (3) and (4) with $0 \leq \bar{y}_i \leq 1$ in polynomial time. After obtaining the fractional solution $\bar{y}_i$, we randomly set y_i to be 1 with probability $\bar{y}_i$.

Lemma 4. *Assume that* $\frac{1}{2}(1-\delta-3\epsilon')|Y'_{opt}|\epsilon'^2 \geq 2\log r$ *and* $\frac{1}{t} = \epsilon'$. *With probability at least* $1-\frac{1}{r}$, *we can get a pair of subsets* $X_A \subseteq X'$ *and* $Y_A \subseteq Y'$ *(an integer solution) by randomized rounding according to the probability* $\bar{y}_i$ *such that* X_A *and* Y_A *form a quasi-*$(\delta+4\epsilon')$*-biclique with* $|X_A|+|Y_A| \geq (1-\delta-4\epsilon')|Y'_{opt}$.

A standard method in [10] can give a de-randomized algorithm.

When $\frac{1}{2}(1-\delta-3\epsilon')|Y'_{opt}|)\epsilon'^2 < 2\log r$, we can enumerate all possible subsets of size $(1-\delta-3\epsilon')|Y'_{opt}|$ in Y' in polynomial time to get the desired solution.

Case 1.2. $|X'_{opt}| < \frac{r(r-1)}{\epsilon'}$: In this case, X'_{opt} and Y'_{opt} form the desired quasi-δ-biclique. Instead of selecting r vertices in X', we select $|X'_{opt}|$ vertices in X'. Though we do not know the value of $|x'_{opt}|$, we can guess the value for $|x'_{opt}| = 1, 2, \ldots, \frac{r(r-1)}{\epsilon'}$. We also solve the integer linear programming (3) and (4) in the same way as in Case 1.1. The algorithm for Case 1 is given in Figure 1.

Theorem 1. *Assume* $|X'_{opt}| \leq \epsilon|Y'_{opt}$. *We set* $\frac{1}{t} = \epsilon'$ *in the algorithm. With probability at least* $1-\frac{1}{r}$, *Algorithm 1 finds a quasi-*$(\delta+4\epsilon')$*-biclique* $X_A \subseteq X$ *and* $Y_A \subseteq Y$ *with* $|X_A|+|Y_A| \geq (1-\delta-4\epsilon')(|X_{opt}|+|Y_{opt}|)(1-\epsilon')/(1+\epsilon)$ *in time* $O((|X||Y|)^{\lceil \delta t \rceil}[|X||Y||Y'|^{\frac{4\log r}{\epsilon'^2}} + |X'|^{\frac{r(r-1)}{\epsilon'}} \frac{r(r-1)}{\epsilon'}(|X|+|Y|)^3)])$.

Case 2: $|X'_{opt}| \geq \epsilon|Y'_{opt}|$. In this case, we have $|X'_{opt}| = \Omega(|X'|)$ and $|Y'_{opt}| = \Omega(|Y'|)$. We will use a quadratic programming approach to solve the problem. We can formulate the quasi-biclique problem for the bipartite graph $G' = (X' \cup Y', E')$ into the following quadratic programming problem.

Quadratic Programming Formulation

Let x_i and y_j be 0/1 variables, where $x_i = 1$ indicates that vertex v_i in X' is in the quasi-biclique and $y_j = 1$ indicates that vertex u_j in Y' is in the quasi-biclique. Define $e_{i,j} = 1$ if $(v_i, u_j) \in E'$ and $e_{i,j} = 0$ otherwise. Let c_1 and c_2 be two integers representing the sizes of X'_{opt} and Y'_{opt}, respectively. We can guess the values of c_1 and c_2 in polynomial time though we do not know c_1 and c_2. We have the following inequalities:

$$y_i(\sum_{j=1}^{|X'|} e_{i,j}x_j) \geq (1-\delta-\frac{1}{t})y_i c_1 \text{ for } i=1,2,\ldots,|Y'| \quad (5)$$

Algorithm 1: Algorithm for Solving Case 1: $|X'_{opt}| \leq \epsilon |Y'_{opt}|$.

Input: a bipartite graph $G = (X \cup Y, E)$, a real number $0 \leq \delta \leq 0.5$, a number $t > 0$, a number $\epsilon > 0$, and a number $\epsilon' > 0$.

Let $k = \lceil \delta t \rceil$.
for any $v_1, v_2, \ldots, v_k \in X$ and any $u_1, u_2, \ldots, u_k \in Y$ **do**
  Set $X' = \cup_{i=1}^{k} D(v_i, Y)$ and $Y' = \cup_{i=1}^{k} D(u_i, X)$.
  $r == \frac{2}{\epsilon'^2} \log(\frac{1}{\epsilon'})$
  Guess $|X'_{opt}|$ and $|Y'_{opt}$ assuming $|X'_{opt}| \leq \epsilon |Y'_{opt}|$.
  if $\frac{1}{2}(1 - \delta - 3\epsilon')|Y'_{opt}|)\epsilon'^2 < 2 \log r)$ **then** enumerate all possible subsets of size $(1 - \delta - 3\epsilon')|Y'_{opt}|$ in Y' in polynomial time to get the desired solution.
  if $\frac{1}{2}(1 - \delta - 3\epsilon')|Y'_{opt}|)\epsilon'^2 > 2 \log r)$ **then**
   for $i = r, r + 1, \ldots \frac{r(r-1)}{\epsilon'}$ **do**
    for every i-elements subset $X_i = \{x_1, x_2, \ldots, x_i\}$ **do**
     give a fractinal solution $\bar{y}_i$ for (3) and (4).
     randomly set $y_i = 1$ with probability $\bar{y}_i$.
Output a $\delta + \frac{1}{t} + 4\epsilon'$ quasi-biclique with the biggest $|X_A| + |Y_A|$.

Fig. 1. The algorithm for solving Case 1

$$x_i(\sum_{j=1}^{|Y'|} e_{i,j} y_j) \geq (1 - \delta - \frac{1}{t}) x_i c_2 \text{ for } i = 1, 2, \ldots, |X'| \tag{6}$$

$$\sum_{i=1}^{|Y'|} y_i = c_1, \tag{7}$$

$$\sum_{i=1}^{|X'|} x_i = c_2. \tag{8}$$

(5) and (6) indicate that $x_i > 0$ and $y_i > 0$ imply that $\sum_{j=1}^{|X'|} e_{i,j} x_j \geq (1 - \delta - \frac{1}{t}) c_1$ and $\sum_{j=1}^{|Y'|} e_{i,j} y_) \geq (1 - \delta - \frac{1}{t}) c_2$, respectively.

Let $\hat{x_i}$ and $\hat{y_j}$ be the 0/1 integer solution for the quadratic programming problem (5)-(8). Let $\hat{r_i} = \sum_{j=1}^{|X'|} e_{i,j} \hat{x_j}$ and $\hat{s_i} = \sum_{j=1}^{|Y'|} e_{i,j} \hat{y_j}$. To deal with the quadratic programming problem, the key idea here is to estimate the values of $\hat{r_i}$ and $\hat{s_j}$. If we know the values of $\hat{r_i}$ and $\hat{y_j}$, then (5) and (6) become

$$y_i \hat{r_i} \geq y_i c_1 (1 - \delta - \frac{1}{t}) \text{ for } i = 1, 2, \ldots, |Y'| \tag{9}$$

$$x_i \hat{s_i} \geq x_i c_2 (1 - \delta - \frac{1}{t}) \text{ for } i = 1, 2, \ldots, |X'|, \tag{10}$$

where $\hat{r_i}$ and $\hat{s_i}$ in (9) and (10) are constants and the quadratic inequalities become linear inequalities.

Estimating $\hat{r_i}$ and $\hat{s_i}$

The approach for giving a good estimation of $\hat{r_i}$ and $\hat{s_i}$ is to randomly and independently select a subset $B_{X'}$ of $O(\log(|X'_{opt}|))$ vertices and a subset $B_{Y'}$

of $O(\log(|Y'_{opt}|))$ vertices in X'_{opt} and Y'_{opt}, respectively. Let $c_1 = |X'_{opt}|$ and $c_2 = |Y'_{opt}|$. We do not know c_1 and c_2, but we can guess them in $O(|X'| \times |Y'|)$ time. Then we can use $\frac{c_1}{k} \sum_{v_j \in B_{X'}} e_{i,j}$ and $\frac{c_2}{k} \sum_{u_j \in B_{Y'}} e_{i,j}$ to estimate $\hat{r_i}$ and $\hat{s_i}$, respectively. Since we do not know X'_{opt} and Y'_{opt}, it is not easy to randomly and independently select vertices from X'_{opt} and Y'_{opt}. We develop a method to randomly select $p \times \log|Y'|$ vertices in Y'_{opt} from Y' when Y'_{opt} is not known. Here p is a constant to be determined later.

Finding $p \log|Y'|$ vertices in Y'_{opt} when Y'_{opt} is not known
Let $|Y'| = c|Y'_{opt}|$. The idea here is to randomly and independently select a subset B of $(c+1) \times p \times \log|Y'|$ vertices from Y' and enumerate all size $p \times \log|Y'|$ subsets of B in time $C^{p\log|Y'|}_{p(c+1)\log|Y'|} \leq O(|Y'|^{p(c+1)})$. We can show that with high probability, we can get a set of $p \log|Y'|$ vertices randomly and independently selected from Y'_{opt}.

Lemma 5. *With probability at least* $1-|Y'|^{-\frac{p}{2c^2(c+1)}}$, *$B$ contains a size $p\log|Y'|$ subset of Y'_{opt}.*

Let $B_{X'}$ and $B_{Y'}$ be the sets of randomly and independently selected vertices in X'_{opt} and Y'_{opt}. Let $|B_{X'}| = p_1 \log|X'|$ and $|B_{Y'}| = p_2 \log|Y'|$. We define $\bar{r_i} = \sum_{v_j \in B_{X'}} e_{i,j}$ and $\bar{s_i} = \sum_{u_j \in B_{Y'}} e_{i,j}$. The following lemma shows that $\frac{c_1}{|B_{X'}|}\bar{r_i}$ and $\frac{c_2}{|B_{Y'}|}\bar{s_i}$ are good approximations of $\hat{r_i}$ and $\hat{s_i}$.

Lemma 6. *With probability at least* $1-2|Y'||X'|^{-\frac{\epsilon^2}{3}p_1} - 2|X'||Y'|^{-\frac{\epsilon^2}{3}p_2}$, *for any $i = 1, 2, \ldots, |X'|$ and $j = 1, 2, \ldots, |Y'|$,*

$$(1-\epsilon)\hat{r_i} \leq \frac{c_1}{|B_{X'}|}\bar{r_i} \leq (1+\epsilon)\hat{r_i}$$

and

$$(1-\epsilon)\hat{s_j} \leq \frac{c_2}{|B_{Y'}|}\bar{s_j} \leq (1+\epsilon)\hat{s_j}.$$

Now, we set $r_i = \frac{c_1}{|B_{X'}|}\bar{r_i}$ and $s_i = \frac{c_2}{|B_{Y'}|}\bar{s_j}$. We consider the following linear programming problem.

$$y_i r_i \geq y_i c_1 (1-\epsilon)(1-\delta) \text{ for } i = 1, 2, \ldots, m, \tag{11}$$

$$x_i s_i \geq x_i c_2 (1-\epsilon)(1-\delta) \text{ for } i = 1, 2, \ldots, m, \tag{12}$$

$$\sum_{i=1}^{|Y'|} y_i = c_1, \tag{13}$$

$$\sum_{i=1}^{|X'|} x_i = c_2 \tag{14}$$

$$\sum_{j=1}^{|X'|} e_{i,j} x_j \geq \frac{r_i}{1+\epsilon} \tag{15}$$

Algorithm 2: Algorithm for Soving Case 2: $|X'_{opt}| > \epsilon|Y'_{opt}|$.

Input: a bipartite graph $G = (X \cup Y, E)$, a real number $0 \leq \delta \leq 0.5$, a number $t > 0$ and a number $\epsilon > 0$.

0. Let $k = \lceil \delta t \rceil$, $p_1 = p_2 = \frac{5}{\epsilon^2}$, $c_x = k(1 + \frac{1}{\epsilon})$ and $c_y = 2k$.
1. **for** any $v_1, v_2, \ldots, v_k \in X$ and any $u_1, u_2, \ldots, u_k \in Y$ **do**
2. Set $X' = \cup_{i=1}^{k} D(v_i, Y)$ and $Y' = \cup_{i=1}^{k} D(u_i, X)$.
3 Randomly and independently select a set $S_{X'}$ of $(c_x + 1)p_1 \log |X'|$ vertices in X' and a set $S_{Y'}$ of $(c_y + 1)p_2 \log |Y'|$ vertices in Y'.
4 **for** any size $p_1 \log |X'|$ subsset $B_{X'}$ of $S_{X'}$ and size $p_2 \log |X'|$ subset $B_{Y'}$ of $S_{Y'}$ **do**
   (a) $\bar{r_i} = \frac{c_1}{|B_{X'}|} \sum_{v_i \in |B|_{X'}} e_{i,j}$
   (b) $\bar{s_i} = \frac{c_2}{|B_{Y'}|} \sum_{u_i \in |B|_{Y'}} e_{i,j}$
   (c) Get a fractional solution x'_i and y'_i for $x_i \in X'$ and $y_i \in Y'$ of (11)-(16)
   (d) do randomrized rouding according to x'_i and y'_i
   (e) $X_A = \{v_i | x_i = 1\}$ and $Y_A = \{u_i | y_i = 1\}$
5. Output a $\delta + \frac{1}{t} + 4\epsilon$ quasi-biclique with the biggest $|X_A| + |Y_A|$.

Fig. 2. The algorithm for Case 2

$$\sum_{j=1}^{|Y'|} e_{i,j} y_j \geq \frac{s_i}{1+\epsilon}. \tag{16}$$

The term $(1 - \epsilon)$ in (11) and (12) ensures that the quadratic programming problem has a solution when the estimated values of r_i and s_i are smaller than $\hat{r_i}$ and $\hat{s_i}$. Similarly, the term $(1 + \epsilon)$ in (15) and (16) ensures that the quadratic programming problem has a solution when the estimated values of r_i and s_i are bigger than $\hat{r_i}$ and $\hat{s_i}$.

Randomized Rounding

Let x'_i and y'_j be a fractional solution for (11) -(16). In order to get a 0/1 solution, we randomly set x_i and y_j to be 1 using the fractional solution as the probability. That is, we randomly set x_i and y_j to be 1's with probability x'_i and y'_i, respectively. (Otherwise, x_i and y_j will be 0.)

Lemma 7. *With probability* $1 - 2exp(-\frac{1}{3}|X'|\epsilon^2) - 2exp(-\frac{1}{3}|Y'|\epsilon^2) - |Y'| exp(-\frac{1}{2}|X'|\epsilon^2) - |X'|exp(-\frac{1}{2}|Y'|\epsilon^2)$, *we can find a subset* $\hat{X} \subseteq X'$ *and a subset* $\hat{Y} \subseteq Y'$ *with* $(1 - \epsilon)c_1 \leq |X'| \leq (1 + \epsilon)c_1$ *and* $(1 - \epsilon)c_2 \leq |Y'| \leq (1 + \epsilon)c_2$ *such that for any* $x \in \hat{X}$, $d(x, Y') \geq (1 - \delta - 4\epsilon)|\hat{Y}|$ *and for any* $y \in \hat{Y}$, $d(y, X) \geq (1 - \delta - 4\epsilon)|\hat{X}|$.

The complete algorithm for Case 2 is given in Figure 2. Let $k = \lceil \delta t \rceil$ as defined in Lemma 2. Here c_x , c_y are set to be $k(1 + \frac{1}{\epsilon})$ and $2k$, respectively. $p_1 = p_2 = \frac{5}{\epsilon^2}$.

Theorem 2. *With probability at least* $1-o(1)$, *Algorithm 2 finds a quasi-*$(\delta+4\epsilon+\frac{1}{t})$*-biclique of size* $(1-\frac{1}{t}-\epsilon)(|X_{opt}|+|Y_{opt}|)$ *in* $O((k \times \frac{1}{\epsilon^2}|X||Y|)^{\lceil \delta t \rceil}(|X|^{\frac{5}{\epsilon^2}k(1+\frac{1}{\epsilon})} + |Y|^{\frac{5}{\epsilon^2}2k})(|X| + |Y|^3))$ *time.*

We can derandomize the algorithm to get a polynomial time deterministic algorithm. Step 3 can be derandomized by using the standared. For instance, instead of randomly and independently choosing $p_1 \log(|X'|)$ and $p_2 \log(|Y'|)$ vertices from X' and Y', we can pick the vertices encountered on a randomwalk of the same length on a constant degree expander [13]. Obviously, the number of such random walks on a constant degree expander is polynomial. Thus, by enumerating all random walks of length $p_1 \log(|X'|)$ and $p_2 \log(|Y'|)$, we have a polynomial time deterministic algorithm. (Also see Arora et al. [11]).

Step 4 (d) can be derandomized by using Raghavan's conditional probabilities method [12]. From Case 1 and Case 2, we can immediately obtain the following theorem.

Theorem 3. *There exists a polynomial time approximation scheme that outputs a quasi-biclique $X_A \subseteq X$ and $Y_A \subseteq Y$ with $|X_A| \geq (1-\epsilon)|X_{opt}|$ and $|Y_A| \geq (1-\epsilon)|Y_{opt}|$ such that any vertex $x \in X_A$ is incident to at least $(1-\delta-\epsilon)|Y_A|$ vertices in Y_A and any vertex $y \in Y_A$ is incident to at least $(1-\delta-\epsilon)|X_A|$ vertices in X_A for any $\epsilon > 0$, where X_{opt} and Y_{opt} form the optimal solution.*

Acknowledgements. This work is fully supported by a grant from the Research Grants Council of the Hong Kong Special Administrative Region, China [Project No. CityU 121207]. The author would like to thank referees for help suggestions.

References

1. Andreopoulos, B., An, A., Wang, X., Faloutsos, M., Schroeder, M.: Clustering by common friends finds locally significant proteins mediating modules. Bioinformatics 23(9), 1124–1131 (2007)
2. Liu, X., Li, J., Wang, L.: Quasi-bicliques: Complexity and Binding Pairs. In: Hu, X., Wang, J. (eds.) COCOON 2008. LNCS, vol. 5092, pp. 255–264. Springer, Heidelberg (2008)
3. Thomas, A., Cannings, R., Monk, N.A.M., Cannings, C.: On the structure of protein-protein interaction networks. Biochemical Society Transanctions 31(6), 1491–1496 (2003)
4. Bu, D., Zhao, Y., Cai, L., Xue, H., Zhu, X., Lu, H., Zhang, J., Sun, S., Ling, L., Zhang, N., Li, G., Chen, R.: Topological structure analysis of the protein-protein interaction network in budding yeast. Nucleic Acids Research 31(9), 2443–2450 (2003)
5. Hishigaki, H., Nakai, K., Ono, T., Tanigami, A., Takagi, T.: Assessment of prediction accuracy of protein function from protein-protein interaction data. Yeast 18(6), 523–531 (2001)
6. Morrison, J.L., Breitling, R., Higham, D.J., Gilbert, D.R.: A lock-and-key model for protein-protein interactions. Bioinformatics 22(16), 2012–2019 (2006)
7. Li, H., Li, J., Wong, L.: Discovering motif pairs at interaction sites from protein sequences on a proteome-wide scale. Bioinformatics 22(8), 989–996 (2006)
8. Liu, X., Li, J., Wang, L.: Modeling Protein Interacting Groups by Quasi-bicliques: Complexity, Algorithm and Application. IEEE/ACM Trans. Comput. Biology Bioinform. (to appear)
9. Lonardia, S., Szpankowskib, W., Yang, Q.: Finding biclusters by random projections. Theoretical Computer Science 368, 271–280 (2006)

10. Li, M., Ma, B., Wang, L.: On the closest string and substring problems. Journal of the ACM 49(2), 157–171 (2002)
11. Arora, S., Karger, D., Karpinski, M.: Polynomial time approximation schemes for dense instances of NP-hard problems. In: Proceedings of the 27th Annual ACM Symposium on Theory of Computing, pp. 284–293. ACM, New York (1995)
12. Raghavan, P.: Probabilistic construction of deterministic algorithms: Approximate packing integer programs. JCSS 37(2), 130–143 (1988)
13. Gillman, D.: A Chernoff bound for randomwalks on expanders. In: Proceedings of the 34th Annual Symposium on Foundations of Computer Science, pp. 680–691. IEEE Computer Society Press, Los Alamitos (1993)
14. Pietrokovski, S.: Searching databases of conserved sequence regions by aligning protein multiple-alignments. Nucleic Acids Research 24, 3836–3845 (1996)
15. Finn, R.D., Marshall, M., Bateman, A.: iPfam: visualization of protein-protein interactions in PDB at domain and amino acid resolutions. Bioinformatics 21(3), 410–412 (2005)
16. Gordeliy, V.I., Labahn, J., Moukhametzianov, R., Efremov, R., Granzin, J., Schlesinger, R., Büldt, G., Savopol, T., Scheidig, A.J., Klare, J.P., Engelhard, M.: Molecular basis of transmembrane signalling by sensory rhodopsin II-transducer complex. Nature 419(6906), 484–487 (2002)
17. Bergo, V.B., Spudich, E.N., Rothschild, K.J., Spudich, J.L.: Photoactivation perturbs the membrane-embedded contacts between sensory rhodopsin II and its transducer. Journal of Biological Chemistry 280(31), 28365–28369 (2005)
18. Inoue, K., Sasaki, J., Spudich, J.L., Terazima, M.: Laser-induced transient grating analysis of dynamics of interaction between sensory rhodopsin II D75N and the HtrII transducer. Biophysical Journal 92(6), 2028–2040 (2007)
19. Sudo, Y., Furutani, Y., Spudich, J.L., Kandori, H.: Early photocycle structural changes in a bacteriorhodopsin mutant engineered to transmit photosensory signals. Journal of Biological Chemistry 282(21), 15550–15558 (2007)
20. Martz, E.: Protein Explorer: easy yet powerful macromolecular visualization. Trends in Biochemical Sciences 27(3), 107–109 (2002)
21. Motwani, R., Raghavan, P.: Randomized Algorithms. Cambridge Univ. Press, Cambridge (1995)

Fast Coupled Path Planning: From Pseudo-Polynomial to Polynomial*

Yunlong Liu[1] and Xiaodong Wu[1,2]

[1] Department of Electrical and Computer Engineering, the University of Iowa
[2] Department of Radiation Oncology, The University of Iowa

Abstract. The coupled path planning (CPP) problem models the motion paths of the leaves of a multileaf collimator for optimally reproducing the prescribed dose in intensity-modulated radiation therapy (IMRT). Two versions of the CPP problem, unconstrained and constrained CPPs, are studied based on whether specifying the starting and ending positions of the leave paths. By exploring the underlying properties of the problem such as submodularity and L$^\natural$-convexity, we solve both CPP problems in polynomial time using the path-end hopping, local searching and proximity scaling techniques, improving current best known pseudo-polynomial time algorithms. Our algorithms are simple and easy to be implemented. Experimental results on real medical data showed that our CPP algorithms outperformed previous best-known algorithm by at least one order of magnitude.

Keywords: Coupled path planning, IMRT, submodularity, L$^\natural$-convexity.

1 Introduction

In this paper, we study an interesting geometric optimization problem, called *coupled path planning* (CPP). Given a non-negative integer reference function f defined on the integer set $\{1, 2, \ldots, n\}$, and three positive integer parameters H, c and Δ, the problem is defined on a uniform grid R_g of size $n \times H$ such that the length of each grid edge is one unit. The goal is to seek two paths $\mathbf{x}_l$ and $\mathbf{x}_r$ in R_g, called the *left* and *right* paths, which are (1) *xy-monotone*: monotone with respect to both the x-axis and the y-axis; (2) *c-steep*: every vertical segment of the paths is of length at least c and every horizontal segment is of unit length; (3) *non-crossing*: two paths $\mathbf{x}_l$ and $\mathbf{x}_r$ do not get across each other; (4) *Δ-error*: the vertical distance between $\mathbf{x}_l$ and $\mathbf{x}_r$ at each column i approximates $f(i)$ within the error bound Δ (i.e., $|\mathbf{x}_l(i) - \mathbf{x}_r(i) - f(i)| \leq \Delta$; and (5) *optimized*: the total sum of the errors (i.e., $\sum_{i=1}^{n} |\mathbf{x}_l(i) - \mathbf{x}_r(i) - f(i)|$) is minimized.

We consider two versions of the CPP problem: (1) The *unconstrained CPP* (UCPP), where the starting and ending positions of the sought paths are not given as part of the input; (2) the *constrained CPP* (CCPP), in which the starting and ending positions of the sought paths are fixed.

* This research was supported in part by the NSF grants CCF-0830402 and CCF-0844765, and the NIH grant K25-CA123112.

M.T. Thai and S. Sahni (Eds.): COCOON 2010, LNCS 6196, pp. 419–428, 2010.

Application Background. The CPP problems arise in a modern cancel therapy technique named *Intensity-Modulated Radiation Therapy* (IMRT), To use this technique effectively, the prescribed radiation dose distributions, namely *intensity maps* (IMs), should be delivered accurately and efficiently. An IM is normally specified by a set of nonnegative integers on a uniform 2D grid. The number in a grid cell indicates the amount of radiation to be delivered to the corresponding body region.

Multileaf collimator (MLC) [16] is one of the most advanced tools for IM delivery in IMRT. An MLC consists of a number of pairs of tungsten alloy leaves of the same rectangular shape and size. The leaves can move left and right (or up and down) to form a rectilinear region, called an *MLC-aperture*. The cross-section of a radiation beam is shaped by an MLC-aperture.

Two radiation delivery methods are commonly used in IMRT [16], called *static* and *dynamic leaf sequencing*. In both leaf sequencing methods, the key problem is to determine a *delivery plan* for a given IM, i.e. a description of how to position the MLC leaves and maintain the on/off states of the radiation to deliver the IM. Two key criteria among others are used to measure the quality of an IMRT delivery plan: (1) *Delivery time* (the efficiency): The reduction of delivery time is beneficial in reducing the intra-fraction uncertainties that occur with organ motions and in minimizing the loss of biological effectiveness [8], thus improving the efficacy of IMRT. (2) *Delivery error* (the accuracy): Various factors may cause the discrepancy between the prescribed radiation dose and the actually delivered dose.

As in Ref. [5], the unconstrained CPP problem models the delivery method of dynamic leaf sequencing. Note that an IM may need to be delivered with a number of MLC leaf pairs, with each pair delivering one row of the IM. Some MLCs (e.g., the Varian MLCs) have the capacity of delivering each IM row independently with one leaf pair. Thus, it suffices to consider an optimal way to deliver one IM row. In the CPP problem, $f(\cdot)$ is the intensity profile (reference function) specifying one row of a given IM, and the two output paths, $\mathbf{x}_l$ and $\mathbf{x}_r$, are the moving trajectories of the two leaf-ends of the MLC leaf pair, i.e., the leaf-end positions (the x-coordinates) at any unit time of the delivery (the y-coordinates). Each MLC leaf has a maximum moving velocity. It takes at lease c units of time (i.e., c vertical edges in R_g) for one leaf-end moves forward one unit length (i.e., one horizontal edge in R_g). Thus, the paths have to be c-steep for some $c > 0$. The vertical distance between two paths, $\mathbf{x}_l$ and $\mathbf{x}_r$, on the i-th column of R_g is the delivery time duration (the delivered intensity level) that the i-th entry of the IM row is exposed to irradiation. To ensure the treatment quality, the delivered intensity level should be close enough to the prescribed intensity level (say, within an error range of Δ). The total error over all columns of R_g gives the delivery error incurred to the IM row specified by $f(\cdot)$. Hence, the CPP problem seeks to deliver one IM row in H units of delivery time while minimizing the total delivery error.

The constrained CPP problem (CCPP) is a key to a newly emerging IMRT delivery technique called *arc-modulated radiation therapy* (AMRT) [17].

Related Work. IMRT delivery technique has been intensively studied by researchers in several different research fields, such as in medical physics [7,14], in operation research [1,2] and in computation geometry [6,11]. Most of those algorithms are designated for static leaf sequencing. Dynamic leaf sequencing algorithms were developed mainly by medical physicists [15,4], aiming to an exact delivery of a given IM with a minimum delivery time. Kamath *et al.* developed an efficient algorithm to exactly deliver an IM with DMLC while minimizing the number of monitor units [11]. Chen *et al.* first modeled the CPP problem in [5], and provided a unified approach for both the UCPP and CCPP problems based on some interesting geometric techniques. Their algorithm runs in a pseudo-polynomial $\mathcal{O}(nH\Delta)$ time and has a pseudo-polynomial space complexity $\mathcal{O}(nH\Delta)$.

2 Modeling the Coupled Path Planning Problems

In this section, we model the coupled path planning problem as an integer programming problem. Recall that the input function $f(\cdot)$ is defined on the integer set $\{1, 2, \ldots, n\}$, denoted by $[1, n]$. Two coupled c-steep paths are represented by two integer vectors, $\mathbf{x}_l = (x_l(1), x_l(2), \ldots, x_l(n))$ and $\mathbf{x}_r = (x_r(1), x_r(2), \ldots, x_r(n))$, with each element representing a horizontal edge on the path. For the sake of convenience, let $\mathbf{x} = (\mathbf{x}_l, \mathbf{x}_r)$ be a feasible solution to the CCP problem. We use the following integer programming to model the problem.

$$\min\ \mathcal{E}(\mathbf{x}) = \sum_{i=1}^{n} |x_l(i) - x_r(i) - f(i)|,\ \mathbf{x} \in \mathbb{Z}_+^{2n}$$

$$\begin{aligned}
\text{s.t.}\quad & H \geq x_l(i) \geq 0, H \geq x_r(i) \geq 0 && i \in [1, n] \quad \text{(1a)}\\
& x_l(i) - x_r(i) \geq 0 && i \in [1, n] \quad \text{(1b)}\\
& x_l(i+1) - x_l(i) \geq 0, x_r(i+1) - x_r(i) \geq 0 && i \in [1, n] \quad \text{(1c)}\\
& x_l(i+1) - x_l(i) \geq c \text{ or } x_l(i+1) = 0 \text{ or } x_l(i) = H && i \in [1, n) \quad \text{(1d)}\\
& x_r(i+1) - x_r(i) \geq c \text{ or } x_r(i+1) = 0 \text{ or } x_r(i) = H && i \in [1, n) \quad \text{(1e)}\\
& f(i) + \Delta \geq x_l(i) - x_r(i) \geq f(i) - \Delta && i \in [1, n] \quad \text{(1f)}
\end{aligned}$$

The integer programming above obviously models the unconstrained CPP problem. In addition, if the four integer path-end parameters, $l_{start}, r_{start}, l_{end}$, and r_{end}, are given (i.e., the left path starts at $(0, l_{start})$ and ends at (H, l_{end}), and the right path starts at $(0, r_{start})$ and ends at (H, r_{end})), then the problem becomes a constrained CPP problem. The path ending constraints can be modeled, as follows.

$$\begin{aligned}
& x_l(i) = 0,\ 1 \leq i < l_{start};\ x_r(i) = 0,\ 1 \leq i < r_{start}\\
& c \leq x_l(i) \leq H - c,\ l_{start} \leq i < l_{end};\ c \leq x_r(i) \leq H - c,\ r_{start} \leq i < r_{end} \quad \text{(1g)}\\
& x_l(i) = H,\ l_{end} \leq i \leq n;\ x_r(i) = H,\ r_{end} \leq i \leq n
\end{aligned}$$

Definitions and notations. Two CCPP problem instances are said to be *associated* iff they have the same inputs except that the path-end parameters are different. One UCPP and one CCPP problem instances are said to be *associated* iff they have the same inputs except that the CCPP instance has specified path-end parameters. For two different associated CCPP instances, $\mathcal{C}$ and $\mathcal{C}'$, with path-end parameters $(l_{start}, r_{start}, l_{end}, r_{end})$ and $(l'_{start}, r'_{start}, l'_{end}, r'_{end})$, respectively, if $0 \leq l_{start} - l'_{start}, r_{start} - r'_{start}, l_{end} - l'_{end}, r_{end} - r'_{end} \leq 1$, then we say $\mathcal{C}'$ is *left adjacent* to $\mathcal{C}$, and $\mathcal{C}$ is *right adjacent* to $\mathcal{C}'$. We introduce ordering relations $\succeq, \preceq$ and $=$ to represent element-wise less than or equal to, greater than or equal to, and equal to, respectively. Also we have two element-wise operations, join($\vee$) and meet($\wedge$), to represent element-wise maximums and minimums of two vectors.

3 Solving the Unconstrained Coupled Path Planning (UCPP) Problem by Exploring Submodularity

In this section, we present our UCPP algorithm by solving $\mathcal{O}(n)$ instances of the constrained coupled path planning (CCPP) problems. Recall that in an CCPP problem instance, the four path-end parameters, $l_{start}, l_{end}, r_{start}, r_{end}$, are specified. A straightforward way to solve the UCPP problem is to compute all $O(n^4)$ constrained CPP instances. However, we can do much better by exploring the submodularity property of the UCPP problem, leading to a so-called *path-end hopping algorithm* for solving the UCPP problem, which reduces the number of CCPP instances need to be solved to $\mathcal{O}(n)$.

Algorithm 1. Path-end hopping

```
Find the minimal feasible solution x° such that any feasible solution x ⪰ x° ;
Solve the CCPP instance C° corresponding to x° to obtain the optimizer x̃;
repeat
    x* ← x̃, and denote by C̃ the CCPP instance corresponding to x̃;
    if none of the CCPP instances left adjacent to C̃ is feasible then
        return x*;
    Solve all CCPP instances left adjacent to C̃ and let x̃ denote the optimizer
    to the CCPP instance which has the smallest objective value;
until E(x̃) > E(x*) ;
return x*;
```

The basic idea of the path-end hopping algorithm is to perform a gradient-descent like search strategy on the non-convex solution space. Start with the minimal feasible solution $\mathbf{x}^\circ$ (i.e., any feasible solution $\mathbf{x} \succeq \mathbf{x}^\circ$). Repeatedly search for a better solution in the CCPP instances left adjacent to the current one. Each CCPP instance has no more than 15 CCPP instances left adjacent to it. In each iteration, the path-end parameters, $l_{start}, l_{end}, r_{start}, r_{end}$, of the newly found CCPP instance with a better solution are non-increasing and at least one path-end parameter decreases by 1. Thus, the total number of CCPP instances need to be solved before the algorithm terminates is bounded by $\mathcal{O}(n)$. Note that Lines 5 and 6 deal with the boundary condition. Hence, in the path-end hopping algorithm, we need to solve $\mathcal{O}(n)$ CCPP instances.

For a given UCPP problem instance $\mathcal{U}$, denote by $\check{\mathbf{x}}^*$ the maximal optimizer of the instance, that is, any optimizer $\mathbf{x}^*$ of $\mathcal{U}$ having $\mathbf{x}^* \preceq \check{\mathbf{x}}^*$. To prove the correctness of the path-end hopping algorithm, we show (1) the maximal optimizer $\check{\mathbf{x}}^*$ is always on the pathway of the path-end hopping process; and (2) when the algorithm terminates, $\mathbf{x}^*$ returned is an optimizer of $\mathcal{U}$. Denote by $\mathbb{D}$ the solution space of $\mathcal{U}$. We first show the submodularity of the UCPP problem.

Lemma 1 (Submodularity). *For any UCPP instance, if* $\mathbf{x}, \mathbf{y} \in \mathbb{D}$, *we have* $\mathbf{x} \vee \mathbf{y}, \mathbf{x} \wedge \mathbf{y} \in \mathbb{D}$, *and* $\mathcal{E}(\mathbf{x} \vee \mathbf{y}) + \mathcal{E}(\mathbf{x} \wedge \mathbf{y}) \leq \mathcal{E}(\mathbf{x}) + \mathcal{E}(\mathbf{y})$.

The proof of Lemma 1 is straightforward. The existence of the minimal feasible solution and the maximal optimizer $\check{\mathbf{x}}^*$ of $\mathcal{U}$ is a direct result of Lemma 1.

Corollary 1. *For any instance of UCPP* $\mathcal{U}$ *with nonempty solution space* $\mathbb{D}$, *(1) there exists a unique minimal feasible solution* $\mathbf{x}^\circ$ *such that* $\forall \mathbf{x} \in \mathbb{D}, \mathbf{x}^\circ \preceq \mathbf{x}$. *(2) There exists a unique maximal optimizer* $\check{\mathbf{x}}^*$ *such that for any optimizer* $\mathbf{x}^*$ *of* $\mathcal{U}$, $\check{\mathbf{x}}^* \succeq \mathbf{x}^*$.

We are now ready to show that the maximal optimizer $\check{\mathbf{x}}^*$ is always on the pathway of the path-end hopping process.

Lemma 2. *Given a UCPP instance* $\mathcal{U}$, *denote by* $\check{\mathbf{x}}^*$ *its maximal optimizer. Suppose* $\widetilde{\mathbf{x}} \preceq \check{\mathbf{x}}^*$ *is an optimizer of an associated CCPP instance* $\mathcal{C}$ *of* $\mathcal{U}$. *(1) There exists an optimizer* $\widetilde{\mathbf{x}}'$ *which has the smallest objective value among those optimizers of the CCPP instances left adjacent to* $\mathcal{C}$, *such that* $\widetilde{\mathbf{x}}' \succeq \widetilde{\mathbf{x}}$. *(2) If* $\mathcal{E}(\widetilde{\mathbf{x}}') \leq \mathcal{E}(\widetilde{\mathbf{x}})$, *then* $\check{\mathbf{x}}^* \succeq \widetilde{\mathbf{x}}'$.

Proof. Based on Lemma 1, it is not difficult to see the existence of $\widetilde{\mathbf{x}}'$ such that $\widetilde{\mathbf{x}}' \succeq \widetilde{\mathbf{x}}$ if the left adjacent CCPP instances of $\mathcal{C}$ exist. Based on submodularity and the fact that $\check{\mathbf{x}}^*$ is optimal, it is easy to show $\mathcal{E}(\check{\mathbf{x}}^* \wedge \widetilde{\mathbf{x}}') \leq \mathcal{E}(\widetilde{\mathbf{x}}')$.

Case 1): $\mathcal{E}(\check{\mathbf{x}}^* \wedge \widetilde{\mathbf{x}}') = \mathcal{E}(\widetilde{\mathbf{x}}')$. We then have $\mathcal{E}(\check{\mathbf{x}}^* \vee \widetilde{\mathbf{x}}') \leq \mathcal{E}(\check{\mathbf{x}}^*)$, which indicates that $\check{\mathbf{x}}^* \vee \widetilde{\mathbf{x}}'$ is an optimizer of $\mathcal{U}$ and $\check{\mathbf{x}}^* \succeq \widetilde{\mathbf{x}}'$.

Case 2): $\mathcal{E}(\check{\mathbf{x}}^* \wedge \widetilde{\mathbf{x}}') < \mathcal{E}(\widetilde{\mathbf{x}}')$. Note that all path-end parameters of the corresponding CCPP instance $\mathcal{C}'$ of $\widetilde{\mathbf{x}}'$ are no larger than those of $\mathcal{C}$, and $\widetilde{\mathbf{x}}' \succeq \widetilde{\mathbf{x}}$ and $\check{\mathbf{x}}^* \succeq \widetilde{\mathbf{x}}$. We have $\widetilde{\mathbf{x}} \preceq \check{\mathbf{x}}^* \wedge \widetilde{\mathbf{x}}' \preceq \widetilde{\mathbf{x}}'$. Since $\widetilde{\mathbf{x}}'$ is the best solution among all CCPP instances left adjacent to $\mathcal{C}$, $\check{\mathbf{x}}^* \wedge \widetilde{\mathbf{x}}'$ is a feasible solution to $\mathcal{C}$. Then, it is obvious that $\mathcal{E}(\widetilde{\mathbf{x}}) \leq \mathcal{E}(\check{\mathbf{x}}^* \wedge \widetilde{\mathbf{x}}') < \mathcal{E}(\widetilde{\mathbf{x}}')$, which contradicts to our assumption $\mathcal{E}(\widetilde{\mathbf{x}}') \leq \mathcal{E}(\widetilde{\mathbf{x}})$. Thus, the second case ($\mathcal{E}(\check{\mathbf{x}}^* \wedge \widetilde{\mathbf{x}}') < \mathcal{E}(\widetilde{\mathbf{x}}')$) is impossible. Hence, $\check{\mathbf{x}}^* \succeq \widetilde{\mathbf{x}}'$, which proves the lemma. □

Note that in the path-end hopping algorithm, we start with the minimal feasible solution $\mathbf{x}^\circ$. Thus, the optimizer $\widetilde{\mathbf{x}}$ of the CCPP instance corresponding to $\mathbf{x}^\circ$ has the following relation with the maximal UCPP optimizer $\check{\mathbf{x}}^*$: $\check{\mathbf{x}}^* \succeq \widetilde{\mathbf{x}}$. Without loss of generality, we assume that current CCPP instance considered in the algorithm is $\mathcal{C}$ and its optimizer $\widetilde{\mathbf{x}} \preceq \check{\mathbf{x}}^*$. Lemma 2 shows that $\check{\mathbf{x}}^*$ is always on the pathway of our gradient-descent like searching process if the best CCPP optimizer $\widetilde{\mathbf{x}}'$ to the left adjacent CCPP instances of $\mathcal{C}$ (i.e., $\widetilde{\mathbf{x}}'$ has the smallest objective value among the optimizers of those CCPP instances), has an objective value no larger than that of $\widetilde{\mathbf{x}}$ (i.e., $\mathcal{E}(\widetilde{\mathbf{x}}') \leq \mathcal{E}(\widetilde{\mathbf{x}})$).

We next show the existence of such a solution $\widetilde{\mathbf{x}}'$. The key idea is to find an appropriate integer $\alpha > 0$ to show that $\widetilde{\mathbf{x}}'$ can be constructed from a UCPP optimizer $\mathbf{x}^*$ and $\widetilde{\mathbf{x}} \preceq \mathbf{x}^*$, more precisely, from $(\mathbf{x}^* - \alpha\mathbf{1}) \vee \widetilde{\mathbf{x}}$ or $\mathbf{x}^* \wedge (\widetilde{\mathbf{x}} + \alpha\mathbf{1})$. First, we need the following lemma.

Lemma 3. *For any instance of UCPP, given $\mathbf{x}, \mathbf{y} \in \mathbb{D}$, if there exists an $\alpha \in \mathbb{Z}_+$, such that $((\mathbf{x} - \alpha\mathbf{1}) \vee \mathbf{y}), (\mathbf{x} \wedge (\mathbf{y} + \alpha\mathbf{1})) \in \mathbb{D}$, then*

$$\mathcal{E}(\mathbf{x}) + \mathcal{E}(\mathbf{y}) \geq \mathcal{E}((\mathbf{x} - \alpha\mathbf{1}) \vee \mathbf{y}) + \mathcal{E}(\mathbf{x} \wedge (\mathbf{y} + \alpha\mathbf{1})) \tag{2}$$

Lemma 4. *For a given UCPP instance $\mathcal{U}$, let $\mathbf{x}^*$ be an optimizer. Assume $\mathbf{x} \preceq \mathbf{x}^*$, is an optimizer of an associated CCPP instance $\mathcal{C}$ of $\mathcal{U}$, but $\mathbf{x}$ is not an optimizer of $\mathcal{U}$. Then there must exist a CCPP instance $\mathcal{C}'$ that is left adjacent to $\mathcal{C}$ and has the optimum no larger than that of $\mathcal{C}$.*

Sketch of the proof. Our goal is to find an appropriate integer $\alpha > 0$, such that (1) $((\mathbf{x}^* - \alpha\mathbf{1}) \vee \mathbf{x})$ and $(\mathbf{x}^* \wedge (\mathbf{x} + \alpha\mathbf{1}))$ are feasible solutions to $\mathcal{U}$; and (2) at least one of the following claims holds: (i) the corresponding CCPP instance of $((\mathbf{x}^* - \alpha\mathbf{1}) \vee \mathbf{x})$ is left adjacent to $\mathcal{C}$ and (ii) the corresponding CCPP instance of $(\mathbf{x}^* \wedge (\mathbf{x} + \alpha\mathbf{1}))$ is right adjacent to $\mathcal{C}^*$.

It is not hard to verify the existence of such α with $\alpha^* = \max\{\mathbf{x}_l^*(l_{start} - 1) - c, \mathbf{x}_r^*(r_{start} - 1) - c, H - c - \mathbf{x}_l(l_{end}^* + 1), H - c - \mathbf{x}_r(r_{end}^* + 1)\}$.

From the selection of α^*, the two paths $((\mathbf{x}^* - \alpha\mathbf{1}) \vee \mathbf{x})$ and $\mathbf{x}$ start at the positions whose differences are no larger than 1, and end at the same positions. Since $\mathbf{x}^*$ is an optimizer of $\mathcal{U}$ and $(\mathbf{x}^* \wedge (\mathbf{x} + \alpha^*\mathbf{1}))$ is a feasible solution, based on Lemma 3, we have $\mathcal{E}((\mathbf{x}^* - \alpha^*\mathbf{1}) \vee \mathbf{x}) \leq \mathcal{E}(\mathbf{x})$. If the corresponding CCPP instance $\mathcal{C}'$ of $((\mathbf{x}^* - \alpha^*\mathbf{1}) \vee \mathbf{x})$ is left adjacent to $\mathcal{C}$ (whose optimizer is $\mathbf{x}$), then $\mathcal{C}'$ has the optimum no larger than that of $\mathcal{C}$. Otherwise, $((\mathbf{x}^* - \alpha^*\mathbf{1}) \vee \mathbf{x})$ must be a solution to $\mathcal{C}$. Since $\mathbf{x}$ is an optimizer to $\mathcal{C}$, based on Lemma 3, $((\mathbf{x} + \alpha^*\mathbf{1}) \wedge \mathbf{x}^*)$ must also be an optimizer to $\mathcal{U}$. Thus, we can apply the procedure mentioned above again until we get a solution better than or equal to $\mathbf{x}$ whose corresponding CCPP instance is left adjacent to $\mathcal{C}$. This proves the lemma. □

Based on Lemmas 2 and 4, we prove that the maximal optimizer $\check{\mathbf{x}}^*$ is always on the pathway of the path-end hopping process. In addition, Lemma 4 also shows that, if we are unable to find any solution to the left adjacent CCPP instances of $\mathcal{C}$ with a better or equal objective value compared with that of the optimizer $\mathbf{x}$ of $\mathcal{C}$, then $\mathbf{x}$ is an optimizer of the UCPP instance $\mathcal{U}$. If there is no left adjacent CCPP instance of $\mathbf{x}$, from Lemmas 2 and 4, $\mathbf{x}$ is certainly an optimizer of $\mathcal{U}$. Hence, we can conclude that the Algorithm 1 is correct.

Lemma 5. *The UCPP problem can be solved by computing $\mathcal{O}(n)$ associated CCPP instances.*

4 Solving the Constrained Couple Path Planning (CCPP) by Exploring L$^\natural$-Convexity

In this section, we present our $\mathcal{O}(n^2 \log H)$ time algorithm for solving the CCPP problem by using local searching and the proximity scaling technique. The key idea is to exploring the L$^\natural$-convexity and total unimodularity of the problem.

Local searching. For a given CCPP instance, the path-end parameters are fixed. Considering Eqn (1g), the nonlinear constraints (Eqns (1d,1e) can be replaced by the following linear constraints.

$$x_l(i+1)-x_l(i) \geq c \;\; i \in [l_{start}, l_{end}), \quad x_r(i+1)-x_r(i) \geq c \;\; i \in [r_{start}, r_{end}) \quad (3)$$

Thus, the domain $\mathbb{D}$ of the CCPP problem is convex. In fact, we can further verify that $\mathbb{D}$ is L$^\natural$-convex (i.e., $\forall \mathbf{x}, \mathbf{y} \in \mathbb{D}$ and $\alpha \in \mathbb{Z}^+$, $((\mathbf{x}-\alpha\mathbf{1})\vee\mathbf{y}), (\mathbf{x}\wedge(\mathbf{y}+\alpha\mathbf{1})) \in \mathbb{D}$).

Lemma 6. *The objective function of CCPP is L$^\natural$-convex.*

A good property of an L$^\natural$-convex function is that $\mathbf{x}^*$ is an optimizer iff for all $\mathbf{x}$ satisfies $\mathbf{x}^* - \mathbf{1} \preceq \mathbf{x} \preceq \mathbf{x}^* + \mathbf{1}$, $\mathcal{E}(\mathbf{x}^*) \leq \mathcal{E}(\mathbf{x})$ [3,12], which indicates the effectiveness of local searching. We adopt Kolmogorov and Shioura's framework [12]. For a CCPP solution $\mathbf{x}^\circ$, define the *upper strip* (resp., *lower strip*) of $\mathbf{x}^\circ$ as $\{\mathbf{x} \mid \mathbf{x}^\circ \preceq \mathbf{x} \preceq \mathbf{x}^\circ + \mathbf{1}\}$ (resp., $\{\mathbf{x} \mid \mathbf{x}^\circ \succeq \mathbf{x} \succeq \mathbf{x}^\circ - \mathbf{1}\}$). The key idea is, as follows. Starting from an initial solution $\mathbf{x}^\circ$, keep on searching the upper strip and the lower strip of $\mathbf{x}^\circ$ to find a better solution. We call the search on the upper strip or the lower strip as a local search step. Kolmogorov and Shioura proved that after $\mathcal{O}(H)$ local search steps, the algorithm can find the optimal solution to the CCPP problem.

For the local search step, if we use Kolmogorov and Shioura's minimum s-t cut based algorithm, it takes $\mathcal{O}(n^2 \log n)$ time. However, we can do much better by formulating the local search step as solving a shortest path in a layered directed acyclic graph (DAG) to achieve a linear $\mathcal{O}(n)$ time algorithm.

Suppose we want to search the upper strip of $\mathbf{x}^\circ$. For the left (resp., right) path $\mathbf{x}_l$ (resp., $\mathbf{x}_r$), we need to determine which of $\mathbf{x}_l^\circ(i)$ and $\mathbf{x}_l^\circ(i)+1$ (resp., $\mathbf{x}_r^\circ(i)$ and $\mathbf{x}_r^\circ(i)+1$) should be on the path for each column i. We construct a DAG $G = (V, E)$, as follows. For each column i, we map each possible value of the pair $(\mathbf{x}_l(i), \mathbf{x}_r(i))$ to a vertex in the layer i, denote by L_i, of the graph. Thus, each layer L_i has four vertices, denote by $L_i(0,0), L_i(0,1), L_i(1,0)$, and $L_i(1,1)$, representing $(\mathbf{x}_l^\circ(i), \mathbf{x}_r^\circ(i))$, $(\mathbf{x}_l^\circ(i), \mathbf{x}_r^\circ(i)+1)$, $(\mathbf{x}_l^\circ(i)+1, \mathbf{x}_r^\circ(i))$, and $(\mathbf{x}_l^\circ(i)+1, \mathbf{x}_r^\circ(i)+1)$, respectively. For each vertex v on Layer L_i, representing $(\mathbf{x}_l(i), \mathbf{x}_r(i))$, the weight of the vertex is assigned the error $|\mathbf{x}_l(i) - \mathbf{x}_r(i) - f(i)|$. Directed edges are put in from vertices in L_i to those in L_{i+1} $(i = 1, 2, \ldots, n-1)$ if all constraints are satisfied. Two dummy vertices s and t are introduced with s connecting to each vertex in L_1 and each vertex in L_n connecting to t. We thus finish the construction of G.

It is not difficult to see that a shortest s-t path in G actually defines a coupled path in the upper strip of $\mathbf{x}^\circ$ with minimum objective value. Note that G has

$4n+2$ vertices and no more than $16n+8$ edges. Thus, it takes $\mathcal{O}(n)$ time to compute a shortest s-t path in G by dynamic programming. The search on the lower strip of $\mathbf{x}^\circ$ can be done in a similar way. Hence, it takes a linear $\mathcal{O}(n)$ time to solving the local search step.

It is already known that the local searching algorithm makes $\mathcal{O}(H)$ calls to the local search step [12]. Thus our CCPP algorithm runs in $\mathcal{O}(nH)$ time, improving Chen *et al.*'s algorithm [5] by a factor of $\mathcal{O}(\Delta)$, which could be as large as H.

Lemma 7. *The CCPP problem can be solved in $\mathcal{O}(nH)$ time.*

However, our algorithm is still pseudo-polynomial due to the factor H. To further eliminate the factor H, we use the proximity scaling technique.

Proximity scaling. The proximity scaling technique was used to solving convex separable integer programming problems on linear constraints [9]. To solve an integer programming (IP) problem, a sequence of scaled versions of the integer programming problem (IP-s) needs to be solved.

Lemma 8 ([9]). *Given an integer programming problem with a convex separable objective function and a constraint matrix with absolute value of subdeterminants (determinant of any square submatrix) bounded by $\Delta_{\mathbf{A}}$, then (1) For each optimal solution $\hat{\mathbf{y}}$ for (IP-s), there exists an optimal solution $\mathbf{z}^*$ for (IP) such that $||s\hat{\mathbf{y}} - \mathbf{z}^*||_\infty \leq ns\Delta_{\mathbf{A}}$; (2) And for each optimal solution $\hat{\mathbf{z}}$ for (IP), there exists an optimal solution $\mathbf{y}^*$ for (IP-s) such that $||\hat{\mathbf{z}} - s\mathbf{y}^*||_\infty \leq ns\Delta_{\mathbf{A}}$.*

However, the objective function $\mathcal{E}(\mathbf{x}) = \sum_{i=1}^{n} |\mathbf{x}_l(i) - \mathbf{x}_r(i) - f(i)|$ is actual a non-separable convex function, which is generally NP-hard[10]. Fortunately, for our CCPP problem, we can perform a linear transform ($\alpha(i) = \mathbf{x}_r(i)$, $\delta(i) = \mathbf{x}_l(i) - \mathbf{x}_r(i)$) to make it become separable convex. Then our objective becomes $\mathcal{E}(\mathbf{x}) = \sum_{i=1}^{n} |\delta(i) - f(i)|$ and is separable convex.

Next, let us consider the subdeterminants of the constraint matrix $\bar{\mathbf{A}}$, which is now with respect to variable $\alpha(i)$ and $\delta(i)$ after performing the transformation.

Lemma 9. *The largest absolute value of the subdeterminants, $\Delta_{\bar{\mathbf{A}}}$, of the constraint matrix $\bar{\mathbf{A}}$ after the linear transformation is no larger than 1.*

Proof. Note that every entry in $\bar{\mathbf{A}}$ is ± 1, or 0. We can safely remove those constraints with a single variable such as $\alpha(i) \geq c$ since they do not affect the absolute value of the determinants, and we can also safely remove one of those two linearly dependent constraints such as $\alpha(i) + \delta(i) \geq c$ and $-\alpha(i) - \delta(i) \geq c - H$, since the removal does not affect the bounds of the subdeterminants. The resulting matrix is denoted by $\widetilde{\mathbf{A}}$, which has the same bound of absolute value of subdeterminants as $\bar{\mathbf{A}}$. Furthermore, $\widetilde{\mathbf{A}}$ consists of simple repetitions of a basic block (or with some rows omitted depends on the position i) since the constraint inequalities repeat for each position i.

Notice that each basic block of $\widetilde{\mathbf{A}}$ overlaps with only two basic blocks immediately before and after it; and each overlapping involves at most 2 columns. If a submatrix has determinant with absolute value larger than 1, it should include at least 2 columns corresponding to non-zero entries for a specified row. Moreover, no row in $\widetilde{\mathbf{A}}$ contains more than 4 non-zero entries. Thus, the largest absolute value of the subdeterminants of $\widetilde{\mathbf{A}}$ can be computed by examining all possible subdeterminants of only 3 successive basic blocks. By computing, the largest absolute value of the subdeterminants of the matrix consisting of 3 successive basic blocks is 1. This proves Lemma 9. □

After performing the transformation, we have an convex separable IP problem: $\mathcal{E}(\bar{\mathbf{x}}) = \sum_{i=1}^{n} |\delta(i) - f(i)|$ subject to $\bar{\mathbf{A}}\bar{\mathbf{x}} \geq \bar{\mathbf{b}}$, where $\bar{\mathbf{x}}$ is an integer vector consisting of all $\alpha(i)$ and $\delta(i)$. If we obtain an optimal solution $\mathbf{x}^s$ to this IP problem with the scaling factor of s (denoted by IP-s), i.e. with additional constraints that $\alpha(i)$ and $\delta(i)$ are multiples of s, based on Lemma 8, there exists an optimizer $\mathbf{x}$ to the original IP that is within a distance $2ns$ away from $\mathbf{x}^s$, since the number of variables is $2n$. That means in the next iteration, we only need to search the strip $\mathbb{S} = \{\mathbf{x} \mid \mathbf{x}^s - (2ns)\mathbf{1} \preceq \mathbf{x} \preceq \mathbf{x}^s + (2ns)\mathbf{1}\}$ for the optimizer of the original IP. Thus, by setting s initially to $\lceil \frac{H}{16n} \rceil$, and reducing both s and the total height of the strips by half at each iteration, we will reach $s = 1$ at get the optimizer in $\mathcal{O}(\log H)$ iterations. Note that each IP-s problem during the scaling process is a CCPP problem with $H = \mathcal{O}(n)$, which can be solved in $\mathcal{O}(n^2)$ time by Lemma 7. Thus, the running time of this proximity scaling algorithm is $\mathcal{O}(n^2 \log H)$.

Theorem 1. *Given an instance of the CCPP, an optimal solution can be obtained in* $\mathcal{O}(n^2 \log H)$ *time.*

Theorem 2. *Given an instance of the UCPP, an optimal solution can be obtained in* $\mathcal{O}(n^3 \log H)$ *time.*

5 Experiments and Discussion

We implemented both our algorithms and Chen *et al.*'s algorithm [5] for the purpose of comparison. We experimented on 20 synthesized data and 30 clinical data from the Department of Radiation Oncology, the University of Iowa.

Our algorithm produced solutions with the same quality as those produced by Chen *et al.*'s since both algorithms guarantee optimal solutions. Compared with Spirou and Chui's algorithm [15], which is a widely used dynamic leaf sequencing algorithm, the treatment time can be reduced by $10\% \sim 45\%$.

As to the average execution time, for typical clinical data with integer precision (a real number is rounded to its nearest integer) , our algorithm ran in less than 1 ms. Chen *et al.*'s algorithm took about 20 ms. The parameters H and Δ have significant impacts on Chen *et al.*'s algorithm, while not affect our algorithm too much. We tested both algorithms on the same data with a higher precision (i.e. scaled up by 10 before rounding). It took Chen *et al.*'s algorithm, about 1 second to find the optimal UCPP solution, while our algorithm still terminated within 1 ms. For even higher precision, Chen *et al.*'s algorithm sometimes failed due to the memory limit since its memory complexity is $\mathcal{O}(nH\Delta)$. Our algorithm worked well with much less memory requirement $\mathcal{O}(n)$.

References

1. Ahuja, R., Hamacher, H.: A network flow algorithm to minimize beam-on time for unconstrained multileaf collimator problems in cancer radiation therapy. Networks 45(1), 36–41 (2005)
2. Baatar, D., Hamacher, H., Ehrgott, M., Woeginger, G.: Decomposition of integer matrices and multileaf collimator sequencing. Discrete Applied Mathematics 152(1-3), 6–34 (2005)
3. Bioucas-Dias, J., Valadão, G.: Phase unwrapping via graph cuts. IEEE Transactions on Image processing 16(3), 698 (2007)
4. Bortfeld, T., Kahler, D., Waldron, T., Boyer, A.: X-ray field compensation with multileaf collimators. International journal of radiation oncology, biology, physics 28(3), 723–730 (1994)
5. Chen, D.Z., Luan, S., Wang, C.: Coupled path planning, region optimization, and applications in intensity-modulated radiation therapy. In: Halperin, D., Mehlhorn, K. (eds.) Esa 2008. LNCS, vol. 5193, pp. 271–283. Springer, Heidelberg (2008)
6. Chen, D., Hu, X., Luan, S., Naqvi, S., Wang, C., Yu, C., Fleischer, R.: Generalized geometric approaches for leaf sequencing problems in radiation therapy. International Journal of Computational Geometry and Applications 16(2-3), 175–204 (2006)
7. Converyt, D., Rosenbloom, M.: The generation of intensity-modulated fields for conformal I radiotherapy by dynamic collimation. Phys. Med. 37(6), 1359–1374 (1992)
8. Fowler, J., Welsh, J., Howard, S.: Loss of biological effect in prolonged fraction delivery. International journal of radiation oncology, biology, physics 59(1), 242–249 (2004)
9. Hochbaum, D.S.: An efficient algorithm for image segmentation, markov random fields and related problems. J. ACM 48(4), 686–701 (2001)
10. Hochbaum, D.: Complexity and algorithms for convex network optimization and other nonlinear problems. 4OR: A Quarterly Journal of Operations Research 3(3), 171–216 (2005)
11. Kamath, S., Sahni, S., Palta, J., Ranka, S.: Algorithms for optimal sequencing of dynamic multileaf collimators. Physics in Medicine and Biology 49(1), 33–54 (2004)
12. Kolmogorov, V., Shioura, A.: New Algorithms for the Dual of the Convex Cost Network Flow Problem with Application to Computer Vision. Mathematical Programming (2007) (submitted)
13. Murota, K.: Discrete Convex Analysis: Monographs on Discrete Mathematics and Applications 10. In: Society for Industrial and Applied Mathematics, Philadelphia, PA,USA (2003)
14. Siochi, R.: Minimizing static intensity modulation delivery time using an intensity solid paradigm. International journal of radiation oncology, biology, physics 43(3), 671–680 (1999)
15. Spirou, S., Chui, C.: Generation of arbitrary intensity profiles by dynamic jaws or multileaf collimators. Medical Physics 21, 1031 (1994)
16. Webb, S.: Intensity-modulated radiation therapy. Inst. of Physics Pub. Inc. (2001)
17. Yu, C.X., Luan, S., Wang, C., Chen, D.Z.: Single arc dose painting: An efficient method of precision radiation therapy. In: Provisional patent application. University of Maryland (2006)

Constant Time Approximation Scheme for Largest Well Predicted Subset*

Bin Fu[1] and Lusheng Wang[2]

[1] Department of Computer Science, University of Texas Pan American Edinburg, TX 78539, USA
binfu@cs.panam.edu

[2] Department of Computer Science, City University of Hong Kong, Hong Kong
lwang@cs.cityu.edu.hk

Abstract. The largest well predicted subset problem is formulated for comparison of two predicted 3D protein structures from the same sequence. Given two ordered point sets $A = \{a_1, \cdots, a_n\}$ and $B = \{b_1, b_2, \cdots b_n\}$ containing n points, and a threshold d, the *largest well predicted subset* problem is to find the rigid transformation T for a largest subset B_{opt} of B such that the distance between a_i and $T(b_i)$ is at most d for every b_i in B_{opt}. A meaningful prediction requires that the size of B_{opt} is at least αn for some constant α [8]. We use LWPS(A, B, d, α) to denote the largest well predicted subset problem with meaningful prediction. An $(1 + \delta_1, 1 - \delta_2)$-approximation for LWPS$(A, B, d, \alpha)$ is to find a transformation T to bring a subset $B' \subseteq B$ of size at least $(1 - \delta_2)|B_{opt}|$ such that for each $b_i \in B'$, the Euclidean distance between the two points distance$(a_i, T(b_i)) \leq (1 + \delta_1)d$. We develop a constant time $(1 + \delta_1, 1 - \delta_2)$-approximation algorithm for LWPS(A, B, d, α) for arbitrary positive constants δ_1 and δ_2.

1 Introduction

3D protein structure prediction is a very important problem in Structural bioinformatics. Many tools have been developed to predict 3D protein structures from their sequences. Evaluating the quality of those softwares in predicting the 3D protein structures is becoming more sand more important. In addition to the popular root mean square deviation (RMSD) by Arun et al [2], some other approaches have been proposed such as MaxSub by Siew et al [10], Global Distance Test, Local/Global Alignment by Zemla [11], and TAMScore by Zhang and Sholnick [12]. A comprehensive survey about those measurements was given by Lancia and Istrail [6]. One of the interesting problems along this line is to find a largest well-predicted subset– a maximum subset of the model which matches the native structure. This problem was first formulated by Siew et al [10]. A protein backbone structure can be characterized by an ordered point

* This research is supported by NSF Career Award 0845376, and Research Grants Council of the Hong Kong Special Administrative Region, China [CityU 121207].

M.T. Thai and S. Sahni (Eds.): COCOON 2010, LNCS 6196, pp. 429–438, 2010.

set $A = \{a_1, \cdots, a_n\}$. A predicted model $B = \{b_1, b_2, \cdots b_n\}$ of the same protein sequence also consists of an ordered set of n points. Given two ordered point sets $A = \{a_1, \cdots, a_n\}$ and $B = \{b_1, b_2, \cdots b_n\}$ containing n points, and a threshold d, the *largest well predicted subset* problem is to find the rigid transformation T for a largest subset B_{opt} of B such that the distance between a_i and $T(b_i)$ is at most d for every b_i in B_{opt}. To evaluate the tools for 3D protein structure prediction, it expects that the predicted 3D structures should have sufficient similarity to the real structure. A meaningful prediction requires that the size of B_{opt} is at least αn for some constant α [8]. We use LWPS(A, B, d, α) to denote the largest well predicted subset problem with meaningful prediction. A $(1 + \delta)$-distance approximation for the largest well-predicted subset problem is to find a transformation T to bring a subset $B' \subseteq B$ of size at least $|B_{opt}|$ such that for each $b_i \in B'$, distance$(a_i, T(b_i)) \leq (1 + \delta)d$, where $\delta > 0$ is a given constant. An $(1 + \delta_1, 1 - \delta_2)$ approximation for the largest well-predicted subset problem is to find a transformation T to bring a subset $B' \subseteq B$ of size at least $(1 - \delta_2)|B_{opt}|$ such that for each $b_i \in B'$, distance$(a_i, T(b_i)) \leq (1 + \delta_1)d$, where $\delta_1 > 0$ and $\delta_2 > 0$ are some given constants. The largest well-predicted subset problem can be solved in $O(n^7)$ time [1,3]. A $(1 + \delta)$-distance approximation algorithm was recently developed by Li et al [8] and the running time was $O(n^3 \log n/\delta^5)$. Li et al [8] also developed an $O(n(\log n)^2/\delta^5)$ time randomized algorithm for the largest well predicted subset problem with meaningful prediction. The geometric pattern pattern matching problems were studied in the computational geometric (e.g. [4,5]), but there are very few of them running sublinear time.

We develop a randomized constant time approximation algorithm to give a $(1 + \delta_1, 1 - \delta_2)$-approximation for the largest well-predicted subset problem. We also study a closely related problem, the *bottleneck distance* problem, where we are given two ordered point sets $A = \{a_1, \cdots, a_n\}$ and $B = \{b_1, b_2, \cdots b_n\}$ containing n points and the problem is to find the smallest d_{opt} such that there exists a rigid transformation T with $distance(a_i, T(b_i)) \leq d_{opt}$ for every point $b_i \in B$. We use $BD(A, B)$ to denote the bottleneck distance problem. A $(1 + \delta)$-approximation for the bottleneck distance problem is to find a transformation T, such that for each $b_i \in B$, distance$(a_i, T(b_i)) \leq (1 + \delta)d_{opt}$, where δ is a constant. For an arbitrary constant δ, we obtain a linear $O(n/\delta^6)$ time $(1 + \delta)$-algorithm for the bottleneck distance problem. The existing algorithms for both problems require super-linear time [8].

2 Preliminaries

Given a set of indices $M_0 \subseteq \{1, 2, \ldots, n\}$ and an ordered point set $P = \{p_1, \cdots, p_n\}$ in three dimensional Euclidean space R^3. M_{opt} is the set of indices for the points in B_{opt}.

For a point p and a straight line L in R^3, distance(p, L) is the perpendicular distance from p to L. We use $L(p_1, p_2)$ to represent the straight line through two points p_1 and p_2 in R^3. Denote $P(p_1, p_2, p_3)$ to be the plane through points p_1 and p_2 and p_3 in R^3. It is unique when p_1, p_2 and p_3 are not in the same line. For two straight lines L_1 and L_2, the distance between L_1 and L_2 is distance$(L_1, L_2) = \min_{p \in L_1}$ distance(p, L_2)).

For a point p and a real $r > 0$, define ball$(p, r) = \{q : \text{distance}(p, q) \leq r\}$ to be all the points in the ball centered at p with radius r, and define sphere$(p, r) = \{q : \text{distance}(p, q) = r\}$ to be the set of all points on the sphere centered at p with radius r. If L is a line, define circle(p, r, L) to be the circle of the intersection of sphere $sphere(p, r)$ and the plane P such that p is on P and L is perpendicular to P.

Theorem 1 ([9]). *Let $X_1, \cdots, X_n$ be n independent random 0-1 variables, where X_i takes 1 with probability p_i. Let $X = \sum_{i=1}^{n} X_i$, and $\mu = E[X]$. Then for any $\delta > 0$, 1) $\Pr(X < (1-\delta)\mu) < e^{-\frac{1}{2}\mu\delta^2}$, and 2) $\Pr(X > (1+\delta)\mu) < \left[\frac{e^{\delta}}{(1+\delta)^{(1+\delta)}}\right]^{\mu}$.*

Corollary 1 ([7]). *Let $X_1, \cdots, X_n$ be n independent random 0-1 variables and $X = \sum_{i=1}^{n} X_i$. Then 1) If X_i takes 1 with probability at most p, then for any $\frac{1}{3} > \epsilon > 0$, $\Pr(X > pn + \epsilon n) < e^{-\frac{1}{3}n\epsilon^2}$, and 2) If X_i takes 1 with probability at least p, then for any $\frac{1}{3} > \epsilon > 0$, $\Pr(X < pn - \epsilon n) < e^{-\frac{1}{2}n\epsilon^2}$.*

3 Randomized Approximation Scheme for LWPS

In this section, we present a randomized $(1+\delta_1, 1-\delta_2)$-approximation algorithm for $LWPS(A, B, d, \alpha)$. The basic idea of our algorithm is to find three points in the 3D structure B such that $\{b_{i_1}, b_{i_2}, b_{i_3}\} \subseteq B_{opt}$ and use the three points to construct a rigid transformation. In order to guarantee the approximation ratio, we require that the three points are far away. For a set of ordered points $B = \{b_1, \cdots, b_n\}$, a *frame* of B is a triple of three points $B' = (b_{i_1}, b_{i_2}, b_{i_3})$ such that $b_{i_1}, b_{i_2}, b_{i_3}$ are in B, distance$(b_{i_1}, b_{i_2}) = \max_{j=1,\cdots,n}\{\text{distance}(b_{i_1}, b_j)\}$ and distance$(b_{i_3}, L(b_{i_1}, b_{i_2})) = \max_{j=1,\cdots,n}\{\text{distance}(b_j, L(b_{i_1}, b_{i_2}))\}$. If we try all possible B_{i_2} and B_{i_3} to get a frame of B, it takes $O(n^2)$ time. Here we use a weaker condition. We want to find $(b_{i_1}, b_{i_2}, b_{i_3})$ from B_{opt} such that distance(b_{i_1}, b_{i_2}) is not less than distance(b_{i_1}, b_j) for at least $(1-\delta_2)|B_{opt}|$ b_j's in B_{opt} and distance$(b_{i_3}, L(b_{i_1}, b_{i_2}))$ is not less than distance$(b_k, L(b_{i_1}, b_{i_2}))$ for at least $(1-\delta_2)|B_{opt}|$ b_k's in B_{opt}. We use a random sampling method to get the three desired points in B. We also use a random sampling approach to estimate the size of B_{opt}.

Rigid transformations based on three points: We can obtain a rigid transformation based on the three points b_{i_1}, b_{i_2} and b_{i_3} in B in three steps. Let $P = (p_1, p_2, p_3)$ be a triple of three points in a structure and $Q = (q_1, q_2, q_3)$ be a triple three points in 3D space. Assume that we want a rigid transformation $T(P, Q)$ that moves the whole structure such that p_1, p_2 and p_3 in the structure are *near* the three points q_1, q_2 and q_3 in 3D space, respectively. The transformation $T(P, Q)$ is a composition of three rigid transformations. The first is a translation that moves p_1 to q_1. The other two are rotations around some fixed axes. The formal definition is as follows:

1. For points q_1 and q_2, define $\text{tran}(q_1, q_2)$ to be the translation that moves point q_1 to point q_2.
2. For three points q_1, q_2, and q_3, let L_O be a line going through q_1 and perpendicular to the plane $P(q_1, q_2, q_3)$. We define $\text{rot1}(q_1, q_2, q_3)$ to be the transformation that uses L_O as the axis to rotate such that line $L(q_1, q_3)$ becomes identical to line $L(q_1, q_2)$. Note that after the transformation, the two points q_3 and q_2 are not necessarily identical since $distance(q_1, q_3)$ is not necessarily equal to $distance(q_1, q_2)$.
3. For three points q_1, q_2, q_3, and q_4, define $\text{rot2}(q_1, q_2, q_3, q_4)$ to be the transformation that uses $L(q_1, q_2)$ as the axis to rotate such that plane $P(q_1, q_2, q_4)$ becomes identical to plane $P(q_1, q_2, q_3)$. Again, after the transformation, the two points q_3 and q_4 may not be identical.
4. For two triples of points $P = (p_1, p_2, p_3)$ and $Q = (q_1, q_2, q_3)$, define $\text{move}(P, Q) = T_3 \circ T_2 \circ T_1$, where $T_1 = \text{tran}(p_1, q_1)$, $T_2 = \text{rot1}(p_1, p_2, T_1(q_2))$, $T_3 = \text{rot2}(p_1, p_2, p_3, T_2(T_1(q_3)))$, and $\circ$ is the composition operator of two transformations such that $(T_2 \circ T_1)(p) = T_2(T_1(p))$.

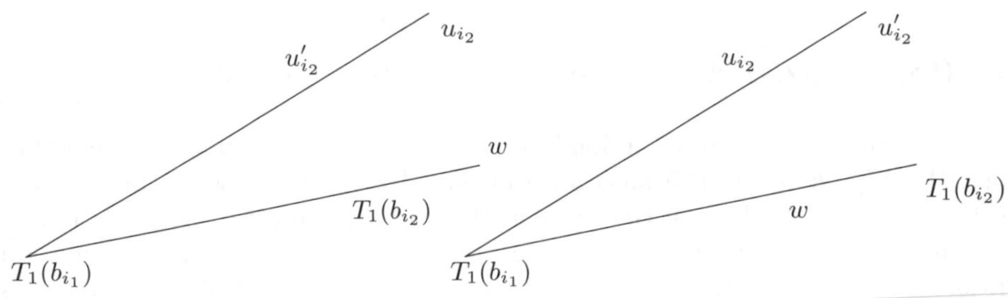

Fig. 1. Left: $\text{distance}(T_1(b_{i_1}), T_1(b_{i_2})) \leq \text{distance}(T_1(b_{i_1}), u_{i_2})$. Right: $\text{distance}(T_1(b_{i_1}), T_1(b_{i_2})) > \text{distance}(T_1(b_{i_1}), u_{i_2})$.

Lemma 1. *Let $B = \{b_1, \cdots, b_n\}$ be a set of points. Let $(b_{i_1}, b_{i_2}, b_{i_3})$ be a frame of B. Let u_{i_1}, u_{i_2} and u_{i_3} be three points such that $\text{distance}(u_{i_1}, b_{i_1}) = d_1$, $\text{distance}(u_{i_2}, T_1(b_{i_2})) = d_2$, and $\text{distance}(u_{i_3}, T_2(T_1(b_{i_3}))) = d_3$, where $T_1 = \text{tran}(u_{i_1}, b_{i_1})$ and $T_2 = \text{rot1}(u_{i_1}, u_{i_2}, T_1(b_{i_2}))$, and $T_3 = \text{rot2}(u_{i_1}, u_{i_2}, u_{i_3}, T_2(T_1(b_{i_3})))$. Let $T' = T_3 \circ T_2 \circ T_1$. Then 1) Point b_{i_1} moves at most d_1 distance by T_1, point $T_1(b_{i_2})$ moves at most $2d_2$ distance by T_2, and point $T_2(T_1(b_{i_3}))$ moves at most $2d_3$ distance by T_3; 2) Each point in B moves at most d_1 distance by T_1, each point in $T_1(B)$ moves at most $2d_2$ distance by T_2, and each point in $T_2(T_1(B))$ moves at most $2d_3$ distance by T_3; and 3) Each point in B is moved at most $d_1 + 2d_2 + 2d_3$ distance by T'.*

Proof. Statement i. We discuss the three transformations T_1, T_2, and T_3 one by one.

Transformation T_1**:** Since $T_1(b_{i_1}) = u_{i_1}$, the transformation moves b_{i_1} to u_{i_1}. Therefore, every point in B is moved the same distance d_1 via the translation T_1.

Transformation T_2: Select a point w alone the line $L(T_1(b_{i_1}), T_1(b_{i_2}))$ such that distance$(T_1(b_{i_1}), w)$ = distance$(T_1(b_{i_1}), u_{i_2})$, and w and $T_1(b_{i_2})$ are in the same side of the line $L(T_1(b_{i_1}), T_1(b_{i_2}))$, which is partitioned into two sides by point $T_1(b_{i_1})$. See Figure 1 for examples. There are two cases that are discussed below:

Case 1: distance$(T_1(b_{i_1}), T_1(b_{i_2})) \leq$ distance$(T_1(b_{i_1}), u_{i_2})$ (see the left of Figure 1).

By triangular inequality, we have the inequalities below distance$(T_1(b_{i_1}), T_1(b_{i_2}))$ + distance$(T_1(b_{i_2}), u_{i_2}) \geq$ distance$(T_1(b_{i_1}), u_{i_2})$ = distance$(T_1(b_{i_1}), w)$ = distance$(T_1(b_{i_1}), T_1(b_{i_2}))$+distance$(T_1(b_{i_2}), w)$. Therefore, distance$(T_1(b_{i_2}), w) \leq$ distance$(T_1(b_{i_2}), u_{i_2}) \leq d_2$. By triangular inequality, we have distance$(w, u_{i_2}) \leq$ distance$(w, T_1(b_{i_2}))$+distance$(T_1(b_{i_2}), u_{i_2}) \leq 2d_2$. We transform line $L(T_1(b_{i_1}), w)$ to line $L(T_1(b_{i_1}), u_{i_2})$ by rotating around the center $T_1(b_{i_1})$ so that w is moved to u_{i_2}. Therefore, w moves at most $2d_2$ distance, and $T_1(b_{i_2})$ also moves at most $2d_2$ distance since distance$(T_1(b_{i_1}), T_1(b_{i_2})) \leq$ distance$(T_1(b_{i_1}), u_{i_2})$ = distance$(T_1(b_{i_1}), w)$.

Case 2: distance$(T_1(b_{i_1}), T_1(b_{i_2})) >$ distance$(T_1(b_{i_1}), u_{i_2})$ (see the right of Figure 1).

Let u'_{i_2} be the point in the line $L(T_1(b_{i_1}), T_1(b_{i_2}))$ with distance$(T_1(b_{i_1}), u'_{i_2})$ = distance$(T_1(b_{i_1}), T_1(b_{i_2}))$. It is easy to see that distance$(w, T_1(b_{i_2}))$ = distance$(u_{i_2}, u'_{i_2}) \leq$ distance$(T_1(b_{i_2}), u_{i_2}) \leq d_2$ because distance$(T_1(b_{i_1}), u'_{i_2})$ = distance$(T_1(b_{i_1}), T_1(b_{i_2}))$ and distance$(T_1(b_{i_1}), u_{i_2})$ = distance$(T_1(b_{i_1}), w)$. Therefore, distance$(T_1(b_{i_2}), u'_{i_2}) \leq$ distance(u_{i_2}, u'_{i_2}) + distance$(u_{i_2}, T_1(b_{i_2})) \leq 2d_2$. Therefore, $T_1(b_{i_2})$ moves at most $2d_2$ distance by T_2.

Transformation T_3: For each point $q \in R^3$, let L_q be the straight line that is parallel to line $L(T_2(T_1(b_{i_1})), T_2(T_1(b_{i_2})))$ and through point q. Let L'_q be the straight line that is perpendicular to $L(T_2(T_1(b_{i_1})), T_2(T_1(b_{i_2})))$, goes through q, and also intersects line $L(T_2(T_1(b_{i_1})), T_2(T_1(b_{i_2})))$.

Select a point w' alone the line $L'_{T_2(T_1(b_{i_3}))}$ with distance$(w', L(T_2(T_1(b_{i_1})), T_2(T_1(b_{i_2}))))$ = distance$(u_{i_3}, L(T_2(T_1(b_{i_1})), T_2(T_1(b_{i_2}))))$. We rotate the around axis line $L(T_2(T_1(b_{i_1})), T_2(T_1(b_{i_2})))$ to move $L_{w'}$ to $L_{u_{i_3}}$. See Figure 2 for examples. There are two cases that are discussed below:

Case 1: distance$(T_2(T_1(b_{i_3})), L(T_2(T_1(b_{i_1})), T_2(T_1(b_{i_2})))) \leq$ distance$(u_{i_3}, L(T_2(T_1(b_{i_1})), T_2(T_1(b_{i_2}))))$ (see the left of Figure 2). We have distance$(u_{i_3}, T_2(T_1(b_{i_3}))))$ + distance$(T_2(T_1(b_{i_3}))), L(T_2(T_1(b_{i_1})), T_2(T_1(b_{i_2})))) \geq$ distance$(u_{i_3}, L(T_2(T_1(b_{i_1})), T_2(T_1(b_{i_2}))))$ = distance$(w', L(T_2(T_1(b_{i_1})), T_2(T_1(b_{i_2}))))$ = distance$(w', T_2(T_1(b_{i_3}))))$ + distance$(T_2(T_1(b_{i_3}))), L(T_2(T_1(b_{i_1})), T_2(T_1(b_{i_2}))))$. Therefore, distance$(w', T_2(T_1(b_{i_3})))) \leq$ distance$(u_{i_3}, T_2(T_1(b_{i_3})))) \leq d_3$.

The distance between the two lines $L_{w'}$ and $L_{u_{i_3}}$ is at most distance$(L_{w'}, L_{u_{i_3}}) \leq$ distance$(w', u_{i_3}) \leq$ distance$(w', T_2(T_1(b_{i_3})))$ + distance$(T_2(T_1(b_{i_3})), u_{i_3}) \leq d_3 + d_3 = 2d_3$. Therefore, w' moves at most $2d_3$, and $T_3(T_1(b_{i_2}))$ moves at most the distance that w' moves. Therefore, $T_2(T_1(b_{i_3}))$ moves at most $2d_3$ via T_3 which has $L(T_2(T_1(b_{i_1})), T_2(T_1(b_{i_2}))$ as axis of rotation.

Case 2: distance$(T_2(T_1(b_{i_3})), L(T_2(T_1(b_{i_1})), T_2(T_1(b_{i_2})))) >$ distance$(u_{i_3}, L(T_2(T_1(b_{i_1})), T_2(T_1(b_{i_2}))))$ (see the right of Figure 2).

We show that $T_2(T_1(b_{i_3}))$ moves at most $2d_3$ via T_3 which has $L(T_2(T_1(b_{i_1})), T_2(T_1(b_{i_2}))$ as axis of the rotation. Let u'_{i3} be the point in the line $L'_{u_{i3}}$ with distance$(u'_{i_3}, L(T_2(T_1(b_{i_1})), T_2(T_1(b_{i_2})))) =$ distance$(T_2(T_1(b_{i_3})), L(T_2(T_1(b_{i_1})), T_2(T_1(b_{i_2}))))$.

$$\begin{aligned}
&\text{distance}(T_2(T_1(b_{i_3})), u_{i_3}) + \text{distance}(u_{i_3}, L(T_2(T_1(b_{i_1})), T_2(T_1(b_{i_2})))\\
&\geq \text{distance}(T_2(T_1(b_{i_3})), L(T_2(T_1(b_{i_1})), T_2(T_1(b_{i_2}))))\\
&= \text{distance}(T_2(T_1(b_{i_3})), w') + \text{distance}(w', L(T_2(T_1(b_{i_1})), T_2(T_1(b_{i_2}))))\\
&= \text{distance}(T_2(T_1(b_{i_3}))), w') + \text{distance}(u_{i_3}, L(T_2(T_1(b_{i_1})), T_2(T_1(b_{i_2})))).
\end{aligned}$$

Therefore, distance$(w', T_2(T_1(b_{i_3})))) \leq$ distance$(u_{i_3}, T_2(T_1(b_{i_3})))) \leq d_3$. Since distance$(w', T_2(T_1(b_{i_3})))) =$ distance(u'_{i_3}, u_{i_3}), we have distance$(u'_{i_3}, u_{i_3}) \leq d_3$.

Therefore, distance$(L_{T_2(T_1(b_{i_3}))}, L_{u'_3}) \leq$ distance$(T_2(T_1(b_{i_3})), u'_{i_3}) \leq$ distance$(T_2(T_1(b_{i_3})), u_{i_3}) +$ distance$(u_{i_3}, u'_{i_3}) \leq 2d_3$. Therefore, $T_2(T_1(b_{i_3}))$ moves at most $2d_3$ distance via rotation T_3 which rotates around the axis line $L(T_2(T_1(b_{i_1})), T_2(T_1(b_{i_2})))$ and moves $L_{T_2(T_1(b_{i_3}))}$ to $L_{u_{i_3}}$.

Statement ii. The transformation T_1 is a translation. Every point moves the same distance d_1 as b_{i_1} does. The rotation T_2 has $T_1(b_{i_1})$ as center. Every point in $T_1(B)$ moves at most that of $T_1(b_{i_2})$ since distance$(T_1(b_j), T_1(b_{i_2}) \leq$ distance$(T_1(b_{i_1}), T_1(b_{i_2}))$ for every point $b_j \in T_1(B)$. The rotation T_3 has $L(T_2(T_1(b_{i_1})), T_2(T_1(b_{i_2}))$ as axis. Every point in $T_2(T_1(B))$ moves at most that of $T_2(T_1(b_{i_3}))$ since distance$(T_2(T_1(b_j)), L(T_2(T_1(b_{i_1})), T_2(T_1(b_{i_2}))) \leq$ distance$(T_2(T_1(b_{i_3})), L(T_2(T_1(b_{i_1})), T_2(T_1(b_{i_2})))$ for every point $b_j \in B$).

Statement iii. By statement ii, the three phases with translation T_1, rotation T_2, and rotation T_3 make every point in B move at most $d_1 + 2d_2 + 2d_3$ distance in total. □

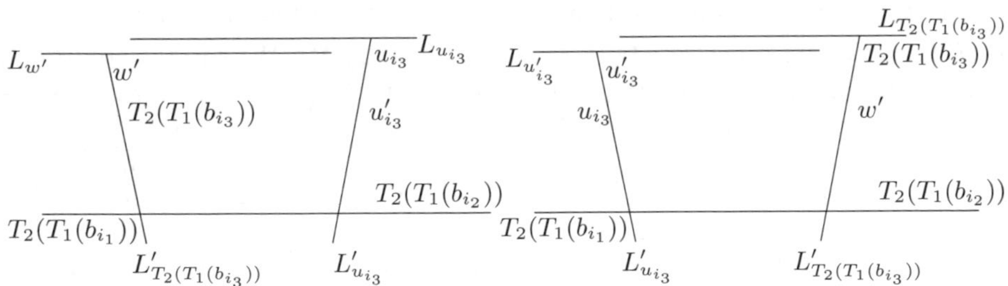

Fig. 2. Left: distance$(T_2(T_1(b_{i_3}))), L(T_2(T_1(b_{i_1})), T_2(T_1(b_{i_2})))) \leq$ distance$(u_{i_3}, L(T_2(T_1(b_{i_1})), T_2(T_1(b_{i_2}))))$. Right: distance$(T_2(T_1(b_{i_3}))), L(T_2(T_1(b_{i_1})), T_2(T_1(b_{i_2})))) >$ distance$(u_{i_3}, L(T_2(T_1(b_{i_1})), T_2(T_1(b_{i_2}))))$.

Definition 1. *For a positive real number t, R^3 can be partitioned into t-grids of side length t, and each (it, jt, kt) is called a t-grid point, where i, j, k are arbitrary integers (t is the size of the grid closely related to the approximation ratio δ_1).*

Definition 2. *For a set of three points $Q = \{q_1, q_2, q_3\}$ and $t > 0$, $U = (u_1, u_2, u_3)$ is a valid t-grid triple for Q if u_1, u_2 and u_3 are t-grid points and there exists a rigid transformation T such that $T(q_i)$ and u_i are in the same cube of the t-grid for $i = 1, 2, 3$.*

Lemma 2. *Assume that $A = \{p_1, \cdots, p_n\}$ and $B = \{b_1, \cdots, b_n\}$ are two ordered point sets and there exists a rigid transformation T with* $\text{distance}(p_i, T(b_i)) \le d_0$ *for $i = 1, \cdots, n$. Let $B' = (b_{i_1}, b_{i_2}, b_{i_3})$ be a frame of B, and ϵ be a real number in $(0, 1)$. Assume that real number d satisfies $d_0 \le d \le cd_0$ for a constant c. Then 1) There exists a valid ϵd-grid triple $U = (u_{i_1}, u_{i_2}, u_{i_3})$ for B' with $u_{i_j} \in \text{ball}(p_{i_j}, d + 2\sqrt{3}\epsilon d)$ for $j = 1, 2, 3$ such that* $\text{distance}(p_i, T'(b_i)) \le d_0 + c_0\epsilon d$ *for $i = 1, \cdots, n$, where $c_0 = \frac{5\sqrt{3}}{2}$ and $T' = \text{move}((u_{i_1}, u_{i_2}, u_{i_3}), B')$; and 2) For any constant $c_1 > 0$, there are at most $\frac{c_2}{\epsilon^6}$ valid ϵd-grid triples $(u_{i_1}, u_{i_2}, u_{i_3})$ for B' with $u_{i_j} \in \text{ball}(p_{i_j}, d + c_1\epsilon d)$ for $j = 1, 2, 3$, where c_2 is a constant for a fixed c_1.*

Proof. Statement 1. Assume that d_0 is the distance threshold and T is the rigid body transformation such that $\text{distance}(p_i, T(b_i)) \le d_0$ for $i = 1, \cdots, n$. We follow Lemma 1, and let $U = \{u_{i_1}, u_{i_2}, u_{i_3}\}$, where u_{i_1} is the ϵd-grid point closest to $T(b_{i_1})$, u_{i_2} is the ϵd-grid point closest to $T_1(T(b_{i_2}))$, u_{i_3} be the ϵd-grid point closest to $T_2(T_1(T(b_{i_3})))$, and T_1, T_2, T_3 are defined as those in Lemma 1. We have that $d_1, d_2, d_3 \le \frac{\sqrt{3}}{2}\epsilon d$, where d_1 is the distance between $T(b_{i_1})$ and u_{i_1}, d_2 is the distance between $T_1(T(b_{i_2}))$ and u_{i_2}, and d_3 is the distance between $T_2(T_1(T(b_{i_3})))$ and u_{i_3}.

By Lemma 1, the transformation $T' = \text{move}(U, B')$ moves every point in B by at most $d_1 + 2d_2 + 2d_3 \le \frac{5\sqrt{3}}{2}\epsilon d = c_0\epsilon d$ distance. Therefore, $\text{distance}(p_i, T'(b_i)) \le d_0 + c_0\epsilon d$ for $i = 1, \cdots, n$.

Statement 2. Let $r_1 = d + c_1\epsilon d$. We first fix the position of u_{i_1} that is in the same ϵd-grid as $T_1(b_{i_1})$. Since u_{i_1} is at an ϵd-grid point and there are $O((\frac{1}{\epsilon})^3)$ ϵd-grids in $\text{ball}(p_{i_1}, r_1)$, there are $O((\frac{1}{\epsilon})^3)$ possible positions for u_{i_1}.

The number of ϵd-grids is $O((\frac{1}{\epsilon})^2)$ touching the intersection of $\text{sphere}(T_1(b_{i_1}), \text{distance}(b_{i_1}, b_{i_2}))$ and $\text{ball}(p_{i_2}, r_1)$. Therefore, there are $O((\frac{1}{\epsilon})^2)$ possible positions for u_{i_2}.

For a point w, let L_w be the line through point w, perpendicular to $L(u_{i_1}, u_{i_2})$, and intersecting with $L(u_{i_1}, u_{i_2})$. Let v_{i_3} be the intersection of $L_{T_2(T_1(b_{i_3}))}$ and $L(u_{i_1}, u_{i_2})$.

The transformation T_3 rotates around axis line $L(u_{i_1}, u_{i_2})$. We just check the number of ϵh-grid points nearby $\text{ball}(p_{i_3}, r_1)$ under this kind of rotations. Since there are $O((\frac{1}{\epsilon}))$ cubes of size $(\epsilon d)^3$ in the intersection of $\text{circle}(v_{i_3}, \text{distance}(T_2(T_1(b_{i_3})), L(u_{i_1}, u_{i_2})), L(u_{i_1}, u_{i_2}))$ and $\text{ball}(p_{i_3}, r_1)$. So, the number of positions for u_{i_3} is $O(\frac{1}{\epsilon})$. Therefore, the total number of cases of $(u_{i_1}, u_{i_2}, u_{i_3})$ is $O(\frac{1}{\epsilon^3} \cdot \frac{1}{\epsilon^2} \cdot \frac{1}{\epsilon}) = O(\frac{1}{\epsilon^6})$. □

ApproximationSchemeLWPS(A, B, d, α)
Input: an ordered set of points $A = \{a_1, \cdots, a_n\}$, an ordered set of points $B = \{b_1, \cdots, b_n\}$, and a distance threshold $d > 0$;

Let δ be selected $\frac{\delta_2}{100}$, and $\epsilon = \frac{\delta_1}{c_0}$ (c_0 is defined in Lemma 2).
Let $k_1 = \lceil \frac{h}{\alpha} \rceil$, $k_2 = \lceil \frac{h}{\alpha\delta} \rceil$, and $k_3 = \lceil \frac{h}{\alpha\delta} \rceil$.
Let $m = \left\lceil \frac{300(h+\ln(\frac{8c_2h^3}{\delta^2\alpha^3\epsilon^6}))}{(\delta_2\alpha)^2} \right\rceil$, where c_2 is a constant in statement ii of Lemma 2.
Select G_1 to be a list of k_1 random elements from $\{1, 2, \cdots, n\}$.
Select G_2 to be a list of k_2 random elements from $\{1, 2, \cdots, n\}$.
Select G_3 to be a list of k_3 random elements from $\{1, 2, \cdots, n\}$.
For each (i_1, i_2, i_3) with i_j from G_j for $j = 1, 2, 3$.
  For each $(u_{i_1}, u_{i_2}, u_{i_3})$ with each ϵd-grid point u_{i_j} in ball$(p_{i_j}, d + 2\sqrt{3}\epsilon d)$)
    Let $T' = $ move$((u_{i_1}, u_{i_2}, u_{i_3}), (b_{i_1}, b_{i_2}, b_{i_3}))$.
    Select M to be a list of m random elements from $\{1, 2, \cdots, n\}$.
    Computer $M_{T'} = \{i$ is from M : distance$(a_i, T'(b_i)) \le (1+\delta)d\}$.
Output $(T', M_{T'}, \frac{|M_{T'}|}{m})$ such that $|M_{T'}|$ is the largest.
End of ApproximationSchemeLWPS

Theorem 2. *Assume δ_1 and δ_2 are parameters in $(0,1)$ and h is an arbitrary constant at least 1. Then with probability at least $1 - 5e^{-h}$, the algorithm ApproximationSchemeLWPS(A, B, d, α) runs in $O(\frac{h^3(h+\ln\max(\frac{1}{\alpha}, \frac{1}{\delta_1}, \frac{1}{\delta_2}))}{\alpha^5\delta_1^6\delta_2^4})$ time and outputs an $(1+\delta_1, 1-\delta_2)$-approximation for a* LWPS(A, B, d, α) *input.*

Proof. In the analysis below, we consider each list G_i to be an ordered multiplicative set of indices in $\{1, 2, \cdots, n\}$.

Claim 1. With probability at most $(1-\alpha)^{k_1}$, G_1 and M_{opt} have no common element.

Proof. Assume that M_{opt} is an optimal solution. We have $|M_{opt}| \ge \alpha n$. Since G_1 has k_1 random indices, the probability is at most $(1-\alpha)^{k_1}$ that none of the indices in G_1 is in M_{opt}. □

Claim 2. Assume that i is an index in M_{opt}. Then with probability at most $(1-\alpha\delta)^{k_2}$, G_2 does not contain an index i_2 that satisfies $i_2 \in M_{opt}$ and $|\{k : k \in M_{opt}$ *and* distance$(b_i, b_k) \le$ distance$(b_i, b_{i_2})\}| \ge (1-\delta)|M_{opt}|$.

Proof. Assume that M_{opt} is an optimal solution. We have $|M_{opt}| \ge \alpha n$. There are at least $\alpha\delta n$ indices i_2 that satisfy the condition $|\{k : k \in M_{opt}$ *and* distance$(b_i, b_k$ $) \le$ distance$(b_i, b_{i_2})\}| \le (1-\delta)|M_{opt}|$. Since set G_2 has k_2 random indices, the probability is at most $(1-\alpha\delta)^{k_2}$ that each index i_2 in G_2 does not satisfy the condition $|\{k : k \in M_{opt}$ *and* distance$(b_i, b_k) \le$ distance$(b_i, b_{i_2})\}| \ge (1-\delta)|M_{opt}|$. □

Claim 3. Assume that L is a line in R^3. With probability at most $(1-\alpha\delta)^{k_3}$ that G_3 does not contain i_3 that satisfies $i_3 \in M_{opt}$ and $|\{k : k \in M_{opt}$ *and* distance$(b_k, L) \le$ distance$(b_{i_3}, L)\}| \ge (1-\delta)|M_{opt}|$.

Proof. Assume that M_{opt} is an optimal solution. We have $|M_{opt}| \ge \alpha n$. There are at least $\alpha\delta n$ indices i_3 that satisfy the condition $|\{k : k \in M_{opt}$ *and* distance$(b_k,$ $L) \le$ distance$(b_{i_3}, L)\}| \ge (1-\delta)|M_{opt}|$.

Set G_3 has k_3 random indices, the probability is at most $(1-\alpha\delta)^{k_3}$ that all the indices i_3 in G_3 does not satisfy the condition $|\{k : k \in M_{opt}$ *and* distance$(b_k, L) \le$ distance$(b_{i_3}, L)\}| \ge (1-\delta)|M_{opt}|$. □

Let L be the line $L(b_{i_1}, b_{i_2})$. By Claims 1, 2, and 3, with probability at most $(1-\alpha)^{k_1} + (1-\alpha\delta)^{k_2} + (1-\alpha\delta)^{k_0} \le e^{-\alpha k_1} + e^{-\alpha\delta k_2} + e^{-\alpha\delta k_3} \le 3e^{-h}$, we get $G_1 \times G_2 \times G_3$ that does not contain a tuple (i_1, i_2, i_3) such that 1) i_1, i_2, i_3 are all in M_{opt}, 2) $|\{k : k \in M_{opt}$ *and* distance$(b_{i_1}, b_k) \le$ distance$(b_{i_1}, b_{i_2}\}| \ge (1-\delta)|M_{opt}|$, and 3) $|\{k : k \in M_{opt}$ *and* distance$(b_k, L) \le$ distance$(b_{i_3}, L)\}| \ge (1-\delta)|M_{opt}|$.

Assume that (i_1, i_2, i_3) satisfy that 1) i_1, i_2, i_3 are all in M_{opt}, 2) $|\{k : k \in M_{opt}$ *and* distance$(b_{i_1}, b_k) \le$ distance$(b_{i_1}, b_{i_2}\}| \ge (1-\delta)|M_{opt}|$, and 3) $|\{k : k \in M_{opt}$ *and* distance$(b_k, L) \le$ distance$(b_{i_3}, L)\}| \ge (1-\delta)|M_{opt}|$.

Let M_0 be a subset of M_{opt} such that 1)distance$(b_{i_1}, b_k) \le$ distance$(b_{i_1}, b_{i_2}\}$ and distance$(b_k, L(b_{i_1}, b_{i_2})) \le$ distance$(b_{i_3}, L(b_{i_1}, b_{i_2}))$ for each $k \in M_0$, and 2)

$$|M_0| \ge (1-\delta)|M_{opt}|. \tag{1}$$

Therefore, (i_1, i_2, i_3) is a frame of M_0. By Lemma 2, there is a tuple $(u_{i_1}, u_{i_2}, u_{i_3})$ of ϵd-grid points from ball$(a_{i_1}, d + 2\sqrt{3}\epsilon d) \times$ ball$(a_{i_2}, d + 2\sqrt{3}\epsilon d) \times$ ball$(a_{i_3}, d + 2\sqrt{3}\epsilon d))$ such distance$(a_i, T_0'(b_i)) \le (1 + c_0\epsilon)d = (1+\delta_1)d$ for each $i \in M_0$, where $T_0' =$ move$((u_{i_1}, u_{i_2}, u_{i_3}), (b_{i_1}, b_{i_2}, b_{i_3}))$.

Let $p = \frac{|\{i : i \in M_0, \text{ and distance}(a_i, T_0'(b_i)) \le (1+\delta_1)d\}|}{n}$. By inequality (1),

$$p \ge \frac{(1-\delta)|M_{opt}|}{n} \ge (1-\delta)\alpha \ge \frac{\alpha}{2}. \tag{2}$$

By Corollary 1, we have probability $\Pr(\frac{|M_{T_0'}|}{m} < p - \delta p) = \Pr(|M_{T_0'}| < pm - \delta pm) < e^{-\frac{1}{3}m(\delta p)^2} \le e^{-\frac{1}{3}m(\delta\alpha)^2} = e^{-h}$. By inequality (2),

$$p - \delta p = (1-\delta)p \ge (1-\delta)(1-\delta)\frac{|M_{opt}|}{n} \ge (1 - \frac{\delta_2}{10})\frac{|M_{opt}|}{n} = (1 - \frac{\delta_2}{10})\mu,$$

where $\mu = \frac{|M_{opt}|}{n}$. Therefore, $\Pr(\frac{|M_{T_0'}|}{m} < (1 - \frac{\delta_2}{10})\mu) \le e^{-h}$.

For each transformation T' generated by the algorithm, let $p_{T'}$ be the probability $p_{T'} = \frac{|\{i : \text{distance}(a_i, T'(b_i)) \le (1+\delta_1))d\}|}{n}$. By Corollary 1, with probability at most $e^{-\frac{1}{3}(\frac{\delta_2}{10}\mu)^2 m}$, we have $\frac{|M_{T'}|}{m} \ge p_{T'} + \frac{\delta_2}{10}\mu$. The size of $G_1 \times G_2 \times G_3$ is $C_1 = k_1k_2k_3 = \lceil\frac{h}{\alpha}\rceil \cdot \lceil\frac{h}{\alpha\delta}\rceil \cdot \lceil\frac{h}{\alpha\delta}\rceil \le (\frac{8h^3}{\alpha^3\delta^2})$. For each $(a_{i_1}, a_{i_2}, a_{i_3})$ in $G_1 \times G_2 \times G_3$, the number of transformation T' processed by the algorithm is $C_2 \le (\frac{c_2}{\delta_1^6})$ (by Lemma 2), where c_2 is a constant. Therefore, there are at most C_1C_2 transformations T' generated by the algorithm. Therefore, with probability at most $(C_1C_2)e^{-\frac{1}{3}(\frac{\delta_2}{10}\mu)^2 m} \le (C_1C_2)e^{-\frac{1}{3}(\frac{\delta_2}{10}\alpha)^2 m} \le e^{-h}$, we have $\frac{|M_{T'}|}{m} \ge p_{T'} + \frac{\delta_2}{10}\mu$ for some T' generated by the algorithm.

Assume that $(T^*, M_{T^*}, \frac{|M_{T^*}|}{m})$ is an output. We have $\frac{|M_{T^*}|}{m} \ge \frac{|M_{T_0'}|}{m}$ since $|M_{T^*}|$ is the largest among all T' generated by the algorithm and T_0 is one of those T's. If $p_{T^*} + \frac{\delta_2}{10}\mu \geq \frac{|M_{T^*}|}{m} \geq \frac{|M_{T'_0}|}{m} \geq p - \delta p$, then $p_{T^*} \geq (p - \delta p) - \frac{\delta_2}{10}\mu \geq (1 - \frac{\delta_2}{10})\mu - \frac{\delta_2}{10}\mu = (1 - \frac{\delta_2}{5})\mu > (1 - \delta_2)\mu$. With probability at most $e^{-h} + e^{-h} = 2e^{-h}$, either $p_{T^*} + \frac{\delta_2}{10}\mu < \frac{|M_{T^*}|}{m}$ or $\frac{|M_{T'_0}|}{m} < p - \delta$. Therefore, with probability at least $1 - 3e^{-h} - 2e^{-h} = 1 - 5e^{-h}$, it outputs an $(1 + \delta_1, 1 - \delta_2)$ approximation.

As we mentioned before, the size of $G_1 \times G_2 \times G_3$ is $C_1 = O(\frac{h^3}{\alpha^3\delta_2^2})$ and for each $(a_{i_1}, a_{i_2}, a_{i_3})$ in $G_1 \times G_2 \times G_3$, the number of transformation T' processed by the algorithm is $C_2 = O(\frac{1}{\delta_1^6})$. The total computational time is $O(C_1 C_2 m) = O((\frac{h^3}{\alpha^3\delta_2^2}) \cdot (\frac{1}{\delta_1^6}) \cdot (\frac{(h+\ln\max(\frac{1}{\alpha}, \frac{1}{\epsilon}, \frac{1}{\delta})}{(\delta_2 \alpha c_1)^2}) = O(\frac{h^3(h+\ln\max(\frac{1}{\alpha}, \frac{1}{\delta_1}, \frac{1}{\delta_2}))}{\alpha^5\delta_1^6\delta_2^4})$. □

Theorem 3. *1) There is an $O(\frac{n}{\epsilon^6})$ time algorithm that gives an $(1+\epsilon)$- approximation for the bottleneck distance problem; and 2) there is no $O(1)$-approximation randomized algorithm for the bottleneck distance problem* BD(A, B) *that queries at most $o(n)$ input points, where both A and B have n points.*

References

1. Ambühl, C., Chakraborty, S., Gärtner, B.: Computing largest common point sets under approximate congruence. In: Paterson, M. (ed.) ESA 2000. LNCS, vol. 1879, pp. 52–63. Springer, Heidelberg (2000)
2. Arun, K.S., Huang, T.S., Blostein, S.D.: Least-squares fitting of two 3-d point sets. IEEE Trans. Pattern Anal. Mach. Intell. 9(5), 698–700 (1987)
3. Choi, V., Goyal, N.: A combinatorial shape matching algorithm for rigid protein docking. In: Sahinalp, S.C., Muthukrishnan, S.M., Dogrusoz, U. (eds.) CPM 2004. LNCS, vol. 3109, pp. 285–296. Springer, Heidelberg (2004)
4. Goodrich, M., Mitchell, J., Orletsky, M.: Practical methods for approximate geometric pattern matching under rigid motions. In: SOCG 1994, pp. 103–112 (1994)
5. Indyk, P., Motwani, R.: Geometric matching under noise: Combinatorial bounds and algorithms. In: SODA 1995, pp. 457–465 (1999)
6. Lancia, G., Istrail, S.: Protein structure comparison: Algorithms and applications. Mathemat. Methods for Protein Struct. Analysis and Design, 1–33 (2003)
7. Li, M., Ma, B., Wang, L.: On the closest string and substring problems. Journal of the ACM 49(2), 157–171 (2002)
8. Li, S.C., Bu, D., Xu, J., Li, M.: Finding largest well-predicted subset of protein structure models. In: Ferragina, P., Landau, G.M. (eds.) CPM 2008. LNCS, vol. 5029, pp. 44–55. Springer, Heidelberg (2008)
9. Motwani, R., Raghavan, P.: Randomized Algorithms. Cambridge University Press, Cambridge (2000)
10. Siew, N., Elofsson, A., Rychlewski, L., Fischer, D.: Maxsub: an automated measure for the assessment of protein structure prediction quality. Bioinformatics 16(9), 776–785 (2000)
11. Zemla, A.: LGA: a method for folding 3d similarities in protein structures. Nucl. Acids Res. 31(13), 3370–3374 (2003)
12. Zhang, Y., Skolnick, J.: Scoring function for automated assessment of protein struct. template quality. Proteins: Struct., Funct., and Bioinf. 57, 702–710 (2004)

On Sorting Permutations by Double-Cut-and-Joins

Xin Chen*

Division of Mathematical Sciences
School of Physical and Mathematical Sciences
Nanyang Technological University, Singapore
chenxin@ntu.edu.sg

Abstract. The problem of *sorting permutations by double-cut-and-joins* (SBD) arises when we perform the double-cut-and-join (DCJ) operations on pairs of unichromosomal genomes without the gene strandedness information. In this paper we show it is a NP-hard problem by reduction to an equivalent previously-known problem, called *breakpoint graph decomposition* (BGD), which calls for a largest collection of edge-disjoint alternating cycles in a breakpoint graph. To obtain a better approximation algorithm for the SBD problem, we made a suitable modification to Lin and Jiang's algorithm which was initially proposed to approximate the BGD problem, and then carried out a rigorous performance analysis via fractional linear programming. The approximation ratio thus achieved for the SBD problem is $\frac{17}{12} + \epsilon \approx 1.4167 + \epsilon$, for any positive ϵ.

1 Introduction

The first problem in the study of genome rearrangements is probably to compute the genomic distance between two genomes based on their gene orders. The genomic distance is generally defined as the minimum number of operations required to transform one genome to another, and the complexity of computing it may largely depend on the choice of operations and on the representation of genomes as well.

Double-cut-and-join (DCJ) is an operation that cuts a chromosome in two places and joins the four ends of the cut in a new way. It was first introduced in [15] and later refined in [3], to unify all the classical genome rearrangement operations including inversions, transpositions, translocations, block-interchanges, fissions and fusions. A simple formula exists for the genomic distance by the DCJ operations, which can be computed in linear time for pairs of genomes when the strandedness of genes have become available.

However, genetic maps produced from many experimental studies such as recombination analysis and physical imaging generally do not specify the strandedness of genes or markers. For instance, the Gramene database

* Supported in part by the Singapore NRF grant NRF2007IDM-IDM002-010 and MOE AcRF Tier 1 grant RG78/08.

M.T. Thai and S. Sahni (Eds.): COCOON 2010, LNCS 6196, pp. 439–448, 2010.

(`http://www.gramene.org`) contains a variety of such maps of the rice, maize, oats and other cereal genomes. As a result, a genome can only be represented as an *unsigned* permutation. When we perform the DCJ operations on pairs of such genomes, a new combinatorial problem naturally arises, which we called *sorting permutations by double-cut-and-joins* or SBD for short. It is specifically defined as the problem of finding the minimum number of DCJs required to transform an unsigned permutation into the identity permutation.

In this paper we study the SBD problem, and for the sake of simplicity, we focus our study on unichromosomal genomes only. We first show that the minimum number of DCJs required to sort an unsigned permutation π, denoted by $d(\pi)$, is equal to the number of breakpoints $b(\pi)$ subtracted by the maximum number $c(\pi)$ of edge-disjoint alternating cycles in the breakpoint graph $G(\pi)$; that is, $d(\pi) = b(\pi) - c(\pi)$. It turns out that the SBD problem is indeed equivalent to a previously known NP-hard problem called *breakpoint graph decomposition* (BGD) [10], implying that the SBD problem is NP-hard as well.

Bafna and Pevzner (1996) [2] presented the first approximation algorithm for the BGD problem with performance ratio $\frac{7}{4}$. It was subsequently improved to $\frac{3}{2}$ and $\frac{33}{23} + \epsilon \approx 1.4348 + \epsilon$, for any positive ϵ, due to Christie (1998) [6] and Caprara and Rizzi (2002) [5], respectively. At present, the best known approximation ratio achievable for BGD (and hence for SBD) is $\frac{5073-15\sqrt{1201}}{3208} + \epsilon \approx 1.4193 + \epsilon$, for any positive ϵ, due to Lin and Jiang (2004) [11]. To further improve the approximation, we present in Section 4.2 a suitable modification on Lin and Jiang's approximation algorithm, and then carry out a rigorous performance analysis via fractional linear programming to obtain a better approximation ratio of $\frac{17}{12} + \epsilon \approx 1.4167 + \epsilon$, for any positive ϵ.

2 Preliminaries

2.1 Breakpoint Graph Decomposition

Let $\pi = \pi_1\pi_2\cdots\pi_n$ be a (unsigned) permutation on $\{1, 2, \cdots, n\}$. We extend a permutation π by adding $\pi_0 = 0$ and $\pi_{n+1} = n + 1$. A pair of consecutive elements π_i and π_{i+1} of π is called an *adjacency* if $|\pi_i - \pi_{i+1}| = 1$ and otherwise, a *breakpoint*. Let $b(\pi)$ denote the number of breakpoints in π. Define the *inverse permutation* π^{-1} of π to be $\pi^{-1}_{\pi_i} := i$, for all $i = 0, \ldots, n + 1$.

The notion of the *breakpoint graph* was first introduced in 1993 by Banfa and Pevzner to study the problem of *sorting by reversals* [1]. Given a permutation π, the breakpoint graph is an edge-colored graph $G(\pi)$ with $n + 2$ vertices π_0, π_1, π_2, $\cdots$, π_n, π_{n+1}. Two vertices π_i and π_j are jointed by a *black* edge if they form a breakpoint in π, or by a *gray* edge if they form a breakpoint in π^{-1}. A cycle in $G(\pi)$ is called *alternating* if the colors of every two consecutive edges are distinct. Henceforth, all cycles referred to in the breakpoint graph will be alternating cycles. The *length* of a cycle is the number of black edges that it contains, and a cycle of length l is called a l-cycle. The breakpoint graph $G(\pi)$ can always be decomposed into a maximum number of edge-disjoint alternating cycles, and this maximum number is denoted by $c(\pi)$. A slightly different definition

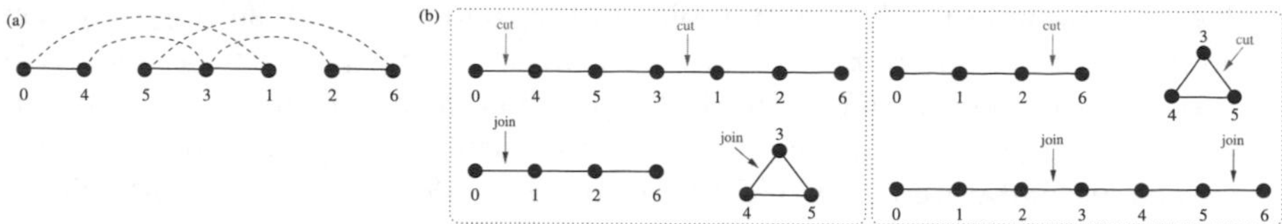

Fig. 1. (a) The breakpoint graph of the permutation $\pi = 4\ 5\ 3\ 1\ 2$, for which $b(\pi) = 4$ and $c(\pi) = 2$. (b) Two double-cut-and-join operations that sort the permutation π. Note that a circular immediate (3 4 5) is created after the first DCJ operation.

of breakpoint graph can also be seen in many studies (e.g. [8]). It allows edges to join not only breakpoints but also adjacencies so that 1-cycles may occur in a breakpoint graph. Let $G^{'}(\pi)$ denote the breakpoint graph to be constructed with this definition and $c^{'}(\pi)$ the number of edge-disjoint alternating cycles in a maximum decomposition of $G^{'}(\pi)$. It is not hard (though not trivial) to see that there always exists a maximum cycle decomposition that retains all the 1-cycles in $G^{'}(\pi)$. Therefore, we can have

$$b(\pi) - c(\pi) = n + 1 - c^{'}(\pi). \tag{1}$$

Unless otherwise stated, we will use the first definition of breakpoint graph in the rest of the paper.

The problem of *breakpoint graph decomposition* is aimed at finding $b(\pi) - c(\pi)$. It was initially introduced to help solve the problem of sorting by reversals [1,5,11]. Since $b(\pi)$ is given with a permutation π, it becomes equivalent to decomposing $G(\pi)$ into a maximum number of edge-disjoint alternating cycles.

The concept of breakpoint graph extends naturally to *signed* permutations. A signed permutation $\overrightarrow{\pi}$, in which each element $\overrightarrow{\pi_i}$ is a signed integer, can be transformed into a unsigned permutation π by replacing $+i$ by $2i-1\ \ 2i$ and $-i$ by $2i\ \ 2i-1$, respectively. Then, define $b(\overrightarrow{\pi}) := b(\pi)$ and $c(\overrightarrow{\pi}) := c(\pi)$.

3 Sorting by Double-Cut-and-Joins

In order to unify all the classical genome rearrangement events, Yancopoulos et al (2005) [15] introduced a new edit operation called double-cut-and-join (DCJ). It generally cuts a chromosome in two places and joins the four ends of the cut in a new way. For example, when a chromosome is given as a permutation π, a DCJ operation ρ_1 that acts on two consecutive pairs $\pi_i\pi_{i+1}$ and $\pi_j\pi_{j+1}$ $(i < j)$ of π will cut both $\pi_i\pi_{i+1}$ and $\pi_j\pi_{j+1}$ and joins either $\pi_i\pi_j$ and $\pi_{i+1}\pi_{j+1}$, or $\pi_i\pi_{j+1}$ and $\pi_{i+1}\pi_j$ to create two new consecutive pairs. If one chooses to join $\pi_i\pi_j$ and $\pi_{i+1}\pi_{j+1}$, it simulates an *inversion* operation, resulting in a new permutation $\pi \cdot \rho_1 = \pi_1 \cdots \pi_i\pi_j \cdots \pi_{i+1}\pi_{j+1} \cdots \pi_n$. On the other hand, to join $\pi_i\pi_{j+1}$ and $\pi_{i+1}\pi_j$, a *circular immediate* $\pi_{i+1} \cdots \pi_j$ is then generated (see Figure 1b). In this case, we may absorb the circular immediate back into the permutation by another DCJ operation ρ_2 that cuts two consecutive pairs $\pi_k\pi_{k+1}$ (assume $k < i$)

and $\pi_l\pi_{l+1}$ (assume $i < l < j$) and joins $\pi_k\pi_{l+1}$ and $\pi_l\pi_{k+1}$, resulting in a new permutation $\pi \cdot \rho_1 \cdot \rho_2 = \pi_1 \cdots \pi_k\pi_{l+1} \cdots \pi_j\pi_{i+1} \cdots \pi_l\pi_{k+1} \cdots \pi_i\pi_{j+1} \cdots \pi_n$. We can see that the composition of two DCJ operations ρ_1 and ρ_2 indeed simulates a block-interchange operation that acts on two segments $\pi_{k+1} \cdots \pi_i$ and $\pi_{l+1} \cdots \pi_j$.

Given a permutation π, the problem of *sorting by double-cut-and-joints* (SBD) is defined to find a sequence of DCJ operations such that $\pi \cdot \rho_1 \cdot \rho_2 \cdots \rho_t$ produces the identity permutation and t is minimum. Let $d(\pi) := t$. In cases of signed permutations $\vec{\pi}$, we already know the following results.

Theorem 1 ([15]). *Let $\vec{\pi}$ be a signed permutation on $\{1, 2, \cdots, n\}$. Then, $d(\vec{\pi}) = b(\vec{\pi}) - c(\vec{\pi})$. Moreover, the optimal sequence of DCJ operations that transform $\vec{\pi}$ into the identity permutation can be found in linear time.*

However, to the best of our knowledge, the problem of sorting unsigned permutations by double-cut-and-joints has not been discussed in the literature so that very little is known about it. In the below, we first show that $d(\pi)$ can still be computed using the same formula as given in the previous theorem.

Theorem 2. *Let π be a unsigned permutation on $\{1, 2, \cdots, n\}$. Then, $d(\pi) = b(\pi) - c(\pi)$.* (Proof omitted due to space limitations)

From the above theorem we can easily see that the SBD problem is indeed equivalent to the problem of breakpoint graph decomposition (BGD), which is already shown to be NP-hard [10].

Theorem 3. *The SBD problem is NP-hard.*

4 Approximation of SBD

4.1 Definitions and Graph-Theoretic Background

Let c_i denote the number of i-cycles in a (fixed) maximum cycle decomposition of $G(\pi)$, for $i \geq 2$. Therefore, $c(\pi) = \sum_{i \geq 2} c_i$. Moreover, let c_2^* denote the maximum number of edge-disjoint 2-cycles in $G(\pi)$, and c_3^* the maximum number of edge-disjoint cycles of length no more than 3. Obviously, $c_2 \leq c_2^*$ and $c_2 + c_3 \leq c_3^*$.

Finding edge-disjoint 2-cycles. To find a collection of edge-disjoint 2-cycles, one may construct a graph $G_2(\pi)$ whose vertices represent all the 2-cycles in $G(\pi)$ and edges joins two 2-cycles that share some edge in $G(\pi)$. An *independent set* in a graph is a subset of vertices in which no two are adjacent, and a *maximum independent set* is an independent set of maximum cardinality. Observe that a maximum independent set of $G_2(\pi)$ gives a largest collection of edge-disjoint 2-cycles with cardinality c_2^*.

Unfortunately, the problem of maximum independent set is NP-hard [9]. In order to obtain a good approximation, Caprara and Rizzi [5] showed a way to reduce the graph $G_2(\pi)$ to another graph $G_2'(\pi)$ with the maximum degree (at most) 4, such that a maximum independent set of $G_2'(\pi)$ is also a maximum independent set of $G_2(\pi)$. By the best-known approximation algorithm (denoted as Approx-MIS) for the maximum independent set problem restricted to graphs with the bounded maximum degree [4], one can obtain the following lemma.

Lemma 1 ([5]). *The problem of finding a largest collection of edge-disjoint 2-cycles in breakpoint graph $G(\pi)$ can be approximated with ratio $\frac{5}{7} - \epsilon$ in polynomial time, for any positive ϵ.*

Finding edge-disjoint 2-cycles and 3-cycles. To find a collection of edge-disjoint cycles of length no more than 3, one may construct a collection C of subsets of the base set S, where S is comprised of all the edges in $G(\pi)$ and each subset of C is comprised of edges from a 2-cycle or 3-cycle in $G(\pi)$. A *set packing* is a sub-collection of pairwise disjoint subsets in C, and a *maximum set packing* is a set packing with maximum cardinality. Observe that a maximum set packing of C gives a largest sub-collection of 2-cycles and 3-cycles with cardinality c_3^*.

The problem of maximum set packing is NP-hard [9] too. Since every subset in C has size at most 6, the problem reduces to the k-set packing problem, where $k = 6$. By the best-known approximation algorithm (denoted as Approx-MSP) for the k-set packing problem [7], one can obtain the following lemma.

Lemma 2 ([11]). *The problem of finding a largest collection of edge-disjoint cycles of length at most 3 in breakpoint graph $G(\pi)$ can be approximated with ratio $\frac{3}{8}$ in polynomial time.*

Local improvement search. The local improvement search is a technique often used to solve many hard combinatorial optimization problems [4,7]. To facilitate our subsequent algorithmic performance analysis, it is briefly introduced below in the context of the k-set packing problem.

Let $I = (S, C)$ be an instance of the k-set packing problem, where S is a base set and C is a collection of subsets of S. We call C' a *feasible* solution of I if it is a sub-collection of disjoint subsets in C. C' is further said to be *maximal* if there is no subset $C_i \in C \setminus C'$ such that $C' \cup \{C_i\}$ is still feasible. Even when a set packing C' is maximal already, it might happen that there exist two subsets $C_{i_1}, C_{i_2} \in C \setminus C'$ and a subset $C_j \in C'$ such that $C' \cup \{C_{i_1}, C_{i_2}\} \setminus C_j$ is feasible. In this case, we may say subsets C_{i_1} and C_{i_2} *improve* C' as its size is growing. If, on the other hand, there is no such pair of subsets improving C', then C' is said to be *2-maximal*.

A local improvement search procedure that finds a 2-maximal set packing for an instance $I = (S, C)$ of the k-set packing problem may start with an arbitrary (possibly empty) set packing C', and then repeatedly update C' by replacing one subset in C' with two subsets in $C \setminus C'$ or adding one subset of $C \setminus C'$ into C' until no more update is possible while the feasibility of C' is not violated.

Lin and Jiang's algorithm. The currently best-known ratio to approximate $b(\pi) - c(\pi)$ is due to Lin and Jiang [11]. Their algorithm is summarized below.

Lin and Jiang's Algorithm

1. Compute a collection P of 2-cycles by algorithm Approx-MIS.
2. Compute a 2-maximal collection Q of cycles of length at most 3 by improving P.
3. Compute a collection R of cycles of length at most 3 by algorithm Approx-MSP.
4. Output the larger collection of Q and R.

Basically, Lin and Jiang's algorithm first runs Approx-MIS and Approx-MSP to obtain two lower bounds of $(\frac{5}{7}-\epsilon)c_2$ and $\frac{3}{8}(c_2+c_3)$ on the number of edge-disjoint cycles, respectively, and then a 2-maximal improvement procedure on the collection of edge-disjoint 2-cycles. The latter leads to a better size guarantee by incorporating a balancing argument, as stated in the following lemma.

Lemma 3 ([11]). *The resulting 2-maximal collection of edge-disjoint cycles either improves the lower bound from $(\frac{5}{7}-\epsilon)c_2$ to $\left(\frac{\sqrt{1201}+89}{168}-\epsilon\right)c_2$ or improves the lower bound from $\frac{3}{8}(c_2+c_3)$ to $\frac{37-\sqrt{1201}}{6}(c_2+c_3)$, but not both.*

With the above lemma, Lin and Jiang (2004) successfully showed the following performance ratio for approximating $b(\pi)-c(\pi)$.

Theorem 4 ([11]). *The problem of minimizing $b(\pi)-c(\pi)$ can be approximated within ratio $\frac{5073-15\sqrt{1201}}{3208}+\epsilon \approx 1.4193+\epsilon$ in polynomial time, for any positive ϵ.*

4.2 A Better Performance Guarantee

In the previous section, we introduced the best-to-date approximation algorithm for minimizing $b(\pi)-c(\pi)$. In this section, we make a suitable modification on this algorithm and then perform a rigorous performance analysis via fractional linear programming to achieve a better approximation ratio.

The modified algorithm. We modified Lin and Jiang's algorithm mainly by removing a computation-intensive step that employs the Approx-MSP algorithm to compute a collection R of cycles of length at most 3 (i.e., step 3 as seen above). The modified algorithm is summarized blow.

<u>The Modified Algorithm</u>

1. Compute a collection P of 2-cycles by algorithm Approx-MIS.
2. Compute a 2-maximal collection Q of cycles of length at most 3 by improving P.
3. Output the collection Q.

Let Q_2 and Q_3 be the sub-collections of 2-cycles and 3-cycles in Q, respectively. Obviously, $Q=Q_2\cup Q_3$ and $Q_2\bigcap Q_3=\emptyset$. Furthermore, let C_2 and C_3 be the sub-collections of 2-cycles and 3-cycles in a (fixed) maximum cycle decomposition of $G(\pi)$, respectively. In the following, we denote the sizes of the above mentioned collections by their respective lower case letters. For short, we simplify the notations $b(\pi)$ and $c(\pi)$ to b and c, respectively.

Lemma 4. $2(c_2+c_3)\le 5q_2+7q_3$.

Proof. Notice that $C_2\cup C_3$ is a (not necessarily maximal) set packing of 2-cycles and 3-cycles. Let S_1 be the collection of cycles in $C_2\cup C_3$ that each intersects exactly one cycle in Q, and let $S_2 = C_2 \cup C_3 \setminus S_1$. Obviously, $s_1 + s_2 = c_2 + c_3$. Since Q is a 2-maximal set packing of 2-cycles and 3-cycles, every cycle in $C_2 \cup C_3$ shall intersect at least one cycle in Q, thereby implying that every cycle in S_2 shall intersect at least two cycles in Q. Moreover, no two cycles in S_1 can intersect a same cycle in Q because otherwise they would further improve Q. Therefore, $s_1 \leq q = q_2 + q_3$. Further notice that every 2-cycle in Q_2 can intersect at most 4 cycles in $C_2 \cup C_3$, and that every 3-cycles in Q_3 can intersect at most 6 cycles in $C_2 \cup C_3$. Therefore, $s_1 + 2s_2 \leq 4q_2 + 6q_3$. Combining the above two inequations yields

$$2s_1 + 2s_2 \leq 5q_2 + 7q_3, \tag{2}$$

which hence establishes the lemma. ■

Lemma 5. $2p \leq 2q_2 + q_3$.

Proof. Notice that Q is initialized as P in our modified algorithm. Therefore, $2p \leq 2q_2 + q_3$ is true at the beginning of the local search for a 2-maximal collection. During the local search process, any improvement by replacing one set with two sets or adding one set into Q will never lower the value of $2q_2 + q_3$. Therefore, $2p \leq 2q_2 + q_3$ remains true at the end of local search. ■

Performance analysis via fractional linear programming. Let C' be a collection of edge-disjoint alternating cycles in which Q_2 and Q_3 are respectively the sub-collections of 2-cycles and 3-cycles output from the above algorithm. It is obvious that $q_2 + q_3 \leq c'$ and $c_2 + c_3 \leq c$. Moreover, $c \leq c_2 + c_3 + \frac{b - 2c_2 - 3c_3}{4}$, which follows from the fact that every i-cycle contains at least 4 black edges for all $i \geq 4$. Then, finding the worst-case ratio between the sizes of the approximate solution C' and some (fixed) optimal solution C is reduced to solving the following optimization problem.

$$\begin{aligned}
\max\ & \frac{b - c'}{b - c} \\
\text{subject to}\ & c_2 + c_3 \leq c, \\
& c \leq c_2 + c_3 + \tfrac{b - 2c_2 - 3c_3}{4}, \\
& (\tfrac{5}{7} - \epsilon)c_2 \leq p, \\
& 2(c_2 + c_3) \leq 5q_2 + 7q_3, \\
& 2p \leq 2q_2 + q_3, \\
& q_2 + q_3 \leq c', \\
& b \geq 1, \\
& c_2,\ c_3,\ p,\ q_2,\ q_3 \geq 0.
\end{aligned}$$

This is a fractional linear programming problem — the generalization of a linear programming problem in which the objective function is the ratio of two linear functions. Notice that the value of c' is bounded from the bottom and the value of c is bounded from the above. Therefore, the maximum is attained when we substitute c' by $q_2 + q_3$ and c by $c_2 + c_3 + \frac{b - 2c_2 - 3c_3}{4}$. That is,

$$\begin{aligned}\max\ & \frac{b - q_2 - q_3}{\frac{3}{4}b - \frac{1}{2}c_2 - \frac{1}{4}c_3} \\ \text{subject to}\ & 2c_2 + 3c_3 \le b, \\ & (\tfrac{5}{7} - \epsilon)c_2 \le p, \\ & 2(c_2 + c_3) \le 5q_2 + 7q_3, \\ & 2p \le 2q_2 + q_3, \\ & b \ge 1, \\ & c_2,\ c_3,\ p,\ q_2,\ q_3 \ge 0.\end{aligned}$$

Consider two cases when $c_2+2c_3 < 1$ and when $c_2+2c_3 \ge 1$, respectively. When $c_2 + 2c_3 < 1$, we have $c_2 = c_3 = 0$ because both c_2 and c_3 are nonnegative integers. The maximum objective value $\frac{4}{3}$ is then attained when $q_2 = q_3 = 0$. When $c_2 + 2c_3 \ge 1$, we further have $2c_2 + 3c_3 \ge 1$. Then, the maximum is attained when $b = 2c_2+3c_3$. Therefore, the worst-case approximation ratio shall be equal to $\max\left\{\frac{4}{3}, \mathcal{R}(\epsilon)\right\}$, where $\mathcal{R}(\epsilon)$ denotes the maximum objective value of the following fractional linear programming problem.

$$\begin{aligned}\max\ & \frac{2c_2 + 3c_3 - q_2 - q_3}{c_2 + 2c_3} \\ \text{subject to}\ & (\tfrac{5}{7} - \epsilon)c_2 \le p, \\ & 2(c_2 + c_3) \le 5q_2 + 7q_3, \\ & 2p \le 2q_2 + q_3, \\ & c_2 + 2c_3 \ge 1, \\ & c_2,\ c_3,\ p,\ q_2,\ q_3 \ge 0.\end{aligned}$$

Because its objective function is bounded both from below and from above, we may transform it to an equivalent linear programming problem.

$$\begin{aligned}\max\ & 2c_2 + 3c_3 - q_2 - q_3 \\ \text{subject to}\ & (\tfrac{5}{7} - \epsilon)c_2 \le p, \\ & 2(c_2 + c_3) \le 5q_2 + 7q_3, \\ & 2p \le 2q_2 + q_3, \\ & c_2 + 2c_3 = 1, \\ & c_2,\ c_3,\ p,\ q_2,\ q_3 \ge 0.\end{aligned}$$

Because p appears only in the first and third constraints, we may remove the variable p by combining these two constraints. Moreover, the fourth equality constraint allows us to remove the variable c_3.

$$\begin{aligned}\max\ & \frac{3}{2} + \frac{1}{2}c_2 - q_2 - q_3 \\ \text{subject to}\ & (\tfrac{5}{7} - \epsilon)c_2 \le q_2 + \tfrac{1}{2}q_3, \\ & c_2 + 1 \le 5q_2 + 7q_3, \\ & c_2 \le 1, \\ & c_2,\ q_2,\ q_3 \ge 0.\end{aligned}$$

It is well known that, for a linear programming problem, the maximum is attained in an extreme point of the feasible region. It turns out that, for the above particular linear programming problem, the maximum value $\mathcal{R}(\epsilon)$ is attained when

$$c_2 = \frac{7}{18-35\epsilon}, \quad q_2 = \frac{5-7\epsilon}{18-35\epsilon}, \quad \text{and} \quad q_3 = 0,$$

such that

$$\mathcal{R}(\epsilon) = \frac{3}{2} - \frac{3-14\epsilon}{36-70\epsilon}.$$

Since $\mathcal{R}(\epsilon)$ is differentiable when $\epsilon \le \frac{1}{2}$, we know from calculus that

$$\mathcal{R}(\epsilon) \le \frac{17}{12} + \epsilon, \qquad 0 \le \epsilon < \frac{1}{2}.$$

Finally, notice that $\frac{17}{12} + \epsilon \ge \frac{4}{3}$, where $\frac{4}{3}$ is the approximation ratio that we can achieve for the instances with $c_2 = c_3 = 0$. Putting them together, we have proved the following theorem, which states an improved approximation ratio for approximating $b(\pi) - c(\pi)$.

Theorem 5. *The problem of minimizing $b(\pi)-c(\pi)$ can be approximated within ratio $\frac{17}{12} + \epsilon \approx 1.4167 + \epsilon$ in polynomial time, for any positive ϵ.*

Along with Theorem 2, the above theorem yields one of our main results below.

Theorem 6. *The SBD problem can be approximated within ratio $\frac{17}{12} + \epsilon \approx 1.4167 + \epsilon$ in polynomial time, for any positive ϵ.*

5 Conclusions

Since the double-cut-and-join (DCJ) operation was first introduced in 2005, a variety of genome rearrangement analyses have been carried out under the DCJ context, such as estimating true evolutionary distances [12], genome halving [13], genome aliquoting [14], and finding genome median [16]. All these analyses are based on a common assumption that the strandedness of genes of interest shall already become available.

In this paper, we studied the problem of sorting permutations by double-cut-and-joints (SBD), which naturally arises when the DCJ operations are performed on pairs of unichromosomal genomes without the gene strandedness information. We first showed that the SBD problem can be equivalently reduced to a previously-known NP-hard problem called breakpoint graph decomposition, thereby implying that the SBD problem is also NP-hard. This result contrasts with our intuition that computing the rearrangement distance by the DCJ operations is always very easy, as exemplified in sorting signed permutations [15]. To achieve a better approximation to the SBD problem, we made a suitable modification to Lin and Jiang's algorithm, which was initially proposed to approximate the BGD problem, and carried out a rigorous performance analysis based on fractional linear programming. The final approximation ratio achieved for the SBD problem is $\frac{17}{12} + \epsilon \approx 1.4167 + \epsilon$, improving over the previously known approximation ratio $\frac{5073-15\sqrt{1201}}{3208} + \epsilon \approx 1.4193 + \epsilon$, for any positive ϵ.

References

1. Bafna, V., Pevzner, P.A.: Genome rearrangements and sorting by reversals. In: 34th IEEE Annual Symposium on Foundations of Computer Science, pp. 148–157 (1993)
2. Bafna, V., Pevzner, P.A.: Genome rearrangements and sorting by reversals. SIAM J. Comput. 25(2), 272–289 (1996)
3. Bergeron, A., Mixtacki, J., Stoye, J.: A unifying view of genome rearrangements. In: Bücher, P., Moret, B.M.E. (eds.) WABI 2006. LNCS (LNBI), vol. 4175, pp. 163–173. Springer, Heidelberg (2006)
4. Berman, P., Fürer, M.: Approximating maximum independent set in bounded degree graphs. In: The fifth annual ACM-SIAM symposium on Discrete Algorithms, pp. 365–371 (1994)
5. Caprara, A., Rizzi, R.: Improved approximation for breakpoint graph decomposition and sorting by reversals. J. Comb. Optim. 6(2), 157–182 (2002)
6. Christie, D.A.: A 3/2-approximation algorithm for sorting by reversals. In: The ninth annual ACM-SIAM symposium on Discrete Algorithms, pp. 244–252 (1998)
7. Halldórsson, M.M.: Approximating discrete collections via local improvements. In: The sixth annual ACM-SIAM symposium on Discrete Algorithms, pp. 160–169 (1995)
8. Hannenhalli, S., Pevzner, P.A.: To cut.. or not to cut (applications of comparative physical maps in molecular evolution). In: The seventh annual ACM-SIAM symposium on Discrete Algorithms, pp. 304–313 (1996)
9. Karp, R.M.: Reducibility among combinatorial problems. In: Miller, R.E., Thatcher, J.W. (eds.) Complexity of Computer Computations, pp. 85–103 (1972)
10. Kececioglu, J., Sankoff, D.: Exact and approximation algorithms for sorting by reversals, with application to genome rearrangement. Algorithmica 13, 180–210 (1995)
11. Lin, G., Jiang, T.: A further improved approximation algorithm for breakpoint graph decomposition. J. Comb. Optim. 8(2), 183–194 (2004)
12. Lin, Y., Moret, B.: Estimating true evolutionary distances under the DCJ model. Bioinformatics 24(13), 114–122 (2008)
13. Warren, R., Sankoff, D.: Genome halving with double cut and join. In: The 6th Asia-Pacific Bioinformatics Conference, vol. 6, pp. 231–240 (2008)
14. Warren, R., Sankoff, D.: Genome aliquoting with double cut and join. BMC Bioinformatics 10(Suppl. 1), S2 (2009)
15. Yancopoulos, S., Attie, O., Friedberg, R.: Efficient sorting of genomic permutations by translocation, inversion and block interchange. Bioinformatics 21(16), 3340–3346 (2005)
16. Zhang, M., Arndt, W., Tang, J.: An exact solver for the dcj median problem. In: Pacific Symposium on Biocomputing, pp. 138–149 (2009)

A Three-String Approach to the Closest String Problem

Zhi-Zhong Chen[1], Bin Ma[2], and Lusheng Wang[3]

[1] Department of Mathematical Sciences, Tokyo Denki University, Hatoyama, Saitama 350-0394, Japan
zzchen@mail.dendai.ac.jp

[2] School of Computer Science, University of Waterloo, 200 University Ave. W, Waterloo, ON, Canada N2L3G1
binma@uwaterloo.ca

[3] Department of Computer Science, City University of Hong Kong, Tat Chee Avenue, Kowloon, Hong Kong SAR
lwang@cs.cityu.edu.hk

Abstract. Given a set of n strings of length L and a radius d, the closest string problem asks for a new string t_{sol} that is within a Hamming distance of d to each of the given strings. It is known that the problem is NP-hard and its optimization version admits a polynomial time approximation scheme (PTAS). Parameterized algorithms have been then developed to solve the problem when d is small. In this paper, with a new approach (called the *3-string approach*), we first design a parameterized algorithm for binary strings that runs in $O\left(nL + nd^3 6.731^d\right)$ time, while the previous best runs in $O\left(nL + nd8^d\right)$ time. We then extend the algorithm to arbitrary alphabet sizes, obtaining an algorithm that runs in $O\left(nL + nd1.612^d \left(\alpha^2 + 1 - 2\alpha^{-1} + \alpha^{-2}\right)^{3d}\right)$ time, where $\alpha = \sqrt[3]{\sqrt{|\Sigma| - 1} + 1}$. This new time bound is better than the previous best for small alphabets, including the very important case where $|\Sigma| = 4$ (i.e., the case of DNA strings).

1 Introduction

An instance of the closest string problem (CSP for short) is a pair $(\mathcal{S}, d)$, where $\mathcal{S}$ is a set of strings of the same length L and d is a nonnegative integer (called the *radius*). The objective is to find a string t_{sol} of length L such that $d(t_{sol}, s) \le d$ for all $s \in \mathcal{S}$. We call t_{sol} a *center string* of radius d for the strings in $\mathcal{S}$. In the optimization version of the problem, only $\mathcal{S}$ is given and the objective is to find the minimum d such that a center string of radius d exists for the strings in $\mathcal{S}$.

The problem finds a variety of applications in bioinformatics, such as universal PCR primer design [4, 11, 20], genetic probe design [11], antisense drug design [3, 11], finding unbiased consensus of a protein family [1], and motif finding [7, 11, 18]. Consequently, the problem has been extensively studied in computational biology [3, 6, 9–12, 14–18].

M.T. Thai and S. Sahni (Eds.): COCOON 2010, LNCS 6196, pp. 449–458, 2010.

The problem is known to be NP-complete [8, 11]. Early attempts to solve this problem mainly focused on approximation algorithms. These include the first non-trivial approximation algorithm with ratio 4/3 [11] and a polynomial time approximation scheme (PTAS) [12]. The time complexity of the PTAS was further improved in [14] and [13]. The main concern of using the PTAS algorithms is that their time complexity is too high. Even with the latest improvement made in [13], the time complexity for achieving an approximation ratio of $1+\epsilon$ is $O(Ln^{O(\epsilon^{-2})})$.

Another approach to the solving of CSP is via parameterized algorithms. A parameterized algorithm computes the exact solution of a problem with time complexity $f(k)\cdot n^c$, where c is a constant, n is the problem size, k is a parameter naturally associated to the input instance, and f is any function [5]. The argument is that if k is typically small for natural instances of the problem, the problem may still be solvable in acceptable time complexity despite that f may be a super-polynomial function.

In fact, for many applications of the closest string problem in bioinformatics, d is relatively small. For example, in the universal PCR primer design problem, d is usually less than 10. For the special case of CSP where $d=1$, Stojanovic *et. al* [17] designed a linear time algorithm. Gramm *et. al* [10] proposed the first parameterized algorithm with time complexity $O\left(nL+n(d+1)^{d+1}\right)$. Ma and Sun [13] gave an algorithm with running time $O\left(nL+nd\cdot 16^d(|\Sigma|-1)^d\right)$, which is the first polynomial time algorithm when d is logarithmic in the input size and the alphabet size is a constant. Wang and Zhu [19] improved the time complexity to $O(nL+nd2^{3.25d}(|\Sigma|-1)^d)$. Chen and Wang [2] further improved the time complexity to $O\left(nL+nd\left(\sqrt{2}|\Sigma|+\sqrt[4]{8}\left(\sqrt{2}+1\right)\left(1+\sqrt{|\Sigma|-1}\right)-2\sqrt{2}\right)^d\right)$ for non-binary strings and to $O\left(nL+nd8^d\right)$ for binary strings. Independently, Zhao and Zhang [21] gave an algorithm running in $O\left(nL+nd\left(2|\Sigma|+4\sqrt{|\Sigma|-1}\right)^d\right)$ time. Note that the algorithm in [2] outperforms the algorithm in [21] for any kind of strings.

In this paper, we introduce a new approach (called the *3-string approach*) and use it to design new parameterized algorithms for the problem. Roughly speaking, with this approach, our algorithm starts by carefully selecting three of the input strings and using them to guess a portion of the output center string. In contrast, all previous algorithms were based on the *2-string approach*, with which the algorithms start by carefully selecting two of the input strings and using them to guess a portion of the output center string. Intuitively speaking, the 3-string approach is better, because it enables the algorithm to guess a larger portion of the output center string in the beginning. The new parameterized algorithm for binary strings runs in $O\left(nL+nd^3 6.731^d\right)$ time, while the previous best runs in $O\left(nL+nd8^d\right)$ time. The new time complexity is asymptotically better. We want to emphasize that $O\left(nL+nd8^d\right)$ seems to be the best possible time complexity achievable by algorithms based on the 2-string approach. We then extend the algorithm to arbitrary strings, obtaining an algorithm that runs in

$O\left(nL + nd1.612^d\left(\alpha^2 + 1 - 2\alpha^{-1} + \alpha^{-2}\right)^{3d}\right)$ time, where $\alpha = \sqrt[3]{\sqrt{|\Sigma| - 1} + 1}$. In particular, in the very important case where $|\Sigma| = 4$ (i.e., the case of DNA strings), our algorithm runs in $O\left(nL + nd13.591^d\right)$ time, while the previous best runs in $O\left(nL + nd13.921^d\right)$ time.

The remainder is organized as follows. Section 2 defines a few notations frequently used in the paper. Section 3 reviews the algorithm in [2], which will be helpful for the presentation of the new algorithm. Section 4 details our algorithm for binary strings. Section 5 then extends the algorithm to general alphabets. All proofs are omitted due to page limit and will be included in the journal version of the paper.

2 Notations

Throughout this paper, Σ denotes a fixed alphabet and a string always means one over Σ. For each positive integer k, $[1..k]$ denotes the set $\{1, 2, \ldots, k\}$. For a string s, $|s|$ denotes the length of s. For each $i \in [1..|s|]$, $s[i]$ denotes the letter of s at its i-th position. Thus, $s = s[1]s[2]\ldots s[|s|]$. A position set of a string s is a subset of $[1..|s|]$. For two strings s and t of the same length, $d(s,t)$ denotes their Hamming distance. For a binary string s, $\bar{s}$ denotes the complement string of s, where $\bar{s}[i] \neq s[i]$ for every $i \in [1..|s|]$.

Two strings s and t of the same length L *agree* (respectively, *differ*) *at a position* $i \in [1..L]$ if $s[i] = t[i]$ (respectively, $s[i] \neq t[i]$). The *position set where s and t agree* (respectively, *differ*) is the set of all positions $i \in [1..L]$ where s and t agree (respectively, differ). The following special notations will be very useful. For two or more strings $s_1, \ldots, s_h$ of the same length, $\{s_1 \equiv s_2 \equiv \cdots \equiv s_h\}$ denotes the position set where s_i and s_j agree for all pairs (i,j) with $1 \le i < j \le h$, while $\{s_1 \not\equiv s_2 \not\equiv \cdots \not\equiv s_h\}$ denotes the position set where s_i and s_j differ for *all* pairs (i,j) with $1 \le i < j \le h$. Moreover, for a sequence $s_1, \ldots, s_h, t_1, \ldots, t_k, u_1, \ldots, u_\ell$ of strings of the same length with $h \ge 2$, $k \ge 1$, and $\ell \ge 0$, $\{s_1 \equiv s_2 \equiv \cdots \equiv s_h \not\equiv t_1 \not\equiv t_2 \not\equiv \cdots \not\equiv t_k \equiv u_1 \equiv u_2 \equiv \cdots \equiv u_\ell\}$ denotes $\{s_1 \equiv s_2 \equiv \cdots \equiv s_h\} \cap \{s_h \not\equiv t_1 \not\equiv t_2 \not\equiv \cdots \not\equiv t_k\} \cap \{t_k \equiv u_1 \equiv u_2 \equiv \cdots \equiv u_\ell\}$.

Another useful concept is that of a *partial string*, which is a string whose letters are only known at its certain positions. If s is a string of length L and P is a position set of s, then $s|_P$ denotes the partial string of length L such that $s|_P[i] = s[i]$ for each position $i \in P$ but $s|_P[j]$ is unknown for each position $j \in [1..L] \setminus P$. Let t be another string of length L. For a subset P of $[1..L]$, the *distance* between $s|_P$ and $t|_P$ is $|\{i \in P \mid s[i] \neq t[i]\}|$ and is denoted by $d(s|_P, t|_P)$. For two disjoint position sets P and Q of s, $s|_P + t|_Q$ denotes the partial string $r|_{P \cup Q}$ such that $r|_{P\cup Q}[i] = s[i]$ if $i \in P$, and $r|_{P\cup Q}[i] = t[i]$ otherwise.

At last, when an algorithm exhaustively tries all possibilities to find the right choice, we say that the algorithm *guesses* the right choice.

3 Previous Algorithms and a New Lemma

In this section we familiarize the readers with the basic ideas in the previously known parameterized algorithms for CSP, as well as introduce a technical lemma that is needed in this paper. All previously known algorithms use the bounded search tree approach for parameterized algorithm design. We explain the ideas based on the algorithm given in [2]. We call the approach used in previous algorithms the *2-string approach* in contrast to the 3-string approach introduced in this paper.

Let $(\mathcal{S}, d)$ be an instance of CSP. Let $s_1, s_2, \ldots, s_n$ be the strings in $\mathcal{S}$, L be the length of each string in $\mathcal{S}$, and t_{sol} be any solution to $(\mathcal{S}, d)$. The idea is to start with a candidate string t with $d(t, t_{sol}) \leq d$. Using some strategies, the algorithm guesses the letters of t_{sol} at some positions (by trying all legible choices), and modify the letters of t to those of t_{sol} at the positions. This procedure is applied iteratively to eventually change t to t_{sol}. The "guessing" causes the necessity of using a search tree, whose size is related to (1) the number of choices to guess from in each iteration (the degree of each tree node) and (2) the total number of iterations (the height of the tree).

At the beginning of each iteration, the algorithm knows the set $\mathcal{S}$ of input strings, the radius d, the current candidate string t, and the remaining distance $b = d(t, t_{sol})$. In addition, if a position in $[1..L]$ has been considered for modification in a previous iteration, it is not helpful to modify it again in the current iteration. Thus, we further record a position set P, at which no further modification is allowed. This gives an *extended closest string problem* (ECSP for short) formerly defined in [2]. An instance of ECSP is a quintuple $\langle \mathcal{S}, d; t, P, b \rangle$, as defined above. A solution of the instance is a string t_{sol} of length L such that $t_{sol}|_P = t|_P$, $d(t_{sol}, t) \leq b$, and $d(t_{sol}, s) \leq d$ for every string $s \in \mathcal{S}$. Obviously, to solve CSP for a given instance $\langle \mathcal{S}, d \rangle$, it suffices to solve ECSP for the instance $\langle \mathcal{S} \setminus \{t\}, d; t, \emptyset, d \rangle$, where t is an arbitrary string in $\mathcal{S}$ and $\emptyset$ is the empty set.

The difference between the algorithms in [10], [13] and [2] exists in the guessing strategy in each iteration. In each iteration, suppose t is not a solution yet, then there must be a string s such that $|\{s \not\equiv t\}| > d$. Since $d(t, t_{sol}) \leq d$, for each subset R of $\{s \not\equiv t\}$ with $|R| = d+1$, there must be at least one position $i \in R$ with $t_{sol}[i] = s[i]$. The algorithm in [10] first chooses an arbitrary subset R of $\{s \not\equiv t\}$ with $|R| = d + 1$, then simply guesses one position $i \in R$, and further changes $t[i]$ to $s[i]$. This reduces b by one. Thus, the degree of the search tree is $d + 1$ and the height of the tree is d. The search tree size is hence bounded by $(d + 1)^{d+1} / d$.

The algorithm in [13] guesses the partial string $t_{sol}|_{\{s \not\equiv t\}}$ (by carefully enumerating all legible choices) and changes t to $t|_{\{s \equiv t\}} + t_{sol}|_{\{s \not\equiv t\}}$. This gives a much greater degree of the search tree than the algorithm in [10]. However, it was shown in [13] that this strategy at least halves the parameter b for the next iteration. Thus, the height of the tree is at most $O(\log d)$. The search tree size can then be bounded by $(O(|\Sigma|))^d$, which is polynomial when $d = O(\log(nL))$ and $|\Sigma|$ is a constant.

The guessing strategy was further refined in [2] as follows. Recall that P is the set of positions of t that have been modified in previous iterations. Suppose there are k positions in $\{s \not\equiv t\} \setminus P$ where t_{sol} and t differ. Out of these k positions, suppose there are $c \leq k$ positions where t_{sol} is different from both t and s. Therefore, at each of the c positions, we need to guess the letter of t_{sol} from

$|\Sigma| - 2$ choices. But at the other $k - c$ positions, we do not need to guess and can simply let t_{sol} be equal to s. So, when c is small, the degree of the search tree node is reduced. On the other hand, when c is large, the following lemma proved in [2] shows that b is greatly reduced for the next iteration, yielding a smaller search tree height. With this lemma, a further improved time complexity is proved in [2]. The algorithm is given in Figure 1.

Lemma 1. *[2] Let $\langle \mathcal{S}, d; t, P, b\rangle$ be an instance of ECSP with a solution t_{sol}. Suppose that s is a string in $\mathcal{S}$ with $d(t, s) = d + \ell > d$. Let k be the number of positions in $\{s \not\equiv t\} \setminus P$ where t_{sol} is different from t. Let c be the number of positions in $\{s \not\equiv t\} \setminus P$ where t_{sol} is different from both t and s. Let b' be the number of positions in $[1..L] \setminus (P \cup \{s \not\equiv t\})$ where t_{sol} is different from t. Then, $b' \le b - k$ and $b' \le k - \ell - c$. Consequently, $b' \le \frac{b-\ell-c}{2}$.*

The 2-String Algorithm

Input: An instance $\langle \mathcal{S}, d; t, P, b\rangle$ of ECSP.
Output: A solution to $\langle \mathcal{S}, d; t, P, b\rangle$ if one exists, or NULL otherwise.

1. If there is no $s \in \mathcal{S}$ with $d(t, s) > d$, then output t and halt.
2. If $d = b$, then find a string $s \in \mathcal{S}$ such that $d(t, s)$ is maximized over all strings in $\mathcal{S}$; otherwise, find an arbitrary string $s \in \mathcal{S}$ such that $d(t, s) > d$.
3. Let $\ell = d(t, s) - d$ and $R = \{s \not\equiv t\} \setminus P$.
4. If $\ell > \min\{b, |R|\}$, then return NULL.
5. *Guess* $t_{sol}|_R$ by the following steps:
  5.1 *Guess* two sets X and Y such that $Y \subseteq X \subseteq R$, $\ell \le |X| \le b$, and $|Y| \le |X| - \ell$.
  5.2 For each $i \in Y$, *guess* a letter z_i different from both $s[i]$ and $t[i]$. Let the partial string $\hat{s}|_Y$ be such that $\hat{s}|_Y[i] = z_i$ for all $i \in Y$.
  5.3 Let $t_{sol}|_R = \hat{s}|_Y + s|_{X\setminus Y} + t|_{R\setminus X}$.
6. Let $t' = t_{sol}|_R + t|_{[1..|t|]\setminus R}$ and $b' = \min\{b - |X|, |X| - \ell - |Y|\}$.
7. Solve $\langle \mathcal{S} - \{s\}, d; t', P \cup R, b'\rangle$ recursively.
8. Return NULL.

Fig. 1. The algorithm given in [2]

The execution of the 2-string algorithm on input $\langle \mathcal{S}, d; t, P, b\rangle$ can be modeled by a tree $\mathcal{T}$ in which the root corresponds to $\langle \mathcal{S}, d; t, P, b\rangle$, each other node corresponds to a recursive call, and a recursive call A is a child of another call B if and only if B calls A directly. We call $\mathcal{T}$ the *search tree* on input $\langle \mathcal{S}, d; t, P, b\rangle$. By the construction of the algorithm, each non-leaf node in $\mathcal{T}$ has at least two children. Thus, the number of nodes in $\mathcal{T}$ is at most twice the number of leaves in $\mathcal{T}$. Consequently, we can focus on how to bound the number of leaves in $\mathcal{T}$. For convenience, we define the *size* of $\mathcal{T}$ to be the number of its leaves. The *depth* of a node u in $\mathcal{T}$ is the distance between the root and u in $\mathcal{T}$. In particular, the depth of the root is 0. The *depth* of $\mathcal{T}$ is the maximum depth of a node in $\mathcal{T}$.

During the execution of the algorithm on input $\langle \mathcal{S}, d; t, P, b\rangle$, d does not change but the other parameters may change. We use $\mathcal{S}_u$, t_u, P_u, and b_u to denote the values of $\mathcal{S}$, t, P, and b when the algorithm enters the node u (i.e., makes the recursive call corresponding to u). Moreover, we use s_u and ℓ_u to denote the string s and the integer ℓ computed in Steps 2 and 3 of the algorithm at node u. Let $T(d, b_u)$ denote the size of the subtree rooted at u.

Lemma 2. *[2]* $T(d, b_u) \leq \binom{\lfloor \frac{2d - d(t_u, t_r) + \ell_r + b_u}{2} \rfloor}{b_u} \left(|\Sigma| + 2\sqrt{|\Sigma| - 1}\right)^{b_u}$ *for each descendant* u *of* r *in* $\mathcal{T}$.

Using Lemma 2, we can prove a new lemma for the 2-string algorithm:

Lemma 3. $T(d, b_u) \leq \binom{d - (2^{h-1} - 1)b_u}{b_u} \left(|\Sigma| + 2\sqrt{|\Sigma| - 1}\right)^{b_u}$ *for each node* u *at depth* $h \geq 2$ *in* $\mathcal{T}$.

4 The 3-String Algorithm for the Binary Case

In addition to Lemma 3, the improvements made in this paper mainly come from a new strategy to collapse the first two levels (the root and its children) of the search tree into a single level. We call this new approach the *3-string approach.* In this section, we demonstrate the approach for the binary-alphabet case because of its simplicity. In Section 5, we will extend it to the general case.

Given an instance $(\mathcal{S}, d)$ of CSP such that there are at most two strings in $\mathcal{S}$, we can solve it trivially in linear time. So, we hereafter assume that each given instance $(\mathcal{S}, d)$ of the problem satisfies that $|\mathcal{S}| \geq 3$.

4.1 First Trick: *Guessing* Ahead

Let us briefly go through the first two levels of the search tree of the 2-string algorithm. The algorithm starts by initializing t_r to be an arbitrary string in $\mathcal{S}$. At the root r, it finds a string $s_r \in \mathcal{S}$ that maximizes $d(t_r, s_r)$. It then uses s_r to modify t_r and further enters a child node u of r (i.e., makes a recursive call). Note that t_r has become t_u at u. Suppose that the subtree $\mathcal{T}_u$ of $\mathcal{T}$ rooted at u contains a solution t_{sol}. The algorithm then finds a string $s_u \in \mathcal{S}$ such that $d(t_u, s_u) > d$ in Step 2. Note that $P_r = \{t_r \not\equiv s_r\}$ and $P_r \setminus P_u = \{t_r \equiv s_r \not\equiv s_u\}$.

The main idea of the 3-string approach is that we *guess* s_u at the very beginning of the algorithm, instead of finding it in the second-level recursion. This will immediately increase the time complexity by a factor of $n - 2$ because there are $n - 2$ choices of s_u. However, with all of the three strings t_r, s_r, and s_u in hand, we will be able to *guess* $t_{sol}|_{P_u}$ easier, which leads to a better time complexity. The trade-off is a good one when d is large. In fact, we do not even need to trade off. In Subsection 4.2, we will introduce another trick to get rid of this factor of $n - 2$.

Lemma 4. *Let* $(\mathcal{S}, d)$ *be an instance of the binary case of CSP. Suppose that* t_r, s_r, *and* s_u *are three strings in* $\mathcal{S}$ *and* t_{sol} *is a solution of* $(\mathcal{S}, d)$. *Let* $P_r = \{t_r \not\equiv s_r\}$, $R_u = \{t_r \equiv s_r \not\equiv s_u\}$, $P' = \{t_{sol} \not\equiv s_u\} \cap P_r$, $R' = \{t_{sol} \not\equiv t_r\} \cap R_u$, *and* $B = \{t_r \equiv s_r \equiv s_u \not\equiv t_{sol}\}$. *Then,* $|P_r| + |P'| + |R_u| + |R'| \leq 3d - 3|B|$.

Lemma 4 suggests that we can construct $t_{sol}|_{P_r \cup R_u}$ by guessing P' and R', then use $\bar{s}_u|_{P'} + s_u|_{P_r \setminus P'} + \bar{t}_r|_{R'} + t_r|_{R_u \setminus R'}$. For positions in B, we can call the 2-string algorithm to solve it (recursively). Because of the bound $|P_r| + |P'| + |R_u| + |R'| \leq 3d - 3|B|$, we either have a smaller number of choices for guessing P' and R', or have a smaller $|B|$ which makes the recursive call of the 2-string algorithm easier. Likely this will lead to a more efficient algorithm than doing the 2-string algorithm from the beginning. We detail the 3-string algorithm for the binary case of CSP in Figure 2.

To analyze the complexity of the algorithm, we need a simple lemma:

Lemma 5. $\binom{p}{k} \leq \left(\frac{1+\sqrt{5}}{2}\right)^{k+p}$. *For* $p \geq 3k$, $\binom{p}{k} \leq \left(\sqrt[4]{6.75}\right)^{k+p}$.

Using Lemma 5, we can now prove our first main theorem:

The 3-String Algorithm for the Binary Case

Input: An instance $(\mathcal{S}, d)$ of the binary case of CSP.
Output: A solution to $(\mathcal{S}, d)$ if one exists, or NULL otherwise.

1. Select an arbitrary string $t_r \in \mathcal{S}$.
2. If there is no $s \in \mathcal{S}$ with $d(t_r, s) > d$, then output t_r and halt.
3. Find a string $s_r \in \mathcal{S}$ such that $d(t_r, s_r)$ is maximized.
4. Let $P_r = \{t_r \not\equiv s_r\}$. If $|P_r| > 2d$, then output NULL and halt.
5. *Guess* a string $s_u \in \mathcal{S} \setminus \{t_r, s_r\}$.
6. *Guess* a subset P' of P_r with $|P'| \leq d$.
7. Let $t' = \bar{s}_u|_{P'} + s_u|_{P_r \setminus P'} + t_r|_{[1..|t_r|] \setminus P_r}$.
8. If there is no $s \in \mathcal{S}$ with $d(t', s) > d$, then output t' and halt.
9. If $d(t', t_r) \leq d$, $d(t', s_r) \leq d$, and $d(t', s_u) > d$, then perform the following:
 9.1. *Guess* a subset R' of R_u such that $|R'| \leq 3d - |P_r| - |R_u| - |P'|$, where $R_u = \{t_r \equiv s_r \not\equiv s_u\}$. (*Comment:* The upper bound on $|R'|$ used in this step comes from Lemma 4.)
 9.2. Let $t = t'|_{P_r} + \bar{t}_r|_{R'} + t_r|_{[1..|t_r|] \setminus (P_r \cup R')}$.
 9.3. If $d(t, t_r) \leq d$, $d(t, s_r) \leq d$, and $d(t, s_u) \leq d$, then do the following:
  9.3.1. Compute $\ell_r = d(t_r, s_r) - d$, $k_1 = d(t_r, t')$, $b_1 = \min\{d - k_1, k_1 - \ell_r\}$, $k_2 = |R'|$, $\ell_2 = d(t', s_u) - d$, and $b_2 = \min\{b_1 - k_2, k_2 - \ell_2, (3d - |P_r| - |R_u| - |P'| - |R'|)/3\}$. (*Comment:* Obviously, $b_1 = \min\{d - k_1, k_1 - \ell_r\}$ mimics the computation of b' in Step 6 of the 2-string algorithm on input $\langle \mathcal{S} \setminus \{t_r\}, d; t_r, \emptyset, d \rangle$, while $b_2 \leq \min\{b_1 - k_2, k_2 - \ell_2\}$ mimics the computation of b' in Step 6 of the 2-string algorithm on input $\langle \mathcal{S} \setminus \{t_r, s_r\}, d; t', P_r, b_1 \rangle$. Moreover, $b_2 \leq (3d - |P_r| - |R_u| - |P'| - |R'|)/3$ follows from Lemma 4.)
  9.3.2. Call the 2-string algorithm to solve $\langle \mathcal{S} \setminus \{t_r, s_r, s_u\}, d; t, P_r \cup R_u, b_2 \rangle$.
10. Output NULL and halt.

Fig. 2. The 3-string algorithm for the binary case

Theorem 1. *The algorithm in Figure 2 runs in* $O\left(nL + n^2d^2 6.731^d\right)$ *time.*

4.2 Second Trick: Avoiding Guessing s_u

The crux is to modify Step 5 of the algorithm in Subsection 4.1 as follows:

5. Find a string $s_u \in \mathcal{S} \setminus \{t_r, s_r\}$ such that $|\{t_r \equiv s_r \not\equiv s_u\}|$ is maximized.

The problem caused by this change is that in Step 9 of the algorithm, we cannot guarantee $d(t', s_u) > d$ any more. Thus, if $d(t', s_u) \leq d$, we want to replace s_u by a new string $\tilde{s}_u$ such that $d(t', \tilde{s}_u) > d$. More specifically, Step 9 of the algorithm in Subsection 4.1 is replaced by the following:

9. If $d(t', t_r) \leq d$ and $d(t', s_r) \leq d$, then perform the following steps:

9.0. If $d(t', s_u) \leq d$, then select an arbitrary string $\tilde{s}_u \in \mathcal{S} \setminus \{t_r, s_r, s_u\}$ with $d(t', \tilde{s}_u) > d$, and further let s_u refer to the same string as $\tilde{s}_u$ does. (*Comment:* Since $\max\{d(t', t_r), d(t', s_r), d(t', s_u)\} \leq d$ but t' is not a solution, $\tilde{s}_u$ must exist.)

9.1 – 9.3. Same as those in the algorithm in Subsection 4.1, respectively.

The key point here is the following lemma:

Lemma 6. *Let t_{sol} be a solution of $(\mathcal{S}, d)$. Consider the time point where the refined algorithm just selected $\tilde{s}_u$ in Step 9.0 but has not let s_u refer to the same string as $\tilde{s}_u$ does. Then, $|P'| < d(t'|_{P_r}, \tilde{s}_u|_{P_r})$. Moreover, $|P_r| + |P'| + |\tilde{R}_u| + |R'| \leq 3d - 3|\tilde{B}|$, where $\tilde{R}_u = \{t_r \equiv s_r \not\equiv \tilde{s}_u\}$, $R' = \{t_{sol} \not\equiv t_r\} \cap \tilde{R}_u$, and $\tilde{B} = \{t_r \equiv s_r \equiv \tilde{s}_u \not\equiv t_{sol}\}$.*

Theorem 2. *The refined 3-string algorithm solves the binary case of CSP in* $O\left(nL + nd^3 6.731^d\right)$ *time.*

The previously best time complexity for the binary case is $O\left(nL + nd8^d\right)$ [2, 21]. Clearly the new algorithm is better when d is large.

5 Extension to Arbitrary Alphabets

There are two main ideas behind the algorithm in Section 4. One is to use Lemma 4 to obtain a better bound on $|B|$. The other is to obtain $t_{sol}|_{P_r}$ from $s_u|_{P_r}$ by modifying $s_u|_{P'}$ (instead of obtaining $t_{sol}|_{P_r}$ from $t_r|_{P_r}$ by modifying $t_r|_{P'}$ as in the 2-string algorithm). It is easy to show that Lemma 4 still holds for arbitrary alphabets. However, the following lemma is stronger than Lemma 4:

Lemma 7. *Suppose that t_r, s_r, and s_u are three strings of the same length L and t_{sol} is another string of length L with $d(t_{sol}, t_r) \leq d$, $d(t_{sol}, s_r) \leq d$, and $d(t_{sol}, s_u) \leq d$. Let $P_r = \{t_r \not\equiv s_r\}$, $R_u = \{t_r \equiv s_r \not\equiv s_u\}$, $P' = \{t_{sol} \not\equiv s_u\} \cap P_r$, $R' = \{t_{sol} \not\equiv t_r\} \cap R_u$, $B = \{t_r \equiv s_r \equiv s_u \not\equiv t_{sol}\}$, $C_1 = \{s_u \equiv t_r \not\equiv s_r \not\equiv t_{sol}\}$, $C_2 = \{s_u \equiv s_r \not\equiv t_r \not\equiv t_{sol}\}$, $C_3 = \{t_r \not\equiv s_r \not\equiv s_u \not\equiv t_{sol}\}$, $C_4 = \{s_u \equiv t_{sol} \not\equiv t_r \not\equiv s_r\}$, and $C_5 = \{t_r \equiv s_r \not\equiv s_u \not\equiv t_{sol}\}$. Then, $|P_r| + |P'| + |R_u| + |R'| \leq 3d - 3|B| - \sum_{i=1}^{5} |C_i|$.*

- $A_1 = \{t_r \equiv s_u \equiv t_{sol} \not\equiv s_r\}$.
- $A_2 = \{t_r \equiv s_u \not\equiv s_r \equiv t_{sol}\}$.
- $A_3 = \{t_r \equiv t_{sol} \not\equiv s_r \equiv s_u\}$.
- $A_4 = \{t_r \equiv t_{sol} \not\equiv s_r \not\equiv s_u\}$.
- $A_5 = \{s_r \equiv t_{sol} \not\equiv t_r \not\equiv s_u\}$.
- $A_6 = \{s_r \equiv s_u \equiv t_{sol} \not\equiv t_r\}$.
- $A_7 = \{t_r \equiv s_r \equiv t_{sol} \not\equiv s_u\}$.
- $A_8 = \{t_r \equiv s_r \not\equiv s_u \equiv t_{sol}\}$.
- $X_1 = A_2 \cup C_1$.
- $X_2 = C_2 \cup A_3$.
- If $|A_4| \leq |C_4|$, then $X_3 = A_4 \cup A_5 \cup C_3$; otherwise, $X_3 = A_5 \cup C_3 \cup C_4$.

As in the binary case, to compute t_{sol}, our algorithm will first use $t_r|_{P_r \cup R_u}$, $s_r|_{P_r \cup R_u}$, and $s_u|_{P_r \cup R_u}$ to compute $t_{sol}|_{P_r \cup R_u}$, and then call the 2-string algorithm to compute $t_{sol}|_{[1..|t_{sol}|] \setminus (P_r \cup R_u)}$. We here explain how to compute $t_{sol}|_{P_r \cup R_u}$. If we know A_1 through A_8, then we can use the three strings t_r, s_r, and s_u to figure out $t_{sol}|_A$, where $A = \bigcup_{i=1}^{8} A_i$. Unfortunately, we do not know A_1 through A_8. Our idea is then to *guess* X_1, X_2, X_3, and R'. In this way, since we know $A_1 \cup X_1 = \{t_r \equiv s_u \not\equiv s_r\}$, $A_6 \cup X_2 = \{s_r \equiv s_u \not\equiv t_r\}$, and either $C_4 \cup X_3 = \{t_r \not\equiv s_r \not\equiv s_u\}$ or $A_4 \cup X_3 = \{t_r \not\equiv s_r \not\equiv s_u\}$, we can find out A_1, A_6, A_7, and either C_4 or A_4. Using X_1, X_2, R', and X_3, we can then *guess* C_1, C_2, C_5, and either $C_3 \cup A_4$ or $C_3 \cup C_4$. Now, we know A_2, A_3, A_5, and A_8. Note that for each position i in one of the four guessed sets C_1, C_2, C_5, and either $C_3 \cup A_4$ or $C_3 \cup C_4$, we can *guess* $t_{sol}[i]$ among only $|\Sigma| - 2$ choices.

6 Conclusion

We have proposed a new 3-string technique for designing more efficient parameterized algorithm for the closest string problem. The time complexity for binary case of the problem is improved from the previously best known $O(nL + nd8^d)$ to $O(nL + nd^3 6.731^d)$. The technique also works for larger alphabet size but our analysis only showed improvement when the alphabet size is no more than 4.

The results are more of theoretical interest at this moment: the new algorithm is faster only for relatively large d (say, $d > 20$), while the current implementations of the parameterized algorithms typically can only solve instances with $d < 15$ in minutes. However, the 3-string technique can potentially be combined with other techniques to develop significantly more efficient computer programs.

Acknowledgment

ZZC is supported by the Grant-in-Aid for Scientific Research of the MEXT of Japan (Grant No. 20500021). BM is supported by a China 863 Program (2008AA02Z313) and NSERC (RGPIN 238748-2006). LW is fully supported by a grant from City University of Hong Kong (Project No. 7002452).

References

1. Ben-Dor, A., Lancia, G., Perone, J., Ravi, R.: Banishing bias from consensus sequences. In: Hein, J., Apostolico, A. (eds.) CPM 1997. LNCS, vol. 1264, pp. 247–261. Springer, Heidelberg (1997)

2. Chen, Z.Z., Wang, L.: Article 3A fast exact algorithm for the closest substring problem and its application to the planted (l, d)-motif model. In: TCBB (2009) (submitted for publication)
3. Deng, X., Li, G., Li, Z., Ma, B., Wang, L.: Genetic design of drugs without side-effects. SIAM Journal on Computing 32(4), 1073–1090 (2003)
4. Dopazo, J., Rodríguez, A., Sáiz, J.C., Sobrino, F.: Design of primers for PCR amplification of highly variable genomes. CABIOS 9, 123–125 (1993)
5. Downey, R.G., Fellows, M.R.: Parameterized complexity. Monographs in Computer Science. Springer, New York (1999)
6. Evans, P.A., Smith, A.D.: Complexity of approximating closest substring problems. In: Lingas, A., Nilsson, B.J. (eds.) FCT 2003. LNCS, vol. 2751, pp. 210–221. Springer, Heidelberg (2003)
7. Fellows, M.R., Gramm, J., Niedermeier, R.: On the parameterized intractability of motif search problems. Combinatorica 26(2), 141–167 (2006)
8. Frances, M., Litman, A.: On covering problems of codes. Theoretical Computer Science 30, 113–119 (1997)
9. Gramm, J., Guo, J., Niedermeier, R.: On exact and approximation algorithms for distinguishing substring selection. In: Lingas, A., Nilsson, B.J. (eds.) FCT 2003. LNCS, vol. 2751, pp. 159–209. Springer, Heidelberg (2003)
10. Gramm, J., Niedermeier, R., Rossmanith, P.: Fixed-parameter algorithms for closest string and related problems. Algorithmica 37, 25–42 (2003)
11. Lanctot, K., Li, M., Ma, B., Wang, S., Zhang, L.: Distinguishing string search problems. In: SODA 1999, pp. 633–642 (1999)
12. Li, M., Ma, B., Wang, L.: On the closest string and substring problems. Journal of the ACM 49(2), 157–171 (2002)
13. Ma, B., Sun, X.: More efficient algorithms for closest string and substring problems. In: RECOMB 2008, pp. 396–409 (2008)
14. Marx, D.: The closest substring problem with small distances. In: FOCS 2005, pp. 63–72 (2005)
15. Meneses, C.N., Lu, Z., Oliveira, C.A.S., Pardalos, P.M.: Optimal solutions for the closest-string problem via integer programming. INFORMS Journal on Computing (2004)
16. Nicolas, F., Rivals, E.: Complexities of the centre and median string problems. In: Proceedings of the 14th Annual Symposium on Combinatorial Pattern Matching, pp. 315–327 (2003)
17. Stojanovic, N., Berman, P., Gumucio, D., Hardison, R., Miller, W.: A linear-time algorithm for the 1-mismatch problem. In: Proceedings of the 5th International Workshop on Algorithms and Data Structures, pp. 126–135 (1997)
18. Wang, L., Dong, L.: Randomized algorithms for motif detection. Journal of Bioinformatics and Computational Biology 3(5), 1039–1052 (2005)
19. Wang, L., Zhu, B.: Efficient algorithms for the closest string and distinguishing string selection problems. In: The Third International Frontiers of Algorithmics Workshop, pp. 261–270 (2009)
20. Wang, Y., Chen, W., Li, X., Cheng, B.: Degenerated primer design to amplify the heavy chain variable region from immunoglobulin cDNA. BMC Bioinformatics 7(Suppl. 4), S9 (2006)
21. Zhao, R., Zhang, N.: A more efficient closest string algorithm. In: 2nd International Conference on Bioinformatics and Computational Biology (to appear, 2010)

A $2k$ Kernel for the Cluster Editing Problem*

Jianer Chen and Jie Meng

Department of Computer Science and Engineering
Texas A&M University
College Station, TX 77843, USA
{chen,jmeng}@cse.tamu.edu

Abstract. The CLUSTER EDITING problem for a given graph G and a given parameter k asks if one can apply at most k edge insertion/deletion operations on G so that the resulting graph is a union of disjoint cliques. The problem has attracted much attention because of its applications in bioinformatics. In this paper, we present a polynomial time kernelization algorithm for the problem that produces a kernel of size bounded by $2k$, improving the previously best kernel of size $4k$ for the problem.

1 Introduction

The CLUSTER EDITING problem is formulated as follows: given a graph G and a parameter k, is it possible to apply at most k edge insertion/deletion operations so that the resulting graph becomes a union of disjoint cliques?

The CLUSTER EDITING problem arises from many application areas [13]. In particular, it has been studied by a number of research groups in biological research [3,4,15]. An example of this line of research is the analysis of gene expression data, in which a critical step is to identify the groups of genes that manifest similar expression patterns. The corresponding problem in algorithmic research is the GENE CLUSTERING problem [15]. An instance of the GENE CLUSTERING problem consists of a set of genes, and a measure of similarity of genes. A threshold can be used to differentiate the similarity of the genes. The goal is to partition the genes into *clusters* that achieve both *homogeneity* (genes in the same cluster are highly similar) and *separation* (genes from different clusters have low similarity) criteria.

Therefore, when an instance of the GENE CLUSTERING problem is given, and a measure threshold is provided, we can represent the instance as a graph G, whose vertices correspond to the genes and whose edges correspond to the high similarity between the genes. Ideally, if the similarity measure is perfect and the measure threshold is precise, the graph G should be a union of disjoint cliques. Unfortunately, the nature of biological research provides biological data by which the graph G can only be "close" to a union of disjoint cliques. This motivates the algorithmic research of the CLUSTER EDITING problem, which tries to "correct"

* This work was supported in part by the USA National Science Foundation under the Grants CCF-0830455 and CCF-0917288.

M.T. Thai and S. Sahni (Eds.): COCOON 2010, LNCS 6196, pp. 459–468, 2010.

a small number k of similarity pairs (i.e., apply a small number k of edge insertion/deletion operations on the graph G) so that the resulting graph becomes a union of disjoint cliques. There has been extensive algorithmic research on the CLUSTER EDITING problem. The optimization version of the problem was first studied by Ben-dor, Shamir and Yakhini [1]. Shamir, Sharan, and Tsur [16,17] proved that the problem is NP-hard. Approximation algorithms for the problem have been studied. Currently the best polynomial time approximation algorithm for the problem has an approximation ratio 2.5 [18]. It is also known that the problem is APX-complete, thus it is unlikely that the problem has a polynomial time approximation scheme [2].

Given the fact that the parameter k is small in the applications of bioinformatics, research on parameterized algorithms and complexity for the CLUSTER EDITING problem has become active recently [6,7,8,9,10,12,17]. The first parameterized algorithm of time $O(2.27^k + n^3)$ for the CLUSTER EDITING problem was developed in [9], which was later improved to $O(1.92^k + n^3)$ [8]. A research direction closely related to the parameterized algorithms is the study of the kernelization of the problem. We say that the problem CLUSTER EDITING *has a kernel of size* $g(k)$ if there is a polynomial-time algorithm that reduces an instance (G, k) of the problem to an equivalent instance (G', k') where the graph G' has at most $g(k)$ vertices and $k' \leq k$. Gramm *et al* [9] showed that the CLUSTER EDITING problem has a kernel of size $2k^2 + k$. Fellows [6] announced an improved kernel of size $24k$ for the problem, and conjectured that a kernel of size bounded by $6k$ for the problem should exist. The conjecture was confirmed later in [7]. The kernel size for the CLUSTER EDITING problem was further improved to $4k$ by Guo [10] based on the idea of *critical cliques* [5,14].

In this paper, we develop a new polynomial-time kernelization algorithm that provides a kernel of size $2k$ for the CLUSTER EDITING problem, improving the previously best result. Because of the space limit, proofs of some of the theorems and lemmas are omitted and will be given in the complete version of the paper.

2 Reduction Rules 1-4

We start with necessary definitions. A *clique* K in a graph G is a subgraph of G that is a complete graph. A *disjoint clique* is a clique K in which no vertex is adjacent to any vertex not in K. For a vertex v, denote by $N(v)$ the set of vertices that are adjacent to v. For a subset S of vertices, denote by $G[S]$ the subgraph of G that is induced by S, by $N(S)$ the set of vertices that are not in S but adjacent to some vertex in S, i.e., $N(S) = \bigcup_{v \in S} N(v) - S$, and by $N_2(S)$ the neighbors of $N(S)$ that are not in $S \cup N(S)$, i.e., $N_2(S) = N(N(S)) - (S \cup N(S))$.

Definition 1. A *critical clique* K in a graph G is a clique such that for all vertices u and v in K, $N(v) - K = N(u) - K$, and K is maximal under this property.

It has been proved [14] that every vertex in a graph G belongs to a unique critical clique. Therefore, the vertices of the graph G are uniquely partitioned into groups such that each group induces a critical clique. The *critical clique graph* G_c of the graph G is defined as follows. Vertices of G_c correspond to critical cliques in

G, and two vertices in G_c are adjacent if the union of the corresponding critical cliques in G induces a larger clique in G. It is known [11] that for a given graph G, the critical clique graph G_c of G can be constructed in linear time. For a critical clique K, in case there is no confusion, we also denote by K the vertex set of the critical clique.

A *solution* to a graph G for the CLUSTER EDITING problem is a sequence of edge insertion/deletion operations that converts G into a collection of disjoint cliques. The solution to the graph G can be represented by a partition $\mathcal{P} = \{C_1, C_2, \ldots, C_h\}$ of the vertex set of G, where each vertex subset C_i (called a *cluster* of $\mathcal{P}$) becomes a disjoint clique after the edge insertion/deletion operations of the solution. An *optimal solution* to G is a solution that uses the minimum number of edge insertion/deletion operations.

Proposition 2. ([10]) *Let K be a critical clique in a graph G. Then in any optimal solution $\mathcal{P}$ to G, K is entirely contained in a single cluster in $\mathcal{P}$.*

Let S be a vertex subset in a graph G. By *making S a disjoint clique*, we mean to perform the following edge operations: adding edges between pairs of vertices in S that are not adjacent, and deleting edges that are between a vertex in S and a vertex not in S. The analysis of the current paper shows that in many cases, an optimal solution to a graph G for the CLUSTER EDITING problem will make $K \cup N(K)$ a disjoint clique for a critical clique K in G. Motivated by this, we introduce the following definition.

Definition 3. Let K be a critical clique in a graph G and let $v \in N(K)$. The *editing degree* $p_K(v)$ of v with respect to K is defined to be the number of vertex pairs $\{v, w_1\}$, where $w_1 \in N(K) - \{v\}$ and $w_1 \notin N(v)$, plus the number of edges $[v, w_2]$, where $w_2 \notin K \cup N(K)$.

Now we are ready to describe our reduction rules. Let (G, k) be an instance of the CLUSTER EDITING problem, and let K be a critical clique in G.

Reduction Rules 1-4

Rule 1 if $|K| > k$, then make $K \cup N(K)$ a disjoint clique, remove $K \cup N(K)$ from G, and decrease k accordingly;

Rule 2 if $|K| \geq |N(K)|$ and $|K| + |N(K)| > \sum_{v \in N(K)} p_K(v)$, then make $K \cup N(K)$ a disjoint clique, remove $K \cup N(K)$ from G, and decrease k accordingly;

Rule 3 if $|K| < |N(K)|$ and $|K| + |N(K)| > \sum_{v \in N(K)} p_K(v)$, and if there is no vertex $u \in N_2(K)$ with $|N(u) \cap N(K)| > (|K| + |N(K)|)/2$, then make $K \cup N(K)$ a disjoint clique, remove $K \cup N(K)$ from G, and decrease k accordingly;

Rule 4 if $|K| < |N(K)|$ and $|K| + |N(K)| > \sum_{v \in N(K)} p_K(v)$, and if there is a vertex $u \in N_2(K)$ with $|N(u) \cap N(K)| > (|K| + |N(K)|)/2$, then insert necessary edges among vertices in $N(K)$ to make $K \cup N(K)$ a clique, remove edges between $N(K)$ and $N_2(K) - \{u\}$, and decrease k accordingly.

In the remaining of this section, we verify that the above rules are all "safe", i.e., the edge operations applied by each of rules are entirely contained in an optimal solution to the graph G for the CLUSTER EDITING problem.

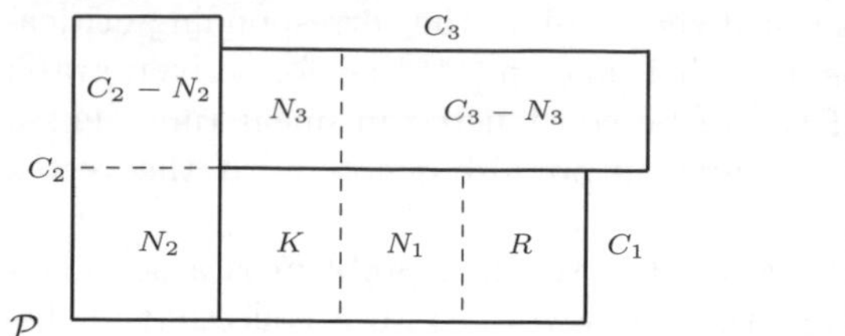

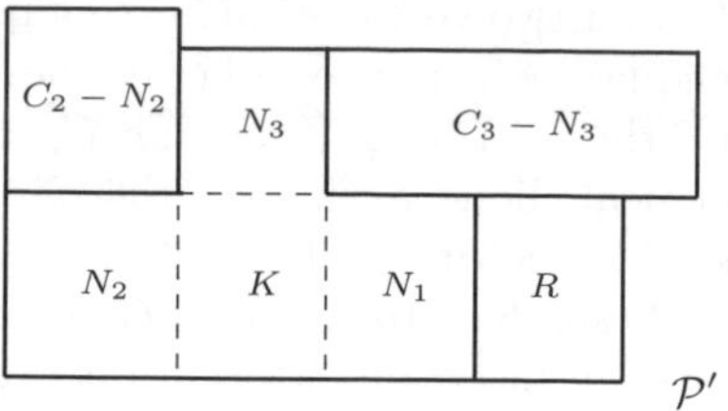

Fig. 1. The critical clique K and the solutions $\mathcal{P}$ and $\mathcal{P}'$

Lemma 4. *Rule 1 is safe.*

Let K be a critical clique in the graph G, and let $\mathcal{P} = \{C_1, C_2, \ldots, C_h\}$ be an optimal solution for G, where C_i, $1 \leq i \leq h$, are the clusters in $\mathcal{P}$. By Proposition 2 and without loss of generality, we can assume $K \subseteq C_1$. Let $N_i = C_i \cap N(K)$ for $1 \leq i \leq h$. Note that some N_i can be empty. Let $R = C_1 - K - N_1$ (see the left figure in Figure 1 for an illustration, where $h = 3$).
We define another solution $\mathcal{P}'$ for G based on the above notations: $\mathcal{P}' = \{K \cup N(K), R, C_2 - N_2, \ldots, C_h - N_h\}$ (see the right figure in Figure 1), and will compare the number of edge operations of the solutions $\mathcal{P}$ and $\mathcal{P}'$.

Besides the edge operations that are common to $\mathcal{P}$ and $\mathcal{P}'$, the solution $\mathcal{P}$ has the following edge operations that are not in $\mathcal{P}'$:

$\mathcal{P}$-operation

(P1) inserting missing edges between K and R;
(P2) deleting all edges between N_i and K for $i \geq 2$;
(P3) inserting missing edges between N_1 and R.
(P4) inserting missing edges between N_i and $C_i - N_i$ for $i \geq 2$;
(P5) deleting all edges between N_i and N_j, for $i \neq j$, $1 \leq i, j \leq h$,

while the solution $\mathcal{P}'$ has the following edge operations that are not in $\mathcal{P}$:

$\mathcal{P}'$-operation

(P1') deleting all edges between N_1 and R;
(P2') inserting missing edges between N_i and N_j, for $i \neq j$, $1 \leq i, j \leq h$;
(P3') deleting all edges between N_i and $C_i - N_i$, for $i \geq 2$.

Lemma 5. *Let G be a graph and let K is a critical clique in G with $|K| \geq |N(K)|$, and for all $v \in N(K)$, $p_K(v) \leq |K|$. Then there is an optimal solution to G that has $K \cup N(K)$ as a cluster.*

Note that for each vertex v in $N(K)$, where K is a critical clique, we can always assume that $p_K(v) \geq 1$. In fact, if $p_K(v) = 0$, then $N(K)$ would consist of a single critical clique and $K \cup N(K)$ would make a disjoint clique in the graph G. Thus, in this case, we can directly reduce the problem instance (G, k) to the smaller instance $(G - (K \cup N(K)), k)$.

Corollary 6. *Rule 2 is safe.*

Lemma 7. *Let K be a critical clique with $|K| < |N(K)|$ and $\sum_{v \in N(K)} p_K(v) < |K| + |N(K)|$. There is an optimal solution $\mathcal{P}_0$ such that $K \cup N(K)$ is entirely contained in a single cluster in $\mathcal{P}_0$.*

Proof. Following the notations in Figure 1, let $\mathcal{P} = \{C_1, C_2, \ldots, C_h\}$ be an optimal solution for the graph G, and let $\mathcal{P}' = \{K \cup N(K), R, C_2 - N_2, \ldots, C_h - N_h\}$. If $N(K) = N_1$, or if $\mathcal{P}'$ is an optimal solution to G, then the lemma is proved. Moreover, if $|N_1| = 0$, then all the $|K| \cdot |N(K)|$ edges between K and $N(K)$ should be deleted by the $\mathcal{P}$-operation set (P2). Thus, the solution $\mathcal{P}$ contains at least $|K| \cdot |N(K)|$ edge operations that are not in $\mathcal{P}'$. On the other hand, because $|N(K)| > |K| \geq 1$ and $\sum_{v \in N(K)} p_K(v) < |K| + |N(K)|$, we have

$$|K| \cdot |N(K)| \geq |K| + |N(K)| - 1 \geq \sum_{v \in N(K)} p_K(v).$$

Since $\sum_{v \in N(K)} p_K(v)$ is an upper bound on the total number of edge operations in the $\mathcal{P}'$-operation sets (P1')-(P3'), the solution $\mathcal{P}'$ contains at most $|K| \cdot |N(K)|$ edge operations that are not in $\mathcal{P}$. But this will imply again that $\mathcal{P}'$ is an optimal solution, thus again proves the lemma.

Thus, in the following discussion, we can assume without loss of generality that $|N(K) - N_1| > 0$, that $\mathcal{P}'$ is not an optimal solution, and that $|N_1| > 0$.

By the lemma assumption $\sum_{v \in N(K)} p_K(v) \leq |K| + |N(K)| - 1$ and by the fact that $p_K(v) \geq 1$ for all $v \in N(K)$, we have

$$\begin{aligned} \sum_{v \in N(K) - N_1} p_K(v) &= \sum_{v \in N(K)} p_K(v) - \sum_{v \in N_1} p_K(v) \\ &\leq (|K| + |N(K)| - 1) - |N_1| = |K| + |N(K) - N_1| - 1. \quad (1) \end{aligned}$$

Note that $\sum_{v \in N(K) - N_1} p_K(v)$ is an upper bound on the total number of missing edges between all N_i and N_j for $i \neq j$ and $1 \leq i, j \leq h$. Since the total number of vertex pairs (v, v') where $v \in N_i$ and $v' \in N_j$ for $i \neq j$ and $0 \leq i, j \leq h$, is not smaller than $|N_1| \cdot |N(K) - N_1|$, we conclude that the total number of existing edges between all N_i and N_j for $i \neq j$ and $1 \leq i, j \leq h$ is at least

$$|N_1| \cdot |N(K) - N_1| - \sum_{v \in N(K) - N_1} p_K(v).$$

Combining this with the inequality (1), we derive that there are at least

$$(|N_1| - 1)(|N(K) - N_1|) - |K| + 1 \quad (2)$$

edges between all N_i and N_j for $i \neq j$ and $1 \leq i, j \leq h$. Note that (2) provides a lower bound on the number of edge deletions by the $\mathcal{P}$-operation set (P5).

The number of edge insertions by the $\mathcal{P}$-operation set (P1) is $|K| \cdot |R|$, and the number of edge deletions by the $\mathcal{P}$-operation set (P2) is $|K| \cdot |N(K) - N_1|$. Combining these with (2), we conclude that the solution $\mathcal{P}$ contains at least

$$(|K| + |N_1| - 1)(|N(K) - N_1|) + |K| \cdot (|R| - 1) + 1 \quad (3)$$

edge operations that are not in the solution $\mathcal{P}'$.

On the other hand, the total number of edge operations in the solution $\mathcal{P}'$ but not in the solution $\mathcal{P}$ is bounded by $\sum_{v \in N(K)} p_K(v)$. By our assumption, $\mathcal{P}'$ is not an optimal solution to G. Thus, we must have

$$(|K|+|N_1|-1)(|N(K)-N_1|)+|K|\cdot(|R|-1)+1 < \sum_{v \in N(K)} p_K(v) \leq |K|+|N(K)|-1.$$

From this inequality, we obtain

$$\begin{aligned} &|K|(|N(K)-N_1|+|R|-2)+|N(K)|(|N(K)-N_1|-1) \\ &< |N(K)-N_1|^2+|N(K)-N_1|-2. \end{aligned} \quad (4)$$

If $|R| > 0$, then since $|N_1| > 0$, we have

$$\begin{aligned} &|K|(|N(K)-N_1|-1)+(|N(K)-N_1|+1)(|N(K)-N_1|-1) \\ &\leq |K|(|N(K)-N_1|-1)+(|N(K)-N_1|+|N_1|)(|N(K)-N_1|-1) \\ &\leq |K|(|N(K)-N_1|+|R|-2)+|N(K)|(|N(K)-N_1|-1) \\ &< |N(K)-N_1|^2+|N(K)-N_1|-2. \end{aligned}$$

The last inequality has used the inequality (4). However, This cannot be true unless $|N(k)-N_1|=0$, a contradiction to our assumption $|N(k)-N_1|>0$.

Thus, we must have $|R| = 0$. As we have analyzed above, the solution $\mathcal{P}$ contains at least $(|K|+|N_1|)(|N(K)-N_1|)-(|K|+(|N(K)-N_1|)-1)$ edge operations that are not in the solution $\mathcal{P}'$ (see (3) and note $|R|=0$). On the other hand, the total number of edge operations in the $\mathcal{P}'$-operation sets (P2') and (P3') is bounded by $\sum_{v \in N(K)-N_1} p_K(v) \leq |K|+|N(K)-N_1|-1$ (see the inequality (1)). Because $|R| = 0$, the $\mathcal{P}'$-operation set (P1') is empty. Thus, the total number of edge operations in $\mathcal{P}'$ that are not in $\mathcal{P}$ is bounded by $|K|+|N(K)-N_1|-1$. Since $\mathcal{P}'$ is not an optimal solution, we must have

$$(|K|+|N_1|)(|N(K)-N_1|)-(|K|+|N(K)-N_1|-1) < |K|+|N(K)-N_1|-1. \quad (5)$$

By our assumption, $|N(K)-N_1|>0$. If $|N(K)-N_1|=1$, then the inequality (5) gives $|N_1|<|K|$, which implies $|N(K)|=|N_1|+|N(K)-N_1|=|N_1|+1 \leq |K|$, contradicting the assumption $|N(K)|>|K|$. If $|N(K)-N_1| \geq 2$, then from the inequality (5) and the assumption $|N_1|>0$, we would have $|K|\cdot(|N(K)-N_1|-2) < |N(K)-N_1|-2$, which implies $|K|<1$ and is again impossible.

The above contradiction verifies that either $\mathcal{P}'$ is an optimal solution that has $K \cup N(K)$ as a cluster, or the optimal solution $\mathcal{P}$ has a cluster (i.e., C_1) that contains $K \cup N(K)$. The lemma now follows directly. □

In fact, we can derive a result that is stronger and more precise than Lemma 7.

Lemma 8. *Let K be a critical clique with $|K| < |N(K)|$ and $\sum_{v \in N(K)} p_K(v) < |K|+|N(K)|$. Then there is at most one vertex u in $N_2(K)$ such that $|N(u) \cap N(K)| > (|K|+|N(K)|)/2$. Moreover, there is an optimal solution $\mathcal{P}$ that either has $K \cup N(K)$ as a cluster, or has a cluster that consists of $K \cup N(K)$ plus a single vertex u in $N_2(K)$ with $|N(u) \cap N(K)| > (|K|+|N(K)|)/2$.*

Corollary 9. *Rule 3 and Rule 4 are safe.*

Proof. By the conditions in the rules, $|K| < |N(K)|$ and $\sum_{v \in N(K)} p_K(v) < |K| + |N(K)|$. By lemma 7, there is an optimal solution $\mathcal{P}$ in which a cluster C_1 contains $K \cup N(K)$. Moreover, by Lemma 8, the cluster C_1 either is $K \cup N(K)$ or consists of $K \cup N(K)$ plus a vertex u in $N_2(K)$. In case C_1 is $K \cup N(K) \cup \{u\}$, the vertex u must satisfy the condition $|N(u) \cap N(K)| > (|K| + |N(K)|)/2$.

Therefore, if no vertex u in $N_2(K)$ satisfies $|N(u) \cap N(K)| > (|K|+|N(K)|)/2$, $K \cup N(K)$ must is a cluster in $\mathcal{P}$. This verifies that Rule 3 is safe.

If there is a vertex u in $N_2(K)$ satisfying $|N(u) \cap N(K)| > (|K| + |N(K)|)/2$, then by Lemma 8, there is only one such a vertex, and the cluster C_1 must be either $K \cup N(K)$ or $K \cup N(K) \cup \{u\}$. In either case, the optimal solution $\mathcal{P}$ has to insert edges among the vertices in $N(K)$ to make $K \cup N(K)$ a clique, and to remove edges between $N(K)$ and $N_2(K) - \{u\}$. Therefore, Rule 4 is safe. □

3 The Pendulum Algorithm and Reduction Rule 5

Let (G, k) be an instance of the CLUSTER EDITING problem. A reduction rule *cannot further reduce the size of* (G, k) if the reduction rule can neither reduce the number of vertices in the graph G nor decrease the parameter value k.

Lemma 10. *Let (G, k) be an instance of* CLUSTER EDITING *such that none of the Reduction Rules 1-4 can further reduce the size of (G, k), and let K be a critical clique in G satisfying $|K| < |N(K)|$ and $\sum_{v \in N(K)} p_K(v) < |K|+|N(K)|$. Then $N_2(K)$ consists of a single vertex u and $N(K)$ is a single critical clique.*

Lemma 10 suggests an interesting special class of instances of the CLUSTER EDITING problem, for which we develop an algorithm, the *Pendulum Algorithm*, that deals with this special case. The algorithm only requires the conditions that $N(K)$ be a single critical clique and that $N_2(K)$ consist of a single vertex, and does not directly depend on the conditions stated in Rules 1-4. Therefore, the algorithm is also of its independent interest. The algorithm is given in Figure 2.

Algorithm **Pendulum Algorithm**

INPUT: (G, k, K): *(G, k) is an instance of* CLUSTER EDITING, *K is a critical clique in G such that $N(K)$ is a critical clique and $N_2(K) = \{u\}$*

OUTPUT: *a reduced equivalent instance (G', k') of* CLUSTER EDITING

1. if $|K| \geq |N(K)|$, then make $K \cup N(K)$ a disjoint clique;
   $G' = G - (K \cup N(K))$; $k' = k - |N(K)|$;
2. if $|K| < |N(K)|$, then pick any set U of $|K|$ vertices in $N(K)$;
   $G' = G - (K \cup U)$; $k' = k - |K|$;
3. return (G', k').

Fig. 2. The Pendulum Algorithm

Lemma 11. *The Pendulum Algorithm is correct.*

Lemmas 10 and 11 suggest that we add another reduction rule to the CLUSTER EDITING problem. Let (G, k) be an instance of the CLUSTER EDITING problem, and let K be a critical clique in G.

Reduction Rule 5

Rule 5 if $|K| < |N(K)|$ and $|K| + |N(K)| > \sum_{v \in N(K)} p_K(v)$, and if none of the Reduction Rules 1-4 can further reduce the size of the instance (G, k), then call the Pendulum Algorithm on (G, k, K).

Corollary 12. *Rule 5 is safe.*

Proof. Under the conditions in Rule 5, by Lemma 10, the set $N_2(K)$ consists of a single vertex u and $N(K)$ is a single critical clique. Therefore, (G, k, K) is a valid input to the Pendulum Algorithm. By Lemma 11, calling the Pendulum Algorithm on (G, k, K) will further reduce the size of the instance (G, k) and produce a reduced equivalent instance. □

4 Final Conclusion: The Kernelization Algorithm

Our kernelization algorithm for the CLUSTER EDITING problem is as follows:

Kernelization
Input: an instance (G, k) of the CLUSTER EDITING problem.
1. $(G', k') = (G, k)$;
2. **while** any of the Rules 1–5 can further reduce the size of (G', k')
 do apply the rule and replace (G', k') by the reduced instance;
3. return (G', k').

By Lemma 4, and Corollaries 6, 9, and 12, in the instance (G', k') returned by the algorithm **Kernelization**, the graph G' has a solution of k' edge operations if and only if in the original instance (G, k) the graph G has a solution of k edge operations. Moreover, none of the Reduction Rules 1–5 can further reduce the size of the instance (G', k') returned by the algorithm **Kernelization**.

Theorem 13. *Let (G, k) be an instance of the* CLUSTER EDITING *problem such that none of the Reduction Rules 1–5 can further reduce the size of (G, k). If there is a solution of no more than k edge operations for the graph G, then G contains at most $2k$ vertices.*

Proof. Let $\mathcal{P}$ be an optimal solution for the graph G such that $\mathcal{P}$ contains no more than k edge operations. We say that a vertex v in G is *touched* (by $\mathcal{P}$) if the solution $\mathcal{P}$ inserted or deleted at least one edge incident to v. The vertex v is *untouched* otherwise. For all vertices v in G, denote by $p_O(v)$ the number of edges incident to v that are inserted/deleted by the solution $\mathcal{P}$.

We divide the clusters in the solution $\mathcal{P}$ into two sub-collections: $\mathcal{P}_1$ consists of the clusters in $\mathcal{P}$ in which all vertices are touched, and $\mathcal{P}_2$ consists of the clusters in $\mathcal{P}$ that are not in $\mathcal{P}_1$. Since clusters in $\mathcal{P}_1$ contains only touched vertices and $p_O(v) \geq 1$ for a touched vertex v, we must have

$$\sum_{C \in \mathcal{P}_1} |C| \leq \sum_{C \in \mathcal{P}_1} \sum_{v \in C} p_O(v), \tag{6}$$

where $|C|$ denotes the number of vertices in the cluster C.

Let C be a cluster in $\mathcal{P}_2$, and let K be the set of vertices in C that are untouched, $K \neq \emptyset$. Because C will become a disjoint clique after the edge operations in the solution $\mathcal{P}$ and all vertices in K are untouched by $\mathcal{P}$, the cluster C must be exactly $K \cup N(K)$. In fact, it can be proved that the induced subgraph $G[K]$ is actually a critical clique in G.

Therefore, every cluster C in $\mathcal{P}_2$ can be written as $C = K \cup N(K)$, where K is the set of untouched vertices in C, and K induces a critical clique in G. By the theorem assumption, none of the Reduction Rules 1–5 can further reduce the size of the instance (G, k). In particular, since none of the Reduction Rules 2–5 can further reduce the size of the instance, we have

$$|C| = |K| + |N(K)| \leq \sum_{v \in N(K)} p_K(v) = \sum_{v \in N(K)} p_O(v) = \sum_{v \in C} p_O(v),$$

the equality second to the last holds true because the solution $\mathcal{P}$ makes $K \cup N(K)$ a disjoint clique so by the definition of the editing degree $p_K(v)$ we have $p_O(v) = p_K(v)$ for all vertices v in $N(K)$. The last equality holds true because all vertices in K are untouched so $p_O(v) = 0$ for all vertices v in K. Summarizing this over all clusters in $\mathcal{P}_2$, we get

$$\sum_{C \in \mathcal{P}_2} |C| \leq \sum_{C \in \mathcal{P}_2} \sum_{v \in C} p_O(v). \tag{7}$$

Adding (6) and (7) and letting V be the vertex set of the graph G, we get

$$|V| = \sum_{C \in \mathcal{P}_1} |C| + \sum_{C \in \mathcal{P}_2} |C| \leq \sum_{C \in \mathcal{P}_1} \sum_{v \in C} p_O(v) + \sum_{C \in \mathcal{P}_2} \sum_{v \in C} p_O(v) = \sum_{v \in V} p_O(v). \tag{8}$$

Since each edge operation in the solution $\mathcal{P}$ increases the p_O value by 1 for exactly two vertices in the graph G, the value $\sum_{v \in V} p_O(v)$ is bounded by 2 times the number of edge operations in the solution $\mathcal{P}$. By the assumption, the solution $\mathcal{P}$ contains no more than k edge operations. Therefore, (8) proves that the number of vertices in the graph G is bounded by $2k$. This completes the proof of the theorem. □

The algorithm **Kernelization** and Theorem 13 give our main result of the paper, as follows.

Theorem 14. *There is an $O(m(n + k))$-time kernelization algorithm for the* CLUSTER EDITING *problem, that on an instance (G, k) produces another instance (G', k') such that $k' \leq k$, that G' has at most $2k'$ vertices, and that the graph G has a solution of at most k edge operations if and only if the graph G' has a solution of at most k' edge operations.*

References

1. Ben-dor, A., Shamir, R., Yakhini, Z.: Clustering gene expression patterns. Journal of Computational Biology 6(3/4), 281–297 (1999)
2. Charikar, M., Guruswami, V., Wirth, A.: Clustering with qualitative information. Journal of Computer and System Sciences 71(3), 360–383 (2005)
3. Chen, Z.-Z., Jiang, T., Lin, G.: Computing phylogenetic roots with bounded degrees and errors. SIAM Journal on Computing 32(4), 864–879 (2003)
4. Dehne, F., Langston, M., Luo, X., Pitre, S., Shaw, P., Zhang, Y.: The cluster editing problem: implementations and experiments. In: Bodlaender, H.L., Langston, M.A. (eds.) IWPEC 2006. LNCS, vol. 4169, pp. 13–24. Springer, Heidelberg (2006)
5. Dom, M., Guo, J., Huffner, F., Niedermeier, R.: Extending the tractability border for closest leaf powers. In: Kratsch, D. (ed.) WG 2005. LNCS, vol. 3787, pp. 397–408. Springer, Heidelberg (2005)
6. Fellows, M.: The lost continent of polynomial time: preprocessing and kernelization. In: Bodlaender, H.L., Langston, M.A. (eds.) IWPEC 2006. LNCS, vol. 4169, pp. 276–277. Springer, Heidelberg (2006)
7. Fellows, M., Langston, M., Rosamond, F., Shaw, P.: Efficient parameterized preprocessing for cluster editing. In: Csuhaj-Varjú, E., Ésik, Z. (eds.) FCT 2007. LNCS, vol. 4639, pp. 312–321. Springer, Heidelberg (2007)
8. Gramm, J., Guo, J., Huffner, F., Niedermeier, R.: Automated generation of search tree algorithms for hard graph modification problems. Algorithmica 39(4), 321–347 (2004)
9. Gramm, J., Guo, J., Huffner, F., Niedermeier, R.: Graph-modeled data clustering: exact algorithms for clique generation. Theory of Computing Systems 38(4), 373–392 (2005)
10. Guo, J.: A more effective linear kernelization for cluster edting. Theoretical Computer Science 410(8-10), 718–726 (2009)
11. Hsu, W., Ma, T.: Substitution decomposition on chordal graphs and applications. In: Hsu, W.-L., Lee, R.C.T. (eds.) ISA 1991. LNCS, vol. 557, pp. 52–60. Springer, Heidelberg (1991)
12. Huffner, F., Komusiewicz, C., Moser, H., Niedermeier, R.: Fixed-parameter algorithms for cluster vertex deletion. In: Laber, E.S., Bornstein, C., Nogueira, L.T., Faria, L. (eds.) LATIN 2008. LNCS, vol. 4957, pp. 711–722. Springer, Heidelberg (2008)
13. Jain, A., Dubes, R.: Algorithms for Clustering Data. Prentice Hall, Englewood Cliffs (1988)
14. Lin, G., Kearney, P.E., Jiang, T.: Phylogenetic k-root and steiner k-root. In: Lee, D.T., Teng, S.-H. (eds.) ISAAC 2000. LNCS, vol. 1969, pp. 539–551. Springer, Heidelberg (2000)
15. Shamir, R., Sharan, R.: Algorithmic approaches to clustering gene expression data. In: Jiang, T., Xu, Y., Zhang, M. (eds.) Current Topics in Computational Molecular Biology, pp. 269–299. MIT press, Cambridge (2002)
16. Shamir, R., Sharan, R., Tsur, D.: Cluster graph modification problems. In: Kučera, L. (ed.) WG 2002. LNCS, vol. 2573, pp. 379–390. Springer, Heidelberg (2002)
17. Shamir, R., Sharan, R., Tsur, D.: Cluster graph modification problems. Discrete Applied Mathematics 144, 173–182 (2004)
18. Zuylen, A., Williamson, D.: Deterministic algorithms for rank aggregation and other ranking and clustering problems. In: Kaklamanis, C., Skutella, M. (eds.) WAOA 2007. LNCS, vol. 4927, pp. 260–273. Springer, Heidelberg (2008)

On the Computation of 3D Visibility Skeletons

Sylvain Lazard[1], Christophe Weibel[2], Sue Whitesides[3], and Linqiao Zhang[4]

[1] INRIA Nancy Grand Est, LORIA, Nancy, France
lazard@loria.fr
[2] McGill University, Math Department, Montreal, QC H3A 2K6, Canada
weibel@math.mcgill.ca
[3] University of Victoria, Department of Computer Science, B. C. V8W 3P6, Canada
sue@uvic.ca
[4] McGill University, School of Computer Science, Montreal, QC H3A 2A7, Canada
lzhang15@cs.mcgill.ca

Abstract. The 3D visibility skeleton is a data structure that encodes the global visibility information of a set of 3D objects. While it is useful in answering global visibility queries, its large size often limits its practical use. In this paper, we address this issue by proposing a subset of the visibility skeleton, which is empirically about 25% to 50% of the whole set. We show that the rest of the data structure can be recovered from the subset as needed, partially or completely. The running time complexity, which we analyze in terms of output size, is efficient. We also prove that the subset is minimal in the sense that the complexity bound ceases to hold if the subset is restricted further.

1 Introduction

Problems of 3D visibility arise commonly in areas such as computer graphics, vision, and robotics. Typical problems include computing the view from a given point, computing pairwise visibility information among objects, or computing shadow boundaries cast by a light source. These visibility problems can roughly be classified into point-based or surface-based, depending on the nature of the view points or light source. While point-based visibility problems have been well studied and understood, surface-based visibility problems remain a challenge.

The *visibility complex* is a data structure that encodes global visibility information by partitioning the space of line segments into sets of components, where each component consists of a set of maximal free line segments that share the same occluders, are tangent to the same objects in the scene, and form one connected component. This data structure was initially proposed in 2D by Pocchiola and Vegter [7]. Durand *et al.* extended the visibility complex data structure to 3D and furthermore introduced the 3D visibility skeleton, which essentially consists of the 0D and 1D cells of the visibility complex [5], together with their adjacencies. Thus the 3D visibility skeleton is a simpler, smaller data structure than the full visibility complex. It has the structure of a graph, where vertices represent 0D cells and arcs represent 1D cells.

M.T. Thai and S. Sahni (Eds.): COCOON 2010, LNCS 6196, pp. 469–478, 2010.

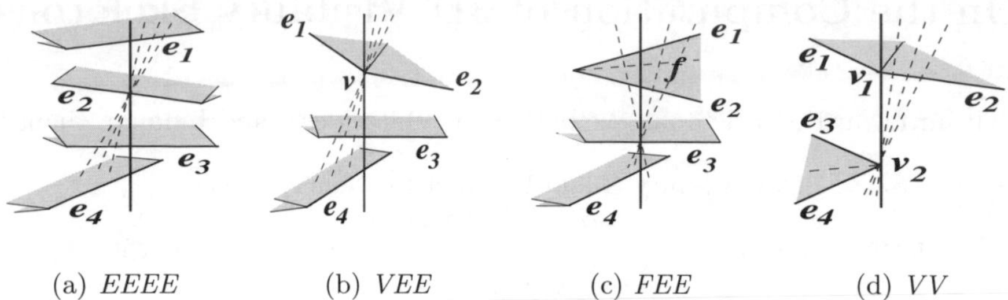

(a) *EEEE* (b) *VEE* (c) *FEE* (d) *VV*

Fig. 1. Some of the 3D visibility skeleton vertices and arcs (dashed lines)

The visibility skeleton has been applied to global illumination computation [5], and has resulted in images with high quality shadow boundaries. However, these applications also show that this data structure, although simplified, can still be enormous and thus has limited practical use.

Various other definitions have been proposed for the 3D visibility skeleton [3,5,6], depending on the notion of visual events, which are based on the context (e.g. smooth manifolds, polyhedral scenes) and the intended use of the data structure (e.g. shadow boundary computation). In this article, we refer to a visibility skeleton based on visual events surfaces for a polytope scene, as defined in Demouth [3].

A visibility skeleton thus defined, with a set of convex disjoint polytopes as input, consists of arcs of so-called type *EEE*, and a subset of arcs of so-called type *VE*, and vertices of so-called types *EEEE*, *VEE*, *FEE*, and a subset of vertices of so-called type *VV* (see Figure 1 and Section 2). For instance, a vertex of type EEEE represents a maximal free line segment that is tangent to four polytopes along four of their edges, and similarly for the other types. Here, the lines supporting the maximal free line segments are required to be tangent to their associated polytopes. Moreover, we only consider among the VV vertices those that correspond to maximal free line segments that lie in some plane tangent to both associated polytopes. This data structure is only a subset of the 3D visibility skeleton defined by Durand [5], which consists of arcs of type *EEE*, *VE* and *FE*, and vertices of types *EEEE*, *VEE*, *FEE*, *VV*, *FvE*, *FF*, *FE* and *FVV*. For convenience, in this paper, we refer to the former definition as a *succinct 3D visibility skeleton*, and to the latter one as a *full 3D visibility skeleton*. We also define vertices contained in the succinct skeleton to be *primary vertices*, and remaining vertices of the full skeleton to be *secondary vertices*.

From the experimental study of Demouth, the size of the succinct 3D visibility skeleton is only about 25% to 50% of the full one [3] (see full paper for details). However, the skeleton vertices and arcs it contains are sufficient to compute the direct shadow boundaries cast by polytopes. While compact and useful on its own, the succinct 3D visibility skeleton does not always contain the necessary visibility information for answering global visibility queries. In particular, arcs *FE* may be needed even though they do not correspond to visual events.

The full visibility skeleton, on the other hand, contains all the necessary information for most visibility queries. However its large size has been seen as an impediment for its practical use [5]. In this paper, we study in detail the 3D visibility skeletons computed from a set of convex disjoint polytopes in general position (defined in Section 2). We prove that knowing the succinct 3D visibility skeleton is sufficient to compute efficiently the secondary vertices; in particular, these computations can be local, that is, only the vertices and arcs of interest need to be computed. The full skeleton can be computed if necessary.

In terms of computing the full skeleton, we prove that, given k disjoint convex polyhedra (henceforth, polytopes) satisfying general position assumptions, with n edges in total, the full visibility skeleton can be computed from the succinct one in $O(p \log p + s \log s)$ time, where p is the number of the primary vertices minus the *EEEE* vertices, and s is the number of secondary vertices. The worst-case size complexity of the primary vertices of interest to us, *EEEE* vertices excluded, is $\Theta(n^2 k)$, and the worst-case size complexity of the secondary vertices is $\Theta(n^2)$ [1]. Thus, in the worst case, $O(p \log p + s \log s)$ is equivalent to $O(n^2 k \log n)$.

There exist various algorithms for computing the secondary vertices. For instance one can use the sweep algorithm described in [1], or one can also compute, in a brute force way, the possibly occluded candidate secondary vertices and perform ray shooting to check for occlusion [4]. The best known worst-case running time is $O(n^2 k \log n)$, obtained by computing $O(n^2)$ candidate secondary vertices in a brute force way and checking for occlusion using the Dobkin-Kirkpatrick hierarchical representation [4], which leads to performing $O(n^2)$ ray shooting queries on each of the k polytopes in $O(\log n)$ time each. Comparatively, the method we propose has the same complexity in the worst case. However, our analysis suggests that our output-dependent method may be much more efficient in many cases. In addition, our method takes as input the primary vertices, whose observed size is, in a random setting, $Ck\sqrt{nk}$ for a small constant C (see [8]), which is much smaller than the worst-case size, that is $\Theta(n^2 k)$.

In the rest of this paper, we provide necessary definitions in Section 2. We introduce the computational relations among the types of visibility skeleton vertices in Section 3. We prove that we can recover the full skeleton from the succinct one in $O(p \log p + s \log s)$ time in Section 4, and show in Section 5 that none of the primary vertex types can be omitted while maintaining the validity of this result. We finally conclude in Section 6.

2 Preliminaries

We introduce the basic definitions we need in this section. We start with some preliminaries in order to explain the types of vertices and arcs of the 3D visibility skeleton of a set of k convex disjoint polytopes in general position. By *general position*, we mean that no edges of different polytopes are parallel, no four points on more than one polytope are coplanar, no line is tangent to more than four polytopes, and no four edges are on a hyperboloid quadratic surface [2].

A *support vertex* of a line is a polytope vertex that lies on the line. A *support edge* of a line is a polytope edge that intersects the line but has no endpoint on it (a support edge intersects the line at only one point of its relative interior). A *support* of a line is one of its support vertices or support edges. A maximal free line segment is a maximal subset of a line that does not intersect the interior of any polytope. The supports of a segment are defined to be the supports of the relative interior of the segment; thus if a maximal free line segment ends at a vertex of a polytope, this vertex is not a support. A *support polytope* of a line is a polytope that a support of the line lies on.

Full 3D visibility skeleton. We introduce this data structure based on the work of Durand [5], and often refer it as the full skeleton. The 3D visibility skeleton is a graph that consists of vertices and arcs. A skeleton vertex is a point in the space of maximal free line segments, and a skeleton arc is a connected sequence of points in the same space, with a skeleton vertex at each extremity. (See Figure 1 for graphical illustrations, and [5] for more details).

There are eight types of skeleton vertices. Note that unless stated otherwise, no two supports come from the same polytope. A skeleton vertex has type *EEEE* if its set of supports consists of four edges; *VEE* if its set of supports consists of a vertex and two edges; *FEE* if its set of supports consists of two edges on one face, and two additional edges; *VV* if its set of supports consists of two vertices; *FF* if its set of supports consists of two edges on one face, and two edges on another face; *FvE* if its set of supports consists of a vertex and an edge on one face, and an edge; *FE* if its set of supports consists of two adjacent vertices of the same polytope; and *FVV* if its set of supports consists of two non-adjacent vertices on the same face of a polytope.

We also consider a special case of *FvE* vertex called an *extremal FvE* vertex, as shown in Figure 2. An extremal *FvE* vertex is also in the plane containing a face incident to the polytope vertex, but does not intersect that face except on the vertex. This ensures that all skeleton arcs have a skeleton vertex at both ends.

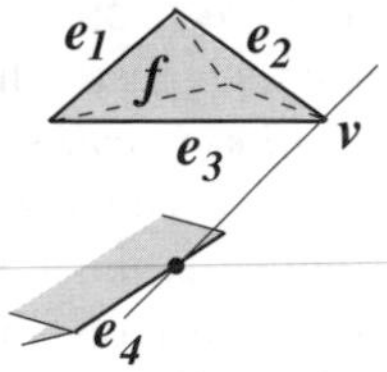

Fig. 2. The extremal case of type *FvE* vertex

There are four types of skeleton arcs. An arc has type: *EEE* if its set of supports consists of three edges; *VE* if its set of supports consists of a vertex and an edge; *FE* if its set of supports consists of two edges on one face, and one additional edge; *FVE* if its set of supports consists of one vertex and one edge on the same face which is not incident to the vertex.

It should be stressed that the maximal free line segment(s) corresponding to a skeleton vertex (arc) is (are) tangent to all its support polytopes.

Succinct 3D visibility skeleton, primary and secondary vertices. We define the *primary vertices* to be the skeleton vertices of types *EEEE*, *VEE*, *FEE*, together with the vertices of type *VV* whose corresponding maximal free line segment lies in a plane tangent to both polytopes; and the *secondary vertices* to be the remaining vertices of type *VV*, and of types *FF*, *FvE*, *FE*, and *FVV*. The succinct 3D visibility skeleton is the subgraph of the full 3D visibility skeleton that contains only primary vertices, and skeleton arcs connecting two primary vertices.

Constraints. Recall that for any skeleton vertex, a support vertex is a polytope vertex that lies on the relative interior of the maximal free line segment, and a support edge is a polytope edge that intersects the maximal free line segment in both of their relative interiors. For any skeleton vertex, we define its *constraints* to be the polytope edges that intersect the maximal free line segment corresponding to the skeleton vertex, and such that the plane containing the edge and the maximal free line segment is tangent to the polytope containing the edge. In other words, the maximal free line segment corresponding to the skeleton vertex can be perturbed so that it intersects the interior of the polytope edge, while remaining tangent to the polytope. We define constraints similarly for arcs. We note that any support edge is a constraint, and that any support vertex is incident to at least two constraints that are not support edges (in the case of extremal *FvE* vertices, there are at least three constraints, and possibly four). A constraint edge that is not a support edge is called a *salient edge.* We note that any skeleton vertex has, under our general position assumption, at least four constraints, and up to seven in some cases. Any skeleton arc has three.

Remark 1. *By continuity, every constraint of a skeleton arc is also a constraint of each of its end vertices.*

3 Computational Relations among the Visibility Skeleton Vertices

Recall from Section 2 that a 3D visibility skeleton vertex corresponds to a maximal free line segment that has 0-degrees of freedom, subject to the condition that the maximal free line segment keeps the same constraints (as defined above).

Remark 2. *We specify that the knowledge of a skeleton vertex includes its corresponding maximal free line segment and its constraint edges, that is, any support polytope edges, any support polytope vertices with incident salient edges, and all support polytopes. The knowledge of a polytope includes cyclic orderings of edges around vertices and faces. Computing a skeleton vertex includes computing its maximal free line segment and all its constraint edges, and similarly for skeleton arcs. Furthermore, we create during preprocessing a binary search data structure on the cyclic orderings of edges around each polytope vertex and face. This is done in $O(n \log n)$, where n is the number of polytope edges in the scene.*

Note that, by Remark 1, and since a skeleton vertex includes the knowledge of its constraint edges, we can determine all skeleton arcs that are incident to a skeleton vertex by removing, in turn, each edge from the list of its constraints. The computation time is constant.

Given a skeleton arc, which includes the knowledge of its constraints and support polytopes, any incident skeleton vertex can be computed without additional knowledge, provided that the skeleton vertex has the same set of support polytopes as the skeleton arc. The computation involves searching edges incident to a polytope vertex to find salient edges, or edges on a polytope face to find the support edges of the incident skeleton vertex. Using the binary search data structure of Remark 2, this requires $O(\log \delta)$ time computation, where δ is the

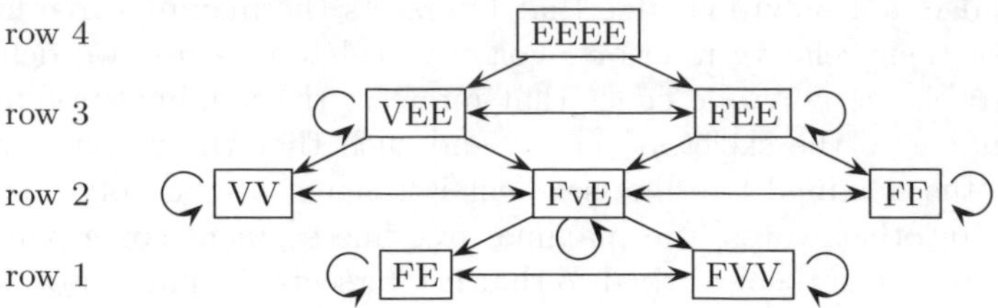

Fig. 3. The possible computational relations among 3D visibility skeleton vertices

maximum degree of a polytope vertex or number of edges on a polytope face. For example, given an *EEE* arc, one can compute the vertices of types *VEE* and *FEE* that are incident to the *EEE* arc in $O(\log \delta)$ time. But, if an *EEE* arc is incident to an *EEEE* vertex, then this *EEEE* vertex cannot be computed directly from the *EEE* arc, since it has an additional, unknown support polytope.

The possible computational relations among skeleton vertex types are summarized in the diagram in Figure 3. The edges in this diagram give all possible pairs of vertex types that can be connected by an arc of the full visibility skeleton. Furthermore, an arrow oriented from one type to another indicates that the set of support polytopes of a vertex of the former type contains that of a vertex of the latter type. Thus, vertices of the latter type can be computed from adjacent vertices of the former type in $O(\log \delta)$ time. Note that skeleton vertices of types appearing in rows 1 through 4 of the diagram in Figure 3 are supported by $1, \ldots, 4$ polytopes, respectively. The computational relations illustrated in Figure 3 are summarized in the next two lemmas, whose proofs are omitted.

Lemma 1. *Let $\mathcal{X}$ and $\mathcal{Y}$ denote vertex types in the full skeleton graph, such that in Figure 3, $\mathcal{X}$ has more support polytopes than $\mathcal{Y}$ (so $\mathcal{X}$ appears in a higher row than $\mathcal{Y}$), and there is an edge directed from $\mathcal{X}$ to $\mathcal{Y}$. Then it is not possible, from a skeleton vertex of type $\mathcal{Y}$, to determine an adjacent skeleton vertex of type $\mathcal{X}$ in time independent of the number of input polytopes, without more knowledge than specified in Remark 2.*

Lemma 2. *If a scene is in general position, then from any skeleton vertex, it is possible to compute each adjacent skeleton vertex in $O(\log \delta)$ time, where δ is the maximum degree of a polytope vertex or number of edges on a polytope face, provided that all support polytopes of the adjacent skeleton vertex are also support polytopes of the given skeleton vertex and of the connecting skeleton arc.*

4 Recovery of the Full Skeleton

We show in this Section that the full visibility skeleton can be recovered from the succinct one in $O(p \log p + s \log s)$ time, where p is the number of primary vertices minus the *EEEE* vertices, and s is the number of secondary vertices.

Recall that the five types of secondary vertices are *FF*, *FvE*, *FVV*, *FE*, and a subset of *VV*. The vertices of type *FVV* and *FE* are easy to find on their own, and can be computed separately (see Remark 3). In this section, we show how to compute vertices of type *FvE*, *FF*, and the subset of *VV* that belongs to the secondary vertices. For this, we explore the subgraph of the full visibility skeleton consisting of *VE* and *FE* arcs and their incident vertices, that is, the *VEE*, *FEE*, *VV*, *FvE* and *FF* vertices. We call this subgraph the *partial graph.*

We first prove that all connected components of the partial graph contain at least a primary vertex of type *VV*, *VEE* or *FEE*. This allows us to find all *FvE*, *FF* and the remaining *VV* vertices by simple graph exploration, examining vertices adjacent to those we have already computed.

To prove that all connected components contain a primary vertex of type *VV*, *VEE* or *FEE*, we proceed as follows. If a pair P_i, P_j of polytopes supports a skeleton vertex, we define G_{ij} to be the subgraph consisting of all skeleton vertices and arcs that have P_i and P_j as supports. The following theorem shows that each connected component of each subgraph G_{ij} contains a primary vertex of the specified types. This allows us to enumerate each subgraph separately starting from these primary vertices.

Theorem 1. *Let P_i and P_j be a pair of polytopes that supports a skeleton vertex. Then each connected component of the subgraph G_{ij} defined as above contains a primary vertex.*

Proof sketch. The proof uses an optimization concept. For each subgraph G_{ij}, we define an *objective function* on the maximum free line segments corresponding to the vertices and arcs of the subgraph. We then prove that each local minimum of each connected component of the subgraph is a primary vertex of type *VV*, *VEE*, or *FEE*. As the subgraph is finite, it must contain such a primary vertex. □

We now show how to explore the partial graph, in order to compute the secondary vertices of types *VV*, *FvE* and *FF* from primary vertices of types *VEE*, *FEE* and *VV*. To explore the partial graph efficiently, we compute the secondary vertices of type *VV*, *FvE* or *FF* that occur in a certain sequence of *VE* or *FE* arcs. In what follows, we will show that this can be done in time $O(p' \log p + s')$, where p' and s' are the number, in the considered sequence, of primary and secondary vertices respectively, and p is the total number of primary vertices that are not *EEEE* in the succinct visibility skeleton.

We consider *VE* arcs and *FE* arcs separately. We define a *sequence of VE arcs* to be a maximal path of *VE* arcs that all share the same support edge and support vertex. We define a *sequence of FE arcs* to be a maximal path of *FE* arcs that share the same support polytope face, and that are tangent to the same other polytope. Note that our definition does not require a sequence of *FE* arcs to be supported by the same edge of the other polytope. The position in these sequences of skeleton vertices of different types is explained in the two following lemmas. The proofs follow from the properties of skeleton vertices.

Lemma 3. *Any sequence of* VE *arcs has a* VV *vertex, a* VEE *vertex, or a non-extremal* FvE *vertex at each extremity, and arcs in the sequence are separated by extremal* FvE *vertices or* VEE *vertices.*

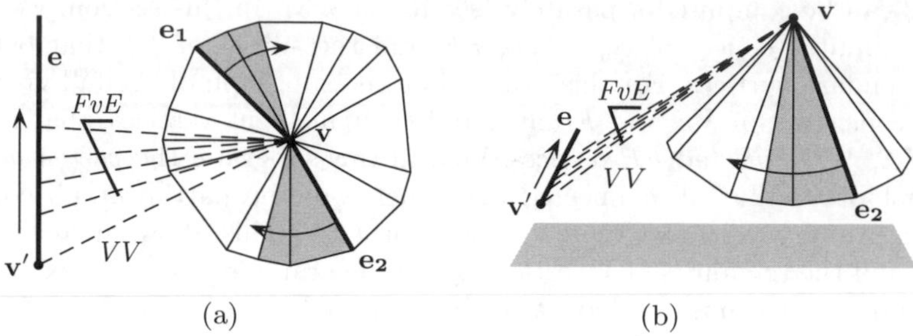

Fig. 4. (a) Bird's eye view and (b) 3D view of extremal *FvE* vertices whose supports are polytope edge **e** and two sequences of faces incident to **v**, starting from $\mathbf{e_1}$ and $\mathbf{e_2}$, which create *VE* arcs with $\mathbf{v}'$

Lemma 4. *Any sequence of* FE *arcs has an extremal* FvE *vertex or* FEE *vertex at each extremity, and arcs in the sequence are separated by non-extremal* FvE *vertices,* FEE *vertices or* FF *vertices.*

The next two lemmas state that any sequence of arcs can be completely explored starting from its primary vertices. (Recall the definitions of p' and s' from above.)

Lemma 5. *Any sequence of* VE *arcs can be computed in* $O(p' \log p + s' + \log \delta)$ *time if all the primary vertices are known, or, when the sequence contains no primary vertex, if one secondary vertex is known.*

Lemma 6. *Any sequence of* FE *arcs can be computed in* $O(p' \log p + s' + \log \delta)$ *if all the primary vertices are known, or, when the sequence contains no primary vertex, if one secondary vertex is known.*

Sketch of proofs. First, if the list of primary vertices in the succinct visibility skeleton is sorted by their supports, we can find the primary vertices in the considered sequence in $O(p' \log p)$. We then show that secondary vertices can be computed in linear time in their number. This is mostly done by enumerating edges incident to a polytope vertex or face in a proper way, as illustrated in Figure 4. The proof mostly consists of case analysis, and is omitted.

If an extremity of a sequence is a secondary vertex, we need to find the constraints of that vertex among the edges incident to a polytope vertex or to a face, and this can be done in $O(\log \delta)$ time for each sequence, using the binary search data structure from Remark 2. □

We describe now our exploration procedure in detail. We compute the secondary vertices by exploring the (unknown) partial graph. That is, we first examine each primary vertex of type *VV*, *VEE* or *FEE*, and find all secondary vertices of type *VV*, *FvE* and *FF* on adjacent sequences of arcs. We keep a list of discovered secondary vertices, and check before adding any new one whether it is already there. We then examine recursively each vertex in that list, looking again for secondary vertices on adjacent sequences of arcs, which are added to the end of the list on the condition that they are not yet there. In this sense, we are treating the list like a queue. To search the list efficiently, we order it (lexicographically for example), and keep track of the queuing order by adding to each vertex a pointer to the next one to be examined. Checking whether a vertex is already in the list is then done in logarithmic time.

Since any vertex is adjacent to a constant number of arcs, the search for secondary vertices can be done in $O(p \log p + s'(\log s' + \log \delta))$ time. Each of the s' secondary vertices found during the search is computed in $O(\log \delta)$ time and added to the list of secondary vertices in $O(\log s')$ time.

Remark 3. *The* FE *vertices correspond to edges of polytopes and can be computed by simple enumeration. Furthermore,* FVV *vertices correspond to diagonals of faces of polytopes, and can also be found by simple enumeration.*

Theorem 1 and Lemmas 5 and 6 yield the following result:

Theorem 2. *Given the succinct visibility skeleton, the full visibility skeleton can be computed from the succinct one in $O(p \log p + s \log s)$ time, where p is the number of primary vertices of type* VV, VEE *and* FEE*; s is the number of secondary vertices of type* VV, FvE, FF, FE, *and* FVV.

The proof is omitted. Note that the complexity of the computation also covers the preprocessing of binary search data structures defined in Remark 2.

We finally note that, if desired, the graph exploration method presented above can be applied to a subset of the input polytopes. In this case, we first find all the primary vertices that are related to the polytopes of the subset, and then apply the graph exploration to these primary vertices only.

5 Tightness of the Succinct Skeleton

In this section, we show, mostly by examples, that Theorem 2 is tight in the sense that if any one of the primary vertex types *EEEE*, *VEE*, *FEE*, or *VV*, is regarded instead as a secondary vertex type, and thus excluded from the succinct skeleton, then Theorem 2 no longer holds.

Type *EEEE*. Any vertex of type *EEEE* requires four support polytopes, but no skeleton arc has four support polytopes, by the general position assumption. Hence, by Lemma 1, vertices of type *EEEE* must be regarded as primary.

Types *VEE* and *FEE*. When three input polytopes are not the support polytopes of any *EEEE* vertex, then the vertices of types *VEE* and *FEE* that have supports on the three polytopes cannot be computed from any *EEEE* vertex.

Moreover, some scenes may generate vertices of type *VEE* but no vertices of type *FEE*, or vertices of type *FEE* but no vertices of type *VEE*. Hence, by Lemma 1, vertices of type *VEE* and *FEE* cannot be dropped.

Type *VV*. When two input polytopes are not the support polytopes of any *EEEE*, *VEE* or *FEE* vertex, then the vertices of type *VV* that have supports on the two polytopes cannot be computed from any *EEEE*, *VEE* or *FEE* vertex. Moreover, when two polytopes resemble two nearly flat tetrahedra and face each other, then the only primary vertices they admit arc of type *VV*. Therefore, the primary vertices of type *VV* must be regarded as primary.

6 Discussion and Conclusion

We have presented a method to recover a full visibility skeleton, either partial or complete, from a succinct one. The full visibility skeleton is the 0D and 1D cells of the 3D visibility complex of polytopes, whereas the succinct one is defined by visual event surfaces, and is a subset of the full one. Recovering the full skeleton mainly consists of computing the secondary vertices of type *FvE*, *FF*, and *VV* (whose supports do not lie on a plane that is tangent to both support polytopes), from the primary vertices of type *VEE*, *FEE*, and *VV* (whose supports lie on a plane that is tangent to both support polytopes).

Given k polytopes with n edges in total, the running time of our method is, in the worst case, $O(p \log p + s \log s)$, where p is the number of primary vertices except type *EEEE*, and s is the number of secondary vertices. In the worst case, p and s are of size $O(n^2 k)$ and $O(n^2)$ respectively, which gives a worst-case total complexity of $O(n^2 k \log n)$. This worst-case complexity is the same as the best previously known algorithm, which consists in computing $O(n^2)$ candidate secondary vertices in a brute-force way and checking the occlusion with each of the k polytopes in $O(\log n)$ time each (using the Dobkin-Kirkpatrick hierarchical representation [4]). Our output-dependent running time analysis suggests that our method may be much more efficient than previous methods in many cases.

An interesting subject for future research is to generalize our results to other types of input, such as intersecting polytopes and non-convex polyhedra, and to consider objects in arbitrary, possibly degenerate, positions.

References

1. Brönnimann, H., Devillers, O., Dujmović, V., Everett, H., Glisse, M., Goaoc, X., Lazard, S., Na, H.-S., Whitesides, S.: Lines and free line segments tangent to arbitrary three-dimensional convex polyhedra. SIAM Journal on Computing 37(2), 522–551 (2007)
2. Brönnimann, H., Everett, H., Lazard, S., Sottile, F., Whitesides, S.: Transversals to line segments in three-dimensional space. Discrete and Computational Geometry 34(3), 381–390 (2005)
3. Demouth, J.: Événements visuels et limites d'ombres. PhD thesis, Univ. Nancy 2 (2008)
4. Dobkin, D.P., Kirkpatrick, D.G.: Determining the separation of preprocessed polyhedra: a unified approach. In: Kirchner, H., Wechler, W. (eds.) ALP 1990. LNCS, vol. 463, pp. 400–413. Springer, Heidelberg (1990)
5. Durand, F.: Visibilité tridimensionnelle: étude analytique et applications. PhD thesis, Université Joseph Fourier - Grenoble I (1999)
6. Koenderink, J., van Doorn, A.: The singularities of the visual mapping. Biological Cybernetics 24, 51–59 (1976)
7. Pocchiola, M., Vegter, G.: The visibility complex. International Journal of Computational Geometry and Applications 6(3), 279–308 (1996); Proceedings of the 9th ACM Annual Symposium on Computational Geometry (SoCG '93)
8. Zhang, L., Everett, H., Lazard, S., Weibel, C., Whitesides, S.: On the size of the 3D visibility skeleton: experimental results. In: Halperin, D., Mehlhorn, K. (eds.) Esa 2008. LNCS, vol. 5193, pp. 805–816. Springer, Heidelberg (2008)

The Violation Heap: A Relaxed Fibonacci-Like Heap

Amr Elmasry*

Max-Planck Institut für Informatik
Saarbrücken, Germany
elmasry@mpi-inf.mpg.de

Abstract. We give a priority queue that achieves the same amortized bounds as Fibonacci heaps. Namely, find-min requires $O(1)$ worst-case time, insert, meld and decrease-key require $O(1)$ amortized time, and delete-min requires $O(\log n)$ amortized time. Our structure is simple and promises an efficient practical behavior when compared to other known Fibonacci-like heaps. The main idea behind our construction is to propagate rank updates instead of performing cascaded cuts following a decrease-key operation, allowing for a relaxed structure.

1 Introduction

The binomial queue [23] is a basic structure that supports the operations: find-min in $O(1)$ worst-case time, insert and meld in $O(1)$ amortized time, decrease-key and delete-min in $O(\log n)$ worst-case time. It can also be extended to support insert in $O(1)$ worst-case time. Being so natural, simple and efficient, binomial queues do exist in most introductory textbooks for algorithms and data structures; see for example [3].

Realizing that many important network-optimization and other algorithms can be efficiently implemented using a heap that better supports decrease-key, and that improving the bound for decrease-key is theoretically possible, Fredman and Tarjan [11] introduced Fibonacci heaps supporting the operations: find-min in $O(1)$ worst-case time, insert, meld and decrease-key in $O(1)$ amortized time, and delete-min in $O(\log n)$ amortized time. Using Fibonacci heaps the asymptotic time bounds for many algorithms have been improved; see [11].

Around the same time, Fredman et al. introduced the pairing heaps [10], a self-adjusting alternative to Fibonacci heaps. They only established $O(\log n)$ amortized time bound for all operations. Stasko and Vitter [21] improved the amortized bound for insert to $O(1)$. They also conducted experiments showing that pairing heaps are more efficient in practice than Fibonacci heaps and than other known heap structures, even for applications requiring many decrease-key operations! More experiments were also conducted [17] illustrating the practical efficiency of pairing heaps. The bounds for the standard implementation were

* Supported by Alexander von Humboldt Fellowship.

M.T. Thai and S. Sahni (Eds.): COCOON 2010, LNCS 6196, pp. 479–488, 2010.

later improved by Iacono [14] to: $O(1)$ per insert, and zero cost per meld. However, Fredman [9] showed that pairing heaps are not theoretically as efficient as Fibonacci heaps by giving a lower bound of $\Omega(\log\log n)$, and precluded the possibility of achieving $O(1)$ decrease-key unless every node carries $\Omega(\log\log n)$ information bits. Later, Pettie [19] improved the analysis for the decrease-key operation to achieve $O(2^{2\sqrt{\log\log n}})$ amortized bound. Recently, Elmasry [5] introduced a variant with $O(\log\log n)$ amortized bound per decrease-key.

Towards a heap that achieves good worst-case time bounds, Driscoll et al. [4] introduced the relaxed heaps. The rank-relaxed heaps achieve the same amortized bounds as Fibonacci heaps, and were easily extended to the run-relaxed heaps that achieve the bounds in the worst case except for meld. Relaxed heaps are good candidates for applications with possible parallel computations. Still, relaxed heaps are not practically efficient and are more complicated than Fibonacci heaps (they are not actually relaxed from this prospective). Another priority queue that achieves the same worst-case time bounds as run-relaxed heaps is the fat heap [16]. Incorporating $O(1)$ worst-case meld to the repertoire of operations, Brodal [2] introduced a priority queue that achieves the same bounds as the Fibonacci heaps, but in the worst-case sense. Brodal's structure is impractical and even more complicated than relaxed heaps.

Several attempts were made [13,15,18,22] to come up with a priority queue that is theoretically as efficient as Fibonacci heaps without sacrificing the practicality. Among those, we find thin heaps [15] the most natural and promising. In spite of being able to improve the space requirements by getting rid of the parent pointers [15,22], or equivalently by using a binary-tree implementation [13,18], the practicality issue is not resolved yet (or at least that is our impression!).

In this paper we claim that we resolved this issue by introducing a priority queue that we call the *violation heap*. Violation heaps have the same amortized bounds as Fibonacci heaps, and are expected to perform in practice in a more efficient manner than other Fibonacci-like heaps and compete with pairing heaps. Our amortized bounds are: $O(1)$ per find-min, insert, meld and decrease-key, and $O(\log n)$ per delete-min. In contrary to other Fibonacci-like heaps, while in agreement with pairing heaps, the *degree* (number of children) of a node in the violation heaps is not necessarily logarithmic in the size of the subtree of this node; in fact there is no bound on node degrees, allowing what we call a *relaxed structure*. (Still, for the purpose of the analysis, the number of children of a node is amortized towards both the delete-min and decrease costs.)

For Fibonacci heaps and thin heaps the degree of a node is bounded by $\approx 1.44\lg n$, and for 2-3 heaps [22] and thick heaps [15] the bound is $\lg n$. Ensuring a larger degree bound (or even no bound) is not necessarily a disadvantage though; it only indicates how much relaxed the structure is. The reason is that a tighter degree bound may require more effort to restrict the structure to such bound. As an alibi, no such degree bound is guaranteed for pairing heaps [10]. Similar arguments can be mentioned about the bound on the height of a splay tree [20] versus that of an AVL [1] tree. In addition, one can resort to known techniques

[7,8] to reduce the number of comparisons performed in a delete-min operation almost by a factor of two.

In the next section we give our motivation: why there is a need for a new structure, what our objectives are, and how to achieve them. Then we introduce the data structure: design, operations and time bounds. Finally, we conclude the paper with some comments.

2 Motivation

In this section we argue why we need a new Fibonacci-like heap structure. We start with the drawbacks of other such heaps. Then, we summarize our objectives and the features required for a better heap structure. We end the section with ideas that lead to the violation heaps.

Drawbacks of Other Structures

The pairing heap [10] is the most efficient among other Fibonacci-like heaps from the practical point of view [17,21]. Still, it is theoretically inefficient according to the following fact [9].

– The amortized cost per decrease-key is not a constant.

In contrary to pairing heaps, all known heaps achieving a constant amortized cost per decrease-key [4,11,13,15,18,22] impose the following constraint, which makes them practically inefficient.

– Every subtree has to permanently maintain some balancing constraint to ensure that its size is exponential with respect to its height.

The Fibonacci heaps [11] have the following drawbacks.

– Every node has a parent pointer, adding up to four pointers per node.
– A subtree is cut and its rank decreases when its root loses two of its children. If the subtrees of these two children are small in size compared to the other children, the cascaded cut is immature. The reason is that the size of this cut-subtree is still exponential in its rank. This results in practical inefficiency as we are unnecessarily losing previously-gained information.
– The worst-case cost per decrease-key can be $\Theta(n)$; see [3, exercise 20.4-1].

Being able to remedy the drawbacks of Fibonacci heaps, other heap structures have not yet achieved the goal though! The trend of imposing more-restricted balancing constraints on every subtree would result in the following pitfalls that accompany a decrease-key operation [13,15,18,22].

– More cuts resulting in the loss of gained information.
– More checks among several cases resulting in performance slow down.

The rank-relaxed heaps [4] is even more restricted, allowing for no structural violations but for only a logarithmic number of heap-order violations. This requires even more case-based checks accompanying decrease-key operations.

Objectives for a New Design

To avoid the above drawbacks, we need a heap with the following properties.

- No parent pointers.
- No cascaded cuts, and in accordance allowing a relaxed structure.
- No upward propagation for rank updates following a rank update of a node that is neither the last nor second-last child.
- Fewer case-based checks following a decrease-key operation.

Insights

The main intuition behind our design is to allow structural violations resulting from decrease-key operations, and only record the amount of violations in every subtree within the root node of this subtree. We rely on the following three ideas:

The first idea, which kills two birds by one stone, is to only consider violations in the first two children of a node (one child is not enough). As in [15], we utilize the unused left pointer of the last child to point to the parent. This makes it easy to convey such violations to a parent node without having a parent pointer.

The second idea, which we have recently used to improve the amortized cost of the decrease-key operation for pairing heaps to $O(\log \log n)$ [5], is not to cut the whole subtree of a node whose key is decreased. Instead, we replace such subtree with the subtree of the last child. The expectation is that the size of a last-child's subtree constitutes a constant fraction of that of the parent, and hence the resulting structural degradation will be smoother. We also reduce the lost information resulting from the cut by keeping a good portion of the subtree.

The third idea is about how to record the violations. As a compensation for the structural violation it has done, a decrease-key operation can pay $1/2$ credit to its parent, $1/4$ credit to its grandparent, and in general $1/2^i$ to its i-th ancestor. Once the credits in a node sum up to at least 1, we declare its subtree as violating. Later, the fix of this violation can be charged to this credit. As long as the credits on a node are still less than 1, its subtree is maintaining a good structure. The question is how to implement this idea! The details follow.

3 The Violation Heaps

Structure

Similar to Fibonacci heaps, 2-3 heaps and thin heaps, the violation heap is a set of heap-ordered node-disjoint multiary trees. The children of a node are ordered according to the time when they are linked to the parent. Every node has three pointers and one integer in addition to its key, utilized as follows.

a. A singly-linked circular list of tree roots, with a root of minimum key first.
b. A doubly-linked list of children for each node, with a pointer to its last child.
c. For each last child, the unused pointer points to its parent.
d. An integer for each node resembling its *rank* (see Section 3 for how to maintain the rank field and for the properties sustained by the rank values).

We maintain the invariant that the size s_z of the subtree of a node z is exponential with respect to its rank r_z, but not with respect to its degree d_z. The *violation* v_z of a node z indicates how bad the structure of its subtree is, and is defined in terms of r_z and d_z as

$$v_z = max(0, d_z/2 - r_z)$$

We emphasize that we only store r_z, but neither d_z nor v_z. The notion of violations is only used in the analysis, but not in the actual implementation.

In the sequel, we call a node *active* if it is one of the last two children of its parent. We maintain the rank of a node z, in accordance with Lemma 2, by updating r_z once the rank of any of its active children r_{z1} and r_{z2} decreases. We use the following formula:

$$r_z \leftarrow \lceil (r_{z1} + r_{z2})/2 \rceil + 1. \tag{1}$$

Evaluating this formula requires an integer addition, a right shift, and one or two increments. If z has one or no children, the rank of a missing child is -1.

The following primitive is used to consolidate the trees of the heap.

3-way-join$(z, z1, z2)$ (the presumption is that $r_z = r_{z1} = r_{z2}$):
Assume w.l.o.g. that z's value is not larger than that of $z1$ and $z2$. Ensure that the active child of z with the larger rank is the last child. Make $z1$ and $z2$ the last two children of z by linking both subtrees to z, and increment r_z.

Operations

- *find-min(h)*: Return the first root of h.

- *insert(x,h)*: A single node x is inserted into the root list of h. The rank of x is initially set to zero. If the key of x is smaller than the minimum of h, then x is inserted in the first position, otherwise in the second position.

- *meld*(h_1, h_2): The root lists of h_1 and h_2 are combined in a new list whose first root is the smaller between the minimums of the two heaps.

- *decrease-key(δ,x,h)*: Subtract δ from the key of x. If x is a root, stop after making it the first root if its new value is smaller than the minimum. If x is an active node whose new value is not smaller than its parent, stop.
  Otherwise, cut the subtree of x and glue in its position the subtree with the larger rank between its active children. Recalculate the rank of x using (1). Promote x's subtree as a tree in h, and make x the first root if its new value is smaller than the minimum. Propagate rank updates by traversing the path of ancestors of x's old position, as long as the visited node is active and as long as its recalculated rank using (1) is smaller than its old rank.

– *delete-min(h)*: Remove from h the first root and make each of its subtrees a tree in h. Repeatedly 3-way-join trees of equal rank until no three trees of the same rank remain. As for Fibonacci heaps [11], this is done in $O(1)$ time per tree using a temporary array indexed by rank values. Finally, the root with the new minimum value is moved to the first position in the root list.

Analysis

First, we point out that 3-way-join is favorable to the normal join for our case.

Lemma 1. *No extra violation units are added to the nodes of the heap as a result of a 3-way-join.*

Proof. Consider the 3-way-join$(z, z1, z2)$ operation, assuming that the value of z is not larger than the values of $z1$ and $z2$. When z gains two extra children, r_z is incremented ensuring that $d_z/2 - r_z$ does not change. □

Next, we show that the assigned rank values fulfill the requirements.

Lemma 2. *The following relations are maintained for every node z.*

$$r_z \begin{cases} = 0 & \text{if } z \text{ has no children,} \\ \leq \lceil (r_{z1} - 1)/2 \rceil + 1 & \text{if } z \text{ has one child } z1, \\ \leq \lceil (r_{z1} + r_{z2})/2 \rceil + 1 & \text{if the active children of } z \text{ are } z1 \text{ and } z2. \end{cases}$$

Proof. When a node z is inserted r_z is set to 0. When a subtree of a node x is cut by a decrease-key operation, the ranks of the ancestors of x are updated by traversing the affected path upwards. As long as the violation of the nodes on the path is to be increased (rank should decrease), the decrease-key operation resumes the upward traversal and decreases the rank. Once the rank of a node is not to be changed, the rank of the other nodes along the path are already valid. For the case when the new rank is more than the old value no updates are done.

When an active child vanishes, as a result of repeated decrease-key operations, another child (if there exist any) becomes active. This would make the inequalities given in the statement of the lemma satisfied as strict "<", and the recalculated rank is now larger than the rank stored in the parent. In such case, no rank updates are done and the upward traversal is terminated.

Consider the 3-way-join$(z, z1, z2)$, assuming that the value of z is not larger than the values of $z1$ and $z2$. Before the join, $r_z = r_{z1} = r_{z2}$. After the join, $z1$ and $z2$ become the active children of z implying that $r_z \leftarrow \lceil (r_{z1} + r_{z2})/2 \rceil + 1$. This explains the validity of incrementing r_z after the join.

If z has one or no children, the lemma analogously follows by assuming the rank of a missing child to be -1. □

The following two lemmas are used in the proof of Lemma 5 and Theorem 1.

Lemma 3. *Consider any node z. Let $z1$ and $z2$ be the active children of z, such that $r_{z1} \geq r_{z2}$. Then, either*

a. $r_{z1} \geq r_z$, *or*
b. $r_{z1} = r_z - 1$ *and* $r_{z2} = r_z - 1$ *or* $r_z - 2$.

Proof. Using $r_z \leq \lceil (r_{z1}+r_{z2})/2 \rceil + 1$ from Lemma 2, then $r_z \leq r_{z1}+1$. Consider the case when $r_{z1} = r_z - 1$. It follows that $r_z \leq \lceil (r_z - 1 + r_{z2})/2 \rceil + 1$, which implies $r_{z2} \geq r_z - 2$. But $r_{z2} \leq r_{z1}$, indicating that r_{z2} equals $r_z - 1$ or $r_z - 2$. □

Lemma 4. *The rank of any node can be decreased by at most* 1 *when propagating rank updates within a decrease-key operation.*

Proof. When a decrease-key is performed on a node z, the subtree of $z1$ is promoted in its position. From lemma 3, $r_{z1} \geq r_z - 1$ implying that we now have a node that may be less in rank but by at most 1, and the lemma holds. Propagating the rank updates using (1) would result in a decrease of 1 in the rank of the parent node, if there is any decrease at all. □

The above lemma is important and worth commenting. A crucial observation is that the parity of the ranks of the active children of a node affects the possibility of updating its rank. From Lemma 2, if the sum of the ranks of the two active children is even and one of them is decremented, then the rank of the parent is not to be changed. If this sum is odd and one of the two ranks is decremented, then the rank of the parent may be decremented. Another consequence is that the sum of the violation units added to the heap nodes is bounded above by the number of rank-update steps performed within the decrease-key operations.

Now, we prove the *structural* lemma, illustrating that the size of a subtree is exponential with respect to its rank.

Lemma 5. *Let* s_z *be the size of the subtree of any node* z, *and* r_z *be its rank. Then* $s_z \geq F_{r_z}$, *where* F_i *is the ith Fibonacci number.*

Proof. If z has no children, then $r_z = 0$ and the lemma holds. If z has one child that is a leaf, then $r_z = 1$ and $s_z = 2$ and the lemma also holds. If z has one child whose rank $r_{z1} > 0$, then $r_z \leq \lceil (r_{z1} - 1)/2 \rceil + 1$ implies $r_{z1} \geq r_z$. Using induction, $s_z > F_{r_{z1}} \geq F_{r_z}$. If z has at least two children, while $r_{z1} \geq r_z$ the bound follows as above by induction. Using Lemma 3, the possibility left is that $r_{z1} = r_z - 1$ and $r_{z2} \geq r_z - 2$. Again using induction, then

$$s_z > F_{r_{z1}} + F_{r_{z2}} \geq F_{r_z - 1} + F_{r_z - 2} = F_{r_z}.$$

□

Corollary 1. *For any node* z, $r_z = O(\log s_z)$.

Finally, we prove the main theorem concerning the time bounds for the operations of the violation heaps.

Theorem 1. *The violation heaps require* $O(1)$ *amortized cost per find-min, insert, meld, and decrease-key; and* $O(\log n)$ *amortized cost per delete-min.*

Proof. An active node is called *critical* if the sum of the ranks of its active children is odd. As a consequence of Lemma 4, the decrease-key terminates the upward traversal for rank updates when it reaches

a. a nonactive node, or
b. an active noncritical node (which becomes critical), or
c. a critical node whose rank is not to be changed (the current value is not larger than the recalculated value).

Let γ be the number of critical nodes in the heap. Let ϑ be the sum of violation units on the nodes of the heap, i.e. $\vartheta = \sum_{\forall z}(v_z = d_z/2 - r_z)$. Let τ be the number of trees in the heap. Let Δ_z be the number of decrease-key operations, since the 3-way-join in which z was linked to its current parent, which have terminated at one of the two nonactive siblings of z that were active just before the join. Let $\Delta = \sum_{\forall z} \Delta_z$. We use the potential function

$$P = 3\gamma + 2\vartheta + \tau + \Delta.$$

The actual time used by find-min, insert, and meld is $O(1)$. Both find-min and meld do not change the potential, and an insertion increases τ by 1. It follows that the amortized cost of these operations is $O(1)$.

We call the path traversed by the decrease-key operation to perform rank updates the *critical path.* The decrease-key uses $O(1)$ time plus the time it traverses the critical path. The crucial idea is that every node on this path was critical and becomes noncritical after the traversal. In addition, the increase in violation units precisely equals the number of such nodes. Let k be the number of nodes on a critical path. Then the actual time used by the decrease-key operation is $O(k)$. On the other hand, γ decreases by at least $k - 2$ (the cut node itself may also become a critical root), ϑ increases by k, τ increases by 1, and Δ may increase by 2 (in that case γ decreases by $k - 1$). Accordingly, the change in potential is at most $-3(k - 2) + 2k + 1 = -k + 7$. These released k credits pay for the $O(k)$ work done by the decrease-key, which only pays the 7 credits.

When the subtree of a node z is cut and becomes a tree root after losing its last child $z1$ (which is glued in its position), the rank of z is updated and may decrease, raising the need for more violation units. In such case, we claim that the increase in 2ϑ is at most the decrease in Δ plus 2; these 2 credits are also paid for by the decrease-key operation. To prove our claim, consider the moment following a 3-way-join linking $z1$ to z. Let r be the rank of $z1$, then the rank of z is $r + 1$. From Lemma 3, the rank of its third-to-last child (whose rank is enforced to be larger than that of the fourth-to-last child) is at least $r - 1$. When z is cut and loses its last child $z1$, the third-to-last child of z before the cut now becomes active with rank say r'. If $r' \geq r$, the rank of z does not decrease and we are done. Otherwise, the decrease in the rank of z, which equals the increase in ϑ, is at most $(r - r' + 1)/2$. On the other hand, Δ_{z1} must have been at least $r - r' - 1$ and is now reset to zero. Then, the increase in potential is at most $2(r - r' + 1)/2 - (r - r' - 1) = 2$, and the claim follows.

The preceding two paragraphs imply that the amortized cost for the decrease-key operation is $O(1)$.

The 3-way-join uses $O(1)$ time, and decreases τ by 2 resulting in a release of 2 potential credits, which pay for the work. Note that the normal join may result in the root of the joined tree being critical, requiring more credits (that is the reason we resort to 3-way-join). In addition, from Lemma 1, no violations are added accompanying the 3-way-join.

Consider a delete-min operation for node z. When the children of z become new trees, τ increases by d_z and ϑ decreases by $v_z \geq d_z/2 - r_z$. Therefore, the change in potential is at most $2r_z$, which is $O(\log n)$ by Lemma 5. Finding the new minimum requires traversing the roots surviving the consolidation, whose count is at most twice the distinct ranks because at most two roots per rank remain. It follows that the amortized cost for the delete-min operation is $O(\log n)$. □

4 Comments

We have given a priority queue that performs the same functions as a Fibonacci heap and with the same running-time asymptotic bounds. The main feature of our priority queue is that it allows any possible structure, with no structural restrictions soever, but only records the structure violations. Our priority queue uses three pointers and an integer per node. Following a decrease-key operation, we retain as much information as possible by not performing cascaded cuts and by keeping a big chunk of the subtree of the decreased node in its place. We expect our priority queue to be practically efficient. Experimental results still need to be conducted to support our intuitive claims.

In comparison with other Fibonacci-like heaps, a drawback of the violation heap is that it cannot be implemented on a pointer machine. Indeed, we need the power of a RAM to recalculate the ranks using formula (1), which involves integer addition and bit operations.

Independently, and after the first technical-report version of this paper [6] (which is to-an-extent different from the current version) was archived, the idea of performing rank updates instead of cascaded cuts appeared in [12]. The structure in [12], which is called rank-pairing heaps, relies on half-ordered half trees and not on multiary trees. Also, the way in which the rank updating mechanism is performed in [12] is different from that in violation heaps.

References

1. Adelson-Velskii, G., Landis, E.: On an information organization algorithm. Doklady Akademia Nauk SSSR 146, 263–266 (1962)
2. Brodal, G.S.: Worst-case efficient priority queues. In: Proceedings of the 7th ACM-SIAM symp. on Disc. Algorithms, pp. 52–58 (1996)
3. Cormen, T., Leiserson, C., Rivest, R., Stein, C.: Introduction to Algorithms, 2nd edn. The MIT Press, Cambridge (2001)
4. Driscoll, J., Gabow, H., Shrairman, R., Tarjan, R.E.: Relaxed heaps: An alternative to Fibonacci heaps with applications to parallel computation. ACM Comm. 31, 1343–1354 (1988)
5. Elmasry, A.: Pairing heaps with O(loglogn) decrease cost. In: Proceedings of the 20th ACM-SIAM symp. on Disc. Algorithms, pp. 471–476 (2009)

6. Elmasry, A.: Violation heaps: an alternative to Fibonacci heaps. CoRR abs/0812.2851 (2008)
7. Elmasry, A.: Layered heaps. In: Hagerup, T., Katajainen, J. (eds.) SWAT 2004. LNCS, vol. 3111, pp. 212–222. Springer, Heidelberg (2004)
8. Elmasry, A., Jensen, C., Katajainen, J.: Two-tier relaxed heaps. Acta Informatica 45, 193–210 (2008)
9. Fredman, M.L.: On the efficiency of pairing heaps and related data structures. J. ACM 46(4), 473–501 (1999)
10. Fredman, M.L., Sedgewick, R., Sleator, D.D., Tarjan, R.E.: The pairing heap: a new form of self_adjusting heap. Algorithmica 1(1), 111–129 (1986)
11. Fredman, M.L., Tarjan, R.E.: Fibonacci heaps and their uses in improved network optimization algorithms. J. ACM 34, 596–615 (1987)
12. Haeupler, B., Sen, S., Tarjan, R.E.: Rank-Pairing Heaps. In: Fiat, A., Sanders, P. (eds.) ESA 2009. LNCS, vol. 5757, pp. 659–670. Springer, Heidelberg (2009)
13. Høyer, P.: A general technique for implementation of efficient priority queues. In: Proceedings of the 3rd Israel Symp. on the Theory of Computing Systems, pp. 57–66 (1995)
14. Iacono, J.: Improved upper bounds for pairing heaps. In: Halldórsson, M.M. (ed.) SWAT 2000. LNCS, vol. 1851, pp. 32–45. Springer, Heidelberg (2000)
15. Kaplan, H., Tarjan, R.E.: Thin heaps, thick heaps. ACM Trans. on Algorithms, Article 3 4(1) (2008)
16. Kaplan, H., Shafrir, N., Tarjan, R.E.: Meldable heaps and boolean union-find. In: Proceedings of the ACM Symp. on Theory of Computing, pp. 573–582 (2002)
17. Moret, B., Shapiro, H.: An empirical assessment of algorithms for constructing a minimum spanning tree. DIMACS Monographs in Disc. Math. and Theoretical Comp. Science 15, 99–117 (1994)
18. Peterson, G.: A balanced tree scheme for meldable heaps with updates. Technical Report GIT-ICS-87-23, School of Information and Comp. Science, Georgia Institute of Technology (1987)
19. Pettie, S.: Towards a final analysis of pairing heaps. In: Proceedings of the 46th IEEE Symp. on Foundations of Comp. Science, pp. 174–183 (2005)
20. Sleator, D.D., Tarjan, R.E.: Self-adjusting binary search trees. J. ACM 32(3), 652–686 (1985)
21. Stasko, J., Vitter, J.S.: Pairing heaps: experiments and analysis. ACM Comm. 30(3), 234–249 (1987)
22. Takaoka, T.: Theory of 2-3 heaps. Disc. Appl. Math. 126(1), 115–128 (2003)
23. Vuillemin, J.: A data structure for manipulating priority queues. ACM Comm. 21, 309–314 (1978)

Threshold Rules for Online Sample Selection

Eric Bach*, Shuchi Chawla**, and Seeun Umboh**

Computer Sciences Dept., University of Wisconsin, Madison
bach@cs.wisc.edu, shuchi@cs.wisc.edu, seeun@cs.wisc.edu

Abstract. We consider the following sample selection problem. We observe in an online fashion a sequence of samples, each endowed by a quality. Our goal is to either select or reject each sample, so as to maximize the aggregate quality of the subsample selected so far. There is a natural trade-off here between the rate of selection and the aggregate quality of the subsample. We show that for a number of such problems extremely simple and oblivious "threshold rules" for selection achieve optimal tradeoffs between rate of selection and aggregate quality in a probabilistic sense. In some cases we show that the same threshold rule is optimal for a large class of quality distributions and is thus oblivious in a strong sense.

1 Introduction

Imagine a heterogeneous sequence of samples from an array of sensors, having different utilities reflecting their accuracy, quality, or applicability to the task at hand. We wish to discard all but the most relevant or useful samples. Further suppose that selection is performed online — every time we receive a new sample we must make an irrevocable decision to keep it or discard it. What rules can we use for sample selection? There is a tradeoff here: while we want to retain only the most useful samples, we may not want to be overly selective and discard a large fraction. So we could either fix a rate of selection (the number of examples we want to retain as a function of the number we see) and ask for the best quality subsample, or fix a desirable level of quality as a function of the size of the subsample and ask to achieve this with the fewest samples rejected.

An example of online sample selection is the following "hiring" process that has been studied previously. Imagine that a company wishing to grow interviews candidates to observe their qualifications, work ethic, compatibility with the existing workforce, etc. How should the company make hiring decisions so as to obtain the higest quality workforce? As with the sensors setting, there is no single correct answer here—again there is a trade-off between being overly selective and growing fast. Broder et al. [6] studied this hiring problem in a

* Sponsored in part by NSF award CCF-0635355 and ARO grant W911NF-0910439.
** Supported in part by NSF award CCF-0643763 and an Alfred P. Sloan fellowship.

M.T. Thai and S. Sahni (Eds.): COCOON 2010, LNCS 6196, pp. 489–499, 2010.

simple setting where each candidate's quality is a one-dimensional random variable and the company wants to maximize the average or median quality of its workforce.

In general performing such selection tasks may require complicated rules that depend on the samples seen so far. Our main contribution is to show that in a number of settings an extremely simple class of rules that we call "threshold rules" is close to optimal on average (within constant factors).

Specifically, suppose that each sample is endowed with a "quality", which is a random variable drawn from a known distribution. We are interested in maximizing the aggregate quality of a set of samples, which is a numerical function of the individual qualities. Suppose that we want to select a subset of n samples out of a total of T seen. Let $Q^*_{T,n}$ denote the maximum aggregate quality that can be achieves by picking the best n out of the T samples. Our goal is to design an online selection rule that approximates $Q^*_{T,n}$ in expectation over the T samples. We use two measures of approximation — the ratio of the expected quality achieved by the offline optimum to that achieved by the online selection rule, $E[Q^*_{T,n}]/E[Q_{T,n}]$, and the expectation of the ratio of the qualities of the two rules, $E[Q^*_{T,n}/Q_{T,n}]$. Here the expectations are taken over the distribution from which the sample is drawn. The approximation ratios are always at least 1 and our goal is to show that they are bounded from above by a constant independent of n. In this case we say that the corresponding selection rule is optimal.

To put this in context, consider the setting studied by Broder et al. [6]. Each sample is associated with a quality in the range $[0, 1]$, and the goal is to maximize the average quality of the subsample we pick. Broder et al. show (implicitly) that if the quality is distributed uniformly in $[0, 1]$ a natural *select above the mean* rule is optimal to within constant factors with respect to the optimal offline algorithm that has the same selection rate as the rule. The same observation holds also for the *select above the median* rule. Both of these rules are adaptive in the sense that the next selection decision depends on the samples seen so far. In more general settings, adaptive rules of this kind can require unbounded space to store information about samples seen previously. For example, consider the following 2-dimensional skyline problem: each sample is a point in a unit square; the quality of a single point (x, y) is the area of its "shadow" $[0, x] \times [0, y]$, and the quality of a set of points is the area of the collective shadows of all the points; the goal is to pick a subsample with the largest shadow. In this case, a natural selection rule is to select a sample if it falls out of the shadow of the previously seen points. However implementing this rule requires remembering on average $O(\log n)$ samples out of n samples seen [4]. We therefore study non-adaptive selection rules.

We focus in particular on so-called "threshold rules" for selection. A threshold rule specifies a criterion or "threshold" that a candidate must satisfy to get selected. Most crucially, the threshold is determined *a priori* given a desired selection rate; it depends only on the number of samples picked so far and is otherwise independent of the samples seen or picked. Threshold rules are extremely simple oblivious rules and can, in particular, be "hard-wired" into the selection process. This suggests the following natural questions. When are threshold rules optimal for online selection problems? Does the answer depend on the desired rate of selection? We answer these questions in three different settings in this paper.

The first setting we study is a single-dimensional-quality setting similar to Broder et al.'s model. In this setting, we study threshold rules of the form *"Pick the next sample whose quality exceeds $f(i)$"* where i is the number of samples picked so far. We show that for a large class of functions f these rules give constant factor approximations. Interestingly, our threshold rules are optimal in an almost distribution-independent way. In particular, every rule f in the aforementioned class is simultaneously constant-factor optimal with respect to any "power law" distribution, and the approximation factor is independent of the parameters of the distribution. In contrast, Broder et al.'s results hold only for the uniform distribution[1].

In the second setting, samples are nodes in a rooted infinite-depth tree. Each node is said to cover all the nodes on the unique path from the root to itself. The quality of a collection of nodes is the total number of distinct nodes that they collectively cover. This is different from the first setting in that the quality defines only a partial order over the samples. Once again, we study threshold rules of the form *"Pick the next sample whose quality exceeds $f(i)$"* and show that they are constant factor optimal.

Our third setting is a generalization of the skyline problem described previously. Specifically, consider a domain X with a probability measure μ and a partial ordering $\prec$ over it. For an element $x \in X$, the "shadow" or "downward closure" of x is the set of all the points that it dominates in this partial ordering, $\mathcal{D}(x) = \{y : y \prec x\}$; likewise the shadow of a subset $S \subseteq X$ is $\mathcal{D}(S) = \cup_{x \in S}\mathcal{D}(x)$. Once again, as in the second setting, we can define the coverage of a single sample to be the measure of all the points in its shadow. However, unlike the tree setting, here it is usually easy to obtain a constant factor approximation to coverage—the maximum coverage achievable is 1 (i.e. the measure of the entire universe), whereas in many cases (e.g. for the uniform distribution over the unit square) a single random sample can in expectation obtain constant coverage. We therefore measure the quality of a subsample $S \subset X$ by its "gap", $\mathrm{Gap}(S) = 1 - \mu(\mathcal{D}(S))$. In this setting, rules that place a threshold on the quality of the next sample to be selected are not constant-factor optimal. Instead, we study threshold rules of the form *"Pick the next sample x for which $\mu(\mathcal{U}(x))$ is at most $f(i)$"*, where $\mathcal{U}(x) = \{y : x \prec y\}$ is the set of all elements that dominate x, or the "upward closure" of x, and show that these rules obtain constant factor approximations.

[1] While Broder et al.'s result can be extended to any arbitrary distribution via a standard tranformation from one space to another, the resulting selection rule becomes distribution dependent, e.g., "select above the mean" is no longer "select above the mean" w.r.t. the other distribution upon applying the transformation.

1.1 Related Work

As mentioned earlier, our work is inspired by and extends the work of Broder et al. [6]. Our third setting is related to the skyline problem that has been studied extensively in online settings by the database community (see, for example, [1] and references therein). Online sample selection is closely related to secretary problems, however there are some key differences. In secretary problems (see, e.g., [7,8,11]) there is typically a fixed bound on the desired number of hires. In our setting the selection process is ongoing and we must pick more and more samples as time passes. This makes the tradeoff between the rate of hiring and the rate of improvement of quality interesting.

Finally, our setting falls within the purview of online algorithms [5]; however, in a departure from standard literature on that topic, we measure the performance of selection rules by their *expected* competitive ratio, and not worst-case competitive ratio. In most cases we show that the competitive ratio is small with very high probability. Crucially, the strategies we study are not only online but also non-adaptive, or oblivious of the samples seen so far. In this sense, our model is closer in spirit to work on oblivious algorithms (see, e.g., [10,2,9]). Oblivious algorithms are highly desirable in practical settings because the rules can be hard-wired into the selection process, making them very easy to implement. The caveat is that for many optimization problems oblivious algorithms do not provide good approximations. Surprisingly, we show that in many scenarios related to sample selection, obliviousness has only a small cost.

2 Models and Results

Let X be a domain with probability measure μ over it. A *threshold rule* $\mathcal{X}$ is specified by a sequence of subsets of X indexed by $\mathbb{N}$: $X = \mathcal{X}_0 \supseteq \mathcal{X}_1 \supseteq \mathcal{X}_2 \supseteq \cdots \supseteq \mathcal{X}_n \supseteq \cdots$. A sample is selected if it belongs to $\mathcal{X}_i$ where i is the number of samples previously selected.

Let $\mathcal{T}$ be an infinite sequence of samples drawn i.i.d. according to μ. Let $\mathcal{T}^{\mathcal{X}}(n)$ denote the prefix of $\mathcal{T}$ such that the last sample on this prefix is the nth sample chosen by the threshold rule $\mathcal{X}$; let $T_n^{\mathcal{X}}$ denote the length of this prefix. We drop the superscript and the subscript when they are clear from the context. The "selection overhead" of a threshold rule as a function of n is the expected waiting time to select n samples, or $E[T_n^{\mathcal{X}}]$, where the expectation is over $\mathcal{T}$.

Let Q be a function denoting "quality". Thus $Q(x)$ denotes the quality of a sample x and $Q(S)$ the aggregate quality of a set $S \subset X$ of samples. $Q(x)$ is a random variable and we assume that it is drawn from a known distribution. Let $Q^*_{\mathcal{T},n}$ denote the quality of an optimal subset of n out of a set $\mathcal{T}$ of samples with respect to measure Q. We use Q^*_n as shorthand for $Q^*_{\mathcal{T}^{\mathcal{X}}(n),n}$ where $\mathcal{X}$ is clear from the context. Let $Q^{\mathcal{X}}_{\mathcal{T}^{\mathcal{X}}(n),n}$ (Q_n for short) denote the quality of a sample of size n selected by threshold rule $\mathcal{X}$ with respect to measure Q.

We look at both maximization and minimization problems. For maximization problems we say that a threshold rule $\mathcal{X}$ achieves a *competitive ratio of* α *in expectation* with respect to Q if for all n,

$$E_{\mathcal{T}\sim\mu}\left[\frac{Q_n^*}{Q_n}\right] \leq \alpha$$

Likewise, $\mathcal{X}$ α*-approximates expected quality* with respect to Q if for all n,

$$\frac{E_{\mathcal{T}\sim\mu}[Q_n^*]}{E_{\mathcal{T}\sim\mu}[Q_n]} \leq \alpha$$

For minimization problems, the ratios are defined similarly:

$$\text{Exp. comp. ratio} = \max_n E_{\mathcal{T}\sim\mu}\left[\frac{Q_n}{Q_n^*}\right]; \text{Approx. to exp. quality} = \max_n \frac{E_{\mathcal{T}\sim\mu}[Q_n]}{E_{\mathcal{T}\sim\mu}[Q_n^*]}$$

We now describe the specific settings we study and the results we obtain.

Model 1: Unit interval (Section 3). Our first setting is the one-dimensional setting studied by Broder et al. [6]. Specifically, each sample is associated with a quality drawn from a distribution over the unit line. Our measure of success is the mean quality of the subsample we select. Note that in the context of approximately optimal selection rules this is a weak notion of success. For example when μ is the uniform distribution over $[0,1]$, even in the absence of any selection rule we can achieve a mean quality of $1/2$, while the maximum achievable is 1. So instead of approximately maximizing the mean quality, we approximately minimize the *mean quality gap*—$\text{Gap}(S) = 1 - (\sum_{x\in S} x)/|S|$—of the subsample.

We focus on *power-law* distributions on the unit line, i.e. distributions with c.d.f. $\mu(1-x) = 1-x^k$ for some constant k, and study threshold rules of the form $\mathcal{X}_i = \{x : x \geq 1 - c_i\}$ where $c_i = \Omega(1/\text{poly}(i))$. We show that these threshold rules are constant factor optimal simultaneously for *any* power-law distribution. Remarkably, this gives an optimal selection algorithm that is oblivious of even the underlying distribution. Formally we obtain the following result.

Theorem 1. *For the unit line equipped with a power-law distribution, any threshold rule* $\mathcal{X}_i = \{x : x \geq 1 - c_i\}$*, where* $c_i = 1/i^\alpha$ *with* $0 \leq \alpha < 1$ *for all* i*, achieves an* $O(1)$ *approximation to the expected gap, where the constant in the* $O(1)$ *depends only on* α *and not on the parameters of the distribution.*

Dominance and shadow. For the next two settings, we need some additional definitions. Let $\prec$ be a partial order over the universe X. As defined earlier, the shadow of an element $x \in X$ is the set of all the points that it dominates, $\mathcal{D}(x) = \{y : y \prec x\}$; likewise the shadow of a sample $S \subseteq X$ is $\mathcal{D}(S) = \cup_{x\in S}\mathcal{D}(x)$. Let $\mathcal{U}(x) = \{y : x \prec y\}$ be the set of points that shadow x; $\mathcal{U}(S)$ for a set S is defined similarly. Note that $\mathcal{U}(x)$ is a subset of $X \setminus \mathcal{D}(x)$ and $\mu(\mathcal{U}(x))$ is the probability that a random sample covers x.

Model 2: Random tree setting (Section 4). While the previous setting was in a continuous domain, next we consider a discrete setting, where the goal is to maximize the cardinality of the shadow set. Specifically, our universe X is the set of all nodes in an rooted infinite-depth binary tree. The following random process generates samples. Let $0 < p < 1$. We start at the root and move left or right at every step with equal probability. At every step, with probability p we terminate the process and output the current node. A node x in the tree dominates another node y if and only if y lies on the unique path from the root to x. For a set S of nodes, we define coverage as $\text{Cover}(S) = |\mathcal{D}(S)|$. Note, that unlike in the previous setting, there is no notion of a gap in this setting.

Once again the threshold rules we consider here are based on sequences of integers $\{c_i\}$. For any such sequence, we define $\mathcal{X}_i = \{x : |\mathcal{D}(x)| \geq c_i\}$. We show that constant-factor optimality can be achieved with exponential or smaller selection overheads.

Theorem 2. *For the binary tree model described above, any threshold rule based on a sequence $\{c_i\}$ with $c_i = O(poly(i))$ achieves an $O(1)$ competitive ratio in expectation with respect to coverage, as well as an $O(1)$ approximation to the expected coverage.*

Model 3: Skyline problem (Section 5). Finally, we consider another continuous domain that is a generalization of the skyline problem mentioned previously. We are interested in selecting a set of samples with a large shadow. Specifically, we define the "gap" of S to be $\text{Gap}(S) = 1 - \mu(\mathcal{D}(S))$. Our goal is to minimize the gap.

We show that a natural class of threshold rules obtains near-optimal gaps in this setting. Recall that $\mathcal{U}(x)$ for an element $x \in X$ denotes the set of elements that dominate x. We consider threshold rules of the form $\mathcal{X}_i = \{x \in X : \mu(\mathcal{U}(x)) \leq c_i\}$ for some sequence of numbers $\{c_i\}$. We require the following continuity assumption on the measure μ.

Definition 1. (Measure continuity) *For all $x \in X$ and $c \in [0, \mu(\mathcal{U}(x))]$, there exists an element $y \in \mathcal{U}(x)$ such that $\mu(\mathcal{U}(y)) = c$. Furthermore, there exist elements $\underline{x}, \overline{x} \in X$ with $\mathcal{U}(\underline{x}) = X$ and $\mathcal{U}(\overline{x}) = \emptyset$.*

Measure continuity ensures that the sets $\mathcal{X}_i$ are all non-empty and proper subsets of each other.

Theorem 3. *For the skyline setting with an arbitrary measure satisfying measure continuity, any threshold rule based on a sequence $\{c_i\}$ with $c_i = i^{-(1/2-\Omega(1))}$ achieves a $1 + o(1)$ competitive ratio in expectation with respect to the gap.*

We note that the class of functions c_i specified in the above theorem includes all functions for which $1/c_i$ grows subpolynomially. In particular, this includes threshold rules with selection overheads that are slightly superlinear.

For the special case of the skyline setting over a two-dimensional unit square $[0, 1]^2$ bestowed with a product distribution and the usual precedence ordering—$(x_1, y_1) \prec (x_2, y_2)$ if and only if $x_1 \leq x_2$ and $y_1 \leq y_2$—we are able to obtain a stronger result that guarantees constant-factor optimality for any polynomial selection overhead:

Theorem 4. *For the skyline setting on the unit square with any product distribution, any threshold rule based on a sequence $\{c_i\}$ with $c_i = \Omega(1/poly(i))$ achieves a $1 + o(1)$ competitive ratio in expectation with respect to the gap.*

In the following sections some proofs are skipped for brevity. The interested reader can find them in the full version of this paper [3].

3 Sample Selection in One Dimension

We will now prove Theorem 1. For a (random) variable $x \in [0, 1]$, let $\overline{x}$ denote its complement $1 - x$. For a cumulative distribution μ with domain $[0, 1]$, we use $\overline{\mu}$ to denote the cumulative distribution for the complementary random variable: $\overline{\mu}(x) = 1 - \mu(1 - x)$.

Let Y denote a draw from the power-law distribution μ, and Y_n denote the (random) quality of the nth sample selected by $\mathcal{X}$. Note that since μ is a power-law distribution, $\overline{Y}_i$ is statistically identical to $c_i\overline{Y}$. Then, the mean quality gap of the first n selected samples is given by $\text{Gap}_n = \frac{1}{n}\sum_{i=1}^{n} \overline{Y}_i = \frac{1}{n}\sum_{i=1}^{n} c_i\overline{Y}$, and, by the linearity of expectation we have $E[\text{Gap}_n] = \frac{E[\overline{Y}]}{n}\sum_{i=0}^{n-1} c_i$.

On the other hand, the expected mean gap of the largest n out of T samples drawn from a distribution with $\overline{\mu}(x) = x^k$ is equal to $\frac{1}{1+1/k}\left(\frac{n}{T+1}\right)^{1/k}$.

To relate these two expressions, we note that the expected selection overhead of $\mathcal{X}$ is given by $E[T_n] = \sum_{i=1}^{n} 1/\overline{\mu}(c_i) = \sum_{i=1}^{n} 1/c_i^k$, and use this to prove the following two lemmas. In fact, we bound the expected online gap against the expected gap of an optimal offline algorithm that is allowed to see not just T_n but $T_{n+1} - 1$ samples. The two lemmas together imply Theorem 1.

Lemma 1. *For selection thresholds $c_i = 1/i^\alpha$ with $0 \le \alpha < 1$, we have*

$$E[Gap_n] \le \frac{1}{1-\alpha}\left(\frac{n}{E[T_n]}\right)^{1/k} .$$

Proof. The proof follows by an application of the Euler-Maclaurin formula.

$$E[\text{Gap}_n] \cdot (E[T_n])^{1/k} = \left(\frac{E[\overline{X}]}{n}\sum_{i=1}^{n} i^{-\alpha}\right)\left(\sum_{i=1}^{n} i^{\alpha k}\right)^{1/k}$$
$$\approx \frac{E[\overline{X}]}{n} \cdot \frac{n^{1-\alpha}}{1-\alpha} \cdot \left(\frac{n^{\alpha k+1}}{\alpha k+1}\right)^{1/k} \le \frac{n^{1/k}}{1-\alpha} .$$

Lemma 2. *For c_i satisfying $c_i = 1/i^\alpha$ with $0 \le \alpha < 1$ for all i, we have*

$$E[Gap^*_{n+1}] \ge \frac{1}{16}\left(\frac{n}{E[T_n]}\right)^{1/k} .$$

Proof. By Markov's inequality we have

$$E[\text{Gap}_n^*] \geq \frac{1}{2}E[\text{Gap}_n^* : T_{n+1} \leq 2E[T_{n+1}]] \geq \frac{1}{2}\frac{1}{(1+1/k)} \cdot \left(\frac{n}{2E[T_{n+1}]}\right)^{1/k}$$
$$\approx \left(\frac{1}{2}\right)^{1+1/k}\left(\frac{1}{1+1/k}\right)\left(\frac{n}{E[T_n]}\right)^{1/k}\left(1+\frac{1}{n}\right)^{-\alpha} \geq \frac{1}{16}\left(\frac{n}{E[T_n]}\right)^{1/k} .$$

4 Sample Selection in Binary Trees

We prove Theorem 2 in two parts: (1) the "fast-growing thresholds" case, that is, $c_i = O(\text{poly}(i))$ and $c_i \geq \log i$ for all i, and, (2) the "slow-growing thresholds" case, that is, $c_i \leq c_{i/2} + O(1)$ for all i. We defer the proof of the slow-growing case to the full paper.

We begin with some notation and observations. Recall that for a node x in the tree $\mathcal{D}(x)$ denotes both the unique path from the root to x as well as the set of nodes covered by x (the shadow of x). Let $\mathcal{D}_k(x)$ be the kth node on $\mathcal{D}(x)$, $\mathcal{D}_{\leq k}$ be the first k nodes of $\mathcal{D}(x)$, and $\mathcal{D}_{\geq k} = \mathcal{D}(x) \setminus \mathcal{D}_{<k}$.

We say that a set of n paths associated with nodes $x_1, \ldots, x_n$ is *independent* at level k if $| \cup_{i=1}^n \mathcal{D}_k(x_i)| = n$. That is, no two paths share the same vertex at level k, and are disjoint after level k. Note that if a set of n paths $\{\mathcal{D}(x_i)\}$, of length $\geq k'$ each, is independent at level $k < k'$, then

$$|\mathcal{D}(\{x_1,\ldots,x_n\})| = \left|\bigcup_{i=1}^n \mathcal{D}(x_i)\right| \geq \left|\bigcup_{i=1}^n \mathcal{D}_{\geq k}(x_i)\right| \geq n(k'-k) .$$

Let $f(i) = c_i - \log i$. We will first obtain an upper bound on Cover_n^*. Let S_n be the n selected nodes, O_n the optimal set of n paths, and R_n the paths that are rejected and are not covered by S_n.

$$\begin{aligned}\text{Cover}_n^* &= |(\mathcal{D}(O_n) \cap \mathcal{D}(R_n)) \cup (\mathcal{D}(O_n) \cap \mathcal{D}(S_n))| \\ &\leq |\mathcal{D}(O_n) \cap \mathcal{D}(R_n)| + |\mathcal{D}(S_n)| \leq (2n + nf(n)) + \text{Cover}_n ,\end{aligned}$$

Here the last inequality follows by noting that $\mathcal{D}(O_n) \cap \mathcal{D}(R_n)$ forms a binary tree with at most $2n$ vertices in the first $\log n$ levels, and at most $nf(n)$ other vertices since it is the union of n paths of length at most $\log n + f(n)$.

Next we obtain a lower bound on Cover_n. Consider the last $n/2$ selected nodes $s_{n/2+1}, \ldots, s_n$. By definition, $|\mathcal{D}(s_{n/2+i})| \geq c_{n/2} = \log n/2 + f(n/2)$. Let $N = |\cup_{i=1}^{n/2} \mathcal{D}_{\log n/2}(s_{n/2+i})|$ be the number of paths $\mathcal{D}(s_{n/2+i})$ that are independent at level $\log n/2$. Since $\mathcal{D}_{\log n/2}(s_{n/2+i})$ chooses from each of the $n/2$ nodes at level $\log n/2$ equiprobably, N has the same distribution as the number of occupied bins when $n/2$ balls are thrown into $n/2$ bins uniformly at random. The expected number of unoccupied bins is $n/2e$. By Markov's inequality, with probability at least 1/2, the number of empty bins is at most $2\frac{n}{2e}$. So, we have that $\Pr[N \geq \frac{n}{2}(1 - 2/e)] \geq 1/2$. Thus, we have

$$\begin{aligned} E\left[\frac{\text{Cover}_n^*}{\text{Cover}_n}\right] &\leq E\left[\frac{\text{Cover}_n^*}{\text{Cover}_n} : N \geq \frac{n}{2}(1-2/e)\right] + \Pr[N \leq \frac{n}{2}(1-2/e)] \\ &\leq E\left[\frac{n(2+f(n)) + \text{Cover}_n}{\text{Cover}_n} : N \geq \frac{n}{2}(1-2/e)\right] + \frac{1}{2} \\ &\leq \frac{3}{2} + \frac{2(2+f(n))}{(1-2/e)f(n/2)} = O(1) \ , \end{aligned}$$

where the constant depends on $f(n)$.

A bound on the approximation for expected coverage can be obtained similarly.

5 Sample Selection in the Skyline Model

In this section we prove Theorem 4. The proof of Theorem 3 is similar and we defer it to the full version. For simplicity of exposition, we first consider the case where μ is the uniform measure, and then describe how to extend the argument to arbitrary product distributions.

Let S_n denote the set of samples selected by the (implicit) threshold rule out of the set T_n of samples seen. Let $R_n = T_n \setminus S_n$ denote the samples rejected by the threshold rule, and O_n denote an optimal subset of T_n of size n. Note that all points in O_n must be undominated by other points in T_n. Let $\mathcal{E}_n$ denote the event that O_n contains a point in R_n, implying that there is a point in $R_n \setminus \mathcal{D}(S_n)$. It is immediate that $\text{Gap}_n \neq \text{Gap}_n^*$ if and only if $\mathcal{E}_n$ happens. In particular, using the fact that the Gap_n^* is always at least c_n, we can relate the expected competitive ratio to the probability of $\mathcal{E}_n$ as follows.

Lemma 3. *For the skyline model with an arbitrary distribution and $\mathcal{E}_n$ as defined above, we have*

$$E_\mu\left[\frac{Gap_n}{Gap_n^*}\right] \leq 1 + \frac{1}{c_n}\Pr[\mathcal{E}_n] \ .$$

Next we show that $\mathcal{E}_n$ happens with very low probability, that is, with high probability every sample in R_n is dominated by some sample in S_n. First, noting that for the uniform measure, $\mu(\mathcal{X}_n) = c_n(1 + \ln 1/c_n)$ for all n, we obtain the following lemma.

Lemma 4. *Let α be a constant satisfying $\frac{c_i}{c_{i/2}} \geq \alpha$ for large enough i. Then with probability $1 - o(c_n)$, $\alpha n/4$ of the samples in S_n belong to $\mathcal{X}_n$.*

The next lemma shows that given a sufficient number of samples in S_n belonging to $\mathcal{X}_n$, with high probability R_n is dominated by these samples. For this lemma let $\mathcal{E}_n'$ denote the event that at least one point in R_n is not dominated by $S_n \cap \mathcal{X}_n$ and let $z = |S_n \cap \mathcal{X}_n|$.

Lemma 5. *Conditioned on z,*

$$\Pr[\mathcal{E}'_n] \leq \exp\left\{-\frac{zc_n}{\mu(\mathcal{X}_n)}\right\} \cdot E[|R_n|] \ .$$

Finally, armed with the fact that $E[|R_n|] \leq E[T_n] = \sum_{i=1}^{n} 1/\mu(\mathcal{X}_i) \leq n/c_n$, we are ready to prove Theorem 4.
Proof of Theorem 4: Using Lemma 3, we have that

$$E_\mu[\text{Gap}_n/\text{Gap}_n^*] \leq 1 + \frac{1}{c_n}(\Pr[\mathcal{E}_n \,|\, z \geq \alpha(n/4)] + \Pr[z < \alpha(n/4)]) \ ,$$

where $z = |S_n \cap \mathcal{X}_n|$. By Lemma 4, the second term in the parentheses is $o(c_n)$. Event $\mathcal{E}_n$ implies event $\mathcal{E}'_n$. So applying Lemma 5 we get

$$\Pr[\mathcal{E}_n : z \geq \alpha(n/4)] \leq \exp\left\{-\frac{zc_n}{\mu(\mathcal{X}_n)}\right\} \cdot E[|R_n|] \leq \exp\{-\alpha(n/4)\} \cdot \frac{n}{c_n} = o(c_n) \ ,$$

since $c_i = \Omega(1/\text{poly}(i))$.

For general product distributions, for a point $(a, b) \in X$, let $\mu(a, b) = \mu^x(a) \cdot \mu^y(b)$ for one-dimensional measures μ^x and μ^y. Lemma 3 holds as before. To bound the probability of $\mathcal{E}_n$, we give a reduction from the product measure setting to the uniform measure setting. In particular, consider mapping X into $X' = [0,1] \times [0,1]$ by mapping a point $(a, b) \in X$ to $(\mu^x(a), \mu^y(b)) \in X'$. Then it is easy to see that $\mathcal{X}_i$ in X gets mapped to $\mathcal{X}_i$ in X' for the same sequence $\{c_i\}$. Then, the probability of the event $\mathcal{E}_n$ under the transformation remains the same as before, and is once again $o(c_n)$. □

References

1. Atallah, M.J., Qi, Y.: Computing all skyline probabilities for uncertain data. In: Proceedings of the 28th ACM SIGMOD-SIGACT-SIGART Symposium on Principles of Database Systems (PODS), pp. 279–287. ACM, New York (2009)
2. Azar, Y., Cohen, E., Fiat, A., Kaplan, H., Räcke, H.: Optimal oblivious routing in polynomial time. J. Comput. Syst. Sci. 69(3), 383–394 (2004)
3. Bach, E., Chawla, S., Umboh, S.: Threshold rules for online sample selection. CoRR abs/1002.5034 (2010)
4. Bentley, J.L., Kung, H.T., Schkolnick, M., Thompson, C.D.: On the average number of maxima in a set of vectors and applications. J. ACM 25(4), 536–543 (1978)
5. Borodin, A., El-Yaniv, R.: Online Computation and Competitive Analysis. Cambridge University Press, Cambridge (1998)
6. Broder, A.Z., Kirsch, A., Kumar, R., Mitzenmacher, M., Upfal, E., Vassilvitskii, S.: The hiring problem and lake wobegon strategies. In: Proceedings of the 19th annual ACM-SIAM Symposium on Discrete Algorithms (SODA), pp. 1184–1193. Society for Industrial and Applied Mathematics, Philadelphia (2008)
7. Ferguson, T.: Who solved the secretary problem? Statistical Science 4(3), 282–289 (1989)

8. Freeman, P.: The secretary problem and its extensions: A review. International Statistical Review/Revue Internationale de Statistique 51(2), 189–206 (1983)
9. Gupta, A., Hajiaghayi, M.T., Räcke, H.: Oblivious network design. In: Proceedings of the 17th Annual ACM-SIAM Symposium on Discrete Algorithms (SODA), pp. 970–979. ACM, New York (2006)
10. Jia, L., Lin, G., Noubir, G., Rajaraman, R., Sundaram, R.: Universal approximations for tsp, steiner tree, and set cover. In: Proceedings of the 37th Annual ACM Symposium on Theory of Computing (STOC), pp. 386–395. ACM Press, New York (2005)
11. Samuels, S.: Secretary problems. In: Handbook of Sequential Analysis, pp. 381–405. Marcel Dekker, Inc., New York (1991)

Heterogeneous Subset Sampling*

Meng-Tsung Tsai, Da-Wei Wang, Churn-Jung Liau, Tsan-sheng Hsu**

Institute of Information Science
Academia Sinica, Taipei 115, Taiwan
{mttsai,wdw,liaucj,tshsu}@iis.sinica.edu.tw

Abstract. In this paper, we consider the problem of heterogeneous subset sampling. Each element in a domain set has a different probability of being included in a sample, which is a subset of the domain set. Drawing a sample from a domain set of size n takes $O(n)$ time if a Naive algorithm is employed. We propose a Hybrid algorithm that requires $O(n)$ preprocessing time and $O(n)$ extra space. On average, it draws a sample in $O(1 + n\sqrt{p^*})$ time, where p^* is $\min(p_\mu, 1 - p_\mu)$ and p_μ denotes the mean of inclusion probabilities. In the worst case, it takes $O(n)$ time to draw a sample. In addition to the theoretical analysis, we evaluate the performance of the Hybrid algorithm via experiments and present an application for particle-based simulations of the spread of a disease.

1 Introduction

Problem HSS. Heterogeneous subset sampling involves drawing several samples from a domain set of size n. Without loss of generality, let the domain set D be $\{1, \ldots, n\}$ in which each element i is associated with an inclusion probability p_i. For each drawn sample, which is a subset of the domain set, the probability of element i being included is equal to the given inclusion probability p_i. As it is necessary to draw several samples from the exemplar application in this paper, and potential new applications, we need to devise an efficient method to achieve the task. Complexity is analyzed under the computation model of *random access machine*, which supports operations on floating-point numbers with a precision of d digits. In addition, we assume it takes $O(1)$ time to calculate $\ln x$ with a precision of d digits and to generate a variate with a precision of d digits from the standard uniform distribution $U(0, 1)$.

The homogeneous case, in which all p_i are identical, is almost equivalent to generating variates from a binomial distribution. However, only a few works, such as [14], have considered the heterogeneous case and they focus on the size of the drawn sample rather than the elements included in it. Although the approach calculates the size of the drawn sample, it does not provide an efficient reduction

* Extended from the technical report TR-IIS-10-002, IIS, Academia Sinica, Taiwan. http://www.iis.sinica.edu.tw/page/library/TechReport/tr2010/tr10002.pdf
Supported in part by National Science Council (Taiwan) Grants NSC 97-2221-E-001-011-MY3.

** Corresponding author.

M.T. Thai and S. Sahni (Eds.): COCOON 2010, LNCS 6196, pp. 500–509, 2010.

method for generating the included elements. An intuitive way to achieve reduction is to sample with replacement, but the HSS problem is a case of sampling without replacement. Trivially applying the method of sampling with replacement to the HSS problem may need infinite computation time in the worst case. With regard to the elements included in a sample, to the best of our knowledge, only the Naive algorithm introduced in Section 2.1 is known [6] [13]. The computational cost of solving the HSS problem with the Naive algorithm is $O(n)$ time for each sample. There is no preprocessing time.

In this paper, we consider the distribution of inclusion probabilities, which may be biased in many applications. For instance, in a particle-based simulation of the spread of a disease, the mean of the distribution is biased toward being small [6] [15]. We propose the Hybrid algorithm, which imposes a tighter bound on the computation time as a trade-off for more preprocessing time and extra space. The algorithm requires $O(n)$ preprocessing time and $O(n)$ extra space. To draw each sample, it takes $O(1 + n\sqrt{p^*})$ time on average and $O(n)$ in the worst case, where p^* is $\min(p_\mu, 1 - p_\mu)$ and p_μ denotes the mean of the inclusion probabilities. The Hybrid algorithm is more efficient than the Naive algorithm when p_μ deviates from 1/2 significantly. Particle-based simulation on the spread of a disease is a particular case that can cope with the requirement for speedup. We consider this case in detail in Section 3.

The remainder of the paper is organized as follows. In Section 2, we introduce the proposed Hybrid algorithm, provide the proof of its correctness, and discuss its complexity. In Section 3, we evaluate the performance of the Hybrid algorithm via experiments and present an application for particle-based simulation of the spread of a disease. Then, in Section 4, we summarize our conclusions.

2 Algorithms

In this section, we introduce three algorithms for solving the HSS problem. The first two algorithms are the building blocks for the third one.

2.1 Naive Algorithm

Algorithm 1. Naive algorithm

```
input : a domain set D = {1, ..., n} in which each element i is associated with
        an inclusion probability p_i
output: S, a drawn sample

S ← φ

foreach i ∈ D do
    t ← U(0, 1)
    if t < p_i then
        S ← S ∪ {i}

return S
```

Algorithm 1 details the steps of a Naive algorithm for the HSS problem. To determine whether an element i should be included in a drawn sample, the algorithm compares its inclusion probability p_i with a variate generated from $U(0,1)$. Because the size of the domain set D is n, the total cost is $O(n)$, which would be the lower bound of the HSS problem if it were necessary to make the decisions one by one.

2.2 Sieve Algorithm

Algorithm 2. Sieve algorithm

```
input  : a domain set D = {1, ..., n} in which each element i is associated
         with an inclusion probability p_i
output : a drawn sample S

p_M ← max(p_1, ..., p_n)
k ← B(n, p_M)
R ← SWOR(k, D)
S ← φ
foreach i ∈ R do
    t ← U(0, 1)
    if t < p_i/p_M then
        S ← S ∪ {i}
return S
```

Procedure SWOR(k, X)

```
built-in: an array E of size n, where E[i] = i for all 1 ≤ i ≤ n
input   : an integer k, where k ≤ |X| ≤ n and a set X = {X_1, ..., X_|X|}
output  : R, a randomly selected subset of X, whose size is k
R ← φ
for i = 1 to k do
    t ← U(0, 1)
    t ← min(⌊t(n − i + 1)⌋, n − i)
    exchange E[i + t] with E[i]
    R ← R ∪ {X_E[i]}
re-swap array E into the built-in state
return R
```

The rationale for the proposed Sieve algorithm, Algorithm 2, is that instead of making a decision about each element in the domain set, we consider a smaller subset selected by a designed mechanism. Then, we only make decisions about the elements in the subset.

The designed mechanism is based on a simple idea. Specifically, for each element i in the domain set, we decouple the decision about it being included in a drawn sample into two decisions, A_i and B_i. For each pair of decisions A_i and B_i, we assign inclusion probabilities to it so that $\Pr(A_i) \times \Pr(B_i)$ is equal to p_i and $\Pr(A_i)$ is the same for all elements. Hence, the original decision is included if and only if both decoupled decisions are included. Since it is meaningful to make decision B_i if and only if the outcome of A_i is included, the decision A_i acts like a sieve by removing unnecessary corresponding decision B_i. To guarantee the success of the decoupling, we let $\Pr(A_i)$ be equal to p_M, the maximum of all p_i's.

One advantage of the decoupling procedure is that the outcome of a decision B_i is meaningful if and only if the corresponding decision A_i is included. Therefore, the outcomes of all A_i identify a subset of all B_i to be determined. Another advantage is that all A_i form a homogeneous case of the HSS problem which can be solved efficiently with two building blocks, namely, Procedure SWOR(k, D) and a binomial sampling from $B(n, p_M)$. Procedure SWOR(k, X) randomly selects a subset of X from all subsets of size k. This is a classic result reported in [11]. $B(n, p)$ denotes the binomial distribution of n trials with success rate p.

We verify the equivalence of the Sieve and Naive algorithms by proving the following assumptions, which are obviously true for the Naive algorithm. First, the probability of an element being included in a drawn sample is equal to the given inclusion probability. Second, the events of the elements being included in a drawn sample are mutually independent. To be noticed that, the lemmas we claim relies on the given $U(0,1)$ source is a true random number generator and no precision error occurs during the calculation. The proofs of the above assumptions for the Sieve algorithm are given in Lemmas 1 and 2 respectively.

Lemma 1. *The probability p'_i of element i being included in a drawn sample is equal to p_i.*

Proof.

$$\begin{aligned} p'_i &= \sum_{x=0}^{n} \Pr((k \leftarrow \Pr(B(n, p_M)) = x) \Pr(i \in \texttt{SWOR}(k, D)) \Pr(B_i) \\ &= \sum_{x=0}^{n} \binom{n}{x} p_M^x (1 - p_M)^{n-x} \left(\frac{x}{n}\right) \left(\frac{p_i}{p_M}\right) = p_i \end{aligned}$$

□

Lemma 2. *$EV_1 \dots EV_n$ are mutually independent, where EV_i denotes the event of element i being included in a drawn sample; that is,*

$$\Pr\left(\bigcap_{i \in R} (EV_i)\right) = \prod_{i \in R} \Pr(EV_i), \textit{ for all } R \subset \{1, \dots, n\} .$$

Proof.

$$\Pr\left(\bigcap_{i\in R}(EV_i)\right)$$
$$= \sum_{x=|R|}^{n} \Pr((k \leftarrow B(n,p_M)) = x) \Pr(R \subset \texttt{SWOR}(k,D)) \prod_{i\in R} \Pr(B_i)$$
$$= \sum_{x=|R|}^{n} \binom{n}{x} p_M^x (1-p_M)^{n-x} \frac{\binom{n-|R|}{x-|R|}}{\binom{n}{x}} \prod_{i\in R} \frac{\Pr(EV_i)}{p_M} = \prod_{i\in R} \Pr(EV_i) \qquad \square$$

Procedure SWOR(k, X) requires $O(n)$ time to initialize the built-in array E and $O(n)$ space to accommodate it. Each invocation of Procedure SWOR(k, X) executes a loop of k iterations (Lines 2-6). Then, the procedure re-swaps the array E into the built-in state, where there are at most $2k$ differences between the two status (Line 7). Therefore, the time complexity of each invocation of Procedure SWOR(k, X) is $O(k + 1)$.

The Sieve algorithm requires $O(n)$ time to calculate p_M (Line 1). This step is ignored in subsequent invocations under the same problem setting. For each invocation, a binomial variate k is generated (Line 2) and Procedure SWOR(k, D) is invoked once (Line 3), after which a loop of k iterations is executed. Let $\mathfrak{C}(B(n,p))$ denote the cost of generating a variate that follows a binomial distribution $B(n,p)$. Then, the following lemma can be derived.

Lemma 3. *Solving the HSS problem with the Sieve algorithm requires $O(n)$ preprocessing time and $O(n)$ extra space; and the time complexity of drawing each sample is*

$$O(\mathfrak{C}(\mathbf{B}(n,p_M)) + k + 1) . \tag{1}$$

Several works focus on binomial random variate generation, e.g., [3] [4] [8] [9] [12]. In [8], the authors report the results of comparing a number of algorithms in experiments. Provided the generated variate is k, Algorithm BG [3] needs $O(k+1)$ computation time for each sampling and Algorithm BALIAS [8] is $O(1)$ fast but requires $O(n)$ preprocessing time and $O(n)$ extra space. In [8], they also conclude that the proposed Algorithm BTPE is the most effective when μ is moderate or large, where μ is the product of the parameters n and p in $B(n,p)$. For cases where μ is small, Algorithm BINV [8] dominates. Since there is no difference between the binomial variate generation algorithms in our theoretical analysis whose complexity can be bounded by $O(k + 1)$ and preprocessing time can be bounded by $O(n)$, we let $\mathfrak{C}(B(n,p))$ be $O(k + 1)$ without specifying which efficient binomial variate generator should be used. In [8], the authors conclude that the combination of Algorithms BINV and BTPE is the fastest experimentally. Hence, we adopt this combination to run our experiments, which we discuss in Section 3. To be noticed, $\ln x$ operation is needed by several binomial samplings [3] [8].

By replacing $\mathfrak{C}(B(n, p_M))$ in Eq.(1) with $O(k+1)$, the time complexity of the Sieve algorithm becomes $O(k+1)$. Due to k is expected np_M and at most n, the Sieve algorithm draws a sample in $O(1+np_M)$ time on average and in $O(n)$ time in the worst case.

Remark 1. The Sieve algorithm is much more efficient than the Naive algorithm when p_M deviates from 1 significantly. Although the algorithms exhibit the same asymptotic behavior when p_M is close to 1, the Sieve algorithm is less efficient because the constant factor hidden in the asymptotic time complexity of the Naive algorithm is smaller.

2.3 Hybrid Algorithm

Algorithm 3. Hybrid algorithm

input : a domain set $D = \{1, \ldots, n\}$ in which element i is associated with an inclusion probability p_i
output : S, a drawn sample

$p_\mu \leftarrow (p_1 + \ldots + p_n)/n$
$p_{thres} \leftarrow \sqrt{p_\mu}$
$X \leftarrow \{i : p_i \leq p_{thres}, i \in D\}$
$Y \leftarrow D - X$
$S \leftarrow sieve(X) \cup naive(Y)$
return S

To leverage the advantages of the Naive and Sieve algorithms, Algorithm 3, the Hybrid algorithm, divides the original domain set D into two disjoint domain sets X and Y. The division is made so that X contains all elements whose inclusion probability is less than a calculated threshold $p_{thres} = \sqrt{p_\mu}$, and Y contains the remaining elements. Then, the Hybrid algorithm applies the Sieve algorithm and the Naive algorithm to X and Y respectively. Therefore, the equivalence of the Hybrid algorithm and the Naive algorithm is verified by the equivalence of the Sieve algorithm and the Naive algorithm, as discussed in the previous subsection. Let $sieve(D)$ and $naive(D)$ represent solving the HSS problem in the domain set D with the Sieve algorithm and the Naive algorithm respectively. Clearly, the following lemma is derived.

Lemma 4. *Solving the HSS problem with the Hybrid algorithm requires $O(n)$ preprocessing time and $O(n)$ extra space; the time complexity of drawing each sample is*

$$O(\mathfrak{C}(B(n, p_{thres})) + k + 1 + |Y|) . \tag{2}$$

Remark 2. Because $\sum_{i \in Y} p_i \leq \sum_{i \in D} p_i = np_\mu$ and $\sqrt{p_\mu} \leq \min_{i \in Y} p_i$, $O(|Y|) = O(n\sqrt{p_\mu})$; similar to Eq. (1), Eq. (2) can be rewritten as $O(1+n\sqrt{p_\mu})$ on average.

Remark 3. When $1 - p_M < p_M$, it is easier to draw the complement of a sample. In Algorithm 3, let $p_i = 1 - p_i$ for all $i \in D$ and Line 5 be $S = D - (sieve(X) \cup naive(Y))$, by Lemma 4, the average time complexity to draw each sample is $O(1 + n\sqrt{1 - p_\mu})$ if Line 5 can be achieved in average $O(k + 1)$-time after $O(n)$ preprocessing time.

Procedure `Complement(`D`,`X`)` /*$X \subset D$*/

```
Initialization(D):          Deletion(i):                     Recovery(X):
for i ∈ D ∪ {0} do          next[prev[i]] ← next[i]          for i ∈ X do
    prev[i] ← i − 1         prev[next[i]] ← prev[i]              next[i − 1] ← i
    crnt[i] ← i                                                  prev[i + 1] ← i
    next[i] ← i + 1
```

As shown in Procedure Complement(D, X), the initialization requires $O(n)$ preprocessing time and the deletion of set X and recover to the initial state take $O(k + 1)$ time. Therefore, the assumption in Remark 3 holds. Combined the results of Remarks 2 and 3, the Hybrid algorithm solves HSS problem in average complexity $O(1 + n\sqrt{p^*})$ where $p^* = \min(p_\mu, 1 - p_\mu)$. More details of the used data structure can be found in Chapter 10.3 of Book [2]. For the ease to compare, the complexities of the introduced algorithms are listed in Table 1.

Table 1. Complexity comparison

Algorithm	Preprocessing Time	Time to Draw Each Sample	Extra Space
Naive	–	avg: $O(n)$ worst: $O(n)$	–
Sieve	$O(n)$	avg: $O(1 + np_M)$ worst: $O(n)$	$O(n)$
Hybrid	$O(n)$	avg: $O(1 + n\sqrt{p^*})$ worst: $O(n)$	$O(n)$

3 Experiment Results

In this section, we compare the performance of the Naive and Hybrid algorithms via experiments. Then, we demonstrate how the Hybrid algorithm can improve the performance of a practical application substantially. The implementation of the combined BTPE and BINV algorithms [8] was downloaded from *GSL*[1]. All experiments were run on a workstation with Intel Xeon 3.2GHz processors.

3.1 Experiments

We conducted a number of experiments to assess the speedup and relative error of the Hybrid algorithm in solving the HSS problem. Let $\mathcal{E}_N(n, p_\mu)$ and $\mathcal{E}_H(n, p_\mu)$ be an experiment that draws 100 samples with Naive and Hybrid algorithms respectively, where $|D| = n$ and $\sum_{i \in D} p_i / n = p_\mu$. In addition, let $\mathcal{S}(n, p_\mu)$ denote the ratio of the running time of $\mathcal{E}_N(n, p_\mu)$ to that of $\mathcal{E}_H(n, p_\mu)$; note that the preprocessing time is included. Let $\mathcal{A}_X(n, p_\mu)$ be the average size of the samples drawn in the experiment $\mathcal{E}_X(n, p_\mu)$; and let $\mathcal{R}(n, p_\mu)$ denote the

[1] GNU Scientific Library `http://www.gnu.org/software/gsl/`

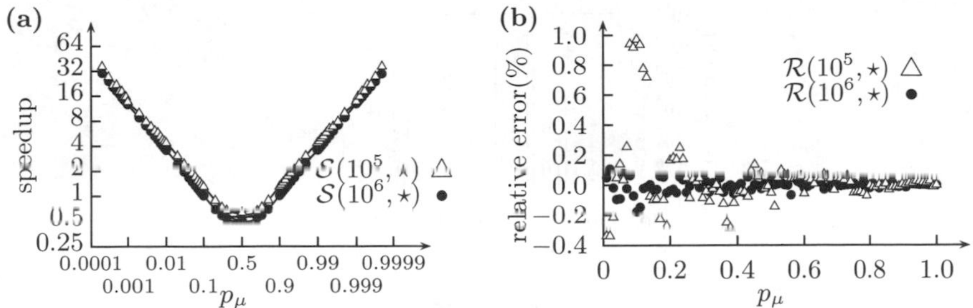

Fig. 1. Comparison of the Hybrid algorithm and the Naive algorithm

relative error $(\mathcal{A}_H(n,p_\mu) - \mathcal{A}_N(n,p_\mu))/\mathcal{A}_N(n,p_\mu)$. Figure 1(a) illustrates the speedup factor. Clearly, the speedup is substantial when p_μ deviates from $1/2$ significantly. The difference between the curves of $\mathcal{S}(10^5,\star)$ and $\mathcal{S}(10^6,\star)$ is caused by the preprocessing time. Figure 1(b) illustrates the relative error of the two algorithms. The distributions of the algorithms are similar; because, in both scenarios, the relative error is bounded in the interval $[-0.4\%, 1\%]$ and converges to the interval $[-0.2\%, 0.2\%]$. Since $\mathcal{A}_N(n,p_\mu)$ is not fixed, the relative error declines as p_μ increases.

3.2 A Practical Application of the Hybrid Algorithm

Particle-based simulation models are now being used in computational epidemiology [6] in addition to traditional SIR (Susceptible, Infector, and Remove/ Recovered) differential-equation-based approaches, such as [1] [5] [7] and [10]. There are two key reasons for this development [6]. The first is that the SIR model can best describe the dynamics of an epidemic when the number of infected persons is large, rather than in the initial or final stages of a disease outbreak. However, the initial stage is crucial because some intervention methods could be applied to prevent or slow down the transmission of the disease at this point. The second reason is that a particle-based simulation model provides more opportunities to fine tune the values of the features considered by the model.

Algorithm 4. Particle-based simulation model.

```
foreach time period T do
    foreach infector I do
        foreach susceptible individual S that had contact with I during T do
            if the infection between I and S takes place then
                change the susceptible status of S
```

Algorithm 4 provides a high-level description of the particle-based simulation model of disease transmission. The most inner loop can be thought as a case of the HSS problem. The most time-consuming part is determining which possible infections have taken place. For each possible infection between an infector I and a susceptible individual S, the occurrence rate p may vary because of a number

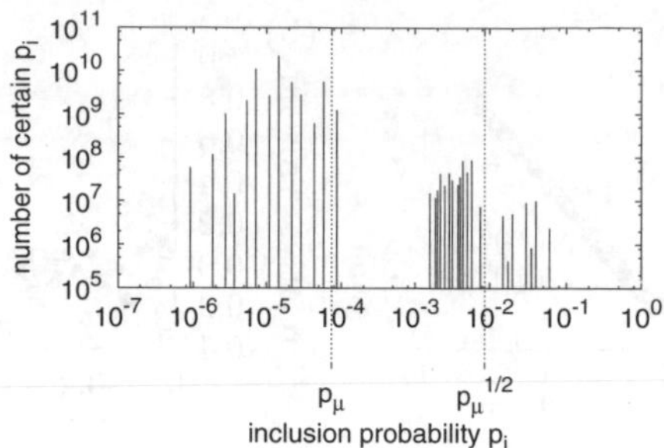

Fig. 2. Histogram of p_i in the disease transmission model

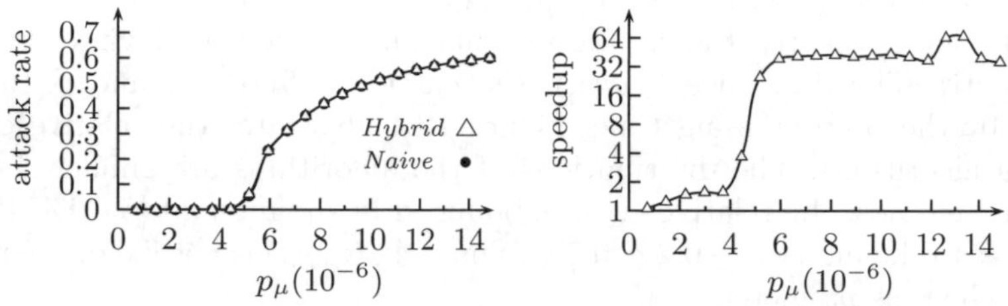

Fig. 3. Simulation results of the Hybrid and Naive algorithms

of factors, such as the nature of the disease, the intervention methods applied to it, and the closeness of I and S. In other words, p varies between different (I, S) pairs and also for the same pair in different time periods.

Following the approach in [6], we construct a simulation model based on Taiwan's census data and apply the Hybrid algorithm to it [15]. As shown in Figure 2, the value of p is usually quite small in the simulation model of disease transmission, as a person seldom has close contact with a large number of family members or friends, compared to the number of casual contacts encountered daily. Figure 2 also shows that setting $\sqrt{p_\mu}$ as p_{thres} is not appropriate. In practice, to maximize the speedup, p_{thres} can be adjusted dynamically based on the distribution of inclusion probabilities. Figure 3 compares the simulated results, the attack rates, and the differences between the computational cost of the two algorithms. Based on the figure, we conclude that the Hybrid algorithm can improve the performance of particle-based disease simulation substantially, without changing the behavior of the model.

4 Concluding Remarks

In this paper, we propose an algorithm, called the Hybrid algorithm, to solve the problem of heterogeneous subset sampling. We prove the correctness of the algorithm and show that time complexity is $O(1 + n\sqrt{p^*})$. In addition, we evaluate the performance of the algorithm via experiments and demonstrate its efficacy by a practical application. The experiment results show that the speedup is substantial when p_μ deviates from 1/2. Moreover, the exemplar application demonstrates that the Hybrid algorithm can be used effectively to simulate particle-based spread of a disease, without changing the behavior of the model. We believe the proposed Hybrid algorithm can be applied to many other problems. In this paper, we focus on dividing the domain set into two groups and solving each of them with a proper algorithm. This method can be extended to improve the performance further. In a future work, we will divide the domain set into $O(\log n)$ groups, where the i-th group consists of the elements, whose inclusion probabilities are in the interval $[2^{1-i}, 2^{-i})$ for all $1 \leq i \leq \log n$, and the remaining elements are contained in the last group. Solving each group with the Sieve algorithm, the average complexity to draw each sample is $O(\log n + np_\mu)$. We will also consider how to dynamically divide the domain set according to the distribution of the inclusion probabilities.

Acknowledgements. We sincerely thank an anonymous reviewer for very helpful comments, one of which is to point out a possible future extension.

References

1. Anderson, R.M., May, R.M.: Infectious diseases of humans. Oxford Press, Oxford (1991)
2. Cormen, T.H., Leiserson, C.E., Rivest, R.L., Stein, C.: Introduction to algorithms, 2nd edn. The MIT Press, Cambridge (2001)
3. Devroye, L.: Generating the maximum of independent identically distributed random variables. Computers and Mathematics with Applications 6, 305–315 (1980)
4. Devroye, L.: Non-Uniform Random Variate Generation. Springer, Heidelberg (April 1986)
5. Evachev, L.A., Longini, I.M.: A mathematical model for the global spread of influenza. Math. Biosci. 75, 3–22 (1985)
6. Germann, T.C., Kadau, K., Longini, I.M., Macken, C.A.: Mitigation strategies for pandemic influenza in the United States. PNAS 103(15), 5935–5940 (2006)
7. Hufnagel, L., Brockmann, D., Geisel, T.: Forecast and control of epidemics in a globalized world. Proc. Natl. Acad. Sci. USA 101, 15124–15129 (2004)
8. Kachitvichyanukul, V., Schmeiser, B.W.: Binomial random variate generation. Commun. ACM 31(2), 216–222 (1988)
9. Kemp, C.D.: A modal method for generating binomial variables. Communications in Statistics Theory and Methods 15, 805–813 (1986)
10. Kermack, W.O., McKendrick, A.G.: A contribution to the mathematical theory of epidemics. Proc. R. Soc. London Ser. A 115, 700–721 (1927)
11. Knuth, D.E.: The Art of Computer Programming. Seminumerical Algorithms, vol. II. Addison-Wesley, Reading (1969)
12. Relles, D.A.: A simple algorithm for generating binomial random variables when n is large. Journal of the American Statistical Association 67(339), 612–613 (1972)
13. Särndal, C.-E., Swensson, B., Wretman, J.: Model assisted survey sampling. Springer Series in Statistics (2003)
14. Serfling, R.J.: Some elementary results on Poisson approximation in a sequence of Bernoulli trials. Society for Industrial and Applied Mathematics 20(3), 567–579 (1978)
15. Tsai, M.-T., Chern, T.-C., Chuang, J.-H., Hsueh, C.-W., Kuo, H.-S., Liau, C.-J., Riley, S., Shen, B.-J., Wang, D.-W., Shen, C.-h., Hsu, T.-s.: Efficient simulation of the spatial transmission dynamics of influenza. PLoS Curr Influenza (2010)

Identity-Based Authenticated Asymmetric Group Key Agreement Protocol

Lei Zhang[1], Qianhong Wu[1,2], Bo Qin[1,3], and Josep Domingo-Ferrer[1]

[1] Universitat Rovira i Virgili, Dept. of Comp. Eng. and Maths
UNESCO Chair in Data Privacy, Tarragona, Catalonia
[2] Wuhan University, School of Computer
Key Lab. of Aerospace Information Security and Trusted Computing
Ministry of Education, China
[3] Xi'an University of Technology, School of Science, Dept. of Maths, China
{lei.zhang,qianhong.wu,bo.qin,josep.domingo}@urv.cat

Abstract. In identity-based public-key cryptography, an entity's public key can be easily derived from its identity. The direct derivation of public keys in identity-based public-key cryptography eliminates the need for certificates and solves certain public key management problems in traditional public-key cryptosystems. Recently, the notion of *asymmetric group key agreement* was introduced, in which the group members merely negotiate a common encryption key which is accessible to any entity, but they hold respective secret decryption keys. In this paper, we first propose a security model for identity-based authenticated asymmetric group key agreement (IB-AAGKA) protocols. We then propose an IB-AAGKA protocol which is proven secure under the Bilinear Diffie-Hellman Exponent assumption.

Keywords: Identity-Based Public-Key Cryptography, Group Key Agreement, Asymmetric Group Key Agreement, Bilinear Map.

1 Introduction

Group Key Agreement (GKA) protocols are widely employed in many modern collaborative and distributed applications. Their main goal is to implement secure broadcast channels. In the conventional GKA definition, a group of members interact over an open network to establish a common secret key to be used to achieve secure broadcast. This secret key is shared by all group members. A limitation of conventional GKA systems is that only group members are allowed to broadcast to other group members. However, in practice, anyone is likely to be a potential sender, just as anyone can encrypt a message in public-key encryption. Observing this fact, Wu *et al.* [17] introduced the notion of Asymmetric Group Key Agreement (AGKA). By their definition, instead of a common secret key, the group members merely negotiate a common encryption key which is accessible to any entity, but they hold respective secret decryption keys.

Motivation. In the real world, sometimes the bandwidth is not critical for GKA protocols but the round efficiency is. One-round key agreement protocols have

M.T. Thai and S. Sahni (Eds.): COCOON 2010, LNCS 6196, pp. 510–519, 2010.

several advantages [11,17] over key agreement protocols in two or more rounds. For instance, imagine a group of friends who wish to share their personal documents via the open network. If a two-round key agreement protocol is employed to establish a secure channel, all friends should be online at the same time. However, if this group of friends live in different time zones, it is difficult for them to be online concurrently.

A trivial way to achieve one round AGKA is for each member in the group to publish a public key and reserve the respective secret key. To send a message to this group, a sender separately encrypts for each member and generates the group ciphertext by concatenating all the individual ciphertexts. However, this trivial solution leads to a long ciphertext and forces the sender to store all the public keys of the group members. Instead of this solution, Wu *et al.* [17] proposed a one-round AGKA protocol from *scratch*, which means that the protocol participants do not hold any secret values prior to the execution of the protocol.

Though the protocols from scratch are efficient, they are only secure against passive adversaries who just eavesdrop the communication channel. Active adversaries are more powerful since they are assumed to have a complete control of the communication channel. They have the ability to relay, delay, modify, interleave or delete the message flows during the execution of the protocol. In particular, an active adversary is able to mount well-known man-in-the-middle attacks. Hence, it is vital for an AGKA protocol to withstand the attacks from active adversaries. This calls for authenticated key agreement protocols.

An authenticated key agreement protocol is a key agreement protocol which aims to ensure that no entities aside from the intended ones can possibly compute the session key agreed. Authenticated key agreement protocols may be designed under different public-key cryptosystems. A number of key agreement protocols have been proposed under the traditional PKI-based public-key paradigm. In that paradigm, key agreement protocols rely on the entities obtaining each other's certificates, extracting each other's public keys, checking certificate chains (which may involve many signature verifications) and finally generating a shared session key. Furthermore, the management of public-key certificates requires a large amount of computation, storage, and communication. To eliminate such costs, Identity-Based Public Key Cryptography (IB-PKC) was introduced by Shamir [15] in 1984. The main feature of IB-PKC is that the public key of an entity can be easily derived from its identity, such as its telephone number or email address; the corresponding private key can only be derived by a trusted Private Key Generator (PKG) who owns the *master secret* of the system.

Contribution. In this paper, we first specify a security model that an Identity-Based Authenticated Asymmetric Group Key Agreement (IB-AAGKA) protocol should satisfy. Our model allows an adversary to adaptively choose his targets, and it captures the IB version of the (modified) common security requirements (*e.g.*, *secrecy*, *known-key security* and *forward secrecy*), which are usually considered in GKA protocols. These newly defined security requirements are described as follows:

- Secrecy requires that only the legitimate participants (group members) can read the messages encrypted by the negotiated public key.
- Known-key security means that, if an adversary learns the group encryption/decryption keys of other sessions, he cannot compute subsequent group decryption keys.
- Forward secrecy ensures that the disclosure of long-term private keys of group members must not compromise the secrecy of the decryption keys established in earlier protocol runs. Specifically, we say a key agreement protocol offers *perfect forward secrecy* if the long-term private keys of all the group members involved may be compromised without compromising any group decryption key previously established by these group members. We say a key agreement offers *partial forward secrecy* if compromise of the long-term keys of one or more specific group members does not compromise the group decryption keys established by these group members.

We also propose a non-trivial one round IB-AAGKA protocol satisfying our security requirements, which we prove in the random oracle model.

Our protocol is based on a specific identity-based multi-signature scheme which we call identity-based batch multi-signature (IB-B-MS), which may itself be interesting in its own right. Our scheme allows x signers to sign t messages in such a way that the length of the batch multi-signature consists only of $t+1$ group elements. Furthermore, the batch multi-signature can be separated into t individual multi-signatures.

Related Work. Since Diffie and Hellman published their solution to key agreement [8], much attention has been paid to this primitive. Joux [11] was the first who extended key agreement to three parties. We notice that both the Diffie-Hellman and Joux protocols are one-round key agreement protocols. However, when the protocol participants are more than three, it seems knotty to construct key agreement protocols without additional rounds. Over the years, many attempts have been made at extending the Diffie-Hellman and Joux protocols to n parties. Among them, the Burmester-Desmedt protocol [5] is one of the best-known. This protocol requires two rounds and is the most efficient existing GKA protocol in round efficiency without constraints on n.

For a key agreement protocol to be usable in open networks, it should be resistant against active adversaries. However, the basic Diffie-Hellman and Joux protocols as well as the Burmester-Desmedt protocol do not authenticate the communication entities. Hence they are not suited for hostile networks where man-in-the-middle attacks may happen. Several protocols have been proposed to add authentication [7,14]; among those, the GKA protocol in [7] is based on IB-PKC. This protocol refers to Katz and Yung's results [12] for an authenticated version and requires two rounds.

The paradigm of provable security subsumes an abstract formalization that considers the protocol environment and identifies its security goals. Bresson *et al.* [4] were the first to formalize the security model for group key agreement protocols. Later, this model was refined by Bresson *et al.* [2,3] and some variants [12,13] of it appeared. These models are widely accepted in proving the security of GKA protocols. In this paper, we will borrow some ideas from these models to define the security model for IB-AAGKA protocols.

2 Bilinear Maps

Let $\mathbb{G}_1$ and $\mathbb{G}_2$ be two multiplicative groups of prime order q, and g be a generator of $\mathbb{G}_1$. A map $\hat{e}: \mathbb{G}_1 \times \mathbb{G}_1 \rightarrow \mathbb{G}_2$ is called a bilinear map if it satisfies the following properties: (1) Bilinearity: $\hat{e}(g^\beta, g^\gamma) = \hat{e}(g, g)^{\beta\gamma}$ for all $\beta, \gamma \in \mathbb{Z}_q^*$; (2) Non-degeneracy: there exists $u, v \in \mathbb{G}_1$ such that $\hat{e}(u, v) \neq 1$; (3) Computability: there exists an efficient algorithm to compute $\hat{e}(u, v)$ for any $u, v \in \mathbb{G}_1$.

The security of our protocol is based on the hardness of following problems.

Computational Diffie-Hellman (CDH) Problem: Given g, g^α, g^β for unknown $\alpha, \beta \in \mathbb{Z}_q$, compute $g^{\alpha\beta}$.

k-Bilinear Diffie-Hellman Exponent (BDHE) Problem: [1] Given g, h, and $y_i = g^{\alpha^i}$ in $\mathbb{G}_1$ for $i = 1, 2, ..., k, k+2, ..., 2k$ as input, compute $\hat{e}(g, h)^{\alpha^k}$.

3 Security Model

The first security model for AGKA protocols was presented by Wu *et al.* [17]. We note that the security model in [17] only considers passive attackers. In the sequel, we will extend this model to capture the ability of active attackers and integrate the notion of IB-PKC.

3.1 Participants and Notations

Let $\mathbb{P}$ be a set with polynomial-size of potential protocol participants. Each participant in $\mathbb{P}$ has an identity and a private key. Any subset $\mathbb{U} = \{\mathcal{U}_1, ..., \mathcal{U}_n\} \subseteq \mathbb{P}$ may decide at any point to establish a confidential channel among them. We use $\Pi_{\mathcal{U}_i}^\pi$ to represent instance π of participant $\mathcal{U}_i$ involved with partner participants $\{\mathcal{U}_1, ..., \mathcal{U}_{i-1}, \mathcal{U}_i+1, \mathcal{U}_n\}$ in a session. Each instance $\Pi_{\mathcal{U}_i}^\pi$ holds the variables $\mathsf{pid}_{\mathcal{U}_i}^\pi$, $\mathsf{sid}_{\mathcal{U}_i}^\pi$, $\mathsf{ms}_{\mathcal{U}_i}^\pi$, $\mathsf{ek}_{\mathcal{U}_i}^\pi$, $\mathsf{dk}_{\mathcal{U}_i}^\pi$ and $\mathsf{state}_{\mathcal{U}_i}^\pi$ which are defined below:

- $\mathsf{pid}_{\mathcal{U}_i}^\pi$ is the *partner ID* of instance $\Pi_{\mathcal{U}_i}^\pi$. It is a set containing the identities of the participants in the group with whom $\Pi_{\mathcal{U}_i}^\pi$ intends to establish a session key including $\mathcal{U}_i$ itself. For simplicity, we assume that the identities in $\mathsf{pid}_{\mathcal{U}_i}^\pi$ are lexicographically ordered.
- $\mathsf{sid}_{\mathcal{U}_i}^\pi$ is the *session ID* of instance $\Pi_{\mathcal{U}_i}^\pi$. We follow [13] in assuming that unique session IDs are provided by some higher-level protocol when the group key-exchange protocol is first initiated. Therefore, all members taking part in a given execution of a protocol will have the same session ID.
- $\mathsf{ms}_{\mathcal{U}_i}^\pi$ is the concatenation of all messages sent and received by $\Pi_{\mathcal{U}_i}^\pi$ during its execution, where the messages are ordered by round, and within each round lexicographically by the identities of the purported senders.
- $\mathsf{ek}_{\mathcal{U}_i}^\pi$ is the encryption key held by $\Pi_{\mathcal{U}_i}^\pi$.
- $\mathsf{dk}_{\mathcal{U}_i}^\pi$ is the decryption key held by $\Pi_{\mathcal{U}_i}^\pi$.

– $\mathsf{state}_{\mathcal{U}_i}^{\pi}$ represents the current (internal) state of instance $\Pi_{\mathcal{U}_i}^{\pi}$. When an instance has *terminated*, it is done sending and receiving messages. We say that an IB-AAGKA protocol has been *successfully terminated* (accepted) in the instance $\Pi_{\mathcal{U}_i}^{\pi}$ if it possesses $\mathsf{ek}_{\mathcal{U}_i}^{\pi}(\neq null)$, $\mathsf{dk}_{\mathcal{U}_i}^{\pi}(\neq null)$, $\mathsf{pid}_{\mathcal{U}_i}^{\pi}$ and $\mathsf{sid}_{\mathcal{U}_i}^{\pi}$.

Definition 1 (Partnering). *We say instances $\Pi_{\mathcal{U}_i}^{\pi}$ and $\Pi_{\mathcal{U}_j}^{\pi'}$ (with $i \neq j$) are partnered iff (1) they are successfully terminated; (2) $\mathsf{pid}_{\mathcal{U}_i}^{\pi} = \mathsf{pid}_{\mathcal{U}_j}^{\pi'}$; and (3) $\mathsf{sid}_{\mathcal{U}_i}^{\pi} = \mathsf{sid}_{\mathcal{U}_j}^{\pi'}$.*

3.2 The Model

In GKA protocols, secrecy is the core security definition. In conventional GKA protocols, secrecy is defined by the indistinguishability of the shared common secret key from a random string in the secret key space. However, in our IB-AAGKA, what is negotiated is only a common public encryption key while the group members' secret decryption keys are different. Observe that both conventional GKAs and our IB-AAGKA have the similar final goal of establishing a confidential broadcast channel among users. Hence, we directly use the confidentiality of the final broadcast channel to define the secrecy of an IB-AAGKA protocol. That is, secrecy is defined by the indistinguishability of a message encrypted under the negotiated public key from a random string in the ciphertext space. Specifically, we use the following game which is run between a challenger $\mathcal{C}$ and an adversary $\mathcal{A}$ who has full control of the network communications to define the security of IB-AAGKA protocols. This game comes as follows.

Initial: At this stage, the challenger $\mathcal{C}$ first runs $\mathsf{Setup}(\ell)$ to generate the system parameters params and $\mathsf{master\text{-}secret}$, then gives the resulting params to the adversary $\mathcal{A}$ while keeping $\mathsf{master\text{-}secret}$ secret.

Training: $\mathcal{C}$ is probed by $\mathcal{A}$ who can make the following queries:

– $\mathsf{Send}(\Pi_{\mathcal{U}_i}^{\pi}, \Delta)$: Send message Δ to instance $\Pi_{\mathcal{U}_i}^{\pi}$, and output the reply generated by this instance. If $\Delta = (\mathsf{sid}, \mathsf{pid})$, this query prompts $\mathcal{U}_i$ to initiate the protocol using session ID sid and partner ID pid. Note that the identity of $\mathcal{U}_i$ should be in pid, and if Δ is incorrect the query returns *null*.
– $\mathsf{Corrupt}(\mathcal{U}_i)$: Output the private key of participant $\mathcal{U}_i$. We will use it to model (partial) forward secrecy.
– $\mathsf{Ek.Reveal}(\Pi_{\mathcal{U}_i}^{\pi})$: Output the encryption key $\mathsf{ek}_{\mathcal{U}_i}^{\pi}$.
– $\mathsf{Dk.Reveal}(\Pi_{\mathcal{U}_i}^{\pi})$: Output the decryption key $\mathsf{dk}_{\mathcal{U}_i}^{\pi}$. We will use it to model *known-key security.*
– $\mathsf{Test}(\Pi_{\mathcal{U}_i}^{\pi})$: At some point, $\mathcal{A}$ returns two messages (m_0, m_1) ($|m_0| = |m_1|$) and an instance $\Pi_{\mathcal{U}_i}^{\pi}$. It is required that $\Pi_{\mathcal{U}_i}^{\pi}$ be fresh (see Definition 2). $\mathcal{C}$ chooses a bit $b \in \{0, 1\}$ uniformly at random, encrypts m_b under $\mathsf{ek}_{\mathcal{U}_i}^{\pi}$ to produce the ciphertext c, and returns c to $\mathcal{A}$. Notice that $\mathcal{A}$ can submit this query only once, and we will use this query to model *secrecy.*

Response: $\mathcal{A}$ returns a bit b'. We say that $\mathcal{A}$ wins if $b' = b$. $\mathcal{A}$'s advantage is defined to be $\mathsf{Adv}(\mathcal{A}) = |2\Pr[b = b'] - 1|$.

Definition 2 (Freshness). *An instance $\Pi^{\pi}_{\mathcal{U}_i}$ is fresh if none of the following happens:*

1. *At some point, $\mathcal{A}$ queried $\mathsf{Dk.Reveal}(\Pi^{\pi}_{\mathcal{U}_i})$ or $\mathsf{Dk.Reveal}(\Pi^{\pi'}_{\mathcal{U}_j})$, where $\Pi^{\pi'}_{\mathcal{U}_j}$ is partnered with $\Pi^{\pi}_{\mathcal{U}_i}$.*
2. *A query $\mathsf{Corrupt}(\mathcal{U}_i)$ was asked before a query of the form $\mathsf{Send}(\Pi^{\pi}_{\mathcal{U}_i}, \Delta)$.*
3. *All the private keys of the participants with $\mathsf{sid}^{\pi}_{\mathcal{U}_i}$ are corrupted*[1].

Definition 3. *An IB-AAGKA protocol is said to be secure against semantically indistinguishable chosen-identity and plaintext attacks (Ind-ID-CPA), if no randomized polynomial-time adversary has a non-negligible advantage in the above game. In other words, any randomized polynomial-time Ind-ID-CPA adversary $\mathcal{A}$ has an advantage $\mathsf{Adv}(\mathcal{A}) = |2\Pr[b = b'] - 1|$ that is negligible.*

Here we only consider security against chosen-plaintext attacks (CPA) for our IB-ASGKA protocol. To achieve security against chosen-ciphertext attacks (CCA), there are some generic approaches that convert a CPA secure encryption scheme into a CCA secure one, such as the Fujisaki-Okamoto conversion [9].

4 Building Block

We propose the signature scheme which will be used in our IB-AAGKA protocol. Our scheme can be viewed as a special identity-based multi-signature scheme which we call identity-based batch multi-signature (IB-B-MS) scheme. In our scheme, each signer will use a single random value to generate t signatures on t different messages. This way, the resulting signature (referred to as batch signature) on t messages of a signer only consists of $t+1$ group elements. Furthermore, our scheme allows signatures on the same message from x signers to be aggregated into an IB-B-MS of $t+1$ group elements. We notice that, when $t = 1$, our scheme degenerates into the Gentry-Ramzan multi-signature [10].

4.1 Definition

An IB-B-MS scheme consists of the following five algorithms:

- $\mathsf{BM.Setup}$: This algorithm takes as input a security parameter ℓ to generate a master-secret and a list of system parameters.
- $\mathsf{BM.Extract}$: This algorithm takes as input an entity's identity ID_i, and the master-secret to produce the entity's private key.
- Sign: On input t messages, a signer's identity ID_i and private key s_i, this algorithm outputs a batch signature.
- $\mathsf{Aggregate}$: On input a collection of x batch signatures on t messages from x signers, this algorithm outputs a batch multi-signature.
- $\mathsf{BM.Verify}$: This algorithm is used to check the validity of a batch multi-signature. It outputs "all valid" if the batch multi-signature is valid; otherwise, it outputs an index set, which means that the multi-signatures on the messages with indices in that set are invalid.

[1] Since we do not allow $\mathcal{A}$ to corrupt all the participants in the same session, our game captures partial forward secrecy.

4.2 The Model

The security of an IB-B-MS scheme is modeled via the following game between a challenger $\mathcal{C}$ and an adversary $\mathcal{A}$.

Initial: $\mathcal{C}$ first runs BM.Setup to obtain a master-secret and params, then sends params to the adversary $\mathcal{A}$ while keeping the master-secret secret.

Training: The adversary $\mathcal{A}$ can perform a polynomially bounded number of the following types of queries in an adaptive manner.

- Extract: $\mathcal{A}$ can request the private key of an entity with identity ID_i. In response, $\mathcal{C}$ outputs the private key of this entity.
- Sign: $\mathcal{A}$ can request an entity's batch signature on n messages. On receiving such a query, $\mathcal{C}$ outputs a batch signature on those messages.

Forgery: $\mathcal{A}$ outputs a set of x entities whose identities form the set $\mathbb{L}_{ID}^* = \{ID_1^*, ..., ID_x^*\}$, a message m^* and a multi-signature σ^*. We say that $\mathcal{A}$ wins the above game if the following conditions are satisfied:

1. σ^* is a valid multi-signature on message m^* under identities $\{ID_1^*, ..., ID_x^*\}$.
2. At least one of the identities in $\mathbb{L}_{ID}^*$ has never been submitted during the BM.Extract queries and m^* together with that identity is not involved in the Sign queries.

Definition 4. *An IB-B-MS scheme is existentially unforgeable under adaptively chosen-message attack if and only if the success probability of any polynomially bounded adversary in the above game is negligible.*

4.3 The Scheme

The construction comes as follows.

- BM.Setup: On input a security parameter ℓ, the KGC chooses two cyclic multiplicative groups $\mathbb{G}_1$ and $\mathbb{G}_2$ with prime order q, where $\mathbb{G}_1$ is generated by g and there exists a bilinear map $\hat{e} : \mathbb{G}_1 \times \mathbb{G}_1 \longrightarrow \mathbb{G}_2$. The KGC also chooses a random $\kappa \in \mathbb{Z}_q^*$ as the master-secret and sets $g_1 = g^\kappa$, and chooses cryptographic hash functions $H_1, H_2 : \{0,1\}^* \longrightarrow \mathbb{G}_1$. The system parameter list is params $= (\mathbb{G}_1, \mathbb{G}_2, \hat{e}, g, g_1, H_1, H_2)$.
- BM.Extract: This algorithm takes as input κ and an entity's identity ID_i. It generates the private key $s_i = id_i^\kappa$ for the entity, where $id_i = H_1(ID_i)$.
- Sign: To sign t messages $m_1, ..., m_t$, a signer with identity ID_i and private key s_i performs the following steps:
  1. Choose a random $\eta_i \in \mathbb{Z}_q^*$ and compute $r_i = g^{\eta_i}$.
  2. For $1 \le j \le t$, compute $f_j = H_2(m_j)$, $z_{i,j} = s_i f_j^{\eta_i}$.
  3. Output the batch signature $\sigma_i = (r_i, z_{i,1}, ..., z_{i,t})$.
- Aggregate: Anyone can aggregate a collection of signatures $\{\sigma_i = (r_i, z_{i,1}, ..., z_{i,t})\}_{1 \le i \le x}$ on the messages $\{m_j\}_{1 \le j \le t}$ from x signers into a batch multi-signature. In particular, $\{\sigma_i = (r_i, z_{i,1}, ..., z_{i,t})\}_{1 \le i \le x}$ can be aggregated into $(w, d_1, ..., d_t)$, where $w = \prod_{i=1}^{x} r_i$, $d_j = \prod_{i=1}^{x} z_{i,j}$.

- BM.Verify: To check the validity of the above batch multi-signature $(w, d_1, \ldots, d_t)$, the verifier computes $Q = \hat{e}(\prod_{i=1}^{s} H_1(ID_i), g_1)$ and for $1 \leq j \leq t$ checks $\hat{e}(d_j, g) \stackrel{?}{=} \hat{e}(f_j, w) \cdot Q$. If all the equations hold, the verifier outputs "all valid"; otherwise, it outputs an index set $\mathbb{I}$, which means the multi-signatures with indices in that set are invalid.

In the full version [18], we prove the following theorem.

Theorem 1. *Suppose an adversary $\mathcal{A}$ who asks at most q_{H_1} times H_1 queries, q_{H_2} times H_2 queries, q_e times* Extract *queries, q_s times* Sign *queries with maximal message size N, and wins the game in Section 4.2 with advantage* Adv($\mathcal{A}$) *in time τ. Then there exists an algorithm to solve the CDH problem with advantage $\frac{4}{(q_e+q_s+x+1)^2 e^2}$* Adv($\mathcal{A}$) *in time $\tau + \mathcal{O}(2q_{H_1} + q_{H_2} + 4Nq_s)\tau_{G_1}$.*

5 The IB-AAGKA Protocol

In this section, we propose our IB-AAGKA protocol. In the sequel, we will consider a group of n participants who wish to establish a broadcast channel.

- Setup: The same as BM.Setup, except that an identity-based signature scheme and a cryptographic hash function $H_3 : \mathbb{G}_2 \longrightarrow \{0,1\}^{\varsigma}$ are chosen, where ς is the bit-length of plaintexts.
- Extract: The same as BM.Extract.
- Agreement: A protocol participant $\mathcal{U}_i$, whose identity is ID_i and private key is s_i, performs the following steps:
  1. Choose a random $\eta_i \in \mathbb{Z}_q^*$, compute $r_i = g^{\eta_i}$.
  2. For $1 \leq j \leq n$, compute $f_j = H_2(j)$, $z_{i,j} = s_i f_j^{\eta_i}$.
  3. Publish $\sigma_i = (r_i, \varrho_i, \{z_{i,j}\}_{j \in \{1,\ldots,n\}, j \neq i})$, where ϱ_i is the identity-based signature on r_i. To keep the whole protocol efficient, one may choose an identity-based signature scheme that supports batch verification [6] to generate ϱ_i.
- Enc.Key.Gen: To get the group encryption key, an entity first checks the n message-signature pairs $(r_1, \varrho_1), \ldots, (r_n, \varrho_n)$. If all of these signatures are valid, then the entity computes $w = \prod_{i=1}^{n} r_i, Q = \hat{e}(\prod_{i=1}^{n} H_1(ID_i), g_1)$, and sets the group encryption key as (w, Q).
- Dec.Key.Gen: Each participant $\mathcal{U}_i$ checks the n message-signature pairs $(r_1, \varrho_1), \ldots, (r_n, \varrho_n)$. If all of these signatures are valid, $\mathcal{U}_i$ computes $d_i = \prod_{j=1}^{n} z_{j,i}$, and checks $\hat{e}(d_i, g) \stackrel{?}{=} \hat{e}(f_i, w) \cdot Q$. If the equation holds, $\mathcal{U}_i$ accepts d_i as the group decryption key; otherwise, it aborts.
- Enc: For a plaintext m, an entity generates the ciphertext as follows:
  1. Select $\rho \in \mathbb{Z}_q^*$, compute $c_1 = g^{\rho}, c_2 = w^{\rho}, c_3 = m \oplus H_3(Q^{\rho})$.
  2. Output the ciphertext $c = (c_1, c_2, c_3)$.
- Dec: To decrypt the ciphertext $c = (c_1, c_2, c_3)$, $\mathcal{U}_i$, whose group decryption key is d_i, computes $m = c_3 \oplus H_3(\hat{e}(d_i, c_1)\hat{e}(f_i^{-1}, c_2))$.

In the full version [18], we prove the following theorem.

Theorem 2. *Suppose an adversary $\mathcal{A}$ who asks at most q_{H_1} times H_1 queries, q_{H_2} times H_2 queries, q_{H_3} times H_3 queries, q_c times* Corrupt *queries, q_s times* Send *queries, q_{er} times* Ek.Reveal *queries and q_{dr} times* Dk.Reveal *queries, and wins the game with advantage* $\mathsf{Adv}(\mathcal{A})$ *in time τ. Then there exists an algorithm to solve the k-BDHE problem with advantage $\frac{4(1-k\mathsf{Adv}_{sig}(\mathcal{A}))}{q_{H_3}(q_c+q_{dr}+k+1)^2e^2}\mathsf{Adv}(\mathcal{A})$ in time $\tau+\mathcal{O}(q_{er})\tau_{\hat{e}}+\mathcal{O}(2q_{H_1}+q_{H_2}+kq_s)\tau_{G_1}$, where $\mathsf{Adv}_{sig}(\mathcal{A})$ is the advantage for $\mathcal{A}$ to forge a valid identity-based signature in time τ, $\tau_{\hat{e}}$ is the time to compute a pairing and τ_{G_1} is the time to compute a scalar multiplication in $\mathbb{G}_1$.*

6 Conclusion

We have defined a security model for IB-AAGKA protocols and proposed a one-round IB-AAGKA protocol from bilinear maps based on the k-BDHE assumption in the random oracle model. The new protocol allows an adversary to adaptively choose his targets, and it offers the key secrecy, known-key security and partial forward secrecy properties. This design is also readily adaptable to implement broadcast encryption in newly emerging vehicular ad hoc networks [16] to provide secure value-added services.

Acknowledgments and Disclaimer

This work is supported by the Spanish Government through projects TSI2007-65406-C03-01 "E-AEGIS" and CONSOLIDER INGENIO 2010 CSD2007-00004 "ARES", by the Catalan Government under 2009 SGR 1135, and by the Chinese NSF projects 60970114, 60970115 and 60970116. The last author is partially supported as an ICREA-Acadèmia researcher by the Catalan Govt. The views of the authors with the UNESCO Chair in Data Privacy do not necessarily reflect the position of UNESCO nor commit that organization.

References

1. Boneh, D., Boyen, X., Goh, E.-J.: Hierarchical Identity Based Encryption with Constant Size Ciphertext. In: Cramer, R. (ed.) EUROCRYPT 2005. LNCS, vol. 3494, pp. 440–456. Springer, Heidelberg (2005)
2. Bresson, E., Chevassut, O., Pointcheval, D.: Provably Authenticated Group Diffie - Hellman Key Exchange - The Dynamic Case. In: Boyd, C. (ed.) ASIACRYPT 2001. LNCS, vol. 2248, pp. 290–309. Springer, Heidelberg (2001)
3. Bresson, E., Chevassut, O., Pointcheval, D.: Dynamic Group Diffie-Hellman Key Exchange under Standard Assumptions. In: Knudsen, L.R. (ed.) EUROCRYPT 2002. LNCS, vol. 2332, pp. 321–336. Springer, Heidelberg (2002)
4. Bresson, E., Chevassut, O., Pointcheval, D., Quisquater, J.J.: Provably Authenticated Group Diffie-Hellman Key Exchange. In: Samarati, P. (ed.) ACM CCS 2001, pp. 255–264. ACM Press, New York (2001)
5. Burmester, M., Desmedt, Y.G.: A Secure and Efficient Conference Key Distribution System. In: De Santis, A. (ed.) EUROCRYPT 1994. LNCS, vol. 950, pp. 275–286. Springer, Heidelberg (1995)

6. Camenisch, J., Hohenberger, S., Pedersen, M.: Batch Verification of Short Signatures. In: Naor, M. (ed.) EUROCRYPT 2007. LNCS, vol. 4515, pp. 246–263. Springer, Heidelberg (2007)
7. Choi, K.Y., Hwang, J.Y., Lee, D.H.: Efficient ID-Based Group Key Agreement with Bilinear Maps. In: Bao, F., Deng, R., Zhou, J. (eds.) PKC 2004. LNCS, vol. 2947, pp. 130–144. Springer, Heidelberg (2004)
8. Diffie, W., Hellman, M.: New Directions in Cryptography. IEEE Transactions on Information Theory 22(6), 644–654 (1976)
9. Fujisaki, E., Okamoto, T.: Secure Integration of Asymmetric and Symmetric Encryption Schemes. In: Wiener, M. (ed.) CRYPTO 1999. LNCS, vol. 1666, pp. 537–554. Springer, Heidelberg (1999)
10. Gentry, C., Ramzan, Z.: Identity-Based Aggregate Signatures. In: Yung, M., Dodis, Y., Kiayias, A., Malkin, T.G. (eds.) PKC 2006. LNCS, vol. 3958, pp. 257–273. Springer, Heidelberg (2006)
11. Joux, A.: A One Round Protocol for Tripartite Diffie-Hellman. J. of Cryptology 17, 263–276 (2004)
12. Katz, J., Yung, M.: Scalable Protocols for Authenticated Group Key Exchange. In: Boneh, D. (ed.) CRYPTO 2003. LNCS, vol. 2729, pp. 110–125. Springer, Heidelberg (2003)
13. Katz, J., Shin, J.S.: Modeling Insider Attacks on Group Key-Exchange Protocols. In: Proceedings of the 12th ACM Conference on Computer and Communications Security-CCS 2005, pp. 180–189. ACM, New York (2005)
14. Kim, H., Lee, S., Lee, D.H.: Constant-Round Authenticated Group Key Exchange for Dynamic Groups. In: Lee, P.J. (ed.) ASIACRYPT 2004. LNCS, vol. 3329, pp. 245–259. Springer, Heidelberg (2004)
15. Shamir, A.: Identity-Based Cryptosystems and Signature Schemes. In: Blakely, G.R., Chaum, D. (eds.) CRYPTO 1984. LNCS, vol. 196, pp. 47–53. Springer, Heidelberg (1985)
16. Wu, Q., Domingo-Ferrer, J., Gonzalez-Nicolas, U.: Balanced Trustworthiness, Safety, and Privacy in Vehicle-to-Vehicle Communications. IEEE Trans. on Veh. Tech. 59(2), 559–573 (2010)
17. Wu, Q., Mu, Y., Susilo, W., Qin, B., Domingo-Ferrer, J.: Asymmetric Group Key Agreement. In: Joux, A. (ed.) EUROCRYPT 2009. LNCS, vol. 5479, pp. 153–170. Springer, Heidelberg (2010)
18. Zhang, L., Wu, Q., Qin, B., Domingo-Ferrer, J.: Identity-Based Authenticated Asymmetric Group Key Agreement Protocol. The full version (2010), http://eprint.iacr.org/2010/209

Zero-Knowledge Argument for Simultaneous Discrete Logarithms

Sherman S.M. Chow[1], Changshe Ma[2], and Jian Weng[3,4,5]

[1] Department of Computer Science
Courant Institute of Mathematical Sciences
New York University, NY 10012, USA
schow@cs.nyu.edu

[2] School of Computer, South China Normal University, Guangzhou, 510631, China
changshema@gmail.com

[3] Department of Computer Science, Jinan University, Guangzhou 510632, China

[4] State Key Laboratory of Information Security
Institute of Software, Chinese Academy of Sciences, Beijing 100080, China

[5] State Key Laboratory of Networking and Switching Technology
Beijing University of Posts and Telecommunications, Beijing 100876, China
cryptjweng@gmail.com

Abstract. In Crypto'92, Chaum and Pedersen introduced a widely-used protocol (CP protocol for short) for proving the equality of two discrete logarithms (EQDL) with unconditional soundness, which plays a central role in DL-based cryptography. Somewhat surprisingly, the CP protocol has never been improved for nearly two decades since its advent. We note that the CP protocol is usually used as a non-interactive proof by using the Fiat-Shamir heuristic, which inevitably relies on the random oracle model (ROM) and assumes that the adversary is computationally bounded. In this paper, we present an EQDL protocol in the ROM which saves ≈40% of the computational cost and ≈33% of the prover's uploading bandwidth. Our idea can be naturally extended for simultaneously showing the equality of n discrete logarithms with $O(1)$-size commitment, in contrast to the n-element adaption of the CP protocol which requires $O(n)$-size. This improvement benefits a variety of interesting cryptosystems, ranging from signatures and anonymous credential systems, to verifiable secret sharing and threshold cryptosystems. As an example, we present a signature scheme that only takes one (offline) exponentiation to sign, without utilizing pairing, relying on the standard decisional Diffie-Hellman assumption.

Keywords: Equality of discrete logarithms, simultaneous discrete logarithms, honest-verifier zero-knowledge argument, online/offline signature, Diffie-Hellman problem.

1 Introduction

The protocol for proving the equality of two discrete logarithms plays a central role in cryptography. Such a protocol, executed by a prover $\mathcal{P}$ and a verifier $\mathcal{V}$,

M.T. Thai and S. Sahni (Eds.): COCOON 2010, LNCS 6196, pp. 520–529, 2010.

can be roughly explained as follows: Suppose $\mathbb{G}$ is a cyclic group with prime order q, and g and h are two random elements chosen from $\mathbb{G}$. $\mathcal{P}$ knows $x \in \mathbb{Z}_q$ such that $y_1 = g^x$ and $y_2 = h^x$. After executing the protocol, $\mathcal{P}$ must convince $\mathcal{V}$ that y_1 and y_2 indeed have the same exponent with respect to the base g and h respectively, i.e., $\log_g y_1 = \log_h y_2$. The idea can be naturally extended to the *simultaneous discrete logarithm* problem, where one wants to show that $y_0 = g_0^x, \cdots, y_{n-1} = g_{n-1}^x$ when given n generators $g_0, g_1, \cdots, g_{n-1}$ of group $\mathbb{G}$.

Chaum and Pedersen [1] introduced a famous protocol (hereinafter referred as CP protocol) for proving the equality of two discrete logarithms, which has been widely used to design a variety of cryptosystems. Examples include but not limited to digital signatures [2, 3], verifiable secret sharing [4], threshold cryptosystems [5, 6], fair exchange protocols [7], pseudonym [8], electronic payment [9], electronic voting[10, 11], etc. Any improvement on the CP protocol is influential since it benefits all these cryptosystems accordingly.

Somewhat surprisingly, the CP protocol has never been improved for nearly two decades since its advent. A natural question to ask is whether we can improve the CP protocol. In this paper, we answer the question motivating the above question instead – can we have a more efficient protocol which "benefits" the cryptosystems utilizing the CP protocol?

To answer our question, we investigate how the CP protocol is typically used and what kinds of its properties have been utilized in cryptographic or security applications. In many cases, interactive proof is undesirable, if not impossible. Examples include digital signatures which provide non-repudiation property and multi-signatures [12, 13] which require all signers to be online if it is done in an interactive setting. So the CP protocol is turned into a non-interactive proof by the Fiat-Shamir heuristic [14], where the prover "challenges" herself by treating the output of the hash function as the challenge given by the verifier. This inevitably relies on the random oracle model [15] (ROM). On the other hand, the CP protocol can be proven secure without relying on the ROM.

As a zero-knowledge proof system, the CP protocol provides unconditional soundness. Roughly speaking, this (soundness) means the prover cannot cheat without being caught in most cases, without any condition imposed on the prover, in particular, the prover may possess unlimited computational power. For real-world applications, we may feel comfortable to assume that the computational abilities of an adversary are bounded, and hence a zero-knowledge *argument*, a system with computational soundness, is sufficient.

1.1 Our Contributions

Based on the above two simple yet important observations, we propose a protocol which gives a zero-knowledge argument system for simultaneous discrete logarithm. In the CP protocol, the prover $\mathcal{P}$ firstly generates two group elements as the commitments. We *carefully* integrate these two elements together in our protocol, which gives a higher performance in terms of both computational cost and communication overhead. More concretely, our protocol offers about 40% computational cost saving, and saves one group element computation overhead.

We remark that it is not entirely trivial to combine two commitments into a single element. We show that a class of trial attempts is insecure.

Our protocol can be naturally extended for simultaneous discrete logarithm for n elements where n is an integer greater than 2. Compared with the n-element adaption of the CP protocol which uses $O(n)$ commitments, we achieve $O(1)$ size.

Finally, we use our protocol to construct an efficient signature scheme which only takes one offline exponentiation and no online exponentiation to sign, under the standard decisional Diffie-Hellman assumption, without utilizing pairing.

2 Preliminaries

2.1 Notations and Complexity Problems

We explain some notations used in the rest of this paper. For a prime q, $\mathbb{Z}_q$ denotes the set $\{0, 1, 2, \cdots, q-1\}$, and $\mathbb{Z}_q^*$ denotes $\mathbb{Z}_q \backslash \{0\}$. For a finite set S, $x \in_R S$ means choosing an element x uniformly at random from S. We use $\mathbb{G}$ to denote a cyclic group with prime order q and $\ell_q = |\mathbb{G}|$ to denote the bit-length of the binary representation of an element in $\mathbb{G}$. We use PPT to mean probabilistic polynomial time.

Definition 1. *The discrete logarithm (DL) problem is, given $g, y \in \mathbb{G}$, to compute $x \in \mathbb{Z}_q$ such that $y = g^x$. We say that an algorithm $\mathcal{B}$ can solve the DL problem in group $\mathbb{G}$ with advantage ϵ if* $\Pr[\mathcal{B}(g, y) = x | g \in_R \mathbb{G}, x \in_R \mathbb{Z}_q, y = g^x] \geqslant \epsilon$.

Definition 2. *The computational Diffie-Hellman (CDH) problem is that, given $g, g^a, g^b \in \mathbb{G}$ with unknown $a, b \in \mathbb{Z}_q$, to compute g^{ab}. We say that an algorithm $\mathcal{B}$ has advantage ϵ in solving CDH problem in group $\mathbb{G}$, if* $\Pr[\mathcal{B}(g, g^a, g^b) = g^{ab} | g \in_R \mathbb{G}, a, b \in_R \mathbb{Z}_q] \geqslant \epsilon$.

Definition 3. *The decisional Diffie-Hellman (DDH) problem is, given g, g^a, g^b, g^c, where $a, b \in_R \mathbb{Z}_q^*$, to tell if $c = ab$ or $c \in_R \mathbb{Z}_q^*$. A tuple (g, g^a, g^b, g^c) is called a "DDH tuple" if $ab = c \mod q$ holds. We say that an algorithm $\mathcal{B}$ has advantage ϵ in solving the DDH problem in $\mathbb{G}$ if* $|\Pr[\mathcal{B}(g, g^a, g^b, g^{ab}) = 1 | g \in_R \mathbb{G}, a, b \in_R \mathbb{Z}_q, c = ab] - \Pr[\mathcal{B}(g, g^a, g^b, g^c) = 1 | g \in_R \mathbb{G}, a, b, c \in_R \mathbb{Z}_q]| \geqslant \epsilon$.

We say that the (t, ϵ)-DL/CDH/DDH assumption holds if no PPT algorithm can solve the respective problem with an advantage at least ϵ within time t.

2.2 Zero-Knowledge Proof for Equality of Discrete Logarithms

The security notions for honest-verifier zero-knowledge protocols for proving the equality of two logarithms are reviewed as follows.

Definition 4. *A protocol for proving the equality of two discrete logarithms is said to be secure, if it simultaneously satisfies the following three properties:*

- ***Completeness.*** *For any honest prover $\mathcal{P}$, an honest verifier $\mathcal{V}$ accepts with overwhelming probability.*
- ***Soundness.*** *An algorithm $\mathcal{A}$ (t, ϵ)-breaks the soundness of the protocol if $\mathcal{A}$ can cheat, without conducting any active attacks, an honest verifier $\mathcal{V}$ (i.e., $\mathcal{A}$ can make $\mathcal{V}$ to accept its proof whereas $\log_g y_1 \neq \log_h y_2$) with at least success probability ϵ within time t, where the probability is taken over the random coins consumed by $\mathcal{A}$ and $\mathcal{V}$*
- ***(Honest-Verifier) Zero-Knowledge.*** *We say that the protocol is statistical (or computational) zero-knowledge, if for every PPT (honest) verifier $\mathcal{V}$, there exists a PPT machine $\mathcal{S}$ (also called simulator), which is not allowed to interact with the prover $\mathcal{P}$, such that the following two distributions are statistically (or computationally) indistinguishable:*
  - *the output of $\mathcal{V}$ after interacting with a prover $\mathcal{P}$ on the common input of public parameters $(\mathbb{G}, g, h, y_1, y_2)$, and*
  - *the output of $\mathcal{S}$ with only the input of public parameters $(\mathbb{G}, g, h, y_1, y_2)$.*

Remark: In our protocol, the public parameters should also contain the description of the hash function H. Our proof for soundness requires modeling H as a random oracle.

2.3 Digital Signature and Its Existential Unforgeability

A signature scheme is defined by the following triple of algorithms:

- $\mathsf{KeyGen}(1^\ell)$: it takes a security parameter 1^ℓ and outputs a private/public key pair (sk, pk).
- $\mathsf{Sign}(sk, m)$: it outputs a signature σ taking a signing key sk and a message m as inputs.
- $\mathsf{Verify}(pk, \sigma, m)$: this verification algorithm takes as input a public key pk, a signature σ and a message m. It outputs 1 if and only if σ is a valid signature on m under pk, 0 otherwise.

Security is modeled by the existential unforgeability under adaptive chosen message attack (EUF-CMA), defined via the following game between a forger $\mathcal{F}$ and a challenger $\mathcal{C}$:

Setup. The challenger $\mathcal{C}$ runs algorithm $\mathsf{KeyGen}(1^\ell)$ to generate a key pair (pk, sk). It then forwards pk to $\mathcal{F}$, keeping sk to himself.

Signing queries. For each query, $\mathcal{F}$ adaptively issues a message m, the challenger runs $\mathsf{Sign}(sk, m)$ to get a signature σ, which is returned to $\mathcal{F}$.

Forge. $\mathcal{F}$ outputs a message/signature pair (m^*, σ^*) where $\mathsf{Verify}(pk, \sigma^*, m^*) = 1$.

$\mathcal{F}$ wins if m^* did not appear in the signing queries. $\mathcal{F}$'s advantage is defined by $\mathrm{Adv} = \Pr[\mathcal{F} \text{ wins}]$, where the probability is taken over the random coins consumed by $\mathcal{C}$ and $\mathcal{F}$.

Definition 5. *We say that a signature scheme is (t, q_s, ϵ)-EUF-CMA secure, if for any t-time forger $\mathcal{F}$ who asking at most q_s signing queries, his advantage is less than ϵ.*

3 Zero-Knowledge Argument for Simultaneous Discrete Logarithms

3.1 Review of Chaum-Pedersen Protocol

Let $\mathbb{G}$ be a cyclic group of prime order q, and let g, h be two random generators of $\mathbb{G}$. The prover $\mathcal{P}$ picks $x \in \mathbb{Z}_q$, and publishes $y_1 = g^x$ and $y_2 = h^x$. To convince the verifier $\mathcal{V}$ that $\log_g y_1 = \log_h y_2$, the CP protocol [1] proceeds as in Fig. 1. It is well known that the CP protocol is an honest-verifier zero-knowledge protocol [16].

1. $\mathcal{P}$ picks $k \in_R \mathbb{Z}_q$ and sends the pair of commitments $(a_1, a_2) = (g^k, h^k)$ to $\mathcal{V}$.
2. Upon receiving (a_1, a_2), $\mathcal{V}$ sends the challenge $c \in \mathbb{Z}_q^*$ to $\mathcal{P}$.
3. $\mathcal{P}$ sends the response $s = k - cx \mod q$ to $\mathcal{V}$.
4. $\mathcal{V}$ accepts if and only if both $a_1 = g^s y_1^c$ and $a_2 = h^s y_2^c$ hold.

Fig. 1. Chaum-Pedersen Protocol for proving $y_1 = g^x$ and $y_2 = h^x$

3.2 Our Proposed Protocol

Let $H : \mathbb{G}^2 \to \mathbb{Z}_q^*$ be a cryptographic hash function (in the proof of soundness, we shall model it as a random oracle). We define the commitment as $v = (g^z h)^k$ where $z = H(y_1, y_2)$. We assume that the discrete logarithm $\log_g h$ is unknown to all participants. This can be easily ensured in practice by requiring h to be the output of a hash function (which again is modeled as a random oracle) with a public seed and a public input. For example, one may set h to be $PRF_s(g)$ where PRF is a pseudorandom function taking the public seed s. Detailed protocol is shown in Fig. 2.

Use of $H()$. It is not entirely trivial to integrate two elements into a single element. The use of $H()$ is crucial for the security. In particular, if $H()$ is not used (as if it always outputs 1), or if y_2 is not required as part of the input of $H()$, e.g. the commitment is computed as $(g^{z'} h)^k$ where $z' = H(g, h, y_1)$ instead, a malicious prover $\mathcal{P}$ can cheat a verifier $\mathcal{V}$ to accept.

Concretely, the malicious prover $\mathcal{P}$ first picks $x, x' \in_R \mathbb{Z}_q^*$, publishes $y_1 = g^x$ and $y_2 = y_1^{-z'}(g^{z'} h)^{x'}$ where $z' = H(g, h, y_1)$, and then interacts with $\mathcal{V}$ as follows: $\mathcal{P}$ sends the commitment $v = (g^{z'} h)^k$ where $k \in_R \mathbb{Z}_q^*$ to $\mathcal{V}$; then responds with $s = k - x'c \mod q$ upon received a challenge $c \in_R \mathbb{Z}_q^*$. Since $(g^{z'} h)^s (y_1^{z'} y_2)^c = (g^{z'} h)^s (y_1^{z'} \cdot y_1^{-z'} (g^{z'} h)^{x'})^c = (g^{z'} h)^{s+x'c} = (g^{z'} h)^k = v$, but $\log_g y_1 \neq \log_h y_2$ with overwhelming probability. So, it fails to provide soundness.

Finally, the range of $H()$ should be large (exponential in the security parameter) even if y_2 is part of the input. Otherwise, a cheating prover can just try all possible z' in the small range of $H()$, construct $y_2 = y_1^{-z'}(g^{z'} h)^{x'}$ and see if $z' = H(y_1, y_2)$ by coincidence. When the range of $H()$ is small, the last equality happens with non-negligible probability, and this exhaustive search is efficient.

1. $\mathcal{P}$ picks $k \in_R \mathbb{Z}_q^*$ and sends the commitment $v = (g^z h)^k$ to $\mathcal{V}$, where $z = H(y_1, y_2)$.
2. Upon receiving v, $\mathcal{V}$ sends the challenge $c \in_R \mathbb{Z}_q^*$ to $\mathcal{P}$.
3. $\mathcal{P}$ sends the response $s = k - cx \mod q$ to $\mathcal{V}$.
4. $\mathcal{V}$ accepts if and only if $v = (g^z h)^s (y_1^z y_2)^c$ holds, where $z = H(y_1, y_2)$.

Fig. 2. Our Protocol for $ZPK[(x) : y_1 = g^x \wedge y_2 = h^x]$

The security of our protocol can be ensured by the following theorem, its proof can be found in the full version.

Theorem 1. *In the random oracle model, our protocol is secure according to Definition 4, under the DL assumption in group* $\mathbb{G}$.

Extension to n-element. Our protocol presented in Fig. 2 can be naturally extended to show the equality of n discrete logarithms, i.e., $\log_{g_0} y_0 = \log_{g_1} y_1 = \cdots = \log_{g_{n-1}} y_{n-1} = x$. Extended protocol is given in Fig. 3. In the full version, we will discuss how to modify the proof of Theorem 1 to fit for our extended protocol. Our protocol can be seen as borrowing some techniques from the multi-signature scheme in [12].

Efficiency. We exploit the fact that an n-element multi-exponentiation (a computation of the form $a_1^{x_1} \cdots a_n^{x_n}$) can be performed significantly more efficiently than n different simple exponentiations using the windows methods in [17–20]. A multi-exponentiation requires a number of multiplications and squarings. For simplicity, we assume multiplication and squaring have the same complexity, and we use t_{m} to denote the computational cost of one modular multiplication in group $\mathbb{G}$. Let $\ell = |\mathbb{Z}_q|$. Setting $\ell = 160$ and the window size w to be 1, the average costs of one exponentiation, 2-element multi-exponentiation and 4-element multi-exponentiation in $\mathbb{G}$ are about $t_{1\mathrm{xp}} = 1.488\ell t_{\mathrm{m}}$, $t_{2\mathrm{xp}} = 1.739\ell t_{\mathrm{m}}$, and $t_{4\mathrm{xp}} = 1.863\ell t_{\mathrm{m}}$ respectively. For n-element multi-exponentiation, its cost increases with n and converges to $t_{n\mathrm{xp}} = 1.988\ell t_{\mathrm{m}}$ when n reaches 15. We obtain these figures from [20][Theorem 3.4].

1. $\mathcal{P}$ picks $k \in_R \mathbb{Z}_q^*$ and sends the commitment $v = (g_0 g_1^{z_1} \cdots g_{n-1}^{z_{n-1}})^k$ to $\mathcal{V}$, where $z_i = H(i, y_1, \cdots, y_{n-1})$ for $i \in [1, n-1]$.
2. Upon receiving v, $\mathcal{V}$ sends the challenge $c \in_R \mathbb{Z}_q^*$ to $\mathcal{P}$.
3. $\mathcal{P}$ sends the response $s = k - cx \mod q$ to $\mathcal{V}$.
4. $\mathcal{V}$ accepts if and only if $v = (g_0 g_1^{z_1} \cdots g_{n-1}^{z_{n-1}})^s (y_0 y_1^{z_1} \cdots y_{n-1}^{z_{n-1}})^c$ holds, where $z_i = H(i, y_1, \cdots, y_{n-1})$ for $i \in [1, n-1]$.

Fig. 3. Our Protocol for proving $y_0 = g_0^x, \cdots, y_{n-1} = g_{n-1}^x$

Table 1. Comparisons between Our Protocol and Chaum-Pedersen Protocol

Protocol	**Chaum-Pedersen [1]**		**Our Protocol**	
Number of DLs	2	n	2	n
Prover's Operation	$2t_{1\text{xp}} = 2.976\ell t_{\text{m}}$	$n \cdot 1.488\ell t_{\text{m}}$	$t_{2\text{xp}} = 1.739\ell t_{\text{m}}$	$t_{n\text{xp}} = 1.988\ell t_{\text{m}}$
Verifier's Operation	$2t_{2\text{xp}} = 3.478\ell t_{\text{m}}$	$n \cdot 1.739\ell t_{\text{m}}$	$t_{4\text{xp}} = 1.863\ell t_{\text{m}}$	$t_{n\text{xp}} = 1.988\ell t_{\text{m}}$
Bandwidth	$2\|\mathbb{G}\| + 2\ell$	$n\|\mathbb{G}\| + 2\ell$	$1\|\mathbb{G}\| + 2\ell$	
Model	Standard		Random Oracle	
Soundness	Unconditional		Computational	

An efficiency comparison between our protocol and the CP protocol is given in Table 1. For the equality of 2 discrete logarithms, our protocol offers about 40% computational cost saving, and saves one group element in terms of the communication overhead. The savings are more for $n > 2$, which basically reduces the computation and the bandwidth requirements from $O(n)$ to $O(1)$.

4 An Efficient Signature Scheme

The well-known CP protocol has been widely used as a building block in many cryptosystems. For examples, applying the Fiat-Shamir transformation technique [14] on the CP protocol, several signature schemes have been constructed [2, 3]. We use our protocol to construct an efficient signature scheme based on the observation made in [2]. The public key contains (g, h, y_1, y_2) and a signature is simply a proof such that the public key is a DDH tuple.

Let $\mathbb{G}$ be a group with ℓ_q-bit prime order q where ℓ_q is the security parameter. Let g be a random generator of $\mathbb{G}$, and let $H_1 : \{0,1\}^* \to \mathbb{Z}_q^*$ and $H_2 : \{0,1\}^* \to \mathbb{Z}_q^*$ be two cryptographic hash functions (which are modeled by random oracles in the security proof). Our DDH-based signature scheme S2 is specified as below:

KeyGen. Pick $x \in_R \mathbb{Z}_q^*, h \in_R \mathbb{G}$, output x as the private key and $(h, y_1 = g^x, y_2 = h^x)$ as the public key.

Sign. Given the private key $x \in \mathbb{Z}_q^*$, the corresponding public key (y_1, y_2), and a message $m \in \{0,1\}^*$, the signer generates a signature as below:
1. compute $z = H_1(y_1, y_2)$; (pre-computable from the public key)
2. compute $u = g^z h$; (pre-computable from the public key too)
3. pick $k \in_R \mathbb{Z}_q^*$, and compute $v = u^k = (g^z h)^k$; (independent of m)
4. output $(e = H_2(m, y_1, y_2, v), s = k - xe \mod q)$.

Sign is derived from our protocol using the Fiat-Shamir heuristic: $v = (g^z h)^k$ is the commitment, e is the challenge, and $s = k - xe$ is the response.

Verify. Given the public key (y_1, y_2), and a signature (e, s) on message m, compute $z = H_1(y_1, y_2)$, and $v' = (g^z h)^s (y_1^z y_2)^e$; return 1 if $e = H_2(m, y_1, y_2, v')$, 0 otherwise.

For a valid signature (e, s) on m, $v' = (g^z h)^s (y_1^z y_2)^e = (g^z)^{s+xe} h^{s+xe} = (g^z h)^k = v$.

Verifier Pre-computation. For a given signer, observe that the values $z, u = g^z h$ and $w = y_1^z y_2$ are invariants for every signature verification, and hence they can be pre-computed. This saves Verify from one 4-element multi-exponentiation to one 2-element multi-exponentiation, and is helpful in the situation where the verifier receives signatures from the same signer more than once.

Signer Pre-computation. The values $z = H_1(y_1, y_2)$ and $u = g^z h$ can be pre-computed without the private key x. Sign only takes one online exponentiation.

Offline/Online Signing. A signing algorithm can be split into offline and online stages. Offline computation refers to computation that may be performed before the message to be signed is known, which possibly includes pre-selecting random numbers for a probabilistic signature scheme. In contrast, online computation must be done after the message is known.

Our Sign algorithm can be naturally split. The offline stage generates the offline token (k, v). The online stage outputs the signature (e, s) by one hash function evaluation and one modular multiplication, which is nearly the least one may expect for a hash-based signature.

Security. The security of our scheme S2 can be ensured by the following theorem. Its detailed proof can be found in the full version.

Theorem 2. *Our signature scheme S2 is EUF-CMA secure in the random oracle model, under the DDH assumption in group $\mathbb{G}$. Concretely, if there exists a t-time adversary $\mathcal{F}$ that has a non-negligible advantage ϵ against the EUF-CMA security of our scheme S2, making at most q_S signing queries, q_{H_1} queries on hash function H_1, and at most q_{H_2} queries on hash function H_2, then there exists an algorithm $\mathcal{B}$ that can solve the DDH problem with*

$$\text{advantage } \epsilon' \geqslant \epsilon - \eta - \frac{\eta + q_{H_2} + q_s}{2^{\ell_q}} \text{ within time } t' \leq t + (q_s + 1)t_{2\text{xp}},$$

where η denotes the success probability to break the soundness of our protocol in Fig. 2, and $t_{2\text{xp}}$ denotes the computational cost of a 2-element multi-exponentiation in $\mathbb{G}$.

Efficiency Comparisons. In [2], Katz and Wang presented a signature scheme (denoted by KW-DDH) with tight security reduction to the DDH assumption. Table 2 presents a detailed comparison between KW-DDH and S2. Both the computational cost for signing and verification of our DDH-based scheme are reduced to 50%, where the comparison is made by instantiating both DDH-based schemes using the same security parameter. Even without signer-specific pre-computation by the verifier, verification only takes $1.863\ell t_{\text{m}}$ which remains competitive. We remark that the aim of [2] is to achieve a tight security reduction. To compensate for the tightness, we need to use a larger security parameter, which results in a larger signature size. However, the 50% saving in computational cost means that our scheme can afford a larger security parameter without incurring too much extra computation time.

Table 2. Efficiency Comparisons between Different Signature Schemes

Scheme	Sign	Verify	Signature Size	Public Key Size
KW-DDH [2]	$2t_{1\text{xp}} = 2.976\ell t_{\text{m}}$	$2t_{2\text{xp}} = 3.478\ell t_{\text{m}}$	$2\lvert\mathbb{Z}_q\rvert$	$3\lvert\mathbb{G}\rvert$
Our Scheme S2	$t_{1\text{xp}} = 1.488\ell t_{\text{m}}$	$t_{2\text{xp}} = 1.739\ell t_{\text{m}}$	$2\lvert\mathbb{Z}_q\rvert$	$3\lvert\mathbb{G}\rvert$

5 Conclusion

We investigated the protocol for the simultaneous discrete logarithms problem – a fundamental building block appears in many cryptosystems of different flavors. We noted that such a protocol is usually used in a non-interactive form by applying the Fiat-Shamir heuristics which inevitably relies on the random oracle model. The adversary is also assumed to be computationally bounded in these applications. We thus proposed an efficient zero-knowledge argument system in the random oracle model which gives a better performance in terms of both the computational cost and the bandwidth. We believe that our protocol can also be used to improve other cryptographic cryptosystems. In particular, we proposed a signature scheme that only takes one (offline) exponentiation to sign, without pairing, based on the standard decisional Diffie-Hellman assumption.

We remark that our technique can be generalized for proving linear relations other than the simple equality. It may be interesting to see if other zero-knowledge argument systems can be speeded up using a similar technique.

References

1. Chaum, D., Pedersen, T.P.: Wallet Databases with Observers. In: Brickell, E.F. (ed.) CRYPTO 1992. LNCS, vol. 740, pp. 89–105. Springer, Heidelberg (1993)
2. Katz, J., Wang, N.: Efficiency Improvements for Signature Schemes with Tight Security Reductions. In: Jajodia, S., Atluri, V., Jaeger, T. (eds.) Computer and Communications Security, pp. 155–164. ACM, New York (2003)
3. Chevallier-Mames, B.: An Efficient CDH-based Signature Scheme with a Tight Security Reduction. In: Shoup, V. (ed.) CRYPTO 2005. Chevallier-Mames, B, vol. 3621, pp. 511–526. Springer, Heidelberg (2005)
4. Stadler, M.: Publicly Verifiable Secret Sharing. In: Maurer, U.M. (ed.) EUROCRYPT 1996. LNCS, vol. 1070, pp. 190–199. Springer, Heidelberg (1996)
5. Shoup, V.: Practical Threshold Signatures. In: Preneel, B. (ed.) EUROCRYPT 2000. LNCS, vol. 1807, pp. 207–220. Springer, Heidelberg (2000)
6. Shoup, V., Gennaro, R.: Securing Threshold Cryptosystems against Chosen Ciphertext Attack. J. Cryptology 15(2), 75–96 (2002)
7. Ateniese, G.: Verifiable Encryption of Digital Signatures and Applications. ACM Transactions on Information and System Security (TISSEC) 7(1), 1–20 (2004)
8. Lysyanskaya, A., Rivest, R.L., Sahai, A., Wolf, S.: Pseudonym Systems. In: Heys, H.M., Adams, C.M. (eds.) SAC 1999. LNCS, vol. 1758, pp. 184–199. Springer, Heidelberg (2000)
9. Camenisch, J., Maurer, U.M., Stadler, M.: Digital Payment Systems with Passive Anonymity-Revoking Trustees. Journal of Computer Security 5(1), 69–90 (1997)

10. Cramer, R., Gennaro, R., Schoenmakers, B.: A Secure and Optimally Efficient Multi-Authority Election Scheme. In: Fumy, W. (ed.) EUROCRYPT 1997. LNCS, vol. 1233, pp. 103–118. Springer, Heidelberg (1997)
11. Chow, S.S.M., Liu, J.K., Wong, D.S.: Robust Receipt-Free Election System with Ballot Secrecy and Verifiability. In: NDSS, The Internet Society (2008)
12. Bellare, M., Neven, G.: Multi-Signatures in the Plain Public-Key Model and A General Forking Lemma. In: Juels, A., Wright, R.N., di Vimercati, S.D.C. (eds.) Computer and Communications Security, pp. 390–399. ACM, New York (2006)
13. Ma, C., Weng, J., Li, Y., Deng, R.H.: Efficient Discrete Logarithm based Multi-Signature Scheme in the Plain Public Key Model. Des. Codes Cryptography 54(2), 121–133 (2010)
14. Fiat, A., Shamir, A.: How to Prove Yourself: Practical Solutions to Identification and Signature Problems. In: Odlyzko, A.M. (ed.) CRYPTO 1986. LNCS, vol. 263, pp. 186–194. Springer, Heidelberg (1987)
15. Bellare, M., Rogaway, P.: Random Oracles are Practical: A Paradigm for Designing Efficient Protocols. Computer and Communications Security, 62–73 (1993)
16. Goldreich, O.: Foundations of Cryptography. Basic Tools, vol. 1. Cambridge University Press, Cambridge (2001)
17. Dimitrov, V.S., Jullien, G.A., Miller, W.C.: Complexity and Fast Algorithms for Multiexponentiations. IEEE Trans. Computers 49(2), 141–147 (2000)
18. Möller, B.: Algorithms for Multi-exponentiation. In: Vaudenay, S., Youssef, A.M. (eds.) SAC 2001. LNCS, vol. 2259, pp. 165–180. Springer, Heidelberg (2001)
19. Menezes, A.J., van Oorschot, P.C., Vanstone, S.A.: Handbook of Applied Cryptography. CRC Press, Boca Raton (2001)
20. Avanzi, R.M.: On Multi-Exponentiation in Cryptography. Cryptology ePrint Archive, Report 2002/154 (2002)

Directed Figure Codes: Decidability Frontier

Michał Kolarz

Institute of Computer Science, Jagiellonian University, Łojasiewicza 6,
30-348 Kraków, Poland
mkol@smp.if.uj.edu.pl

Abstract. We consider directed figures defined as labelled polyominoes with designated start and end points, with a partial catenation. As we are interested in codicity verification for sets of figures, we show that it is decidable for sets of figures with parallel translation vectors, which is one of the decidability border cases in this setting. We give a constructive proof which leads to a straightforward algorithm.

1 Introduction

Our research concentrates on two-dimensional codes which are a natural extension of variable-length word codes, i.e., subsets X of a monoid such that every product of the elements decomposes uniquely over X. The interest in two-dimensional codes is natural in the context of various disciplines that use picture encodings, e.g. [1]. The classical variable-length word codes have been studied by many authors and the literature on this subject is abundant, including e.g. the well-known monograph [2]. On the other hand, the results on codicity in other spaces are rare. Some authors have extended word codes to trees (see [3]) or polyominoes (e.g. [4,5]); decidability of the codicity testing problem varies in these cases. It is undecidable in the case of both standard and labelled polyominoes (cf. [6,7]) and decidable for directed figures with a merging function (see a recent paper [8]). Some authors also study the border between decidability and undecidability of codicity testing (e.g. [9]) for standard codes. Here we explore this field too in the context of two-dimensional structures.

In the present paper we study pictures composed of labelled polyominoes, equipped with start and end points; we call them directed figures. Catenation of directed figures is a partial operation, not defined in case of overlapping domains. We extend the results of [10], proving decidability of codicity testing for figures with parallel translation vectors. This is an important border case that has been left open in [10].

Section 2 introduces the definitions. Section 3 recalls some of the results that we are going to extend in this paper. Then we proceed to the main result in Sect. 4. The codicity verification algorithm, presented in Sect. 5, summarizes our results. We conclude with final remarks in Sect. 6.

M.T. Thai and S. Sahni (Eds.): COCOON 2010, LNCS 6196, pp. 530–539, 2010.

2 Definitions

Throughout this paper Σ is a finite, non-empty alphabet.

For $u = (u_x, u_y) \in \mathbb{Z}^2$, a *translation* in $\mathbb{Z}^2$ by vector u is denoted by tr_u, i.e.,

$$\mathrm{tr}_u : \mathbb{Z}^2 \ni (x, y) \mapsto (x + u_x, y + u_y) \in \mathbb{Z}^2.$$

For a set $V \subset \mathbb{Z}^2$ and an arbitrary function $f : V \to \Sigma$ it obviously induces two mappings

$$\mathrm{tr}_u : P(\mathbb{Z}^2) \ni V \mapsto \{\mathrm{tr}_u(v) \mid v \in V\} \in P(\mathbb{Z}^2), \quad (1)$$

$$\mathrm{tr}_u : \Sigma^V \ni f \mapsto f \circ \mathrm{tr}_{-u} \in \Sigma^{\mathrm{tr}_u(V)}. \quad (2)$$

A *directed figure* over Σ (or simply *figure*) is a quadruple $F = (D, b, e, l)$, where $D \subseteq \mathbb{Z}^2$ is finite and non-empty, $b, e \in \mathbb{Z}$ and $l : D \to \Sigma$. Here $D =: \mathrm{dom}(F)$ is a *domain*, $b =: \mathrm{begin}(F)$ and $e =: \mathrm{end}(F)$ are *begin* and *end points*, $\mathrm{end}(F) - \mathrm{begin}(F) =: \mathrm{tran}(F)$ is a *translation vector*, and $l =: \mathrm{label}(F)$ is a *labeling function* of F. In addition we define the *empty directed figure* ε as $(\emptyset, (0,0), (0,0), \emptyset)$. Figures are considered up to translation; it is the relative position of figures that is important.

A set of all figures over Σ is denoted as $\Sigma^{\diamond}$.

Example 1. A directed figure and its graphical representation (where a circle marks the start point and a diamond marks the end point of the figure).

$$(\{(0,0), (1,0), (1,1)\}, (0,0), (2,1), \{(0,0) \mapsto a, (1,0) \mapsto b, (1,1) \mapsto c\})$$

Let $x = (D_x, b_x, e_x, l_x)$ and $y = (D_y, b_y, e_y, l_y)$ be figures.
If $D_x \cap \mathrm{tr}_{x_e - y_b}(D_y) = \emptyset$, *catenation* of x and y is defined as

$$xy = (D_x \cup \mathrm{tr}_{x_e - y_b}(D_y), b_x, \mathrm{tr}_{x_e - y_b}(e_y), l),$$

where

$$l(z) = \begin{cases} l_x(z) & \text{for } z \in D_x, \\ \mathrm{tr}_{x_e - y_b}(l_y)(z) & \text{for } z \in \mathrm{tr}_{x_e - y_b}(D_y). \end{cases}$$

If $D_x \cap \mathrm{tr}_{x_e - y_b}(D_y) \neq \emptyset$, catenation of x and y is not defined.

Example 2. Two figures and their catenation.

$x =$ $\quad y =$ $\quad xy =$

Example 3. Catenation of the following figures is not defined, since there is a conflict (point labeled by b in the first figure overlaps with point labeled by c in the second figure).

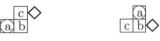

Abusing the notation, we will also write $X^\diamond$ to denote the set of all figures that can be composed by catenation from figures of $X \subseteq \Sigma^\diamond$.

A set $X \subseteq \Sigma^\diamond$ is a *code* if for any $x \in X^\diamond$ there exists only one sequence $x_1, \ldots, x_k \in X$ such that $x = x_1 \cdots x_k$

Example 4. { a, a } $\subseteq \{a\}^\diamond$ is a code.

Example 5. $\{w =$ a a, $x =$ a a, $y =$ a a, $z =$ a a $\} \subseteq \{a\}^\diamond$ is not a code since $wx = yz =$ a a a a .

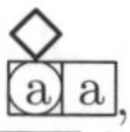

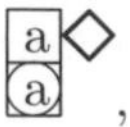

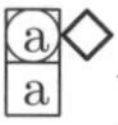

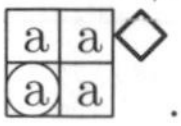

3 Previous Results

In our recent work ([10]), we showed that if a given set of figures $X \in \Sigma^\diamond$ contains only figures with translation vectors lying in an open half-plane, i.e.:

$$\exists \tau \in \mathbb{Z}^2 \; \forall x \in X : \mathrm{tran}(x) \cdot \tau > 0,$$

it is decidable whether X is a code. We constructed a verification algorithm for this case.

On the other hand, the problem whether a given set of figures is a code is undecidable in general, even when restricted to figures with connected domains. This was obtained by a reduction from the Post Correspondence Problem, using a technique introduced in [5]. The proof gives one of the border cases for decidability w.r.t. translation vectors: when they lie in a closed half-plane, the problem of codicity verification becomes undecidable. Moreover, a simple modification of the proof leads to the conclusion that for any non-parallel vectors $u, v \in \mathbb{Z}^2$ there exist constants $a, b \in \mathbb{Z}$ such that codicity problem remains undecidable even for sets of figures with translation vectors restricted to au, $-au$ and bv.

Hence, a very natural question is the decidability in the case of all translation vectors being parallel. Of course, in that case, the problem becomes one-dimentional, but it cannot be easily transformed to any known words problem. This leads us to the main result of this paper.

4 Main Result

In this section we prove that testing whether a given finite set of figures with parallel translation vectors is a code is decidable. We present a constructive proof, which gives a verification algorithm for this case.

Theorem 1. *For sets of figures with parallel translation vectors, it is decidable whether a given set is a code.*

Let $X \subseteq \Sigma^\diamond$ be finite and non-empty, and without loss of generality assume

$$\forall x \in X : \mathrm{begin}(x) = (0, 0) \ .$$

We assume that translation vectors of elements of X are parallel, i.e. there exists a shortest vector $\tau \in \mathbb{Z}^2$ (fixed throughout this paper) such that

$$\forall x \in X : \mathrm{tran}(x) \in \mathbb{Z}\tau = \{j\tau \mid j \in \mathbb{Z}\} \ .$$

In particular, if $(t_1, t_2) = \mathrm{tran}(x)$ (for some $x \in X$, such that $\mathrm{tran}(x) \neq (0,0)$), then τ is one of the following vectors:

$$(t_1/\gcd(|t_1|,|t_2|)\ ,\ t_2/\gcd(|t_1|,|t_2|)), \tag{3}$$
$$(-t_1/\gcd(|t_1|,|t_2|)\ ,\ -t_2/\gcd(|t_1|,|t_2|)), \tag{4}$$

where gcd denotes greatest common divisor. If all translation vectors of elements of X are $(0,0)$, then decidability problem is trivial, since then X is not a code if and only if there exist $x, y \in X$ such that $\mathrm{dom}(x) \cap \mathrm{dom}(y) = \emptyset$ - then $xy = yx$; this is obviously decidable.

We define the following *bounding areas*:

$$\begin{aligned}
\mathrm{B_L} &:= \{u \in \mathbb{Z}^2 \mid 0 > u \cdot \tau\}, \\
\mathrm{B_0} &:= \{u \in \mathbb{Z}^2 \mid 0 \leq u \cdot \tau < \tau \cdot \tau\}, \\
\mathrm{B_R} &:= \{u \in \mathbb{Z}^2 \mid \tau \cdot \tau \leq u \cdot \tau\} \ .
\end{aligned}$$

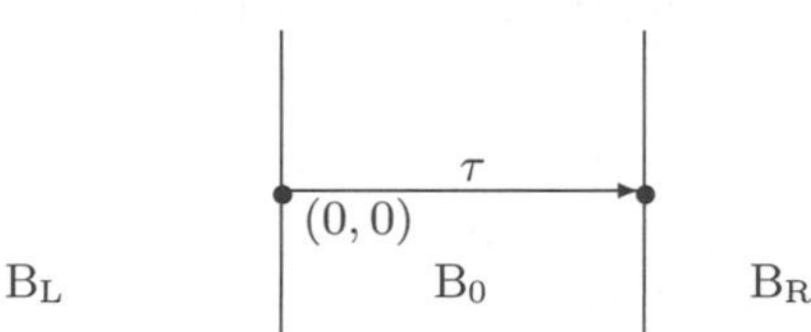

Fig. 1. Bounding areas $\mathrm{B_L}$, $\mathrm{B_0}$ and $\mathrm{B_R}$

For a non-empty figure $x \in \Sigma^\diamond$, a *bounding hulls* of x is a set:

$$\begin{aligned}
\mathrm{hull}(x) &:= \bigcup_{n=m\ldots M} \mathrm{tr}_{n\tau}(\mathrm{B_0}), \\
\mathrm{hull}^*(x) &:= \bigcup_{n=-M\ldots -m} \mathrm{tr}_{n\tau}(\mathrm{B_0}),
\end{aligned}$$

where

$$\begin{aligned}
m &= \min\{n \in \mathbb{Z} \mid \mathrm{tr}_{n\tau}(\mathrm{B_0}) \cap (\mathrm{dom}(x) \cup \{\mathrm{begin}(x), \mathrm{end}(x)\}) \neq \emptyset\}, \\
M &= \max\{n \in \mathbb{Z} \mid \mathrm{tr}_{n\tau}(\mathrm{B_0}) \cap (\mathrm{dom}(x) \cup \{\mathrm{begin}(x), \mathrm{end}(x)\}) \neq \emptyset\} \ .
\end{aligned}$$

In addition, for an empty figure $\mathrm{hull}(\varepsilon) := \emptyset$ and $\mathrm{hull}^*(\varepsilon) := \emptyset$.

4.1 Starting Configurations

Our goal is either to find a figure $x \in X^\diamond$ that has two different factorizations over elements of X, or to show that such a figure does not exist. If it exists, without loss of generality we can assume that it has the following two different x– and y–*factorizations*:

$$x = \dot{x}_1\ddot{x}_1 \cdots \ddot{x}_{k-1}\dot{x}_k\ddot{x}_k = \dot{y}_1\ddot{y}_1 \cdots \ddot{y}_{l-1}\dot{y}_l\ddot{y}_l$$

where $\dot{x}_1 \neq \dot{y}_1$, $\text{begin}(\dot{x}_1) = \text{begin}(\dot{y}_1) = (0,0)$ and for $i \in \{1,\ldots,k\}$ and $j \in \{1,\ldots,l\}$ we have:

$$\begin{array}{ll}\dot{x}_i \in X & \text{and } \text{hull}(\dot{x}_i) \cap \text{B}_0 \neq \emptyset, \\ \ddot{x}_i \in X^\diamond \cup \{\varepsilon\} & \text{and } \text{hull}(\ddot{x}_i) \cap \text{B}_0 = \emptyset, \\ \dot{y}_j \in X & \text{and } \text{hull}(\dot{y}_j) \cap \text{B}_0 \neq \emptyset, \\ \ddot{y}_j \in X^\diamond \cup \{\varepsilon\} & \text{and } \text{hull}(\ddot{y}_j) \cap \text{B}_0 = \emptyset \quad . \end{array}$$

Observe that the following conditions for x–factorization are satisfied for $i \in \{1,\ldots,k-1\}$:

- if $\text{end}(\dot{x}_i) \in \text{B}_\text{L}$, then $\text{begin}(\dot{x}_{i+1}) \in \text{B}_\text{L}$,
- if $\text{end}(\dot{x}_i) = (0,0)$, then $\ddot{x}_i = \varepsilon$ and $\text{begin}(\dot{x}_{i+1}) = (0,0)$,
- if $\text{end}(\dot{x}_i) \in \text{B}_\text{R}$, then $\text{begin}(\dot{x}_{i+1}) \in \text{B}_\text{R}$,

they are trivial implications of the assumption that $\text{hull}(\ddot{x}_i) \cap \text{B}_0 = \emptyset$ and the fact that $\dot{x}_i$ must be somehow linked with $\dot{x}_{i+1}$. Similar conditions are satisfied for the y–factorization. In addition, the x–factorization must match the y–factorization,i.e.:

- if $\text{end}(\dot{x}_k) \in \text{B}_\text{L}$, then $\text{end}(\dot{y}_l) \in \text{B}_\text{L}$,
- if $\text{end}(\dot{x}_k) = (0,0)$, then $\text{end}(\dot{y}_l) = (0,0)$,
- if $\text{end}(\dot{x}_k) \in \text{B}_\text{R}$, then $\text{end}(\dot{y}_l) \in \text{B}_\text{R}$.

Also, it is obvious that

$$\bigcup_{i=1\ldots k} \text{dom}(\dot{x}_i) \cap \text{B}_0 = \bigcup_{i=1\ldots l} \text{dom}(\dot{y}_i) \cap \text{B}_0, \tag{5}$$

$$\bigcup_{i=1\ldots k} \text{label}(\dot{x}_i) \mid_{\text{B}_0} = \bigcup_{i=1\ldots l} \text{label}(\dot{y}_i) \mid_{\text{B}_0} \quad . \tag{6}$$

Now we consider all possible pairs of sequences $((\dot{x}_i)_i, (\dot{y}_j)_j)$ satisfying the above conditions. Note that equality of such sequences is considered not up to translation: relative position of sequences' elements is important. Such a pair will be called a *starting configuration.* Observe that there can be only a finite number of such configurations, since

$$\bigcup_{i=1\ldots k} \text{dom}(\dot{x}_i) \subseteq \bigcup_{x\in X} (\text{hull}(x) \cup \text{hull}^*(x)),$$

$$\bigcup_{i=1\ldots l} \text{dom}(\dot{y}_i) \subseteq \bigcup_{x\in X} (\text{hull}(x) \cup \text{hull}^*(x))$$

and the set on the right side is bounded in the direction of τ. Also note that if there is no starting configuration for X, then obviously X is a code.

4.2 Left and Right Configurations

We consider independently all starting configurations constructed for X. Since (5) and (6) hold, we can now forget the labelling of B_0. From a starting configuration $((\dot{x}_i)_{i=1}^k, (\dot{y}_j)_{j=1}^l)$ we construct L– and R–*configurations* (left and right configurations)

$$\mathrm{C_R} = ((D^x_\mathrm{R}, l^x_\mathrm{R}, \mathrm{EB}^x_\mathrm{R}), (D^y_\mathrm{R}, l^y_\mathrm{R}, \mathrm{EB}^y_\mathrm{R})),$$
$$\mathrm{C_L} = ((D^x_\mathrm{L}, l^x_\mathrm{L}, \mathrm{EB}^x_\mathrm{L}), (D^y_\mathrm{L}, l^y_\mathrm{L}, \mathrm{EB}^y_\mathrm{L})) \ .$$

First we show a construction for an x–part of configuration:

$$D^x_\mathrm{R} := \bigcup_{i=1\dots k} \mathrm{dom}(\dot{x}_i) \cap \mathrm{B_R} \text{ and } l^x_\mathrm{R} := \bigcup_{i=1\dots k} \mathrm{label}(\dot{x}_i) \mid_{\mathrm{B_R}},$$
$$D^x_\mathrm{L} := \bigcup_{i=1\dots k} \mathrm{dom}(\dot{x}_i) \cap \mathrm{B_L} \text{ and } l^x_\mathrm{L} := \bigcup_{i=1\dots k} \mathrm{label}(\dot{x}_i) \mid_{\mathrm{B_L}}$$

and multisets EB^x_L, EB^x_R are obtained in the following way.
For each $i \in \{1, \dots, k-1\}$:

- if $\mathrm{end}(\dot{x}_i) \in \mathrm{B_L}$, then $(\mathrm{end}(\dot{x}_i), \mathrm{begin}(\dot{x}_{i+1}))$ is added to EB^x_L,
- if $\mathrm{end}(\dot{x}_i) = (0,0)$, then no pair is added to EB^x_L or EB^x_R,
- if $\mathrm{end}(\dot{x}_i) \in \mathrm{B_R}$, then $(\mathrm{end}(\dot{x}_i), \mathrm{begin}(\dot{x}_{i+1}))$ is added to EB^x_R

and

- if $\mathrm{end}(\dot{x}_k) \in \mathrm{B_L}$, then $(\mathrm{end}(\dot{x}_k), \bullet)$ is added to EB^x_L,
- if $\mathrm{end}(\dot{x}_k) = (0,0)$, then no pair is added to EB^x_L or EB^x_R,
- if $\mathrm{end}(\dot{x}_k) \in \mathrm{B_R}$, then $(\mathrm{end}(\dot{x}_1), \bullet)$ is added to EB^x_R.

These multisets keep information on how figures $\dot{x}_i$ and $\dot{x}_{i+1}$ should be linked by $\ddot{x}_i$ factors. The $\bullet$ symbol denotes the end of the whole figure.

A y–part is created in the analogous way.

Example 6. Consider a set containing the following figures (vertical lines separate the figures):

Taking $\tau = (2,1)$, we construct one of possible starting configurations (x–part only). We also show the construction of x–part of L– and R–configurations.

Figure 2 shows the construction. Each image presents a current figure (with bold lines) and its translation vector. Also domain and labeling of all of the previous figures is presented, together with the end point of the previous figure (which is important for the construction). B_0 lies between the slanted lines. Domains and labellings of L– and R–configurations are presented on Fig. 3.

Now let us consider R–configuration only (L–configuration is handled in a similar way). We say that R–configuration $((D^x_\mathrm{R}, l^x_\mathrm{R}, \mathrm{EB}^x_\mathrm{R}), (D^y_\mathrm{R}, l^y_\mathrm{R}, \mathrm{EB}^y_\mathrm{R}))$ is *terminating* if it satisfies the following conditions:

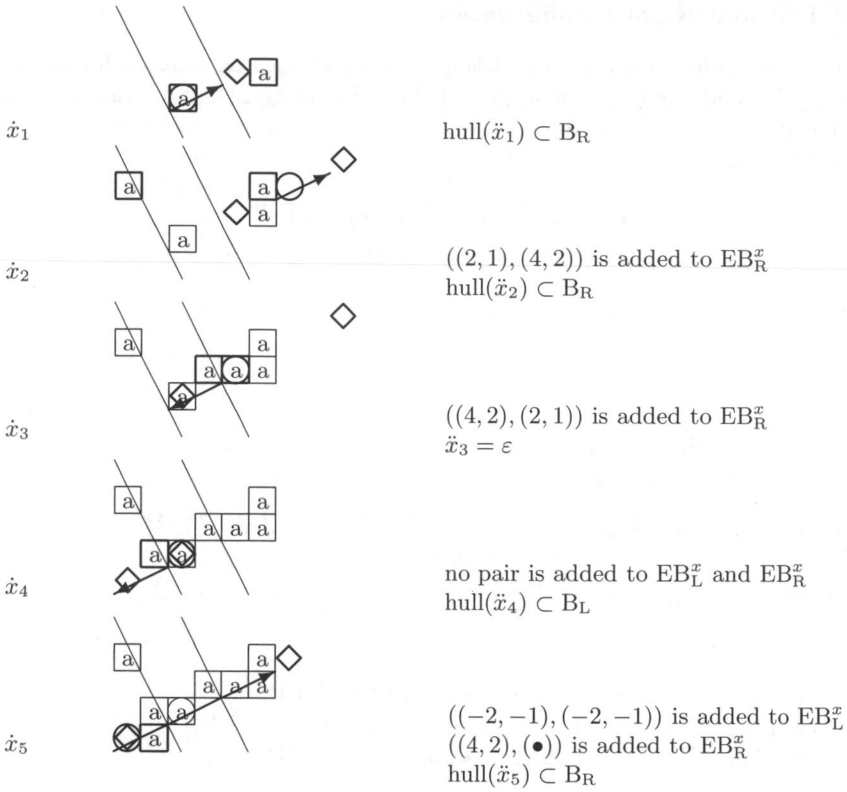

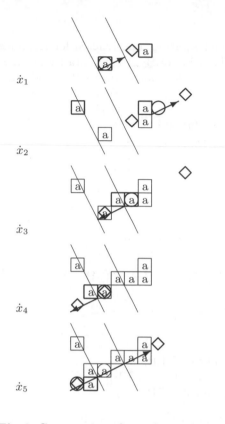

$\dot{x}_1$ hull($\ddot{x}_1$) $\subset$ B$_\mathrm{R}$

$\dot{x}_2$ $((2,1),(4,2))$ is added to EB^x_R
hull($\ddot{x}_2$) $\subset$ B$_\mathrm{R}$

$\dot{x}_3$ $((4,2),(2,1))$ is added to EB^x_R
$\ddot{x}_3 = \varepsilon$

$\dot{x}_4$ no pair is added to EB^x_L and EB^x_R
hull($\ddot{x}_4$) $\subset$ B$_\mathrm{L}$

$\dot{x}_5$ $((-2,-1),(-2,-1))$ is added to EB^x_L
$((4,2),(\bullet))$ is added to EB^x_R
hull($\ddot{x}_5$) $\subset$ B$_\mathrm{R}$

Fig. 2. Construction of sample starting configuration and its L– and R–configurations

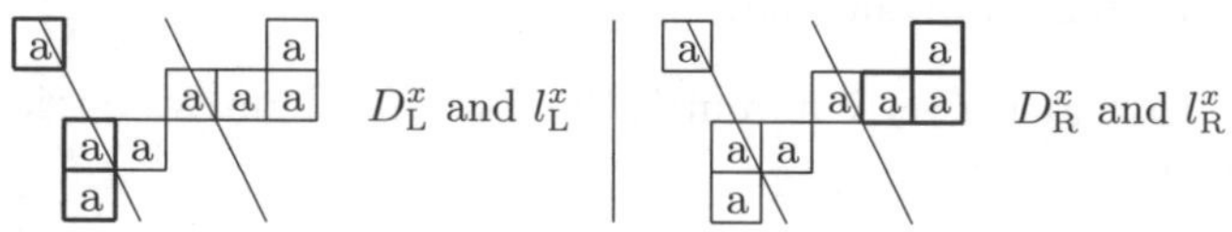

Fig. 3. Domains and labellings of sample L– and R–configurations (respective objects marked with thick lines)

– the domain and labelling of an x–part of a R–configuration match to the domain and labelling of its y–part, i.e.:

$$D^x_\mathrm{R} = D^y_\mathrm{R} \text{ and } l^x_\mathrm{R} = l^y_\mathrm{R},$$

– if a location of the end point of the whole figure is encoded in an R–configuration, then its location is the same in both x– and y–parts, i.e.:

$$\forall e \in \mathbb{Z} : (e, \bullet) \in \mathrm{EB}^x_\mathrm{R} \Leftrightarrow (e, \bullet) \in \mathrm{EB}^y_\mathrm{R},$$

– all points that should be linked together are trivially linked, since they are the same points, i.e.:

$$\forall(e,b) \in \mathrm{EB}^x_{\mathrm{R}} \cup \mathrm{EB}^y_{\mathrm{R}} : e = b \text{ or } b = \bullet \ .$$

Note that if for some starting configuration we obtain terminating L–configuration and terminating R–configuration, then X is not a code.

On the other hand, if we show that for all starting configurations such pair of terminating L– and R–configurations cannot be reached, then X is a code.

4.3 Obtaining New R–Configurations

When a R–configuration derived from a starting configuration is terminating, then we can proceed to analysis of L–configuration. If the R–configuration is not terminating, we must check if adding new figures to it may create terminating configuration.

At the beginning such a derived configuration lies in $\mathrm{B_R}$. For simplicity of notation, we can translate such a configuration by a vector $-\tau$ (translating its all elements; see (1) and (2)).

Now from the given R–configuration we want to obtain new R–configuration by adding new figures from X. In order to obtain a new R–configuration from a given R–configuration, we create the new R–configuration as a copy of the old one. Then zero or more of the following operations must be performed (note that they need not be admissible for an arbitrary R–configuration or we may not need such operations to be performed)

– an x–part operation: add any $x \in X$ for which

$$\mathrm{hull}(x) \cap \mathrm{B_L} = \emptyset, \tag{7}$$

$$\mathrm{hull}(x) \cap \mathrm{B_0} \neq \emptyset, \tag{8}$$

$$\mathrm{dom}(x) \cap D^x_{\mathrm{R}} = \emptyset \tag{9}$$

  to the new configuration, adding its domain and labelling function to the domain and labelling function of the R–configuration, and replacing any pair (e,b) from $\mathrm{EB}^x_{\mathrm{R}}$ in the old configuration with two pairs $(e, \mathrm{begin}(x))$ and $(\mathrm{end}(x), b)$ in the new one,
– an y–part operation: similarly.

In each step of creating the new generation of an R–configuration, we add only figures that change the given R–configuration within $\mathrm{B_0}$ - this implies (8). We add such figures into R–configuration only at that step. In consecutive steps adding such figures is forbidden - this implies (7). At the first step this is a consequence of restrictions for $\ddot{x}_i$ and $\ddot{y}_i$. Condition (9) is obvious. Of course it is possible that given R–configuration is not extendable at all.

After these operations we want that x–part of obtained R–configuration matches to its y–part on $\mathrm{B_0}$, i.e.,

$$D^x_{\mathrm{R}} \cap \mathrm{B_0} = D^y_{\mathrm{R}} \cap \mathrm{B_0} \text{ and } l^x_{\mathrm{R}}\mid_{\mathrm{B_0}} = l^y_{\mathrm{R}}\mid_{\mathrm{B_0}} \ .$$

In addition, for the x–part (similar conditions for the y–part):

- if $((0,0),b) \in \mathrm{EB}^x_{\mathrm{R}}$, then $b = (0,0)$ or $b = \bullet$,
- if $(e,(0,0)) \in \mathrm{EB}^x_{\mathrm{R}}$, then $e = (0,0)$,

and for both parts

- $((0,0),\bullet) \in \mathrm{EB}^x_{\mathrm{R}}$ if and only if $((0,0),\bullet) \in \mathrm{EB}^y_{\mathrm{R}}$.

These conditions are trivial consequences of (7), (8) and (9) on new figures added to R–configuration. Of course it is possible, that one cannot obtain any R–configuration form the old one.

Here, since x–part and y–part of each newly created R–configuration are the same, we now do not have to remember the labelling of B_0. When we forget this information, configurations created lie in B_{R}, so we can translate them by $-\tau$ as previously.

Now observe that all parts of an R–configuration are bounded: domains are contained in the area restricted by the widest hull of elements of X; multisets $\mathrm{EB}^x_{\mathrm{R}}$ and $\mathrm{EB}^y_{\mathrm{R}}$ cannot be infinite, since eventually all points must be linked. There is only finitely many of such configurations. Either we find a terminating R–configuration or we consider all configurations that can be obtained from a given starting configuration performing one or more described steps. □

5 Algorithm

To summarize our considerations, we present an algorithm that answers whether a given set of figures with parallel translation vectors is a code:

1. **input**: $X \subseteq \Sigma^\diamond$ with parallel translation vectors and begin points in $(0,0)$
2. **if for each** $x \in X$ $\mathrm{tran}(x) = (0,0)$
   (a) **if** there exist $x, y \in X$ such that $\mathrm{dom}(x) \cap \mathrm{dom}(y) = \emptyset$
   then return false
   (b) **else return true**
3. **compute** τ *(see (3) or (4))* and set of all starting configurations SC
4. **for each** $c \in$SC
   (a) **compute** sets ALLLC and ALLRC, L– and R–configurations for c (in fact, here both sets are singletons)
   (b) **set** CHECKEDLC=$\emptyset$, NEWLC=$\emptyset$
   (c) **repeat**
      i. **set** ALLLC=ALLLC ∪ NEWLC, NEWLC=$\emptyset$
      ii. **for each** $c_l \in$ALLLC \ CHECKEDLC
         A. **set** CHECKEDLC=CHECKEDLC ∪ $\{c_l\}$
         B. **compute** set TMPLC, L–configurations obtained from c_l
         C. **set** NEWLC=NEWLC ∪ TMPLC
      iii. **if** NEWLC ⊆ ALLLC **then continue** step 4 with next c
      until NEWLC contains a terminating L–configuration
   (d) *repeat the analogous steps 4.b-4.c for R–configurations*
   (e) **return false**
5. **return true**

5 Final Remarks

The algorithm we have obtained has very poor complexity. However it is clear that although the number of all possible configurations (starting configurations, L– and R–configurations) is huge, most of them cannot be obtained for a given set of figures. Thus a careful analysis should provide a much better estimate for the complexity, which is a natural subject for further research. We are also interested in finding other border cases for codicity verification decidability.

References

1. Costagliola, G., Ferrucci, F., Gravino, C.: Adding symbolic information to picture models: definitions and properties. Theoretical Computer Science 337, 51–104 (2005)
2. Berstel, J., Perrin, D.: Theory of Codes. Academic Press, London (1985)
3. Mantaci, S., Restivo, A.: Codes and equations on trees. Theoretical Computer Science 255, 483–509 (2001)
4. Aigrain, P., Beauquier, D.: Polyomino tilings, cellular automata and codicity. Theoretical Computer Science 147, 165–180 (1995)
5. Beauquier, D., Nivat, M.: A codicity undecidable problem in the plane. Theoretical Computer Science 303, 417–430 (2003)
6. Moczurad, M., Moczurad, W.: Decidability of simple brick codes. In: Mathematics and Computer Science III (Algorithms, Trees, Combinatorics and Probabilities). Trends in Mathematics, pp. 541–542. Birkhäuser, Basel (2004)
7. Moczurad, M., Moczurad, W.: Some open problems in decidability of brick (labelled polyomino) codes. In: Chwa, K.-Y., Munro, J.I.J. (eds.) COCOON 2004. LNCS, vol. 3106, pp. 72–81. Springer, Heidelberg (2004)
8. Kolarz, M., Moczurad, W.: Directed figure codes are decidable. Discrete Mathematics and Theoretical Computer Science 11(2), 1–14 (2009)
9. Fernau, H., Reinhardt, K., Staiger, L.: Decidability of code properties. Theoretical Informatics and Applications RAIRO 41, 243–259 (2007)
10. Kolarz, M.: Directed figures - extended models. Theoretical Informatics and Applications RAIRO (2009) (submitted)

Author Index

Adiga, Abhijin 3
Adiwijaya 209
Anzai, Shinya 235
Augustine, John 90
Ausiello, Giorgio 160

Babenko, Maxim 120
Bach, Eric 489
Bayzid, Md. Shamsuzzoha 399
Berman, Piotr 226
Bhowmick, Diptendu 3
Biswas, Sudip 182
Blum, Manuel 1
Bretto, Alain 173
Busaryev, Oleksiy 278

Chandran, L. Sunil 3
Chawla, Shuchi 489
Chen, Jianer 459
Chen, Xin 439
Chen, Xue 308
Chen, Zhi-Zhong 449
Chin, Francis Y.L. 100
Chow, Sherman S.M. 520
Chrobak, Marek 254
Chun, Jinhee 235
Cibulka, Josef 192
Csűrös, Miklós 358

de Berg, Mark 216
Devismes, Stéphane 80
Dey, Tamal K. 278
Djordjevic, Bojan 244
Domingo-Ferrer, Josep 510
Dürr, Christoph 254

Elmasry, Amr 338, 479
Eppstein, David 90
Erickson, Alejandro 288
Estivill-Castro, Vladimir 264

Fernández Anta, Antonio 378
Fernau, Henning 34
Fomin, Fedor V. 34
Franciosa, Paolo G. 160
Friedrich, Tobias 130
Fu, Bin 429
Fukuyama, Shota 348
Fung, Stanley P.Y. 389

Gudmundsson, Joachim 244
Guíñez, Flavio 254
Gusakov, Alexey 120

Hartwig, Michael 318
Healy, Patrick 110
Heednacram, Apichat 264
Holroyd, Alexander 298
Hsu, Tsan-sheng 500
Hu, Guangda 308

Ibarra, Oscar H. 2
Italiano, Giuseppe F. 160

Jain, Rahul 54

Kakugawa, Hirotsugu 80
Kamei, Sayaka 80
Karpinski, Marek 226
Kasai, Ryosei 235
Khosravi, Amirali 216
Klauck, Hartmut 54
Kolarz, Michał 530
Korman, Matias 235
Krebs, Andreas 44
Kynčl, Jan 192

Lazard, Sylvain 469
Lee, D.T. 368
Liau, Churn-Jung 500
Limaye, Nutan 44
Lin, Ching-Chi 368
Lingas, Andrzej 226
Liu, Yunlong 419
Lokshtanov, Daniel 199
Lozano, Antoni 254
Lu, Chi-Jen 13

Ma, Bin 449
Ma, Changshe 520

Mahajan, Meena 44
Meng, Jie 459
Mészáros, Viola 192
Misra, Neeldhara 199
Mondal, Debajyoti 182
Mosteiro, Miguel A. 378
Musdalifah, S. 209

Nishat, Rahnuma Islam 182
Nussbaum, Doron 140

Philip, Geevarghese 34
Pu, Shuye 140

Qin, Bo 510

Rahman, Md. Saidur 182, 399
Razenshteyn, Ilya 120
Rextin, Aimal 110
Ribichini, Andrea 160
Rosen, Ricky 23
Ruskey, Frank 288, 298

Sack, Jörg-Rüdiger 140
Sauerwald, Thomas 130
Saurabh, Saket 34, 199
Schalekamp, Frans 70
Schurch, Mark 288
Silvestre, Yannick 173
Stolař, Rudolf 192
Sudarsana, I W. 209
Sun, Xiaoming 308
Suraweera, Francis 264
Su, Yu-Hsuan 368

Takhanov, Rustem 328
Tayu, Satoshi 348
Thang, Nguyen Kim 254
Ting, Hing-Fung 100
Tixeuil, Sébastien 80
Tokuyama, Takeshi 235
Tsai, Meng-Tsung 500

Ueno, Shuichi 348
Umboh, Seeun 489
Uno, Takeaki 140

Valtr, Pavel 192
van Zuylen, Anke 60, 70

Wang, Da-Wei 500
Wang, Lusheng 409, 429, 449
Wang, Yusu 278
Weibel, Christophe 469
Weng, Jian 520
Whitesides, Sue 469
Williams, Aaron 298
Woodcock, Jennifer 288
Wortman, Kevin A. 90
Wu, Hsin-Lung 13
Wu, Qianhong 510
Wu, Xiaodong 419

Xiao, Mingyu 150

Yanhaona, Muhammad Nur 399
Yu, Michael 70

Zarrabi-Zadeh, Hamid 140
Zhang, Lei 510
Zhang, Linqiao 469
Zhang, Shengyu 54
Zhang, Yong 100

Printing: Mercedes-Druck, Berlin
Binding: Stein+Lehmann, Berlin